HANDBOOK OF RESEARCH ON FUNCTIONAL MATERIALS

Principles, Capabilities, and Limitations

HANDBOOK OF RESEARCH ON FUNCTIONAL MATERIALS

Principles, Capabilities, and Limitations

Edited by

**Charles Wilkie, PhD, Georges Geuskens, PhD,
and Victor Manuel de Matos Lobo, PhD**

Gennady E. Zaikov, DSc, and A. K. Haghi, PhD
Reviewers and Advisory Board Members

Apple Academic Press

TORONTO   NEW JERSEY

Apple Academic Press Inc. | Apple Academic Press Inc.
3333 Mistwell Crescent | 9 Spinnaker Way
Oakville, ON L6L 0A2 | Waretown, NJ 08758
Canada | USA

© 2014 by Apple Academic Press, Inc.

First issued in paperback 2021

Exclusive worldwide distribution by CRC Press, a member of Taylor & Francis Group

No claim to original U.S. Government works

ISBN 13: 978-1-77463-293-2 (pbk)
ISBN 13: 978-1-926895-65-9 (hbk)

Library of Congress Control Number: 2013954918

Library and Archives Canada Cataloguing in Publication

Handbook of research on functional materials: principles, capabilities, and limitations/ edited by Charles Wilkie, PhD, Georges Geuskens, PhD, and Victor Manuel de Matos Lobo, PhD ; Gennady E. Zaikov, DSc, and A. K. Haghi, PhD, reviewers and advisory board members.

Includes bibliographical references and index.
ISBN 978-1-926895-65-9
1. Materials--Technological innovations--Handbooks, manuals, etc. I. Lobo, Victor M. M., writer of preface, editor of compilation II. Wilkie, C. A. (Charles A.), writer of preface, editor of compilation III. Geuskens, G. (Georges), writer of preface, editor of compilation IV. Title: Research on functional materials.

TA403.6.H35 2014   620.1'1   C2013-907752-9

Apple Academic Press also publishes its books in a variety of electronic formats. Some content that appears in print may not be available in electronic format. For information about Apple Academic Press products, visit our website at **www.appleacademicpress.com** and the CRC Press website at **www.crcpress.com**

ABOUT THE EDITORS

Charles Wilkie, PhD
Professor of Polymer and Organic Chemistry, Marquette University, Milwaukee, Wisconsin, USA

Charles Wilkie, PhD, is the Pfetschinger-Habermann Professor, Polymer and Organic Chemistry at Marquette University in Milwaukee, Wisconsin. He is an editor of *Polymers for Advanced Technologies* and serves on the editorial boards of *Polymer Degradation and Stability* and *Thermochimica Acta*. In 1992, Professor Wilkie was the recipient of the ACS Milwaukee Section Award. He was a Fulbright-Hays scholar at the Universite Libre de Bruxelles in 1991–1992. In 2007, he received the Marquette University Award for Excellence in Research. He is one of the organizers of the American Chemical Society Meeting on fire retardancy, held every four years, and he has taught a number of short courses on fire retardancy.

Georges Geuskens, PhD
Professor Emeritus, Department of Chemistry and Polymers, Universite de Libre de Bruxelles, Belgium

Georges Geuskens, PhD, is Professor Emeritus in the Department of Chemistry and Polymers at the Universite de Libre de Bruxelles in Belgium. Dr. Geuskens is world-renowned scientist in the field of photochemistry and particularly in the field of photodegradation and light stabilization of organic compounds (including oligomers, polymers, composites and nanocomposites). He has published several books and volumes in this area and in general, 200 original papers and reviews.

Victor Manuel de Matos Lobo, PhD
Professor, Coimbra University, Coimbra, Portugal

Victor Manuel de Matos Lobo, PhD, earned his PhD in 1966 from Cambridge University (UK), where he developed a new isothermal diffusion cell. He was the Head for the Department of Chemistry at the University of Coimbra in Portugal for many years. He has been a member of the Portuguese National Council for Education, representing the Portuguese scientific societies. He is a member of the council responsible for accepting scientific journals in SciELO (Scientific Electronic Library Online). He has published around 50 articles in the Portuguese daily press on matters concerning education. He has been the Editor of *Portugaliae Electrochimica Acta* from September 2002. Since 1991, he headed the Portuguese delegation at international meetings of the ISO (International Organization for Standardization) and the CEN (European Committee for Standardization).

REVIEWERS AND ADVISORY BOARD MEMBERS

Gennady E. Zaikov, DSc

Gennady E. Zaikov, DSc, is Head of the Polymer Division at the N. M. Emanuel Institute of Biochemical Physics, Russian Academy of Sciences, Moscow, Russia, and Professor at Moscow State Academy of Fine Chemical Technology, Russia, as well as Professor at Kazan National Research Technological University, Kazan, Russia. He is also a prolific author, researcher, and lecturer. He has received several awards for his work, including the the Russian Federation Scholarship for Outstanding Scientists. He has been a member of many professional organizations and on the editorial boards of many international science journals.

A. K. Haghi, PhD

A. K. Haghi, PhD, holds a BSc in urban and environmental engineering from University of North Carolina (USA); a MSc in mechanical engineering from North Carolina A&T State University (USA); a DEA in applied mechanics, acoustics and materials from Université de Technologie de Compiègne (France); and a PhD in engineering sciences from Université de Franche-Comté (France). He is the author and editor of 65 books as well as 1000 published papers in various journals and conference proceedings. Dr. Haghi has received several grants, consulted for a number of major corporations, and is a frequent speaker to national and international audiences. Since 1983, he served as a professor at several universities. He is currently Editor-in-Chief of the *International Journal of Chemoinformatics and Chemical Engineering* and *Polymers Research Journal* and on the editorial boards of many international journals. He is also a faculty member of University of Guilan (Iran) and a member of the Canadian Research and Development Center of Sciences and Cultures (CRDCSC), Montreal, Quebec, Canada.

CONTENTS

LIST OF CONTRIBUTORS

M. Abbasi
Textile Engineering department, Faculty of Engineering, University of Guilan, Rasht P.O. BOX 3756, Guilan, Iran

I. Afanasov
EFTEK-Bio LTD. 123458, Technopark "Strogino" ul. Tvardovskogo 8, Moscow 123458, Russia

L. F. Akhmetshina
Izhevsk State Technical University, Studencheskaya St. 7, Izhevsk 426069, Russia; OJSC "Izhevsk Electromechanical Plant "Kupol", Pesochnaya St. 3, Izhevsk 426033, Russia, E-mail: techprour@mail.ru

A. Y. Bedanokov
D.I. Mendeleev Russian University for Chemical Technology, Moscow, Russia

A. A. Berlin
N.N. Semenov Institute of chemical Physics, Russian Academy of Sciences, 4 Kosygin str., Moscow 119991, Russia

M. A. Chashkin
Izhevsk State Technical University, Russian Academy of Sciences; Izhevsk Electromechanical Plant, Russian Academy of Sciences; BRHE Centre of Chemical Physics and Mesoscopy, Udmurt Scientific Centre, UD, RAS

K. S. Dibirova
Dagestan State Pedagogical University, Makhachkala 367003, Yaragskii st., 57, Russian Federation

Kazuo Furuya
High Voltage Electron Microscopy Station, National Institute for Materials Science, 3–13 Sakura, Tsukuba 305-0003, Japan

Yanovskii Yurii Grigor'evich
Institute of Applied Mechanics of Russian Academy of Sciences, Leninskii pr., 32 a, Moscow 119991, Russian Federation, E-mail: iam@ipsun.ras.ru

A. K. Haghi
Textile Engineering department, Faculty of Engineering, University of Guilan, Rasht P.O.BOX 3756, Guilan, Iran

Akihisa Inoue
WPI Advanced Institute for Materials Research; Institute for Materials Research, Tohoku University, Sendai 980-8577, Japan

Iordanskii, A. L.
N.N. Semenov Institute of chemical Physics, Russian Academy of Sciences, 4 Kosygin str., Moscow 119991, Russia, Tel. +7-495-939-7434. E mail: aljordan08@gmail.com

V. I. Kodolov
Izhevsk State Technical University, Russian Academy of Sciences; Izhevsk Electromechanical Plant, Russian Academy of Sciences; BRHE Centre of Chemical Physics and Mesoscopy, Udmurt Scientific Centre, UD, RAS

V. I. Kodolov
Studencheskaya St. 7, Izhevsk 426069, Russia; OJSC "Izhevsk Electromechanical Plant "Kupol", Pesochnaya St. 3, Izhevsk 426033, Russia; BRHE Centre of Chemical Physics and Mesoscopy, Udmurt Scientific Centre, UD, RAS

G. V. Kozlov
N.M. Emanuel Institute of Biochemical Physics of Russian Academy of Sciences, Moscow 119334, Kosygin st., 4, Russian Federation

T. Z. Lygina
Central Scientific-Research Institute of Geology Non-Ore Minerals, Zinin Street 4, 420097 Kazan, Russia, E-mail lygina@geolnerud.ru

G. M. Magomedov
Dagestan State Pedagogical University, Makhachkala 367003, Yaragskii st., 57, Russian Federation

O. V. Mikhailov
Kazan National Research Technological University, K. Marx Street 68, 420015 Kazan, Russia, E-mail ovm@kstu.ru

A. K. Mikitaev
Kabardino – Balkarian State University, Nalchik, Russia

M. A. Mikitaev
L.Ya.Karpov Research Institute, Moscow, Russia

M. S. Motlagh
Textile Engineering department, Faculty of Engineering, University of Guilan, Rasht P.O.BOX 3756, Guilan, Iran

V. Mottaghitalab
Textile Engineering department, Faculty of Engineering, University of Guilan, Rasht P.O.BOX 3756, Guilan, Iran

N. I. Naumkina
Central Scientific-Research Institute of Geology Non-Ore Minerals, Zinin Street 4, 420097 Kazan, Russia, E-mail lygina@geolnerud.ru

Yu. V.

Pershin
Izhevsk State Technical University, Russian Academy of Science, Izhevsk, Russia

S. Z. Rogovina
N.N. Semenov Institute of chemical Physics, Russian Academy of Sciences, 4 Kosygin str., Moscow 119991, Russia

M. Sajedi
Textile Engineering department, Faculty of Engineering, University of Guilan, Rasht P.O.BOX 3756, Guilan, Iran

Anamika Singh
Division of Reproductive and Child Health, Indian Council of Medical Research, New Delhi

Rajeev Singh
Division of Reproductive and Child Health, Indian Council of Medical Research, New Delhi

Minghui Song
High Voltage Electron Microscopy Station, National Institute for Materials Science, 3–13 Sakura, Tsukuba 305-0003, Japan

V. V. Trineeva
Izhevsk State Technical University, Russian Academy of Sciences; Izhevsk Electromechanical Plant, Russian Academy of Sciences; BRHE Centre of Chemical Physics and Mesoscopy, Udmurt Scientific Centre, UD, RAS; Institute of Applied Mechanics, Ural Division, Russian Academy of Science

M. A. Vakhrushina
Izhevsk Electromechanical Plant, Russian Academy of Sciences

Yu. M. Vasilchenko
Izhevsk State Technical University, Russian Academy of Sciences; Izhevsk Electromechanical Plant, Russian Academy of Sciences; BRHE Centre of Chemical Physics and Mesoscopy, Udmurt Scientific Centre, UD, RAS

Kozlov Georgii Vladimirovich
Kh.M. Berbekov Kabardino-Balkarian State University, Department of Chemistry, 360004, Nalchik, Chernyshevskogo, 173, Russia, E-mail: I_dolbin@mail.ru

Guoqiang Xie
Institute for Materials Research, Tohoku University, Sendai 980-8577, Japan, Tel.: +81-22-215-2492; E-mail: xiegq@imr.tohoku.ac.jp

G. E. Zaikov
N.M. Emanuel Institute of Biochemical Physics of Russian Academy of Sciences, Moscow 119334, Kosygin st., 4, Russian Federation, E-mail: gezaikov@yahoo.com

A. I. Zakharov
Izhevsk Electromechanical Plant, Russian Academy of Sciences

LIST OF ABBREVIATIONS

ACA	amino caproic acid
ANN	artificial neural network
ANOVA	analysis of variance
BSA	bovine serum albumin
CA	contact angle
CCD	central composite design
CHEC	cold-hardened epoxy composition
CHT	chitosan
CMC	cell membrane complex
CME	clathrin-mediated endocytosis
CNTs	carbon nanotubess
CvME	caveolae-mediated endocytosis
DCM	dichloromethane
DLS	dynamic light scattering
DSSC	dye-sensitized solar cells
EBID	electron-beam-induced deposition
ED	electron diffraction
EDS	energy dispersive spectroscopy
EMD	electron microdiffraction
ER	epoxy resin
ERM	effective reinforcing modulus
EWG	electron-withdrawing groups
HDPE	high-density polyethylene
LMC	low-molecular compound
MFD	mean fiber diameter
MNPs	magnetic nanoparticles
MNSs	magnetically targeted nanosystems
MWCNT	multiwalled carbon nanotube
MWNTs	multiwalled nanotubes
NPT	isothermal–isobaric
NVE	microcanonical
NVT	canonical
PAN	polyacrylonitrile
PCA	polycaproamide
PEO	polyethylene oxide

PEPA	polyethylene polyamine
PVA	polyvinyl alcohol
RDP	radial density profile
RES	reticuloendothelial system
RME	receptor-mediated endocytosis
RMSE	root mean square errors
RSM	response surface methodology
RVE	representative volume element
RVP	radial velocity profile
SAD	selected-area diffraction
SEM	scanning electron microscope
SLN	solid lipid nanoparticles
STM	scanning tunneling
SUSHI	Simulation Utilities for Soft and Hard Interfaces
SWCN	single walled carbon nanotube
SWNTs	single walled nanotubes
TDGL	time-dependent Ginsburg–Landau
TEM	transmission electron microscopy
TFA	triflouroacetic acid
TG	thermogravimetric
μVT	grand canonical

LIST OF SYMBOLS

x_i and x_j	independent variables
a	temperature conductivity coefficient
d	dimension of Euclidean space
D	optical density
F	work angle constant
Y	is the predicted response

GREEK VARIABLES

h	sample thickness
l_0	length of the main chain skeletal bond
M_e	molecular weight of polymer chain
N_A	Avogadro number
R_s	sample heat resistance
S	cross-sectional area of macromolecule
S	sample area
c	heat capacity
η	dynamic viscosity
κ	sphere constant according to the test certificate
λ	wavelength
ν	Poisson's ratio
ρ	density
ρ_1	sphere density according to the test certificate
ρ_2	sample density
r_p	polymer density
s_Y	the temperature dependence of yield stress
τ	sphere movement time
j_{cl}	relative fraction of local order domains (nanoclusters)
χ	a relative fraction of elastically deformed polymer

PREFACE

Functional materials have played a significant part in humans' existence. They have a role in every aspect of modern life, including as health care, food, information technology, transportation, energy industries, etc. The speed of developments within the polymer sector is phenomenal and at same time crucial to meet demands of today's and future life. Specific applications for polymers range from adhesives, coatings, painting, foams and packaging to structural materials, composites, textiles, electronic and optical devices, biomaterials and many other uses in industries and daily life. Functional materials are the basis of natural and synthetic materials. They are macromolecules and, in nature, are the raw material for proteins and nucleic acids, which are essential for human bodies.

Cellulose, wool, natural rubber and synthetic rubber, and plastics are well-known examples of natural and synthetic types. Natural and synthetic polymers play a massive role in everyday life, and a life without functional materials really does not exist. These developments have made polymers such as polyesters contribute enormously to today's modern life. One of the most important applications is the furnishing sector (home, office, cars, aviation industry, etc.), which benefits hugely from the advances in technology. There are a number of requirements for a fabric to function in its chosen end use, for example, resistance to pilling and abrasion, as well as dimensional stability. Polyester is now an important part of the upholstery fabrics. The shortcomings attributed to the fiber in its early days have mostly been overcome. Now it plays a significant part in improving the life-span of a fabric, as well as its dimensional stability, which is due to its heat-setting properties.

About a half century has passed since synthetic leather, a composite material completely different from conventional ones, came to the market. Synthetic leather was originally developed for end-uses such as the upper of shoes. Gradually other uses like clothing steadily increased the production of synthetic leather and suede. Synthetic leathers and suede have a continuous ultrafine porous structure comprising a three-dimensional entangled nonwoven fabric and an elastic material principally made of polyurethane. Polymeric materials consisting of the synthetic leathers are polyamide and polyethylene terephthalate for the fiber and polyurethanes with various soft segments, such as aliphatic polyesters, polyethers and polycarbonates for the matrix.

The introduction of plastics is associated with the twentieth century but the first plastic material, celluloid, was made in 1865. During the 1970s, clothes of polyester became fashionable but by the 1980s synthetics lost the popularity in favor of natural materials. Although people were less enthusiastic about synthetic fabrics for everyday wear, Gore-Tex and other synthetics became popular for outdoor and workout clothing. At the same time as the use of synthetic materials in clothing declined, alternative uses were found. One great example is the use of polyester for making beverage bottles where it replaced glass with its shatterproof properties as a significant property.

In general it can be said that plastics enhance and even preserve life. Kevlar, for instance, when it is used in making canoes for recreation or when used to make a bulletproof vest. Polyester enhances life, such as when this highly nonreactive material is used to make replacement human blood vessels or even replacement skin for burn victims. With all the benefits attributed to plastics, they have their negative side. A genuine environmental problem exists due to the fact that the synthetic polymers do not break down easily compared with the natural polymers, hence the need not only to develop biodegradable plastics, but also to work on more effective means of recycling. A lot of research is needed to study the methods of degradation and stabilization of polymers in order to design polymers according to the end-use.

Among the most important and versatile of the hundreds of commercial plastics is polyethylene. Polyethylene is used in a wide variety of applications because it can be produced in many different forms. The first type to be commercially exploited was called low density polyethylene (LDPE). This polymer is characterised by a large degree of branching, forcing the molecules to pack together rather than loosely forming a low density material. LDPE is soft and pliable and has applications ranging from plastic bags, containers, textiles, and electrical insulation, to coatings for packaging materials.

Another form of polyethylene differing from LDPE in structure is high density polyethylene (HDPE). HDPE demonstrates little or no branching, resulting in the molecules to be tightly packed. HDPE is much more rigid than LDPE and is used in applications where rigidity is important. Major uses of HDPE are plastic tubing, bottles, and bottle caps. Other variations of polyethylene include high and ultra-high molecular mass ones. These types are used in applications where extremely tough and resilient materials are needed.

Natural polymers unlike the synthetic ones do possess very complex structure. Natural polymers such as cellulose, wool, and natural rubber are used in many products in large proportions. Cellulose derivatives are one of the most versatile groups of regenerated materials with various fields of application. Cellulose is found in nature in all forms of plant life, particularly in wood and cotton. The purest form of cellulose is obtained from the seed hairs of the cotton plant which contain up to 95% cellulose. The first cellulose derivatives came to the stage around 1845 when the nitration of starch and paper led to discovery of cellulose nitrate.

In 1865 for the first time a mouldable thermoplastic was made of cellulose nitrate and castor oil.

In 1865 the first acetylation of cellulose was carried out but the first acetylation process for use in industry was announced in 1894. In 1905 an acetylation process was introduced which yielded a cellulose acetate soluble in the cheap solvent, acetone. It was during the First World War when cellulose acetate dope found importance for weather proofing and stiffening the fabric of aircraft wings. There was a large surplus production capacity after the war which led to civilian end uses such as the production of cellulose acetate fibers by 1920s. Cellulose acetate became the main thermoplastic moulding material when the first modern injection moulding machines were designed. Among the cellulose derivatives, cellulose acetates are produced in the largest volume. Cellulose acetate can be made into fibers, transparent films and the less substituted derivatives are true thermoplastics. Cellulose acetates are mouldable and can be fabricated by the conventional processes. They have toughness, good appearance, and are capable of many color variations including white transparency.

New applications are being developed for polymers at a very fast rate all over the world at various research centers. Examples of these include electro active polymers, nano products, robotics, etc. Electro active polymers are special types of materials that can be used for example as artificial muscles and facial parts of robots or even in nano robots. These polymers change their shape when activated by electricity or even by chemicals. They are light weight but can bear a large force that is very useful when being utilized for artificial muscles. Electro active polymers together with nanotubes can produce very strong actuators. Currently research works are carried out to combine various types of electro active polymers with carbon nanotubes to make the optimal actuator. Carbon nanotubes are very strong, elastic, and the conduct electricity. When they are used as an actuator, in combination with an electro active polymer, the contractions of the artificial muscle can be controlled by electricity. Already works are under way to use electro active polymers in space. Various space agencies are investigating the possibility of using these polymers in space. This technology has a lot to offer for the future, and with the ever-increasing work on nanotechnology, electro active materials will play a very important part in modern life.

This book covers a broad range of functional materials and provides industry professionals and researchers in polymer science and technology with a single, comprehensive book summarizing all aspects involved in the functional materials production chain. The book focuses on industrially important materials, analytical techniques, and formulation methods, with chapters covering step-growth, radical, and co-polymerization, crosslinking and grafting, reaction engineering, advanced technology applications, including conjugated, dendritic, and nano-material polymers and emulsions, and characterization methods, which includes spectroscopy, light scattering, and microscopy.

The book introduces current state-of-the-art technology in functional materials with an emphasis on the rapidly growing technologies. It takes a unique approach by presenting specific materials and then progresses into a discussion of the ways in which these materials and processes are integrated into today's functioning manufacturing industry. It follows a more quantitative and design-oriented approach than other texts in the market, helping readers gain a better understanding of important concepts. Readers will also discover how material properties relate to the process variables in a given process as well as how to perform quantitative engineering analysis of manufacturing processes.

— **Charles Wilkie, PhD, Georges Geuskens, PhD,
and Victor Manuel de Matos Lobo, PhD**

CHAPTER 1

A STUDY ON NATURAL HYBRID NANOCOMPOSITES

G. M. MAGOMEDOV, K. S. DIBIROVA, G. V. KOZLOV,
and G. E. ZAIKOV

CONTENTS

1.1 INTRODUCTION

It has been shown that at semicrystalline polymers with devitrificated amorphous phase consideration as natural hybrid nanocomposites their anomalous high reinforcement degree is realized at the expense of crystallites partial recrystallization (mechanical disordering), that means crystalline phase participation in these polymers elastic properties formation. It is obvious, that the proposed mechanism is inapplicable for the description of polymer nanocomposites with inorganic nanofillers.

As it is known [1], semicrystalline polymers, similar to widely applicable polyethylene and polypropylene, at temperatures of the order of room ones have devitrificated amorphous phase. This means that such phase elasticity modulus is small and makes up the value of the order of 10 MPa [2]. At the same time elasticity modulus of the semicrystalline polymers can reach values of ~ 1.0–1.4 GPa and is a comparable one with the corresponding parameter for amorphous glassy polymers. In case of the latters it has been shown [3, 4], that they can be considered as natural nanocomposites, in which local order domains (nanoclusters) serve as nanofiller and as loosely packed matrix of polymer within the framework of the cluster model of polymers amorphous state structure [5] is considered as matrix. In this case elasticity modulus of glassy loosely packed matrix makes up the value of order of 0.8 GPa and a corresponding parameter for polymer (e.g., polycarbonate or polyarylate) ~ 1.6 GPa. In other words, the reinforcement degree of loosely packed matrix by nanoclusters for amorphous glassy polymers is equal to ~ 2, whereas for the semicrystalline polymers it can exceed two orders. By analogy with amorphous [3, 4] and cross-linked [6] polymers semicrystalline polymers can be considered as natural hybrid nanocomposites, in which rubber-like matrix is reinforced by two kinds of nanofiller: nanoclusters (analog of disperse nanofiller with particles size of the order of ~ 1 nm [5]) and crystallites (analog of organoclay with platelets size of the order of ~ 30–50 nm [7]). The clarification of abnormally high reinforcement degree mechanism allows giving an answer to the question, would this mechanism be applicable to polymer nanocomposites, filled with inorganic nanofiller (e.g., organoclay). Therefore, the purpose of the present work is the study of reinforcement mechanism of rubber-like matrix of high-density polyethylene (HDPE) at its consideration as natural hybrid nanocomposite.

1.2 EXPERIMENTAL

The gas-phase HDPE of industrial production of mark HDPE-276, GOST 16338-85 with average weight molecular mass 1.4×10^5 and crystallinity degree 0.723, measured by the sample density, was used.

The testing specimens were prepared by method of casting under pressure on a casting machine Test Samples Molding Apparate RR/TS MP of firm Ray-Ran (Taiwan) at material cylinder temperature 473 K, compression mold temperature 333 K and pressure of blockage 8 MPa.

The impact tests have been performed by using a pendulum impact machine on samples without a notch according to GOST 4746-80, type II, within the testing temperatures range T=213–333 K. Pendulum impact machine was equipped with a piezoelectric load sensor, that allows to determine elasticity modulus E and yield stress σ_Y in impact tests according to the techniques [8] and [9], respectively.

Uniaxial tension mechanical tests have been performed on the samples in the shape of two-sided spade with sizes according to GOST 11262-80. The tests have been conducted on a universal testing apparatus Gotech Testing Machine CT-TCS 2000, production of German Federal Republic, within the testing temperatures range T=293–363 K and strain rate of 2×10^{-3} s^{-1}.

1.3 RESULTS AND DISCUSSION

In Fig. 1, the temperature dependences of elasticity modulus E for the studied HDPE have been adduced. As one can see, at comparable testing temperatures E value in case of quasistatic tests is about twice smaller, than in impact ones. Let us note, that this distinction is not due to tests type. Thus, HDPE with the same crystallinity degree at T=293 K E value can reach 1252 MPa [10]. Let us consider the physical grounds of this discrepancy. The value of fractal dimension d_f of polymer structure, which is its main characteristic, can be determined by several methods application. The first from them uses the following equation [11]:

$$d_f = (d-1)(1+\nu),\tag{1}$$

where d is dimension of Euclidean space, in which a fractal is considered (it is obvious, that in our case d=3), ν is Poisson's ratio, estimated according to the mechanical tests results with the aid of the equation [12]:

$$\frac{\sigma_Y}{E} = \frac{1-2\nu}{6(1+\nu)}\tag{2}$$

The second method assumes the value d_f calculation according to the equation [5]:

$$d_f = d - 6.44 \times 10^{-10}\left(\frac{\varphi_{cl}}{C_\infty S}\right)^{1/2},\tag{3}$$

where φ_{cl} is relative fraction of local order domains (nanoclusters), C_∞ is characteristic ratio, S is cross-sectional area of macromolecule.

For HDPE $C_\infty = 7$ [13], $S = 14.4$ Å^2 [14] and φ_{cl} value can be calculated according to the following percolation relationship [5]:

$$\varphi_{cl} = 0.03(1-K)(T_m - T)^{0.55}, \tag{4}$$

where K is crystallinity degree, T_m and T are melting and testing temperatures, respectively. For HDPE $T_m \approx 400K$ [15].

And at last, for semicrystalline polymers d_f value can be evaluated as follows [16]:

$$d_f = 2 + K \tag{5}$$

In Table 1, the comparison of d_f values, determined by the three indicated methods has been adduced (K change with temperature was estimated according to the data of paper [7]). As one can see, if in case of impact tests the calculation according to all three indicated methods gives coordinated results, then for quasistatic tests estimation according to the Eqs. (1) and (2) gives clearly understated d_f values, especially with appreciation of possible variation of this dimension for nonporous solids ($2 \leq d_f \leq 2.95$ [11]).

TABLE 1   The values of fractal dimension d_f of HDPE structure, calculated by different methods.

Tests type	T, K	d_f the equation (1)	d_f the equation (3)	d_f the equation (5)
	293	2.302	2.800	2.723
	303	2.296	2.796	2.723
Quasistatic	313	2.272	2.802	2.713
	323	2.353	2.801	2.693
	333	2.248	2.799	2.673
	343	2.182	2.799	2.663
	353	2.170	2.800	2.643
	363	2.078	2.808	2.633

TABLE 1 *(Continued)*

Tests type	T, K	d_f the equation (1)	d_f the equation (3)	d_f the equation (5)
	213	2.764	2.734	2.723
Impact	233	2.762	2.741	2.723
	253	2.700	2.727	2.723
	273	2.750	2.756	2.723
	293	2.680	2.729	2.723
	313	2.624	2.743	2.713
	333	2.646	2.766	2.763

Let us consider the causes of the indicated discrepancy in more details. At present it has been assumed [1], that in case of semicrystalline polymers with devitrificated amorphous phase deformation in elasticity region, i.e., at E value determination, the indicated amorphous phase is only deformed, that defines smaller values of both E and d_f. This conclusion is confirmed by disparity between d_f values, calculated on the basis of mechanical characteristics (the Eqs. (1) and (2)) and crystallinity degree (the Eq. (5)). And on the contrary, a good correspondence of d_f values, obtained by the three indicated methods (Table 1), assumes crystalline phase participance at HDPE deformation in elasticity region in case of impact tests (Fig. 1).

In Fig. 2, the temperature dependence of yield stress σ_Y has been adduced for HDPE in case of both types of tests. As one can see, the data for quasistatic and impact tests are described by the same curve. As it has been shown in work [5], σ_Y values are defined by contribution of both nanoclusters, i.e., amorphous phase, and crystallites. Combined consideration of the plots of Figs. 1 and 2 demonstrates, that the distinction of d_f values, determined according to the Eqs. (1), (2) and (5), is only due to the above indicated structural distinction of HDPE deformation in an elasticity region.

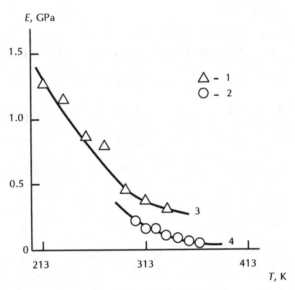

FIGURE 1 The dependences of elasticity modulus E on testing temperature T for HDPE, obtained in impact (1, 3) and quasistatic (2, 4) tests. 1, 2 – the experimental data; 3, 4 – calculation according to the Eq. (12).

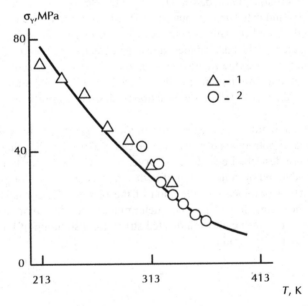

FIGURE 2 The dependence of yield stress σ_Y on testing temperature T for HDPE, obtained in impact (1) and quasistatic (2) tests.

The quantitative evaluation of crystalline regions contribution in HDPE elasticity can be performed within the framework of yield fractal conception [17], according to which the value of Poisson's ration in yield point v_Y can be estimated as follows:

$$v_y = v\chi + 0.5(1 - \chi),\qquad(6)$$

where v is Poisson's ratio in elastic strains region, determining according to the Eq. (2), χ is a relative fraction of elastically deformed polymer.

For amorphous glassy polymers it has been shown that χ value is equal to a relative fraction of loosely packed matrix $\varphi_{l.m.}$. In case of semicrystalline polymers in deformation process partial recrystallization (mechanical disordering) of a crystallites part can be realized, the relative fraction of which χ_{cr} is determined by the following equation [5]:

$$v_y = v\chi + 0.5(1 - \chi)\qquad(7)$$

If to consider HDPE as natural hybrid nanocomposite, then amorphous phase (the indicated nanocomposite matrix) elasticity modulus E_{am} can be determined within the framework of high-elasticity conception, using the known equation [2]:

$$G_{am} = kNT,\qquad(8)$$

where G_{am} is a shear modulus of amorphous phase, k is Boltzmann constant, N is a number of active chains of polymer network.

As it is known [5], in amorphous phase two types of macromolecular entanglements are present: traditional macromolecular "binary hookings" and entanglements, formed by nanoclusters, networks density of which is equal to v_e and v_{cl} respectively. v_e value is determined within the framework of rubber high-elasticity conception [2]:

$$v_e = \frac{\rho_p N_A}{M_e},\qquad(9)$$

where ρ_p is polymer density, N_A is Avogadro number, M_e is molecular weight of polymer chain part between macromolecular "binary hookings," which is equal to 1390 [18] and ρ_p value can be accepted equal to 960 kg/m³ [15].

In its turn, v_{cl} value can be determined according to the following equation [5]:

$$G_{am} = kNT,\qquad(10)$$

where l_0 is length of the main chain skeletal bond, for HDPE equal to 1.54 Å [13].

E_{am} and G_{am} are connected by a simple fractal formula [11]:

$$E_{am} = df G_{am} \qquad (11)$$

Calculation according to the Eqs. (6) and (7) has shown that in case of HDPE quasistatic tests χ_{cr} value is close to zero and in case of impact tests χ_{cr}=0.400–0.146 within the range of T=231–333 K. Now reinforcement degree of HDPE, considered as hybrid nanocomposite, can be expressed as the ratio E/E_{am}. In Fig. 3, the dependence $E/E_{am}(\chi_{cr}^2)$ has been adduced (such form of dependence was chosen for its linearization). As one can see, this linear dependence, demonstrates E/E_{am} (or E) increase at χ_{cr} growth and is described analytically by the following relationship:

$$\frac{E}{E_{am}} = 590 \chi_{cr}^2 , \text{ MPa.} \qquad (12)$$

The comparison of experimental E and calculated according to the Eq. (12) E^T elasticity modulus values for the studied HDPE has been adduced in Fig. 1. As one can see, the good correspondence between theory and experiment is obtained (the average discrepancy between E and E^T does not exceed 6%, i.e., comparable with an error of elasticity modulus experimental determination).

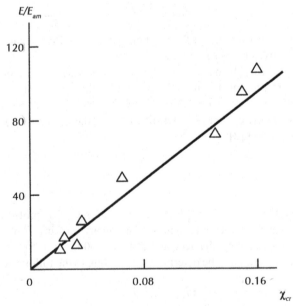

FIGURE 3 The dependence of reinforcement degree E/E_{am} on relative fraction of crystalline phase χ_{cr}, subjecting to partial recrystallization, for HDPE in impact tests

Thus, the performed estimations demonstrated, that high values of reinforcement degree E/E_{am} for semicrystalline polymers, considered as hybrid nanocomposites (in case of studied HDPE E/E_{am} value is varied within the limits of 10–110) were due to recrystallization process (mechanical disordering of crystallites) in elastic deformation process and, as consequence, to contribution of crystalline regions in polymers elastic properties formation. It is obvious, that this mechanism does not work in case of inorganic nanofiller (e.g., organoclay). Besides, a nanofiller (crystallites) is formed spontaneously in a polymer crystallization process, that automatically cancels the problem of its dispersion at large K of the order of 70 mass %, whereas to obtain exfoliated organoclay at contents larger than 3 mass % is difficult [19]. Thus, it is now clear that nanocomposites polymer/organoclay maximum reinforcement degree does not exceed 4 [20].

1.4 CONCLUSIONS

The performed analysis has shown that at the consideration of semicrystalline polymers with devitrificated amorphous phase as natural hybrid nanocomposites their abnormally high reinforcement degree is realized at the expense of crystallites partial recrystallization (mechanical disordering), that means crystalline phase participation in the formation of these polymers elastic properties. It is obvious, that the proposed mechanism is inapplicable for the description of reinforcement of polymer nanocomposites with inorganic nanofiller.

KEYWORDS

- **binary hookings**
- **high-density polyethylene**
- **nanoclusters**
- **semicrystalline polymers**

REFERENCES

1. Narisawa, I.; *Strength of Polymeric Materials*. Chemistry Publishing House (in rus.): Moscow, **1987**, p. 400.

2. Bartenev, G. M.; Frenkel, S. Ya. *Physics of Polymers*. Chemistry Publishing House (in rus.): Moscow, **1990**, p. 432.
3. Kozlov, G. V.; *Recent Patents on Chemical Engineering*, **2011**, *4(1)*, 53–77.
4. Kozlov, G. V.; Mikitaev, A. K.; *Polymers as Natural Nanocomposites: Unrealized Potential*. Lambert Academic Publishing: Saarbrücken, **2010**, p. 323.
5. Kozlov, G. V.; Ovcharenko, E. N.; Mikitaev, A. K.; *Structure of Polymer Amorphous State* (in rus.). RKhTU Publishing House: Moscow, **2009**, p. 392.
6. Magomedov, G. M.; Kozlov, G. V.; Zaikov, G. E.; *Structure and Properties of Cross-Linked Polymers*. A Smithers Group Company: Shawbury, **2011**, p. 492.
7. Tanabe, Y.; Strobl, G. R.; Fisher, E. W.; *Polymer*, **1986**, *27(8)*, 1147–1153.
8. Kozlov, G. V.; Shetov, R. A.; Mikitaev, A. K.; *Russian Polymer Science* (in rus.), **1987**, *29(5)*, 1109–1110.
9. Kozlov, G. V.; Shetov, R. A.; Mikitaev, A. K.; *Russian Polymer Science* (in rus.), **1987**, *29(9)*, 2012–2013.
10. Pegoretti, A.; Dorigato, A.; Penati, A. *EXPRESS Polymer Lett.;* **2007**, *1(3)*, 123–131.
11. Balankin, A. S.; *Synergetics of Deformable Body*. Ministry of Defence SSSR Publishing House: Moscow, **1991**, p. 404.
12. Kozlov, G. V.; Sanditov, D. S.; *Anharmonic Effects and Physical-Mechanical Properties of Polymers*. Nauka (Science) Publishing House (in rus.): Novosibirsk, **1994**, p. 264.
13. Aharoni, S. M.; *Macromolecules*, **1983**, *16(9)*, 1722–1728.
14. Aharoni, S. M.; *Macromolecules*, **1985**, *18(12)*, 2624–2630.
15. Kalinchev, E. L.; Sakovtseva, M. B.; *Properties and Processing of Thermoplastics* (in rus.). Chemistry Publishing House: Leningrad, **1983**, p. 288.
16. Aloev, V. Z.; Kozlov, G. V.; *Physics of Orientation Phenomena in Polymeric Materials* (in rus.). Polygraphservice and, T. Nal'chik, **2002**, p. 288.
17. Balankin, A. S.; Bugrimov, A. L.; *Russian Polymer Science* (in rus.), **1992**, *34(5)*, 129–132.
18. Graessley, W. W.; Edwards, S. F.; *Polymer*, **1981**, *22(10)*, 1329–1334.
19. Miktaev, A. K.; Kozlov, G. V.; Zaikov, G. E.; *Polymer Nanocomposites: Variety of Structural Forms and Applications*. Nova Science Publishers, Inc.: New York, **2008**, p. 319.
20. Liang, Z.-M.; Yin, J.; Wu, J.-H.; Qiu, Z.-X.; He, F.-F.; *Europ. Polymer J.* **2004**, *40(2)*, 307–314.

CHAPTER 2

GREEN NANOFIBERS— PRODUCTION AND LIMITS

A. K. HAGHI and G. E. ZAIKOV

CONTENTS

2.1   INTRODUCTION

Over the recent decades, the fabrication of biodegradable nanofibers for many biomedical applications, such as tissue engineering, drug delivery, wound dressing, enzyme immobilization, and so forth, has been recorded. Nanofiber fabrics have unique characteristics, such as very large surface area, ease of functionalization for various purposes and superior mechanical properties. Electrospinning (Fig. 1) is an important technique that can be used for the production of polymer nanofibers with diameters from several micrometers down to tens of nanometers. In electrospinning, the charged jets of a polymer solution, collected on a target, are created by using an electrostatic force. Many parameters can influence the quality of fibers, including the solution properties (polymer concentration, solvent volatility and solution conductivity), the governing variables (flow rate, voltage and tip-to-collector distance), and the ambient parameters (humidity, solution temperature, air velocity in the electrospinning chamber).

In recent years, scientists have manifested increased interest in electrospinning of natural materials, such as collagen, fibrogen, gelatin, silk, chitin and chitosan, due to their high biocompatible and biodegradable properties. Chitin is the second most abundant natural polymer in the world and chitosan (poly (1-4)-2-amino-2-deoxy-β-D-glucose) is the deacetylated product of chitin[1-12].

Researchers are interested in this natural polymer because of its properties, including solid-state structure and chain conformations in dissolved state.

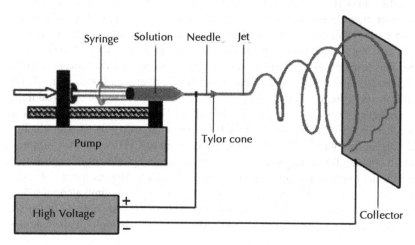

FIGURE 1   A typical image of Electrospinning process.

The physical characteristics of electrospun nanofibers such as fiber diameter depend on various parameters that are mainly divided into three categories: solution properties (solution viscosity, solution concentration, polymer molecular weight, and surface tension), processing conditions (applied voltage, volume flow rate, spinning distance, and needle diameter), and ambient conditions (temperature, humidity, and atmosphere pressure). Numerous applications require nanofibers with desired properties suggesting the importance of the process control. It does not come true unless having a comprehensive outlook of the process and quantitative study of the effects of governing parameters.

Besides physical characteristics, medical scientists showed a remarkable attention to biocompatibility and biodegradability of nanofibers made of biopolymers such as collagen, fibrogen, gelatin, silk, chitin and chitosan. Chitin is the second abundant natural polymer in the world and Chitosan (poly(1–4)-2-amino-2-deoxy-β-D-glucose) is the deacetylated product of chitin. CHT is well known for its biocompatible and biodegradable properties.

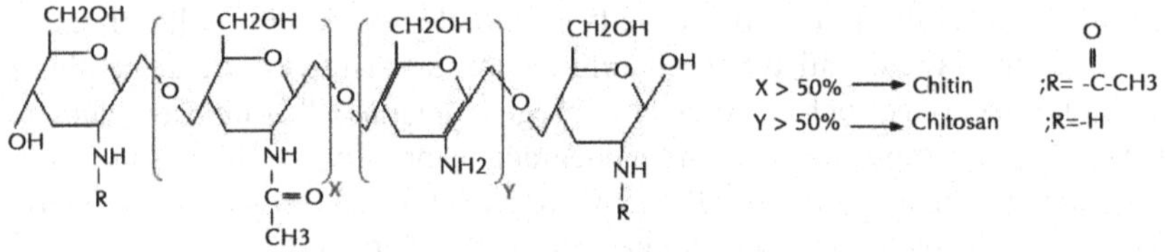

SCHEME 1   Chemical structures of Chitin and Chitosan biopolymers.

Chitosan (CHT) is insoluble in water, alkali, and most mineral acidic systems. However, though its solubility in inorganic acids is quite limited, CHT is in fact soluble in organic acids, such as dilute aqueous acetic, formic, and lactic acids. CHT also has free amino groups, which make it a positively charged polyelectrolyte. This property makes CHT solutions highly viscous and complicates its electrospinning. Furthermore, the formation of strong hydrogen bonds in a 3-D network prevents the movement of polymeric chains exposed to the electrical field.

Different strategies were used for bringing CHT in nanofiber form. The three top most abundant techniques include blending of favorite polymers for electrospinning process with CHT matrix, alkali treatment of CHT backbone to improve electro spinnability through reducing viscosity and employment of concentrated organic acid solution to produce nanofibers by decreasing of surface tension. Electrospinning of polyethylene oxide (PEO)/CHT and polyvinyl alcohol (PVA)/ CHT blended nanofiber are two recent studies based on first strategy. In the second protocol, the molecular weight of CHT decreases through alkali treatment.

Solutions of the treated CHT in aqueous 70–90% acetic acid have been employed to produce nanofibers with appropriate quality and processing stability.

Using concentrated organic acids such as acetic acid and triflouroacetic acid (TFA) with and without dichloromethane (DCM) has been reported exclusively for producing neat CHT nanofibers. They similarly reported the decreasing of surface tension and at the same time enhancement of charge density of CHT solution without significant effect on viscosity. This new method suggests significant influence of the concentrated acid solution on the reducing of the applied field required for electrospinning.

The mechanical and electrical properties of neat CHT electrospun natural nanofiber mat can be improved by addition of the synthetic materials including carbon nanotubes (CNTs). CNTs are one of the key synthetic polymers that were discovered in 1991. CNTs either single walled nanotubes (SWNTs) or multiwalled nanotubes (MWNTs) combine the physical properties of diamond and graphite. They are extremely thermally conductive like diamond and appreciably electrically conductive like graphite. Moreover, the flexibility and exceptional specific surface area to mass ratio can be considered as significant properties of CNTs. The scientists are becoming more interested to CNTs for existence of exclusive properties such as superb conductivity and mechanical strength for various applications. To best of our knowledge, there has been no report on electrospinning of CHT/MWNTs blend, except those ones that use PVA to improve spinnability. Results showed uniform and porous morphology of the electrospun nanofibers. Despite adequate spinnability, total removing of PVA from nanofiber structure to form conductive substrate is not feasible. Moreover, thermal or alkali solution treatment of CHT/PVA/MWNTs nanofibers extremely influence on the structural morphology and mechanical stiffness. The CHT/CNT composite can be produced by the hydrogen bonds due to hydrophilic positively charged polycation of CHT due to amino groups and hydrophobic negatively charged of CNT due to carboxyl and hydroxyl groups.

2.2 EXPERIMENTAL PART

CHT with degree of deacetylation of 85% and molecular weight of 5×10^5 was supplied by Sigma-Aldrich. The MWNTs, supplied by Nutrino, have an average diameter of four nm and the purity of about 98%. All of the other solvents and chemicals were commercially available and used as received without further purification.

2.3 PREPARATION OF CHT-MWNTS DISPERSIONS

A Branson Sonifier 250 operated at 30W used to prepare the MWNTs dispersions in CHT /organic acid (90%w/w acetic acid, 70/30 TFA/DCM) solution based on different protocols. In first approach, 3 mg of as received MWNTs was dispersed into deionized water or DCM using solution sonicating for 10 min (current chapter, sample 1). Different amount of CHT was then added to MWNTs dispersion for preparation of a 8–12 w/w % solution and then sonicated for another 5 min. Figure 2 shows two different protocols used in this study.

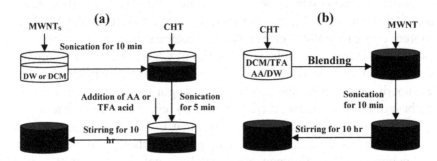

FIGURE 2   Two protocols used in this study for preparation of MWNTs/CHT dispersion (a) Current study (b) (AA/DW abbreviated for acetic acid diluted in de-mineralized water).

In next step, the organic acid solution added to obtain a CHT/MWNT solution with total volume of 5 mL and finally the dispersion was stirred for another 10 h. The sample 2 was prepared using the second technique. Same amount of MWNTs were dispersed in CHT solution, and the blend with total volume of 5 mL were sonicated for 10 min and dispersion was stirred for 10 h.

2.4 ELECTROSPINNING PROCESS

After the preparation of spinning solution, it transferred to a 5 mL syringe and became ready for spinning of nanofibers. The experiments were carried out on a horizontal electrospinning setup shown schematically in Fig. 1. The syringe containing CHT/MWNTs solution was placed on a syringe pump (New Era NE-100) used to dispense the solution at a controlled rate. A high voltage DC power supply (Gamma High Voltage ES-30) employed to generate the required electric field for electrospinning. The positive electrode and the grounding electrode of the high voltage supplier attaches respectively to the syringe needle and flat collector wrapped with aluminum foil where electrospun nanofibers accumulates via an alligator clip to form a nonwoven mat. The voltage and the tip-to-collector distance fixed respectively on 18–24 kV and 4–10 cm. In addition, the electrospinning carried out at room temperature and the aluminum foil removed from the collector.

2.5 MEASUREMENTS AND CHARACTERIZATIONS METHOD

A small piece of mat placed on the sample holder and gold sputter-coated (Bal-Tec). Thereafter, the micrograph of electrospun CHT/MWNTs nanofibers were obtained using scanning electron microscope (SEM, Phillips XL-30). Fourier transform infrared spectra (FT-IR) recorded using a Nicolet 560 spectrometer to investigate the interaction between CHT and MWNT in the range of 800–4000 cm^{-1} under a transmission mode. The size distribution of the dispersed solution was evaluated using dynamic light scattering technique (Zetasizer, Malvern Instruments). The conductivity of nanofiber samples was measured using a home-made four-probe electrical conductivity cell operating at constant humidity. The electrodes were circular pins with separation distance of 0.33 cm and fibers connect to pins by silver paint (SPI). Between the two outer electrodes, a constant DC current applies by Potentiostat/Galvanostat model 363 (Princeton Applied Research). The generated potential difference between the inner electrodes and the current flow between the outer electrodes recorded by digital multimeter 34401A (Agilent). Figure 3 illustrates the experimental setup for conductivity measurement. The conductivity (δ: S/cm) of the nanofiber with rectangular surface can then be calculated according to Eq. (1), which parameters call for length (L:cm), width (W:cm), thickness (t:cm), DC current applied (mA) and the potential drop across the two inner electrodes (mV). All measuring repeated at least five times for each set of samples.

$$\delta = \frac{I \times L}{V \times W \times t} \qquad (1)$$

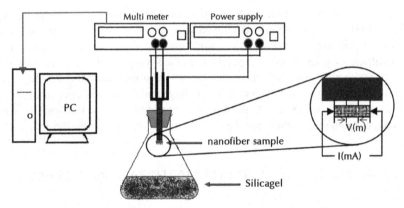

FIGURE 3   The experimental setup for four-probe electrical conductivity measurement of nanofiber thin film.

2.6   RESULTS

Utilization of MWNTs in biopolymer matrix initially requires their homogenous dispersion in a solvent or polymer matrix. Dynamic light scattering (DLS) is a sophisticated technique used for evaluation of particle size distribution. DLS provides many advantages for particle size analysis to measures a large population of particles in a very short time period, with no manipulation of the surrounding medium. DLS of MWNTs dispersions indicate that the hydrodynamic diameter of the nanotube bundles is between 150 and 400 nm after 10 min of sonication for sample 2 (Fig. 4).

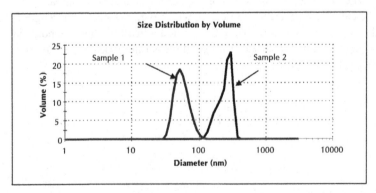

FIGURE 4   Hydrodynamic diameter distributions of MWNT bundles in CHT/acetic acid (1%) solution for different preparation technique

MWNTs bundle in sample 1(different approach but same sonication time compared to sample 2) shows a range of hydrodynamic diameter between 20–100 nm (Fig. 4). The lower range of hydrodynamic diameter for sample 1 correlates to more exfoliated and highly stable nanotubes strands in CHT solution. The higher stability of sample 1 compared to sample 2 over a long period of time was confirmed by solution stability test. The results presented in Fig. 5 indicate that procedure employed for preparation of sample 1 (current chapter) was an effective method for dispersing MWNTs in CHT/acetic acid solution. However, MWNTs bundles in sample 2 showed reagglomeration upon standing after sonication.

Despite the method reported in Ref. [35] neither sedimentation nor aggregation of the MWNTs bundles observed in first sample. Presumably, this behavior in sample 1 can be attributed to contribution of CHT biopolymer to forms an effective barrier against reagglomeration of MWNTs nanoparticles. In fact, using sonication energy, in first step without presence of solvent, make very tiny exfoliated but unstable particle in water as dispersant. Instantaneous addition of acetic acid as solvent and long mixing most likely helps the wrapping of MWNTs strands with CHT polymer chain[13-22].

Figure 6 shows the FT-IR spectra of neat CHT solution and CHT/MWNTs dispersions prepared using strategies explained in experimental part. The interaction between the functional group associated with MWNTs and CHT in dispersed form has been understood through recognition of functional groups. The enhanced peaks at ~1600 cm^{-1} can be attributed to (N-H) band and (C=O) band of amide functional group. However, the intensity of amide group for CHT/MWNTs dispersion increases presumably due to contribution of G band in MWNTs. More interestingly, in this region, the FT-IR spectra of MWNTs-CHT dispersion (sample 1) have been highly intensified compared to sample 2. It correlates to higher chemical interaction between acid functionalized C-C group of MWNTs and amide functional group in CHT.

This probably is the main reason of the higher stability and lower MWNTs dimension demonstrated in Figs. 4 and 5. Moreover, the intensity of protonated secondary amine absorbance at 2400 cm^{-1} for sample 1 prepared by new technique is negligible compared to sample 2 and neat CHT. Furthermore, the peak at 2123 cm^{-1} is a characteristic band of the primary amine salt, which is associated to the interaction between positively charged hydrogen of acetic acid and amino residues of CHT. In addition, the broad peaks at ~3410 cm^{-1} due to the stretching vibration of OH group superimposed on NH stretching bond and broaden due to hydrogen bonds of polysaccharides. The broadest peak of hydrogen bonds observed at 3137–3588 cm^{-1} for MWNTs/CHT dispersion prepared by new technique (sample1).

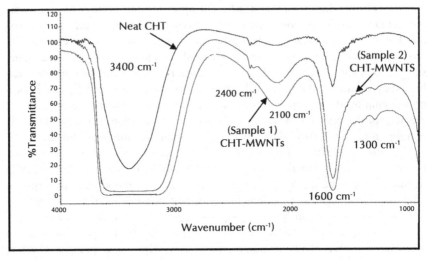

FIGURE 5 FT-IR spectra of CHT-MWCNT in 1% acetic acid with different techniques of dispersion

The different solvents including acetic acid 1–90%, pure formic acid, and TFA/DCM tested for preparation of spinning solution-using protocol explained for sample 1. Upon applying of the high voltage even more than 25 kV, no polymer jet forms using of acetic acid 1–30% and formic acid as the solvent for CHT/MWNT. However, experimental observation shows bead formation when the acetic acid (30–90%) used as the solvent. Therefore, one does not expect the formation of electrospun fiber of CHT/MWNTs using prescribed solvents (data not shown).

Figure 6 shows scanning electronic micrographs of the CHT/MWNTs electrospun nanofibers in different concentration of CHT in TFA/DCM (70:30) solvent. As presented in Fig. 6a, at low concentrations of CHT the beads deposited on the collector and thin fibers coexited among the beads. When the concentration of CHT increases as shown in Figs. 6a–c the bead density decreases. Figure 6c show the homogenous electrospun nanofibers with minimum beads, thin and interconnected fibers. More increasing of concentration of CHT lead to increasing of interconnected fibers at Figs. 6d–e. Figure 7 shows the effect of concentration on average diameter of MWNTs/CHT electrospun nanofibers. Our assessments indicate that the fiber diameter of CHT/MWNTs increases with the increasing of the CHT concentration. Hence, CHT/MWNTs (10 wt%) solution in TFA/DCM (70:30) considered as resulted in optimized concentration. An average diameter of 275 nm (Fig. 6c: diameter distribution, 148–385) investigated for these conditions.

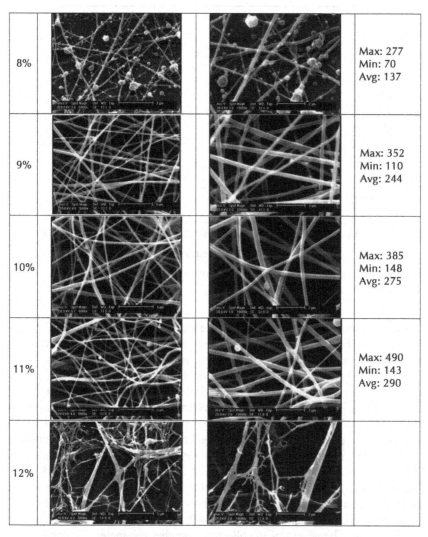

FIGURE 6 Scanning electron micrographs of electrospun nanofibers at different CHT concentration (wt %): (a) 8, (b) 9, (c) 10, (d) 11, (e) 12, 24 kV, 5 cm, TFA/DCM: 70/30, (0.06%wt MWNTs).

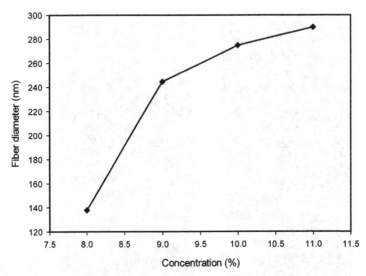

FIGURE 7   The effect of the CHT concentration in CHT/MWNT dispersion on nanofiber diameter.

Figure 7 shows the effect of the CHT concentration in CHT/MWNT dispersion on nanofiber diameter. Figure 8 shows the SEM image of CHT/MWNTs electrospun nanofibers produced in different voltage. In our experiments, 18 kV attains as threshold voltage, where fiber formation occurs.

For lower voltage, the beads and some little fiber deposited on collector (Fig. 8a). As shown in Figs. 8a–d, the beads decrease while voltage increasing from 18 kV to 24 kV. The collected nanofibers by applying 18 kV (9a) and 20 kV (9b) were not quite clear and uniform. The higher the applied voltage, the uniform nanofibers with narrow distribution starts to form. The average diameter of fibers, 22 kV (9c), and 24 kV (9d), respectively, were 204 (79–391), and 275 (148–385). The conductivity measurement given in Table 2 confirms our observation in first set of conductivity data. As can be seen from last row, the amount of electrical conductivity reaches to a maximum level of 9×10^{-5} S/cm at prescribed setup.

The distance between tips to collector is another parameter that controls the fiber diameter and morphology. Fig. 9 shows the change in morphologies of CHT/MWNTs electrospun nanofibers at different distance. When the distance is not long enough, the solvent could not find opportunity for separation. Hence, the interconnected thick nanofiber deposits on the collector (Fig. 9a). However, the adjusting of the distance on 5 cm (Fig. 9b) leads to homogenous nanofibers with negligible beads and interconnected areas. However, the beads increases by increasing of distance of the tip-to-collector as represented from Fig. 9b to Fig.

9f. Also, the results show that the diameter of electrospun fibers decreases by increasing of distance tip to collector in Figs. 9b–d, respectively, 275 (148–385), 170 (98–283), 132 (71–224). A remarkable defects and nonhomogeneity appears for those fibers prepared at a distance of 8 cm (Fig. 9e) and 10 cm (Fig. 10f). However, a 5 cm distance selected as proper amount for CHT/MWNTs electrospinning process.

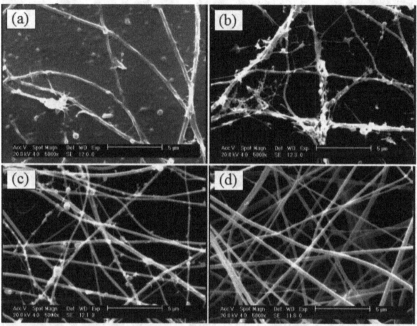

FIGURE 8 Scanning electronic micrographs of electrospun fibers at different voltage (kV): (a) 18, (b) 20, (c) 22, (d) 24, 5 cm, 10 wt%, TFA/DCM: 70/30. (0.06 wt% MWNTs).

The nonhomogeneity and huge bead densities plays as a barrier against electrical current and still a bead free and thin nanofiber mat shows higher conductivity compared to other samples. Experimental framework in this study was based on parameter adjusting for electrospinning of conductive CHT/MWNTs nanofiber. It can be expected that, the addition of nanotubes can boost conductivity and change morphological aspects, which is extremely important for biomedical applications.

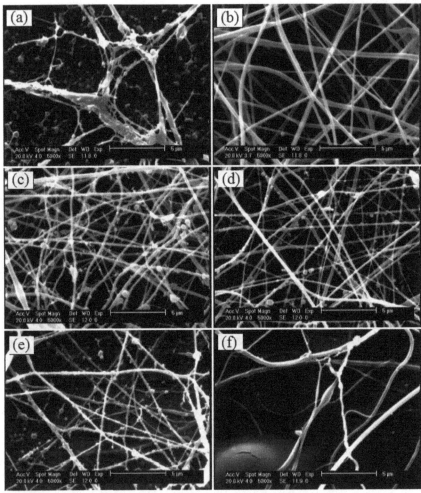

FIGURE 9 Scanning electronic micrographs of electrospun fibers of CHT/MWNT at different tip-to-collector distances (cm): (a) 4, (*b*) 5, (*c*) 6, (d) 7, (e) 8, (f) 10, 24 kV, 10 wt%, TFA/DCM: 70/30.

2.7 SILK FIBERS

In recent years, the electrospinning process has gained much attention because it is an effective method to manufacture ultrafine fibers or fibrous structures of many polymers with diameter in the range from several micrometers down to tens of nanometers. In the electrospinning process, a high voltage is used to create an electrically charged jet of a polymer solution or a molten polymer. This jet is collected on a target as a nonwoven fabric. The jet typically develops a bending instability and then solidifies to form fibers, which measures in the range of nanometers to 1 mm. Because these nanofibers have some useful properties such as high specific surface area and high porosity, they can be used as filters, wound dressings, tissue engineering scaffolds, and so forth.

Recently, the protein-based materials have been interested in biomedical and biotechnological fields. The silk polymer, a representative fibrous protein, has been investigated as one of promizing resources of biotechnology and biomedical materials due to its unique properties. Furthermore, high molecular weight synthetic polypeptides of precisely controlled amino acid composition and sequence have been fabricated using the recombinant tools of molecular biology. These genetic engineered polypeptides have attracted the researcher's attention as a new functional material for biotechnological applications including cellular adhesion promoters, biosensors, and suture materials.

Silks are generally defined as fibrous proteins that are spun into fibers by some Lepidoptera larvae such as silkworms, spiders, scorpions, mites and flies. Silkworm silk has been used as a luxury textile material since 3000 B.C., but it was not until recently that the scientific community realized the tremendous potential of silk as a structural material. While stiff and strong fibers (such as carbon, aramid, glass, etc.) are routinely manufactured nowadays, silk fibers offer a unique combination of strength and ductility which is unrivaled by any other natural or man-made fibers. Silk has excellent Properties such as lightweight (1.3 g/cm³) and high tensile strength (up to 4.8 GPa as the strongest fiber known in nature). Silk is thermally stable up to 250 °C, allowing processing over a wide range of temperatures. The origins of this behavior are to be found in the organization of the silks at the molecular and supramolecular levels, which are able to impart a large load bearing capability together with a damage tolerant response. The analysis of the fracture micro mechanisms in silk may shed light on the relationship between its microstructure and the mechanical properties. The antiparallel β pleated structure of silk gives rise to its structural properties of combined strength and toughness[22-31].

B. Mori silk consists of two types of proteins, fibroin and sericin. Fibroin is the protein that forms the filaments of silkworm silk. Fibroin filaments made up of bundles of nanofibrils with a bundle diameter of around 100 nm. The nanofibrils are oriented parallel to the axis of the fiber, and are thought to interact strongly with each other. Fibroin contains 46% Glycine, 29% Alanine and 12% Serine. Fibroin is a giant molecule (4700 amino acids) comprising a "crystalline" portion of about two-thirds and an "amorphous" region of about one-third. The crystalline portion comprises about 50 repeats of polypeptide of 59 amino acids whose sequence is known: Gly Ala-Gly Ala-Gly Scr-Gly Ala-Ala-Gly (Scr-Gly Ala-Gly

Ala-Gly)s-Tyr. This repeated unit forms, β-sheet and is responsible for the mechanical properties of the fiber.

Silk fibroin (SF) can be prepared in various forms such as gels, powders, fibers, and membranes. Number of researchers has investigated silk-based nanofibers as one of the candidate materials for biomedical applications, because it has several distinctive biological properties including good biocompatibility, good oxygen and water vapor permeability, biodegradability, and minimal inflammatory reaction. Several researchers have also studied processing parameters and morphology of electrospun silk nanofibers using Hexafluoroacetone, Hexafluoro-2-propanol and formic acid as solvents. In all these reports nanofibers with circular cross sections have been observed.

In this section, effects of electrospinning parameters are studied and nanofibers dimensions and morphology are reported. Morphology of fibers diameter of silk precursor was investigated varying concentration, temperature and applied voltage and observation of ribbon like silk nanofibers were reported. Furthermore, a more systematic understanding of the process conditions was studied and a quantitative basis for the relationships between average fiber diameter and electrospinning parameters was established using response surface methodology (RSM), which will provide some basis for the preparation of silk nanofibers with desired properties.

2.7.1   PREPARATION OF REGENERATED SF SOLUTION

Raw silk fibers (B.mori cocoons were obtained from domestic producer, Abrisham Guilan Co., IRAN) were degummed with 2 gr/L Na_2CO_3 solution and 10 gr/L anionic detergent at 100 °C for 1 h and then rinsed with warm distilled water. Degummed silk (SF) was dissolved in a ternary solvent system of $CaCl_2/CH_3CH_2OH/H_2O$ (1:2:8 in molar ratio) at 70 °C for 6 h. After dialysis with cellulose tubular membrane (Bialysis Tubing D9527 Sigma) in H_2O for 3 days, the SF solution was filtered and lyophilized to obtain the regenerated SF sponges.

2.7.2   PREPARATION OF THE SPINNING SOLUTION

SF solutions were prepared by dissolving the regenerated SF sponges in 98% formic acid for 30 min. Concentrations of SF solutions for electrospinning was in the range from 8% to 14% by weight.

2.7.3 ELECTROSPINNING

In the electrospinning process, a high electric potential (Gamma High voltage) was applied to a droplet of SF solution at the tip (0.35 mm inner diameter) of a syringe needle, The electrospun nanofibers were collected on a target plate which was placed at a distance of 10 cm from the syringe tip. The syringe tip and the target plate were enclosed in a chamber for adjusting and controlling the temperature. Schematic diagram of the electrospinning apparatus is shown in Fig. 10. The processing temperature was adjusted at 25, 50 and 75 °C. A high voltage in the range from 10 kV to 20 kV was applied to the droplet of SF solution.

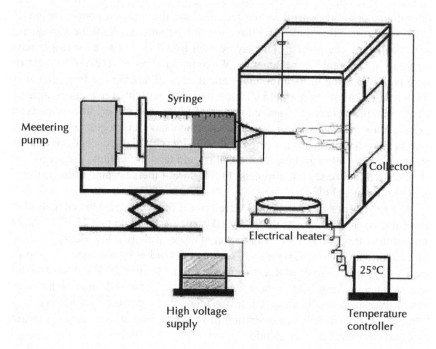

FIGURE 10 Schematic diagram of electrospinning apparatus.

2.7.4 CHARACTERIZATION

Optical microscope (Nikon Microphot-FXA) was used to investigate the macroscopic morphology of electrospun SF fibers. For better resolving power, morphology, surface texture and dimensions of the gold-sputtered electrospun nanofibers were determined using a Philips XL-30 scanning electron microscope. A measurement of about 100 random fibers was used to determine average fiber diameter and their distribution.

2.7.5 RESULTS AND DISCUSSION

2.7.5.1 EFFECT OF SILK CONCENTRATION

One of the most important quantities related with electrospun nanofibers is their diameter. Since nanofibers are resulted from evaporation of polymer jets, the fiber diameters will depend on the jet sizes and the solution concentration. It has been reported that during the traveling of a polymer jet from the syringe tip to the collector, the primary jet may be split into different sizes multiple jets, resulting in different fiber diameters. When no splitting is involved in electrospinning, one of the most important parameters influencing the fiber diameter is Concentration of regenerated silk solution. The jet with a low concentration breaks into droplets readily and a mixture of fibers, bead fibers and droplets as a result of low viscosity is generated. These fibers have an irregular morphology with large variation in size, on the other hand jet with high concentration don't break up but traveled to the grounded target and tend to facilitate the formation of fibers without beads and droplets. In this case, Fibers became more uniform with regular morphology.

At first, a series of experiments were carried out when the silk concentration was varied from 8 to 14% at the 15KV constant electric field and 25 °C constant temperature. Below the silk concentration of 8% as well as at low electric filed in the case of 8% solution, droplets were formed instead of fibers. Figure 11 shows morphology of the obtained fibers from 8% silk solution at 20 KV. The obtained fibers are not uniform. The average fiber diameter is 72 nm and a narrow distribution of fiber diameters is observed. It was found that continuous nanofibers were formed in the above silk concentration of 8% regardless of the applied electric field and electrospinning condition.

In the electrospinning of silk fibroin, when the silk concentration is more than 10%, thin and rod like fibers with diameters range from 60–450 nm were obtained. Figs. 12–14 show the SEM micrographs and diameter distribution of the resulted fibers.

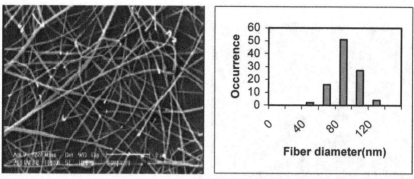

FIGURE 11   SEM micrograph and fiber distribution of 8 wt% of silk at 20 KV and 25 °C.

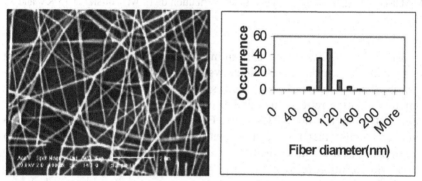

FIGURE 12   SEM micrograph and fiber distribution of 10 wt% of silk at 15 KV and 25 °C.

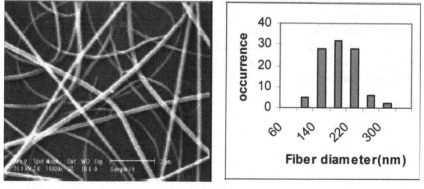

FIGURE 13   SEM micrograph and fiber distribution of 12 wt% of silk at 15 KV and 25 °C.

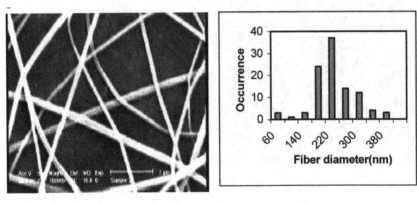

FIGURE 14   SEM micrograph and fiber distribution of 14 wt% of silk at 15 KV and 25 °C.

There is a significant increase in mean fiber diameter with the increasing of the silk concentration, which shows the important role of silk concentration in fiber formation during electrospinning process. Concentration of the polymer solution reflects the number of entanglements of polymer chains in the solution, thus solution viscosity. Experimental observations in electrospinning confirm that for forming fibers, a minimum polymer concentration is required. Below this critical concentration, application of electric field to a polymer solution results electrospraying and formation of droplets to the instability of the ejected jet. As the polymer concentration increased, a mixture of beads and fibers is formed. Further increase in concentration results in formation of continuous fibers as reported in this chapter. It seems that the critical concentration of the silk solution in formic acid for the formation of continuous silk fibers is 10%.

Experimental results in electrospinning showed that with increasing the temperature of electrospinning process, concentration of polymer solution has the same effect on fibers diameter at 25 °C. Figures 15–17 show the SEM micrographs and diameter distribution of the fibers at 50 °C.

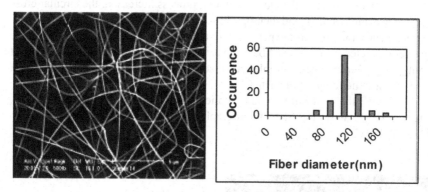

FIGURE 15   SEM micrograph and fiber distribution of 10 wt% of silk at 15 KV and 50 °C.

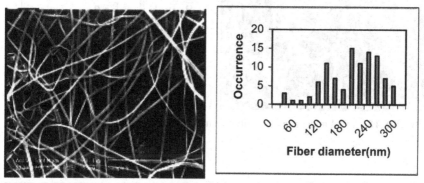

FIGURE 16   SEM micrograph and fiber distribution of 12 wt% of silk at 15 KV and 50 °C.

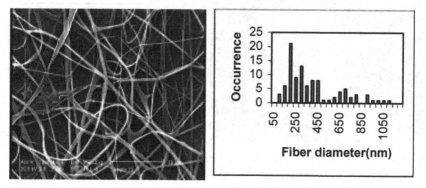

FIGURE 17   SEM micrograph and fiber distribution of 14 wt% of silk at 15 KV and 50 °C.

When the temperature of the electrospinning process increased, the circular cross section became elliptical and then flat, forming a ribbon with a cross-sectional perimeter nearly the same as the perimeter of the jet. Flat and ribbon like fibers have greater diameter than circular fibers. It seems that at high voltage the primary jet splits into different sizes multiple jets, resulting in different fiber diameters that nearly in a same range of diameter.

At 75 °C, the concentration and applied voltage have same effect as 50 °C. Figures 18–20 show the SEM micrographs and diameter distribution of Nano fibers at 75 °C.

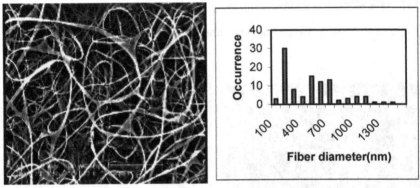

FIGURE 18   SEM micrograph and fiber distribution of 10 wt% of silk at 15 KV and 75 °C.

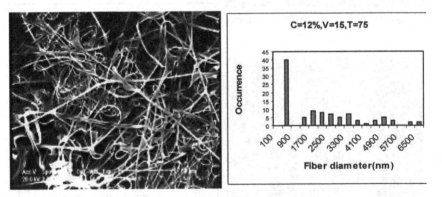

FIGURE 19   SEM micrograph and fiber distribution of 12 wt% of silk at 15 KV and 75 °C.

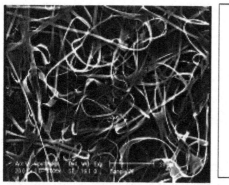 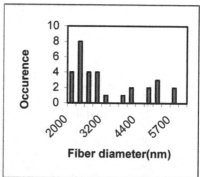

FIGURE 20   SEM micrograph and fiber distribution of 14 wt% of silk at 15 KV and 75 °C.

Figure 21 shows the relationship between mean fiber diameter and SF concentration at different electrospinning temperature. There is a significant increase in mean fiber diameter with increasing of the silk concentration, which shows the important role of silk concentration in fiber formation during electrospinning process. It is well known that the viscosity of polymer solutions is proportional to concentration and polymer molecular weight. For concentrated polymer solution, concentration of the polymer solution reflects the number of entanglements of polymer chains, thus have considerable effects on the solution viscosity. At fixed polymer molecular weight, the higher polymer concentration resulting higher solution viscosity. The jet from low viscosity liquids breaks up into droplets more readily and few fibers are formed, while at high viscosity, electrospinning is prohibit because of the instability flow causes by the high cohesiveness of the solution. Experimental observations in electrospinning confirm that for fiber formation to occur, a minimum polymer concentration is required. Below this critical concentration, application of electric field to a polymer solution results electro spraying and formation of droplets to the instability of the ejected jet. As the polymer concentration increased, a mixture of beads and fibers is formed. Further increase in concentration results in formation of continuous fibers as reported in this chapter. It seems that the critical concentration of the silk solution in formic acid for the formation of continuous silk fibers is 10% when the applied electric field was in the range of 10 to 20 kV.

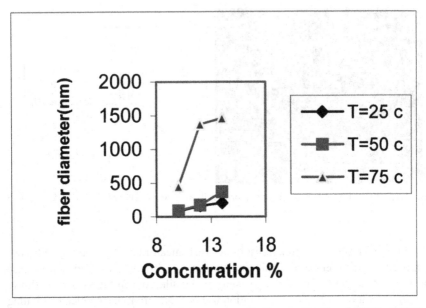

FIGURE 21   Mean fiber diameter of electrospun silk fibers at 25, 50 and 75 °C at15 kV.

2.7.5.2   EFFECT OF ELECTRIC FIELD

It was already reported that the effect of the applied electrospinning voltage is much lower than effect of the solution concentration on the diameter of electrospun fibers. In order to study the effect of the electric field, silk solution with the concentration of 10%, 12%, and 14% were electrospun at 10, 15, and 20 KV at 25 °C. The variation of the mean fiber diameter at different applied voltage for different concentration is shown in **fig. 22** This figure and SEM micrographs show that at all electric fields continuous and uniform fibers with different fiber diameter are formed. At a high solution concentration, Effect of applied voltage is nearly significant. It is suggested that, at this temperature, higher applied voltage causes multiple jets formation, which would provide decrees fiber diameter.

As the results of this finding it seems that electric field shows different effects on the nanofibers morphology. This effect depends on the polymer solution concentration and electrospinning conditions.

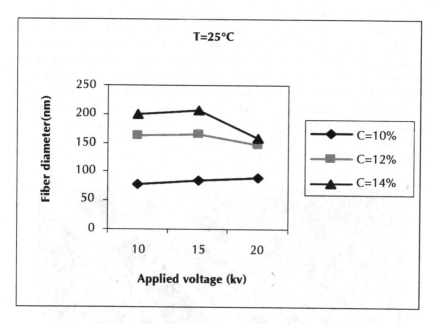

FIGURE 22   Mean fiber diameter of electrospun silk fibers at 10%, 15%, 20% and 25 °C.

2.7.5.3   EFFECT OF ELECTROSPINNING TEMPERATURE

One of the most important quantities related with electrospun nanofibers is their diameter. Since nanofibers are resulted from evaporation of polymer jets, the fiber diameters will depend on the jet sizes. The elongation of the jet and the evaporation of the solvent both change the shape and the charge per unit area carried by the jet. After the skin is formed, the solvent inside the jet escapes and the atmospheric pressure tends to collapse the tube like jet. The circular cross section becomes elliptical and then flat, forming a ribbon-like structure. In this chapter we believe that ribbon-like structure in the electrospinning of SF at higher temperature thought to be related with skin formation at the jets. With increasing the electrospinning temperature, solvent evaporation rate increases, which results in the formation of skin at the jet surface. Non- uniform lateral stresses around the fiber due to the uneven evaporation of solvent and/or striking the target make the nanofibers with circular cross-section to collapse into ribbon shape.

Bending of the electrospun ribbons were observed on the SEM micrographs as a result of the electrically driven bending instability or forces that occurred when the ribbon was stopped on the collector. Another problem that may be occurring in the electrospinning of SF at high temperature is the branching of jets. With increasing the temperature of electrospinning process, the balance between the surface tension and electrical forces can shift so that the shape of a jet becomes unstable. Such an unstable jet can reduce its local charge per unit surface area by ejecting a smaller jet from the surface of the primary jet or by splitting apart into two smaller jets. Branched jets, resulting from the ejection of the smaller jet on the surface of the primary jet were observed in electrospun fibers of SF. The axis of the cones from which the secondary jets originated were at an angle near 90° with respect to the axis of the primary jet. Figure 23 shows the SEM micrographs of flat, ribbon like and branched fibers.

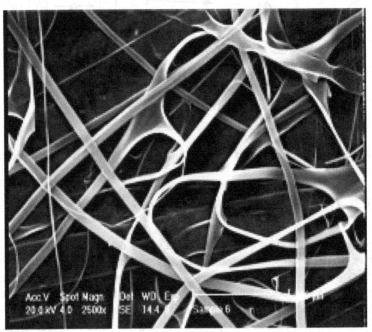

FIGURE 23   SEM micrograph of silk flat, ribbon like and branched nano fibers at high temperature.

In order to study the effect of electrospinning temperature on the morphology and texture of electrospun silk nanofibers, 12% silk solution was electrospun at various temperatures of 25, 50 and 75 °C. Results are shown in Fig. 24. Interestingly, the electrospinning of silk solution showed flat fiber morphology at 50 and 75 °C, whereas circular structure was observed at 25 °C. At 25 °C, the nanofibers with a rounded cross section and a smooth surface were collected on the target. Their diameter showed a size range of approximately 100 to 300 nm with 180 nm being the most frequently occurring. They are within the same range of reported size for electrospun silk nanofibers. With increasing the electrospinning temperature to 50 °C, The morphology of the fibers was slightly changed from circular cross section to ribbon like fibers. Fiber diameter was also increased to a range of approximately 20 to 320 nm with 180 nm the most occurring frequency. At 75 °C, The morphology of the fibers was completely changed to ribbon like structure. Furthermore, fibers dimensions were increased significantly to the range of 500 to 4100 nm with 1100 nm the most occurring frequency.

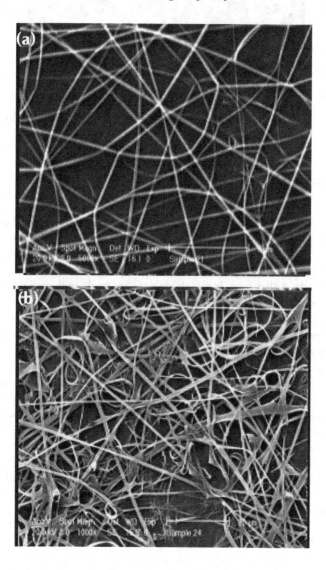

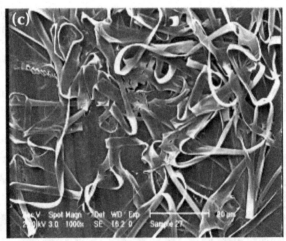

FIGURE 24   SEM micrograph of 12 wt% of silk at 20 KV and (a) 25 °C, (b) 50 °C, (c) 75 °C.

2.8 MODELING

Response surface methodology (RSM) is a collection of mathematical and statistical techniques for empirical model building (Appendix). By careful design of *experiments*, the objective is to optimize a *response* (output variable), which is influenced by several *independent variables* (input variables). An experiment is a series of tests, called *runs*, in which changes are made in the input variables in order to identify the reasons for changes in the output response.

In order to optimize and predict the morphology and average fiber diameter of electrospun silk, design of experiment was employed in the present chapter. Morphology of fibers and distribution of fiber diameter of silk precursor were investigated varying concentration, temperature and applied voltage. A more systematic understanding of these process conditions was obtained and a quantitative basis for the relationships between average fiber diameter and electrospinning parameters was established using response surface methodology (Appendix), which will provide some basis for the preparation of silk nanofibers.

A central composite design was employed to fit a second-order model for three variables. Silk concentration ($\times_1$), applied voltage ($\times_2$), and temperature ($\times_3$) were three independent variables (factors) considered in the preparation of silk nanofibers, while the fibers diameter were dependent variables (response). The actual and corresponding coded values of three factors ($\times_1$, X_2, and X_3) are given in Table 1. The following second-order model in X_1, X_2 and X_3 was fitted using the data in Table 1:

$$Y = \beta_0 + \beta_1 x_1 + \beta_2 x_2 + \beta_3 x_3 + \beta_1 x_1^2 + \beta_2 x_2^2 + \beta_3 x_3^2 + \beta_1 x_1 x_2 + \beta_1 x_1 x_3 + \beta_1 x_2 x_3 + \varepsilon$$

TABLE 1   Central composite design.
Coded values

X_i Independent variables		-1–0 1
X_1 Silk concentration (%)		10–12–14
X_2	applied voltage (KV)	10–15–20
X_3	temperature (°C)	25–50–75

The Minitab and Mathlab programs were used for analysis of this second-order model and for response surface plots (Minitab 11, Mathlab 7).

By Regression analysis, values for coefficients for parameters and P-values (a measure of the statistical significance) are calculated. When P-value is less than 0.05, the factor has significant impact on the average fiber diameter. If P-value is greater than 0.05, the factor has no significant impact on average fiber diameter. And R^2_{adj} (represents the proportion of the total variability that has been explained by the regression model) for regression models were obtained (Table 2) and main effect plots on fiber diameter (Fig. 25) were obtained and reported.

The fitted second-order equation for average fiber diameter is given by:

$$Y = 391 + 311 X_{1-} 164 X_2 + 57 X_{3-} 162 X_1{}^2 + 69 X_2{}^2 + 391 X_3{}^2 - 159 X_1 X_2 + 315 X_1 X_{3-} 144 X_2 X_3 \quad (1)$$

where Y is the average fiber diameter.

TABLE 2   Regression Analysis for the three factors (concentration, applied voltage, temperature) and coefficients of the model in coded unit*.

Variables	Constant		p-value
	β_0	391.3	0.008
x_1	β_1	310.98	0.00
x_2	β_2	−164.0	0.015
x_3	β_3	57.03	0.00
x_1^2	β_{11}	161.8	0.143
x_2^2	β_{22}	68.8	0.516
x_3^2	β_{33}	390.9	0.002

TABLE 2   *(Continued)*

Variables	Constant		*p*-value
x_1x_2	β_{12}	−158.77	0.048
x_1x_3	β_{13}	314.59	0.001
x_2x_3	β_{23}	−144.41	0.069
F	*p* -value	R^2	R^2 (adj)
18.84	0.00	0.907	0.858

* Model: $Y=\beta_0+\beta_1x_1+\beta_2x_2+\beta_3x_3+\beta_{11}x^2_1+\beta_{22}x^2_2+\beta_{332}x^2_3+\beta_{12}x_1x2+\beta_{13}x_1x3+\beta_{13}x_2x3$, where "*y*" is average fiber diameter.

From the *p*-values listed in Table 2, it is obvious that *p*-value of term X_2 is greater than P-values for terms X_1 and X_3. And other P-values for terms related to applied voltage such as, X_2^2, X_1X_2, X_2X_3 are much greater than significance level of 0.05. That is to say, applied voltage has no much significant impact on average fiber diameter and the interactions between concentration and applied voltage, temperature and applied voltage are not significant, either. But *p*-values for term related to X_3 and X_1 are less than 0.05. Therefore, temperature and concentration have significant impact on average fiber diameter. Furthermore, R^2_{adj} is 0.858, That is to say, this model explains 86% of the variability in new data.

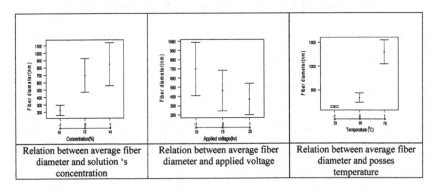

Relation between average fiber diameter and solution 's concentration	Relation between average fiber diameter and applied voltage	Relation between average fiber diameter and posses temperature

FIGURE 25   Effect of electrospinning parameters on silk nanofibers diameter.

2.9   CONCLUDING REMARKS

Several solvents, including acetic acid 1–90%, formic acid and TFA/DCM (70:30), were used for electrospinning of chitosan/carbon nanotube dispersion.

It has been observed that the TFA/DCM (70:30) solvent is the only solvent with a proper reliability for the electrospinnability of chitosan/carbon nanotube. This is a significant improvement in the electrospinning of chitosan/carbon nanotube dispersion. It was also observed that homogenous fibers with an average diameter of 455 nm (306–672) could be prepared with chitosan/carbon nanotube dispersion in TFA/DCM 70:30. In addition, the SEM images showed that the fiber diameter decreased by decreasing voltage and increasing the tip-to-collector distance.

The electrospinning of silk fibroin was processed and the average fiber diameters depend on the electrospinning condition. Morphology of fibers and distribution of diameter were investigated at various concentrations, applied voltages and temperature. The electrospinning temperature and the solution concentration have a significant effect on the morphology of the electrospun silk nanofibers. There effects were explained to be due to the change in the rate of skin formation and the evaporation rate of solvents. To determine the exact mechanism of the conversion of polymer into nanofibers require further theoretical and experimental work.

From the practical view the results of the present chapter can be condensed. Concentration of regenerated silk solution was the most dominant parameter to produce uniform and continuous fibers. The jet with a low concentration breaks into droplets readily and a mixture of fibers and droplets as a result of low viscosity is generated. On the other hand jets with high concentration do not break up but traveled to the target and tend to facilitate the formation of fibers without beads and droplets. In this case, fibers become more uniform with regular morphology. In the electrospinning of silk fibroin, when the silk concentration is more than 10%, thin and rod like fibers with diameters range from 60–450 nm were obtained. Furthermore, in the electrospinning of silk fibroin, when the process temperature is more than 25 °C, flat, ribbon like and branched fibers with diameters range from 60–7000 nm were obtained.

Two-way analysis of variance was carried out at the significant level of 0.05 to study the impact of concentration, applied voltages and temperature on average fiber diameter. It was concluded that concentration of solution and electrospinning temperature were the most significant factors impacting the diameter of fibers. Applied voltage had no significant impact on average fiber diameter. The average fiber diameter increased with polymer concentration and electrospinning temperature according to proposed relationship under the experimental conditions studied in this chapter [1–31].

KEYWORDS

- amorphous region
- chitin
- chitosan
- collagen
- crystalline portion
- fibroin
- gelatin
- silk

REFERENCES

1. Miwa, M.; Nakajima, A.; Fujishima, A.; Hashimoto, K.; Watanabe, T.; *Langmuir,* **2000,** *16,* 5754.
2. Ö ner, D.; McCarthy, T. J. *Langmuir,* **2000,** *16,* 7777.
3. Abdelsalam, M. E.; Bartlett, P. N.; Kelf, T.; Baumberg, J.; *Langmuir,* **2005,** *21,* 1753,
4. A.; Nakajima, K.; Hashimoto, T.; Watanabe, K.; Takai, G.; Yamauchi, A.; Fujishima, *Langmuir,* **2000,** *16,* 7044.
5. Zhong, W.; Liu, S.; Chen, X.; Wang, Y.; Yang, W.; *Macromolecules,* **2006,** *39,* 3224.
6. Shams Nateri, A.; Hasanzadeh, M.; *J. Comput. Theor. Nanosci.;* **2009,** *6,* 1542.
7. Kilic, A.; Oruc, F.; Demir, A.; *Text. Res. J.* **2008,** *78,* 532.
8. Reneker, D. H.; Chun, I.; *Nanotechnology* **1996,** *7,* 216.
9. Shin, Y. M.; Hohman, M. M.; Brenner, M. P.; Rutledge, G. C.; *Polymer,* **2001,** *42,* 9955.
10. Reneker, D. H.; Yarin, A. L.; Fong, H.; Koombhongse, S.; *J. Appl. Phys.;* **2000,** *87,* 4531.
11. Zhang, S.; Shim, W. S.; Kim, J.; *Mater. Design,* **2009,** *30,* 3659.
12. Yördem, O. S.; Papila, M.; Menceloğlu, Y. Z.; *Mater. Design,* **2008,** *29,* 34.
13. Chronakis, I. S.; *J. Mater. Process. Tech.;* **2005,** *167,* 283.
14. Dotti, F.; Varesano, A.; Montarsolo, A.; Aluigi, A.; Tonin, C.; Mazzuchetti, G.; *J. Ind. Text.* **2007,** *37,* 151.
15. Lu, Y.; Jiang, H.; Tu, K.; Wang, L.; *Acta Biomater.* **2009,** *5,* 1562.
16. Lu, H.; Chen, W.; Xing, Y.; Ying, D.; Jiang, B.; *J. Bioact. Compat. Pol.;* **2009,** *24,* 158.
17. Nisbet, D. R.; Forsythe, J. S.; Shen, W.; Finkelstein D. I.; Horne, M. K.; *J. Biomater. Appl.* **2009,** *24,* 7.
18. Ma, Z.; Kotaki, M.; Inai, R.; Ramakrishna, S.; *Tissue Eng.* **2005,** *11,* 101.
19. Hong, K. H.; *Polym. Eng. Sci.;* **2007,** *47,* 43.
20. Zhang, W.; Pintauro, P. N.; *ChemSusChem.;* **2011,** *4,* 1753.

21. Lee, S.; Obendorf, S. K.; *Text. Res. J.;* **2007,** *77,* 696.
22. Myers, R. H.; Montgomery D. C.; Anderson-cook, C. M.; *Response surface methodology: process and product optimization using designed experiments,* 3rd ed.; John Wiley and Sons, USA, **2009.**
23. Gu, S. Y.; Ren J.; Vancso, G. J.; *Eur. Polym. J.* **2005,** *41,* 2559.
24. Dev, V.; Venugopal, R. G.; Senthilkumar, J. R.; Gupta, M. D.; Ramakrishna, S.; *J. Appl. Polym. Sci.;* **2009,** *113,* 3397.
25. Galushkin, A. L.; *Neural networks Theory,* Springer, Moscow Institute of Physics and Technology **2007.**
26. Ma, M.; Mao, Y.; Gupta, M.; Gleason K. K.; Rutledge, G. C.; *Macromolecules,* **2005,** *38,* 9742.
27. Haghi, A. K.; Akbari, M.; *Phys. Status. Solidi. A.;* **2007,** *204,* 1830.
28. Ziabari, M.; Mottaghitalab, V.; Haghi, A. K.; in *Nanofibers: Fabrication, Performance, and Applications,* Chang, W. N.; Nova Science Publishers, USA, **2009.**
29. Ramakrishna, S.; Fujihara, K.; Teo, W. E.; Lim, T. C.; Ma, Z. *An Introduction to Electrospinning and Nanofibers,* World Scientific Publishing, Singapore, **2005.**
30. Zhang, S.; Shim, W. S.; Kim, J.; *Mater. Design,* **2009,** *30,* **3659.**
31. Kasiri, M. B.; Aleboyeh H.; Aleboyeh, A.; *Environ. Sci. Technol.;* **2008,** *42,* 7970.

APPENDIX A

Materials produced from fibers and threads are classified as textile: fabrics, non-woven materials, fur fabric, carpets and rugs, and so forth.

Textile fibers are the main raw material of the textile industry. According to their origin, these fibers are divided into natural and chemical ones.

Natural fibers are plant (cotton, bast fibers), animal (wool, silk) and mineral (asbestos) origin ones.

Chemical fibers are produced from modified natural or synthetic high-molecular substances and are classified as artificial ones obtained by chemical processing of natural raw material, commonly cellulose (viscose, acetate), and synthetic ones obtained from synthetic polymers (nylon-6, polyester, acryl, PVC fibers, etc.).

Textile materials are damaged by microorganisms, insects, rodents and other biodamaging agents. Fibers and fabrics resistance to biodamages primarily depends upon chemical nature of the fibers from which they are made. Most frequently we have to put up with microbiological damages of textile materials based on natural fibers – cotton, linen, and so forth, used by saprophyte microflora. Chemical fibers and fabrics, especially synthetic ones, are higher biologically resistant, but microorganisms-biodegraders can also adapt to them.

Textile material degradation by microorganisms depends on their wear rate, kind and origin, organic composition, temperature and humidity conditions, degree of aeration, and so forth.

With increased humidity and temperature, and restricted air exchange microorganisms damage fibers and fabrics at different stages of their manufacture and application, starting from the primary processing of fibers including spinning, weaving, finishing and storage, transportation and operation of textile materials and articles from them. Fiber and fabric biodamage intensity sharply increases when contacting with the soil and water, specifically in the regions with warm and humid climate.

Textile materials are damaged by bacteria and microscopic fungi. Bacterial degradation of textile materials is more intensive than the fungal one. The damaging bacterium genus are: *Cytophaga, Micrococcus, Bacterium, Bacillus, Cellulobacillus, Pseudomonas, Sarcina*. Among fungi damaging textile materials in the air and in the soil the following are detected: *Aspergillus, Penicillium, Alternaria, Cladosporium, Fusarium, Trichoderma*, and so forth.

Annual losses due to microbiological damaging of fabrics reach hundreds of millions of dollars.

Fiber and fabric biodamaging by microorganisms is usually accompanied by the mass loss and mechanical strength of the material as a result, for example, of fiber degradation by microorganism metabolites: enzymes, organic acids, and so forth.

CURIOUS FACTS

The immediate future of the textile industry belongs to biotechnology. Even today suggestions on the synthesis of various polysaccharides using microbiological methods are present. These methods may be applied to synthesis of fiber-forming monomers and polymers.

Scientists have demonstrated possibilities of microbiological synthesis of some monomers to produce dicarboxylic acids, caprolactam, and so forth. Some kinds of fiber-forming polymers, polyethers, in particular, can also be obtained by microbiological synthesis.

Some kinds of fiber-forming polypeptides have already been obtained by the microbiological synthesis. In some cases, concentration of these products may reach 40% of the biomass weight and they can be used as a perspective raw material for synthetic fibers. Studies in this direction are widely performed in many countries all over the world.

The impact of microorganisms on textile materials that causes their degradation is performed, at least, by two main ways (direct and indirect):

- fungi and bacteria use textile materials as the nutrient source (assimilation);
- textile materials are damaged by microorganism metabolism (degradation).

Biodamages of textile materials induced by microorganisms and their metabolites manifest in coloring (occurrence of spots on textile materials or their coatings), defects (formation of bubbles on colored surfaces of textile materials), bond breaks in fibrous materials, penetration deep inside (penetration of microorganisms into the cavity of the natural fiber), deterioration of mechanical properties (e.g., strength at break reduction), mass loss, a change of chemical properties (cellulose degradation by microorganisms), liberation of volatile substances and changes of other properties.

It is known that when microorganisms completely consume one part of the substrate they then are able to liberate enzymes degrading other components of the culture medium. It is found that, degrading fiber components each group of microorganisms due to their physiological features decomposes some definite part of the fiber, damages it differently and in a different degree. It is found that along with the enzymes, textile materials are also degraded by organic acids produced by microorganisms: lactic, gluconic, acetic, succinic, fumaric, malic, citric, oxalic, and so forth. It is also found that enzymes and organic acids liberated by microorganisms continue degrading textile materials even after microorganisms die. As noted, the content of cellulose, proteins, pectins and alcohol-soluble waxes increases with the fiber damage degree in it; pH increases, and concentration of water-soluble substances increases that, probably, is explained by increased accumulation of metabolites and consumption of nutrients by microorganisms for their vital activity.

The typical feature of textile material damaging by microorganisms is occurrence of honeydew, red violet or olive spots with respect to a pigment produced by microorganisms and fabric color. As microorganism's pigment interacts with the fabric dye, spots of different hue and tints not removed by laundering or by hydrogen peroxide oxidation. They may sometimes be removed by hot treatment in blankly solution. Spot occurrence on textile materials is usually accompanied by a strong musty odor.

CURIOUS FACTS

Textile materials are also furnished by biotechnological methods based on the use of enzymes performing various physicochemical processes. Biotechnologies are mostly full for preparation operations (cloth softening, boil-off and bleaching of cotton fabrics, wool washing) and bleaching effects of jeans and other fabrics, biopolishing, and making articles softer.

Enzymes are used to improve sorption properties of cellulose fibers, to increase specific area and volume of fibers, to remove pectin "companions" of cotton and linen cellulose. Enzymes also hydrolyze ether bonds on the surface of polyether fibers.

Cotton fabric dyeing technologies in the presence of enzymes improving coloristic parameters of prepared textile articles under softer conditions, which reduce pollutants in the sewage. To complete treatment of the fabric surface, that is, removal of surface fiber fibrilla, enzymes are applied. The enzymatic processing also decreases and even eliminates the adverse effect of long wool fiber puncturing. The application of enzymes to treatment of dyed tissues is industrially proven. This processing causes irregular bleaching that adheres the articles fashionable "worn" style.

Temperature and humidity are conditions promoting biodamaging of fibrous materials. The comparison of requirements to biological resistance of textile materials shall be based on the features of every type of fibers, among which mineral fiber are most biologically resistant.

Different microorganism damaging rate for fabrics is due to their different structure. Thinner fabrics with the lower surface density and higher through porosity are subject to the greatest biodamaging, because these properties of the material provide large contact area for microorganisms and allow their easy penetration deep into it. The thread bioresistance increases with the yarning rate.

Cotton fiber is a valuable raw material for the textile industry. Its technological value is due to a complex of properties that should be retained during harvesting, storage, primary and further processing to provide high quality of products. One of the factors providing retention of the primary fiber properties is its resistance to bacteria and fungi impacts. This is tightly associated with the chemical and physical features of cotton fiber structures.

Ripe cotton fiber represents a unit extended plant cell shaped as a flattened tube with a corkscrew waviness. The upper fiber end is cone-shaped and dead-ended. The lower end attached to the seed is a torn open channel.

The cotton fiber structure is formed during its maturation, when cellulose is biosynthesized and its macromolecules are regularly disposed.

The basic elements of cotton fiber morphological structure are known to be the cuticle, primary wall, convoluted layer, secondary wall, and tertiary wall with the central channel. The fiber surface is covered by a thin layer of wax substances – the primary wall (cuticle). This layer is a protective one and possesses rather high chemical resistance.

There is a primary wall under the cuticle consisting of a cellulose framework and fatty-wax-pectin substances. The upper layer of the primary wall is less densely packed as compared with the inner one, due to fiber surface expansion during its growth. Cellulose fibrils in the primary wall are not regularly oriented.

The primary layer covers the convoluted layer having structure different from the secondary wall layers. It is more resistant to dissolution, as compared with the main cellulose mass.

The secondary cotton fabric wall is more homogeneous and contains the greatest amount of cellulose. It consists of densely packed, concurrently oriented cellulose fibrils composed in thin layers. The fibril layers are spiral-shaped twisted around the fiber axis. There are just few micropores in this layer.

The tertiary wall is an area adjacent to the fiber channel. Some authors think that the tertiary wall contains many pores and consists of weakly ordered cellulose fibrils and plenty of protein admixtures, protoplasm, and pectin substances.

The fiber channel is filled with protoplasm residues, which are proteins, and contains various mineral salts and a complex of microelements. For the mature fiber, channel cross-section is 4–8% of total cross-section.

Cotton fiber is formed during maturation. This includes not only cellulose biosynthesis, but also ordering of the cellulose macromolecules shaped as chains and formed by repetitive units consisting two β-D-glucose residues bound by glucosidic bonds.

Among all plant fibers, cotton contains the maximum amount of cellulose (95–96%).

The morphological structural unit of cellulose is a cluster of macromolecules – a fibril 1.0–1.5 μm long and 8–15 nm thick, rather than an individual molecule. Cellulose fibers consist of fibril clusters uniform oriented along the fiber or at some angle to it.

It is known that cellulose consists not only of crystalline areas – micelles where molecule chains are concurrently oriented and bound by intermolecular forces, but also of amorphous areas.

Amorphous areas in the cellulose fiber are responsible for the finest capillaries formation, that is, a "submicroscopic" space inside the cellulose structure is formed. The presence of the submicroscopic system of capillaries in the cellulose fibers is of paramount significance, because it is the channel of chemical reactions by which water-soluble reagents penetrate deep in the cellulose structure. More active hydroxyl groups interacting with various substances are also disposed here.

It is known that hydrogen bonds are present between hydroxyl groups of cellulose molecules in the crystalline areas. Hydroxyl groups of the amorphous area may occur free of weakly bound and, as a consequence, they are accessible for sorption. These hydroxyl groups represent active sorption centers able to attract water.

Among all plant fibers, cotton has the highest quantity of cellulose (95–96%). Along with cellulose, the fibers contain some fatty, wax, coloring mineral substances (4–5%). Cellulose concomitant substances are disposed between macromolecule clusters and fibrils. Raw cotton contains mineral substances (K, Na, Ca, Mg) that promote mold growth and also contains microelements (Fe, Cu, Zn) stimulating growth of microorganisms. Moreover, it contains sulfates, phosphorus, glucose, glycidols and nitrogenous substances, which also stimulate growth of microbes. Differences in their concentrations are one of the reasons for different aggressiveness of microorganisms in relation to the cotton fiber.

The presence of cellulose, pectin, nitrogen containing and other organic substances in the cotton fiber, as well as its hygroscopicity makes it a good culture medium for abundant microflora.

Cotton is infected by microorganisms during harvesting, transportation and storage. When machine harvested, the raw cotton is clogged by various admixtures. It obtains multiple fractures of leaves and cottonseed hulls with humidity higher than of the fibers. Such admixtures create a humid macrozone around, where microorganisms intensively propagate. Fiber humidity above 9% is the favorable condition for cotton fiber degradation by microorganisms.

It is found that cotton fiber damage rate directly in hulls may reach 42–59%; hence, the fiber damage rate depends on a number of factors, for example, cultivation conditions, harvesting period, type of selection, and so forth.

Cotton fiber maturity is characterized by filling with cellulose. As the fiber becomes more mature, its strength, elasticity and coloring value increase.

Low quality cotton having higher humidity is damaged by microorganisms to a greater extent. The fifths fibers contain microorganisms 3–5 times more than the first quality fibers. When cotton hulls open, the quantity of microorganisms sharply increases in them, because along with dust wind brings fungus spores and bacteria to the fibers.

Cotton is most seriously damaged during storage: in compartments with high humidity up to 24% of cotton is damaged. Cotton storage in compact bales covered by tarpaulin is of high danger, especially after rains. For instance, after one and half months of such storage the fiber is damaged by 50% or higher.

Cotton microflora remains active under conditions of the spinning industry. As a result, the initial damage degree of cotton significantly increases.

CURIOUS FACTS

On some textile enterprises of Ivanovskaya Oblast, sickness cases of spinning-preparatory workshops were observed. When processing biologically contaminated cotton, plenty of dust particles with microorganisms present on them are liberated to the air. This can affect the health status of the employees.

Under natural conditions, cotton products are widely used in contact with the soil (fabrics for tents) receiving damage both from the inside and the outside. The main role here is played by cellulose degrading bacteria and fungi.

It was considered over a number of years that the main role in cotton fiber damage is played by cellulose degrading microorganisms. Not denying participation of cellulose degrading bacteria and fungi in damaging of the cotton fiber, it is noted that a group of bacteria with yellow pigmented mucoid colonies representing epi-phytic microflora always present on the cotton plant dominate in the process of the fiber degradation. Nonsporeforming epiphytic bacteria inhabiting in the cotton plants penetrate from their seeds into the fiber channel and begin developing there. Using chemical substances of the channel, these microorganisms then permeate into the submicroscopic space of the tertiary wall primarily consuming pectins of the walls and proteins of the channels.

Enzymes and metabolites produced by microorganisms induce hydrolysis of cellulose macromolecules, increasing damage of internal areas of the fiber. Thus, the fiber delivered to processing factories may already be significantly damaged by microorganisms that inevitably affects production of raw yarn, fabric, and so forth.

One hundred and thirty five strains of fungi of different genus capable of damaging cotton fibers are currently determined. It is found that the population of phytopathogenic fungi is much lower than that of cellulose degraders: *Chaetomium globosum, Aspergillus flavus, Aspergillus niger, Rhizopus nigricans, Trichothecium roseum*. These species significantly deteriorate the raw cotton condition, sharply reduce spinning properties of the fiber, in particular.

It is also found that the following species of fungi are usually present on cotton fibers: *Mucor* (consumes water-soluble substances), *Aspergillus* and *Penicillium* (consumes insoluble compounds), *Chaetomium, Trichoderma*, and so forth. (degrade cellulose). This points to the fact that some species of mold fungi induce the real fiber decomposition that shall be distinguished from simple surface growth of microorganisms. For example, *Mucor* fungi incapable of inducing cellulose degradation may actively vegetate on the yarn finish. Along with fungi, bacteria are always present on the raw cotton, most represented by *Bacillus* and *Pseudomonas* genus species.

Figures A1 and A2 present surface micrographs of the first and fifth quality grade primary cotton fibers.

Figure A3, a micrograph, shows the cotton fiber surface after the impact of spontaneous microflora during 7 days. It is observed that bacterial cells are accumulated in places of fiber damage, at clearly noticeable cracks. Figure 4 shows the first quality cotton fiber surface after impact of *Aspergillus niger* culture during 14 days. On the fiber surface mycelium is observed. Figures A5 and A6 show photos of cotton fiber surfaces infected by *Bac. subtilis* (14-day exposure): bacterial cells on first quality fibers are separated by the surface, forming no conglomerates, whereas on the fifth quality fibers conglomerates are observed that indicates their activity (Fig. A7).

Academician A.A. Imshenetsky has demonstrated that aerobic cellulose bacteria are able to propagate under increased humidity, whereas fungi propagate at lower humidity. Textile products are destroyed by fungi at their humidity about 10%, whereas bacteria destroy them at humidity level of, at least, 20%. As a consequence, the main attention at cotton processing to yarn should be paid to struggle against fungi, and at wet spinning and finishing fabrics and knitwear not only fungi, but mostly bacteria should be struggled against.

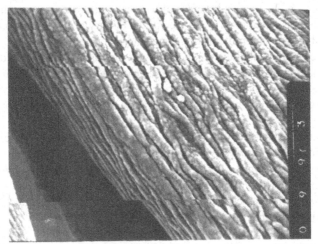

FIGURE A1   The surface of initial first quality cotton specimen (x 4500).

FIGURE A2   The surface of initial fifth quality cotton specimen (x 4500).

FIGURE A3   The first quality cotton, 7-day exposure (spontaneous microflora) (x 10000).

FIGURE A4   The first quality cotton, *Asp. niger* contaminated, 14-day exposure (x 3000).

FIGURE A5   The first quality cotton, *Bacillus subtilis* contaminated, 14-day exposure (x 4500).

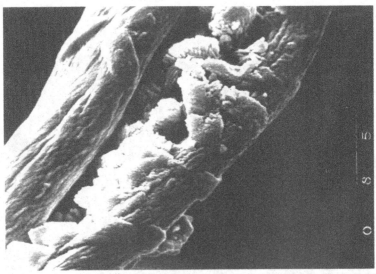

FIGURE A6   The fifth quality cotton, *Bacillus subtilis* contaminated, 14-day exposure (x 4500).

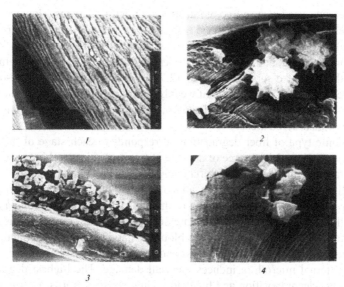

FIGURE A7   Cotton fiber micrographs: *1* – initial fiber (× 4500); *2–4:* fibers damages by different microorganisms: *2 – Aspergillus niger* (× 3000); *3 – Bacillus subtilis* (× 4500); *4 – Pseudomonas fluorescens* (× 10000)

Cotton damage leads to:
- significant decrease of strength of the fibers and articles from them;
- disturbance of technological process (the smallest particles of sticky mucus excreted by some species of bacteria and fungi become the reason for sticking executive parts of machines);
- abruptness increase;
- waste volume increase.

Damaging of cotton fibers, fabrics and textile products by microorganisms is primarily accompanied by occurrence of colored yellow, orange, red, violet, and so forth, spots and then by putrefactive odor, and, finally, the product loses strength and degrades.

The effect of microorganisms results in noticeable changes in chemical composition and physical structure of cotton fibers.

As found by electron microscopy, cotton fiber degradation by enzymes is most intensive in the zones of lower fibril structure density.

In the cotton fiber damaged by microorganisms, cellulose concentration decreases by 7.5%, pectin substances – by 60.7%, hemicellulose – by 20%, and noncellulose polysaccharides content also decreases. Cellulose biostability increases with its crystallinity degree and macromolecule orientation, as well as with hydroxyl group replacement by other functional groups. Microscopic fungi and bacteria are able to degrade cellulose and as a result glucose is accumulated in the medium, used as a source of nutrition by microorganisms. However, some part of cellulose is not destroyed and completely preserves its primary structure.

Cellulose of undamaged cotton fiber has 76.5% of well-ordered area, 7.8% of weakly ordered area, and 15.7% of disordered area. Microbiological degradation reduces the part of disordered area to 12.7%, whereas the part of well-ordered area increases to 80.4%. The ratio of weakly ordered area changes insignificantly. This goes to prove that the order degree of cotton cellulose increases due to destruction of disordered areas.

A definite type of fiber degradation corresponds to each stage of the cotton fiber damage. The initial degree of damage is manifested in streakiness, when the fiber surface obtains cracks of different length and width due to its wall break.

Swellings are formed resulting abundant accumulation of microorganisms and their metabolites in a definite part of the fiber. They may be accompanied by fiber wall break induced by biomass pressure. In this case, microorganisms and their metabolites splay out that causes blobs formation from the fiber and breaks in the yarn, as well as irregular fineness and strength.

The external microflora induces the wall damage. The highest degradation stage is fiber decomposition and breakdown into separate fibrils. Hence, perfect fiber structure is absent in this case.

In all cases of damage, a high amount of fungal mycelium may be present on the fiber surface, which hyphae penetrate through the fiber or wrap about it thus preventing spinning and coloring of textile materials.

Enzymatic activity of fungi is manifested in strictly defined places of cellulose microfibrils, and the strength loss rate depends on both external climate conditions and contamination conditions. Cotton fabrics inoculated by the microscopic fungus *Aspergillus niger* under laboratory conditions at a temperature +29°C lose 66% of the initial strength 2–3 weeks after contamination, whereas inoculation by *Chaetomium globosum* induces 98.7% loss of strength, that is, completely destroys the material.

The same fabric exposed to soil at +29°C during 6 days loses 92% of the initial strength.

And cotton fabric exposed to seawater for 65 days loses up to 90% of strength.

A1   BAST FIBER BIODAMADING

Fibers produced from stalks, leaves or fruit covers of plants are called bast fibers. Hemp stalks give strong, coarse fibers – the hemp used for packing cloth and ropes. Coarse technical fibers: jute, ambary, ramie, and so forth, are produced from stalks of cognominal plants. Among all bast fibers, linen ones are most widely used.

The linen complex fiber, from which yarn and fabrics are manufactured, represents a batch of agglutinated filaments (plant cells) stretched and arrow-headed. The linen filament represents a plant cell with thick walls, narrow channel and knee-shaped nodes called shifts. Shifts are traces of fractures or bends of the fiber occurred during growth, and especially during mechanical treatment. Fiber ends are arrow-shaped, and the channel is closed. The cross-section represents an irregular polygon with five or six edges and a channel in the center. Coarser fibers have oval cross-section with wider and slightly flattened channel.

Complex fibers consist of filament batches (15–30 pieces in a batch) linked by middle lamellae. Middle lamellae consist of various substances: pectins, lignin, hemicellulose, and so forth.

Bast fibers contain a bit lower amount of cellulose (about 70%) than cotton ones. Moreover, they contain such components as lignin (10%), wax and trace amounts of antibiotics, some of which increase biostability of the fiber. The presence of lignins induces coarsening (lignifications) of plant cells that promotes the loss of softness, flexibility, elasticity, and increased friability of fibers.

The main method for fiber separation from the flax is microbiological one, in which vital activity of pectin degrading microorganisms degrade pectins linking bast batches to the stalk tissues. After that the fiber can be easily detached by mechanical processing.

Microorganisms affect straw either at its spreading directly at the farm that lasts 20–30 days or at its retting at a flax-processing plant where retting lasts 2–4 days.

In the case of spreading and retting of spread straw by atmospheric fallouts and dew under anaerobic conditions, the main role is played by microscopic fungi. According to data by foreign investigators, the following fungi are the most widespread at straw spreading: *Pullularia* (spires in the stalk bark); *Cladosporium* (forms a velvet taint of olive to dark green color); *Alternaria* (grows through the bark by a flexible colorless chain and unambiguously plays an important role at dew spreading).

The studies indicate that *Cladosporium* fungus is the most active degrader of flax straw pectins.

When retting linen at flax-processing plants, conditions different from spreading are created for microflora. Here flax is submerged to the liquid with low oxygen content due to its displacement from straws by the liquid and consumption by aerobic bacteria, which propagate on easily accessible nutrients extracted from the straw.

These conditions are favorable for multiplication of anaerobic, pectin degrading clostridia related to the group of soil spore bacteria, which includes just few species. Most of them are thermophiles and, therefore, the process takes 2–4 days in the warmed up water; however, at lower temperature (+15...20°C) it takes 10–15 days.

CURIOUS FACTS

In Russia and Check Republic, spreading is the most popular way of processing flax. In Poland, Romania and Hungary, the flax is processed at flax-processing plants by retting, and I Netherlands – by retting and partly by spreading.

The linen fiber obtained by different methods (spread or retted straw) has different spinning properties. The spread straw is now considered to be the best, where the main role in degradation of stalk pectins is played by mold fungi. In production of retted fiber, this role is played by pectin degrading bacteria, some strains of which being able to form an enzyme (cellulose) that degrades cellulose itself. Such impact may be one of the damaging factors in the processes of linen retting. Thus, biostability of the flax depends on the method of fiber production.

The studies show that all kinds of biological treatment increase the quantity of various microbial damages of the fiber. Meanwhile, the spread fiber had lower total number of microscopic damages compared with any other industrial method.

There are other methods for flax production, steaming, for example, that give steamed fiber. It has been found that steamed flax is the most biostable fiber. Possible reasons for so high biostability are high structure ordering of this fiber and high content of modified lignin in it. Moreover, during retting and spreading the fiber is enriched with microorganisms able to degrade cellulose under favorable conditions, whereas steaming sterilizes the fiber.

When exposed to microorganisms, pectins content in the linen fiber decreases by 38%, whereas cellulose content – by 1.2% only. The quantity of wax and ash content of the fiber exposed to microorganisms do not virtually change.

The ordered area share in the linen cellulose is 83.6%, the weakly ordered area – 5.1%, and disordered area – 15.7%. During microbiological degradation the share of disordered areas in the linen cellulose decreases to 7.8%, and the share of ordered areas increases to 86.9%. The share of weakly ordered areas varies insignificantly.

Microbiological damages of linen, jute and other bast fibers and fabrics are manifested by separate staining (occurrence of splotches of color or fiber darkening) and putrefactive odor. On damaged bast fibers, microscopic cross fractures and chips, and microholes and scabs in the fiber walls are observed.

The studies of relative biostability of bast fibers demonstrate that Manilla hemp and jute are most stable, whereas linen and cannabis fibers have the lowest stability.

Natural biostability of bast fibers is generally low and in high humidity and temperature conditions, when exposed to microorganisms, physicochemical and strength indices of both fibers and articles from them rapidly deteriorate. Generally, bast fibers are considered to have virtually the same biostability, as cotton fibers do.

Biostability of cellulose fivers is highly affected by further treatment with finishing solutions (sizing and finishing) containing starch, powder, resins and other substances which confer wearing capacity, wrinkle resistance, fire endurance, and so forth, to textile materials. Many of these substances represent a good culture medium for microorganisms. Therefore, at the stage of yarn and fabric sizing and finishing, the main attention is paid to strict compliance with sanitary and technological measures, which are to prevent fabric infection, by microorganisms and further biodamaging.

A2   BIODAMAGING OF ARTIFICIAL FIBERS

Artificial fibers and fabrics are produced by chemical treatment of natural cellulose obtained from spruce, pine tree and fir. Artificial fibers based on cellulose are viscose, acetate, and so forth, ones. These fibers obtained from natural raw material have higher amorphous structure as compared with high-molecular natural material and, therefore, have lower stability, higher moisture and swelling capacity.

By chemical structure and microbiological stability viscose fibers are similar to common cotton fibers. Biostability of these fibers is low: many cellulosolytic microorganisms are capable of degrading them. Under laboratory conditions, some species of mold fungi shortly (within a month) induces complete degradation of viscose fibers, whereas wool fibers under the same conditions preserve up to 50% of initial stability. For viscose fabrics, the loss of stability induced by soil microorganisms during 12–14 days gives 54 to 76%. These parameters of artificial fibers and fabrics are somewhat higher than for cotton.

Acetate fibers are produced from acetyl cellulose – the product of cellulose etherification by acetic anhydride. Their properties significantly differ from those of viscose fibers and more resemble artificial fibers. For instance, they possess lower moisture retaining property, lesser swelling and loss of strength under wet condition. They are more stable to damaging effect of cellulosolytic enzymes of bacteria and microscopic fungi, because contrary to common cellulose fibers possessing side hydroxyl groups in macromolecules, acetate fiber macromolecules have side acetate groups hindering interaction of macromolecules with enzymes.

Among artificial textile materials of the new generation, textile fibers from bamboo, primarily obtained by Japanese, are highlighted. Bamboo possesses the reference antimicrobial properties due to the presence of "bambocane" substance in the fiber. Bamboo fivers possess extremely porous structure that makes them much more hygroscopic that cotton. Clothes from bamboo fibers struggle against sweat secretion – moisture is immediately absorbed and evaporated by fabric due to presence of pores, and high antimicrobial properties of bamboo prevent perspiration odor.

Various modified viscose fibers, micromodal and modal, for example, produced from beech were not studies for biostability. Information on biostability of artificial fibers produced from lactic casein, soybean protein, maize, peanut and corn is absent.

A3 WOOL FIBER BIODAMAGING

By wool the animal hair is called, widely used in textile and light industry. The structure and chemical composition of the wool fiber significantly differ it from other types of fibers and shows great variety and heterogeneity of properties. Sheep, camel, goat and rabbit wool is used as the raw material.

After thorough cleaning, the wool fiber can be considered virtually consisting of a single protein – keratin. The wool contains the following elements (in %): carbon – 50; hydrogen – 6–7; nitrogen – 15–21; oxygen 21–24; sulfur – 2–5, and other elements.

The chemical feature of wool is high content of various amino acids. It is known that wool is a copolymer of, at least, 17 amino acids, whereas the most of synthetic fibers represent copolymers of two monomers.

Different content of amino acids in wool fibers promotes the features of their chemical properties. Of the great importance is the quantity of cystine containing virtually all sulfur, which is extremely important for the wool fiber properties. The higher sulfur content in the wool is, the better its processing properties are, the higher resistance to chemical and other impacts is and the higher physic-mechanical properties are.

Wool fiber layers, in turn, differ by the sulfur content: it is higher in the cortical layer that in the core.

Among all textile fibers, wool has the most complex structure.

The fine merino wool fiber consists of two layers: external flaky layer or cuticle and internal cortical layer – the cortex. Coarser fibers have the third layer – the core.

The cuticle consists of flattened cells overlapping one another (the flakes) and tightly linked to one another and the cortical layer inside.

Cuticular cells have a membrane, the so-called epicuticle, right around. It is found that epicuticle gives about 2% of the fiber mass. Cuticle cells limited by walls quite tightly adjoin one another, but, nevertheless, there is a thin layer of intercellular protein substance between them, which mass is 3–4% of the fiber mass.

The cortical layer, the cortex, is located under the cuticle and forms the main mass of the fiber and, consequently, defines basic physicomechanical and many other properties of the wool. Cortex is composed of spindle-shaped cells connive to one another. Protein substance is also located between the cells.

The cortical layer cells are composed of densely located cylindrical, thread-like macrofibrils of about 0.05–0.2 µm in diameter. Macrofibrils of the cortical layer are composed of microfibrils with the average diameter of 7–7.5 nm.

Microfibrils, sometimes call the secondary agents, are composed of primary aggregates – protofibrils. Protofibril represents two or three twisted α-spiral chains.

It is suggested [27, 28] that α-spirals are twisted due to periodic repetition of amino acid residues in the chain, hence, side radicals of the same spiral are disposed in the inner space of another α-spiral providing strong interaction, including for the account of hydrophobic bonds, because each seventh residue has a hydrophobic radical.

According to the data by English investigator J.D. Leeder, the wool fiber can be considered as a collection of flaky and cortical cells bound by a cell membrane complex (CMC) which thus forms a uniform continuous phase in the keratin substance of the fiber. This intercellular cement can easily be chemically and microbiologically degraded, that is, a δ-layer about 15 nm thick (CMC or intercellular cement) is located between cells filling in all gaps.

The studies show that the composition of intercellular material between flaky cells may differ from that of the material between cortical cells. In the cuticle-cuticle, cuticle-cortex and cortex-cortex complexes the intercellular "cement" has different chemical compositions.

Although the cell membrane complex gives only 6% of the wool fiber mass, there are proofs that it causes the main effect on many properties of the fiber and fabric. For instance, a suggestion was made that cmC components may affect such mechanical properties, as wear resistance and torsion fatigue, as well as such chemical properties, as resistance to acids, proteolytic enzymes and chemical finishing agents.

The core layer is present in the fibers of coarser wool with the core cell content up to 15%. Disposition and shape of the core layer cells significantly vary with respect to the fiber type. This layer can be continuous (along the whole fiber) or may be separated in sections. The cell carcass of the core layer is composed of protein similar to microfibril cortex protein.

By its chemical composition, wool is a protein substance. The main substance forming wool is keratin – a complex protein containing much sulfur in contrast with other proteins. Keratin is produced during amino acid biosynthesis in the hair bag epidermis in the hide. Keratin structure represents a complex of high-molecular chain batches interacting both laterally and transversally. Along with keratin, wool contains lower amounts of other substances.

Wool keratin reactivity is defined by its primary, secondary and tertiary structures, that is, the structure of the main polypeptide chains, the nature of side radicals and the presence of cross bonds.

Among all amino acids, only cystine forms cross bonds; their presence considerably defines wool insolubility in many reagents. Cystine bond decomposition simplifies wool damaging by sunlight, oxidants and other agents. Cystine contains almost all sulfur present in the wool fibers. Sulfur is very important for the wool quality, because it improves chemical properties, strength and elasticity of fibers.

Along with general regularities in the structure of high-molecular compounds, fibers differ from one another by chemical composition, monomer structure, polymerization degree, orientation, intermolecular bond strength and type, and so forth, that defines different physicomechanical and chemical properties of the fibers.

The main chemical component of wool – keratin, is a nutrition for microorganisms.

Microorganisms may not directly consume proteins. Therefore, they are only consumed by microbes having proteolytic enzymes – exoproteases that are excreted by cells to the environment.

Wool damage may start already before sheep shearing, that is, in the fleece, where favorable nutritive (sebaceous matters, wax, epithelium), temperature, aeration and humidity conditions are formed.

Contrary to microorganisms damaging plant fibers, the wool microflora is versatile, generally represented by species typical of the soil and degrading plant residues.

Initiated in the fleece, wool fiber damages are intensified during its storage, processing and transportation under unfavorable conditions.

Specific epiphytic microflora typical of this particular fiber is always present on its surface. Representatives of this microflora excrete proteolyric enzymes (mostly pepsin), which induce hydrolytic keratin decay by polypeptide bonds to separate amino acids.

Wool is degraded in several stages: first, microorganisms destroy the flaky layer and then penetrate into the cortical layer of the fiber, although the cortical layer itself is not destroyed, because intercellular substance located between the cells is the culture medium. As a result, the fiber structure is disturbed: flakes and cells are not bound yet, the fiber cracks and decays.

The mechanism of wool fiber hydrolysis by microorganisms suggested by American scientist E. Race represents a sequence of transformations: proteins – peptones – polypeptides – water + ammonia + carboxylic acids.

The most active bacteria: *Alkaligenes bookeri, Pseudomonas aeroginosa, Proteus vulgaris, Bacillus agri, B. mycoides, B. mesentericus, B. megatherium, B. subtilis, and microscopic fungi: Aspergillus, Alternaria, Cephalothecium, De-*

matium, Fusarium, Oospora, Penicillium, Trichoderma, were extracted from the wool fiber surface.

However, the dominant role in the wool degradation is played by bacteria. Fungi are less active in degrading wool. Consuming fat and dermal excretion, fungi create conditions for further vital activity of bacteria-degraders. The role of microscopic fungi may also be reduced to splitting the ends of fibers resulting mechanical efforts of growing hyphae. Such splitting allows bacteria to penetrate into the fiber. Fungi weakly use wool as the source of carbon.

In 1960 s, the data on the effect of fat and dirt present on the surface of unclean fibers on the wool biodamaging were published. It is found that unclean wool is damaged much faster than clean one. The presence of fats on unclean wool promotes fungal microflora development.

The activity of microbiological processes developing on the wool depends on mechanical damages of the fiber and preliminary processing of the wool.

It is found that microorganism penetration may happen through fiber cuts or microcracks in the flaky layer. Cracks may be of different origins – mechanical, chemical, and so forth. It is also found that wool subject to intensive mechanical or chemical treatment is easier degraded by microorganisms than untreated one.

For instance, high activity of microorganisms during wool bleaching by hydrogen peroxide in the presence of alkaline agents and on wool washed in the alkaline medium was observed. When wool is treated in a weak acid medium, the activity of microorganisms is abruptly suppressed. This also takes place on the wool colored by chrome and metal-containing dyes. The middle activity of microorganisms is observed on the wool colored by acid dyes.

When impacted by microorganisms, structural changes in the wool are observed: flaky layer damages, its complete exfoliation, and lamination of the cortical layer.

Wool fiber damages can be reduced to several generalized types provided by their structural features:

- channeling and overgrowth – accumulation of bacteria or fungal hyphae and their metabolites on the fiber surface;
- flaky layer damage, local and spread;
- cortical layer lamination to spindle-shaped cells;
- spindle-shaped cell destruction.

Along with the fiber structure damage, some bacteria and fungi decrease its quality by making wool dirty blue or green that may not be removed by water or detergents. Splotches of color also occur on wool, for example, due to the impact of *Pseudomonas aeruginosa bacteria; in this case, color depends on medium pH: green splotches occurred in a weakly alkaline medium, and in weak acid medium they are red. Green splotches may also be caused by development of Dermatophilus congolensis* fungi. Black color of wool is provided by *Pyronellaea glomerata* fungi.

Thus, wool damage reduces its strength, increases waste quantity at combing and imparts undesirable blue, green or dirty color and putrefactive odor.

However, wool is degraded by microorganisms slower than plant fibers.

A4   CHANGES OF STRUCTURE AND PROPERTIES OF WOOL FIBERS BY MICROORGANISMS

To evaluate bacterial contamination of wool fibers, it is suggested to use an index suggested by A.I. Sapozhnikova, which characterizes discoloration rate of resazurin solution, a weak organic dye and currently hydrogen acceptor. It is also indicator of both presence and activity of reductase enzyme.

The method is based on resazurin ability to lose color in the presence of reductase, which is microorganisms' metabolite, due to redox reaction proceeding. This enables judging about quantity of active microorganisms present in the studied objects by solution discoloration degree.

Discoloration of the dye solution was evaluated both visually and spectrophotometrically by optical density value.

Table 1 shows results of visual observations of color transitions and optical density measurements of incubation solutions after posing the reductase test.

As follows from the data obtained, coloration of water extracts smoothly changed from blue-purple for control sterile physiological solution (D = 0.889) to purple for initial wool samples (D_{thin} = 0.821 and D_{coarse} = 0.779), crimson (D_{thin} = 0.657 and D_{coarse} = 0.651) and light crimson at high bacterial contamination (D_{thin} = 0.548 and D_{coarse} = 0.449 and 0.328) depending on bacterial content of the fibers.

TABLE A1   Visual coloration and optical density of incubation solutions with wool fibers at the wavelength λ = 600 nm and different stages of spontaneous microflora development.

Time of microorganism development, days	Thin merino wool		Coarse caracul wool	
	Optical density	Color (visual assessment)	Optical density	Color (visual assessment)
Control, physiological saline	0.889	Blue purple	0.889	Blue purple
0 (init.)	0.821	Purple	0.779	Purple
7	0.712	Purple	0.657	Crimson
14	0.651	Crimson	0.449	Light crimson
28	0.548	Light crimson	0.328	Light crimson

This dependence can be used for evaluation of bacterial contamination degree for wool samples applying color standard scale.

It is found that the impact of microorganisms usually reduces fiber strength, especially for coarse caracul wool: after 28 days of impact strength decreased by 57–65%. The average rate of strength reduction is about 2% per day (Fig. A8).

It is found that after 28 days of exposure, the highest reduction of breaking load of wool fibers is induced by *Bac. subtilis* bacteria.

Figure A9 clearly shows that 14 days after exposure to microorganisms the surface of coarse wool fiber is almost completely covered by bacterial cells. Meanwhile, note also (Fig. A10) that cuticular cells themselves are not damaged, but their bonding is disturbed that grants access to cortical cells for microorganisms. Figure A11 shows the wool fiber decay to separate fibrils caused by microorganisms.

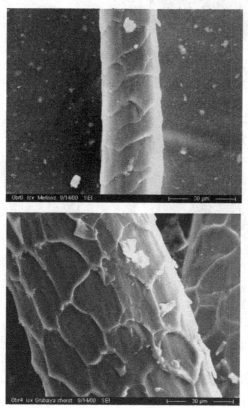

FIGURE A8 Micrographs of initial wool fibers (×1000): a – thin merino wool; *b* – coarse caracul wool

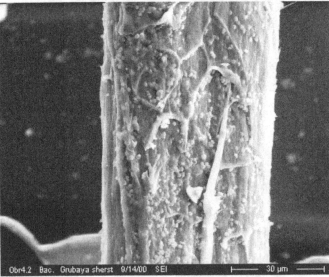

FIGURE A9 Micrographs of thin merino wool (a) and coarse caracul wool (*b*) after 14 days of exposure to *Bac. subtilis* (×1000)

FIGURE A10   Micrographs of flaky layer destruction of thin (a) and coarse (b) wool fibers after 14 days of exposure to spontaneous microflora (×1000)

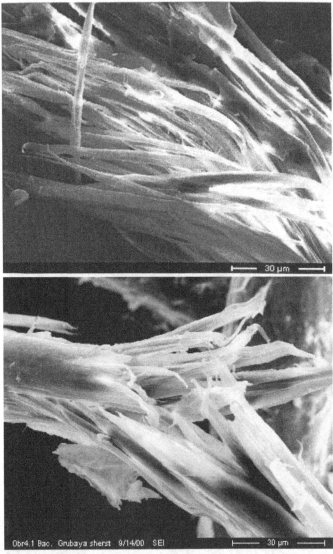

FIGURE A11   Micrographs of thin (a) and coarse (b) wool fiber fibrillation after 28 days of exposure to microorganisms (×1000)

Wool fiber biodegradation changes the important quality indices, such as whiteness and yellowness. This process can be characterized as "yellowing" of wool fibers.

Table A2 shows data obtained on yellowness of thin merino and coarse caracul wool fibers exposed to spontaneous microflora, *Bac. subtilis* bacteria and *Asp. niger* fungi during 7, 14 and 28 days.

TABLE A2   Yellowness of thin merino and coarse caracul wool fibers after different times of exposure to different microorganisms.

Time of exposure to microorganisms, days Type of fibers	Yellowness, %		
	Bac. subtilis	*Asp. niger*	**Spontaneous microflora**
Thin, 0	27.7	27.7	27.7
Thin, 7	36.4	37.3	29.1
Thin, 14	45.4	42.5	34.8
Thin, 28	53.9	48.1	39.9
Coarse 0	39.3	39.3	39.3
Coarse 7	46.8	44.2	41.2
Coarse 14	51.8	46.4	42.1
Coarse 28	57.1	49.5	43.3

In terms of detection of biodegradation mechanism and changes of material properties, determination of relations between changes in properties and structure of fibers affected by microorganisms is of the highest importance. Yellowness increase of wool fibers affected by microorganisms testifies occurrence of additional coloring centers.

For the purpose of elucidating the mechanism of microorganisms' action on the wool fibers that leads to significant changes of properties and structure of the material, the changes of amino acid composition of wool keratin proteins being the nutrition source of microorganisms damaging the material are studied.

Exposure to microorganisms during 28 days leads to wool fiber degradation and, consequently, to noticeable mass reduction of all amino acids in the fiber composition. To the greatest extent, these changes are observed for coarse wool, where total quantity of amino acids is reduced by 10–12 rel. %, and for merino wool slightly lower reduction (4.7 rel. %) is observed.

It should be noted that at comparatively low reduction of the average amount of amino acids in the system (not more than 12 rel. %) all types of wool fibers demonstrate a significant reduction of the quantity of some amino acids, such as serine, cystine, methionine, and so forth (up to 25–33 rel. %).

Analysis of the data obtained testifies that in all types of wool fibers (but to different extent) reduction of quantity of amino acids with disulfide bonds (cystine, methionine) and ones related to polar (hydrophilic) amino acids, including serine, glycine, threonine and tyrosine, is observed. These very amino acids provide hydrogen bonds imparting stability to keratin structure.

In the primary structure of keratin, serine is the N-end group, and Tyrosine is C-end group. In this connection, reduction of the quantity of these amino acids testifies degradation of the primary structure of the protein.

Changes observed in the amino acid composition of wool fiber proteins exposed to spontaneous microflora may testify that microorganisms degrade peptide and disulfide bonds, which provide stability of the primary structure of proteins, and break hydrogen bonds, which play the main role in stabilization of spatial structure of proteins (secondary, tertiary and quaternary).

Very important data were obtained in the study of the wool fiber structure by IR-spectroscopy method. It is found that when microorganisms affect wool fibers, their surface layers demonstrate increasing quantity of hydroxyl groups that indicates accumulation of functional COO-groups, nitrogen concentration in keratin molecule decreases, and protein chain configuration partly changes – β-configuration (stretched chains) transits to α-configuration (a spiral). This transition depends on α- and β-forms ratio in the initial fiber and is more significant for thin fibers, which mostly have β-configuration of chains in the initial fibers.

Thus, it is found that microorganisms generally affect the cell membrane complex and degrade amino acids, such as cystine, methionine, serine, glycine, threonine and tyrosine. Microorganisms reduce breaking load and causes "yellowing" of the wool fibers (Fig. A12).

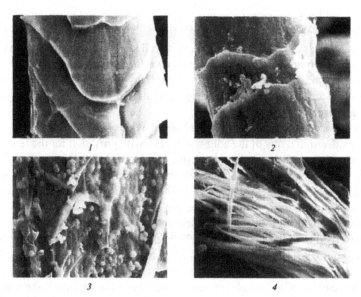

FIGURE A12   Micrographs of wool fibers: 1 – initial undamaged fiber (× 3000); 2, 3 – bacterial cells on the fiber surface (× 3000); 4 – fiber fibrillation after exposure to microorganisms during 4 weeks (× 1000)

A5   BIODAMAGES OF SYNTHETIC FIBERS

Synthetic fibers are principally different from natural and artificial ones by structure and, being an alien substrate for microorganisms, are harder damaged by them. Since occurrence of synthetic fabrics in 1950 s, it is suggested that they are "everlasting" and are not used by microorganisms. However, it has been found with time that, firstly, microorganisms although slower, but yet are capable of colonizing synthetic fabrics and using their carbon in the course of development (i.e. causing biodamage), and secondly, there are both more and less microorganism resistant fabrics among synthetic ones.

Among microorganisms damaging synthetic fibers, Trichoderma genus fungi are identified, at the initial stages developing due to lubricants and finishing agents without fiber damage and then wrap them with mycelium, loosen threads and, hence, reduce fabric strength.

When studying fabrics from nitrone, lavsan, caprone, it has been found that soil fungi and bacteria cause roughly the same effect on characteristics of these fabrics increasing the fiber swelling degree by 20–25%, reducing strength by 10–15% and elongation at break by 15–20%.

Synthetic fibers represent potential source of energy and nutrition for microorganisms. The ability of microorganisms to attach to surfaces of insoluble solids, then using them as the nutritive substrate, is well known. Living cells of microorganisms have complex structure, just on the surface of bacterial cells complexes of proteins, lipids and polysaccharides were found; it contains hydrophilic and hydrophobic areas, various functional groups and mosaic electric charge (at total negative charge of the cells).

The first stage of microorganism interaction with synthetic fibers can be rightfully considered in terms of the adhesion theory with provision for the features of structure and properties of microorganisms as a biological system.

The entire process of microorganism impact of the fiber can conditionally be divided into several stages: attachment to the fiber, growth and multiplication on it and consumption of it, as the nutrition and energy source.

Enzymes excreted by bacteria act just in the vicinity of bacterial membrane. Been adsorbed onto the fiber, living cells attach to the surface and adapt to new living conditions. The ability to be adsorbed onto the surface of synthetic fibers is caused by:

- the features of chemical structure of the fibers. For instance, fibers adsorbing microorganisms are polyamide and polyvinyl alcohol ones; the fiber not adsorbing microorganisms is, for example, ftorin;
- physical structure of the fiber. For example, fibers with smaller linear density, with a lubricant on the surface absorb greater amount of microorganisms;
- the presence of electric charge on the surface, its value and sign. Positively charged chemical fibers adsorb virtually all bacteria, fibers having no electric charge adsorb the majority of bacteria, and negatively charged fibers do not adsorb bacteria.

Supermolecular structure also stipulates the possibility for microorganisms and their metabolites to diffuse inside the internal areas of the fiber. Microorganism assimilation of the fiber starts from the surface, and further degradation processes and their rate are determined by microphysical state of the fiber. Microorganism metabolite penetration into inner areas of the fiber and deep layers of a crystalline material is only possible in the presence of capillaries.

Chemical fiber damaging and degradation starting from the surface are, in many instances, promoted by defects like cracks, chips or hollows which may occur in the course of fiber production and finishing.

Along with physical inhomogeneity, chemical inhomogeneity may promote biodegradation of synthetic fibers. Chemical inhomogeneity occurs during polymer synthesis and its thermal treatments, manifesting itself in different content of monomers and various end groups. The possibility for microorganism metabolites to penetrate inside the structure of synthetic fibers depends on the quantity and accessibility of functional end groups in the polymer, which are abundant in oligomers.

The ability to synthetic fibers to swell also makes penetration of biological agents inside low ordered areas of fibers and weakens intermolecular interactions, off-orientation of macromolecules, and degradation in the amorphous and crystalline zones. Structural changes result in reduction of strength properties of fibers.

Theoretical statements that synthetic fibers with the lower ordered structure and higher content of oligomers possess lower stability to microorganism impact than fibers with highly organized structure and lower content of low-molecular compounds.

Thus, the most rapid occurrence and biodegradation of synthetic fibers are promoted by low ordering and low orientation of macromolecules in the fibers, their low density, low crystallinity and the presence of defects in macro and microstructure of the fibers, pores and cavities in their internal zones.

Carbochain polymer based fibers are higher resistant to microbiological damages. These polymers are: polyolefins, polyvinyl chloride, polyvinyl fluoride, polyacrylonitrile, polyvinyl alcohol. Fibers based on heterochain polymers: polyamide, polyether, polyurethane, and so forth, are less bioresistant.

Comparative soil tests for biostability of artificial and synthetic fibers demonstrate that viscose fiber is completely destroyed on the 17th day of tests; bacterium and fungus colonies occur on lavsan on the 20th day; caprone is overgrown by fungus mycelium on the 30th day. Chlorin and ftorlon have the highest biostability. The initial signs of their biodamage are only observed 3 months after the test initiation.

The studies of nitron, lavsan and capron fabric biostability have found that soil fungi and bacteria cause nearly equal influence on parameters of these fabrics, increasing swelling degree of the fibers by 20–25%, reducing strength by 10–15% and elongation at break by 15–20%. Meanwhile, nitron demonstrated higher biostability, as compared with lavsan and capron.

A6   CHANGES IN STRUCTURE AND PROPERTIES OF POLYAMIDE FIBERS INDUCED BY MICROORGANISMS

In contrast with natural fibers, chemical fibers have no permanent and particular microflora. Therefore, the most widespread species of microorganisms possessing increased adaptability are the main biodegraders of these materials.

Occurrence and progression of biological degradation of polyamide fibers is, in many instances, is induced by their properties and properties of affecting microorganisms, and their species composition. Generally, the species of microorganisms degrading polyamide and other chemical fibers are determined by their operation conditions, which form microflora, and its adaptive abilities.

Polyamide fibers are most frequently used in mixtures with natural fibers. Natural fibers contain specific microflora on the surface and inside. Therefore, capron fibers mixed with cotton, wool or linen are affected by their microflora. It is found that capron fiber degradation by microorganisms obtained from wool is characterized as deep fiber decay; microorganisms extracted from natural silk cause streakiness of capron fibers; microorganisms extracted from cotton cause fading and decomposition; microorganisms extracted from linen cause fading, streakiness and decomposition.

The microorganism interaction with polyamide fibers is most fully studied in the works by scientists. For the purpose of detecting bacteria-degraders of polyamide fibers, microorganisms were extracted from fibers damaged in the medium of active sewage silt, soil, microflora of natural fibers and test-bacteria complex selected as degraders of polyamide materials. Capron fibers were inoculated by extracted cultures of microorganisms and types of damages were reproduced.

Polyamide fiber materials were natural nutritive and energy source for these microorganisms. Therefore, bacterial strains extracted from damaged fibers were different from the initial strains. It is proven that the existence on a new substrate has stipulated changes of intensity and direction of physiological-biochemical processes of bacterial cells, the change of their morphological and culture properties.

Extraction, cultivation and use of such adaptive strains are of both scientific and practical interest. Using bacterial strains adaptive to capron, polyamide production waste, warn products, toxic substances may be used. This allows obtaining of secondary raw materials and solving the problem of environmental protection.

It is proved experimentally that polycaproamide fibers possess high adsorbability, which value depends on the properties of impacting bacteria. For instance, gram-positive, especially sporeforming bacteria *Bacillus subtilis, Bac. mesentericus* (from 84.5 to 99.3% of living cells), are most highly adsorbed, and adsorption of gram-positive bacteria varies significantly.

The extensive research of test-bacterium complex impact was performed. These bacteria were chosen as degraders of polyamide fibers, along with microflora of active sewage silts, linen and jute microorganisms as degraders of a complex capron thread. It is found that the test-bacterium complex injected by the author, after 7 months of exposure, increases biodegradation index to 1.27 that testifies about intensive degradation of the fiber microstructure. The highest degradation of complex polycaproamide (PCA) thread is caused by from active sewage silt microorganisms.

High activity of silt microorganisms and test-bacteria is explained by the fact that they include bacterium strains *Bacillus subtilis* and *Bac. mesentericus*, which according to the data by a number of authors may induce full degradation of caprolactam to amino acids using it, as the source of carbon and nitrogen.

Thus, adaptive forms of microorganisms induce the highest degradation of polyamide fibers.

Polyamide fibers are characterized by physical structure inhomogeneity, which occurs during processing and is associated with differences in crystallinity and orientation of macromolecules determining fiber accessibility for microorganisms and their metabolites penetration. The surface layer is damaged during orientational stretching and, consequently, has lower molecular alignment. That is why the surface layer is most intensively changed by microorganisms. The study of polyamide fiber macrostructure after exposure to microorganisms shows that streakiness and cover damage are the main damages of these fibers.

The studies of supermolecular structure of capron fiber surface show that after exposure to microorganisms the capron fiber cover becomes loose and uneven [39, 43]. Surface supermolecular structure degradation increases with the microbial impact: fibrils and their yarns become split and misaligned both laterally and transversely, multiple defects in the form of pores and cavities, and cracks of various depths are formed.

Chemical inhomogeneity of polyamide fibers also promotes changes in the fiber structure, when exposed to microorganisms. It is found that the polyamide fiber degradation increases with low-molecular compound (LMC) content in them; meanwhile, at the same content of low-molecular compounds, thermally treated fibers were higher biostable, as compared with untreated specimens.

Along with the morphological characters, which characterize biodamaging of the fibers, functional features, such as strength decrease and increase of deformation properties of the fiber, were detected. The greatest strength decrease (by 46.4%) was observed for thermally untreated capron fiber containing 3.4% LMC, and the smallest decrease (by 5%) was observed for the fibers with 3.2% LMC, thermally treated at the optimum time of 5 s.

IR-spectroscopy method was applied to detect polycaproamide fiber damage by microorganisms [39]. It is found that carboxyl and amide groups are accumulated during their biodegradation.

The change of various property indices of polyamide fibers also results from macro, micro and chemical changes.

Of interest are studies of the microorganism effect on polyamide fabric quality. Fabrics (both bleached and colored) from capron monofilament were exposed to a set of test-cultures: *Bac.subtilis, Ps.fluorecsens, Ps.herbicola, Bac.mesentericus.* After 3–9 month exposure, yellowness and dark spots occurred, coloring intensity decreased, and an odor appeared. The optical microscopy studies indicated that all fibers exposed to microorganisms, had damages typical of synthetic fibers – overgrowing, streakiness, bubbles, wall damages. The increasing quantity of biodamages with time results in tensile strength reduction: by 6–8% for capron fibers after 9 months, at inconsiderable change of relative elongation.

It all goes to show that development of microorganisms on polyamide fibrous materials results in changes of fibers morphology, their molecular and supermolecular structure and, as a consequence, reduction of strength properties, color change, and odor.

To clear up the mechanism of polycaproamide fiber degradation by polarographic investigation, a possibility of ε-amino caproic acid accumulation by *Bacillus subtilis k1* culture during degradation of polycaproamide fibers 0.3 and 0.7 tex fineness was studied (fig.A13).

As a source of carbon, ε-amino caproic acid (10 mg/l) or PCA (0.5 g/l) was injected into the mineral medium.

It is found that at PCA fibrous materials exposure to *Bacillus subtilis k1* strain, the maximum quantity of ε-amino caproic acid is liberated on the fifth day: 32 mg/l, 66 mg/l.

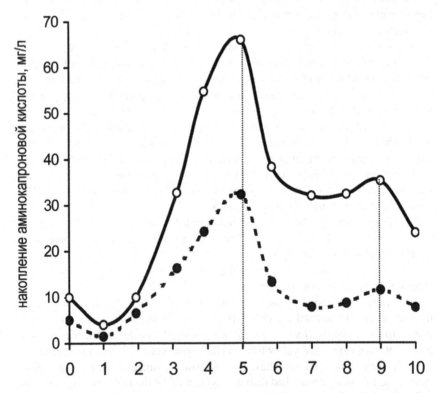

FIGURE A13   The change of ε-amino caproic acid (ACA) concentration during *Bacillus subtilis k1* development on PCA material fibers: – 0.7 tex; – 0.3 tex.

It is known that if the medium includes several substrates metabolized by a particular strain of microorganisms, the substrate providing the maximum culture propagation rate is consumed first. As this substrate is going to exhaust, bacteria subsequently consume other substrates, which provide lower rates of cell multiplication.

In the mineral medium containing chemically pure ε-amino caproic acid (with the initial concentration of 10 mg/L) as the source of carbon, its concentration decreased gradually, and 5 days after it was not detected in the solution.

Thus, basing on the data obtained, one may conclude that, firstly, PCA fibers of 0.3 tex fineness are more accessible for microorganisms; secondly, *Bacillus subtilis k1* strain can be used to use polycaproamide fibrous materials; thirdly, polarographic analysis has proven the mechanism of PCA fibrous material degradation with ε-amino caproic acid liberation. As a consequence, suggested *Bacillus subtilis k1* strain VKM No.V-1676D degrades PCA fibrous materials at the both macro and microstructure levels, with ε-amino caproic acid formation.

A7 METHODS OF TEXTILE MATERIAL PROTECTION AGAINST DAMAGING BY MICROORGANISMS

Imparting antimicrobial properties to textile materials pursues two main aims: protection of the objects contacting with textile materials against actions of microorganisms and pathogenic microflora.

In the first case, we speak about imparting biostability to materials and, as a consequence, about passive protection. The second case concerns creation of conditions for preventive attack of a textile material on pathogenic bacteria and fungi to prevent their impact on the protected object.

The basic method of increasing biostability of textile materials is application of antimicrobial agents (biocides). The requirements to the "ideal" biocide are the following:

• The efficacy against the most widespread microorganisms at minimal concentration and maximal action time;
• Non-toxicity of applied concentrations for people;
• The absence of color and odor;
• Low price and ease of application;
• Retaining of physicomechanical, hygienic and other properties of the product;
• Compatibility with other finishing agents and textile auxiliaries;
• light stability and weather ability.

At any time, nearly every class of chemical compounds was applied to impart textile materials antibacterial or antifungal activity. Today, application of nanotechnologies, specifically injection of silver and iodine nanoparticles, to impart textile antimicrobial properties is of the greatest prospect.

At all times, copper, silver, tin, mercury, and so forth, salts were used to protect fibrous materials against biodamages. Among these biocides, the most widespread are copper salts due to their low cost and comparatively toxicity. The use of zinc salts is limited by their low biocide action, whereas mercury, tin and arsenic salts are highly toxic for the man. However, there are organomercury preparations applied to synthetic and natural fibrous materials used as linings and shoe plates, widely advertised for antibacterial and antifungal finishing.

The data are cited that impregnation of textile materials by a mixture of neomycin with tartaric, propionic, stearic, phthalic and some other acids imparts them the bacteriostatic effect. Acids were dissolved in water, methyl alcohol or butyl alcohol and were sprayed on the material.

The methods of imparting textile materials biostability can be divided into the following groups:

- impregnation by biocides, chemical and physical modification of fibers and threads, which then form a textile material;
- cloth impregnation by antimicrobial agent solutions of emulsions, its chemical modification;
- injection of antimicrobial agents into the binder (at nonwoven material manufacture by the chemical method);
- imparting antimicrobial properties to textile materials during their coloring and finishing;
- application of disinfectants during chemical cleaning or laundry of textile products.

However, impregnation of fibers and cloth does not provide firm attachment of reagents. As a result, the antimicrobial action of such materials is nondurable. The most effective methods of imparting biocide properties to textile materials are those providing chemical bond formation, that is, chemical modification methods. Chemical modification methods for fibrous materials represent processing that leads to clathrate formation, for example, injection of biological active agents into spinning melts or solutions.

At the stage of capron polymerization, an antibacterial organotin compound (tributyltin oxide or hydroxide) is added that retains the antibacterial effect after multiple laundries. Methods of imparting antimicrobial properties to textile materials by injection of nitrofuran compounds into spinning melts with further fixing them at molding in the fine structure of fibers similar to clathrates were designed.

There are data on imparting antimicrobial properties to synthetic materials during oiling. Prior to drafting, fibers are treated by compounds based on oxyquinoline derivatives, by aromatic amines or nitrofuran derivatives. Such fibers possess durable antimicrobial effect.

Nanotechnologies are actively intruded in the light industry, allowing obtaining of materials with antimicrobial properties. The following directions using nanotechnologies, which are now investigated, should be outlined, including creation of new textile materials on the account of:
primarily, the use of textile nonfibers and threads in the materials;
secondly, the use of nanodispersions and nanoemulsions for textile finishing.

It may be said that nanotechnologies allow a significant decrease of expenses at the main production stage, where consumption of raw materials and semiproducts is considerable. For nanoparticles imparting antimicrobial properties, silver, copper, palladium, and so forth, particles are widely used. Silver is the natural antimicrobial agent, which properties are intensified by the nanoscale of particles (the surface area sharply increases), so that such textile is able to kill multiple microorganisms and viruses. In this form, silver also reduces necessity in fabric cleaning, eliminates sweat odor as a result of microorganism development on the human body during wearing.

Properties of materials designed with the use of silver nanoparticles, which prevent multiplication of various microorganisms, may be useful in medicine, for example. The examples are surgical retention sutures, bandages, plasters, surgical boots, medical masks, whites, skullcaps, towels, and so forth. Customers know well sports clothes, prophylactic socks with antimicrobial properties.

Along with chemical fibers and threads, natural ones are treated by nanoparticles. For example, silver and palladium nanoparticles (5–20 nm in diameter) were synthesized in citric acid, which prevented their agglutination, and then natural fibers were dipped in the solution with these negatively charged particles. Nanoparticles imparted antibacterial properties and even ability to purify air from pollutants and allergens to clothes and underwear.

When these products appeared at the world market, disputes about ecological properties and the influence of these technologies on the human organism have arisen. There are no accurate data yet how these developments may affect the human organism. However, it should be noted that some specialists do not recommend everyday use of antibacterial socks, because these antibacterial properties affect the natural skin microflora.

Nanomaterials are primarily hazardous due to their microscopic size. Firstly, owing to small size they are chemically more active because of a great total area of the nanosubstance. As a result, low toxic substance may become extremely toxic. Secondly, chemical properties of the nanosubstance may significantly change due to manifestations of quantum effects that, finally, may make a safe substance extremely hazardous. Thirdly, due to small size, nanoparticles freely permeate through cellular membranes damaging bioplasts and disturbing the cell operation.

Physical modification of fibers or threads is the direct change of their composition (without new chemical formations and transformations), structure (supermolecular and textile), properties, production technology and processing. Modernization of the structure and increase of the fiber crystallinity degree induces biostability increase. However, in contrast with chemical modification, physical modification does not impart antimicrobial properties to the fibers, but may increase biostability.

By no means always textile materials produced completely from antimicrobial fibers are required. Even a small fracture of highly active antimicrobial fiber (e.g., 1/3 or even 1/4) is able to provide sufficient biostability to the entire material. The studies show that antimicrobial fibers were found not only protected themselves against microorganism damage, but also capable of shielding plant fibers from their impact.

Manufacture of antimicrobial nonwoven materials by injection of active microcapsule ingredients into it is of interest. Microcapsules can contain solid particles of microdrops of antimicrobial substances liberated under particular conditions (e.g., by friction, pressure, dissolution of capsule coatings or their biodegradation).

Biostability of fibrous materials may be significantly affected by the dye selection. Dyes possessing antimicrobial activity on the fiber are known: salicylic acid derivatives capable of bonding copper, triphenylmethane, acridic, thiazonic, and so forth, ones. For instance, chromium-containing dyes possess antibacterial action, but resistance to mold fungi is not imparted.

It is known that synthetic fibers dyed by dispersed pigments are more intensively degraded by microorganisms. It is suggested that these pigments make the fiber surface more accessible for bacteria and fungi.

Single bath coloring and bioprotective finishing of textile materials are also applied. A combination of these processes is not only of theoretical interest, but is also perspective in terms of technology and economy.

Processing of textile materials by silicones also imparts antimicrobial properties to these clothes. Some authors state that textile material sizing by water repellents imparts them sufficient antimicrobial activity. Water repellency of materials may reduce the adverse impact of microorganisms, because the quantity of adsorbed moisture is reduced. However, Hydrophobic finishing itself may not fully eliminate the adverse effect of microorganisms. Therefore, antimicrobial properties imparted to some textile materials during silicone finishing may be related to application of metal salts as catalysts, such as copper, chromium and aluminum.

Disinfectants, for example, at laundry, may be applied by the customer himself. The method of sanitizing substance application for carpets, which is spraying or dispensing of a disinfectant on the surface of floor covers during operation is known. The acceptable disinfection level may be obtained during laundry of textile products by such detergents, which may create residual fungal and bacteriostatic activity.

APPENDIX B

Figure B1 shows the scanning electronic micrographs of carbon chitosan/nanotube electrospun fibers, at different chitosan concentrations, in a TFA/DCM (70:30) solvent. As presented in Fig. B1a, at low concentrations of chitosan, the beads deposited on the collector and the thin fibers coexisted. As the concentration of chitosan increased (Figs. B1a–c), the beads decreased significantly. Figure B1c shows homogenous electrospun fibers with minimum beads, thin fibers and interconnected fibers. The increase of chitosan concentration leads to an increase of the interconnected fibers, as shown in Figures B1d-e. The average diameter of chitosan/carbon nanotube fibers increased when increasing the concentration of chitosan (Figs. B1 a–e). Hence, a chitosan/carbon nanotube solution of TFA/DCM (70:30) with 10 wt% of chitosan assured optimized conditions for electrospinning of this solution. When the voltage was low, the beads were deposited on the collector (Fig. B2a). As shown in Fig. B2a–d, the number of beads decreased with increasing voltage from 18 to 24 kV. In our study, the average diameter of fibers prepared by 18 kV was measured as 307 nm. As the applied voltage increased, the average fiber diameters also increased. The average diameter of fibers for 20 kV (2b), 22 kV (2c), and 24 kV (2d), was 308 (194–792), 448 (267–656) and 455 (306–672), respectively.

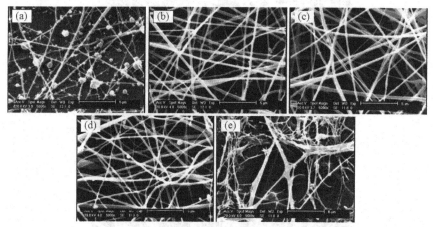

FIGURE B1 Scanning electron micrographs of electrospun fibers at different chitosan concentrations (wt%): (a) 8, (b) 9, (c) 10, (d) 11, (e) 12, 24 kV, 5 cm, TFA/DCM: 70/30.

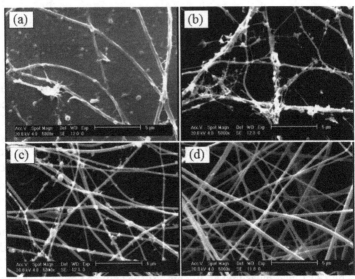

FIGURE B2 Scanning electron micrographs of electrospun fibers at different voltages (kV): (a) 18, (b) 20, (c) 22, (d) 24, 5 cm, 10 wt%, TFA/DCM: 70/30.

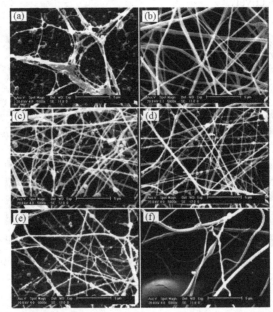

FIGURE B3 Scanning electron micrographs of electrospun fibers of chitosan/carbon nanotubes at different tip-to-collector distances (cm): (a) 4, (b) 5, (c) 6, (d) 7, (e) 8, (f) 10, 24 kV, 10 wt%, TFA/DCM: 70/30.

The morphologies of chitosan/carbon nanotube electrospun fibers at different tip-to-collector distances are presented in Fig. B3. When the tip-to-collector distance was low, a little interconnected fiber (with high fiber diameter) was deposited on the collector (as shown in Fig. B3a). At a 5 cm tip-to-collector distance (Fig. 3b), more homogenous fibers with negligible beads were obtained. However, the beads increased with increasing the tip-to-collector distance (Figs. B3b–f).

Also, our study has demonstrated that the diameter of electrospun fibers decreased by increasing tip-to-collector distance (as shown in Figs. B3b–d; 455 (306–672), 134 (87–163), 107 (71–196)). The fibers prepared within a distance of 8 cm (Fig. B3e) and 10 cm (Fig. B3f) presented defects and nonhomogenous diameter. However, a 5-cm tip-to-collector distance appears to be reliable for electrospinning.

APPENDIX C

In this appendix, effects of four electrospinning parameters, including solution concentration (wt.%), applied voltage (kV), tip to collector distance (cm), and volume flow rate (mL/h), on contact angle (CA) of polyacrylonitrile (PAN) nanofiber mat are presented. To optimize and predict the CA of electrospun fiber mat, RSM and artificial neural network (ANN) are employed and a quantitative relationship between processing variables and CA of electrospun fibers is established.

C1   MEASUREMENT AND CHARACTERIZATION

The morphology of the gold-sputtered electrospun fibers were observed by scanning electron microscope (SEM, Philips XL-30). The average fiber diameter and distribution was determined from selected SEM image by measuring at least 50 random fibers. The wettability of electrospun fiber mat was determined by CA measurement. The CA measurements were carried out using specially arranged microscope equipped with camera and PCTV vision software as shown in Figure 2. The droplet used was distilled water and was 1 μl in volume. The CA experiments were carried out at room temperature and were repeated five times. All contact angles measured within 20 s of placement of the water droplet on the electrospun fiber mat. A typical SEM image of electrospun fiber mat, its corresponding diameter distribution and CA image are shown in Figs. C1 and C2.

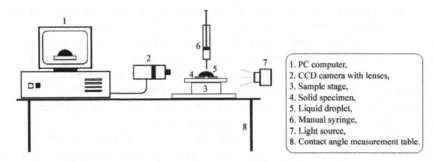

FIGURE C1   Schematic of CA measurement set up.

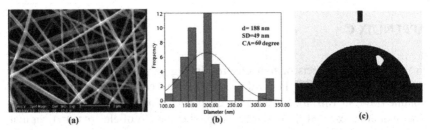

FIGURE C2   A typical (a) SEM image, (b) fiber diameter distribution, and (c) CA of electrospun fiber mat.

C2   RESPONSE SURFACE METHODOLOGY

Response surface methodology (RSM) is a combination of mathematical and statistical techniques used to evaluate the relationship between a set of controllable experimental factors and observed results. This optimization process is used in situations where several input variables influence some output variables (responses) of the system.

In the present study, central composite design (CCD) was employed to establish relationships between four electrospinning parameters and the CA of electrospun fiber mat. The experiment was performed for at least three levels of each factor to fit a quadratic model. Based on preliminary experiments, polymer solution concentration (wt.%), applied voltage (kV), tip to collector distance (cm), and volume flow rate (mL/h) were determined as critical factors with significance effect on CA of electrospun fiber mat. These factors were four independent variables and chosen equally spaced, while the CA of electrospun fiber mat was dependent variable. The values of −1, 0, and 1 are coded variables corresponding to low, intermediate and high levels of each factor respectively. The experimental parameters and their levels for four independent variables are shown in Table C1. The regression analysis of the experimental data was carried out to obtain an empirical model between processing variables. The contour surface plots were obtained using Design-Expert® software.

TABLE C1 Design of experiment (factors and levels).

Factor	Variable	Unit	Factor level		
			−1	0	1
X_1	Solution concentration	(wt.%)	10	12	14
X_2	Applied voltage	(kV)	14	18	22
X_3	Tip to collector distance	(cm)	10	15	20
X_4	Volume flow rate	(mL/h)	2	2.5	3

The quadratic model, Eq. (1) including the linear terms, was fitted to the data.

$$Y = \beta_0 + \sum_{i=1}^{4}\beta_i x_i + \sum_{i=1}^{4}\beta_{ii} x_i^2 + \sum_{i=1}^{3}\sum_{j=2}^{4}\beta_{ij} x_i x_j \tag{C1}$$

where, Y is the predicted response, x_i and x_j are the independent variables, β_0 is a constant, β_i is the linear coefficient, β_{ii} is the squared coefficient, and β_{ij} is the second-order interaction coefficients.

The quality of the fitted polynomial model was expressed by the determination coefficient (R^2) and its statistical significance was performed with the Fisher's statistical test for analysis of variance (ANOVA).

C3 ARTIFICIAL NEURAL NETWORK

Artificial neural network (ANN) is an information processing technique, which is inspired by biological nervous system, composed of simple unit (neurons) operating in parallel. A typical ANN consists of three or more layers, comprizing an input layer, one or more hidden layers and an output layer. Every neuron has connections with every neuron in both the previous and the following layer. The connections between neurons consist of weights and biases. The weights between the neurons play an important role during the training process. Each neuron in hidden layer and output layer has a transfer function to produce an estimate as target. The interconnection weights are adjusted, based on a comparison of the network output (predicted data) and the actual output (target), to minimize the error between the network output and the target.

In this study, feed forward ANN with one hidden layer composed of four neurons was selected. The ANN was trained using back-propagation algorithm. The same experimental data used for each RSM designs were also used as the input variables of the ANN. There are four neurons in the input layer corresponding to four electrospinning parameters and one neuron in the output layer corresponding to CA of electrospun fiber mat. Figure C3 illustrates the topology of ANN used in this investigation.

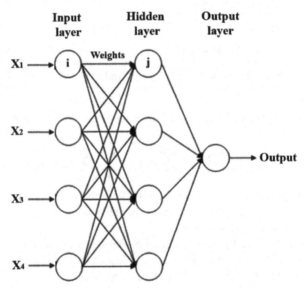

FIGURE C3   The topology of artificial neural network used in this study.

C4   THE ANALYSIS OF VARIANCE (ANOVA)

All 30 experimental runs of CCD were performed according to Table C2. A significance level of 5% was selected; that is, statistical conclusions may be assessed with 95% confidence. In this significance level, the factor has significant effect on CA if the p-value is less than 0.05. On the other hand, when p-value is greater than 0.05, it is concluded the factor has no significant effect on CA.

The results of ANOVA for the CA of electrospun fibers are shown in Table C3. Eq. (2) is the calculated regression equation.

TABLE C2   The actual design of experiments and response.

No.	Electrospinning parameters				Response
	X_1 Concentration	X_2 Voltage	X_3 Distance	X_4 Flow rate	CA (°)
1	10	14	10	2	44±6
2	10	22	10	2	54±7
3	10	14	20	2	61±6
4	10	22	20	2	65±4
5	10	14	10	3	38±5
6	10	22	10	3	49±4
7	10	14	20	3	51±5
8	10	22	20	3	56±5
9	10	18	15	2.5	48±3
10	12	14	15	2.5	30±3
11	12	22	15	2.5	35±5
12	12	18	10	2.5	22±3
13	12	18	20	2.5	30±4
14	12	18	15	2	33±4
15	12	18	15	3	25±3
16	12	18	15	2.5	26±4
17	12	18	15	2.5	29±3
18	12	18	15	2.5	28±5
19	12	18	15	2.5	25±4
20	12	18	15	2.5	24±3
21	12	18	15	2.5	21±3
22	14	14	10	2	31±4
23	14	22	10	2	35±5
24	14	14	20	2	33±6
25	14	22	20	2	37±4
26	14	14	10	3	19±3
27	14	22	10	3	28±3
28	14	14	20	3	39±5
29	14	22	20	3	36±4
30	14	18	15	2.5	20±3

$$CA = 25.80 - 9.89X_1 + 2.17X_2 + 4.33X_3 - 2.33X_4$$
$$-1.63X_1X_2 - 1.63X_1X_3 + 1.63X_1X_4 - 0.88X_2X_3 - 0.63X_2X_4 + 0.37X_3X_4 \quad \text{(C2)}$$
$$+7.90X_1^2 + 6.40X_2^2 - 0.096X_3^2 + 2.90X_4^2$$

TABLE C3   Analysis of variance for the CA of electrospun fiber mat.

Source	SS	DF	MS	F-value	Probe > F	Remarks
Model	4175.07	14	298.22	32.70	<0.0001	Significant
X_1 Concentration	1760.22	1	1760.22	193.01	<0.0001	Significant
X_2 Voltage	84.50	1	84.50	9.27	0.0082	Significant
X_3 Distance	338.00	1	338.00	37.06	<0.0001	Significant
X_4 Flow rate	98.00	1	98.00	10.75	0.0051	Significant
X_1X_2	42.25	1	42.25	4.63	0.0481	Significant
X_1X_3	42.25	1	42.25	4.63	0.0481	Significant
X_1X_4	42.25	1	42.25	4.63	0.0481	Significant
X_2X_3	12.25	1	12.25	1.34	0.2646	
X_2X_4	6.25	1	6.25	0.69	0.4207	Significant
X_3X_4	2.25	1	2.25	0.25	0.6266	
X_1^2	161.84	1	161.84	17.75	0.0008	Significant
X_2^2	106.24	1	106.24	11.65	0.0039	Significant
X_3^2	0.024	1	0.024	0.0026	0.9597	
X_4^2	21.84	1	21.84	2.40	0.1426	
Residual	136.80	15	9.12			
Lack of Fit	95.30	10	9.53	1.15	0.4668	

From the p-values presented in Table C3, it can be concluded that the p-values of terms X_3^2, X_4^2, X_2X_3, X_2X_4 and X_3X_4 is greater than the significance level of 0.05, therefore they have no significant effect on the CA of electrospun fiber mat. Since the above terms had no significant effect on CA of electrospun fiber mat, these terms were removed. The fitted equations in coded unit are given in Eq. (C3).

$$CA = 26.07 - 9.89X_1 + 2.17X_2 + 4.33X_3 - 2.33X_4$$
$$-1.63X_1X_2 - 1.63X_1X_3 + 1.63X_1X_4 \qquad (C3)$$
$$+9.08X_1^2 + 7.58X_2^2$$

Now, all the p-values are less than the significance level of 0.05. Analysis of variance showed that the RSM model was significant ($p<0.0001$), which indicated that the model has a good agreement with experimental data. The determination coefficient (R^2) obtained from regression equation was 0.958.

C5 ARTIFICIAL NEURAL NETWORK

In this section, the best prediction, based on minimum error, was obtained by ANN with one hidden layer. The suitable number of neurons in the hidden layer was determined by changing the number of neurons

The good prediction and minimum error value were obtained with four neurons in the hidden layer. The weights and bias of ANN for CA of electrospun fiber mat are given in Table C4. The R^2 and mean absolute percentage error were 0.965 and 5.94% respectively, which indicates that the model was shows good fitting with experimental data.

TABLE C4   Weights and bias obtained in training ANN.

Hidden layer	Weights	IW_{11}	IW_{12}	IW_{13}	IW_{14}
		1.0610	1.1064	21.4500	3.0700
		IW_{21}	IW_{22}	IW_{23}	IW_{24}
		−0.3346	2.0508	0.2210	−0.2224
		IW_{31}	IW_{32}	IW_{33}	IW_{34}
		−0.6369	−1.1086	−41.5559	0.0030
		IW_{41}	IW_{42}	IW_{43}	IW_{44}
		−0.5038	−0.0354	0.0521	0.9560
	Bias	b_{11}	b_{21}	b_{31}	b_{41}
		−2.5521	−2.0885	−0.0949	1.5478

TABLE C4 *(Continued)*

Output layer	Weights	LW_{11}
		0.5658
		LW_{21}
		0.2580
		LW_{31}
		−0.2759
		LW_{41}
		−0.6657
	Bias	b
		0.7104

C6   EFFECTS OF SIGNIFICANT PARAMETERS ON RESPONSE

The morphology and structure of electrospun fiber mat, such as the nanoscale fibers and interfibrillar distance, increases the surface roughness as well as the fraction of contact area of droplet with the air trapped between fibers. It is proved that the CA decrease with increasing the fiber diameter, therefore the thinner fibers, due to their high surface roughness, have higher CA than the thicker fibers. Hence, we used this fact for comparing CA of electrospun fiber mat. The interaction contour plot for CA of electrospun PAN fiber mat are shown in Figure C4.

As mentioned in the literature, a minimum solution concentration is required to obtain uniform fibers from electrospinning. Below this concentration, polymer chain entanglements are insufficient and a mixture of beads and fibers is obtained. On the other hand, the higher solution concentration would have more polymer chain entanglements and less chain mobility. This causes the hard jet extension and disruption during electrospinning process and producing thicker fibers. Figure C4(a) show the effect of solution concentration and applied voltage at middle level of distance (15 cm) and flow rate (2.5 mL/h) on CA of electrospun fiber mat. It is obvious that at any given voltage, the CA of electrospun fiber mat decrease with increasing the solution concentration.

Figure C4(b) shows the response contour plot of interaction between solution concentration and spinning distance at fixed voltage (18 kV) and flow rate (2.5 mL/h). Increasing the spinning distance causes the CA of electrospun fiber mat to increase. Because of the longer spinning distance could give more time for the solvent to evaporate, increasing the spinning distance will decrease the nanofiber diameter and increase the CA of electrospun fiber mat. As demonstrated in Figure C4(b), low solution concentration cause the increase in CA of electrospun fiber mat at large spinning distance.

The response contour plot in Figure C4(c) represented the CA of electrospun fiber mat at different solution concentration and volume flow rate. Ideally, the volume flow rate must be compatible with the amount of solution removed from the tip of the needle. At low volume flow rates, solvent would have sufficient time to evaporate and thinner fibers were produced, but at high volume flow rate, excess amount of solution fed to the tip of needle and thicker fibers were resulted. Therefore the CA of electrospun fiber mat will be decreased.

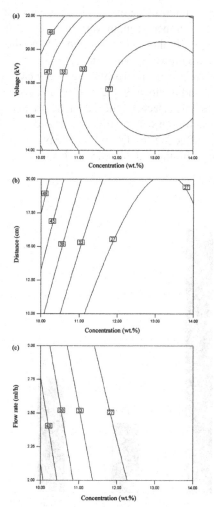

FIGURE C4   Contour plots for contact angle of electrospun fiber mat showing the effect of: (a) solution concentration and applied voltage, (b) solution concentration and spinning distance, (c) solution concentration and volume flow rate.

As shown by Eq. (4), the relative importance (RI) of the various input variables on the output variable can be determined using ANN weight matrix.

$$RI_j = \frac{\sum_{m=1}^{N_h}\left(\left(\left|IW_{jm}\right| \middle/ \sum_{k=1}^{N_i}\left|IW_{km}\right|\right) \times \left|LW_{mn}\right|\right)}{\sum_{k=1}^{N_i}\left\{\sum_{m=1}^{N_h}\left(\left(\left|IW_{km}\right| \middle/ \sum_{k=1}^{N_i}\left|IW_{km}\right|\right) \times \left|LW_{mn}\right|\right)\right\}} \times 100 \quad\quad (C4)$$

where RI_j is the relative importance of the jth input variable on the output variable, N_i and N_h are the number of input variables and neurons in hidden layer, respectively ($N_i = 4$, $N_h = 4$ in this study), IW and LW are the connection weights, and subscript "n" refer to output response (n=1).

The relative importance of electrospinning parameters on the value of CA calculated by Eq. (C4) and is shown in Fig. C5. It can be seen that, all of the input variables have considerable effects on the CA of electrospun fiber mat. Nevertheless, the solution concentration with relative importance of 49.69% is found to be most important factor affecting the CA of electrospun nanofibers. These results are in close agreement with those obtained with RSM.

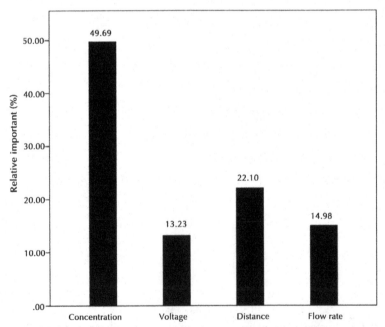

FIGURE C5   Relative importance of electrospinning parameters on the CA of electrospun fiber mat.

C7 OPTIMIZING THE CA OF ELECTROSPUN FIBER MAT

The optimal values of the electrospinning parameters were established from the quadratic form of the RSM. Independent variables (solution concentration, applied voltage, spinning distance, and volume flow rate) were set in range and dependent variable (CA) was fixed at minimum. The optimal conditions in the tested range for minimum CA of electrospun fiber mat are shown in Table C5. This optimum condition was a predicted value, thus to confirm the predictive ability of the RSM model for response, a further electrospinning and CA measurement was carried out according to the optimized conditions and the agreement between predicted and measured responses was verified. Figure C6 shows the SEM, average fiber diameter distribution and corresponding CA image of electrospun fiber mat prepared at optimized conditions.

TABLE C5   Optimum values of the process parameters for minimum CA of electrospun fiber mat.

Solution concentration (wt.%)	Applied voltage (kV)	Spinning distance (cm)	Volume flow rate (mL/h)	Predicted CA (°)	Observed CA (°)
13.2	16.5	10.6	2.5	20	21

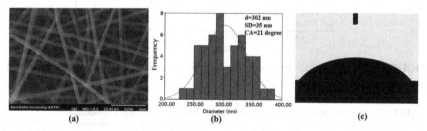

(a)   (b)   (c)

FIGURE C6   (a) SEM image, (b) fiber diameter distribution, and (c) CA of electrospun fiber mat prepared at optimized conditions.

C8 COMPARISON BETWEEN RSM AND ANN MODEL

Table C6 gives the experimental and predicted values for the CA of electrospun fiber mat obtained from RSM as well as ANN model. It is demonstrated that both models performed well and a good determination coefficient was obtained for both RSM and ANN. However, the ANN model shows higher determination coefficient ($R^2=0.965$) than the RSM model ($R^2=0.958$). Moreover, the absolute percentage error in the ANN prediction of CA was found to be around 5.94%, while for the RSM model, it was around 7.83%. Therefore, it can be suggested that the ANN model shows more accurately result than the RSM model. The plot of actual and predicted CA of electrospun fiber mat for RSM and ANN is shown in Figure C7.

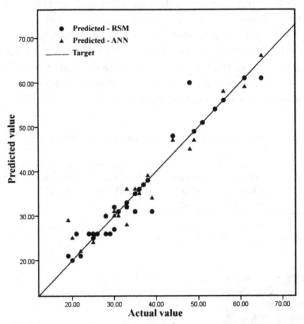

FIGURE C7   Comparison between the actual and predicted contact angle of electrospun nanofiber for RSM and ANN model.

TABLE C6   Experimental and predicted values by RSM and ANN models.

No.	Experimental	Predicted		Absolute error (%)	
		RSM	ANN	RSM	ANN
1	44	47	48	6.41	9.97
2	54	54	54	0.78	0.46
3	61	59	61	3.70	0.42
4	65	66	61	2.06	6.06
5	38	39	38	2.37	0.54
6	49	47	49	5.10	0.68

TABLE C6 *(Continued)*

No.	Experimental	Predicted		Absolute error (%)	
		RSM	**ANN**	**RSM**	**ANN**
7	51	51	51	0.35	0.45
8	56	58	56	4.32	0.17
9	48	45	60	6.17	24.37
10	30	31	27	4.93	9.35
11	35	36	31	2.34	11.15
12	22	22	21	1.18	4.15
13	30	30	32	1.33	6.04
14	33	28	33	13.94	0.60
15	25	24	25	5.04	0.87
16	26	26	26	0.27	1.33
17	29	26	26	10.10	9.16
18	28	26	26	6.89	5.91
19	25	26	26	4.28	5.38
20	24	26	26	8.63	9.77
21	21	26	26	24.14	25.45
22	31	30	31	2.26	0.57
23	35	31	35	10.34	0.66
24	33	36	32	8.18	2.18
25	37	37	37	0.59	0.34
26	19	29	21	52.11	10.23
27	28	30	30	7.07	8.20
28	39	34	31	12.05	21.30
29	36	35	36	1.72	0.04
30	20	25	20	26.30	2.27
R^2		0.958	0.965		
Mean absolute error (%)				7.83	5.94

APPENDIX D

Variables that potentially can alter the electrospinning process (Fig. D1) are large. Hence, investigating all of them in the framework of one single research would almost be impossible. However, some of these parameters can be held constant during experimentation. For instance, performing the experiments in a controlled environmental condition, which is concerned in this study, the ambient parameters (i.e. temperature, air pressure, and humidity) are kept unchanged. Solution viscosity is affected by polymer molecular weight, solution concentration, and temperature. For a particular polymer (constant molecular weight) at a fixed temperature, solution concentration would be the only factor influencing the viscosity. In this circumstance, the effect of viscosity could be determined by the solution concentration. Therefore, there would be no need for viscosity to be considered as a separate parameter.

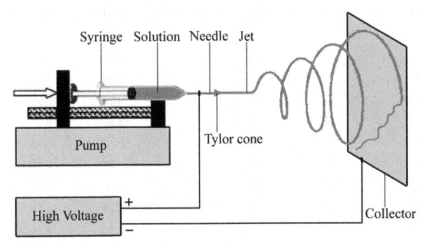

FIGURE D1   A typical image of Electrospinning process.

In this regard, solution concentration ©, spinning distance (d), applied voltage (V), and volume flow rate (Q) were selected to be the most influential parameters. The next step is to choose the ranges over which these factors are varied. Process knowledge, which is a combination of practical experience and theoretical understanding, is required to fulfill this step. The aim is here to find an appropriate range for each parameter where dry, bead-free, stable, and continuous fibers without breaking up to droplets are obtained. This goal could be achieved by conducting a set of preliminary experiments while having the previous works in mind along with using the reported relationships.

The relationship between intrinsic viscosity ($[\eta]$) and molecular weight (M) is given by the well-known Mark-Houwink-Sakurada equation as follows:

$$[\eta] = KM^a \tag{D1}$$

where K and a are constants for a particular polymer-solvent pair at a given temperature. Polymer chain entanglements in a solution can be expressed in terms of Berry number (B), which is a dimensionless parameter and defined as the product of intrinsic viscosity and polymer concentration ($B=[\eta]C$). For each molecular weight, there is a lower critical concentration at which the polymer solution cannot be electrospun.

As for determining the appropriate range of applied voltage, referring to previous works, it was observed that the changes of voltage lay between 5 kV to 25 kV depending on experimental conditions; voltages above 25 kV were rarely used. Afterwards, a series of experiments were carried out to obtain the desired voltage domain. At V<10 kV, the voltage was too low to spin fibers and $10\ kV \leq V < 15\ kV$ resulted in formation of fibers and droplets; in addition, electrospinning was impeded at high concentrations. In this regard, $15\ kV \leq V \leq 25\ kV$ was selected to be the desired domain for applied voltage.

The use of 5 cm – 20 cm for spinning distance was reported in the literature. Short distances are suitable for highly evaporative solvents whereas it results in wet coagulated fibers for nonvolatile solvents due to insufficient evaporation time. Afterwards, this was proved by experimental observations and $10\ cm \leq d \leq 20\ cm$ was considered as the effective range for spinning distance.

Few researchers have addressed the effect of volume flow rate. Therefore in this case, the attention was focused on experimental observations. At Q<0.2 mL/h, in most cases especially at high polymer concentrations, the fiber formation was hindered due to insufficient supply of solution to the tip of the syringe needle. Whereas, excessive feed of solution at Q>0.4 mL/h incurred formation of droplets along with fibers. As a result, $0.2\ mL/h \leq Q \leq 0.4\ mL/h$ was chosen as the favorable range of flow rate in this study.

Consider a process in which several factors affect a response of the system. In this case, a conventional strategy of experimentation, which is extensively used in practice, is the *one-factor-at-a-time* approach. The major disadvantage of this approach is its failure to consider any possible interaction between the factors, say the failure of one factor to produce the same effect on the response at different levels of another factor. For instance, suppose that two factors A and B affect a response. At one level of A, increasing B causes the response to increase, while at the other level of A, the effect of B totally reverses and the response decreases with increasing B. As interactions exist between electrospinning parameters, this approach may not be an appropriate choice for the case of the present chapter. The correct strategy to deal with several factors is to use a full factorial design. In this method, factors are all varied together; therefore all possible combinations of the levels of the factors are investigated. This approach is very efficient, makes the most use of the experimental data and takes into account the interactions between factors.

It is trivial that in order to draw a line at least two points and for a quadratic curve at least three points are required. Hence, three levels were selected for each parameter in this study so that it would be possible to use quadratic models. These levels were chosen equally spaced. A full factorial experimental design with four factors (solution concentration, spinning distance, applied voltage, and flow rate) each at three levels (3^4 design) were employed resulting in 81 treatment combinations. This design is shown in Fig. D 2.

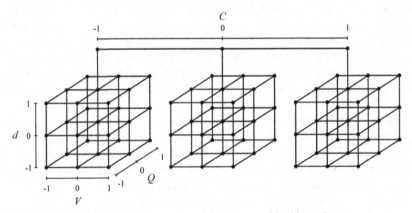

FIGURE D2   A 3^4 full factorial experimental design used in this study.

A −1, 0, and 1 are coded variables corresponding to low, intermediate and high levels of each factor respectively. The coded variables (x_j) were calculated using Eq. (2) from natural variables (ξ_i). The indices 1 to 4 represent solution concentration, spinning distance, applied voltage, and flow rate respectively. In addition to experimental data, 15 treatments inside the design space were selected as test data and used for evaluation of the models. The natural and coded variables for experimental data (numbers 1–81) as well as test data (numbers 82–96) are listed in Table A8 in Appendix.

$$x_j = \frac{\xi_j - [\xi_{hj} + \xi_{lj}]/2}{[\xi_{hj} - \xi_{lj}]/2} \tag{D2}$$

The mechanism of some scientific phenomena has been well understood and models depicting the physical behavior of the system have been drawn in the form of mathematical relationships. However, there are numerous processes at the moment, which have not been, sufficiently been understood to permit the theoretical approach. RSM is a combination of mathematical and statistical techniques useful for empirical modeling and analysis of such systems. The application of RSM is in situations where several input variables are potentially influence some performance measure or quality characteristic of the process – often called responses. The relationship between the response (y) and k input variables (ξ_1, ξ_2, ..., ξ_k) could be expressed in terms of mathematical notations as follows:

$$y = f(\xi_1, \xi_2, ..., \xi_k) \tag{D3}$$

where the true response function f is unknown. It is often convenient to use coded variables (x_1, x_2,.., x_k) instead of natural (input) variables. The response function will then be:

$$y = f(x_1, x_2, ..., x_k) \tag{D4}$$

Since the form of true response function f is unknown, it must be approximated. Therefore, the successful use of RSM is critically dependent upon the choice of appropriate function to approximate f. Low-order polynomials are widely used as approximating functions. First order (linear) models are unable to capture the interaction between parameters, which is a form of curvature in the true response function. Second order (quadratic) models will be likely to perform well in these circumstances. In general, the quadratic model is in the form of:

$$y = \beta_0 + \sum_{j=1}^{k} \beta_j x_j + \sum_{j=1}^{k} \beta_j x_j^2 + \sum_{i<j}\sum_{j=2}^{k} \beta_j x_i x_j + \varepsilon \tag{D5}$$

where ε is the error term in the model. The use of polynomials of higher order is also possible but infrequent. The βs are a set of unknown coefficients needed to be estimated. In order to do that, the first step is to make some observations on the system being studied. The model in Eq. (5) may now be written in matrix notations as:

$$y = X\beta + \varepsilon \tag{D6}$$

where y is the vector of observations, X is the matrix of levels of the variables, β is the vector of unknown coefficients, and ε is the vector of random errors. Afterwards, method of least squares, which minimizes the sum of squares of errors, is employed to find the estimators of the coefficients ($\hat{\beta}$) through:

$$\hat{\beta} = (X'X)^{-1}X'y \tag{D7}$$

The fitted model will then be written as:

$$\hat{y} = X\hat{\beta} \tag{D8}$$

Finally, response surfaces or contour plots are depicted to help visualize the relationship between the response and the variables and see the influence of the parameters. As you might notice, there is a close connection between RSM and linear regression analysis.

After the unknown coefficients (βs) were estimated by least squares method, the quadratic models for the mean fiber diameter (MFD) and standard deviation of fiber diameter (StdFD) in terms of coded variables are written as:

$$
\begin{aligned}
MFD = 282.031 + 34.953x_1 + 5.622x_2 - 2.113x_3 + 9.013x_4 \\
-11.613x_1^2 - 4.304x_2^2 - 15.500x_3^2 \\
-0.414x_4^2 + 12.517x_1x_2 + 4.020x_1x_3 - 0.162x_1x_4 + 20.643x_2x_3 + 0.741x_2x_4 + 0.877x_3x_4
\end{aligned} \tag{D9}
$$

$$
\begin{aligned}
StdFD = 36.1574 + 4.5788x_1 - 1.5536x_2 + 6.4012x_3 + 1.1531x_4 \\
-2.2937x_1^2 - 0.1115x_2^2 - 1.1891x_3^2 + 3.0980x_4^2 \\
-0.2088x_1x_2 + 1.0010x_1x_3 + 2.7978x_1x_4 + 0.1649x_2x_3 - 2.4876x_2x_4 + 1.5182x_3x_4
\end{aligned} \tag{D10}
$$

In the next step, a couple of very important hypothesis-testing procedures were carried out to measure the usefulness of the models presented here. First, the test for significance of the model was performed to determine whether there is a subset of variables, which contributes significantly in representing the response variations. The appropriate hypotheses are:

$$
\begin{aligned}
H_0 : \beta_1 = \beta_2 = \cdots = \beta_k \\
H_1 : \beta_j \neq 0 \quad \text{for at least one } j
\end{aligned} \tag{D11}
$$

The F statistics (the result of dividing the factor mean square by the error mean square) of this test along with the p-values (a measure of statistical significance, the smallest level of significance for which the null hypothesis is rejected) for both models are shown in Table D1.

TABLE D1 Summary of the results from statistical analysis of the models.

	F	p-value	R^2	R^2_{adj}	R^2_{pred}
MFD	106.02	0.000	95.74%	94.84%	93.48%
StdFD	42.05	0.000	89.92%	87.78%	84.83%

The p-values of the models are very small (almost zero), therefore it could be concluded that the null hypothesis is rejected in both cases suggesting that there are some significant terms in each model. There are also included in Table D1, the values of R^2, R^2_{adj}, and R^2_{pred}. R^2 is a measure for the amount of response variation which is explained by variables and will always increase when a new term is added to the model regardless of whether the inclusion of the additional term is statistically significant or not. R^2_{adj} is the adjusted form of R^2 for the number of terms in the model; therefore it will increase only if the new terms improve the model and decreases if unnecessary terms are added. R^2_{pred} implies how well the model predicts the response for new observations, whereas R^2 and R^2_{adj} indicate how well the model fits the experimental data. The R^2 values demonstrate that 95.74% of MFD and 89.92% of StdFD are explained by the variables. The R^2_{adj} values are 94.84% and 87.78% for MFD and StdFD respectively, which account for the number of terms in the models. Both R^2 and R^2_{adj} values indicate that the models fit the data very well. The slight difference between the values of R^2 and R^2_{adj} suggests that there might be some insignificant terms in the models. Since the R^2_{pred} values are so close to the values of R^2 and R^2_{adj}, models does not appear to be overfit and have very good predictive ability.

The second testing hypothesis is evaluation of individual coefficients, which would be useful for determination of variables in the models. The hypotheses for testing of the significance of any individual coefficient are:

$$H_0 : \beta_j = 0$$
$$H_1 : \beta_j \neq 0 \qquad \text{(D12)}$$

The model might be more efficient with inclusion or perhaps exclusion of one or more variables. Therefore the value of each term in the model is evaluated using this test, and then eliminating the statistically insignificant terms, more efficient models could be obtained. The results of this test for the models of MFD and StdFD are summarized in Table AD2 and Table D3, respectively. T statistic in these tables is a measure of the difference between an observed statistic and its hypothesized population value in units of standard error.

TABLE D2 The test on individual coefficients for the model of mean fiber diameter (MFD).

Term (coded)	Coeff.	T	p-value
Constant	282.031	102.565	0.000
C	34.953	31.136	0.000
d	5.622	5.008	0.000

TABLE D2   *(Continued)*

Term (coded)	Coeff.	T	p-value
V	−2.113	−1.882	0.064
Q	9.013	8.028	0.000
C^2	−11.613	−5.973	0.000
d^2	−4.304	−2.214	0.030
V^2	−15.500	−7.972	0.000
Q^2	−0.414	−0.213	0.832
Cd	12.517	9.104	0.000
CV	4.020	2.924	0.005
CQ	−0.162	−0.118	0.906
dV	20.643	15.015	0.000
dQ	0.741	0.539	0.592
VQ	0.877	0.638	0.526

TABLE D3   The test on individual coefficients for the model of standard deviation of fiber diameter (StdFD).

Term (coded)	Coef	T	p-value
Constant	36.1574	39.381	0.000
C	4.5788	12.216	0.000
D	−1.5536	−4.145	0.000
V	6.4012	17.078	0.000
Q	1.1531	3.076	0.003
C^2	−2.2937	−3.533	0.001
d^2	−0.1115	−0.172	0.864
V^2	−1.1891	−1.832	0.072
Q^2	3.0980	4.772	0.000
Cd	−0.2088	−0.455	0.651

TABLE D3   *(Continued)*

Term (coded)	Coef	T	p-value
CV	1.0010	2.180	0.033
CQ	2.7978	6.095	0.000
dV	0.1649	0.359	0.721
dQ	−2.4876	−5.419	0.000
VQ	1.5182	3.307	0.002

As depicted, the terms related to Q^2, CQ, dQ, and VQ in the model of MFD and related to d^2, CD, and dV in the model of StdFD have very high p-values, therefore they do not contribute significantly in representing the variation of the corresponding response. Eliminating these terms will enhance the efficiency of the models. The new models are then given by recalculating the unknown coefficients in terms of coded variables in Eqs. (D13) and (14), and in terms of natural (uncoded) variables in Eqs. (D15), and (D16).

$$
\begin{aligned}
MFD = {} & 281.755 + 34.953x_1 + 5.622x_2 - 2.113x_3 + 9.013x_4 \\
& -11.613x_1^2 - 4.304x_2^2 - 15.500x_3^2 \\
& +12.517x_1x_2 + 4.020x_1x_3 + 20.643x_2x_3
\end{aligned}
\tag{D13}
$$

$$
\begin{aligned}
StdFD = {} & 36.083 + 4.579x_1 - 1.554x_2 + 6.401x_3 + 1.153x_4 \\
& -2.294x_1^2 - 1.189x_3^2 + 3.098x_4^2 \\
& +1.001x_1x_3 + 2.798x_1x_4 - 2.488x_2x_4 + 1.518x_3x_4
\end{aligned}
\tag{D14}
$$

$$
\begin{aligned}
MFD = {} & 10.3345 + 48.7288C - 22.7420d + 7.9713V + 90.1250Q \\
& -2.9033C^2 - 0.1722d^2 - 0.6120V^2 \\
& +1.2517Cd + 0.4020CV + 0.8257dV
\end{aligned}
\tag{D15}
$$

$$
\begin{aligned}
StdFD = {} & -1.8823 + 7.5590C + 1.1818d + 1.2709V - 300.3410Q \\
& -0.5734C^2 - 0.0476V^2 + 309.7999Q^2 \\
& +0.1001CV + 13.9892CQ - 4.9752dQ + 3.0364VQ
\end{aligned}
\tag{D16}
$$

The results of the test for significance as well as R^2, R^2_{adj}, and R^2_{pred} for the new models are given in Table D4. It is obvious that the p-values for the new models are close to zero indicating the existence of some significant terms in each model. Comparing the results of this table with Table D1, the F statistic increased for the new models, indicating the improvement of the models after eliminating the insignificant terms. Despite the slight decrease in R^2, the values of R^2_{adj}, and R^2_{pred} increased substantially for the new models. As it was mentioned earlier in the chapter, R^2 will always increase with the number of terms in the model. Therefore, the smaller R^2 values were expected for the new models, due to the fewer terms. However, this does not necessarily suggest that the pervious models were more efficient. Looking at the tables, R^2_{adj}, which provides a more useful tool for comparing the explanatory power of models with different number of terms, increased after eliminating the unnecessary variables. Hence, the new models have the ability to better explain the experimental data. Due to higher R^2_{pred}, the new models also have higher prediction ability. In other words, eliminating the insignificant terms results in simpler models, which not only present the experimental data in superior form, but also are more powerful in predicting new conditions.

TABLE D4 Summary of the results from statistical analysis of the models after eliminating the insignificant terms.

	F	p-value	R^2	R^2_{adj}	R^2_{pred}
MFD	155.56	0.000	95.69%	95.08%	94.18%
StdFD	55.61	0.000	89.86%	88.25%	86.02%

The test for individual coefficients was performed again for the new models. The results of this test are summarized in Table A5 and Table A6. This time, as it was anticipated, no terms had higher p-value than expected, which need to be eliminated. Here is another advantage of removing unimportant terms. The values of T statistic increased for the terms already in the models implying that their effects on the response became stronger.

TABLE D5 The test on individual coefficients for the model of mean fiber diameter (MFD) after eliminating the insignificant terms.

Term (coded)	Coeff.	T	p-value
Constant	281.755	118.973	0.000
C	34.953	31.884	0.000
d	5.622	5.128	0.000
V	−2.113	−1.927	0.058
Q	9.013	8.221	0.000
C^2	−11.613	−6.116	0.000
d^2	−4.304	−2.267	0.026

TABLE D5 *(Continued)*

Term (coded)	Coeff.	T	*p*-value
V^2	−15.500	−8.163	0.000
Cd	12.517	9.323	0.000
CV	4.020	2.994	0.004
dV	20.643	15.375	0.000

TABLE D6 The test on individual coefficients for the model of standard deviation of fiber diameter (StdFD) after eliminating the insignificant terms.

Term (coded)	Coef	T	*p*-value
Constant	36.083	45.438	0.000
C	4.579	12.456	0.000
d	−1.554	−4.226	0.000
V	6.401	17.413	0.000
Q	1.153	3.137	0.003
C^2	−2.294	−3.602	0.001
V^2	−1.189	−1.868	0.066
Q^2	3.098	4.866	0.000
CV	1.001	2.223	0.029
CQ	2.798	6.214	0.000
dQ	−2.488	−5.525	0.000
VQ	1.518	3.372	0.001

After developing the relationship between parameters, the test data were used to investigate the prediction ability of the models. Root mean square errors (RMSE) between the calculated responses (C_i) and real responses (R_i) were determined using Eq. (D17) for experimental data as well as test data for the sake of evaluation of both MFD and StdFD models.

$$\text{RMSE} = \sqrt{\frac{\sum_{i=1}^{n}(C_i - R_i)^2}{n}} \tag{D17}$$

CHAPTER 3

POLYMERIC NANOSYSTEMS

A. K. HAGHI and G. E. ZAIKOV

CONTENTS

3.1   INTRODUCTION

The appearance of "nanoscience" and "nanotechnology" stimulated the burst of terms with "nano-" prefix. Historically the term "nanotechnology" appeared before and it was connected with the appearance of possibilities to determine measurable values up to 10^{-9} of known parameters: 10^{-9} m-nm (nanometer), 10^{-9} s-ns (nanosecond), 10^{-9} degree (nanodegree, shift condition). Nanotechnology and molecular nanotechnology comprise the set of technologies connected with transport of atoms and other chemical particles (ions, molecules) at distances contributing the interactions between them with the formation of nanostructures with different nature.

When scanning tunnel microscope was invented there was an opportunity to influence atoms of a substance thus stimulating the work in the field of probe technology, which resulted in substantiation and practical application of nanotechnological methods in 1994. With the help of this technique it is possible to handle single atoms and collect molecules or aggregates of molecules, construct various structures from atoms on a certain substrate (base). Naturally, such a possibility cannot be implemented without preliminary computer designing of so-called "nanostructures architecture." Nanostructures architecture assumes a certain given location of atoms and molecules in space that can be designed on computer and afterwards transferred into technological program of nanotechnological facility.

The term "chemical assembly" appeared together with the development of chemistry and physics of surface after the birth of "electron spectroscopy for chemical analysis."

Chemical assembly comprises the interaction of chemical particles with the surface and "grafting" of functional groups to the surface or interface boundary "gas–solid."

It should be noted that in nanoproduct obtained fullerene C_{60} predominates by its content, represents an enclosed cluster of 60 carbon atoms. This cluster has a stable and symmetrical structure. Furthermore in 1991 nonotubes were discovered. Afterwards the investigations in the field of nanoparticles and nanosystems started spreading all over the world.

The development of these trends predetermined the appearance and development of so-called "nanoscience." In the same way as "nanotechnology" and "nanoscience" is determined as a combination of scientific knowledge from various disciplines, such as physics, chemistry, biology, mathematics, programming, etc., adapted to nanostructures and nanosystems. Nanoscience comprises fundamental and applied scientific knowledge. Therefore, it is possible to speak about nanochemistry, nanometallurgy, nanoelectronics, nanomachine-building, science of nanomaterials and similar disciplines. If we think of the world from the point that the nature is unified and different disciplines in the science were created by people for the convenience of perception and understanding of the world around, the appearing areas of nanoscience closely connected with nanotechnology represent a vast aggregation of disciplines the list of which will still be incomplete analogously to [4]. Let us be restricted to definitions connected with science of nanomaterials.

The notion "science of nanomaterials" assumes scientific knowledge for obtaining, composition, properties and possibilities to apply nanostructures, nanosystems and nanomaterials. A simplified definition of this term can be as follows: material science dealing with materials comprising particles and phases with nanometer dimensions. To determine the existence area for nanostructures and nanosystems it is advisable to find out the difference of these formations from analogous material objects.

From the analysis of literature the following can be summarized: the existence area of nanosystems and nanoparticles with any structure is between the particles of molecular and atomic level determined in pictometers and aggregates of molecules or per molecular formations over micron units. Here it should be mentioned that in polymer chemistry particles with nanometer dimensions belong to the class of per molecular structures, such as globules and fibrils by one of parameters, for example, by diameter or thickness. In chemistry of complex compounds clusters with nanometer dimensions are also known.

The notion "cluster" assumes energy-wise compensated nucleus with a shell, the surface energy of which is rather small, as a result under given conditions the cluster represents a stable formation.

In chemical literature a cluster is equated with a complex compound containing a nucleus and a shell. Usually a nucleus consists of metal atoms combined with metallic bond, and a shell – of ligands. Manganese carbonyls [$(Co)_5$ $MnMn(Co)_5$] and cobalt carbonyls [$Co_6(Co)_{18}$], nickel pentadienyls [$Ni_6(C_5H_6)_6$] belong to elementary clusters.

In recent years the notion "cluster" has got an extended meaning. At the same time the nucleus can contain not only metals or not even contain metals. In some clusters, for instance, carbon ones there is no nucleus at all. In this case their shape can be characterized as a sphere (icosahedron, to be precise) – fullerenes, or as a cylinder – fullerene tubules. Surely a certain force field is formed by atoms on internal walls inside such particles. It can be assumed that electrostatic, electromagnetic and gravitation fields conditioned by corresponding properties of atoms contained in particle shells can be formed inside tubules and fullerenes. If analyze papers recently published, it should be noted that a considerable exceeding of surface size over the volume and, consequently, a relative growth of the surface energy in comparison with the growth of volume and potential energy is the main feature of clusters. If particle dimensions (diameters of "tubes" and "spheres") change from 1 up to several hundred nanometers, they would be called nanoparticles. In some papers the area of nanoclusters existence is within 1–10 nm.

Based on classical definitions, in given paper metal nanoparticles and nano-crystals are referred to as nanoclusters. Apparently, the difference of nanoparticles from other particles (smaller or larger) is determined by their specific character-istics. The search of nanoworld distinctions from atomic-molecular, micro and macroworld can lead to finding analogies and coincidences in colloid chemistry, chemistry of polymers and coordination compounds. First of all it should be noted that nanoparticles usually represent a small collective aggregation of atoms being within the action of adjacent atoms, thus conditioning the shape of nanoparticles. A nanoparticle shape can vary depending upon the nature of adjacent atoms and character of formation medium. Obviously, the properties of separate atoms and molecules (of small size) are determined by their energy and geometry charac-teristics, the determinative role being played by electron properties. In particular, electron interactions determine the geometry of molecules and atomic structures of small size and mobility of these chemical particles in media, as well as their activity or reactivity.

When the number of atoms in chemical particle exceeds 30, a certain stabili-zation of its shape being also conditioned by collective influence of atoms con-stituting the particle is observed. At the same time, the activity of such a particle remains high but the processes with its participation have a directional character. The character of interactions with the surroundings of such structures is deter-mined by their formation mechanism.

During polymerization or copolymerization the influence of macromolecule growth parameters changes with the increase of the number of elementary acts of its growth. After 7–10 acts the shape or geometry of nanoparticles formed be-comes the main determinative factor providing the further growth of macromol-ecule (chain development). A nanoparticle shape is usually determined not only by its structural elements but also by its interactions with surrounding chemical particles.

From the aforesaid it can be concluded that the possibility of self-organization of nanoparticles with the formation of corresponding nanosystems and nanoma-terials is the main distinction of nanoworld from pico, micro, and macroworld. Recently, much attention has been paid to synergetics or the branch of science dealing with self-organization processes since these processes, in many cases, proceed with small energy consumption and, consequently, are more ecologically clear in comparison with existing technological processes.

In turn, nanoparticle dimensions are determined by its formation conditions. When the energy consumed for macroparticle destruction or dispersion over the surface increases, the dimensions of nanomaterials are more likely to decrease. The notion "nanomaterial" is not strictly defined. Several researchers consider nanomaterials to be aggregations of nanocrystals, nanotubes or fullerenes. At the same time there is a lot of information available that nanomaterials can represent materials containing various nanostructures. The most attention researchers pay to metallic nanocrystals. Special attention is paid to metallic nanowires and nano-fibers with different compositions.

Here are some names of nanostructures:

1) fullerenes, 2) gigantic fullerenes, 3) fullerenes filled with metal ions, 4) fullerenes containing metallic nucleus and carbon (or mineral) shell, 5) one-layer nanotubes, 6) multilayer nanotubes, 7) fullerene tubules, 8) "scrolls", 9) conic nanotubes, 10) metal-containing tubules, 11) "onions", 12) "Russian dolls", 13) bamboo-like tubules, 14) "beads", 15) welded nanotubes, 16) bunches of nano-tubes, 17) nanowires, 18) nanofibers, 19) nanoropes, 20) nanosemispheres (nano-cups), 21) nanobands and similar nanostructures, as well as various derivatives from enlisted structures. It is quite possible that a set of such structures and no-tions will be enriched.

In most cases nanoparticles obtained are bodies of rotation or contain parts of bodies of rotation. In natural environment there are minerals containing fullerenes or representing thread-like formations comprising nanometer pores or structures. In the first case we are talking about schungite that is available in quartz rock in unique deposit in Prionezhje. Similar mineral can also be found in the river Lena basin but it consists of micro and macrodimensional cones, spheroids and complex fibers. In the second case we are talking about kerite from pegmatite on Volyn (Ukraine) that consists of polycrystalline fibers, spheres and spirals mostly of micron dimensions, or fibrous vetcillite from the state of Utah (USA); globular anthraxolite and asphaltite.

Diameters of some internal channels are up to 20–50 nm. Such channels can be of interest as nanoreactors for the synthesis of organic, carbon and polymeric substances with relatively low energy consumption. In case of directed location of internal channels in such matrixes and their inner-combinations the spatial struc-tures of certain purpose can be created. Terminology in the field of nanosystems existence is still being developed, but it is already clear that nanoscience obtains qualitatively new knowledge that can find wide application in various areas of human practice thus significantly decreasing the danger of people's activities for themselves and environment.

The system classification by dimensional factor is known, based on which we consider the following:

- microobjects and microparticles 10^{-6}–10^{-3} m in size;
- nanoobjects and nanoparticles 10^{-9}–10^{-6} m in size;
- picoobjects and picoparticles 10^{-12}–10^{-9} m in size.

Assuming that nanoparticle vibration energies correlate with their dimensions and comparing this energy with the corresponding region of electromagnetic waves, we can assert that energy action of nanostructures is within the energy region of chemical reactions. System self-organization refers to synergetics. Quite often, scientists considered that nanotechnology is based on self-organization of metastable systems. As assumed, self-organization can proceed by dissipative (synergetic) and continual (conservative) mechanisms. At the same time, the system can be arranged due to the formation of new stable ("strengthening") phases or due to the growth provision of the existing basic phase. This phenomenon underlies the arising nanochemistry. Below is one of the possible definitions of nanochemistry.

Nanochemistry is a science investigating nanostructures and nanosystems in metastable ("transition") states and processes flowing with them in near-"transition" state or in "transition" state with low activation energies.

To carry out the processes based on the notions of nanochemistry, the directed energy action on the system is required, with the help of chemical particle field as well, for the transition from the prepared near-"transition" state into the process product state (in our case – into nanostructures or nanocomposites). The perspective area of nanochemistry is the chemistry in nanoreactors. Nanoreactors can be compared with specific nanostructures representing limited space regions in which chemical particles orientate creating "transition state" prior to the formation of the desired nanoproduct. Nanoreactors have a definite activity, which predetermines the creation of the corresponding product. When nanosized particles are formed in nanoreactors, their shape and dimensions can be the reflection of shape and dimensions of the nanoreactor.

In the recent years, a lot of scientific information in the field of nanotechnology and science of nanomaterials appeared. Scientists defined the interval from 1 to 1000 nm as the area of nanostructure existence, the main feature of which is to regulate the system self-organization processes. However, scientists limited the upper threshold at 100 nm. At the same time it was not well substantiated and taken from. Now many nanostructures varying in shapes and sizes are known. These nanostructures have sizes that fit into the interval, and are active in the processes of self-organization, and also demonstrate specific properties.

Problems of nanostructure activity and the influence of nanostructure super small quantities on the active media structural changes are explained.

The molecular nanotechnology ideology is analyzed [19]. In accordance with the development tendencies in self-organizing systems under the influence of nanosized excitations the reasons for the generation of self-organization in the range 10^{-6}–10^{-9} m should be determined.

Based on the law of energy conservation the energy of nanoparticle field and electromagnetic waves in the range 1–1000 nm can transfer, thus corresponding to the range of energy change from soft X-ray to near IR radiation. This is the range of energies of chemical reactions and self-organization (structuring) of systems connected with them.

Apparently the wavelengths of nanoparticle oscillations near the equilibrium state are close or correspond to their sizes. Then based on the concepts of ideologists of nanotechnology in material science the definition of nanotechnology can be as follows:

Nanotechnology is a combination of knowledge in the ways and means of conducting processes based on the phenomenon of nano-sized system self-organization and utilization of internal capabilities of the systems that results in decreasing the energy consumption required for obtaining the targeted product while preserving the ecological cleanness of the process.

Nanotechnology expands itself as one of the major area of science [1–25]. An important and exciting aspect of nanomedicine is the use of nanoparticle for target specific drug delivery systems and it allows a new innovative therapeutic approaches. Due to their small size, these drug delivery systems are promising tools in therapeutic approaches such as selective or targeted drug delivery towards a specific tissue or organ. It also enhances drug transport across biological barriers and intracellular drug delivery. Nanotherapeutic agents work successfully against traditional drugs, which exhibits very less bioavailability and are often associated with gastro-intestinal side effects. Nanotherapeutics improves the delivery of drugs that cannot normally be taken orally and it also improves the safety and efficacy of low molecular weight drugs. It also improves the stability and absorption of proteins that normally cannot be taken orally.

3.2   PROPERTIES OF NANOTHERAPEUTICS

A good therapeutic agent or nanoparticle should have ability to: (1) cross one or various biological membranes (e.g., mucosa, epithelium, endothelium) before (2) diffusing through the plasma membrane to (3) finally gain access to the appropriate organelle where the biological target is located.

Different methods of drug delivery are listed below:
1.   Oral Drug Delivery
2.   Injection Based Drug Delivery
3.   Transdermal Drug Delivery
4.   Bone Marrow Infusion
5.   Control Release Systems
6.   Targeted Drug Delivery:
     a. Therapeutical Monoclonal
     b. Antibodies
     c. Liposomes
     d. Microparticles
     e. Modified Blood Cells
     f. Nanoparticles
7.   Implant Drug Delivery System

3.3 METAL-BASED NANOPARTICLES

Metallic nanoparticles are either spherical metal or semiconductor particles with nanometer-sized diameters. Similar to nanoparticles "nanocrystal" (NC) or "nanocrystallite" term is used for nanoparticles made up of semiconductors. Generally semiconductor materials from the groups II-VI (CdS, CdSe, CdTe), III-V (InP, InAs) and IV-VI (PbS, PbSe, PbTe) are of particular interest. Metal Based nanoparticle are used for nanoscale transistors, biological sensors, and next generation photovoltaics [26–49].

3.4 LIPID-BASED NANOPARTICLES

Lipid based nanoparticles are widely used for drug targeting and drug delivery. Solid lipid nanoparticles (SLN) introduced in 1991 represent an alternative carrier system for drugs [4]. Nanoparticles made from solid lipids are attracting major attention as novel colloidal drug carrier for intravenous applications as they have been proposed as an alternative particulate carrier system. SLN are submicron colloidal carriers ranging from 50 to 1000 nm, which are composed of physiological lipid [23–52].

Lipid based system are many used because:

1. Lipids enhance oral bioavailability and reduce plasma profile variability.
2. Better characterization of lipoid excipients.
3. An improved ability to address the key issues of technology transfers and manufactures scale-up.

3.5 POLYMER-BASED NANOPARTICLES

Polymeric nanoparticles are nanoparticles, which are prepared from polymers. Polymeric nanoparticles forms (1) the micronization of a material into nanoparticles and (2) the stabilization of the resultant nanoparticles. As for the micronization, one can start with either small monomers or a bulk polymer. The drug is dissolved, entrapped, encapsulated or attached to a nanoparticles and one can obtain different nanoparticles, nanospheres or nanocapsules according to methods of preparation. Gums, Gelatin Sodium alginate Albumin are used for polymer based drug delivery. Polymeric nanoparticles are prepared by Cellulosics, Poly(2-hydroxy ethyl methacrylate), Poly(N-vinyl pyrrolidone), Poly(vinyl alcohol), Poly(methyl methacrylate), Poly(acrylic acid), Polyacrylamide, Poly(ethylene-covinyl acetate) like polymeric materials. Polymer used in drug delivery must have following qualities like it should be chemically inert, nontoxic and free of leachable impurities [39–60].

3.6   BIOLOGICAL NANOPARTICLES

For proper interaction with molecular targets, a biological or molecular coating or layer is required which act as a bioinorganic interface, and it should be attached to the nanoparticle for proper interaction. These biological coating may be antibodies, biopolymers like collagen. Layers of small molecules that make the nanoparticles biocompatible are also used as bioinorganic interface. For detection and analysis these nanoparicles should be optically active and they should either fluoresces or change their color in different intensities of light.

3.7   NANOCARRIERS IN DRUG DELIVERY

Nanoparticles and Liposomes get absorbed by blood streams for its mode of action. For particular function theses nanoparticles must have specific structure and composition. These particles rapidly get cleared from blood stream by macrophages of reticuloendothelial system (RES) [13–28].

Nanoparticles can be delivered to targets by phagocytosis pathway and non-phagocytosis pathway.

3.7.1 PHAGOCYTOSIS PATHWAY

Phagocytosis occurs by profession phagocytes and by nonprofessional phagocytes. A professional phagocyte includes macrophages, monocytes, neutrophils and dendritic cell, while nonprofessional phagocytes are fibroblast, epithelial and endothelial cell.

Phagocytosis takes place by:
1.   Recognition of the opsonized particles in the blood stream.
2.   Adhesion of particles to macrophages
3.   Ingestion of the particle

Opsonization of the nanoparticles is a major step and it takes place before phagocytosis. In this process nanoparticles get tagged by a protein known as *opsonins*. Due to this tagging nanoparticles make a visible complex for macrophages, this whole process takes place in blood streams. These activated particles then get attached to macrophages and this interaction is similar to receptor-ligand interaction [61–80].

As actin is depolymerized from the phagosome, the newly denuded vacuole membrane becomes accessible to early endosomes. Through a series of fusion and fission events, the vacuolar membrane and its contents will mature, fusing with late endosomes and ultimately lysosomes to form a phagolysosome (Fig. 1). The rate of these events depends on the surface properties of the ingested particle, typically from half to several hours. The phagolysosomes become acidified due to the vacuolar proton pump ATPase located in the membrane and acquire many enzymes, including esterases and cathepsins [81–95].

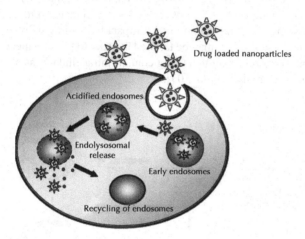

FIGURE 1   Nanoparticles enter the cell through receptor-mediated endocytosis and get localized in the endolysosomal compartment.

3.7.2   NON-PHAGOCYTIC PATHWAY

This is also known as pinocytosis or cell drinking method. It is basically taking of fluids and solutes. Very small nanoparticles get entered in well and absorbed by pinocytosis as phagocytosis is restricted to size of particle. The process of Pinocytosis occurs in specialized cell. It may be clathrin mediated endocytosis, caveolae-mediated endocytosis, macropinocytosis and caveolae-independent endocytosis

3.7.2.1   CLATHRIN-MEDIATED ENDOCYTOSIS

Endocytosis via clathrin-coated pits, or clathrin-mediated endocytosis (CME), occurs constitutively in all mammalian cells, and fulfills crucial physiological roles, including nutrient uptake and intracellular communication. For most cell types, CME serves as the main mechanism of internalization for macromolecules and plasma membrane constituents. CME via specific receptor-ligand interaction is the best described mechanism, to the extent that it was previously referred to as "receptor-mediated endocytosis" (RME). However, it is now clear that alternative nonspecific endocytosis via clathrin-coated pits also exists (as well as receptor-mediated but clathrin-independent endocytosis). Notably, the CME, either receptor-dependent or independent, causes the endocytosed material to end up in degradative lysosomes. This has an important impact in the drug delivery field since the drug-loaded nanocarriers may be tailored in order to become metabolized into the lysosomes, thus releasing their drug content intracellularly as a consequence of lysosomal biodegradation [96–103].

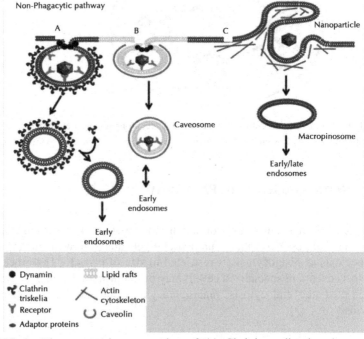

FIGURE 2 Diagrammamtic presentation of (a) Clathrin-mediated endocytosis (b) Caveolae-mediated endocytosis (© Macropinocytosis).

3.7.2.2 CAVEOLAE-MEDIATED ENDOCYTOSIS

Although CME is the predominant endocytosis mechanism in most cells, alternative pathways have been more recently identified, caveolae-mediated endocytosis (CvME) being the major one. Caveolae are characteristic flask-shaped membrane invaginations, having a size generally reported in the lower end of the 50–100 nm range, typically 50–80 nm. They are lined by caveolin, a dimeric protein, and enriched with cholesterol and sphingolipids (Fig. 2). Caveolae are particularly abundant in endothelial cells, where they can constitute 10–20% of the cell surface, but also smooth muscle cells and fibroblasts. CvMEs are involved in endocytosis and trancytosis of various proteins; they also constitute a port of entry for viruses (typically the SV40 virus) and receive increasing attention for drug delivery applications using nanocarriers.

Unlike CME, CvME is a highly regulated process involving complex signaling, which may be driven by the cargo itself. After binding to the cell surface, particles move along the plasma membrane to caveolae invaginations, where they may be maintained through receptor-ligand interactions. Fission of the caveolae from the membrane, mediated by the GTPase dynamin, then generates the cytosolic caveolar vesicle, which does not contain any enzymatic cocktail. Even this pathway is employed by many pathogens to escape degradation by lysosomal enzymes. The use of nanocarriers exploiting CvME may therefore be advantageous to by-pass the lysosomal degradation pathway when the carried drug(e.g., peptides, proteins, nucleic acids, etc.) is highly sensitive to enzymes. On the whole, the uptake kinetics of CvME is known to occur at a much slower rate than that of CME. Ligands known to be internalized by CvME include folic acid, albumin and cholesterol.

3.7.2.3 MACROPINOCYTOSIS

Macropinocytosis is another type of clathrin-independent endocytosis pathway, occurring in many cells, including macrophages. It occurs via formation of actin-driven membrane protusions, similarly to phagocytosis. However, in this case, the protusions do not zipper up along the ligand-coated particle; instead, they collapse onto and fuse with the plasma membrane. This generates large endocytic vesicles, called macropinosomes, which sample the extracellular milieu and have a size generally bigger than 1 lm (and sometimes as large as 5 lm). The intracellular fate of macropinosomes varies depending on the cell type, but in most cases, they acidify and shrink. They may eventually fuse with lysosomal compartments or recycle their content to the surface (Fig. 2). Macropinosomes have not been reported to contain any specific coating, nor do they concentrate receptors.

This endocytic pathway does not seem to display any selectivity, but is involved, among others, in the uptake of drug nanocarriers [80–103].

3.7.2.4   MAGNETIC NANOPARTICLES (MNPS)

Magnetic nanoparticles (MNPs) have many applications in different areas of biology and medicine. MNPs are used for hyperthermia, magnetic resonance imaging, immunoassay, purification of biologic fluids, cell and molecular separation, tissue engineering. The design of magnetically targeted nanosystems (MNSs) for a smart delivery of drugs to target cells is a promising direction of nanobiotechnology. They traditionally consist on one or more magnetic cores and biological or synthetic molecules, which serve as a basis for polyfunctional coatings on MNPs surface. The coatings of MNSs should meet several important requirements. They should be biocompatible, protect magnetic cores from influence of biological liquids, prevent MNSs agglomeration in dispersion, provide MNSs localization in biological targets and homogeneity of MNSs sizes. The coatings must be fixed on MNPs surface and contain therapeutic products (drugs or genes) and biovectors for recognition by biological systems. The model, which is often used when MNSs are developed, is presented in Fig. 3.

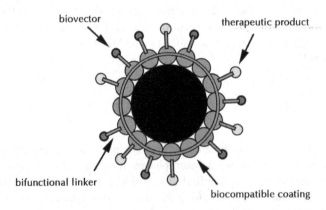

FIGURE 3   The classical scheme of magnetically targeted nanosystem for a smart delivery of therapeutic products.

Proteins are promising materials for creation of coatings on MNPs for biology and medicine. When proteins are used as components of coatings it is of the first importance that they keep their functional activity. Protein binding on MNPs surface is a difficult scientific task. Traditionally bifunctional linkers (glutaralde-hyde, carbodiimide) are used for protein cross-linking on the surface of MNPs and modification of coatings by therapeutic products and biovectors. Research-ers modified MNPs surface with aminosilanes and performed protein molecules attachment using glutaraldehyde. In the issue bovine serum albumin (BSA) was adsorbed on MNPs surface in the presence of carbodiimide. These works revealed several disadvantages of this way of protein fixing which make it unpromising. Some of them are clusters formation as a result of linking of protein molecules ad-sorbed on different MNPs, desorption of proteins from MNSs surface as a result of incomplete linking, uncontrollable linking of proteins in solution (Fig. 4). The creation of stable protein coatings with retention of native properties of molecules still is an important biomedical problem.

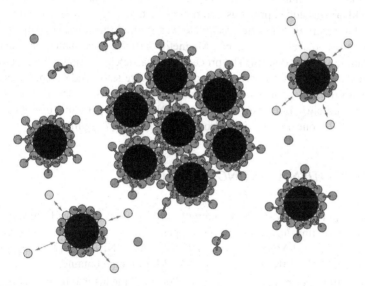

FIGURE 4 Nonselective linking of proteins on MNPs surface by bifunctional linkers leading to clusters formation and desorption of proteins from nanoparticles surface.

It is known that proteins can be chemically modified in the presence of free radicals with formation of cross-links. The goals of the work were to create stable protein coating on the surface of individual MNPs using a fundamen-tally novel approach based on the ability of proteins to form interchain covalent bonds under the action of free radicals and estimate activity of proteins in the coating.

3.8 EXPERIMENTAL

3.8.1 MAGNETIC SORBENT SYNTHESIS

Nanoparticles of magnetite Fe_3O_4 were synthesized by coprecipitation of ferrous and ferric salts in water solution at 4 °C and in the alkaline medium:

$$Fe^{2+} + 2Fe^{3+} + 8OH- \rightarrow Fe_3O_4\downarrow + 4H_2O$$

1.4 g of $FeSO_4\cdot7H_2O$ and 2.4 g of $FeCl_3\cdot6H_2O$ were dissolved in 50 ml of distilled water so that molar ratio of Fe^{2+}/Fe^{3+} was equal to 1:2. After filtration of the solution 10 mL of 25 mass% NH_4OH was added to it on a magnetic stirrer. 2.4 g of PEG 2 kDa) was added previously in order to reduce the growth of nanoparticles during the reaction. After the precipitate was formed the solution (with 150 mL of water) was placed on a magnet. Magnetic particles precipitated on it and supernatant liquid was deleted. The procedure of particles washing was repeated for 15 times until neutral pH was obtained. MNPs were stabilized by double electric layer with the use of disperser. To create the double electric layer 30 ml of 0.1 M phosphate-citric buffer solution (0.05 M NaCl) with pH value of 4 was introduced. MNPs concentration in hydrosol was equal to 37 mg/mL.

3.8.2 PROTEIN COATINGS FORMATION

Bovine serum albumin and thrombin with activity of 92 units per 1 mg were used for protein coating formation. Several types of reaction mixtures were created: "A1-MNP-0", "A2-MNP-0", "A1-MNP-1", "A2-MNP-1", "A2-MNP-1-acid", "T1-MNP-0", "T1-MNP-0" and "T1-0-0." All of them contained:

(1) 2.80 mL of protein solution ("A1" or "A2" means that there is BSA solution with concentration of 1 mg/mL or 2 mg/mL in 0.05 M phosphate buffer with pH 6.5 (0.15 M NaCl) in the reaction mixture; "T1" means that there is thrombin solution with concentration of 1 mg/mL in 0.15M NaCl with pH 7.3);

(2) 0.35 mL of 0.1 M phosphate-citric buffer solution (0.05 M NaCl) or MNPs hydrosol ("MNP" in the name of reaction mixture means that it contains MNPs);

(3) 0.05 mL of distilled water or 3 mass% H_2O_2 solution ("0" or "1" in the reaction mixture names correspondingly).

Hydrogen peroxide interacts with ferrous ion on MNPs surface with formation of hydroxyl-radicals by Fenton reaction:

$$Fe^{2+} + H_2O_2 \rightarrow Fe^{3+} + OH^{\cdot} + OH^-$$

"A2-MNP-1-acid" is a reaction mixture, containing 10 µL of ascorbic acid with concentration of 152 mg/mL. Ascorbic acid is known to form free radicals in reaction with H_2O_2 and generate free radicals in solution but not only on MNPs surface.

The sizes of MNPs, proteins and MNPs in adsorption layer were analyzed using dynamic light scattering (Zetasizer Nano S "Malvern", England) with detection angle of 173° at temperature 25 °C.

3.8.3   STUDY OF PROTEINS ADSORPTION ON MNPS

The study of proteins adsorption on MNPs was performed using ESR-spectroscopy of spin labels. The stable nitroxide radical used as spin label is presented in Fig. 5. Spin labels technique allows studying adsorption of macromolecules on nano-sized magnetic particles in dispersion without complicated separation processes of solution components. The principle of quantitative evaluation of adsorption is the following. Influence of local fields of MNPs on spectra of radicals in solution depends on the distance between MNPs and radicals. If this distance is lower than 40 nm for magnetite nanoparticles with the average size of 17 nm ESR spectra lines of the radicals broaden strongly and their intensity decreases to zero. The decreasing of the spectrum intensity is proportional to the part of radicals, which are located inside the layer of 40 nm in thickness around MNP. The same happens with spin labels covalently bound to protein macromolecules. An intensity of spin labels spectra decreases as a result of adsorption of macromolecules on MNPs (Fig. 6). We have shown that spin labels technique can be used for the study of adsorption value, adsorption kinetics, calculation of average number of molecules in adsorption layer and adsorption layer thickness, concurrent adsorption of macromolecules.

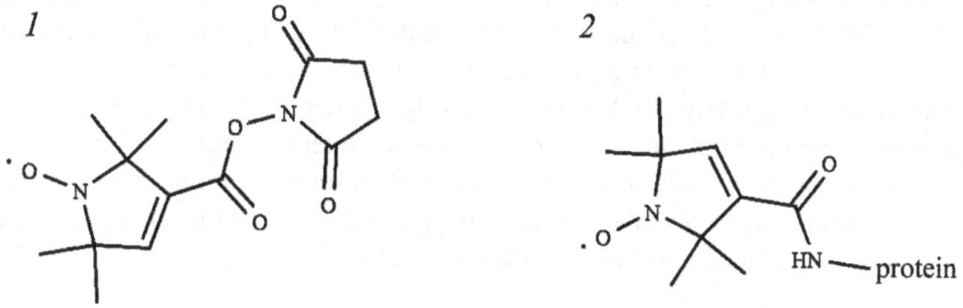

FIGURE 5   The stable nitroxide radical used for labeling of macromolecules containing aminogroups (*1*) and spin label attached to protein macromolecule (*2*).

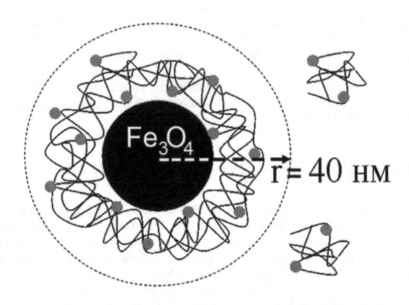

FIGURE 6   Magnetic nanoparticles and spin-labeled macromolecules in solution.

The reaction between the radical and protein macromolecules was conducted at room temperature. 25 μL of radical solution in 96% ethanol with concentration of 2.57 mg/mL was added to 1 mL of protein solution. The solution was incubated for 6 h and dialyzed. The portion of adsorbed protein was calculated from intensity of the low-field line of nitroxide radical triplet I_{+1}.

The method of ferromagnetic resonance was also used to study adsorption layer formation.

The spectra of the radicals and magnetic nanoparticles were recorded at room temperature using spectrometer at a microwave power of 5 mW, modulation frequency 100 kHz and amplitude 1 G. The first derivative of the resonance absorption curve was detected. The samples were placed into the cavity of the spectrometer in a quartz flat cell. Magnesium oxide powder containing Mn^{2+} ions was used as an external standard in ESR experiments. Average amount of spin labels on protein macromolecules reached 1 per 4–5 albumin macromolecules and 1 per 2–3 thrombin macromolecules. Rotational correlation times of labels were evaluated as well as a fraction of labels with slow motion.

3.8.4 COATING STABILITY ANALYSIS AND ANALYSIS OF SELECTIVITY OF FREE RADICAL PROCESS

In our previous works it was shown that fibrinogen (FG) adsorbed on MNPs surface forms thick coating and micron-sized structures. Also FG demonstrates an ability to replace BSA previously adsorbed on MNPs surface. This was proved by complex study of systems containing MNPs, spin-labeled BSA and FG with spin labels technique and ferromagnetic resonance. The property of FG to replace BSA from MNPs surface was used in this work for estimating BSA coating stability. 0.25 mL of FG solution with concentration of 4 mg/mL in 0.05 M phosphate buffer with pH 6.5 was added to 1 mL of the samples "A1-MNP-0", "A2-MNP-0", "A1-MNP-1", "A2-MNP-1." The clusters formation was observed by dynamic light scattering.

The samples "A2-MNP-0", "A2-MNP-1", "T1-MNP-0", "T1-MNP-1" were centrifugated at 120,000 g during 1 h. On these conditions MNPs precipitate, but macromolecules physically adsorbed on MNPs remain in supernatant liquid. The precipitates containing MNPs and protein fixed on MNPs surface were dissolved in buffer solution with subsequent evaluation of the amount of protein by Bradford colorimetric method. Spectrophotometer was used as well.

Free radical modification of proteins in supernatant liquids of "A2-MNP-0", "A2-MNP-1" and the additional sample "A2-MNP-1-acid" were analyzed by IR-spectroscopy using FTIR-spectrometer with DTGS-detector with 2 cm^{-1} resolution. Comparison of "A2-MNP-0", "A2-MNP-1" and "A2-MNP-1-acid" helps to reveal the selectivity of free radical process in "A2-MNP-1."

3.8.5 ENZYME ACTIVITY ESTIMATION

Estimation of enzyme activity of protein fixed on MNPs surface was performed on the example of thrombin. This protein is a key enzyme of blood clotting system, which catalyzes the process of conversion of fibrinogen to fibrin. Thrombin may lose its activity as a result of free radical modification and the rate of the enzyme reaction may decrease. So estimation of enzyme activity of thrombin cross-linked on MNPs surface during free radical modification was performed by comparison of the rates of conversion of fibrinogen to fibrin under the influence of thrombin contained in reaction mixtures. 0.15 mL of the samples "T1-MNP-0", "T1-MNP-1" and "T1–0-0" was added to 1.4 mL of FG solution with concentration of 4 mg/mL. Kinetics of fibrin formation was studied by Rayleigh light scattering on spectrometer with multibit 64-channel correlator.

3.9   RESULTS AND DISCUSSION

ESR spectra of spin labels covalently bound to BSA and thrombin macromolecules (Fig. 7) allow obtaining information about their microenvironment. The
spectrum of spin labels bound to BSA is a superposition of narrow and wide lines
characterized by rotational correlation times of 10^{-9} s and 2×10^{-8} s respectively.
This is an evidence of existence of two main regions of spin labels localization on
BSA macromolecules. The portion of labels with slow motion is about 70%. So a
considerable part of labels are situated in internal areas of macromolecules with
high microviscosity. The labels covalently bound to thrombin macromolecules
are characterized by one rotational correlation time of 0.26 ns. These labels are
situated in areas with equal microviscosity.

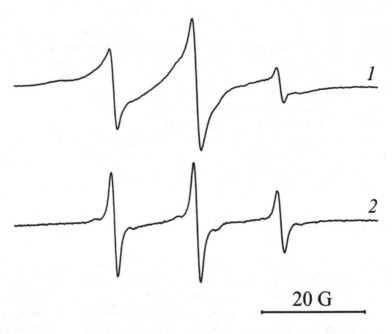

FIGURE 7   ESR spectra of spin labels on BSA (*1*) and thrombin (*2*) at 25 °C.

The signal intensity of spin-labeled macromolecules decreased after introduction of magnetic nanoparticles into the solution that testifies to the protein adsorption on MNPs (Fig. 8). Spectra of the samples "A1-MNP-0" and "T1-MNP-0"
consist of nitroxide radical triplet, the third line of sextet of Mn^{2+} (the external
standard) and ferromagnetic resonance spectrum of MNPs. Rotational correlation

time of spin labels does not change after MNPs addition. The dependences of spectra lines intensity for spin-labeled BSA and thrombin in the presence of MNPs on incubation time are shown in Table 1. Signal intensity of spin-labeled BSA changes insignificantly. These changes correspond to adsorption of approximately 12% of BSA after the sample incubation for 100 min. The study of adsorption kinetics allows establishing that adsorption equilibrium in "T1-MNP-0" takes place when the incubation time equals to 80 min and approximately 41% of thrombin is adsorbed. The value of adsorption A may be estimated using the data on the portion of macromolecules adsorbed and specific surface area calculated from MNPs density (5200 mg/m^3), concentration and size. Hence BSA adsorption equals to 0.35 mg/m^2 after 100 min incubation. The dependence of thrombin adsorption value on incubation time is shown in Fig. 9. Thrombin adsorption equals to 1.20 mg/m^2 after 80 min incubation.

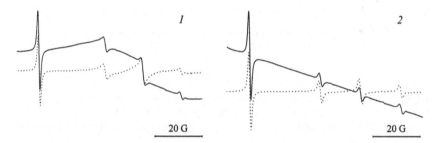

FIGURE 8   ESR spectra of spin labels on BSA (*1*) and thrombin (2) macromolecules before (dotted line) and 75 min after (solid line) MNPs addition to protein solution at 25 °C. External standard – MgO powder containing Mn^{2+}.

TABLE 1   The dependence of relative intensity of low-field line of triplet I_{+1} of nitroxide radical covalently bound to BSA and thrombin macromolecules, and the portion N of the protein adsorbed on incubation time t of the samples "A1-MNP-0" and "T1-MNP-0."

	Spin-labeled BSA		Spin-labeled thrombin	
t, min.	I_{+1}, rel. units	N,%	I_{+1}, rel. units	N,%
0	0.230 ± 0.012	0 ± 5	0.25 ± 0.01	0 ± 4
15	—	—	0.17 ± 0.01	32 ± 4
35	0.205 ± 0.012	9 ± 5	0.16 ± 0.01	36 ± 4
75	0.207 ± 0.012	10 ± 5	0.15 ± 0.01	40 ± 4
95	—	—	0.15 ± 0.01	40 ± 4
120	0.200 ± 0.012	13 ± 5	0.14 ± 0.01	44 ± 4

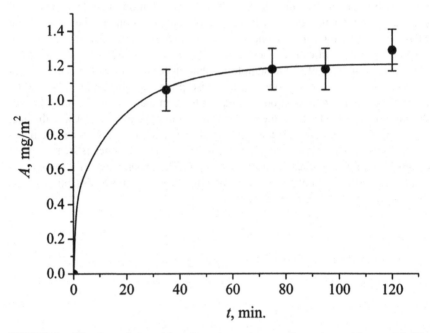

FIGURE 9 Kinetics of thrombin adsorption on magnetite nanoparticles at 25 °C. Concentration of thrombin in the sample is 0.9 mg/mL, MNPs – 4.0 mg/mL.

The FMR spectra of the samples "A1-MNP-0", "T1-MNP-0" and MNPs are characterized by different position in magnetic field (Fig. 10). The center of the spectrum of MNPs is 3254 G, while the center of "A1-MNP-0" and "T1-MNP-0" spectra is 3253 G and 3449 G, respectively. Resonance conditions for magnetic nanoparticles in magnetic field of spectrometer include a parameter of the shift of FMR spectrum $|M_1| = \frac{3}{2}|H_1|$, where H_1 is a local field created by MNPs in linear aggregates, which form in spectrometer field. $H_1 = 2\sum_1^{\infty} \frac{2\mu}{(nD)^3}$, where D is a distance between MNPs in linear aggregates, μ is MNPs magnetic moment, n is a number of MNPs in aggregate. Coating formation and the thickness of adsorption layer influence on the distance between nanoparticles decrease dipole interactions and particles ability to aggregate. As a result the center of FMR spectrum moves to higher fields. This phenomenon of FMR spectrum center shift we observed in the system "A1-MNP-0" after FG addition. The spectrum of MNPs with thick coating becomes similar to FMR spectra of isolated MNPs. So the similar center positions of FMR spectra of MNPs without coating and MNPs in BSA coating point to a very thin coating and low adsorption of protein in this case. In contrast according to FMR center position the thrombin coating on MNPs is thicker than albumin coating. This result is consistent with the data obtained by ESR spectroscopy.

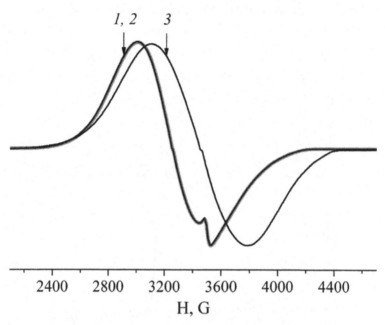

FIGURE 10 FMR spectra of MNPs (*1*), MNPs in the mixture with BSA (the sample "A1-MNP-0") after incubation time of 120 min (*2*) and MNPs in the mixture with thrombin (the sample "T1-MNP-0") after incubation time of 120 min (*3*).

FG ability to replace BSA in adsorption layer on MNPs surface is demonstrated in Fig. 11. Initially there is bimodal volume distribution of particles over sizes in the sample "A2-MNP-0" that can be explained by existence of free (unabsorbed) BSA and MNPs in BSA coating. After FG addition the distribution changes. Micron-sized clusters form in the sample that proves FG adsorption on MNPs [18]. In the case of "A2-MNP-1" volume distribution is also bimodal. The peak of MNPs in BSA coating is characterized by particle size of maximal contribution to the distribution of ~23 nm. This size is identical to MNPs in BSA coating in the sample "A2-MNP-0." It proves that H_2O_2 addition does not lead to uncontrollable linking of protein macromolecules in solution or cluster formation. Since MNPs size is 17 nm, the thickness of adsorption layer on MNPs is approximately 3 nm.

After FG addition to "A2-MNP-1" micron-sized clusters do not form. So adsorption BSA layer formed in the presence of H_2O_2 keeps stability. This stability can be explained by formation of covalent bonds between protein macromole-cules [13] in adsorption layer as a result of free radicals generation on MNPs surface. Stability of BSA coating on MNPs was demonstrated for the samples "A1-MNP-1" and "A2-MNP-1" incubated for more than 100 min before FG addition. Clusters are shown to appear if the incubation time is insufficient.

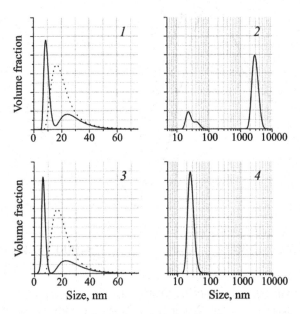

FIGURE 11   Volume distributions of particles in sizes in systems without (*1, 2*) and with (*3, 4*) H_2O_2 ("A2-MNP-0", "A2-MNP-1") incubated for 2 h before (*1, 3*) and 20 min after (*2, 4*) FG addition. Dotted line is the volume distribution of nanoparticles in sizes in dispersion.

The precipitates obtained by ultracentrifugation of "A2-MNP-0", "A2-MNP-1", "T1-MNP-0" and "T1-MNP-1" were dissolved in buffer solution. The amount of protein in precipitates was evaluated by colorimetric method (Table 2). The results showed that precipitates of systems with H_2O_2 contained more protein than the same systems without H_2O_2. Therefore in the samples containing H_2O_2 the significant part of protein molecules does not leave MNPs surface when centrifuged while in the samples "A2-MNP-0" and "T1-MNP-0" the most of protein molecules leaves the surface. This indicates the stability of adsorption layer formed in the presence of free radical generation initiator and proves crosslinks formation.

TABLE 2   The amount of protein in precipitates after centrifugation of the samples "A2-MNP-0", "A2-MNP-1", "T1-MNP-0" and "T1-MNP-1" of 3.2 mL in volume.

Sample name	Amount of protein in precipitates, mg
"A2-MNP-0"	0.05
"A2-MNP-1"	0.45
"T1-MNP-0"	0.15
"T1-MNP-1"	1.05

Analysis of content of supernatant liquids obtained after ultracentrifugation of reaction systems containing MNPs and BSA that differed by H_2O_2 and ascorbic acid presence ("A2-MNP-0", "A2-MNP-1" and "A2-MNP-1-acid") allows evaluating the scale of free radical processes in the presence of H_2O_2. As it was above-mentioned in the presence of ascorbic acid free radicals generate not only on MNPs surface but also in solution. So both molecules on the surface and free molecules in solution can undergo free radical modification in this case. From Fig. 12, we can see that the IR-spectrum of "A2-MNP-1-acid" differs from the spectra of "A2-MNP-0" and "A2-MNP-1", while the spectra of "A2-MNP-0" and "A2-MNP-1" almost have no differences. The IR-spectra differ in the region of 1200–800 cm^{-1}. The changes in this area are explained by free radical oxidation of amino acid residues of methionine, tryptophane, histidine, cysteine, phenylalanine. These residues are sulfur-containing and cyclic ones which are the most sensitive to free radical oxidation. The absence of differences in "A2-MNP-0" and "A2-MNP-1" proves that cross-linking of protein molecules in the presence of H_2O_2 is selective and takes place only on MNPs surfaces.

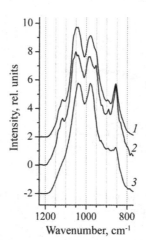

FIGURE 12   IR-spectra of supernatant solutions obtained after centrifugation of the samples "A2-MNP-0" (*1*), "A2-MNP-1" (*2*) and "A2-MNP-1-acid" (*3*).

When proteins are used as components of coating on MNPs for biology and medicine their functional activity retaining is very important. Proteins fixed on MNPs can lose their activity as a result of adsorption on MNPs or free radical modification, which is cross-linking, and oxidation but it was shown that they do not lose it. Estimation of enzyme activity of thrombin cross-linked on MNPs surface was performed by comparison of the rates of conversion of fibrinogen to fibrin under the influence of thrombin contained in reaction mixtures "T1-MNP-0", "T1-MNP-1" and "T1-0-0." The curves for the samples containing thrombin and MNPs, which differ by the presence of H_2O_2, had no fundamental differences that illustrate preservation of enzyme activity of thrombin during free radical cross-linking on MNPs surface. Fibrin gel was formed during ~15 min in both cases. Rayleigh light scattering intensity was low when "T1-0-0" was used and small fibrin particles were formed in this case. The reason of this phenomenon is autolysis (self-digestion) of thrombin. Enzyme activity of thrombin, one of serine proteinases, decreases spontaneously in solution. So the proteins can keep their activity longer when adsorbed on MNPs. This way, the method of free radical cross-linking of proteins seems promising for enzyme immobilization (Fig. 13).

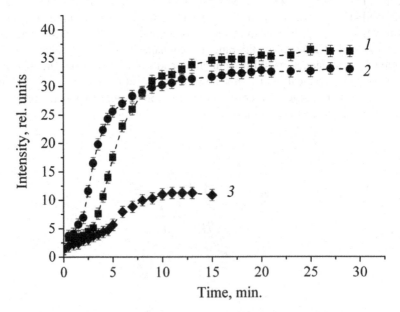

FIGURE 13 Kinetics curves of growth of Rayleigh light scattering intensity in the process of fibrin gel formation in the presence of "T1-MNP-0" (*1*), "T1-MNP-1" (*2*) and "T1-0-0" (*3*).

3.10 TRANSFER OF HEAT IN A NANOSYSTEM

Heat and mass transfer in wet nanostructure are coupled in a complicated way. The structure of the solid matrix varies widely in shape. There is, in general, a distribution of void sizes, and the nanostructures may also be locally irregular. Energy transport in such a medium occurs by conduction in all of the phases. Mass transport occurs within voids of the medium. In an unsaturated state these voids are partially filled with a liquid, whereas the rest of the voids contain some gas. It is a common misapprehension that nonhygroscopic fibers (i.e., those of low intrinsic for moisture vapor) will automatically produce a hydrophobic fabric. The major significance of the fine geometry of a nanostructure in contributing to resistance to water penetration can be stated in the following manner:

For instance the requirements of a water repellent fabric are (*a*) that the nanofibers shall be spaced uniformly and as far apart as possible and (*b*) that they should be held so as to prevent their ends drawing together. In the meantime, wetting takes place more readily on surfaces of high fiber density and in a fabric where there are regions of high fiber density such as yarns, the peripheries of the yarns will be the first areas to wet out and when the peripheries are wetted, water can pass unhindered through the nanofabric.

For thermal analysis of wet nanofabrics, the liquid is water and the gas is air. Evaporation or condensation occurs at the interface between the water and air so that the air is mixed with water vapor. A flow of the mixture of air and vapor may be caused by external forces, for instance, by an imposed pressure difference. The vapor will also move relative to the gas by diffusion from regions where the partial pressure of the vapor is higher to those where it is lower.

Again, heat transfer by conduction, convection, and radiation and moisture transfer by vapor diffusion are the most important mechanisms in very cool or warm environments from the skin.

Meanwhile, nanotextile manufacturing involves a crucial energy-intensive drying stage at the end of the process to remove moisture left from dye setting. Determining drying characteristics for nanotextiles, such as temperature levels, transition times, total drying times and evaporation rates, etc. is vitally important so as to optimize the drying stage. In general, drying means to make free or relatively free from a liquid. We define it more narrowly as the vaporization and removal of water from textiles.

3.10.1 HEAT

When a wet nanofabric is subjected to thermal drying two processes occur simultaneously, namely:

(a) Transfer of heat to raise the wet nanofabric temperature and to evaporate the moisture content.
(b) Transfer of mass in the form of internal moisture to the surface of the nanofabric and its subsequent evaporation.

The rate at which drying is accomplished is governed by the rate at which these two processes proceed. Heat is a form of energy that can across the boundary of a system. Heat can, therefore, be defined as "the form of energy that is transferred between a nanosystem and its surroundings as a result of a temperature difference." There can only be a transfer of energy across the boundary in the form of heat if there is a temperature difference between the system and its surroundings. Conversely, if the nanosystem and surroundings are at the same temperature there is no heat transfer across the boundary.

Strictly speaking, the term "heat" is a name given to the particular form of energy crossing the boundary. However, heat is more usually referred to in thermodynamics through the term "heat transfer", which is consistent with the ability of heat to raise or lower the energy within a nanosystem.

There are three modes of heat transfer:
(1) Convection;
(2) Conduction; and
(3) Radiation.

All these three modes are different. Convection relies on movement of a fluid. Conduction relies on transfer of energy between molecules within a solid or fluid. Radiation is a form of electromagnetic energy transmission and is independent of any substance between the emitter and receiver of such energy. However, all three modes of heat transfer rely on a temperature difference for the transfer of energy to take place.

The greater the temperature difference the more rapidly will the heat be transferred. Conversely, the lower the temperature difference, the slower will be the rate at which heat is transferred. When discussing the modes of heat transfer it is the rate of heat transfer Q that defines the characteristics rather than the quantity of heat.

As it was mentioned earlier, there are three modes of heat transfer, convection, conduction and radiation. Although two, or even all three, modes of heat transfer may be combined in any particular thermodynamic situation, the three are quite different and will be introduced separately.

The coupled heat and liquid moisture transport of nano-porous material has wide industrial applications in textile engineering and functional design of apparel products. Heat transfer mechanisms in nano-porous textiles include conduction by the solid material of fibers, conduction by intervening air, radiation, and convection. Meanwhile, liquid and moisture transfer mechanisms include vapor diffusion in the void space and moisture sorption by the fiber, evaporation, and capillary effects. Water vapor moves through textiles as a result of water vapor concentration differences. Nanofibers absorb water vapor due to their internal chemical compositions and structures. The flow of liquid moisture through the textiles is caused by fiber-liquid molecular attraction at the surface of fiber materials, which is determined mainly by surface tension and effective capillary pore distribution and pathways. Evaporation and/or condensation take place, depending on the temperature and moisture distributions. The heat transfer process is coupled with the moisture transfer processes with phase changes such as moisture sorption and evaporation.

Mass transfer in the drying of a wet nanofabric will depend on two mechanisms: movement of moisture within the fabric which will be a function of the internal physical nature of the solid and its moisture content; and the movement of water vapor from the material surface as a result of water vapor from the material surface as a result of external conditions of temperature, air humidity and flow, area of exposed surface and supernatant pressure.

3.10.2 CONVECTION HEAT TRANSFER

A very common method of removing water from textiles is convective drying. Convection is a mode of heat transfer that takes place as a result of motion within a fluid. If the fluid, starts at a constant temperature and the surface is suddenly increased in temperature to above that of the fluid, there will be convective heat transfer from the surface to the fluid as a result of the temperature difference. Under these conditions the temperature difference causing the heat transfer can be defined as:

$$\Delta T = \text{(surface temperature)} - \text{(mean fluid temperature)}$$

Using this definition of the temperature difference, the rate of heat transfer due to convection can be evaluated using Newton's law of cooling:

$$Q = h_c A \Delta T \tag{1}$$

where A is the heat transfer surface area and is the coefficient of heat transfer from the surface to the fluid, referred to as the "convective heat transfer coefficient."

The units of the convective heat transfer coefficient can be determined from the units of other variables:

$$Q = h_c A \Delta T$$
$$W = (h_c) m^2 K$$

So the units of are.

The relationship given in Eq. (1) is also true for the situation where a surface is being heated due to the fluid having higher temperature than the surface. However, in this case the direction of heat transfer is from the fluid to the surface and the temperature difference will now be

(Mean fluid temperature) – (Surface temperature)

The relative temperatures of the surface and fluid determine the direction of heat transfer and the rate at which heat transfer take place.

As given in Eq. (1), the rate of heat transfer is not only determined by the temperature difference but also by the convective heat transfer coefficient. This is not a constant but varies quite widely depending on the properties of the fluid and the behavior of the flow. The value of must depend on the thermal capacity of the fluid particle considered, that is, for the particle. The higher the density and of the fluid the better the convective heat transfer.

Two common heat transfer fluids are air and water, due to their widespread availability. Water is approximately 800 times denser than air and also has a higher value of. If the argument given above is valid then water has a higher thermal capacity than air and should have a better convective heat transfer performance.

3.10.3 CONDUCTION HEAT TRANSFER

If a fluid could be kept stationary there would be no convection taking place. However, it would still be possible to transfer heat by means of conduction. Conduction depends on the transfer of energy from one molecule to another within the heat transfer medium and, in this sense, thermal conduction is analogous to electrical conduction.

Conduction can occur within both solids and fluids. The rate of heat transfer depends on a physical property of the particular solid of fluid, termed its thermal conductivity k, and the temperature gradient across the medium. The thermal conductivity is defined as the measure of the rate of heat transfer across a unit width of material, for a unit cross-sectional area and for a unit difference in temperature.

From the definition of thermal conductivity k it can be shown that the rate of heat transfer is given by the relationship:

$$Q = \frac{kA\Delta T}{x}$$

ΔT is the temperature difference $T_1 - T_2$, defined by the temperature on the either side of the porous surface. The units of thermal conductivity can be determined from the units of the other variables:

$$Q = kA\Delta T / x$$
$$W = (k)m^2 K / m$$

The unit of k is $W / m^2 K / m$.

3.10.4 RADIATION HEAT TRANSFER

The third mode of heat transfer, radiation, does not depend on any medium for its transmission. In fact, it takes place most freely when there is a perfect vacuum between the emitter and the receiver of such energy. This is proved daily by the transfer of energy from the sun to the earth across the intervening space.

Radiation is a form of electromagnetic energy transmission and takes place between all matters providing that it is at a temperature above absolute zero. Infrared radiation form just part of the overall electromagnetic spectrum. Radiation is energy emitted by the electrons vibrating in the molecules at the surface of a body. The amount of energy that can be transferred depends on the absolute temperature of the body and the radiant properties of the nanosurface.

A body that has a surface that will absorb all the radiant energy it receives is an ideal radiator, termed a "black body". Such a body will not only absorb radiation at a maximum level but will also emit radiation at a maximum level. However, in practice, bodies do not have the surface characteristics of a black body and will always absorb, or emit, radiant energy at a lower level than a black body.

It is possible to define how much of the radiant energy will be absorbed, or emitted, by a particular surface by the use of a correction factor, known as the "emissivity" and given the symbol ε. The emissivity of a surface is the measure of the actual amount of radiant energy that can be absorbed, compared to a black body. Similarly, the emissivity defines the radiant energy emitted from a surface compared to a black body. A black body would, therefore, by definition, have an emissivity ε of 1. It should be noted that the value of emissivity is influenced more by the nature of texture of clothes, than its color. The practice of wearing white clothes in preference to dark clothes in order to keep cool on a hot summer's day is not necessarily valid. The amount of radiant energy absorbed is more a function of the texture of the clothes rather than the color.

Since World War II, there have been major developments in the use of microwaves for heating applications. After this time it was realized that microwaves had the potential to provide rapid, energy-efficient heating of materials. These main applications of microwave heating today include food processing, wood drying, plastic and rubber treating as well as curing and preheating of ceramics. Broadly speaking, microwave radiation is the term associated with any electromagnetic radiation in the microwave frequency range of 300 MHz-300 Ghz. Domestic and industrial microwave ovens generally operate at a frequency of 2.45 Ghz corresponding to a wavelength of 12.2 cm. However, not all materials can be heated rapidly by microwaves. Materials may be classified into three groups, that is, conductors' insulators and absorbers. Materials that absorb microwave radiation are called dielectrics, thus, microwave heating is also referred to as dielectric heating. Dielectrics have two important properties:

(1) They have very few charge carriers. When an external electric field is applied there is very little change carried through the material matrix.

(2) The molecules or atoms comprising the dielectric exhibit a dipole movement distance. An example of this is the stereochemistry of covalent bonds in a water molecule, giving the water molecule a dipole movement. Water is the typical case of nonsymmetric molecule. Dipoles may be a natural feature of the dielectric or they may be induced. Distortion of the electron cloud around nonpolar molecules or atoms through the presence of an external electric field can induce a temporary dipole movement. This movement generates friction inside the dielectric and the energy is dissipated subsequently as heat.

The interaction of dielectric materials with electromagnetic radiation in the microwave range results in energy absorbance. The ability of a material to absorb energy while in a microwave cavity is related to the loss tangent of the material.

This depends on the relaxation times of the molecules in the material, which, in turn, depends on the nature of the functional groups and the volume of the molecule. Generally, the dielectric properties of a material are related to temperature, moisture content, density and material geometry.

An important characteristic of microwave heating is the phenomenon of "hot spot" formation, whereby regions of very high temperature form due to nonuniform heating. This thermal instability arises because of the nonlinear dependence of the electromagnetic and thermal properties of material on temperature. The formation of standing waves within the microwave cavity results in some regions being exposed to higher energy than others.

Microwave energy is extremely efficient in the selective heating of materials as no energy is wasted in "bulk heating" the sample. This is a clear advantage that microwave heating has over conventional methods. Microwave heating processes are currently undergoing investigation for application in a number of fields where the advantages of microwave energy may lead to significant savings in energy consumption, process time and environmental remediation.

Compared with conventional heating techniques, microwave heating has the following additional advantages:

– higher heating rates;
– no direct contact between the heating source and the heated material;
– selective heating may be achieved;
– greater control of the heating or drying process.

3.10.5 COMBINED HEAT TRANSFER COEFFICIENT

For most practical situations, heat transfer relies on two, or even all three modes occurring together. For such situations, it is inconvenient to analyze each mode separately. Therefore, it is useful to derive an overall heat transfer coefficient that will combine the effect of each mode within a general situation. The heat transfer in moist fabrics takes place through three modes, conduction, radiation, and the process of distillation. With a dry fabric, only conduction and radiation are present.

3.10.6 POROSITY AND PORE SIZE DISTRIBUTION IN NANOFABRIC

The amount of porosity, that is, the volume fraction of voids within the nanofabric, determines the capacity of a nanofabric to hold water; the greater the porosity, the more water the fabric can hold. Porosity is obtained by dividing the total volume of water extruded from fabric sample by the volume of the sample:

Porosity = volume of water/volume of fabric

= (volume of water per gram sample)(density of sample)

It should be noted that most of water is stored between the yarns rather than within them. In the other words, all the water can be accommodated by the pores within the yarns, and it seems likely that the water is chiefly located there. It should be noted that pores of different sizes are distributed within a nanofabric. By a porous medium we mean a material contained a solid matrix with an interconnected void. The interconnectedness of the pores allows the flow of fluid through the fabric. In the simple situation ("single phase flow") the pores is saturated by a single fluid. In "two-phase flow" a liquid and a gas share the pore space.

The usual way of driving the laws governing the macroscopic variables are to begin with standard equations obeyed by the fluid and to obtain the macroscopic equations by averaging over volumes or areas contained many pores.

In defining porosity we may assume that all the pore space is connected. If in fact we have to deal with a fabric in which some of the pore space is disconnected from the reminder, then we have to introduce an "effective porosity", defined as the ratio of the connected pore to total volume.

A further complication arises in forced convection in fabric which is a porous medium. There may be significant thermal dispersion, i.e., heat transfer due to hydrodynamic mixing of the fluid at the pore scale. In addition to the molecular diffusion of heat, there is mixing due to the nature of the fabric.

3.10.7 THERMAL EQUILIBRIUM AND COMFORT

Some of the issues of clothing comfort that are most readily involve the mechanisms by which clothing materials influence heat and moisture transfer from skin to the environment. Heat flow by conduction, convection, and radiation and moisture transfer by vapor diffusion are the most important mechanisms in very cool or warm environments from the skin

It has been recognized that the moisture transport process in clothing under a humidity transient is one of the most important factors influencing the dynamic comfort of a wearer in practical wear situations. However, the moisture transport process is hardly a single process since it is coupled with the heat-transfer process under dynamic conditions. Some materials will posses properties promoting rapid capillary and diffusion movement of moisture to the surface and the controlling factor will be the rate at which surface evaporation can be secured. In the initial stages of drying materials of high moisture content, also it is important to obtain the highest possible rate of nanosurface evaporation. This surface evaporation is essentially the diffusion of vapor from the surface of the fabric to the surrounding atmosphere through a relatively stationary film of air in contact with its surface. This air film, in addition to presenting a resistance to the vapor flow, is itself a heat insulates. The thickness of this film rapidly decreases with increase in the velocity of the air in contact with it while never actually disappearing. The inner film of air in contact with the wet fabric remains saturated with vapor so long as the fabric surface has free moisture present. This result in a vapor pressure gradient through the film from the wetted solid surface to the outer air and, with large air movements, the rate of moisture diffusion through the air film will be considerable. The rate of diffusion, and hence evaporation of the moisture will be directly proportional to the exposed area of the fabric, inversely proportional to the film thickness and directly proportional to the inner film surface and the partial pressure of the water vapor in the surrounding air. It is of importance to note at this point that, since the layer of air film in contact with the wetted fabric undergoing drying remains saturated at the temperature of the area of contact, the temperature of the fabric surface while still possessing free moisture will lie very close to wet-bulb temperature of the air.

3.10.8 MOISTURE IN NANOFIBERS

The amount of moisture that a fiber can take up varies markedly, at low relative humidifies, below 0.35, water is adsorbed mono-molecularly by many natural fibers. From thermodynamic reasoning, we expect the movement of water through a single fiber to occur at a rate that depends on the chemical potential gradient. Meanwhile, moisture has a profound effect on the physical properties of many fibers. Hygroscopic fibers will swell as moisture is absorbed and shrink as it is driven off. Very wet fabrics lose the moisture trapped between the threads first, and only when the threads themselves dry out will shrinkage begin. The change in volume on shrinkage is normally assumed to be linear with moisture content. With hydrophilic materials moisture is found to reduce stiffness and increase creep, probably as a result of plasticization. Variations in moisture content can enhance creep. To describe movement of moisture at equilibrium relative humidity below unity, the idea of absorptive diffusion can be applied. Only those molecules with kinetic energies greater than the activation energy of the moisture fiber bonds can migrate from one site to another. The driving force for absorptive diffusion is considered to be the spreading pressure, which acts over molecular surfaces in two-dimensional geometry and is similar to the vapor pressure, which acts over three-dimensional spaces.

3.11 CONCLUSION

Now-a-days n-number of drugs are available in market, so nanotherapeutics is particularly very important to minimize health hazards caused by drugs. The most important aspect of nanotherapeutics is, it should be noninvasive and target oriented, it should have very less or no side effects. In-spite of using nonmaterial one can create nanodevices for better functioning. As most of the peptides and proteins have short half-lives so frequent dosing or injections are required, which is not applicable for many situations by using nanotherapeutics one can solve these problems. Thus, nanotherapeutics is expanding day by day in drug delivery system.

The novel method of fixation of proteins on MNPs proposed in the work was successfully realized on the example of albumin and thrombin. The blood plasma proteins are characterized by a high biocompatibility and allow decreasing toxicity of nanoparticles administered into organism. The method is based on the ability of proteins to form interchain covalent bonds under the action of free radicals. The reaction mixture for stable coatings obtaining should consist on protein solution, nanoparticles containing metals of variable valence (for example, Fe, Cu, Cr) and water-soluble initiator of free radicals generation. In this work albumin and thrombin were used for coating being formed on magnetite nanoparticles. Hydrogen peroxide served as initiator. By the set of physical (ESR-spectroscopy, ferromagnetic resonance, dynamic and Rayleigh light scattering, IR-spectroscopy) and biochemical methods it was proved that the coatings obtained are stable and formed on individual nanoparticles because free radical processes are localized strictly in the adsorption layer. The free radical linking of thrombin on the surface of nanoparticles has been shown to almost completely keep native properties of the protein molecules. Since the method provides enzyme activity and formation of thin stable protein layers on individual nanopaticles it can be successfully used for various biomedical goals concerning a smart delivery of therapeutic products and biologically active substances (including enzymes). It reveals principally novel technologies of one-step creation of biocompatible magnetically targeted nanosystems with multiprotein polyfunctional coatings, which meet all the requirements and contain both biovectors and therapeutic products (Fig. 14).

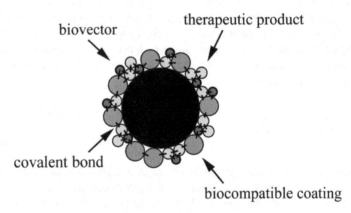

FIGURE 14   The scheme of magnetically targeted nanosystem for a smart delivery of therapeutic products based on the free radical protein cross-linking.

For Heat flow analysis of wet porous nanosystems, the liquid is water and the gas is air. Evaporation or condensation occurs at the interface between the water and air so that the air is mixed with water vapor. A flow of the mixture of air and vapor may be caused by external forces, for instance, by an imposed pressure difference. The vapor will also move relative to the gas by diffusion from regions where the partial pressure of the vapor is higher to those where it is lower.

Heat flow in porous nanosystems is the study of energy movement in the form of heat which occurs in many types of processes. The transfer of heat in porous nanostructure fabrics occurs from the high to the low temperature regions. Therefore, a temperature gradient has to exist between the two regions for heat transfer to happen. It can be done by conduction (within one porous solid or between two porous solids in contact), by convection (between two fluids or a fluid and a porous solid in direct contact with the fluid), by radiation (transmission by electromagnetic waves through space) or by combination of the above three methods.

KEYWORDS

- black body
- chemical assembly
- cluster
- convective heat transfer
- electron spectroscopy for chemical analysis
- fibrinogen
- nanostructures architecture
- science of nanomaterials

REFERENCES

1. Wu, Y.; Clark, R. L.; *Journal of Colloid and Interface Science*, **2007**, *310, 2*, 529.
2. Jaworek, A.; Sobczyk, A. T.; *Journal of Electrostatics*, **2008**, *66(3–4)*, 197.
3. Koombhongse, S.; Liu, W.; Reneker, D. H.; *Journal of Polymer Science Part B: Polymer Physics*, **2001**, *39(21)*, 2598.
4. Yu, Q. Z.; Li, Y.; Wang, M.; Chen, H. Z.; *Chinese Chemical Letters*, **2008**, *19(2)*, 223.
5. Luo, C. J.; Nangrejo, M.; Edirizinghe, M.; *Polymer*, **2010**, *51(7)*, 1654.
6. Zheng, J.; He, A.; Li, J.; Xu, J.; Han, C. C.; *Polymer*, **2006**, *47(20)*, 7095.
7. Leong, M. F.; Chian, K. S.; Mhaisalkar, P. S.; Ong, W. F.; Ratner, B. D.; *Journal of Biomedical Materials Research. Part, A.;* **2009**, *89(4)*, 1040.
8. Servoli, E.; Ruffo, G. A.; Migliaresi, C. *Polymer*, **2010**, *51(11)*, 2337.
9. Dayal, P.; Kyu, T.;*Journal of Applied Physics*, **2006**, *100(4)*, 043512.
10. Megelski, S.; Stephens, J. S.; Chase, D. B.; Rabolt, J. F.; *Macromolecules*, **2002**, *35(22)*, 8456.
11. Rangkupan, R.; Reneker, D. H.; *Journal of Metals, Materials and Minerals*, **2003**, *12(2)*, 81.
12. Dalton, P. D.; Klinkhammer, K.; Salber, J.; Klee, D.; Möller, M.; *Biomacromolecules*, **2006**, *7(3)*, 686.

13. Shin, J. W.; Shin, H. J.; Heo, S. J.; Lee, Y. J.; Hwang, Y. M.; Kim, D. H.; Kim, J. H.; J. W.; Shin in *Proceedings of the 3rd Kuala Lumpur International Conference on Biomedical Engineering (IFMBE)*, Kuala Lumpur, Malaysia, **2006**, p. 692.
14. Bonani, W.; Maniglio, D.; Motta, A.; Tan, W.; Migliaresi, C.; *Journal of Biomedical Materials Research Part B: Applied Biomaterials*, **2011**, *96(2)*, 276.
15. Lin, T.; Wang, H.; Wang, X. *Advanced Materials*, **2005**, *17(22)*, 2699.
16. Gupta, P.; Wilkes, G. L.; *Polymer*, **2003**, *44(20)*, 6353.
17. Jiang, H.; Hu, Y.; Li, Y.; Zhao, P.; Zhu, K.; Chen, W. *Journal of Controlled Release*, **2005**, *108*, 2–3, 237.
18. Díaz, J. E.; Fernández-Nieves, A.; Barrero, A.; Márquez, M.; Loscertales, I. G.; *Journal of Physics: Conference Series*, **2008**, *127*, *1*, 012008.63 *Advanced Electrospinning Setups and Special Fibre and Mesh Morphologies*.
19. Stankus, J. J.; Guan, J.; Fujimoto, K.; Wagner, W. R.; *Biomaterials*, **2006**, *27(5)*, 735.
20. Brauker, J. H.; Carr-Brendel, V. E.; Martinson, L. A.; Crudele, J.; Johnston, W. D.; Johnson, R. C.; *Journal of Biomedical Materials Research*, **1995**, *29(12)*, 1517.
21. Lu, B.; Wang, Y.; Liu, Y.; Duan, H.; Zhou, J.; Zhang, Z.; Wang, Y.; Li, X.; Wang, W.; Lan, W.; Xie, E. *Small*, **2010**, *6(15)*, 1612.
22. Fang, D.; Hsiao, B. S.; Chu, B. *Polymer Preprints*, **2003**, *44(2)*, 59.
23. Zhou, F-L.; Gong, R-H.; Porat, I. *Journal of Applied Polymer Science*, **2010**, *115*, 5, 2591.
24. Theron, S. A.; Yarin, A. L.; Zussman, E.; Kroll, E. *Polymer*, **2005**, *46(9)*, 2889.
25. Varesano, A.; Carletto, R. A.; Mazzuchetti, G.; *Journal of Materials Processing Technology*, **2009**, *209(11)*, 5178.
26. Kim, G.; Cho, Y-S.; Kim, W. D.; *European Polymer Journal*, **2006**, *42(9)*, 2031.
27. Yamashita, Y.; Ko, F.; Miyake, H.; Higashiyama, A.; *Sen'i Gakkaishi*, **2008**, *64(1)*, 24.61 *Advanced Electrospinning Setups and Special Fibre and Mesh Morphologies*
28. Zhou, F-L.; Gong, R-H.; Porat, I. *Polymer Engineering and Science*, **2009**, *49(12)*, 2475.
29. Kumar, A.; Wei, M.; Barry, C.; Chen, J.; Mead, J. *Macromolecular Materials and Engineering*, **2010**, *295*, 8, 701.
30. Dosunmu, O. O.; Chase, G. G.; Kataphinan, W.; Reneker, D. H.; *Nanotechnology*, **2006**, *17(4)*, 1123.
31. Badrossamay, M. R.; McIlwee, H. A.; Goss, J. A.; Parker, K. K.; *Nano Letters*, **2010**, *10(6)*, 2257.
32. Srivastava, Y.; Marquez, M.; Thorsen, T.; *Journal of Applied Polymer Science*, **2007**, *106(5)*, 3171.
33. Jirsak, O.; Sanetrnik, F.; Lukas, D.; Kotek, V.; Martinova, L.; Chaloupek, J.; Inventors, Technicka Universita V Liberci, assignee; US Patent 7585437B2, **2009**.
34. Lukas, D.; Sarkar, A.; Pokorny, P. *Journal of Applied Physics*, **2008**, *103(8)*, 084309.
35. Yarin, A.; Zussman, E. *Polymer*, **2004**, *45(9)*, 2977.
36. Thoppey, N. M.; Bochinski, J. R.; Clarke, L. I.; Gorga, R. E.; *Polymer*, **2010**, *51(21)*, 4928.
37. Tang, S.; Zeng, Y.; Wang, X. *Polymer Engineering and Science*, **2010**, *50(11)*, 2252.
38. Cengiz, F.; Dao, T. A.; Jirsak, O. *Polymer Engineering and Science*, **2010**, *50(5)*, 936.
39. Jirsak, O.; Syzel, P.; Sanetrnik, F.; Hruza, J.; Chaloupek, J. *Journal of Nanomaterials*, **2010**, 2010, 1.

40. Wang, X.; Niu, H.; Lin, T.; Wang, X.; *Polymer Engineering and Science*, **2009**, *49(8)*, 1582.
41. Liu, Y.; He, J-H.; Yu, J-Y. *Journal of Physics: Conference Series*, **2008**, *96*, 012001.
42. Liu, Y.; Dong, L.; Fan, J.; Wang, R.; Yu, J-Y. *Journal of Applied Polymer Science*, **2011**, *120, 1*, 592.62 *Electrospinning for Advanced Biomedical Applications and Therapies*
43. Salem, D. R.; in *Nanofibers and Nanotechnology in Textiles*, Brown, P. J.; Stevens, K. (eds).; CRC Press, Boca Raton, FL, USA, **2007**, p.1.
44. Kelly, A. J.; *Journal of Aerosol Science*, **1994**, *25(6)*, 1159.
45. Sun, D.; Chang, C.; Li, S.; Lin, L. *Nano Letters*, **2006**, *6(4)*, 839.
46. Chang, C.; Limkrailassiri, K.; Lin, L. *Applied Physics Letters*, **2008**, *93(12)*, 123111.
47. Levit, N.; Tepper, G. *The Journal of Supercritical Fluids*, **2004**, *31(3)*, 329.
48. Larrondo, L.; Manley, R. S. J.; *Journal of Polymer Science. Part B.; Polymer Physics*, **1981**, *19(6)*, 909.
49. Liu, Y.; Wang, X.; Li, H.; Yan, H.; Yang, W. *Society of Plastics Engineers Plastics Research Online*, **2010**, 10.1002/spepro.003055.
50. Dalton, P. D.; Grafahrend, D.; Klinkhammer, K.; Klee, D.; Möller, M.*Polymer*, **2007**, *48(23)*, 6823.
51. Hou, Q.; Grijpma, D. W.; Feijen, J. *Biomaterials*, **2003**, *24(11)*, 1937.
52. Nam, J.; Huang, Y.; Agarwal, S.; Lannutti, J. *Tissue Engineering*, **2007**, *13(9)*, 2249.
53. Kim, T. G.; Chung, H. J.; Park, T. G.; *Acta Biomaterialia*, **2008**, *4(6)*, 1611.
54. Baker, B. M.; Gee, A. O.; Metter, R. B.; Nathan, A. S.; Marklein, R. A.; Burdick, J. A.; Mauck, R. L.; *Biomaterials*, **2008**, *29(15)*, 2348.
55. Whited, B. M.; Whitney, J. R.; Hofmann, M. C.; Xu, Y.; Rylander, M. N.; *Biomaterials*, **2011**, *32(9)*, 2294.
56. Gentsch, R.; Boyzen, B.; Lankenau, A.; Börner, H. G.; *Macromolecular Rapid Communications*, **2010**, *31(1)*, 59.
57. Soliman, S.; Pagliari, S.; Rinaldi, A.; Forte, G.; Fiaccavento, R.; Pagliari, F.; Franzese, O.; Minieri, M.; Di Nardo, P.; Licoccia, S.; Traversa, E. *Acta Biomaterialia*, **2010**, *6(4)*, 1227.
58. Santos, M. I.; K.; Tuzlakoglu, S.; Fuchs, Gomes, M. E.; K.; Peters, Unger, R. E.; E.; Piskin, Reis, R. L.; Kirkpatrick, C. J.; *Biomaterials*, **2008**, *29(32)*, 4306.
59. Tuzlakoglu, K.; Bolgen, N.;Salgado, A. J.; Gomes, M. E.; Piskin, E.; Reis, R. L.; *Journal of Materials Science: Materials in Medicine*, **2005**, *16(12)*, 1099.65 *Advanced Electrospinning Setups and Special Fibre and Mesh Morphologies*
60. Thorvaldsson, A.; Stenhamre, H.; Gatenholm, P.; Walkenström, P.; *Biomacromolecules*, **2008**, *9(3)*, 1044.
61. Smit, E.; Büttner, U.; Sanderson, R. D.; *Polymer*, **2005**, *46(8)*, 2419.
62. Khil, M-S.; Bhattarai, S. R.; Kim, H-Y.; Kim S-Z.; Lee, K-H.; *Journal of Biomedical Materials Research. Part, B.; Applied Biomaterials*, **2005**, *72(1)*, 117.
63. Teo, W. E.; Ramakrishna, S. *Nanotechnology*, **2006**, *17(14)*, R89.64 *Electrospinning for Advanced Biomedical Applications and Therapies*
64. Simonet, M.; Schneider, O. D.; Neuenschwander, P.; Stark, W. J.; *Polymer Engineering and Science Engineering*, **2007**, *47(12)*, 2020.
65. Leong, M. F.; Rasheed, M. Z.; Lim, T. C.; Chian, K. S.; *Journal of Biomedical Materials Research Part, A.;* **2009**, *91(1)*, 231.

66. Schneider, O. D.; Weber, F.; Brunner, T. J.; Loher, S.; Ehrbar, M.; Schmidlin, P. R.; Stark, W. J.; *Acta Biomaterialia*, **2009**, *5(5)*, 1775.
67. Nam, Y. S.; Yoon, J. J.; Park, T. G.; *Journal of Biomedical Materials Research*, **2000**, *53(1)*, 1.
68. Lee, Y. H.; Lee, J. H.; An, I-G.; Kim, C.; Lee, D. S.; Lee, Y. K.; Nam, J-D.; *Biomaterials*, **2005**, *26(16)*, 3165.
69. Curtis, S. G.; Gadegaard, A. N.; Dalby, M. J.; Riehle, M. O.; Wilkinson, D. W.; Aitchison, C. G. *IEEE Transactions in Nanobioscience*, **2004**, *3(1)*, 61.
70. Wang, Y.; Bella, E.; Lee, S. D.; Migliaresi, C. C.; Pelcastre, L.; Schwartz, Z.; Boyan, B. D.; Motta, A. *Biomaterials*, **2010**, *31(17)*, 4672.
71. Kim, G.; Son, J.; Park, S.; Kim, W. *Macromolecular Rapid Communications*, **2008**, *29(19)*, 1577.
72. Ahn, S. H.; Koh, Y. H.; Kim, G. H.; *Journal of Micromechanics and Microengineering*, **2010**, *20(6)*, 065015.
73. Yoon, H.; Kim, G. *Journal of Pharmaceutical Sciences*, **2011**, *100(2)*, 424.
74. Tayalia, P.; Mendonca, C. R.; Baldacchini, T.; Mooney, D. J.; E Mazur, *Advanced Materials*, **2008**, *20(23)*, 4494.
75. Sachlos, E.; Czernuszka, J. T.; *European Cells and Materials*, **2003**, *5*, 29.
76. Mota, C.; Puppi, D.; Dinucci, D.; Errico, C.; Bártolo, P.; Chiellini, F. *Materials*, **2011**, *4(3)*, 527.
77. Wang, L.; Pai, C. -L.; Boyce, M. C.; Rutledge, G. C.; *Applied Physics Letters*, **2009**, *94(15)*, 151916.
78. Casper, C. L.; Stephens, J. S.; Tassi, N. G.; Chase, D. B.; Rabolt, J. F.; *Macromolecules*, **2004**, *37(2)*, 573.
79. Rosso, F.; Giordano, A.; Barbarisi, M.; Barbarisi, A. *Journal of Cellular Physiology*, **2004**, *199, 2*, 174.
80. Boyan, B. D.; Lossdörfer, S.; Wang, L.; Zhao, G.; Lohmann, C. H.; Cochran, D. L.; Z Schwartz, *European Cells and Materials*, **2003**, *6*, 22.
81. Chung, T-W.; Liu, D-Z.; Wang, S-Y.; Wang, S-S.; *Biomaterials*, **2003**, *24(25)*, 4655.
82. Moroni, L.; Licht, R.; de Boer, J.; de Wijn, J. R.; van Blitterswijk, C. A. *Biomaterials*, **2006**, *27(28)*, 4911.
83. Truong, Y. B.; Glattauer, V.; Lang, G.; Hands, K.; Kyratzis, I. L.; Werkmeister, J. A.; J. Ramshaw, A. M.; *Biomedical Materials*, **2010**, *5(2)*, 25005.
84. Lee, K. Y.; Yuk, S. H.; *Progress in Polymer Science*, **2007**, *32(7)*, 669.
85. Tao, S. L.; Desai, T. A.; *Advanced Drug Delivery Reviews*, **2003**, *55(3)*, 315.66 *Electrospinning for Advanced Biomedical Applications and Therapies*
86. Moroni, L.; de Wijn, J. R.; van Blitterswijk, C. A. *Journal of Biomaterials Science. Polymer Edition*, **2008**, *19(5)*, 543.
87. Luginbuehl, V.; Meinel, L.; Merkle, H. P.; B Gander, *European Journal of Pharmaceutics and Biopharmaceutics*, **2004**, *58(2)*, 197.
88. Willerth, S. M.; Sakiyama-Elbert, S. E.; *Advanced Drug Delivery Reviews*, **2007**, *59*, 4–5, 325.
89. Biondi, M.; Ungaro, F.; Quaglia, F.; Netti, P. A.; *Advanced Drug Delivery Reviews*, **2008**, *60(2)*, 229.
90. Munir, M. M.; Suryamas, A. B.; Iskandar, F.; Okuyama, K. *Polymer*, **2009**, *50(20)*, 4935.

91. Qi, Z.; Yu, H.; Chen, Y.; Zhu, M. *Materials Letters*, **2009**, *63*, 3–4, 415.
92. Luna-Bárcenas, G.; Kanakia, S. K.; Sanchez, I. C.; Johnston, K. P.; *Polymer*, **1995**, *36(16)*, 3173.67 *Advanced Electrospinning Setups and Special Fibre and Mesh Morphologies*
93. Liu, J.; Shen, Z.; Lee, S-H.; Marquez, M.; McHugh, M. A. *The Journal of Supercritical Fluids*, **2010**, *53*, 1–3, 142.
94. McCann, J. T.; Marquez, M.; Xia, Y. *Journal of the American Chemical Society*, **2006**, *128*, *5*, 1436.
95. You, Y.; Youk, J. H.; Lee, S. W.; Min, B-M.; Lee, S. J.; Park, W. H.; *Materials Letters*, **2006**, *60(6)*, 757.
96. Lyoo, W. S.; Youk, J. H.; Lee, S. W.; Park, W. H.; *Materials Letters*, **2005**, *59(28)*, 3558.
97. Chakraborty, S.; Liao, I-C.; Adler, A.; Leong, K. W.; *Advanced Drug Delivery Reviews*, **2009**, *61(12)*, 1043.
98. Srikar, R.; Yarin, A. L.; Megaridis, C. M.; Bazilevsky, A. V.; Kelley, E. *Langmuir*, **2008**, *24(3)*, 965.
99. Gandhi, M.; Srikar, R.; Yarin, A. L.; Megaridis, C. M.; Gemeinhart, R. A.; *Molecular Pharmaceutics*, **2009**, *6(2)*, 641.
100. Wang, M.; Yu, J. H.; Kaplan, D. L.; Rutledge, G. C.; *Macromolecules*, **2006**, *39(3)*, 1102.
101. Gupta, A. K.; Gupta, M.; *Biomaterials* **2005**, *26*, 3995
102. Vatta, L. L.; Sanderson, D. R.; Koch, K. R.; *Pure Appl. Chem.;* **2006**, *78*, 1793.
103. Laurent, S.; Forge, D.; Port, M.; Roch, A.; Robic, C.; Elst, L. V.; Muller, R. N.; *Chem. Rev.;* **2008**, *108*, 2064.

ADDITIONAL READINGS

Deitzel, J.; Kleinmeyer, J. D.; Hirvonen, J. K.; Beck Tan, N. C.; *Polymer*, **2001**, *42(19)*, 8163.

Hong, J. K.; Madihally, S. V.; *Tissue Engineering*, **2011**, *17(2)*, 125.

Huang, Z-M.; Zhang, Y-Z.; Kotaki, M.; Ramakrishna, S. *Composites Science and Technology*, **2003**, *63(15)*, 2223.

Katta, P.; Alessandro, M.; Ramsier, R. D.; Chase, G. G.; *Nano Letters*, **2004**, *4*, *11*, 2215.

Li, D.; Wang, Y.; Xia, Y. *Advanced Materials*, **2004**, *16(4)*, 361.

Li, D.; Wang, Y.; Xia, Y. *Nano Letters*, **2003**, *3(8)*, 1167.

Teo, W. E.; Kotaki, M.; Mo, X. M. Ramakrishna, S. *Nanotechnology*, **2005**, *16(6)*, 918.

Theron, A.; Zussman, E.; Yarin, A. L.; *Nanotechnology*, **2001**, *12(3)*, 384.

Thorvaldsson, A.; Engström, J.; Gatenholm, P.; Walkenström, P. *Journal of Applied Polymer Science*, **2010**, *118(1)*, 511.

Volpato, F. Z.; S. Ramos, L. F.; Motta, A.; Migliaresi, C. *Journal of Bioactive and Compatible Polymers*, **2011**, *26(1)*, 35.

CHAPTER 4

CARBON NANOTUBES AND RELATED STRUCTURES

A. K. HAGHI and G. E. ZAIKOV

CONTENTS

4.1 INTRODUCTION

The appearance of "nanoscience" and "nanotechnology" stimulated the burst of terms with "nano-" prefix. Historically the term "nanotechnology" appeared before and it was connected with the appearance of possibilities to determine measurable values up to 10^{-9} of known parameters: 10^{-9} m-nm (nanometer), 10^{-9} s-ns (nanosecond), 10^{-9} degree (nanodegree, shift condition). Nanotechnology and molecular nanotechnology comprise the set of technologies connected with transport of atoms and other chemical particles (ions, molecules) at distances contributing the interactions between them with the formation of nanostructures with different nature. Researchers showed the possibility to develop technologies on nanometer level, Eric Drexler is considered to be the founder and ideologist of nanotechnology. When scanning tunnel microscope was invented there was an opportunity to influence atoms of a substance thus stimulating the work in the field of probe technology, which resulted in substantiation and practical application of nanotechnological methods in 1994. With the help of this technique it is possible to handle single atoms and collect molecules or aggregates of molecules, construct various structures from atoms on a certain substrate (base). Naturally, such a possibility cannot be implemented without preliminary computer designing of so-called "nanostructures architecture." Nanostructures architecture assumes a certain given location of atoms and molecules in space that can be designed on computer and afterwards transferred into technological program of nanotechnological facility.

The notion "science of nanomaterials" assumes scientific knowledge for obtaining, composition, properties and possibilities to apply nanostructures, nanosystems and nanomaterials. A simplified definition of this term can be as follows: material science dealing with materials comprizing particles and phases with nanometer dimensions. To determine the existence area for nanostructures and nanosystems it is advisable to find out the difference of these formations from analogous material objects.

From the analysis of literature the following can be summarized: the existence area of nanosystems and nanoparticles with any structure is between the particles of molecular and atomic level determined in picometers and aggregates of molecules or permolecular formations over micron units. Here, it should be mentioned that in polymer chemistry particles with nanometer dimensions belong to the class of permolecular structures, such as globules and fibrils by one of parameters, for example, by diameter or thickness. In chemistry of complex compounds clusters with nanometer dimensions are also known.

In 1991, Japanese researcher Idzhima was studying the sediments formed at the cathode during the spray of graphite in an electric arc. His attention was attracted by the unusual structure of the sediment consisting of microscopic fibers and filaments. Measurements made with an electron microscope showed that the diameter of these filaments does not exceed a few nanometers and a length of one to several microns.

Having managed to cut a thin tube along the longitudinal axis, the researchers found that it consists of one or more layers, each representing a hexagonal grid of graphite, which is based on hexagon with vertices located at the corners of the carbon atoms. In all cases, the distance between the layers is equal to 0.34 nm that is the same as that between the layers in crystalline graphite.

Typically, the upper ends of tubes are closed by multilayer hemispherical caps, each layer is composed of hexagons and pentagons, reminiscent of the structure of half a fullerene molecule.

The extended structure consisting of rolled hexagonal grids with carbon atoms at the nodes are called nanotubes.

Lattice structure of diamond and graphite are shown in Fig. 1 Graphite crystals are built of planes parallel to each other, in which carbon atoms are arranged at the corners of regular hexagons. Each intermediate plane is shifted somewhat toward the neighboring planes, as shown in the Fig. 1.

The elementary cell of the diamond crystal represents a tetrahedron, with carbon atoms in its center and four vertices. Atoms located at the vertices of a tetrahedron form a center of the new tetrahedron, and thus, are also surrounded by four atoms each, etc. All the carbon atoms in the crystal lattice are located at equal distance (0.154 nm) from each other.

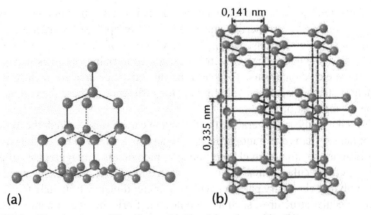

FIGURE 1   The structure of the diamond lattice (a) and graphite (b).

Nanotubes are rolled into a cylinder (hollow tube) graphite plane, which is lined with regular hexagons with carbon atoms at the vertices of a diameter of several nanometers. Nanotubes can consist of one layer of atoms – single-wall nanotubes SWNT and represent a number of "nested" one into another layer pipes – multiwalled nanotubes – MWNT.

Nanostructures can be built not only from individual atoms or single molecules, but the molecular blocks. Such blocks or elements to create nanostructures are graphene, carbon nanotubes and fullerenes.

4.1.1 GRAPHENE

Graphene is a single flat sheet, consisting of carbon atoms linked together and forming a grid, each cell is like a bee's honeycombs (Fig. 2). The distance between adjacent carbon atoms in graphene is about 0.14 nm.

Graphite, from which slates of usual pencils are made, is a pile of graphene sheets (Fig. 3). Graphenes in graphite is very poorly connected and can slide relative to each other. So, if you conduct the graphite on paper, then after separating graphene from sheet the graphite remains on paper. This explains why graphite can write.

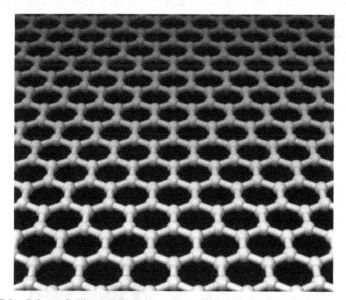

FIGURE 2 Schematic illustration of the graphene. Light balls – the carbon atoms, and the rods between them – the connections that hold the atoms in the graphene sheet.

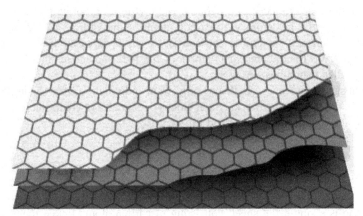

FIGURE 3   Schematic illustration of the three sheets of graphene, which are one above the other in graphite

4.1.2   CARBON NANOTUBES

Many perspective directions in nanotechnology are associated with carbon nanotubes.

Carbon nanotubes are a carcass structure or a giant molecule consisting only of carbon atoms.

Carbon nanotube is easy to imagine, if we imagine that we fold up one of the molecular layers of graphite – graphene (Fig. 4).

FIGURE 4   Carbon nanotubes.

The way of folding nanotubes – the angle between the directions of nanotube axis relative to the axis of symmetry of graphene (the folding angle) – largely determines its properties.

Of course, no one produces nanotubes, folding it from a graphite sheet. Nanotubes formed themselves, for example, on the surface of carbon electrodes during arc discharge between them. At discharge, the carbon atoms evaporate from the surface, and connect with each other to form nanotubes of all kinds – single, multilayered and with different angles of twist (Fig. 5).

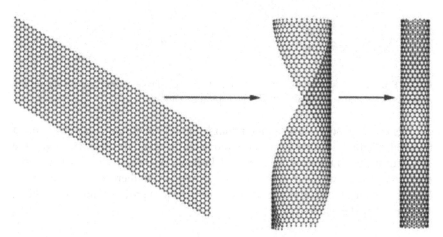

FIGURE 5 One way of imaginary making nanotube (right) from the molecular layer of graphite (left).

The diameter of nanotubes is usually about 1 nm and their length is a thousand times more, amounting to about 40 microns. They grow on the cathode in perpendicular direction to surface of the butt. The so-called self-assembly of carbon nanotubes from carbon atoms occurs. Depending on the angle of folding of the nanotube they can have conductivity as high as that of metals, and they can have properties of semiconductors.

Carbon nanotubes are stronger than graphite, although made of the same carbon atoms, because carbon atoms in graphite are located in the sheets. And everyone knows that sheet of paper folded into a tube is much more difficult to bend and break than a regular sheet. That's why carbon nanotubes are strong. Nanotubes can be used as a very strong microscopic rods and filaments, as Young's modulus of single-walled nanotube reaches values of the order of 1–5 TPa, which is much more than steel! Therefore, the thread made of nanotubes, the thickness of a human hair is capable to hold down hundreds of kilos of cargo.

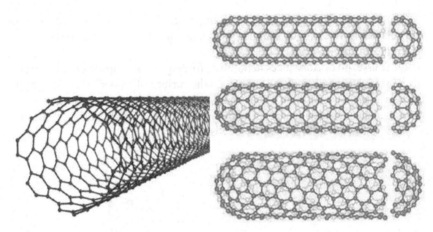

FIGURE 6   Left – schematic representation of a single-layer carbon nanotubes, on the right (top to bottom) – two-ply, straight and spiral nanotubes.

It is true that at present the maximum length of nanotubes is usually about a hundred microns – which is certainly too small for everyday use. However, the length of the nanotubes obtained in the laboratory is gradually increasing – now scientists have come close to the millimeter border. So there is every reason to hope that in the near future, scientists will learn how to grow a nanotube length in centimeters and even meters.

4.1.3   FULLERENES

The carbon atoms, evaporated from a heated graphite surface, connecting with each other, can form not only nanotube, but also other molecules, which are closed convex polyhedra, for example, in the form of a sphere or ellipsoid. In these molecules, the carbon atoms are located at the vertices of regular hexagons and pentagons, which make up the surface of a sphere or ellipsoid.

The molecules of the symmetrical and the most studied fullerene consisting of 60 carbon atoms (C_{60}), form a polyhedron consisting of 20 hexagons and 12 pentagons and resembles a soccer ball (Fig. 7). The diameter of the fullerene C_{60} is about 1 nm.

FIGURE 7   Schematic representation of the fullerene C_{60}.

4.1.4   CLASSIFICATION OF NANOTUBES

The main classification of nanotubes is conducted by the number of constituent layers.

Single-walled nanotubes – the simplest form of nanotubes. Most of them have a diameter of about 1 nm in length, which can be many thousands of times more. The structure of the nanotubes can be represented as a "wrap" a hexagonal network of graphite (graphene), which is based on hexagon with vertices located at the corners of the carbon atoms in a seamless cylinder. The upper ends of the tubes are closed by hemispherical caps, each layer is composed of hexa – and pentagons, reminiscent of the structure of half of a fullerene molecule. The distance d between adjacent carbon atoms in the nanotube is approximately equal to $d = 0,15$ nm (Fig. 8).

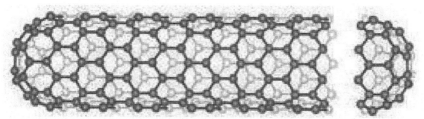

FIGURE 8   Graphical representation of single-walled nanotube.

Multi-walled nanotubes consist of several layers of graphene stacked in the shape of the tube. The distance between the layers is equal to 0.34 nm, that is the same as that between the layers in crystalline graphite (Fig. 9).

Carbon nanotubes could find applications in numerous areas:
- additives in polymers;
- catalysts (autoelectronic emission for cathode ray lighting elements, planar panel of displays, gas discharge tubes in telecom networks);
- absorption and screening of electromagnetic waves;
- transformation of energy;
- anodes in lithium batteries;
- keeping of hydrogen;
- composites (filler or coating);
- nanosondes;
- sensors;
- strengthening of composites;
- supercapacitors.

FIGURE 9   Graphic representation of a multiwalled nanotube.

More than a decade, carbon nanotubes, despite their impressive performance characteristics have been used, in most cases, for scientific research.

To date, the most developed production of nanotubes has Asia, the production capacity that is 2–3 times higher than in North America and Europe combined. Is dominated by Japan, which is a leader in the production of MWNT. Manufacturing North America, mainly focused on the SWNT. Growing at an accelerated rate production in China and South Korea. In the coming years, China will surpass the level of production of the U.S. and Japan, and by 2013, a major supplier of all types of nanotubes, according to experts, could be South Korea.

4.1.5 CHIRALITY

Chirality is a set of two integer positive indices (n, m), which determines how the graphite planes folds and how many elementary cells of graphite at the same time fold to obtain the nanotube.

From the value of parameters (n, m) are distinguished as direct (achiral) high-symmetry carbon nanotubes.

o   armchair $n = m$
o   zigzag $m = 0$ or $n = 0$
o   helical (chiral) nanotube

In Fig. 10a is shown a schematic image of the atomic structure of graphite plane – graphene, and shown how a nanotube can be obtained from it. The nanotube is fold up with the vector connecting two atoms on a graphite sheet. The cylinder is obtained by folding this sheet so that were combined the beginning and end of the vector. That is, to obtain a carbon nanotube from a graphene sheet, it should turn so that the lattice vector $\overline{R}$ has a circumference of the nanotube in Fig. 10b. This vector can be expressed in terms of the basis vectors of the elementary cell graphene sheet $\vec{R} = n\vec{r_1} + m\vec{r_2}$. Vector $\overline{R}$, which is often referred to simply by a pair of indices (n, m), called the chiral vector. It is assumed that $n > m$. Each pair of numbers (n, m) represents the possible structure of the nanotube.

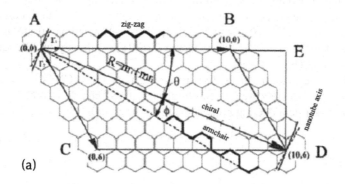

(a)

FIGURE 10   *(Continued)*

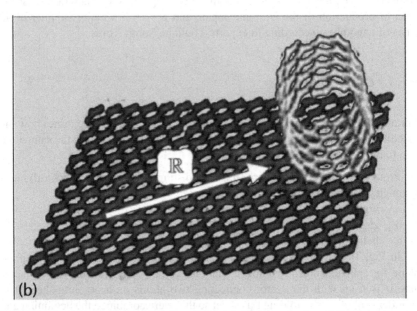

(b)

FIGURE 10   Schematic representation of the atomic structure of graphite plane.

In other words the chirality of the nanotubes (n, m) indicates the coordinates of the hexagon, which as a result of folding the plane has to be coincide with a hexagon, located at the beginning of coordinates (Fig. 11).

Many of the properties of nanotubes (e.g., zonal structure or space group of symmetry) strongly depend on the value of the chiral vector. Chirality indicates what property has a nanotube – a semiconductor or metallicheskm. For example, a nanotube (10,10) in the elementary cell contains 40 atoms and is the type of metal, whereas the nanotube (10, 9) has already in 1084 and is a semiconductor (Fig. 12).

If the difference $n - m$ is divisible by 3, then these CNTs have metallic properties. Semimetals are all achiral tubes such as "chair". In other cases, the CNTs show semiconducting properties. Just type chair CNTs $(n = m)$ are strictly metal.

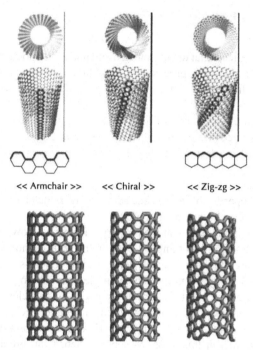

FIGURE 11 Single-walled carbon nanotubes of different chirality (in the direction of convolution). Left to right: the zigzag (16,0), armchair (8,8) and chiral (10,6) carbon nanotubes.

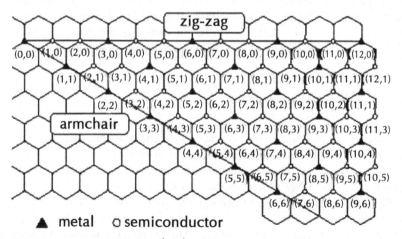

FIGURE 12 The scheme of indices (n, m) of lattice vector $\bar{R}$ tubes having semiconductor and metallic properties.

4.2   EFFECTIVE PARAMETERS

In addition to size effects that occur in micro and nano, we should note the following factors that determine the processes in low-dimensional systems:
- surface roughness (resistance to flow effects, interactions with the particles, etc.);
- dissolved gases (formation of bubbles, sticking to the surface, etc.);
- chemical surface properties (chemical reactions, etc.);
- hydrophobic – hydrophilic of surface;
- contaminants;
- heating due to uncontrollable processes;
- electrical properties of the surface (double layer, the change of surface and volume charge, charge transfer, etc.).

Viscous forces in the fluid can lead to large dispersion flow along the axis of motion. They have a significant impact, both on the scale of individual molecules, and the scale of microflows – near the borders of the liquid-solid (beyond a few molecular layers), during the motion on a complex and heterogeneous borders.

Influence of the effect of boundary regions on the particles and fluxes have been observed experimentally in the range of molecular thicknesses up to hundreds of nanometers. If the surface has a superhydrophobic properties, this range can extend to the micron thickness. Molecular theory can predict the effect of hydrophobic surfaces in the system only up to tens of nanometers.

Fluids, the flow of liquid or gas, have properties that vary continuously under the action of external forces. In the presence of fluid shear forces are small in magnitude, leads large changes in the relative position of the element of fluid. In contrast, changes in the relative positions of atoms in solids remain small under the action of any small external force. Termination of action of the external forces on the fluid does not necessarily lead to the restoration of its initial form.

4.2.1   CAPILLARY EFFECTS

To observe the capillary effects, you must open the nanotube, that is, remove the upper part – lids. Fortunately, this operation is quite simple.

The first study of capillary phenomena have shown that there is a relationship between the magnitude of surface tension and the possibility of its being drawn into the channel of the nanotube. It was found that the liquid penetrates into the channel of the nanotube, if its surface tension is not higher than 200 mN/m. Therefore, for the entry of any substance into the nanotube using solvents having a low surface tension. For example concentrated nitric acid with surface tension of 43 mN/m is used to inject certain metals into the channel of a nanotube. Then annealing is conducted at 4000 °C for 4 h in an atmosphere of hydrogen, which leads to the recovery of the metal. Thus, the obtained nanotubes containing nickel, cobalt and iron.

Along with the metals carbon nanotubes can be filled with gaseous substances, such as hydrogen in molecular form. This ability is of great practical importance, since opening the ability to safely store hydrogen, which can be used as a clean fuel in internal combustion engines.

4.2.2 SPECIFIC ELECTRICAL RESISTANCE OF CARBON NANOTUBES

Due to small size of carbon nanotubes only in 1996 they succeeded to directly measure their electrical resistivity ρ . The results of direct measurements showed that the resistivity of the nanotubes could be varied within wide limits to 0.8 ohm/ cm. The minimum value is lower than that of graphite. Most of the nanotubes have metallic conductivity, and the smaller shows properties of a semiconductor with a bandgap of 0.1 to 0.3 eV.

The resistance of single-walled nanotube is independent of its length, because of this it is convenient to use for the connection of logic elements in microelectronic devices. The permissible current density in carbon nanotubes is much greater than in metallic wires of the same cross section, and one hundred times better achievement for superconductors.

The results of the study of emission properties of the material, where the nanotubes were oriented perpendicular to the substrate, have been very interesting for practical use. Attained values of the emission current density of the order of 0.5 mA/mm^2 . The value obtained is in good agreement with the Fowler-Nordheim expression.

The most effective and common way to control microflow substances is *electrokinetic* and *hydraulic.* At the same time the most technologically advanced and automated considered electrokinetic.

Charges transfer in mixtures occurs as a result of the directed motion of charge carriers – ions. There are different mechanisms of such transfer, but usually are *convection, migration and diffusion.*

Convection is called mass transfer the macroscopic flow. *Migration* – the movement of charged particles by electrostatic fields. The velocity of the ions depends on field strength. In microfluidics a special role is played *electrokinetic processes* that can be divided into four types: *electro-osmosis, electrophoresis, streaming potential and sedimentation potential.* These processes can be qualitatively described as follows:

(a) *electro-osmosis* – the movement of the fluid volume in response to the applied electric field in the channel of the electrical double layers on its wetted surfaces.

(b) *Electrophoresis* – the forced motion of charged particles or molecules, in mixture with the acting electric field.

(c) *Streamy potential* – the electric potential, which is distributed through a channel with charged walls, in the case when the fluid moves under the action of pressure forces. Joule electric current associated with the effect of charge transfer is flowing stream.

(d) *The potential of sedimentation* – an electric potential is created when charged particles are in motion relative to a constant fluid. The driving force for this effect – usually gravity.

In general, for the microchannel cross-section S amount of introduced probe (when entering electrokinetic method) depends on the applied voltage U, time t during which the received power, and mobility of the sample components μ:

$$Q = \frac{\mu S U t}{L} \cdot c$$

where;

c – probe concentration in the mixture,

L – the channel length.

Amount of injected substance is determined by the electrophoretic and total electroosmotic mobilities μ.

In the hydrodynamic mode of entry by the pressure difference in the channel or capillary of circular cross section, the volume of injected probe V_c:

$$V_c = \frac{4}{128} \cdot \frac{\Delta p \pi d t}{\eta L}$$

where;

Δp – pressure differential,

d – diameter of the channel,

η – viscosity.

In the simulation of processes in micron-sized systems the following basic principles are fundamental:

1. hypothesis of *laminar* flow (sometimes is taken for granted when it comes to microfluidics);

2. continuum hypothesis (detection limits of applicability);

3. laws of formation of the velocity profile, mass transfer, the distribution of electric and thermal fields;

4. boundary conditions associated with the geometry of structural elements (walls of channels, mixers zone flows, etc.).

Since we consider the physical and chemical transport processes of matter and energy, mathematical models, most of them have the form of systems of differential equations of second order partial derivatives. Methods for solving such equations are analytical (Fourier and its modifications, such as the method of Greenberg, Galerkin, in some cases, the method of d'Alembert and the Green's functions, the Laplace operator method, etc.) or numerical (explicit or, more effectively, implicit finite difference schemes) – traditional. The development involves, basically, numerical methods and follows the path of saving computing resources, and increasing the speed of modern computers.

Laminar flow – a condition in which the particle velocity in the liquid flow is not a random function of time. The small size of the microchannels (typical dimensions of 5 to 300 microns) and low surface roughness create good conditions for the establishment of laminar flow. Traditionally, the image of the nature of the flow gives the dimensionless characteristic numbers: the Reynolds number and Darcy's friction factor.

In the motion of fluids in channels the turbulent regime is rarely achieved. At the same time, the movement of gases is usually turbulent.

Although the liquid – are quantized in the length scale of intermolecular distances (about 0.3 nm to 3 nm in liquids and for gases), they are assumed to be continuous in most cases, microfluidics. Continuum hypothesis (continuity, continuum) suggests that the macroscopic properties of fluids consisting of molecules, the same as if the fluid was completely continuous (structurally homogeneous). Physical characteristics: mass, momentum and energy associated with the volume of fluid containing a sufficiently large number of molecules must be taken as the sum of all the relevant characteristics of the molecules.

Continuum hypothesis leads to the concept of fluid particles. In contrast to the ideal of a point particle in ordinary mechanics, in fluid mechanics, particle in the fluid has a finite size.

At the atomic scale we would see large fluctuations due to the molecular structure of fluids, but if we the increase the sample size, we reach a level where it is possible to obtain stable measurements. This volume of probe must contain a sufficiently large number of molecules to obtain reliable reproducible signal with small statistical fluctuations. For example, if we determine the required volume as a cube with sides of 10 nm, this volume contains some of the molecules and determines the level of fluctuations of the order of 0.5%.

The most important position in need of verification is to analyze the admissibility of mass transfer on the basis of the continuum model that can be used instead of the concentration dependence of the statistical analysis of the ensemble of individual particles. The position of the continuum model is considered as a necessary condition for microfluidics.

The applicability of the hypothesis is based on comparison of free path length of a particle λ in a liquid with a characteristic geometric size d. The ratio of these lengths – the Knudsen number: $Kn = \lambda / d$. Based on estimates of the Knudsen number defined two important statements:

a) $Kn < 10^{-3}$ – justified hypothesis of a continuous medium,

b) $Kn < 10^{-1}$ – allowed the use of adhesion of particles to the solid walls of the channel.

Wording of the last condition can also be varied: both in form $U = 0$ and in a more complex form, associated with shear stresses. The calculation of λ can be carried out as $\lambda \approx \sqrt[3]{\overline{V} / Na}$, where;

$\overline{V}$ – molar volume,

Na – Avogadro's number.

Under certain geometrical approximations of the particles of substance free path length can be calculated as $\lambda \approx 1 / \left(\sqrt{2} \pi r_S^2 Na \right)$, if used instead r_S Stokes radius, as a consequence of the spherical approximation of the particle. On the other hand, for a rigid model of the molecule r_S should be replaced by the characteristic size of the particles R_g – the radius of inertia, calculated as $R_g = n_i \cdot \delta_i / \sqrt{6}$. Here δ_i – the length of a fragment of the chain (link), n_i – the number of links.

Of course, the continuum hypothesis is not acceptable when the system under consideration is close to the molecular scale. This happens in nanoliquid, such as liquid transport through nano-pores in cell membranes or artificially made nano-channels.

In contrast to the continuum hypothesis, the essence of modeling the molecular dynamics method is as follows. We consider a large ensemble of particles, which simulate atoms or molecules, i.e., all atoms are material points. It is believed that the particles interact with each other and, moreover, may be subject to external influence. Interatomic forces are represented in the form of the classical potential force (the gradient of the potential energy of the system).

The interaction between atoms is described by means of van der Waals forces (intermolecular forces), mathematically expressed by the Lennard-Jones potential:

$$V(r) = \frac{Ae^{-\sigma r}}{r} - \frac{C_6}{r^6}$$

where;

A and C_6 – some coefficients depending on the structure of the atom or molecule,

σ — the smallest possible distance between the molecules.

In the case of two isolated molecules at a distance of r_0 the interaction force is zero, that is, the repulsive forces balance attractive forces. When $r > r_0$ the resultant force is the force of gravity, which increases in magnitude, reaching a maximum at $r = r_m$ and then decreases. When $r < r_0$ – a repulsive force. Molecule in the field of these forces has potential energy $V(r)$, which is connected with the force of $f(r)$ by the differential equation

$$dV = -f(r)dr$$

At the point $r = r_0$, $f(r) = 0$, $V(r)$ reaches an extremum (minimum).

The chart of such a potential is shown in Fig. 13. The upper (positive) half-axis r corresponds to the repulsion of the molecules, the lower (negative) half-plane – their attraction. We can say simply: at short distances the molecules mainly repel each, on the long – draw each other. Based on this hypothesis, and now an obvious fact, the van der Waals received his equation of state for real gases.

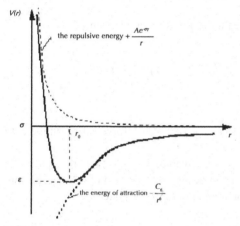

FIGURE 13 The chart of the potential energy of intermolecular interaction.

The exponential summand in the expression for the potential describing the repulsion of the molecules at small distances, often approximated as

$$\frac{Ae^{-\sigma r}}{r} \approx \frac{C_{12}}{r^{12}}$$

In this case we obtain the Lennard-Jones potential:

$$V(r) = \frac{C_{12}}{r^{12}} - \frac{C_6}{r^6} \tag{1}$$

The interaction between carbon atoms is described by the potential

$$V_{CC}\left(r\right)=K\left(r-b\right)^{2},$$

where,

K is – constant tension (compression) connection,

$b = 1,4A$ – the equilibrium length of connection,

r – current length of the connection.

The interaction between the carbon atom and hydrogen molecule is described by the Lennard-Jones

$$V(r)=4\varepsilon\left[\left(\frac{\sigma}{r}\right)^{12}-\left(\frac{\sigma}{r}\right)^{6}\right]$$

For all particles (Fig. 15) the equations of motion are written:

$$m\frac{d^{2}\overline{r_{i}}}{dt^{2}}=\overline{F}_{T-H_{2}}\left(\overline{r_{i}}\right)+\sum_{j\neq i}\overline{F}_{H_{2}-H_{2}}\left(\overline{r_{i}}-\overline{r_{j}}\right),$$

where,

$\overline{F}_{T-H_{2}}\left(\overline{r}\right)$ – force, acting by the CNT, $\overline{F}_{H_{2}-H_{2}}\left(\overline{r_{i}}-\overline{r_{j}}\right)$ – force acting on the i-th molecule from the j-th molecule

The coordinates of the molecules are distributed regularly in the space, the velocity of the molecules are distributed according to the Maxwell equilibrium distribution function according to the temperature of the system:

$$f\left(u,v,w\right)=\frac{\beta^{3}}{\pi^{3/2}}\exp\left(-\beta^{2}\left(u^{2}+v^{2}+w^{2}\right)\right)\quad\beta=\frac{1}{\sqrt{2RT}}$$

The macroscopic flow parameters are calculated from the distribution of positions and velocities of the molecules:

$$\overline{V}=\left\langle\overline{v}_{i}\right\rangle=\frac{1}{n}\sum_{i}\overline{v}_{i},$$

$$\rho=\frac{nm}{V_{0}},\qquad\frac{3}{2}RT=\frac{1}{2}\left\langle\left|\overline{v}_{i}^{\prime}\right|^{2}\right\rangle.\overline{v}_{i}^{\prime}=\overline{v}_{i}-\overline{V},$$

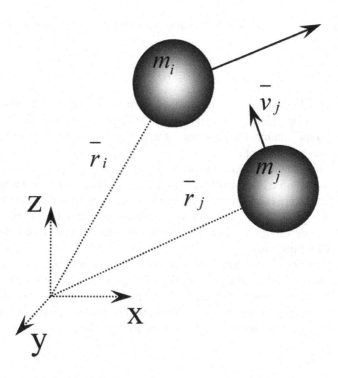

FIGURE 14 Position of Particles.

The resulting system of equations is solved numerically. However, the molecular dynamics method has limitations of applicability:

1. the de Broglie wavelength h/mv (where h – Planck's constant, m – the mass of the particle, v – velocity) of the atom must be much smaller than the interatomic distance;

2. Classical molecular dynamics cannot be applied for modeling systems consisting of light atoms such as helium or hydrogen;

3. at low temperatures, quantum effects become decisive for the consideration of such systems must use quantum chemical methods;

4. necessary that the times at which we consider the behavior of the system were more than the relaxation time of the physical quantities.

In 1873, Van der Waals proposed an equation of state is qualitatively good description of liquid and gaseous systems. It is for one mole (one mole) is:

$$\left(p + \frac{a}{v^2}\right)(v - b) = RT \qquad (2)$$

Note that at $p \gg \dfrac{a}{v^2}$ and $v \gg b$ this equation becomes the equation of state of ideal gas

$$pv = RT \qquad\qquad (3)$$

Van der Waals equation can be obtained from the Clapeyron equation of Mendeleev by an amendment to the magnitude of the pressure a/v^2 and the amendment b to the volume, both constant a and b independent of T and v but dependent on the nature of the gas.

The amendment b takes into account:

1. the volume occupied by the molecules of real gas (in an ideal gas molecules are taken as material points, not occupying any volume);
2. so-called "dead space", in which can not penetrate the molecules of real gas during motion, i.e. volume of gaps between the molecules in their dense packing.

Thus, $b = v_{mon} + v_{3a3}$ (Fig. 16). The amendment a/v^2 takes into account the interaction force between the molecules of real gases. It is the internal pressure, which is determined from the following simple considerations. Two adjacent elements of the gas will react with a force proportional to the product of the quantities of substances enclosed in these elementary volumes.

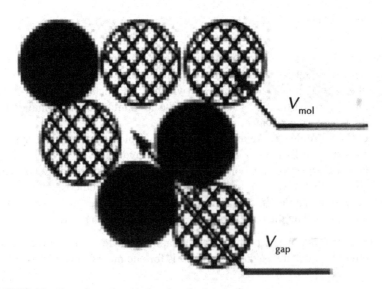

FIGURE 15   Location scheme of molecules in a real gas.

Therefore, the internal pressure P_{BH} is proportional to the square of the concentration n:

$$p_{BH} \sim n^2 \sim \rho^2 \sim \frac{1}{v^2},$$

Where;
ρ – the gas density.
Thus, the total pressure consists of internal and external pressures:

$$p + p_{BH} = p + \frac{a}{v^2}$$

Equation (3) is the most common for an ideal gas. Under normal physical conditions and from Eq. (3) we obtain:

$$R = \frac{R\mu}{\mu} = \frac{8314}{\mu}$$

Knowing $R\mu$ we can find the gas constant for any gas with the help of the value of its molecular mass μ (Table 1):

TABLE 1 The molecular weight of some gases.

Gas	N	Ar	H_2	O_2	CO	CO_2	Ammonia	Air
μ	28	40	2	32	28	44	17	29

For gas mixture with mass M state equation has the form:

$$pv = MR_{cm}T = \frac{8314MT}{\mu_{cm}} \tag{4}$$

where;
R_{cm} – gas constant of the mixture.
The gas mixture can be given by the mass proportions g_i, voluminous r_i or mole fractions n_i respectively, which are defined as the ratio of mass m_i, volume v_i or number of moles N_i of i gas to total mass M, volume v or number of moles N of gas mixture. Mass fraction of component is $g_i = \frac{m_i}{M}$, where $i = 1, n$. It is obvious that $M = \sum_{i=1}^{n} m_i$ and $\sum_{i=1}^{n} g_i = 1$. The volume fraction is $r_i = \frac{v_i}{v_{cm}}$, where v_i – partial volume of component mixtures.

Similarly, we have $\displaystyle\sum_{i=1}^{n} v_i = v_{cm}, \sum_{i=1}^{n} r_i = 1$.

Depending on specificity of tasks the gas constant of the mixture may be determined as follows:

$$R_{cm} = \frac{1}{\displaystyle\sum_{i=1}^{n} r_i R_i^{-1}}; \quad R_{cm} = \frac{1}{\displaystyle\sum_{i=1}^{n} r_i R_i^{-1}}$$

If we know the gas constant R_{cm}, the seeming molecular weight of the mixture is equal to

$$\mu_{cm} = \frac{8314}{R_{cm}} = \frac{8314}{\displaystyle\sum_{i=1}^{n} g_i R_i} = 8314 \sum_{i=1}^{n} r_i R_i^{-1}$$

The pressure of the gas mixture p is equal to the sum of the partial pressures of individual components in the mixture p_i:

$$p = \sum_{i=1}^{n} p_i \tag{5}$$

Partial pressure p_i – pressure that has gas, if it is one at the same temperature fills the whole volume of the mixture ($p_i v_{cm} = RT$).

With various methods of setting the gas mixture partial pressures

$$p_i = p r_i; \; p_i = \frac{p g_i \mu_{cm}}{\mu_i} \tag{6}$$

From the Eq. (6) we see that for the calculation of the partial pressures p_i necessary to know the pressure of the gas mixture, the volume or mass fraction i of the gas component, as well as the molecular weight of the gas mixture μ and the molecular weight of i of gas μ_i.

The relationship between mass and volume fractions are written as follows:

$$g_i = \frac{m_i}{m_{cm}} = \frac{\rho_i v_i}{\rho_{cm} v_{cm}} = \frac{R_{cm}}{R_i} r_i = \frac{\mu_i}{\mu_{cm}} r_i$$

We rewrite Eq. (2) as

$$v^3 - \left(b + \frac{RT}{p}\right) v^2 + \frac{a}{p} v - \frac{ab}{p} = 0 \tag{7}$$

When $p = p_k$ and $T = T_k$, where p_k and T_k – critical pressure and temperature, all three roots of Eq. (7) are equal to the critical volume v_k:

$$v^3 - \left(b + \frac{RT_k}{P_k}\right)v^2 + \frac{a}{p_k}v - \frac{ab}{p_k} = 0 \qquad (8)$$

Because the $v_1 = v_2 = v_3 = v_k$, then Eq. (8) must be identical to the equation:

$$(v - v_1)(v - v_2)(v - v_3) = (v - v_k)^3 = v^3 - 3v^2 v_k + 3vv_k^2 - v_k^3 = 0 \qquad (9)$$

Comparing the coefficients at the equal powers of v in both equations leads to the equalities:

$$b + \frac{RT_k}{p_k} = 3v_k \frac{a}{p_k} = 3v_k^2 \frac{ab}{p_k} = v_k^3 \qquad (10)$$

Hence,

$$a = 3v_k^2 p_k; b = \frac{v_k}{3} \qquad (11)$$

Considering Eq. (10) as equations for the unknowns p_k, v_k, T_k, we obtain:

$$p_k = \frac{a}{27b^2}; v_k = 3b; T_k = \frac{8a}{27bR} \qquad (12)$$

From Eqs. (10) and (11) or (12) we can find the relation:

$$\frac{RT_k}{p_k v_k} = \frac{8}{3} \qquad (13)$$

Instead of the variables p, v, T let's introduce the relationship of these variables to their critical values (leaden dimensionless parameters).

$$\pi = \frac{p}{p_k}; \omega = \frac{v}{V_k}; \tau = \frac{T}{T_k} \qquad (14)$$

Substituting Eqs. (12) and (14) in Eq. (7) and using Eq. (13), we obtain:

$$\left(\pi p_k + \frac{3v_k^2 p_k}{\omega^2 v_k^2}\right)\left(\omega v_k - \frac{v_k}{3}\right) = RT_k \tau,$$

$$\left(\pi+\frac{3}{\omega^2}\right)(3\omega-1)=3\frac{RT_k}{p_k v_k}\,\tau,$$

$$\left(\pi+\frac{3}{\omega^2}\right)(3\omega-1)=8\tau \qquad\qquad (15)$$

4.3   SLIPPAGE OF THE FLUID PARTICLES NEAR THE WALL

Features of the simulation results of Poiseuille flow in the microtubules, when the molecules at the solid wall and the wall atoms at finite temperature of the wall make chaotic motion lies in the fact that in the intermediate range of Knudsen numbers there is slippage of the fluid particles near the wall.

Researchers describe three possible cases: (1) the liquid can be stable (no slippage), (2) slides relative to the wall (with slippage flow), (3) the flow profile is realized; this is when the friction of the wall is completely absent (complete slippage).

In the framework of classical continuum fluid dynamics, according to the Navier boundary condition the velocity slip is proportional to fluid velocity gradient at the wall:

$$v\big|_{y=0}=L_S\,dv/dy\big|_{y=0} \qquad\qquad (16)$$

Here and in Fig. 17 L_S represents the "slip length" and has a dimension of length.

Because of the slippage, the average velocity in the channel $\langle v_{pdf}\rangle$ increases.

In a rectangular channel (of width > height h and viscosity of the fluid η) due to an applied pressure gradient $-dp/dx$ the authors of that article obtained:

$$\langle v_{pdf}\rangle=\frac{h^2}{12\eta}\left(-\frac{dp}{dx}\right)\left(1+\frac{6L_S}{h}\right) \qquad\qquad (17)$$

The results of molecular dynamics simulation for nanosystems with liquid, with characteristic dimensions of the order of the size of the fluid particles, show that a large slippage lengths (of the order of microns) should occur in the carbon nanotubes of nanometer diameter and, consequently, can increase the flow rate by three orders of ($6L_S/h > 1000$). Thus, the flow with slippage is becoming more and more important for hydrodynamic systems of small size.

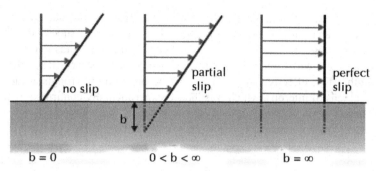

FIGURE 17 Three cases of slip flow past a stationary surface. The slip length b is indicated.

The results of molecular dynamics simulation of unsteady flow of mixtures of water – water vapor, water and nitrogen in a carbon nanotube are reported by many reserchers.

Carbon nanotubes have been considered "zigzag" with chiral vectors (20, 20), (30, 30) and (40, 40), corresponding to pipe diameters of 2.712, 4.068 and 5.424 nm, respectively.

Knowing the value of the flow rate and the system pressure, which varies in the range of 600–800 bars, are high enough to ensure complete filling of the tubes. This pressure can be achieved by the total number of water molecules 736, 904, and 1694, respectively.

The effects of slippage of various liquids on the surface of the nanotube were studied in detail.

The length of slip, can be calculated using the current flow velocity profiles of liquid, shown in Fig. 18, were 11, 13, and 15 nm for the pipes of 2.712, 4.068 and 5.424 nm, respectively. The dotted line marked by theoretical modeling data. The vertical lines indicate the position of the surface of carbon nanotubes.

It was found out that as the diameter decreases, the speed of slippage of particles on the wall of nanotube also decreases. The authors attribute this to the increase of the surface friction.

Experiments with various pressure drops in nanotubes demonstrated slippage of fluid in micro and nanosystems. The most remarkable were the two recent experiments, which were conducted to improve the flow characteristics of carbon nanotubes with the diameters of 2 and 7 nm respectively. In the membranes in which the carbon nanotubes were arranged in parallel, there was a slip of the liquid in the micrometer range. This led to a significant increase in flow rate – up to three – four orders of magnitude.

In the experiments for the water moving in microchannels on smooth hydrophobic surfaces, there are slidings at about 20 nm. If the wall of the channel is not smooth but twisty or rough, and at the same time, hydrophobic, such a structure would lead to an accumulation of air in the cavities and become superhydrophobic (with contact angle greater than 160°). It is believed that this leads to creation of contiguous areas with high and low slippage, which can be described as "effective slip length". This effective length of the slip occurring on the rough surface can be several tens of microns, which was indeed experimentally confirmed by many researchers.

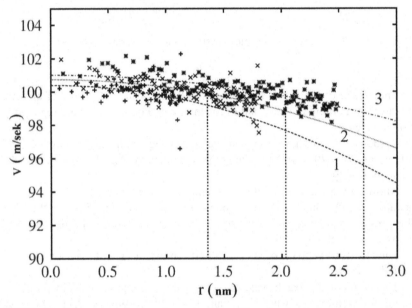

FIGURE 18 Time average streaming velocity profiles of water in a carbon nanotubes of different diameter: 2.712 nm, curve 1; 4.068 nm, curve 2; 5.424 nm, curve 3.

Another possible problem is filling of the hydrophobic systems with liquid. Filling of micron size hydrophobic capillaries is not a big problem, because pressure of less than 1 atm is sufficient. Capillary pressure, however, is inversely proportional to the diameter of the channel, and filling for nanochannels can be very difficult.

4.3.1 THE DENSITY OF THE LIQUID LAYER NEAR A WALL OF CARBON NANOTUBE

Researchers showed radial density profiles of oxygen averaged in time and hydrogen atoms in the "zigzag" carbon nanotube with chiral vector (20, 20) and a radius

$R = 1,356$ nm (Fig. 19). The distribution of molecules in the area near the wall of the carbon nanotube indicated a high-density layer near the wall of the carbon nanotube. Such a pattern indicates the presence of structural heterogeneity of the liquid in the flow of the nanotube. In **Figure (3.3)** $\rho^*(r) = \rho(r)/\rho(0)$, 2,712 nm diameter pipe is completely filled with water molecules at $300^0 K$. The overall density $\rho(0) = 1000 KS / M^3$. The arrows denote the location of distinguishable layers of the water molecules and the vertical line the position of the CNT wall.

The distribution of molecules in the region of $0,95 \le r \le R$ nm indicates a high-density layer near the wall carbon nanotube. Such a pattern indicates the presence of structural heterogeneity of the liquid in the flow of the nanotube.

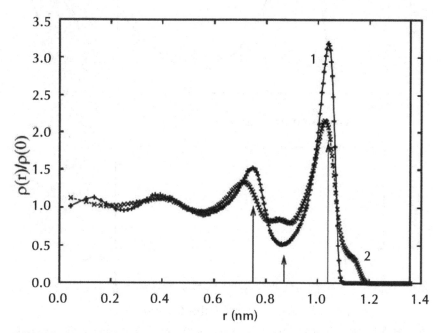

FIGURE 19 Radial density profiles of oxygen (curve 1) and hydrogen (curve 2) atoms for averaged in the time interval.

Many researchers obtained similar results for the flow of the water in the carbon nanotubes using molecular dynamic simulations.

Figure 20 shows the scheme of the initial structure and movement of water molecules in the model carbon nanotube (a), the radial density profile of water molecules inside nanotubes with different radii (b) and the velocity profile of water molecules in a nanotube chirality (60,60) (c).

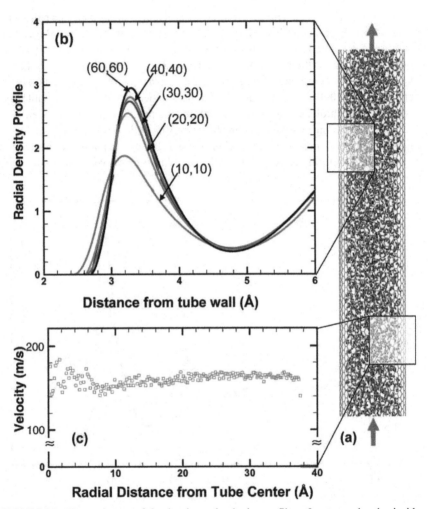

FIGURE 20   Dependences of the density and velocity profiles of water molecules inside the nanotubes with different radii

(a)   Schematic of the initial structure and transport of water molecules in a model CNT.

(b)   The radial density profile (RDP) of water molecules inside CNTs with different radii.

(c)   The representative radial velocity profile (RVP) of water molecules inside a (60,60) nanotube

It is mentioned that the area available for molecules of the liquid is less than the area bounded by a solid wall, primarily due to the van der Waals interactions.

4.3.2 THE EFFECTIVE VISCOSITY OF THE LIQUID IN A NANOTUBE

Scientists showed a significant increase in the effective viscosity of the fluid in the nano volumes compared to its macroscopic value. It was shown that the effective viscosity of the liquid in a nanotube depends on the diameter of the nanotube. The effective viscosity of the liquid in a nanotube is defined as follows.

Let's establish a conformity nanotubes filled with liquid, possibly containing crystallites with the same size tube filled with liquid, considered as a homogeneous medium (i.e., without considering the crystallite structure), in which Poiseuille flow is realized at the same pressure drop and consumption rate. The viscosity of a homogeneous fluid, which ensures the coincidence of these parameters, will be called the effective viscosity of the flow in the nanotube.

Researchers showed that while flowing in the narrow channels of width less than 2 nm, water behaves like a viscous liquid. In the vertical direction water behaves as a rigid body, and in a horizontal direction it maintains its fluidity.

It is known that at large distances the van der Waals interaction has a magnetting tendency and occurs between any molecules like polar as well as nonpolar. At small distances it is compensated by repulsion of electron shells. Van der Waals interaction decreases rapidly with distance. Mutual convergence of the particles under the influence magnetting forces continues until these forces are balanced with the increasing forces of repulsion.

Knowing the deceleration of the flow (Fig. 20) of water a, the authors of the cited article calculates the effective shear stress between the wall of the pipe length l and water molecules by the formula

$$\tau = Nma / (2\pi Rl) \qquad (18)$$

Here, the shear stress is a function of tube radius and flow velocity $\bar{v}$, m – mass of water molecules, the average speed is related to volumetric flow $\bar{v} = Q / (\pi R^2)$.

Denoting n_0 the density of water molecules number, we can calculate the shear stress in the form of:

$$\tau\big|_{r=R} = n_0 mRa / 2 \qquad (19)$$

Figure 21 shows the results of calculations of the authors of the cited article – influence of the size of the tube R_0 on the effective viscosity (squares) and shear stress τ (triangles), when the flow rate is approximately 165 m/sec.

According to classical mechanics of liquid flow at different pressure drops Δp along the tube length l is given by Poiseuille formula:

$$Q_P = \frac{\pi R^4 \Delta p}{8 \eta l} \quad Q_P = \pi R^2 \bar{v} \tag{20}$$

$$\tau = \frac{\Delta p R}{2l} \tag{21}$$

and the effective viscosity of the fluid can be estimated as $\eta = \tau \cdot R / (4\bar{v})$.

The change in the value of shear stress directly causes the dependence of the effective viscosity of the fluid from the pipe size and flow rate. In this case the effective viscosity of the transported fluid can be determined from Eqs. (20) and (21) as:

$$\eta = \frac{\tau \cdot R}{4\bar{v}} \tag{22}$$

With increasing radius τ according to Eq. (21) increases.

Calculations showed that the magnitude of the shear stress τ is relatively small in the range of pipe sizes considered. This indicates that the surface of carbon nanotubes is very smooth and the water molecules can easily slide through it.

In fact, shear stress is primarily due to van der Waals interaction between the solid wall and the water molecules. It is noted that the characteristic distance between the near-wall layer of fluid and pipe wall depends on the equilibrium distance between atoms O and C, the distribution of the atoms of the solid wall and bend of the pipe.

From Fig. 20, we can see, that the effective viscosity η increases by two orders of magnitude when R_0 changes from 0.67 to 5.4 nm. The value of the calculated viscosity of water in the tube (10,10) is $8,5 \times 10^{-8}$ Pa/s, roughly 4 orders of magnitude lower than the viscosity of a large mass of water.

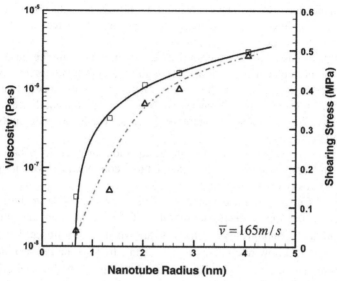

FIGURE 20 Size effect of shearing stress (triangle) and viscosity (square), with $\bar{v}$ =165M/c.

According to Eqs. (20)–(22), the effective viscosity can be calculated as $\eta = \dfrac{\pi R^4 \Delta p}{8QL}$. The results of calculations are shown in Fig. 21.

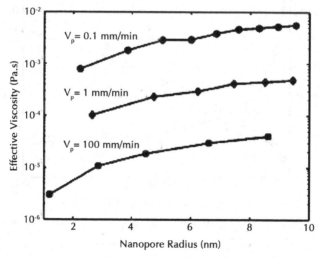

FIGURE 21 Effective viscosity as a function of the nanopore radius and the loading rate.

The dependence of the shear stress on the flow rate is illustrated in Fig. 22. From the example of the tube (20,20) it's clear that τ increases with $\bar{v}$. The growth rate slowed down at higher values $\bar{v}$.

At high speeds $\bar{v}$, while water molecules are moving along the surface of the pipe, the liquid molecules do not have enough time to fully adjust their positions to minimize the free energy of the system. Therefore, the distance between adjacent carbon atoms and water molecules may be less than the equilibrium van der Waals distances. This leads to an increase in van der Waals forces of repulsion and leads to higher shear stress.

Scientists showed that, even though the equation for viscosity is based on the theory of the continuum, it can be extended to a complex flow to determine the effective viscosity of the nanotube.

Figure 22 also shows a dependence η on $\bar{v}$ on the inside of the nanotube (20,20). It is seen that η decreases sharply with increasing flow rate and begins to asymptotically approaches a definite value when $\bar{v} > 150$ m/sec. For the current pipe size and flow rate ranges $\eta \sim 1/\sqrt{\bar{v}}$, this trend is the result of addiction $\tau - \bar{v}$, contained the same Fig. 22. According to Fig. 21, high-speed effects are negligible.

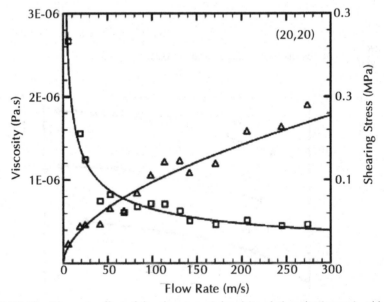

FIGURE 22   Flow rate effect of shearing stress (triangle) and viscosity (square), with R_0 =1.336 HM.

One can easily see that the dependence of viscosity on the size and speed is consistent qualitatively with the results of molecular dynamic simulations. In all studied cases, the viscosity is much smaller than its macroscopic analogy. When the radius of the pores varies from about 1 nm to 10 nm, the value of the effective viscosity increases by an order of magnitude respectively. A more significant change occurs when the speed increases from 0.1 mm/min up to 100 mm/min. This results in a change in the value of viscosity η, respectively, by 3–4 orders. The discrepancy between simulation and test data can be associated with differences in the structure of the nanopores and liquid phase.

Figure 23 shows the viscosity dependence of water, calculated by the method of DM, the diameter of the CNT. The viscosity of water, as shown in the Fig. 23, increases monotonically with increasing diameter of the CNT.

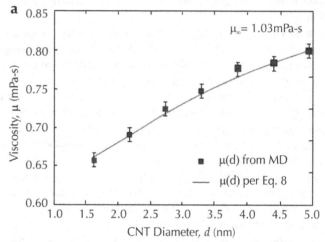

FIGURE 23   Variation of water viscosity with CNT diameter.

4.3.3   ENERGY RELEASE DUE TO THE COLLAPSE OF THE NANOTUBE

Scientists theoretically predicted the existence of a "domino effect" in single-walled carbon nanotube.

Squashing it at one end by two rigid moving to each other by narrow graphene planes (about 0.8 nm in width and 8.5 nm in length), one can observe it rapidly (at a rate exceeding 1 km/s) release its stored energy by collapsing along its length like a row of dominoes. The effect resembles a tube of toothpaste squeezing itself (Fig. 24).

The structure of a single-walled carbon nanotube has two possible stable states: circular or collapsed. Chang realized that for nanotubes wider than 3.5 nm, the circular state stores more potential energy than the collapsed state as a result of van der Waal's forces. He performed molecular dynamics simulations to find out what would happen if one end of a nanotube was rapidly collapsed by clamping it between two graphene bars.

This phenomenon occurs with the release of energy, and thus allows for the first time to talk about carbon nanotubes as energy sources. This effect can also be used as a accelerator of molecules.

The tube does not collapse over its entire length at the same time, but sequentially, one after the other carbon ring, starting from the end, which is tightened (Fig. 24). It happens just like a domino collapses, arranged in a row (this is known as the "domino effect"). Only here the role of bone dominoes is performed by a ring of carbon atoms forming the nanotube, and the nature of this phenomenon is quite different.

Recent studies have shown that nanotubes with diameters ranging from 2 to 6 nm, there are two stable equilibrium states – cylindrical (tube no collapses) and compressed (imploded tube) – with difference values of potential energy, the difference between which and can be used as an energy source.

Researchers found that switching between these two states with the subsequent release of energy occurs in the form of arising domino effect wave. The scientists have shown that such switching is not carried out in carbon nanotubes with diameters of 2 nm and not more, as evident from previous studies, but with a little more, starting from 3.5 nm.

A theoretical study of the "domino effect" was conducted using a special method of classical molecular dynamics, in which the interaction between carbon atoms was described by van der Waals forces (intermolecular forces), mathematically expressed by the Lennard-Jones potential.

The main reason for the observed effect, in author's opinion, is the competition of the potential energy of the van der Waals interactions, which "collapses" the nanotube with the energy of elastic deformation, which seeks to preserve the geometry of carbon atoms, which eventually leads to a bistable (collapsed and no collapsed) configuration of the carbon nanotube.

For small diameter tubes the dominant is the energy of elastic deformation, the cylindrical shape of such a nanotube is stable. For nanotubes with sufficiently large diameter the van der Waals interaction energy is dominant. This means more stability and less compressed nanotube stability, or, as physicists say, metastability (i.e., apparent stability) of its cylindrical shape.

Thus, "domino effect" wave can be produced in a carbon nanotube with a relatively large diameter (more than 3.5 nm, as the author's calculations), because only in such a system the potential energy of collapsing structures may be less than the potential energy of the "normal" nanotube. In other words, the cylindrical

and collapsing structure of a nanotubes with large diameters are, respectively, its
metastable and stable states.

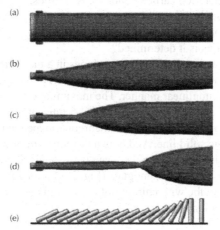

FIGURE 24   "Domino effect" in a carbon nanotube. (a) The initial form of carbon
nanotubes – cylindrical. (b) One end of the tube is squeezed. (c), (d) Propagation of domino
waves – the configuration of the nanotube 15 and 25 picoseconds after the compression
of its end. (e) Schematic illustration of the "domino effect" under the influence of gravity.

Change of the potential energy of a carbon nanotube with a propagating "do-
min effect" wave in it with time is represented as a graph in Fig. 25a.

This chart shows three sections of features in the change of potential energy.
The first (from 0 ps to 10 ps) are composed of elastic strain energy, which appears
due to changes in the curvature of the walls of the nanotubes in the process of collapsing, the energy change of van der Waals interactions occurring between the
opposite walls of the nanotubes, as well as the interaction between the tube walls
and graphene planes, compressing her end.

The second region (from 10 ps to 35 ps) corresponds to the "domino effect"–
"domino effect" wave spreads along the surface of the carbon nanotube. Energetically, it looks like this: at every moment, when carbon ring collapses, some of the
potential energy of the van der Waals is converted to kinetic energy (the rest to the
energy of elastic deformation), which is kind of stimulant to support and "falling
domino" – following the collapse of the rings, which form the nanotube, with
each coagulated ring reducing the total potential energy of the system.

Finally, the third segment (from 35 ps to 45 ps) corresponds to the ended
"domino" process – carbon tube collapsed completely. We emphasize that the
nanotube, which collapsed (as seen from the Fig. 25a), has less potential energy
than it was before beginning of the "domino effect".

In other words, the spread "domino effect" waves – a process that goes with the release of energy: about 0.01 eV per atom of carbon. This is certainly not comparable in any way with the degree of energy yield in nuclear reactions, but the fact of power generation carbon nanotube is obvious.

Later scientists analyzed the kinematic characteristics of the process – what the rate of propagation of the wave of destruction or collapse of carbon nanotubes and their characteristics is it determined?

Calculations show that the wave of dominoes in a tube diameter of 4–5 nm is about 1 km/s (as seen from the Fig. 25b) and depends on its geometry – the diameter and chirality in a nonlinear manner. The maximum effect should be observed in the tube with a diameter slightly less than 4.5 nm – it will be carbon rings to collapse at a speed of 1.28 km/sec. The theoretical dependence obtained by the author, shows the blue solid line. And now an example of how energy is released in such a system with a "domino effect" can be used in nanodevices. The author offers an original way to use – "nanogun" (Fig. 26a). Imagine that at our disposal there is a carbon nanotube with chirality of (55.0) and corresponding to observation of the dominoes diameter.

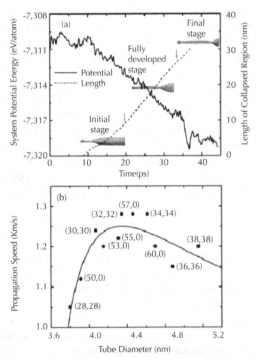

FIGURE 25   (a) Time dependences of potential energy and the length of the collapsed nanotube, (b) The velocity of propagation of the domino wave for carbon nanotubes with different diameters and chirality.

Put a C_{60} fullerene inside a nanotube. A little imagination, and a carbon nanotube can be considered as the gun trunk, and the C_{60} molecule – as its shell.

The molecule located inside will extruded from it into the other, open end under the influence of squeezing nanotube (see fig. 5.6).

The question is, what is the speed of the "core"? Chang estimated that, depending on the initial position of the fullerene molecule when leaving a nanotube, it can reach speeds close to the velocity of "domino effect" waves – about 1 km/s (Fig. 26b). Interestingly, this speed is reached by the "core" for just 2 picoseconds and at a distance of 1 nm. It is easy to calculate that the observed acceleration is of great value $0,5 \cdot 10^{15} \, m/c^2$. For comparison, the speed of bullets in an AK-47 is 1.5 times lower than the rate of fullerene emitted from a gun.

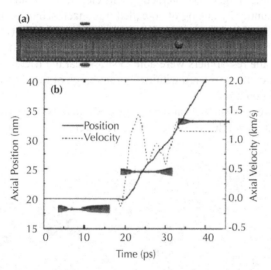

FIGURE 26   Nano cannon scheme acting on the basis of "domino effect" in the incision. (a) Inside a carbon nanotube (55.0) is the fullerene molecule C_{60}. (b) The initial position and velocity of the departure of the "core" (a fullerene molecule), depending on the time. The highest rate of emission of C_{60} (1.13 km /s) comparable to the velocity of the domino wave.

Necessary be noted that the simulation took place in nanogun at assumption of zero temperature in Kelvin. However, this example is not so abstract and may be used in the injecting device.

Thus, for the first time been demonstrated, albeit only in theory, the use of single-walled carbon nanotubes as energy sources.

4.4 FLUID FLOW IN NANOTUBES

The friction of surface against the surface in the absence of the interlayer between the liquid material (so-called dry friction) is created by irregularities in the given surfaces that rub one another, as well as the interaction forces between the particles that make up the surface.

As part of their study, the researchers built a computer model that calculates the friction force between nano-surfaces (Fig. 27). In the model, these surfaces were presented simply as a set of molecules for which forces of intermolecular interactions were calculated.

As a result, scientists were able to establish that the friction force is directly proportional to the number of interacting particles. The researchers propose to consider this quantity by analog of so-called true macroscopic contact area. It is known that the friction force is directly proportional to this area (it should not be confused with common area of the contact surfaces of the bodies).

In addition, the researchers were able to show that the friction surface of the nano-surfaces can be considered within the framework of the classical theories of friction of nonsmooth surfaces.

A literature review shows that nowadays molecular dynamics and mechanics of the continuum in are the main methods of research of fluid flow in nanotubes.

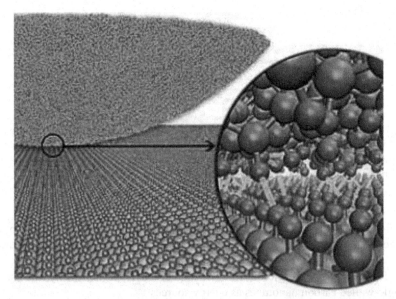

FIGURE 27   Computer model of friction at the nanoscale (Right shows the surfaces of interacting particles).

Although, the method of molecular dynamics simulations is effective, it at the same time requires enormous computing time especially for large systems. Therefore, simulation of large systems is more reasonable to carry out nowadays by the method of continuum mechanics.

The fluid flow in the channel can also be considered in the framework of the continuum hypothesis. The Navier-Stokes equation can be used and the velocity profile can be determined for Poiseuille flow.

However, the water flow by means of pressure differential through the carbon nanotubes with diameters ranging from 1.66 to 4.99 nm is studies using molecular dynamics simulation study. For each nanotube the value enhancement predicted by the theory of liquid flow in the carbon nanotubes is calculated. This formula is defined as a ratio of the observed flow in the experiments to the theoretical values without considering slippage on the model of Hagen-Poiseuille. The calculations showed that the enhancement decreases with increasing diameter of the nanotube.

Important conclusion is that by constructing a functional dependence of the viscosity of the water and length of the slippage on the diameter of carbon nanotubes, the experimental results in the context of continuum fluid mechanics can easily be described. The aforementioned is true even for carbon nanotubes with diameters of less than 1.66 nm.

The theoretical calculations use the following formula for the steady velocity profile of the viscosity η of the fluid particles in the CNT under pressure gradient $\partial p / \partial z$:

$$v(r) = \frac{R^2}{4\eta}\left[1 - \frac{r^2}{R^2} + \frac{2L_S}{R}\right]\frac{\partial p}{\partial z} \qquad (23)$$

The length of the slip, which expresses the speed heterogeneity at the boundary of the solid wall and fluid is defined as:

$$L_S = \frac{v(r)}{dv/dr}\bigg|_{r=R} \qquad (24)$$

Then the volumetric flow rate, taking into account the slip Q_S is defined as:

$$Q_S = \int_0^R 2\pi r \cdot v(r)dr = \frac{\pi\left[(d/2)^4 + 4(d/2)^3 \cdot L_S\right]}{8\eta} \cdot \frac{\partial p}{\partial z} \qquad (25)$$

Equation (4.3) is a modified Hagen-Poiseuille equation, taking into account slippage. In the absence of slip $L_S = 0$ (4.3) coincides with the Hagen-Poiseuille flow, Eq. (20) for the volumetric flow rate without slip Q_P. In some works the parameter enhancement flow ε is also introduced. It is defined as the ratio of the calculated volumetric flow rate of slippage to Q_P (calculated using the effective viscosity and the diameter of the CNT). If the measured flux is modeled using Eq. (25), the degree of enhancement takes the form:

$$\varepsilon = \frac{Q_S}{Q_P} = \left[1 + 8\frac{L_S(d)}{d}\right]\frac{\eta_\infty}{\eta(d)} \tag{26}$$

where $d = 2R$ – diameter of CNT, η_∞ – viscosity of water, $L_S(d)$ – CNT slip length depending on the diameter, $\eta(d)$ – the viscosity of water inside CNTs depending on the diameter.

If $\eta(d)$ finds to be equal to η_∞, then the influence of the effect of slip on ε is significant, if $L_S(d) \geq d$. If $L_S(d) < d$ and $\eta(d) = \eta_\infty$, then there will be no significant difference compared to the Hagen-Poiseuille flow with no slip.

Table 2 shows the experimentally measured values of the enhancement water flow. Enhancement flow factor and the length of the slip were calculated using the equations given above.

TABLE 2   Experimentally measured values of the enhancement water flow.

Nanosystems	Diameter (nm)	Enhancement, ε	Slip length, L_S, (nm)
carbon nanotubes	300–500	1	0
	44	22–34	113–177
carbon nanotubes	7	$10^4 - 10^5$	3900–6800
	1.6	560–9600	140–1400

Figure 28 depicts the change in viscosity of the water and the length of the slip in diameter. As can be seen from the Fig. 28, the dependence of slip length to the diameter of the nanotube is well described by the empirical relation:

$$L_S(d) = L_{S,\infty} + \frac{C}{d^3} \tag{27}$$

where $L_{S,\infty} = 30$ nm – slip length on a plane sheet of graphene, a C – const.

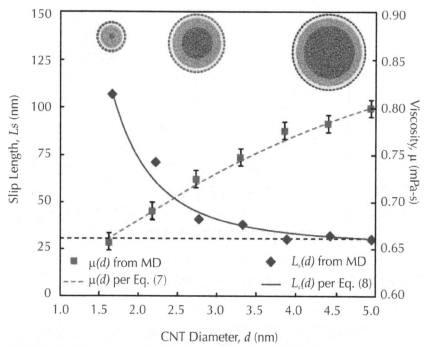

FIGURE 28 Variation of viscosity and slip length with CNT diameter

Figure 29 shows dependence of the enhancement of the flow rate ε on the diameter for all seven CNTs.

There are three important features in the results. First, the enhancement of the flow decreases with increasing diameter of the CNT. Second, with increasing diameter, the value tends to the theoretical value of Eqs. (26) and (27) with a slip $L_{s\infty}$ = 30 nm and the effective viscosity $\mu(d) = \mu_\infty$. The dotted line shown the curve of 15% in the second error in the theoretical data of viscosity and slip length. Third, the change ε in diameter of CNTs cannot be explained only by the slip length.

To determine the dependence the volumetric flow of water from the pressure gradient along the axis of single-walled nanotube with the radii of 1.66, 2.22, 2.77, 3.33, 3.88, 4.44 and 4.99 nm in the method of molecular modeling was used. Snapshot of the water-CNT is shown in Fig. 30.

Figure 30 shows the results of calculations to determine the pressure gradient along the axis of the nanotube with the diameter of 2.77 nm and a length of 20 nm. Change of the density of the liquid in the cross sections was less than 1%.

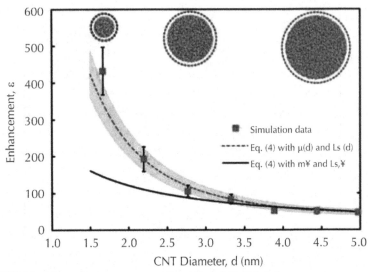

FIGURE 29   Flow enhancement as predicted from MD simulations.

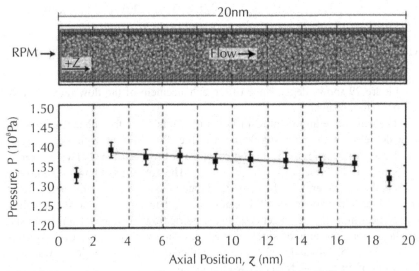

FIGURE 30   Axial pressure gradient inside the 2.77 nm diameter CNT.

Figure 30 shows the dependence of the volumetric flow rate from the pressure gradient for all seven CNTs. The flow rate ranged from 3–14 m/ sec. In the range considered here the pressure gradient $(0-3)\times10^9 \, atm/m$ Q

($pl / sek = 10^{-15} m^3 / sek$) is directly proportional $\partial p / \partial z$. Coordinates of chirality for each CNT are indicated in the figure legend. The linearity of the relations between flow and pressure gradient confirms the validity of calculations of the Eq. (29).

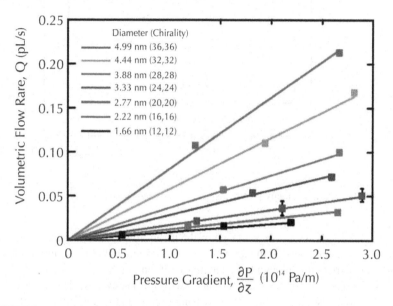

FIGURE 31   Relationship between volumetric liquid flow rate in carbon nanotubes with different diameters and applied pressure gradient

Figure 32 shows the profile of the radial velocity of water particles in the CNT with diameter 2.77 nm. The vertical dotted line at 1.38 nm marked surface of the CNT. It is seen that the velocity profile is close to a parabolic shape.

Researchers consider the flow of water under a pressure gradient in the single-walled nanotubes of "chair" type of smaller radii: 0.83, 0.97, 1.10, 1.25, 1.39 and 1.66 nm.

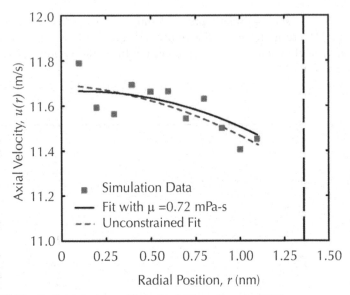

FIGURE 32   Radial velocity profile inside 2.77 nm diameter CNT.

Figure 33 shows the dependence of the mean flow velocity $\bar{v}$ on the applied pressure gradient $\Delta P / L$ in the long nanotubes – 75 nm at 298 K. A similar picture pattern occurs in the tube with the length of 150 nm.

As we can see, there is conformance with the Darcy law, the average flow rate for each CNT increases with increasing pressure gradient. For a fixed value of $\Delta p / L$, however, the average flow rate does not increase monotonically with increasing, diameter of the CNTs, as follows from Poiseuille equation. Instead, when at the same pressure gradient, decrease of the average speed in a CNT with the radius of 0.83 nm to a CNT with the radius of 1.10 nm, similar to the CNTs 1.10 and 1.25 nm, then increases from a CNT with the radius of 1.25 nm to a CNT of 1.66 nm.

The nonlinearity of the relationship between $\bar{v}$ and $\Delta P / L$ are the result of inertia losses (i.e., insignificant losses) in the two boundaries of the CNT. Inertial losses depend on the speed and are caused by a sudden expansion, abbreviations, and other obstructions in the flow.

We note the important conclusion of the work of scientists in which the method of molecular modeling shows that the Eq. (23) (Poiseuille parabola) correctly describes the velocity profile of liquid in a nanotube when the diameter of a flow is 5–10 times more than the diameter of the molecule (≈ 0.17 nm for water).

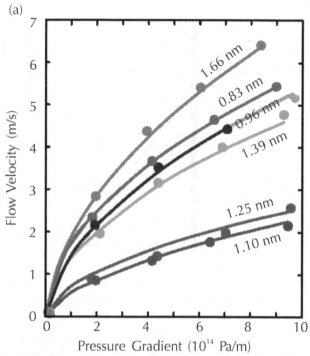

FIGURE 33   Relationship between average flow velocity and applied pressure gradient for the 75 nm long CNTs

In Fig. 34, we can see the effect of slip on the velocity profile at the boundary of radius R of the pipe and fluid. When $L_s = 0$ the fluid velocity at the wall vanishes, and the maximum speed (on the tube axis) exceeds flow speed twice.

The Fig. 34 shows the velocity profiles for Poiseuille flow without slip ($L_s = 0$) and with slippage $L_s = 2R$. The flow rate is normalized to the speed corresponding to the flow without slip. Thick vertical lines indicate the location of the pipe wall. The thick vertical lines indicate the location of the tube wall. As the length of the slip, the flow rate increases, decreases the difference between maximum and minimum values of the velocity and the velocity profile becomes more like a plug.

Velocity of the liquid on a solid surface can also be quantified by the coefficient of slip L_c. The coefficient of slippage – there is a difference between the radial position in which the velocity profile would be zero and the radial position of the solid surface. Slip coefficient is equal to $L_C = \sqrt{R^2 + 2RL_s} = \sqrt{5}R$.

For linear velocity profiles (e.g., Couette flow), the length of the slip and slip rate are equal. These values are different for the Poiseuille flow.

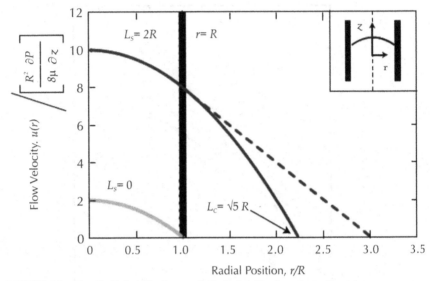

FIGURE 34   No-slip Poiseuille flow and slip Poiseuille flow through a tube.

Figure 35 shows dependence of the volumetric flow rate Q from the pressure gradient $\partial p/\partial z$ in long nanotubes with diameters between 1.66 nm and 6.93 nm. As can be seen in the studied range of the pressure gradient Q is proportional to $\partial p/\partial z$. As in the Poiseuille flow, volumetric flow rate increases monotonically with the diameter of CNT at a fixed pressure gradient. Magnitude of calculations error for all the dependencies are similar to the error for the CNT diameter 4.44 nm, marked in the Fig. 35.

Many researchers considered steady flow of incompressible fluids in a channel width $2h$ under action of the force of gravity ρg or pressure gradient $\partial p/\partial y$, which is described by the Navier-Stokes equations. The velocity profile has a parabolic form:

$$U_y(z) = \frac{\rho g}{2\eta} \cdot \left[(\delta + h)^2 - z^2 \right]$$

where δ – length of the slip, which is equal to the distance from the wall to the point at which the velocity extrapolates to zero.

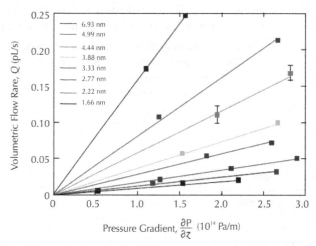

FIGURE 35   Volumetric flow rate in CNTs versus pressure gradient.

4.4.1 SOME OF THE IDEAS AND APPROACHES FOR MODELING IN NANOHYDROMECANICS

Let's consider the fluid flow through the nanotube. Molecules of a substance in a liquid state are very close to each other (Fig. 36).

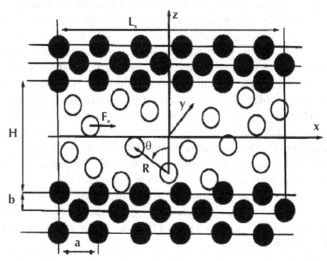

FIGURE 36   The fluid flow through the nanotube.

Most liquid's molecules have a diameter of about 0.1 nm. Each molecule of the fluid is "squeezed" on all sides by neighboring molecules and for a period of time $\left(10^{-10} - 10^{-13} s\right)$ fluctuates around certain equilibrium position, which itself from time to time is shifted in distance commensurating with the size of molecules or the average distance between molecules l_{cp}:

$$l_{cp} \approx \sqrt[3]{\frac{1}{n_0}} = \sqrt[3]{\frac{\mu}{N_A \rho}},$$

where,

n_0 – number of molecules per unit volume of fluid,

N_A – Avogadro's number,

ρ – fluid density,

μ – molar mass.

Estimates show that one cubic of nano water contains about 50 molecules. This gives a basis to describe the mass transfer of liquid in a nanotube-based continuum model. However, the specifics of the complexes, consisting of a finite number of molecules, should be kept in mind. These complexes, called clusters in literatures, are intermediately located between the bulk matter and individual particles – atoms or molecules. The fact of heterogeneity of water is now experimentally established.

There are groups of molecules in liquid—"microcrystals" containing tens or hundreds of molecules. Each microcrystal maintains solid form. These groups of molecules, or "clusters" exist for a short period of time, then break up and are re-created again. Besides, they are constantly moving so that each molecule does not belong at all times to the same group of molecules, or "cluster".

Modeling predicts that gas molecules bounce off the perfectly smooth inner walls of the nanotubes as billiard balls, and water molecules slide over them without stopping. Possible cause of unusually rapid flow of water is maybe due to the small-diameter nanotube molecules move on them orderly, rarely colliding with each other. This "organized" move is much faster than usual chaotic flow. However, while the mechanism of flow of water and gas through the nanotubes is not very clear, and only further experiments and calculations can help understand it.

The model of mass transfer of liquid in a nanotube proposed in this paper is based on the availability of nanoscale crystalline clusters in it.

A similar concept was developed in which the model of structured flow of fluid through the nanotube is considered. It is shown that the flow character in the nanotube depends on the relation between the equilibrium crystallite size and the diameter of the nanotube.

Figure 37 shows the results of calculations by the molecular dynamics of fluid flow in the nanotube in a plane (a) and three-dimensional (b) statement. The figure shows the ordered regions of the liquid.

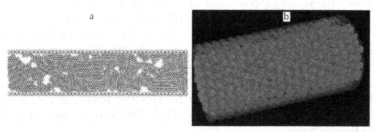

FIGURE 37   The results of calculations of fluid flow in the nanotube.

The typical size of crystallite is 1–2 nm, that is, compared, for example, with a diameter of silica nanotubes of different composition and structure. The flow model proposed in the present chapter is based on the presence of "quasi-solid" phase in the central part of the nanotube and liquid layer, nonautonomous phases.

Consideration of such a structure that is formed when fluid flows through the nanotube is also justified by the aforementioned results of the experimental studies and molecular modeling.

When considering the fluid flow with such structure through the nanotube, we will take into account the aspect ratio of "quasi-solid" phase and the diameter of the nanotube so that a character of the flow is stable and the liquid phase can be regarded as a continuous medium with viscosity η.

Let's establish relationship between the volumetric flow rate of liquid Q flowing from a liquid layer of the nanotube length l, the radius R and the pressure drop $\Delta p/l$, $\Delta p = p - p_0$, where p_0 is the initial pressure in the tube (Fig. 38).

Let R_0 be a radius of the tube from the "quasi-solid" phase, v – velocity of fluid flow through the nanotube.

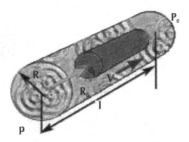

FIGURE 38   Flow through liquid layer of the nanotube.

Structural regime of fluid flow (Fig. 39) implies the existence of the continuous laminar layer of liquid (the liquid layer in the nanotube) along the walls of a pipe. In the central part of a pipe a core of the flow is observed, where the fluid moves, keeping his former structure, that is, as a solid ("quasi-solid" phase in the nanotube). The velocity slip is indicated in Fig. 39a through v_0.

Let's find the velocity profile $v(r)$ in a liquid interlayer $R_0 \leq r \leq R$ of the nanotube. We select a cylinder with radius r and length l in the interlayer, located symmetrically to the centerline of the pipe (see Fig. 39b).

At the steady flow, the sum of all forces acting on all the volumes of fluid with effective viscosity η, is zero.

The following forces are applied on the chosen cylinder: the pressure force and viscous friction force affects the side of the cylinder with radius r, calculated by the Newton formula.

Thus,

$$(p - p_0)\pi r^2 = -\eta \frac{dv}{dr} 2\pi r l \qquad (28)$$

Integrating Eq. (28) between r to R with the boundary conditions $r = R$: $v = v_0$, we obtained a formula to calculate the velocity of the liquid layers located at a distance r from the axis of the tube:

$$v(r) = (p - p_0)\frac{R^2 - r^2}{4\eta l} + v_0 \qquad (29)$$

Maximum speed $v_{\mathcal{A}}$ has the core of the nanotube $0 \leq r \leq R_0$ and is equal to:

$$v_{\mathcal{A}} = (p - p_0)\frac{R^2 - R_0^2}{4\eta l} + v_0 \qquad (30)$$

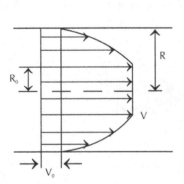 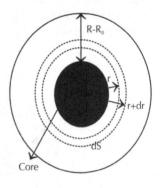

FIGURE 39   Structure of the flow in the nanotube.

Such structure of the liquid flow through nanotubes considering the slip is similar to a behavior of viscoplastic liquids in the tubes. Indeed, as we know, for viscoplastic fluids a characteristic feature is that they are to achieve a certain critical internal shear stresses τ_0 behave like solids, and only when internal stress exceeds a critical value above begin to move as normal fluid. Scientists shown that liquid behave in the nanotube the similar way. A critical pressure drop is also needed to start the flow of liquid in a nanotube.

Structural regime of fluid flow requires existence of continuous laminar layer of liquid along the walls of pipe. In the central part of the pipe is observed flow with core radius R_{00}, where the fluid moves, keeping his former structure, that is, as a solid.

The velocity distribution over the pipe section with radius R of laminar layer of viscoplastic fluid is expressed as follows:

$$v(r) = \frac{\Delta p}{4\eta l}\left(R^2 - r^2\right) - \frac{\tau_0}{\eta}(R - r) \tag{31}$$

The speed of flow core in $0 \le r \le R_{00}$ is equal.

Let's calculate the flow or quantity of fluid flowing through the nanotube cross-section S at a time unit. The liquid flow dQ for the inhomogeneous velocity field flowing from the cylindrical layer of thickness dr, which is located at a distance r from the tube axis is determined from the relation

$$v_{\mathcal{A}} = \frac{\Delta p}{4\eta l}\left(R^2 - R_{00}^2\right) - \frac{\tau_0}{\eta}(R - R_{00})$$

where dS – the area of the cross-section of cylindrical layer (between the dotted lines in Fig. 39).

Let's place Eq. (29) in Eq. (33), integrate over the radius of all sections from R_0 to R and take into account that the fluid flow through the core flow is determined from the relationship $Q_{\mathcal{A}} = \pi R_0^2 v_{\mathcal{A}}$. Then we get the formula for the flow of liquid from the nanotube:

$$Q = \pi R^2 v_0 + Q_P\left[1 - \left(\frac{R_0}{R}\right)^4\right] \tag{34}$$

If $(R_0/R)^4 < 1$ (no nucleus) and $v_0\Delta p R^2/8l\eta \ll 1$ (no slip), then Eq. (34) coincides with Poiseuille Eq. (20). When $R_0 \approx R$ (no of a viscous liquid interlayer in the nanotube), the flow rate Q is equal to volumetric flow $Q \approx \pi R^2 v_0$ of fluid for a uniform field of velocity (full slip).

Accordingly, flow rate of the viscoplastic fluid flowing with a velocity Eq. (29), is equal to:

$$Q = -\frac{\pi R^3 \tau_0}{3\eta}\left[1-\left(\frac{R_{00}}{R}\right)^3\right] + Q_P\left[1-\left(\frac{R_{00}}{R}\right)^4\right] \tag{35}$$

Comparing Eqs. (29)–(32) and (34), (35), we can see that the structure of the flow of the liquid through the nanotubes considering the slippage, is similar to that of the flow of viscoplastic fluid in a pipe of the same radius R.

Given that the size of the central core flow of viscoplastic fluid (radius R_{00}) is defined by:

$$R_{00} = \frac{2\tau_0 l}{\Delta p} \tag{36}$$

for viscoplastic fluid flow we obtain Buckingham formula:

$$Q = Q_P\left[1+\frac{1}{3}\left(\frac{2l\tau_0}{R\Delta p}\right)^4 - \frac{4}{3}\left(\frac{2l\tau_0}{R\Delta p}\right)\right] \tag{37}$$

We'll establish conformity of the pipe, which implements the flow of a viscoplastic fluid with a fluid-filled nanotube, the same size and with the same pressure drop. We say that an effective internal critical shear stress τ_{0ef} of viscoplastic fluid flow, which ensures the coincidence rate with the flow of fluid in the nanotube. Then from Eq. (37) we obtain equation of fourth order to determine τ_{0ef}:

$$\left(\frac{2l\tau_{0ef}}{R\Delta p}\right)^4 - 4\left(\frac{2l\tau_{0ef}}{R\Delta p}\right) = A; A = 3(\varepsilon-1); \varepsilon = Q/Q_P \tag{38}$$

The solution of Eq. (38) can be found, for example, the iteration method of Newton:

$$\bar{\tau}_{0ef\,n} = \bar{\tau}_{0ef\,n-1} - \frac{\bar{\tau}_{0ef\,n-1}^{-4} - 4\bar{\tau}_{0ef\,n-1} - A}{4\bar{\tau}_{0ef\,n-1}^{-3} - 4}\,\bar{\tau}_{0ef} = \frac{2l\tau_{0ef}}{R\Delta p} \tag{39}$$

The first component in Eq. (34) represents the contribution to the fluid flow due to the slippage, and it becomes clear that the slippage significantly enhances the flow rate in the nanotube, when $l\eta v_0 >\approx \Delta p R^2$.

This result is consistent with experimental and theoretical results (which show that water flow in nanochannels can be much higher than under the same conditions, but for the liquid continuum.

In the absence of slippage $\varepsilon = 1$ the Eq. (38) has a trivial solution $\bar{\tau}_{0ef} = 0$.

4.4.2 RESULTS

Let's determine the dependence of the effective critical inner shear stress τ_{0ef} on the radius of the nanotubes, by taking necessary values for calculations $\varepsilon = Q/Q_P$. The results of calculations at $\Delta p / l = 2,1 \cdot 10^{14}$ Pa/m are in the table below:

R, м	τ_{0ef} (Па)	$\varepsilon = Q/Q_P$
$0,8 \times 10^{-9}$	498498	350
$1,11 \times 10^{-9}$	577500	200
$1,385 \times 10^{-9}$	632599	114
$1,665 \times 10^{-9}$	699300	84
$1,94 \times 10^{-9}$	782208	68
$2,22 \times 10^{-9}$	855477	57
$2,495 \times 10^{-9}$	932631	50

Calculations show that the value of effective internal shear stress depends on the size of the nanotube.

Figure 40. shows the dependence τ_{0ef} on the nanotube radius.

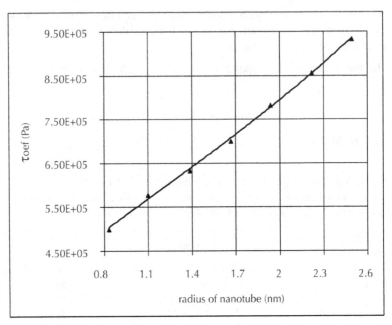

FIGURE 40   Dependence of the effective inner shear stress from the radius of the nanotube.

As you can see, this dependence $\tau_{0ef}(R)$ is almost linear. Within the range of the considered nanotube sizes τ_{0ef} has a relatively low value, which indicates the smoothness of the surface of carbon nanotubes.

4.4.3   THE FLOW OF FLUID WITH AN EMPTY INTERLAYER

The works of many researchers were analyzed in the aforementioned analysis of the structure of liquid flow in carbon nanotubes. The results of the calculations of the cited works showed that during the flow of the liquid particles, an empty layer between the fluid and the nanotube is formed. The area near the walls of the carbon nanotube $R_* \leq r \leq R$ becomes inaccessible for the molecules of the liquid due to van der Waals repulsion forces of the heterogeneous particles of the carbon and water. Moreover, according to the results of man scientists thicknesses of the layers $R_* \leq r \leq R$ regardless of radiuses of the nanotubes are practically identical: $R_* / R \approx 0,8.$

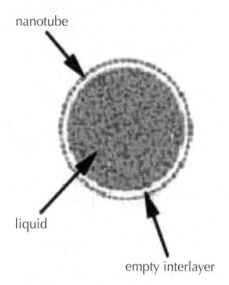

FIGURE 41   The structure of the flow.

A similar result was obtained in by researchers which is an image (Fig. 38) of the configuration of water molecules inside (8, 8) single-walled carbon nanotubes at different temperatures: 298, 325, and 350 ^0K.

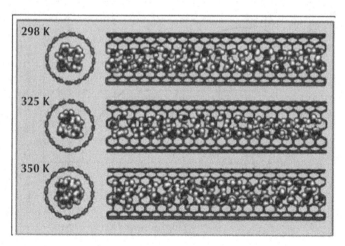

FIGURE 38   The configuration of water molecules inside single-walled carbon nanotubes.

Integrating Eq. (28) between r to R_* at the boundary conditions $r = R_* : v = v_*$, we obtain a formula to calculate the velocity of the liquid layers located at a distance r from the axis of the tube:

$$v(r) = v_* + \frac{2Q_p}{\pi R^2}\left[\left(\frac{R_*}{R}\right)^2 - \left(\frac{r}{R}\right)^2\right] \tag{40}$$

Let's insert Eq. (40) in Eq. (33), integrate over the radius of all sections from 0 to R_*. Then we get a formula for the flow of liquid from the nanotube:

$$Q = Q_P\left[v_*\frac{8l\eta R_*^2}{\Delta p R^4} + \left(\frac{R_*}{R}\right)^4\right] \tag{41}$$

or

$$\varepsilon = \frac{8l\eta v_*}{\Delta p R^2}\left(\frac{R_*}{R}\right)^2 + \left(\frac{R_*}{R}\right)^4$$

from which we can determine the unknown v_*:

$$v_* = \frac{Q_P}{\pi R^2}\left[\varepsilon - \left(\frac{R_*}{R}\right)^4\right]\left(\frac{R}{R_*}\right)^2 \tag{42}$$

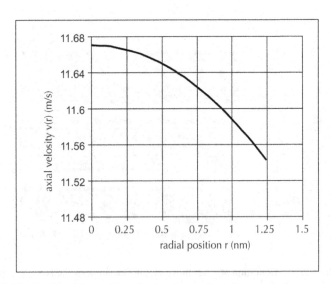

FIGURE 43   Profile of the radial velocity of water in carbon nanotubes.

Figure 43 shows the profile of the radial velocity of water particles in a carbon nanotube with a diameter of 2.77 nm, calculated using the Eq. (44) at $Q_P = 4,75 \cdot 10^{-19} n / m^2$, $\varepsilon = 114$. The velocity at the border v_* is equal to 11,55 m/s. It is seen that the velocity profile is similar to a parabolic shape, and at the same time agrees with the numerical values.

The calculations suggest the following conclusions. Flow of liquid in a nanotube was investigated using synthesis of the methods of the continuum theory and molecular dynamics. Two models are considered. The first is based on the fact that fluid in the nanotube behaves like a viscoplastic. A method of calculating the value of limiting shear stress is proposed, which was dependent on the nanotube radius. A simplified model agrees quite well with the results of the molecular simulations of fluid flow in carbon nanotubes. The second model assumes the existence of an empty interlayer between the liquid molecules and wall of the nanotube. This formulation of the task is based on the results of experimental works known from the literature. The velocity profile of fluid flowing in the nanotube is practically identical to the profile determined by molecular modeling.

As seen from the results of the calculations, the velocity value varies slightly along the radius of the nanotube. Such a velocity distribution of the fluid particles can be explained by the lack of friction between the molecules of the liquid and the wall due to the presence of an empty layer. This leads to an easy slippage of the liquid and, consequently, anomalous increase in flow compared to the Poiseuille flow.

APPENDIX A—MODELING AND SIMULATION TECHNIQUES

A1   MOLECULAR SCALE METHODS

The modeling and simulation methods at molecular level usually employ atoms, molecules or their clusters as the basic units considered. The most popular methods include molecular mechanics (MM), Molecular dynamics (MD) and Monte Carlo (MC) simulation. Modeling of polymer nanocomposites at this scale is predominantly directed toward the thermodynamics and kinetics of the formation, molecular structure and interactions. The diagram in Fig. A1 describes the equation of motion for each method and the typical properties predicted from each of them [1–22]. We introduce here the two widely used molecular scale methods: MD and MC.

A1.1   MOLECULAR DYNAMICS

MD is a computer simulation technique that allows one to predict the time evolution of a system of interacting particles (e.g., atoms, molecules, granules, etc.) and estimate the relevant physical properties [23, 24]. Specifically, it generates such information as atomic positions, velocities and forces from which the macroscopic properties (e.g., pressure, energy, heat capacities) can be derived by means of statistical mechanics. MD simulation usually consists of three constituents: (i) a set of initial conditions (e.g., initial positions and velocities of all particles in the system); (ii) the interaction potentials to represent the forces among all the particles; (iii) the evolution of the system in time by solving a set of classical Newtonian equations of motion for all particles in the system. The equation of motion is generally given by:

$$\vec{F}_i(t) = m_i \frac{d^2 \vec{r}_i}{dt^2} \tag{A1}$$

where $\vec{F}_i(t)$ is the force acting on the ith atom or particle at time t, which is obtained as the negative gradient of the interaction potential U, mi is the atomic mass and $\vec{r}_i$ the atomic position. A physical simulation involves the proper selection of interaction potentials, numerical integration, periodic boundary conditions, and the controls of pressure and temperature to mimicbphysically meaningful thermodynamic ensembles. The interaction potentials together with their parameters, i.e., the so-called force field, describe in detail how the particles in a system interact with each other, i.e., how the potential energy of a system depends on the particle coordinates. Such a force field may be obtained by quantum method (e.g., ab initio), empirical method (e.g., Lennard–Jones, Mores, Born-Mayer) or quantum-empirical method (e.g., embedded atom model, glue model, bond order potential). The criteria for selecting a force field include the accuracy, transferability and computational speed. A typical interaction potential U may consist of a number of bonded and nonbonded interaction terms:

$$U(\vec{r_1},\vec{r_2},\vec{r_3},\dots,\vec{r_n}) = \sum_{i_{bond}}^{N_{bond}} U_{bond}(i_{bond},\vec{r_a},\vec{r_b}) + \sum_{i_{angle}}^{N_{angle}} U_{angle}(i_{angle},\vec{r_a},\vec{r_b},\vec{r_c})$$

$$+ \sum_{i_{torsion}}^{N_{torsion}} U_{torsion}\left(i_{torsion},\vec{r_a},\vec{r_b},\vec{r_c},\vec{r_d}\right)$$

$$+ \sum_{i_{inversion}}^{N_{inversion}} U_{inversion}\left(i_{inversion},\vec{r_a},\vec{r_b},\vec{r_c},\vec{r_d}\right)$$

$$+ \sum_{i=1}^{N-1}\sum_{j>i}^{N} U_{vdw}\left(i,j,\vec{r_a},\vec{r_b}\right)$$

$$+ \sum_{i=1}^{N-1}\sum_{j>i}^{N} U_{electrostatic}\left(i,j,\vec{r_a},\vec{r_b}\right) \tag{A2}$$

The first four terms represent bonded interactions, that is, bond stretching Ubond, bond-angle bend Uangle and dihedral angle torsion Utorsion and inversion interaction Uinversion, while the last two terms are nonbonded interactions, that is, van der Waals energy Uvdw and electrostatic energy Uelectrostatic. In the equation, $\vec{r}_a, \vec{r}_b, \vec{r}_c, \vec{r}_d$ are the positions of the atoms or particles specifically involved in a given interaction; N_{bond}, N_{angle}, $N_{torsion}$ and $N_{inversion}$ stand for the total numbers of these respective interactions in the simulated system; i_{bond}, i_{angle}, $i_{torsion}$ and $i_{inversion}$ uniquely specify an individual interaction of each type; i and j in the van der Waals and electrostatic terms indicate the atoms involved in the interaction. There are many algorithms for integrating the equation of motion using finite difference methods. The algorithms of varlet, velocity varlet, leap-frog and Beeman, are commonly used in MD simulations [23]. All algorithms assume that the atomic position $\vec{r}$, velocities $\vec{v}$ and accelerations $\vec{a}$ can be approximated by a Taylor series expansion:

$$\vec{r}(t + \delta t) = \vec{r}(t) + \vec{v}(t)\delta t + \frac{1}{2}\vec{a}(t)\delta^2 t + \cdots$$

$$\vec{r}(t + \delta t) = \vec{r}(t) + \vec{v}(t)\delta t + \frac{1}{2}\vec{a}(t)\delta^2 t + \cdots \tag{A3}$$

$$\vec{v}(t + \delta t) = \vec{v}(t) + \vec{a}(t)\delta t + \frac{1}{2}\vec{b}(t)\delta^2 t + \cdots$$

$$\vec{v}(t + \delta t) = \vec{v}(t) + \vec{a}(t)\delta t + \frac{1}{2}\vec{b}(t)\delta^2 t + \cdots \qquad (A4)$$

$$\vec{a}(t + \delta t) = \vec{a}(t) + \vec{b}(t)\delta t + \cdots \qquad (A5)$$

Generally speaking, a good integration algorithm should conserve the total energy and momentum and be time-reversible. It should also be easy to implement and computationally efficient, and permit a relatively long time step. The verlet algorithm is probably the most widely used method. It uses the positions $\vec{r}(t)$ and accelerations $\vec{a}(t)$ at time t, and the positions $\vec{r}(t - \delta t)$ from the previous step (t–δ) to calculate the new positions $\vec{r}(t + \delta t)$ at (t+δt), we have:

$$\vec{r}(t + \delta t) = \vec{r}(t) + \vec{v}(t)\delta t + \frac{1}{2}\vec{a}(t)\delta t^2 + \cdots$$

$$\vec{r}(t + \delta t) = \vec{r}(t) + \vec{v}(t)\delta t + \frac{1}{2}\vec{a}(t)\delta t^2 + \cdots \qquad (A6)$$

$$\vec{r}(t - \delta t) = \vec{r}(t) - \vec{v}(t)\delta t + \frac{1}{2}\vec{a}(t)\delta t^2 + \cdots$$

$$\vec{r}(t - \delta t) = \vec{r}(t) - \vec{v}(t)\delta t + \frac{1}{2}\vec{a}(t)\delta t^2 + \cdots \qquad (A7)$$

$$\vec{r}(t + \delta t) = 2\vec{r}(t) - \vec{r}(t - \delta t) + \vec{a}(t)\delta t^2 + \cdots$$

$$\vec{r}(t + \delta t) = 2\vec{r}(t) - \vec{r}(t - \delta t) + \vec{a}(t)\delta t^2 + \cdots \qquad (A8)$$

The velocities at time t and $t + \frac{1}{2\delta t}$ can be respectively estimated.

$$\vec{v}(t) = \frac{\left[\vec{r}(t + \delta t) - \vec{r}(t - \delta t)\right]}{2\delta t} \qquad (A9)$$

$$\vec{v}\left(t + \frac{1}{2\delta t}\right) = \frac{\left[\vec{r}(t + \delta t) - \vec{r}(t - \delta t)\right]}{\delta t}$$

$$\vec{v}\left(t + \frac{1}{2\delta t}\right) = \frac{\left[\vec{r}(t + \delta t) - \vec{r}(t - \delta t)\right]}{\delta t} \tag{A10}$$

MD simulations can be performed in many different ensembles, such as grand canonical (μVT), microcanonical (NVE), canonical (NVT) and isothermal–isobaric (NPT). The constant temperature and pressure can be controlled by adding an appropriate thermostat (e.g., Berendsen, Nose, Nose–Hoover and Nose–Poincare) and barostat (e.g., Andersen, Hoover and Berendsen), respectively. Applying MD into polymer composites allows us to investigate into the effects of fillers on polymer structure and dynamics in the vicinity of polymer–filler interface and also to probe the effects of polymer–filler interactions on the materials properties.

A1.2 MONTE CARLO

MC technique, also called Metropolis method [24], is a stochastic method that uses random numbers to generate a sample population of the system from which one can calculate the properties of interest. A MC simulation usually consists of three typical steps. In the first step, the physical problem under investigation is translated into an analogous probabilistic or statistical model. In the second step, the probabilistic model is solved by a numerical stochastic sampling experiment. In the third step, the obtained data are analyzed by using statistical methods. MC provides only the information on equilibrium properties (e.g., free energy, phase equilibrium), different from MD, which gives nonequilibrium, as well as equilibrium properties. In a NVT ensemble with N atoms, one hypothesizes a new configuration by arbitrarily or systematically moving one atom from position i→j. Due to such atomic movement, one can compute the change in the system Hamiltonian ΔH:

$$\Delta H = H(j)-H(i) \tag{A11}$$

where H(i) and H(j) are the Hamiltonian associated with the original and new configuration, respectively.

This new configuration is then evaluated according to the following rules. If $\Delta H \geq 0$, then the atomic movement would bring the system to a state of lower energy. Hence, the movement is immediately accepted and the displaced atom remains in its new position. If $\Delta H \geq 0$, the move is accepted only with a certain probability $P^{i \rightarrow j}$ which is given by:

$$P^{i \rightarrow j} \propto \exp\left(-\frac{\Delta H}{K_B T}\right) \tag{A12}$$

where K_B is the Boltzmann constant. According to Metropolis et al. [25] one can generate a random number ζ between 0 and 1 and determine the new configuration according to the following rule:

$$\zeta \leq \exp\left(-\frac{\Delta H}{K_B T}\right); \qquad (A13)$$

the move is accepted;

$$\zeta > \exp\left(-\frac{\Delta H}{K_B T}\right); \qquad (A14)$$

the move is not accepted.

If the new configuration is rejected, one counts the original position as a new one and repeats the process by using other arbitrarily chosen atoms. In a μVT ensemble, one hypothesizes a new configuration j by arbitrarily choosing one atom and proposing that it can be exchanged by an atom of a different kind. This procedure affects the chemical composition of the system. Also, the move is accepted with a certain probability. However, one computes the energy change ΔU associated with the change in composition. The new configuration is examined according to the following rules. If $\Delta U \geq 0$, the move of compositional change is accepted. However, if $\Delta U \geq 0$, the move is accepted with a certain probability which is given by:

$$P^{\,i \to j} \propto \exp\left(-\frac{\Delta U}{K_B T}\right) \qquad (A15)$$

where ΔU is the change in the sum of the mixing energy and the chemical potential of the mixture. If the new configuration is rejected one counts the original configuration as a new one and repeats the process by using some other arbitrarily or systematically chosen atoms. In polymer nanocomposites, MC methods have been used to investigate the molecular structure at nanoparticle surface and evaluate the effects of various factors.

A2 MICROSCALE METHODS

The modeling and simulation at microscale aim to bridge molecular methods and continuum methods and avoid their shortcomings. Specifically, in nanoparticle–polymer systems, the study of structural evolution (i.e., dynamics of phase separation) involves the description of bulk flow (i.e., hydrodynamic behavior) and the interactions between nanoparticle and polymer components. Note that hydrodynamic behavior is relatively straightforward to handle by continuum methods but is very difficult and expensive to treat by atomistic methods. In contrast, the interactions between components can be examined at an atomistic level but are usually not straightforward to incorporate at the continuum level. Therefore, various simulation methods have been evaluated and extended to study the microscopic structure and phase separation of these polymer nanocomposites, including BD, DPD, LB, time-dependent Ginsburg–Landau (TDGL) theory, and dynamic DFT. In these methods, a polymer system is usually treated with a field description or microscopic particles that incorporate molecular details implicitly. Therefore, they are able to simulate the phenomena on length and time scales currently inaccessible by the classical MD methods.

A2.1 BROWNIAN DYNAMICS

BD simulation is similar to MD simulations [26]. However, it introduces a few new approximations that allow one to perform simulations on the microsecond timescale whereas MD simulation is known up to a few nanoseconds. In BD the explicit description of solvent molecules used in MD is replaced with an implicit continuum solvent description. Besides, the internal motions of molecules are typically ignored, allowing a much larger time step than that of MD. Therefore, BD is particularly useful for systems where there is a large gap of time scale governing the motion of different components. For example, in polymer–solvent mixture, a short time-step is required to resolve the fast motion of the solvent molecules, whereas the evolution of the slower modes of the system requires a larger time step. However, if the detailed motion of the solvent molecules is concerned, they may be removed from the simulation and their effects on the polymer are represented by dissipative ($-\gamma$P) and random ($\sigma \zeta(t)$) force terms. Thus, the forces in the governing Eq. (16) is replaced by a Langevin equation,

$$F_i(t) = \sum_{i \neq j} F_{ij}^c - \gamma P_i + \sigma \zeta_i(t) \qquad (A16)$$

Where F_{ij}^c is the conservative force of particle j acting on particle i, γ and σ are constants depending on the system, Pi the momentum of particle i, and $\zeta(t)$ a Gaussian random noise term. One consequence of this approximation of the fast degrees of freedom by fluctuating forces is that the energy and momentum are no longer conserved, which implies that the macroscopic behavior of the system will not be hydrodynamic. In addition, the effect of one solute molecule on another through the flow of solvent molecules is neglected. Thus, BD can only reproduce the diffusion properties but not the hydrodynamic flow properties since the simulation does not obey the Navier–Stokes equations.

A2.2 DISSIPATIVE PARTICLE DYNAMICS

DPD was originally developed by Hoogerbrugge and Koelman [27]. It can simulate both Newtonian and nonNewtonian fluids, including polymer melts and blends, on microscopic length and time scales. Like MD and BD, DPD is a particle-based method. However, its basic unit is not a single atom or molecule but a molecular assembly (i.e., a particle). DPD particles are defined by their mass M_i, position r_i and momentum P_i. The interaction force between two DPD particles i and j can be described by a sum of conservative F_{ij}^C, dissipative F_{ij}^D and random forces F_{ij}^R [28–30]:

$$F_{ij} = F_{ij}^C + F_{ij}^D + F_{ij}^R \qquad (A17)$$

While the interaction potentials in MD are high-order polynomials of the distance r_{ij} between two particles, in DPD the potentials are softened so as to approximate the effective potential at microscopic length scales. The form of the conservative force in particular is chosen to decrease linearly with increasing r_{ij}. Beyond a certain cut-off separation r_c, the weight functions and thus the forces are all zero. Because the forces are pair wise and momentum is conserved, the macroscopic behavior directly incorporates Navier–Stokes hydrodynamics. However, energy is not conserved because of the presence of the dissipative and random force terms, which are similar to those of BD, but incorporate the effects of Brownian motion on larger length scales. DPD has several advantages over MD, for example, the hydrodynamic behavior is observed with far fewer particles than required in a MD simulation because of its larger particle size. Besides, its force forms allow larger time steps to be taken than those in MD.

A2.3 LATTICE BOLTZMANN

LB [31] is another microscale method that is suited for the efficient treatment of polymer solution dynamics. It has recently been used to investigate the phase separation of binary fluids in the presence of solid particles. The LB method is originated from lattice gas automaton, which is constructed as a simplified, fictitious molecular dynamic in which space, time and particle velocities are all discrete. A typical lattice gas automaton consists of a regular lattice with particles residing on the nodes. The main feature of the LB method is to replace the particle occupation variables (Boolean variables), by single-particle distribution functions (real variables) and neglect individual particle motion and particle–particle correlations in the kinetic equation. There are several ways to obtain the LB equation from either the discrete velocity model or the Boltzmann kinetic equation, and to derive the macroscopic Navier–Stokes equations from the LB equation. An important advantage of the LB method is that microscopic physical interactions of the fluid particles can be conveniently incorporated into the numerical model. Compared with the Navier– Stokes equations, the LB method can handle the interactions among fluid particles and reproduce the microscale mechanism of hydrodynamic behavior. Therefore it belongs to the MD in nature and bridges the gap between the molecular level and macroscopic level. However, its main disadvantage is that it is typically not guaranteed to be numerically stable and may lead to physically unreasonable results, for instance, in the case of high forcing rate or high interparticle interaction strength.

A2.4   TIME-DEPENDENT GINZBURG–LANDAU METHOD

TDGL is a microscale method for simulating the structural evolution of phase-separation in polymer blends and block copolymers. It is based on the Cahn–Hilliard–Cook (CHC) nonlinear diffusion equation for a binary blend and falls under the more general phase-field and reaction-diffusion models [32–34]. In the TDGL method, a free-energy function is minimized to simulate a temperature quench from the miscible region of the phase diagram to the immiscible region. Thus, the resulting time-dependent structural evolution of the polymer blend can be investigated by solving the TDGL/CHC equation for the time dependence of the local blend concentration. Glotzer and co-workers have discussed and applied this method to polymer blends and particle-filled polymer systems [35]. This model reproduces the growth kinetics of the TDGL model, demonstrating that such quantities are insensitive to the precise form of the double-well potential of the bulk free-energy term. The TDGL and CDM methods have recently been used to investigate the phase-separation of polymer nanocomposites and polymer blends in the presence of nanoparticles [36–40].

A2.5   DYNAMIC DFT METHOD

Dynamic DFT method is usually used to model the dynamic behavior of polymer systems and has been implemented in the software package Mesodyn™ from Accelrys [41]. The DFT models the behavior of polymer fluids by combining

Gaussian mean-field statistics with a TDGL model for the time evolution of conserved order parameters. However, in contrast to traditional phenomenological free-energy expansion methods employed in the TDGL approach, the free energy is not truncated at a certain level, and instead retains the full polymer path integral numerically. At the expense of a more challenging computation, this allows detailed information about a specific polymer system beyond simply the Flory–Huggins parameter and mobilities to be included in the simulation. In addition, viscoelasticity, which is not included in TDGL approaches, is included at the level of the Gaussian chains. A similar DFT approach has been developed by Doi and co-workers [42, 43] and forms the basis for their new software tool Simulation Utilities for Soft and Hard Interfaces (SUSHI), one of a suite of molecular and mesoscale modeling tools (called OCTA) developed for the simulation of polymer materials [44]. The essence of dynamic DFT method is that the instantaneous unique conformation distribution can be obtained from the off-equilibrium density profile by coupling a fictitious external potential to the Hamiltonian. Once such distribution is known, the free energy is then calculated by standard statistical thermodynamics. The driving force for diffusion is obtained from the spatial gradient of the first functional derivative of the free energy with respect to the density. Here, we describe briefly the equations for both polymer and particle in the diblock polymer–particle composites [38].

A3 MESOSCALE AND MACROSCALE METHODS

Despite the importance of understanding the molecular structure and nature of materials, their behavior can be homogenized with respect to different aspects, which can be at different scales. Typically, the observed macroscopic behavior is usually explained by ignoring the discrete atomic and molecular structure and assuming that the material is continuously distributed throughout its volume. The continuum material is thus assumed to have an average density and can be subjected to body forces such as gravity and surface forces. Generally speaking, the macroscale methods (or called continuum methods hereafter) obey the fundamental laws of: (i) continuity, derived from the conservation of mass; (ii) equilibrium, derived from momentum considerations and Newton's second law; (iii) the moment of momentum principle, based on the model that the time rate of change of angular momentum with respect to an arbitrary point is equal to the resultant moment; (iv) conservation of energy, based on the first law of thermodynamics; and (v) conservation of entropy, based on the second law of thermodynamics. These laws provide the basis for the continuum model and must be coupled with the appropriate constitutive equations and the equations of state to provide all the equations necessary for solving a continuum problem. The continuum method relates the deformation of a continuous medium to the external forces acting on the medium and the resulting internal stress and strain. Computational approaches range from simple closed-form analytical expressions to micromechanics and complex structural mechanics calculations based on beam and shell theory. In this section, we introduce some continuum methods that have been used in polymer nanocomposites, including micromechanics models (e.g., Halpin–Tsai model, Mori–Tanaka model), equivalent-continuum model, self-consistent model and finite element analysis.

A4   MICROMECHANICS

Since the assumption of uniformity in continuum mechanics may not hold at the microscale level, micromechanics methods are used to express the continuum quantities associated with an infinitesimal material element in terms of structure and properties of the micro constituents. Thus, a central theme of micromechanics models is the development of a representative volume element (RVE) to statistically represent the local continuum properties. The RVE is constructed to ensure that the length scale is consistent with the smallest constituent that has a first-order effect on the macroscopic behavior. The RVE is then used in a repeating or periodic nature in the full-scale model. The micromechanics method can account for interfaces between constituents, discontinuities, and coupled mechanical and nonmechanical properties. Ourpurpose is to review the micromechanics methods used for polymer nanocomposites. Thus, we only discuss here some important concepts of micromechanics as well as the Halpin–Tsai model and Mori–Tanaka model.

A4.1   BASIC CONCEPTS

When applied to particle reinforced polymer composites, micromechanics models usually follow such basic assumptions as (i) linear elasticity of fillers and polymer matrix; (ii) the fillers are axisymmetric, identical in shape and size, and can be characterized by parameters such as aspect ratio; (iii) well-bonded filler–polymer interface and the ignorance of interfacial slip, filler–polymer debonding or matrix cracking. The first concept is the linear elasticity, i.e., the linear relationship between the total stress and infinitesimal strain tensors for the filler and matrix as expressed by the following constitutive equations:

For filler $\sigma^f = C^f \varepsilon^f$

For matrix $\sigma^m = C^m \varepsilon^m$

where C is the stiffness tensor. The second concept is the average stress and strain. Since the point wise stress field $\sigma(x)$ and the corresponding strain field $\varepsilon(x)$ are usually nonuniform in polymer composites, the volume–average stress $\bar{\sigma}$ and strain $\bar{\varepsilon}$ are then defined over the representative averaging volume V, respectively.

$$\bar{\sigma} = \frac{1}{V} \int \sigma(x) dv$$

$$\bar{\varepsilon} = \frac{1}{V} \int \varepsilon(x) dv$$

Therefore, the average filler and matrix stresses are the averages over the corresponding volumes v_f and v_m, respectively.

$$\bar{\sigma_f} = \frac{1}{V_f} \int \sigma(x) dv$$

$$\bar{\sigma_m} = \frac{1}{V_m} \int \sigma(x) dv$$

The average strains for the fillers and matrix are defined, respectively, as

$$\bar{\varepsilon_f} = \frac{1}{V_f} \int \varepsilon(x) dv$$

$$\bar{\varepsilon_m} = \frac{1}{V_m} \int \varepsilon(x) dv$$

Based on the above definitions, the relationships between the filler and matrix averages and the overall averages can be derived as follows:

$$\bar{\sigma} = \bar{\sigma_f} v_f + \bar{\sigma_m} v_m \tag{A18}$$

$$\bar{\varepsilon} = \bar{\varepsilon_f} v_f + \bar{\varepsilon_m} v_m \tag{A19}$$

where v_f, v_m are the volume fractions of the fillers and matrix, respectively.

The third concept is the average properties of composites, which are actually the main goal of a micromechanics model. The average stiffness of the composite is the tensor C that maps the uniform strain to the average stress.

$$\bar{\sigma} = \bar{\varepsilon} C$$

The average compliance S is defined in the same way:

$$\bar{\varepsilon} = \bar{\sigma}S$$

Another important concept is the strain–concentration and stress–concentration tensors A and B, which are basically the ratios between the average filler strain (or stress) and the corresponding average of the composites.

$$\bar{\varepsilon}_f = \bar{\varepsilon}A$$

$$\bar{\sigma}_f = \bar{\sigma}B$$

Using the above concepts and equations, the average composite stiffness can be obtained from the strain concentration tensor A and the filler and matrix properties:

$$C = C_m + v_f (C_f - C_m)A \qquad (A19)$$

A4.2 HALPIN–TSAI MODEL

The Halpin–Tsai model is a well-known composite theory to predict the stiffness of unidirectional composites as a functional of aspect ratio. In this model, the longitudinal E11 and transverse E22 engineering moduli are expressed in the following general form:

$$\frac{E}{E_m} = \frac{1 + \zeta \eta v_f}{1 - \eta v_f} \qquad (A20)$$

where E and E_m represent the Young's modulus of the composite and matrix, respectively, v_f is the volume fraction of filler, and η is given by:

$$\eta = \frac{\dfrac{E}{E_m} - 1}{\dfrac{E_f}{E_m} + \zeta_f}$$

where E_f represents the Young's modulus of the filler and ζ_f the shape parameter depending on the filler geometry and loading direction. When calculating longitudinal modulus E_{11}, ζ_f is equal to l/t, and when calculating transverse modulus E_{22}, ζ_f is equal to w/t. Here, the parameters of l, w and t are the length, width and thickness of the dispersed fillers, respectively. If $\zeta_f \to 0$, the Halpin–Tsai theory converges to the inverse rule of mixture (lower bound):

$$\frac{1}{E} = \frac{v_f}{E_f} + \frac{1 - v_f}{E_m} \qquad (A21)$$

Conversely, if $\zeta_f \to \infty$, the theory reduces to the rule of mixtures (upper bound),

$$E = E_f v_f + E_m \left(1 - v_f\right) \tag{A22}$$

A4.3 MORI–TANAKA MODEL

The Mori–Tanaka model is derived based on the principles of Eshelby's inclusion model for predicting an elastic stress field in and around ellipsoidal filler in an infinite matrix. The complete analytical solutions for longitudinal E_{11} and transverse E_{22} elastic moduli of an isotropic matrix filled with aligned spherical inclusion are [45, 46]:

$$\frac{E_{11}}{E_m} = \frac{A_0}{A_0 + v_f(A_1 + 2v_0 A_2)} \tag{A23}$$

$$\frac{E_{22}}{E_m} = \frac{2A_0}{2A_0 + v_f(-2A_3 + (1 - v_0 A_4) + (1 + v_0)A_5 A_0)} \tag{A24}$$

where E_m represents the Young's modulus of the matrix, v_f the volume fraction of filler, v_0 the Poisson's ratio of the matrix, parameters, A0, A1, ..., A5 are functions of the Eshelby's tensor and the properties of the filler and the matrix, including Young's modulus, Poisson's ratio, filler concentration and filler aspect ratio [45].

A4.4 EQUIVALENT-CONTINUUM AND SELF-SIMILAR APPROACHES

Numerous micromechanical models have been successfully used to predict the macroscopic behavior of fiber-reinforced composites. However, the direct use of these models for nanotube-reinforced composites is doubtful due to the significant scale difference between nanotube and typical carbon fiber. Recently, two methods have been proposed for modeling the mechanical behavior of single walled carbon nanotube (SWCN) composites: equivalent-continuum approach and self-similar approach [47]. The equivalent-continuum approach was proposed by Odegard, et al. [48]. In this approach, MD was used to model the molecular interactions between SWCN–polymer and a homogeneous equivalent-continuum reinforcing element (e.g., (a) SWCN surrounded (b) polymer) were constructed as shown in Fig. A2. Then, micromechanics are used to determine the effective bulk properties of the equivalent-continuum reinforcing element embedded in a continuous polymer. The equivalent-continuum approach consists of four major steps, as briefly described below.

Step 1: MD simulation is used to generate the equilibrium structure of a SWCN–polymer composite and then to establish the RVE of the molecular model and the equivalent-continuum model.

Step 2: The potential energies of deformation for the molecular model and effective fiber are derived and equated for identical loading conditions. The bonded and nonbonded interactions within a polymer molecule are quantitatively described by MM. For the SWCN/polymer system, the total potential energy U^m of the molecular model is:

$$U^m = \sum U^r(k_r) + \sum U^\theta(k_\theta) + \sum U^{vdw}(k_{vdw}) \tag{A25}$$

where U^r, U^θ and U^{vdw} are the energies associated with covalent bond stretching, bond-angle bending, and van der Waals interactions, respectively. An equivalent-truss model of the RVE is used as an intermediate step to link the molecular and equivalent-continuum models. Each atom in the molecular model is represented by a pin-joint, and each truss element represents an atomic bonded or nonbonded interaction. The potential energy of the truss model is

$$U^t = \sum U^a(E^a) + \sum U^b(E^b) + \sum U^c(E^c) \tag{A26}$$

where U^a, U^b and U^c are the energies associated with truss elements that represent covalent bond stretching, bond-angle bending, and van der Waals interactions, respectively. The energies of each truss element are a function of the Young's modulus, E.

Step 3: A constitutive equation for the effective fiber is established. Since the values of the elastic stiffness tensor components are not known a priori, a set of loading conditions are chosen such that each component is uniquely determined from

$$U^f = U^t = U^m$$

Step 4: Overall constitutive properties of the dilute and unidirectional SWCN/polymer composite are determined with Mori–Tanaka model with the mechanical properties of the effective fiber and the bulk polymer. The layer of polymer molecules that are near the polymer/nanotube interface (Fig. 2) is included in the effective fiber, and it is assumed that the matrix polymer surrounding the effective fiber has mechanical properties equal to those of the bulk polymer. The self-similar approach was proposed by Pipes and Hubert [49], which consists of three major steps:

First, a helical array of SWCNs is assembled. This array is termed as the SWCN nanoarray where 91 SWCNs make up the cross-section of the helical nanoarray. Then, the SWCN nanoarrays is surrounded by a polymer matrix and assembled into a second twisted array, termed as the SWCN nanowire Finally, the SWCN nanowires are further impregnated with a polymer matrix and assembled into the final helical array—the SWCN microfiber. The self-similar geometries described in the nanoarray, nanowire and microfiber (Fig. A3) allow the use of the same mathematical and geometric model for all three geometries [49].

A4.5   FINITE ELEMENT METHOD

FEM is a general numerical method for obtaining approximate solutions in space to initial-value and boundary-value problems including time-dependent process-es. It employs preprocessed mesh generation, which enables the model to fully capture the spatial discontinuities of highly inhomogeneous materials. It also al-lows complex, nonlinear tensile relationships to be incorporated into the analysis. Thus, it has been widely used in mechanical, biological and geological systems. In FEM, the entire domain of interest is spatially discretized into an assembly of simply shaped subdomains (e.g., hexahedra or tetrahedral in three dimensions, and rectangles or triangles in two dimensions) without gaps and without overlaps. The subdomains are interconnected at joints (i.e., nodes). The implementation of FEM includes the important steps shown in Fig. A.1. The energy in FEM is taken from the theory of linear elasticity and thus the input parameters are simply the elastic moduli and the density of the material. Since these parameters are in agreement with the values computed by MD, the simulation is consistent across the scales. More specifically, the total elastic energy in the absence of tractions and body forces within the continuum model is given by [50]:

$$U = U_v + U_k$$

$$U_k = \frac{1}{2} \int dr \, p(r) |\dot{U}_r|^2$$

$$U_v = \frac{1}{2} \int dr \sum_{\mu,v,\lambda,\sigma=1}^{3} \varepsilon_{\mu v}(r) C_{\mu v \lambda \sigma} \lambda_\sigma(r)$$

where U_v is the Hookian potential energy term, which is quadratic in the symmetric strain tensor e, contracted with the elastic constant tensor C. The Greek indices (i.e., m, n, l, s) denote Cartesian directions. The kinetic energy U_k involves the time rate of change of the displacement field $\dot{U}$, and the mass density ρ.

These are fields defined throughout space in the continuum theory. Thus, the total energy of the system is an integral of these quantities over the volume of the sample dυ. The FEM has been incorporated in some commercial software packages and open source codes (e.g., ABAQUS, ANSYS, Palmyra and OOF) and widely used to evaluate the mechanical properties of polymer composites. Some attempts have recently been made to apply the FEM to nanoparticle-reinforced polymer nanocomposites. In order to capture the multiscale material behaviors, efforts are also underway to combine the multiscale models spanning from molecular to macroscopic levels [51, 52].

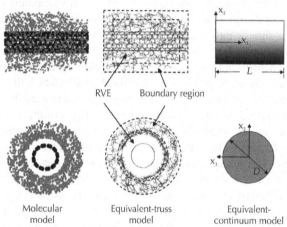

FIGURE A1 Equivalent-continuum modeling of effective fiber.

A5 MULTI SCALE MODELING OF MECHANICAL PROPERTIES

In Odegard's study [48], a method has been presented for linking atomistic simulations of nano-structured materials to continuum models of the corresponding bulk material. For a polymer composite system reinforced with SWNT, the method provides the steps whereby the nanotube, the local polymer near the nanotube, and the nanotube/polymer interface can be modeled as an effective continuum fiber by using an equivalent-continuum model. The effective fiber retains the local molecular structure and bonding information, as defined by molecular dynamics, and serves as a means for linking the equivalent-continuum and micromechanics models. The micromechanics method is then available for the prediction of bulk mechanical properties of SWNT/polymer composites as a function of nanotube size, orientation, and volume fraction. The utility of this method was examined by modeling tow composites that both having a interface. The elastic stiffness constants of the composites were determined for both aligned and three-dimensional randomly oriented nanotubes, as a function of nanotube length and volume fraction. They used Mori– Tanaka model [53] for random and oriented fibers position and compare their model with mechanical properties, the interface between fiber and matrix was assumed perfect. Motivated by micrographs showing that embedded nanotubes often exhibit significant curvature within the polymer, Fisher et al.[54] have developed a model combining finite element results and micromechanical methods (Mori-Tanaka)to determine the effective reinforcing modulus of a wavy embedded nanotube with perfect bonding and random fiber orientation assumption. This effective reinforcing modulus (ERM) is then used within a multiphase micromechanics model to predict the effective modulus of a polymer reinforced with a distribution of wavy nanotubes. We found that even slight nanotube curvature significantly reduces the effective reinforcement when compared to straight nanotubes. These results suggest that nanotube waviness may be an additional mechanism limiting the modulus enhancement of nanotube-reinforced polymers. Bradshaw et al. [55] investigated the degree to which the characteristic waviness of nanotubes embedded in polymers can impact the effective stiffness of these materials. A 3D finite element model of a single infinitely long sinusoidal fiber within an infinite matrix is used to numerically compute the dilute strain concentration tensor. A Mori–Tanaka model uses this tensor to predict the effective modulus of the material with aligned or randomly oriented inclusions. This hybrid finite element micromechanical modeling technique is a powerful extension of general micromechanics modeling and can be applied to any composite microstructure containing nonellipsoidal inclusions. The results demonstrate that nanotube waviness results in a reduction of the effective modulus of the composite relative to straight nanotube reinforcement. The degree of reduction is dependent on the ratio of the sinusoidal wavelength to the nanotube diameter. As this wavelength ratio increases, the effective stiffness of a composite with randomly oriented wavy nanotubes converges to the result obtained with straight nanotube inclusions.

The effective mechanical properties of carbon nanotube-based composites are evaluated by Liu & Chen [56] using a 3-D nanoscale RVE based on 3-D elasticity theory and solved by the finite element method. Formulas to extract the material constants from solutions for the RVE under three loading cases are established using the elasticity. An extended rule of mixtures, which can be used to estimate the Young's modulus in the axial direction of the RVE and to validate the numerical solutions for short CNTs, is also derived using the strength of materials theory. Numerical examples using the FEM to evaluate the effective material constants of a CNT-based composites are presented, which demonstrate that the reinforcing capabilities of the CNTs in a matrix are significant. With only about 2% and 5% volume fractions of the CNTs in a matrix, the stiffness of the composite in the CNT axial direction can increase as many as 0.7 and 9.7 times for the cases of short and long CNT fibers, respectively. These simulation results, which are believed to be the first of its kind for CNT-based composites, are consistent with the experimental results reported in the literature [57–59]. The developed extended rule of mixtures is also found to be quite effective in evaluating the stiffness of the CNT-based composites in the CNT axial direction. Many research issues need to be addressed in the modeling and simulations of CNTs in a matrix material for the development of nanocomposites. Analytical methods and simulation models to extract the mechanical properties of the CNT-based nanocomposites need to be further developed and verified with experimental results. The analytical method and simulation approach developed in this paper are only a preliminary study. Different type of RVEs, load cases and different solution methods should be investigated. Different interface conditions, other than perfect bonding, need to be investigated using different models to more accurately account for the interactions of the CNTs in a matrix material at the nanoscale. Nanoscale interface cracks can be analyzed using simulations to investigate the failure mechanism in nanomaterials. Interactions among a large number of CNTs in a matrix can be simulated if the computing power is available. Single-walled and multiwalled CNTs as reinforcing fibers in a matrix can be studied by simulations to find out their advantages and disadvantages. Finally, large multiscale simulation models for CNT-based composites, which can link the models at the nano, micro and macro scales, need to be developed, with the help of analytical and experimental work [56]. The three RVEs proposed in Ref. [60] and shown in Fig. A2 are relatively simple regarding the models and scales and pictures in Fig. A3 are three loading cases for the cylindrical RVE. However, this is only the first step toward more sophisticated and large-scale simulations of CNT-based composites. As the computing power and confidence in simulations of CNT-based composites increase, large scale 3-D models containing hundreds or even more CNTs, behaving linearly or nonlinearly, with coatings or of different sizes, distributed evenly or randomly, can be employed to investigate the interactions among the CNTs in a matrix and to evaluate the effective material properties. Other numerical methods can also be attempted for the modeling and simulations of CNT-based composites, which may offer some advantages over the FEM approach. For example, the boundary element method, [60, 61], accelerated with the fast multipole techniques, [62, 63], and the mesh free methods [64] may enable one to model an RVE with thousands of CNTs in a matrix on a desktop computer. Analysis of the CNT-based composites using the boundary element method is already underway and will be reported subsequently.

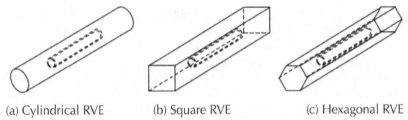

(a) Cylindrical RVE   (b) Square RVE   (c) Hexagonal RVE

FIGURE A2   Three nanoscale representative volume elements for the analysis of CNT-based nanocomposites.

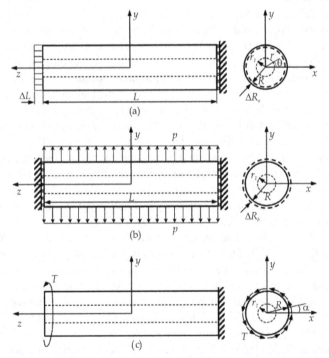

FIGURE A3   Three loading cases for the cylindrical RVE used to evaluate the effective material properties of the CNT-based composites. (a)Under axial stretch DL; (b) under lateral uniform load P; (c) under torsional load T.

The effective mechanical properties of CNT based composites are evaluated using square RVEs based on 3-D elasticity theory and solved by the FEM. Formulas to extract the effective material constants from solutions for the square RVEs under two loading cases are established based on elasticity. Square RVEs with multiple CNTs are also investigated in evaluating the Young's modulus and Poisson's ratios in the transverse plane. Numerical examples using the FEM are presented, which demonstrate that the load-carrying capabilities of the CNTs in a matrix are significant. With the addition of only about 3.6% volume fraction of the CNTs in a matrix, the stiffness of the composite in the CNT axial direction can increase as much as 33% for the case of long CNT fibers [65]. These simulation results are consistent with both the experimental ones reported in the literature [56–59, 66]. It is also found that cylindrical RVEs tend to overestimate the effective Young's moduli due to the fact that they overestimate the volume fractions of the CNTs in a matrix. The square RVEs, although more demanding in modeling and computing, may be the preferred model in future simulations for estimating the effective material constants, especially when multiple CNTs need to be considered. Finally, the rules of mixtures, for both long and short CNT cases, are found to be quite accurate in estimating the effective Young's moduli in the CNT axial direction. This may suggest that 3-D FEM modeling may not be necessary in obtaining the effective material constants in the CNT direction, as in the studies of the conventional fiber reinforced composites. Efforts in comparing the results presented in this paper using the continuum approach directly with the MD simulations are underway. This is feasible now only for a smaller RVE of one CNT embedded in a matrix. In future research, the MD and continuum approach should be integrated in a multiscale modeling and simulation environment for analyzing the CNT-based composites. More efficient models of the CNTs in a matrix also need to be developed, so that a large number of CNTs, in different shapes and forms (curved or twisted), or randomly distributed in a matrix, can be modeled. The ultimate validation of the simulation results should be done with the nanoscale or microscale experiments on the CNT reinforced composites [64].

Griebel and Hamaekers [67] reviewed the basic tools used in computational nanomechanics and materials, including the relevant underlying principles and concepts. These tools range from subatomic ab initio methods to classical molecular dynamics and multiple-scale approaches. The energetic link between the quantum mechanical and classical systems has been discussed, and limitations of the standing alone molecular dynamics simulations have been shown on a series of illustrative examples. The need for multiscale simulation methods to tackle nanoscale aspects of material behavior was therefore emphasized; that was followed by a review and classification of the mainstream and emerging multiscale methods. These simulation methods include the broad areas of quantum mechanics, molecular dynamics and multiple-scale approaches, based on coupling the atomistic and continuum models. They summarize the strengths and limitations of currently available multiple-scale techniques, where the emphasis is made on the latest perspective approaches, such as the bridging scale method, multiscale boundary conditions, and multiscale fluidics. Example problems, in which multiple-scale simulation methods yield equivalent results to full atomistic simulations at fractions of the computational cost, were shown. They compare their results with Odegard, et al. [48], the micromechanic method was BEM Halpin-Tsai Eq. [68] with aligned fiber by perfect bonding.

The solutions of the strain-energy-changes due to a SWNT embedded in an infinite matrix with imperfect fiber bonding are obtained through numerical method by Wan, et al. [69]. A "critical" SWNT fiber length is defined for full load transfer between the SWNT and the matrix, through the evaluation of the strain-energy-changes for different fiber lengths The strain-energy-change is also used to derive the effective longitudinal Young's modulus and effective bulk modulus of the composite, using a dilute solution. The main goal of their research was investigation of strain-energy-change due to inclusion of SWNT using FEM. To achieve full load transfer between the SWNT and the matrix, the length of SWNT fibers should be longer than a "critical" length if no weak interphase exists between the SWNT and the matrix [69].

A hybrid atomistic/continuum mechanics method is established in the Feng, et al. study [70] the deformation and fracture behaviors of carbon nanotubes (CNTs) in composites. The unit cell containing a CNT embedded in a matrix is divided in three regions, which are simulated by the atomic-potential method, the continuum method based on the modified Cauchy–Born rule, and the classical continuum mechanics, respectively. The effect of CNT interaction is taken into account via the Mori–Tanaka effective field method of micromechanics. This method not only can predict the formation of Stone–Wales (5-7-7-5) defects, but also simulate the subsequent deformation and fracture process of CNTs. It is found that the critical strain of defect nucleation in a CNT is sensitive to its chiral angle but not to its diameter. The critical strain of Stone–Wales defect formation of zigzag CNTs is nearly twice that of armchair CNTs. Due to the constraint effect of matrix, the CNTs embedded in a composite are easier to fracture in comparison with those not embedded. With the increase in the Young's modulus of the matrix, the critical breaking strain of CNTs decreases.

Estimation of effective elastic moduli of nanocomposites was performed by the version of effective field method developed in the framework of quasicrystalline approximation when the spatial correlations of inclusion location take particular ellipsoidal forms [71]. The independent justified choice of shapes of inclusions and correlation holes provide the formulae of effective moduli, which are symmetric, completely explicit, and easily to use. The parametric numerical analyzes revealed the most sensitive parameters influencing the effective moduli which are defined by the axial elastic moduli of nanofibers rather than their transversal moduli as well as by the justified choice of correlation holes, concentration and prescribed random orientation of nanofibers [72].

Li and Chou [73, 74] have reported a multiscale modeling of the compressive behavior of carbon nanotube/polymer composites. The nanotube is modeled at the atomistic scale, and the matrix deformation is analyzed by the continuum finite element method. The nanotube and polymer matrix are assumed to be bonded by van der Waals interactions at the interface. The stress distributions at the nanotube/polymer interface under isostrain and isostress loading conditions have been examined. They have used beam elements for SWCNT using molecular structural mechanics, truss rod for vdW links and cubic elements for matrix, the rule of mixture was used as for comparison in this research. The buckling forces of nanotube/polymer composites for different nanotube lengths and diameters are computed. The results indicate that continuous nanotubes can most effectively enhance the composite buckling resistance.

Anumandla and Gibson [75] describes an approximate, yet comprehensive, closed form micromechanics model for estimating the effective elastic modulus of carbon nanotube-reinforced composites. The model incorporates the typically observed nanotube curvature, the nanotube length, and both 1D and 3D random arrangement of the nanotubes. The analytical results obtained from the closed form micromechanics model for nanoscale representative volume elements and results from an equivalent finite element model for effective reinforcing modulus of the nanotube reveal that the reinforcing modulus is strongly dependent on the waviness, wherein, even a slight change in the nanotube curvature can induce a prominent change in the effective reinforcement provided. The micromechanics model is also seen to produce reasonable agreement with experimental data for the effective tensile modulus of composites reinforced with multiwalled nanotubes (MWNTs) and having different MWNT volume fractions.

Effective elastic properties for carbon nanotube reinforced composites are obtained through a variety of micromechanics techniques [76]. Using the in-plane elastic properties of graphene, the effective properties of carbon nanotubes are calculated using a composite cylinders micromechanics technique as a first step in a two-step process. These effective properties are then used in the self-consistent and Mori–Tanaka methods to obtain effective elastic properties of composites consisting of aligned single or multiwalled carbon nanotubes embedded in a polymer matrix. Effective composite properties from these averaging methods are compared to a direct composite cylinders approach extended from the work of Z. Hashin and B. Rosen [77] and R. Christensen and K. Lo [78]. Comparisons with finite element simulations are also performed. The effects of an interphase layer between the nanotubes and the polymer matrix as result of functionalization is also investigated using a multilayer composite cylinders approach. Finally, the modeling of the clustering of nanotubes into bundles due to interatomic forces is accomplished herein using a tessellation method in conjunction with a multiphase Mori–Tanaka technique. In addition to aligned nanotube composites, modeling of the effective elastic properties of randomly dispersed nanotubes into a matrix is performed using the Mori–Tanaka method, and comparisons with experimental data are made.

Selmi, et al. [79] deal with the prediction of the elastic properties of polymer composites reinforced with single walled carbon nanotubes. Our contribution is the investigation of several micromechanical models, while most of the papers on the subject deal with only one approach. They implemented four homogenization schemes, a sequential one and three others based on various extensions of the Mori–Tanaka (M–T) mean-field homogenization model: two-level (M–T/M–T), two-step (M–T/M–T) and two-step (M–T/Voigt). Several composite systems are studied, with various properties of the matrix and the graphene, short or long nanotubes, fully aligned or randomly oriented in 3D or 2D. Validation targets are experimental data or finite element results, either based on a 2D periodic unit cell or a 3D representative volume element. The comparative study showed that there are cases where all micromechanical models give adequate predictions, while for some composite materials and some properties, certain models fail in a rather spectacular fashion. It was found that the two-level (M–T/M–T) homogenization model gives the best predictions in most cases. After the characterization of the discrete nanotube structure using a homogenization method based on energy equivalence, the sequential, the two-step (M–T/M–T), the two-step (M–T/Voigt), the two-level (M–T/M–T) and finite element models were used to predict the elastic properties of SWNT/polymer composites. The data delivered by the micromechanical models are compared against those obtained by finite element analyzes or experiments. For fully aligned, long nanotube polymer composite, it is the sequential and the two-level (M–T/M–T) models, which delivered good predictions. For all composite morphologies (fully aligned, two-dimensional in-plane random orientation, and three-dimensional random orientation), it is the two-level (M–T/M–T) model, which gave good predictions compared to finite element and experimental results in most situations. There are cases where other micromechanical models failed in a spectacular way.

Luo, et al. [80] have used multiscale homogenization (MH) and FEM for wavy and straight SWCNTs, they have compare their results with Mori-Tanaka, Cox, Halpin-Tsai, Fu, et al. [81], Lauke [82]. Trespass, et al. [83] used 3D elastic beam for C-C bond and, 3D space frame for CNT and progressive fracture model for prediction of elastic modulus, they used rule of mixture for compression of their results. Their assumption was embedded a single SWCNT in polymer with Perfect bonding. The multiscale modeling, Monte Carlo, FEM and using equivalent continuum method was used by Spanos and Kontsos [84] and compared with Zhu, et al. [85] and Paiva, et al. [86]'s results.

The effective modulus [87] of CNT/PP composites is evaluated using FEA of a 3D RVE which includes the PP matrix, multiple CNTs and CNT/PP interphase and accounts for poor dispersion and non homogeneous distribution of CNTs within the polymer matrix, weak CNT/polymer interactions, CNT agglomerates of various sizes and CNTs orientation and waviness. Currently, there is no other model, theoretical or numerical, that accounts for all these experimentally observed phenomena and captures their individual and combined effect on the effective modulus of nanocomposites. The model is developed using input obtained from experiments and validated against experimental data. CNT reinforced PP composites manufactured by extrusion and injection molding are characterized in terms of tensile modulus, thickness and stiffness of CNT/PP interphase, size of CNT agglomerates and CNT distribution using tensile testing, AFM and SEM, respectively. It is concluded that CNT agglomeration and waviness are the two dominant factors that hinder the great potential of CNTs as polymer reinforcement. The proposed model provides the upper and lower limit of the modulus of the CNT/PP composites and can be used to guide the manufacturing of composites with engineered properties for targeted applications. CNT agglomeration can be avoided by employing processing techniques such as sonication of CNTs, stirring, calendaring etc., whereas CNT waviness can be eliminated by increasing the injection pressure during molding and mainly by using CNTs with smaller aspect ratio. Increased pressure during molding can also promote the alignment of CNTs along the applied load direction. The 3D modeling capability presented in this study gives an insight on the upper and lower bound of the CNT/PP composites modulus quantitatively by accurately capturing the effect of various processing parameters. It is observed that when all the experimentally observed factors are considered together in the FEA the modulus prediction is in good agreement with the modulus obtained from the experiment. Therefore, it can be concluded that the FEM models proposed in this study by systematically incorporating experimentally observed characteristics can be effectively used for the determination of mechanical properties of nanocomposite materials. Their result is in agreement with the results reported in Ref. [88], The theoretical micromechanical models, shown in Fig. A4, are used to confirm that our FEM model predictions follow the same trend with the one predicted by the models as expected.

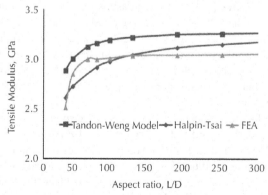

FIGURE A4   Effective modulus of 5 wt.% CNT/PP composites: theoretical models vs. FEA.

For reasons of simplicity and in order to minimize the mesh dependency on the results the hollow CNTs are considered as solid cylinders of circular cross-sectional area with an equivalent average diameter, shown in Fig. A.5calculated by equating the volume of the hollow CNT to the solid one [87].

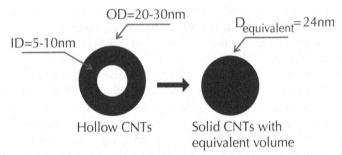

FIGURE A5   Schematic of the CNTs considered for the FEA.

The micromechanical models used for the comparison were Halpin–Tsai (H–T) [89] and Tandon–Weng (T–W) [90] model and the comparison was performed for 5 wt.% CNT/PP. It was noted that the H–T model results to lower modulus compared to FEA because H–T equation does not account for maximum packing fraction and the arrangement of the reinforcement in the composite. A modified H–T model that account for this has been proposed in the literature [91]. The effect of maximum packing fraction and the arrangement of the reinforcement within the composite become less significant at higher aspect ratios [92].

A finite element model of carbon nanotube, interphase and its surrounding polymer is constructed to study the tensile behavior of embedded short carbon nanotubes in polymer matrix in presence of vdW interactions in interphase region by Shokrieh and Rafiee [93]. The interphase is modeled using nonlinear spring elements capturing the force-distance curve of vdW interactions. The constructed model is subjected to tensile loading to extract longitudinal Young's modulus. The obtained results of this work have been compared with the results of previous research of the same authors [94] on long embedded carbon nanotube in polymer matrix. It shows that the capped short carbon nanotubes reinforce polymer matrix less efficient than long CNTs.

Despite the fact that researches have succeeded to grow the length of CNTs up to 4 cm as a world record in US Department of Energy Los Alamos National Laboratory [95] and also there are some evidences on producing CNTs with lengths up to millimeters [96,97], CNTs are commercially available in different lengths ranging from 100 nm to approximately 30 lm in the market based on employed process of growth [98–101]. Chemists at Rice University have identified a chemical process to cut CNTs into short segments [102]. As a consequent, it can be concluded that the SWCNTs with lengths smaller than 1000 nm do not contribute significantly in reinforcing polymer matrix. On the other hand, the efficient length of reinforcement for a CNT with (10, 10) index is about 1.2 lm and short CNT with length of 10.8 lm can play the same role as long CNT reflecting the uppermost value reported in our previous research [94]. Finally, it is shown that the direct use of Halpin–Tsai equation to predict the modulus of SWCNT/ composites overestimates the results. It is also observed that application of previously developed long equivalent fiber stiffness [94] is a good candidate to be used in Halpin–Tsai equations instead of Young's modulus of CNT. Halpin–Tsai equation is not an appropriate model for smaller lengths, since there is not any reinforcement at all for very small lengths.

Earlier, a nano-mechanical model has been developed by Chowdhury et al. [103] to calculate the tensile modulus and the tensile strength of randomly oriented short carbon nanotubes (CNTs) reinforced nanocomposites, considering the statistical variations of diameter and length of the CNTs. According to this model, the entire composite is divided into several composite segments which contain CNTs of almost the same diameter and length. The tensile modulus and tensile strength of the composite are then calculated by the weighted sum of the corresponding modulus and strength of each composite segment. The existing micromechanical approach for modeling the short fiber composites is modified to account for the structure of the CNTs, to calculate the modulus and the strength of each segmented CNT reinforced composites. Multi-walled CNTs with and without intertube bridging (see Fig. A6) have been considered. Statistical variations of the diameter and length of the CNTs are modeled by a normal distribution. Simulation results show that CNTs intertube bridging, length and diameter affect the nanocomposites modulus and strength. Simulation results have been compared with the available experimental results and the comparison concludes that the developed model can be effectively used to predict tensile modulus and tensile strength of CNTs reinforced composites.

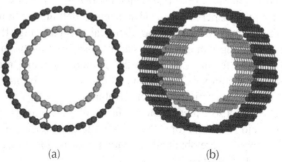

(a)   (b)

FIGURE A6.   Schematic of MWNT with intertube bridging. (a) Top view and (b) oblique view.

The effective elastic properties of carbon nanotube-reinforced polymers have been evaluated by K.I. Tserpes and A. Chanteli [104] as functions of material and geometrical parameters using a homogenized RVE. The RVE consists of the polymer matrix, a multiwalled carbon nanotube (MWCNT) embedded into the matrix and the interface between them. The parameters considered are the nanotube aspect ratio, the nanotube volume fraction as well as the interface stiffness and thickness. For the MWCNT, both isotropic and orthotropic material properties have been considered. Analyzes have been performed by means of a 3D FE model of the RVE. The results indicate a significant effect of nanotube volume fraction. The effect of nanotube aspect ratio appears mainly at low values and diminishes after the value of 20. The interface mostly affects the effective elastic properties at the transverse direction. Having evaluated the effective elastic properties of the MWCNT–polymer at the microscale, the RVE has been used to predict the tensile modulus of a polystirene specimen reinforced by randomly aligned MWCNTs for which experimental data exist in the literature. A very good agreement is obtained between the predicted and experimental tensile moduli of the specimen. The effect of nanotube alignment on the specimen's tensile modulus has been also examined and found to be significant since as misalignment increases the effective tensile modulus decreases radically. The proposed model can be used for the virtual design and optimization of CNT– polymer composites since it has proven capable of assessing the effects of different material and geometrical parameters on the elastic properties of the composite and predicting the tensile modulus of CNT-reinforced polymer specimens [105–144].

APPENDIX B

The notion "cluster" assumes energy-wise compensated nucleus with a shell, the surface energy of which is rather small, as a result under given conditions the cluster represents a stable formation. Nanocomposites of various shapes can be obtained either by the dispersion of a substance in a definite medium, or condensation or synthesis from low-molecular chemical particles (Fig. B1). Thus, the techniques for obtaining nanostructures can be classified by the mechanism of their formation. Other features, by which the methods for nanostructure production can be classified, comprise the variants of nanoproduct formation process by the change of energy consumption. A temperature or an energy factor is usually evident here. Besides, a so-called apparatus factor plays an important role together with the aforesaid features. At present, nanostructural "formations" of various shapes and compositions are obtained in a rather wide region of actions upon the chemical particles and substances.

In chemical literature a cluster is equated with a complex compound containing a nucleus and a shell. Usually a nucleus consists of metal atoms combined with metallic bond, and a shell of ligands. Manganese carbonyls $[(Co)_5 MnMn(Co)_5]$ and cobalt carbonyls $[Co_6(Co)_{18}]$, nickel pentadienyls $[Ni_6(C_5H_6)_6]$ belong to elementary clusters.

In recent years, the notion "cluster" has got an extended meaning. At the same time the nucleus can contain not only metals or not even contain metals. In some clusters, for instance, carbon ones there is no nucleus at all. In this case their shape can be characterized as a sphere (icosahedron, to be precise)—fullerenes, or as a cylinder—fullerene tubules. Surely a certain force field is formed by atoms on internal walls inside such particles. It can be assumed that electrostatic, electromagnetic, and gravitation fields conditioned by corresponding properties of atoms contained in particle shells can be formed inside tubules and fullerenes. If analyze papers recently published, it should be noted that a considerable exceeding of surface size over the volume and, consequently, a relative growth of the surface energy in comparison with the growth of volume and potential energy is the main feature of clusters. If particle dimensions (diameters of "tubes" and "spheres") change from 1 up to several hundred nanometers, they would be called nanoparticles. In some papers, the area of nanoclusters existence is within 1–10 nm.

Based on classical definitions, in given paper, metal nanoparticles and nanocrystals are referred to as nanoclusters. Apparently, the difference of nanoparticles from other particles (smaller or larger) is determined by their specific characteristics. The search of nanoworld distinctions from atomic-molecular, micro and macroworld can lead to finding analogies and coincidences in colloid chemistry, chemistry of polymers, and coordination compounds. Firstly, it should be noted that nanoparticles usually represent a small collective aggregation of atoms being within the action of adjacent atoms, thus conditioning the shape of nanoparticles. A nanoparticle shape can vary depending upon the nature of adjacent atoms and character of formation medium. Obviously, the properties of separate atoms and molecules (of small size) are determined by their energy and geometry characteristics, the determinative role being played by electron properties. In particular, electron interactions determine the geometry of molecules and atomic structures of small size and mobility of these chemical particles in media, as well as their activity or reactivity.

When the number of atoms in chemical particle exceeds 30, a certain stabilization of its shape being also conditioned by collective influence of atoms constituting the particle is observed. Simultaneously, the activity of such a particle remains high but the processes with its participation have a directional character. The character of interactions with the surroundings of such structures is determined by their formation mechanism.

During polymerization or copolymerization the influence of macromolecule growth parameters changes with the increase of the number of elementary acts of its growth. According to scientists, after 7–10 acts the shape or geometry of nanoparticles formed becomes the main determinative factor providing the further growth of macromolecule (chain development). A nanoparticle shape is usually determined not only by its structural elements but also by its interactions with surrounding chemical particles.

Due to the overlapping of different classification features, it is appropriate to present a set of diagrams by main features. For instance, the methods for nanoparticle formation by substance dispersion and chemical particle condensation can be identified with physical and chemical methods, though such decision is incorrect, since substance destruction methods can contain both chemical and physical impacts. In turn, when complex nanostructures are formed from simple ones, both purely physical and chemical factors are possible. However, in the process of substance dispersion high-energy sources, such as electric arc, laser pyrolysis, plasma sources, mechanical crushing or grinding should be applied.

From the aforesaid, it can be concluded that the possibility of self-organization of nanoparticles with the formation of corresponding nanosystems and nanomaterials is the main distinction of nanoworld from pico, micro, and macroworld. Recently, much attention has been paid to synergetics or the branch of science dealing with self-organization processes since these processes, in many cases, proceeds with small energy consumption and, consequently, is more ecologically clear in comparison with existing technological processes.

In turn, nanoparticle dimensions are determined by its formation conditions. When the energy consumed for macroparticle destruction or dispersion over the surface increases, the dimensions of nanomaterials are more likely to decrease. The notion "nanomaterial" is not strictly defined. Several researchers consider nanomaterials to be aggregations of nanocrystals, nanotubes, or fullerenes. Simultaneously, there is a lot of information available that nanomaterials can represent materials containing various nanostructures. The most attention researchers pay to metallic nanocrystals. Special attention is paid to metallic nanowires and nanofibers with different compositions.

Here are some names of nanostructures:

(1) fullerenes, (2) gigantic fullerenes, (3) fullerenes filled with metal ions, (4) fullerenes containing metallic nucleus and carbon (or mineral) shell, (5) one-layer nanotubes, (6) multilayer nanotubes, (7) fullerene tubules, (8) "scrolls", (9) conic nanotubes, (10) metal-containing tubules, (11) "onions", (12) "Russian dolls", (13) bamboo-like tubules, (14) "beads", (15) welded nanotubes, (16) bunches of nanotubes, (17) nanowires, (18) nanofibers, (19) nanoropes, 20) nanosemispheres (nanocups), (21) nanobands and similar nanostructures, as well as various derivatives from enlisted structures. It is quite possible that a set of such structures and notions will be enriched.

In most cases nanoparticles obtained are bodies of rotation or contain parts of bodies of rotation. In natural environment there are minerals containing fullerenes or representing thread-like formations comprizing nanometer pores or structures. In the first case, it is talked about schungite that is available in quartz rock in unique deposit in Prionezhje. Similar mineral can also be found in the river Lena basin, but it consists of micro and macrodimensional cones, spheroids, and complex fibers. In the second case, it is talked about kerite from pegmatite on Volyn (Ukraine) that consists of polycrystalline fibers, spheres, and spirals mostly of micron dimensions, or fibrous vetcillite from the state of Utah (USA); globular anthraxolite and asphaltite.

Diameters of some internal channels are up to 20–50 nm. Such channels can be of interest as nanoreactors for the synthesis of organic, carbon, and polymeric substances with relatively low energy consumption. In case of directed location of internal channels in such matrixes and their inner-combinations the spatial structures of certain purpose can be created. Terminology in the field of nanosystems existence is still being developed, but it is already clear that nanoscience obtains qualitatively new knowledge that can find wide application in various areas of human practice thus, significantly decreasing the danger of people's activities for themselves and environment.

The system classification by dimensional factor is known, based on which we consider the following:

- microobjects and microparticles 10^{-6}–10^{-3} m in size;
- nanoobjects and nanoparticles 10^{-9}–10^{-6} m in size;
- picoobjects and picoparticles 10^{-12}–10^{-9} m in size.

Assuming that nanoparticle vibration energies correlate with their dimensions and comparing this energy with the corresponding region of electromagnetic waves, we can assert that energy action of nanostructures is within the energy region of chemical reactions. System self-organization refers to synergetics. Quite often, especially recently, the papers are published, for example, it is considered that nanotechnology is based on self-organization of metastable systems. As assumed, self-organization can proceed by dissipative (synergetic) and continual (conservative) mechanisms. Simultaneously, the system can be arranged due to the formation of new stable ("strengthening") phases or due to the growth provision of the existing basic phase. This phenomenon underlies the arising nanochemistry. Below is one of the possible definitions of nanochemistry.

Nanochemistry is a science investigating nanostructures and nanosystems in metastable ("transition") states and processes flowing with them in near-"transition" state or in "transition" state with low activation energies.

To carry out the processes based on the notions of nanochemistry, the directed energy action on the system is required, with the help of chemical particle field as well, for the transition from the prepared near-"transition" state into the process product state (in our case–into nanostructures or nanocomposites). The perspective area of nanochemistry is the chemistry in nanoreactors. Nanoreactors can be compared with specific nanostructures representing limited space regions in which chemical particles orientate creating "transition state" prior to the formation of the desired nanoproduct. Nanoreactors have a definite activity, which predetermines the creation of the corresponding product. When nanosized particles are formed in nanoreactors, their shape and dimensions can be the reflection of shape and dimensions of the nanoreactor.

In the last years a lot of scientific information in the field of nanotechnology and science of nanomaterials appeared. Scientists defined the interval from 1 to 1000 nm as the area of nanostructure existence, the main feature of which is to regulate the system self-organization processes. However, later some scientists limited the upper threshold at 100 nm. At the same time, it was not well substantiated. Now many nanostructures varying in shapes and sizes are known. These nanostructures have sizes that fit into the interval, determined by Smally, and are active in the processes of self-organization, and also demonstrate specific properties.

Problems of nanostructure activity and the influence of nanostructure super small quantities on the active media structural changes are explained.

The molecular nanotechnology ideology is analyzed. In accordance with the development tendencies in self-organizing systems under the influence of nanosized excitations the reasons for the generation of self-organization in the range 10^{-6}–10^{-9} m should be determined.

Based on the law of energy conservation the energy of nanoparticle field and electromagnetic waves in the range 1–1000 nm can transfer, thus corresponding to the range of energy change from soft X-ray to near IR radiation. This is the range of energies of chemical reactions and self-organization (structuring) of systems connected with them.

Apparently the wavelengths of nanoparticle oscillations near the equilibrium state are close or correspond to their sizes. Then based on the concepts of ideologists of nanotechnology in material science the definition of nanotechnology can be as follows:

Nanotechnology is a combination of knowledge in the ways and means of conducting processes based on the phenomenon of nano-sized system self-organization and utilization of internal capabilities of the systems that results in decreasing the energy consumption required for obtaining the targeted product while preserving the ecological cleanness of the process.

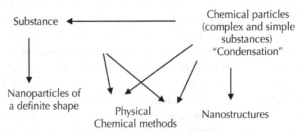

FIGURE B1   Classification diagram of nanostructure formation techniques by the features "dispersion" and "condensation".

At the same time, the conceptions "dispersion" and "condensation" are conditional and can be explained in various ways (Figs. B2 and B3). Among the physical methods of "dispersion" high-temperature methods and methods with relatively low temperatures or high-energy and low-energy ones are distinguished.

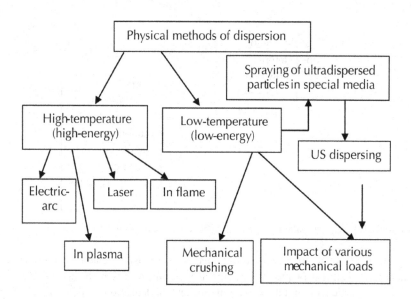

FIGURE B2   Classification diagram of physical methods of dispersion.

Considerably fewer chemical methods of dispersion are known, though, in the physical methods listed chemical processes are surely present, since it is difficult to imagine spraying and grinding of a substance without chemical reactions of destruction. Therefore, it is more appropriate to speak about physical methods of impact upon the substance that lead to their dispersion (decomposition, destruction at high temperatures, radiations or mechanical loads). Then, chemical methods are identified with methods of impact upon the substances of chemical particles and media.

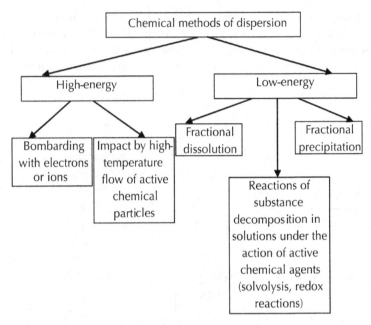

FIGURE B3   Classification diagram of dispersion chemical methods.

Basically nanostructures are obtained from picosized chemical particles by means of chemical or physical-chemical techniques, in which the activity of a chemical particle but not its activation during "condensation" is determinant (Fig. B4). Therefore, the classification diagram of chemical techniques for nanostructure formation under the action of chemical media can be given as follows:

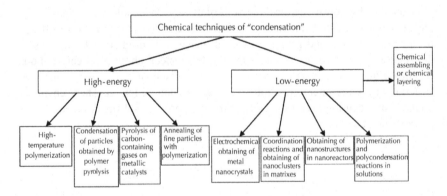

FIGURE B4   Classification diagram of chemical techniques of "condensation".

Actually, such diagrams are approximate in the same way as conditional are separate points of separation and difference between them. Several techniques can be referred to combined or physical methods of impact upon active chemical particles. For instance, CVD method comprises high-energy technique that leads to the formation of gaseous phase that can be attributed to chemical methods of dispersion, and then active chemical particles formed during the pyrolysis "transform" into nanostructures. The polymerization processes of gaseous phase particles are carried out by means of probe technological stations, and this can be referred to the physical techniques of "condensation." So, under the physical methods of formation of various nanostructures, including nanocrystals, nanoclusters, fullerenes, nanotubes, and so on, it means the techniques in which physical impact results in the formation of nanostructures from active picosized chemical particles (Fig. B5).

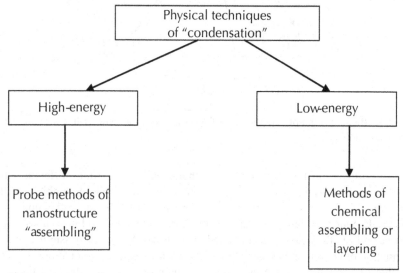

FIGURE B5 Physical impact results in the formation of nanostructures from active picosized chemical particles.

The proposed classification does not reflect multivicissitude of methods for nanostructure formation. Usually nanostructure production comprises the following:
1. preparation of "embryos" or precursors of nanostructures;
2. production of nanostructures;
3. isolation and refining of nanostructures of a definite shape.

The production of nanostructures by various methods, including mechanical ones, for instance, extrusion, grinding and similar operations can proceed in several stages.

Mechanical crushing, combined methods for grinding in media and influence of action power and medium, where the substance is crushed and sprayed, upon the size and shape of nanostructures formed will be discussed in this chapter. At the same time, multistepped combined methods for obtaining nanostructures with different shapes and sizes have been applied in recent years. Ways of substance dispersion by mechanical methods till nanoproducts are obtained (fine powders or ultradispersed particles) can be given in the following diagram (Fig. B6):

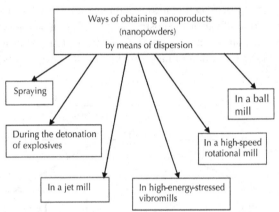

FIGURE B6   Diagram of classification of ways for obtaining powder-like nanoproducts by means of dispersion.

One of the most widely applied methods used for grinding different substances is the crushing and grinding in mills: ball, rod, colloid combined with the action of shock force, abrasion force, centrifugal force and vibration upon the materials. Since, initially, carbon nanostructures, fullerenes and nanotubes were extracted from the carbon dust obtained in electrical charge, the possibility of the formation of corresponding nanostructures when using conventional mechanical methods of action upon substances seems doubtful. However, the investigation of the products obtained after graphite crushing in ball mills allows making the conclusion that the method proposed can be quite competitive high-temperature way for forming nanostructures when the mechanical power provides practically the same conditions as the heat. In this case, it is difficult to imagine a directed action of the combined forces upon the substance. However, at the set speed of the mill drum the directed action of the "stream" of steel balls upon the material can be predicted. As an example, let us give the description of one of such processes during graphite grinding. Carbon nanostructures resembling "onions" by their shape are obtained when grinding graphite in ball mills. The product is ground in planetary ball mill in the atmosphere of pure argon. Mass ratio of steel balls to the powder of pure graphite equals 40. The rotation speed of the drum is 270 rpm. The grinding time changed from 150 to 250 h. It was observed that after 150 h of grinding the nanoproduct obtained resembles by characteristics and appearance the nanoparticles obtained in electric-arc method.

Iron-containing nanostructures were also obtained in planetary ball mills after milling the iron powders in heptane adding oleic acid.

The distinctive feature of the considered technique for producing metal/carbon nanocomposites is a wide application of independent, modern, experimental and theoretical analysis methods to substantiate the proposed technique and investigation of the composites obtained (quantum-chemical calculations, methods of transmission electron microscopy and electron diffraction, method of X-ray photoelectron spectroscopy, X-ray phase analysis, etc.). The technique developed allows synthesizing a wide range of metal/carbon nanocomposites by composition, size and morphology depending on the process conditions. In its application it is possible to use secondary metallurgical and polymer raw materials. Thus, the nanocomposite structure can be adjusted to extend the function of its application without prefunctionalization. Controlling the sizes and shapes of nanostructures by changing the metal-containing phase, to some extent, apply completely new, practicable properties to the materials which sufficiently differ from conventional materials.

The essence of the method consists in coordination interaction of functional groups of polymer and compounds of 3d-metals as a result of grinding of metal-containing and polymer phases Further, the composition obtained undergoes thermolysis following the temperature mode set with the help of thermogravimetric and differential thermal analyzes. At the same time, one observes the polymer carbonization, partial or complete reduction of metal compounds and structuring of carbon material in the form of nanostructures with different shapes and sizes.

Metal/carbon nanocomposite (Me/C) represents metal nanoparticles stabilized in carbon nanofilm structures. In turn, nanofilm structures are formed with carbon amorphous nanofibers associated with metal containing phase. As a result of stabilization and association of metal nanoparticles with carbon phase, the metal chemically active particles are stable in the air and during heating as the strong complex of metal nanoparticles with carbon material matrix is formed.

APPENDIX C

The propagation of ultrashort optical pulses in the array of carbon nanotubes was investigated. The electromagnetic field is considered based on the Maxwell's equations, and the electronic system of carbon nanotubes based on the Boltzmann kinetic equation in the relaxation time approximation. We construct the distribution of the pulse intensity and observed the periodic effect of a partition of pulse maximum into two maxima, and their subsequent merger into a single maximum. The objective of this appendix is to consider the problem of three-dimensional propagation of ultrashort optical pulse in the array of carbon nanotubes. The rapid development of laser and optical technologies, the application of powerful lasers and the unique precision of optical measurements allowed substantial progress in the study of nonlinear phenomena that occur in a wide range of substances with a very different physical properties. This is due both to the rapid progress of computer technology, and with interest in modern physics the study of nonlinear dynamic processes. The object of the study should be as a substance with pronounced nonlinear properties and important from the viewpoint of practical applications. In the last decade, as a matter of researchers are increasingly attracted to carbon nanotubes (CNTs) are unique macromolecular systems. Nanometer diameter and micron length of CNTs make them attractive for use in nano- and microelectronics, as they allow them to be considered the closest in structure to the ideal one-dimensional systems. The study of optical soliton propagation in CNTs is one of the most promising areas of research. Although these papers were predicted by the existence possibility of electromagnetic solutions and dependence of their characteristics on the parameters of CNTs remained a series of questions requiring clarification. This question related to output beyond the one-dimensional approximation, and the question of the propagation of optical pulses with light diffraction, and the scattering problem of the optical pulse in the multidimensional case the inhomogeneities of various types. These structures are localized in three spatial dimensions, will be called "light bullets", because in this problem localization occurs in three dimensions. All these factors make the problem of the study of nonlinear dynamic processes in CNTs relevant for theory and in practice.

C1 MODEL AND BASIC RELATIONS

We assume in the model that the electric field vector $\mathbf{E}(x, y, z, t)$ directed along the tube axis x, and the electromagnetic wave is moving in the transverse direction (Fig. 1). Carbon nanotubes are considered to be ideal to have a structural modification of the "zigzag" and located on the same interatomic distances equal to 0.34 nm [3] to simplify the calculations. Method of packing CNTs in the array is not important, because it does not take into account the interaction between the CNTs, and therefore used a tetragonal packing of single-walled carbon nanotubes in two-dimensional array with space group P42/mc(D_{4h}^9) in Fig. C1.

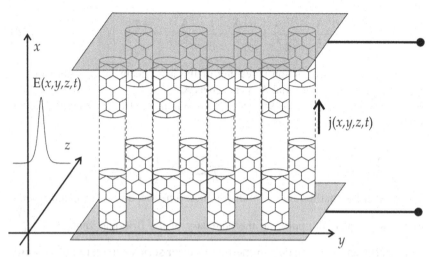

FIGURE C1   The geometry of the problem. Ultrashort optical pulse with the electric field $E(x, y, z, t)$ is directed along the tube axis x, moving in the transverse direction along the axis z.

In this appendix, we ignore the influence of the substrate on which the nanotubes are grown. This can lead particularly to the appearance of surface waves of these system, we will not considered these features in such article.

Electronic structure of "zigzag" carbon nanotubes is characterized by chiral indices $(n, 0)$ and described by the well-known dispersion relation which obtained in the framework of Huckel π-electron approximation:

$$E(\mathbf{p}) = \pm \gamma \sqrt{1 + 4\cos\left(ap_x\right)\cos\left(\pi s / n\right) + 4\cos^2\left(\pi s / n\right)} \qquad (27)$$

where $a = 3d / 2\hbar$, $d = 0.142$ nm is the distance between adjacent carbon atoms in graphene, $\mathbf{p} = (p_x, s)$ is the quasimomentum of the electrons in graphene, p_x is the parallel component of the graphene sheet of the quasimomentum and $s = 1, 2, \ldots,$ n are the quantization numbers of the momentum components depending on the width of the graphene ribbon. Different signs are related to the conductivity band and to the valence band accordingly.

The electromagnetic field pulse is described classically on the basis of Maxwell's equations. We are considering the dielectric and magnetic properties of CNTs and two-dimensionality of the problem the Maxwell equations for the vector potential $\mathbf{A}$ in the gauge $\mathbf{E} = -\dfrac{1}{c}\dfrac{\partial \mathbf{A}}{\partial t}$ will be:

$$\frac{\partial^2 \mathbf{A}}{\partial x^2} + \frac{\partial^2 \mathbf{A}}{\partial y^2} + \frac{\partial^2 \mathbf{A}}{\partial z^2} - \frac{1}{c}\frac{\partial^2 \mathbf{A}}{\partial t^2} = -\frac{4\pi}{c}\mathbf{j} \qquad (28)$$

We ignore the diffraction spreading of the laser beam in the x-direction in Eq. (28). The vector potential $\mathbf{A}$ is chosen as $\mathbf{A} = (Ax\ (x, y, z, t),\ 0,\ 0)$. Using the apparatus shown in Refs. [4–9], Eq. (27), we obtain an effective equation for the vector potential:

$$\frac{\partial^2 A_z}{\partial x^2} + \frac{\partial^2 A_z}{\partial y^2} + \frac{\partial^2 A_z}{\partial z^2} - \frac{1}{c}\frac{\partial^2 A_z}{\partial t^2} + \frac{q}{\pi h}\sum_m c_m \sin\left(\frac{maq}{c}A_z(t)\right) = 0 \qquad (29)$$

where F0 is the equilibrium Fermi distribution function, q is the elementary electron charge, q0 is the pulse on the boundary of the Brillouin zone $q_0 = \frac{2\pi h}{3b}$.

The coefficients a_{ms} are determined by Fourier series expansion of velocity component v_x (s, p) obtained from (1) as $v_x = \partial E(p)/\partial p_x$.

These simplifications have been made: (a) interband transitions are not counted. This assumption imposes a restriction on the maximum frequency of laser pulses, which for the CNT is in the near infrared. (b) the solution for the field component $A_x(t)$ is in the class of rapidly decreasing functions. (c) the relaxation time τ is large enough for the common durations of ultrashort laser pulses.

Equation (3) were solved numerically using the direct difference scheme "cross". Steps to time and coordinate were determined from the standard conditions of stability and decreased successively in 2 times as long as the solution is not changed in the eighth digit after dot. These calculations for the one-dimensional case are presented in Refs. [4–9]. The evolution of the initial conditions considered in time (the variable t). The initial pulse profile has a Gaussian form:

$$E(x, y, z, t) = E_{\max} \exp\left(-\frac{(x - x_0)^2}{\mu_x}\right)\exp\left(-\frac{(y - y_0)^2}{\mu_y}\right)\exp\left(-\frac{(z - z_0)^2}{\mu_z}\right) \qquad (30)$$

where E_{max} is the maximum field amplitude, x_0 is the coordinate of the maximum intensity the axis of x, y_0 is the coordinate of the maximum intensity the axis of y, z_0 is the coordinate of the maximum intensity the axis of z, μ_x, μ_y and μ_z are the parameters defining pulse width to the coordinates x, y and z respectively.

This appendix focused on the results of the impulse, which localized in three spatial dimensions. In the evolution of ultrashort pulse periodically split into two pulses and then combined into one, while the amplitude of the pulses is changed. Figures C2 and C3 shows the results of numerical simulation of the pulse evolution in the three-dimensional CNTs array such as (8.0) and (20.0), respectively.

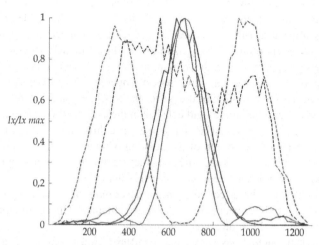

FIGURE C2 Profiles of the intensities I_x/I_{xmax} bullets in the cross section parallel to the plane ZOY and passing through a maximum pulse propagating in the array of CNTs such as (8.0). The ordinate shows the ratio I_x/I_{xmax}. Type of pulse: 1 – initial pulse, 2 – pulse after $t = 0.8 \cdot 10^{-12} s$, 3 – pulse after $t = 1.6 \cdot 10^{-12} s$, 4 – pulse after $t = 2.4 \cdot 10^{-12} s$, 5 – pulse after $t = 2.9 \cdot 10^{-12} s$.

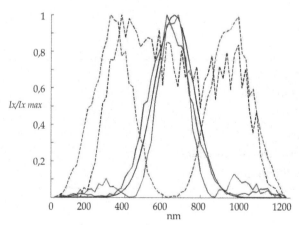

FIGURE C3 Profiles of the intensities I_x/I_{xmax} bullets in the cross section parallel to the plane ZOY and passing through a maximum pulse propagating in the array CNT type (20.0). The ordinate shows the ratio I_x/I_{xmax}. Type of pulse: 1 – initial pulse, 2 – pulse after $t = 0.8 \cdot 10^{-12} s$, 3 – pulse after $t = 1.6 \cdot 10^{-12} s$, 4 – pulse after $t = 2.4 \cdot 10^{-12} s$, 5 – pulse after $t = 2.9 \cdot 10^{-12} s$.

Figures C2 and C3 shows the distribution of intensities $I_x(x, y, z, t) = E_x^2(x, y, z, t)$ at different times. Analysis of the data graphs show that the stable propagation of pulses in the array of CNTs is possible, which are localized in three dimensions. These pulses are often called in literature as "light bullets". The interesting effect of a periodic partition of pulse maximum into two maxima, and their subsequent merger into a single maximum was discovered. The pulse shape varies significantly during the evolution of the pulse. We consider that this effect is akin to the dynamics of internal modes and consists in the excitation of internal vibration modes of the "light bullet" when passing a pulse through the array of CNTs, which leads to periodic separation of the pulse maximum and their interference with the original "bullet." This fact suggests that there may be bumeron analogs in the other highly nonlinear three-dimensional systems.

1. The model and the effective equation describing the dynamics of an ultrashort pulse laser beams in the CNT are obtained. The approximations, which used in this model were formulated.

2. Numerical calculations showed that in three-dimensional case are possible stable nonlinear waves that are localized in three directions of light pulses, which are analogs of "light bullets".

3. The periodic separation of peak pulse into two pulses and subsequent integration occur in the case of scattering when light pulse passing through the array of CNTs. The pulse shape at the same time varies considerably.

KEYWORDS

- domino effect
- Hagen-Poiseuille
- light bullets
- nanogun
- nanostructures architecture
- quasi-solid phase

REFERENCES

1. Iijima, S.; Helical microtubules of graphitic carbon. Nature **1991**, *354*, 6348, 56.
2. Dresselhaus, M. S.; Dresselhaus, G.; Eklund, P. C.;Science of fullerenes and carbon nanotubes. New York: Academic Press; **1996**.

3. Saito, R.; Dresselhaus, G.; Dresselhaus, M. S.; Physical properties of carbon nanotubes. Imperial College Press; **1998**.

4. Harris, P.; J. F.; Carbon nanotubes and related structures: new materials for the 20-first century. Cambridge, United Kingdom: Cambridge University Press; **1999**, 279.

5. Wagner, H. D.; Vaia, R. A.; Nanocomposites: issues at the interface. Mater Today **2004**, *7(11)*, 38.

6. Zeng, Q. H.; Yu, A. B.; Lu, M. G. Q. ultiscale modeling and simulation of polymer nanocomposites, Prog. Polym. Sci. **2008**, *33*, 191–269.

7. Iijima, S.; Helical microtubules of graphitic carbon. Nature **1991**, *354*, 56–58.

8. Bethune, D. S.; Klang, C. H.; De Vries, M. S.; et al. Cobalt-catalyzed growth of carbon nanotubes with single-atomic-layer walls. Nature **1993**, *363*, 605–607.

9. Dresselhaus, M. S.; Dresselhaus, G.; Saito, R.; Physics of carbon nanotubes. Carbon **1995**, *33*, 883–891.

10. Thostenson, E. T.; Ren, Z. F.; Chou, T. W.; Advances in the science and technology of CNTs and their composites: a review. Compos Sci Technol **2001**, *61*, 1899–912.

11. Yakobson, B. I.; Avouris, P.; Mechanical properties of carbon nanotubes. Top Appl Phys **2001**, *80*, 287–327.

12. Ajayan, P. M.; Schadler, L. S.; Braun, P. V.; Nanocomposite science and technology. Weinheim: Wiley-VCH; **2003**, p. 77–80.

13. Li J, Ma, P.; C, Chow, W. S.; To C. K, Tang, B. Z.; Kim J. K.; Correlations between percolation threshold, dispersion state and aspect ratio of carbon nanotube, Adv Funct Mater **2007**, *17*, 3207–15.

14. Ajayan, P. M.; Stephan, O.; Colliex, C.; Trauth, D.; Aligned carbon nanotube arrays formed by cutting a polymer resin-nanotube composite. Science **1994**, *265*, 1212–4.

15. Summary of Searching Results. <http://www.scopus.com> (accessed January 2010).

16. Du, J. H. ; Bai, J.; Cheng, H. M.;The present status and key problems of carbon nanotube based polymer composites. Express Polym Lett **2007**, *1*, 253–73.

17. Lee, J. Y.; Baljon, A. R. C.; Loring, R. F.; Panagiotopoulos, A. Z.; Simulation of polymer melt intercalation in layered nanocomposites. J Chem Phys **1998**, *109*, 10321–30.

18. Smith, G. D.; Bedrov, D.; Li, L.; W, Byutner, O.; A molecular dynamics simulation study of the viscoelastic properties of polymer nanocomposites. J Chem Phys **2002**, *117*, 9478–89.

19. Smith, J. S.; Bedrov, D.; Smith, G. D. A molecular dynamics simulation study of nanoparticle interactions in a model polymer–nanoparticle composite. Compos Sci Technol **2003**, *63*, 1599–605.

20. Zeng, Q. H.; Yu, A. B. ; Lu, G. Q.; Standish, R. K.; Molecular dynamics simulation of organic-inorganic nanocomposites: layering behavior and interlayer structure of organoclays. Chem Mater **2003**, *15*, 4732–8.

21. Vacatello, M.; Predicting the molecular arrangements in polymer-based nanocomposites. Macromol Theory Simul **2003**, *12*, 86–91.

22. Zeng, Q. H.; Yu, A. B.; Lu, G. Q.; Interfacial interactions and structure of polyurethane intercalated nanocomposite. Nanotechnology **2005**, *16*, 2757–63.

23. Allen, M. P.; Tildesley, D. J.; Computer simulation of liquids.Oxford: Clarendon Press; 1989.

24. Frenkel, D.; Smit, B.; Understanding molecular simulation: from algorithms to applications. 2nd Ed. San Diego: Academic Press; 2002.
25. Metropolis, N.; Rosenbluth, A. W.; Marshall, N.; Rosenbluth, M. N.; Teller, A. T.;Equation of state calculations by fast computing machines. J Chem Phys **1953**, *21*, 1087–92.
26. Carmesin, I.; Kremer, K.; The bond fluctuation method: a new effective algorithm for the dynamics of polymers in all spatial dimensions. Macromolecules **1988**, *21*, 2819–23.
27. Hoogerbrugge, P. J.; Koelman, J. M. V. A.; Simulating microscopic hydrodynamic phenomena with dissipative particle dynamics. Europhys Lett **1992**, *19*, 155–60.
28. Gibson, J. B.; Chen, K.; Chynoweth, S.; Simulation of particle adsorption onto a polymer-coated surface using the dissipative particle dynamics method. J Colloid Interface Sci **1998**, *206*, 464–74.
29. Dzwinel V.; Yuen, D. A.; A two-level, discrete particleapproach for large-scale simulation of colloidal aggregates. Int J Mod Phys C **2000**, *11*, 1037–61.
30. Dzwinel, W.; Yuen, D. A.; A two-level, discrete-particle approach for simulating ordered colloidal structures. J Colloid Interface Sci **2000**, *225*, 179–90.
31. Chen, S.; Doolen, G. D.; Lattice Boltzmann method for fluid flows. Annu Rev Fluid Mech **1998**, *30*, 329–64.
32. Cahn, J. W.; On spinodal decomposition. Acta Metall **1961**, *9*, 795–801.
33. Cahn, J. W.; Hilliard, J. E.; Spinodal decomposition: a reprise. Acta Metall **1971**, *19*, 151–61.
34. Cahn, J. W.; Free energy of a nonuniform system. II.; Thermodynamic basis. J Chem Phys **1959**, *30*, 1121–4.
35. Lee, B. P.; Douglas, J. F.; Glotzer, S. C.; Filler-induced composition waves in phase-separating polymer blends. Phys Rev E **1999**, *60*, 5812–22.
36. Ginzburg, V. V.; Qiu, F.; Paniconi, M.; Peng, G. W.; Jasnow, D.; Balazs, A. C.; Simulation of hard particles in a phaseseparating binary mixture. Phys Rev Lett **1999**, *82*, 4026–9.
37. Qiu, F.; Ginzburg, V. V.; Paniconi, M.; Peng, G. W.; Jasnow, D.; Balazs, A. C.; Phase separation under shear of binary mixtures containing hard particles. Langmuir **1999**, *15*, 4952–6.
38. Ginzburg, V. V.; Gibbons, C.; Qiu, F.; Peng, G. W.; Balazs, A. C.; Modeling the dynamic behavior of diblock copolymer/ particle composites. Macromolecules **2000**, *33*, 6140–7.
39. Ginzburg, V. V.; Qiu, F.; Balazs, A. C.; Three-dimensional simulations of diblock copolymer/particle composites.Polymer **2002**, *43*, 461–6.
40. He G, Balazs, A. C.; Modeling the dynamic behavior of mixtures of diblock copolymers and dipolar nanoparticles.J Comput Theor Nanosci **2005**, *2*, 99–107.
41. Altevogt, P.; Ever, O. A.; Fraaije J. G. E. M.; Maurits, N. M.; van Vlimmeren, B. A. C.; The MesoDyn project: software for mesoscale chemical engineering. J Mol Struct **1999**, *463*, 139–43.
42. Kawakatsu, T.; Doi, M.; Hasegawa, A.; Dynamic density functional approach to phase separation dynamics of polymer systems. Int J Mod Phys C **1999**, *10*, 1531–40.

43. Morita, H.; Kawakatsu, T.; Doi, M.; Dynamic density functional study on the structure of thin polymer blend films with a free surface. Macromolecules **2001**, *34*, 8777–83.
44. Doi, M.; OCTA-a free and open platform and softwares of multiscale simulation for soft materials /http://octa.jp/S.2002.
45. Tandon, G. P.; Weng, G. J.; The effect of aspect ratio of inclusions on the elastic properties of unidirectionally aligned composites. Polym Compos **1984**, *5*, 327–33.
46. Fornes, T. D.; Paul, D. R.; Modeling properties of nylon 6/clay nanocomposites using composite theories. Polymer **2003**, *44*, 4993–5013.
47. Odegard, G. M.; Pipes, R. B.; Hubert, P.; Comparison of two models of SWCN polymer composites. Compos Sci Technol **2004**, *64*, 1011–20.
48. Odegard, G. M.; Gates, T. S.; Wise, K. E.; Park, C.; Siochi, E. J.; Constitutive modeling of nanotube-reinforced polymer composites. Compos Sci Technol **2003**, *63*, 1671–87.
49. Pipes, R. B.; Hubert, P.; Helical carbon nanotube arrays: mechanical properties. Compos Sci Technol **2002**, *62*, 419–28.
50. Rudd, R. E.; Broughton, J. Q.; Concurrent coupling of length scales in solid state systems. Phys Stat Sol B **2000**, *217*, 251–91.
51. Starr, F. W.; Glotzer, S. C.; Simulations of filled polymers on multiple length scales, In: Nakatani, A. I.; Hjelm, R. P.; Gerspacher, M.; Krishnamoorti, R.; editors. Filled and nanocomposite polymer materials, Materials research symposium proceedings. Warrendale: Materials Research Society, **2001**, p. KK4.1.1–KK4.1.13.
52. Glotzer, S. C.; Starr, F. W.; Towards multiscale simulations of filled and nanofilled polymers, In: Cummings, P. T.; Westmoreland, P. R.; Carnahan, B.; editors. Foundations of molecular modeling and simulation: Proceedings of the 1st international conference on molecular modeling and simulation. Keystone: American Institute of Chemical Engineers, **2001**, p. 44–53.
53. Mori, T.; Tanaka, K.; Average stress in matrix and average elastic energy of materials with misfitting inclusions. Acta Metallurgica **1973**, *21*, 571–575.
54. Fisher, F. T.; Bradshaw, R. D.; Brinson, L. C.; Fiber waviness in nanotube-reinforced polymer composites—I: modulus predictions using effective nanotube properties, Comp Sci and Tech **2003**, *63*, 1689–1703.
55. Fisher, F. T.; Bradshaw, R. D.; Brinson LC Fiber waviness in nanotube-reinforced polymer composites—II: modeling via numerical approximation of the dilute strain concentration tensor. Comp Sci and Tech **2003**, *63*, 1705–1722.
56. Liu, Y. J.; Chen, X. L.; Evaluations of the effective material properties of carbon nanotube-based composites using a nanoscale representative volume element, Mech of Mat **2003**, *35*, 69–81.
57. Schadler, L. S.; Giannaris, S. C.; Ajayan, P. M.; Load transfer in carbon nanotube epoxy composites. Applied Physics Letters **1998**, *73 (26)*, 3842–3844.
58. Wagner, H. D.; Lourie, O.; Feldman, Y.; Tenne, R.; Stress-induced fragmentation of multiwall carbon nanotubes in a polymer matrix. Applied Physics Letters **1998**, *72 (2)*, 188–190.
59. Qian, D.; Dickey, E. C.; Andrews, R.; Rantell, T.; Load transfer and deformation mechanisms in carbon nanotube polystyrene composites. Applied Physics Letters **2000**, *76 (20)*, 2868–2870.

60. Liu, Y. J.; Chen, X. L.; Modeling and analysis of carbon nanotube-based composites using the FEM BEM.; Submitted to CMES: Computer Modeling in Engineering and Science. **2002.**

61. Liu, Y. J.; Xu, N.; Luo, J. F.; Modeling of inter phases in fiber-reinforced composites under transverse loading using the boundary element method. Journal of Applied Mechanics **2000,** *67 (1),* 41–49.

62. Fu, Y.; Klimkowski, K. J.; Rodin, G. J.; Berger, E.; et al.; A fast solution method for three-dimensional many-particle problems of linear elasticity. International Journal for Numerical Methods in Engineering **1998,** *42,* 1215–1229.

63. Nishimura, N.; Yoshida, K.-I.; Kobayashi, S.; A fast multipole boundary integral equation method for crack problems in 3D Engineering Analysis with Boundary Elements **1999,** *23,* 97–105.

64. Qian, D.; Liu, W. K.; Ruoff, R. S.; Mechanics of C60 in nanotubes. The Journal of Physical Chemistry B **2001,** *105 (44),* 10753–10758.

66. Chen, X. L.; Liu, Y. J.; Square representative volume elements for evaluating the effective material properties of carbon nanotube-based composites, Comput Mater Sci **2004,** *29,* 1–11.

67. C.; Bower, R.; Rosen, L.; Jin, J.; Han, O.; Zhou, Deformation of carbon nanotubes in nanotube–polymer composites, Applied Physics Letters **1999,** *74,* 3317–3319.

68. Gibson, R. F.; Principles of composite material mechanics, CRC Press, 2nd edition. **2007,** 97–134.

69. Wan, H.; Delale, F.; Shen, L.; Effect of CNT length and CNT-matrix interphase in carbon nanotube (CNT) reinforced composites, Mech Res Commun **2005,** *32,* 481–489.

70. Shi, D.; Feng, X.; Jiang, H.; Huang, Y. Y.; Hwang, K.; Multiscale analysis of fracture of carbon nanotubes embedded in composites, Int J of Fract **2005,** *134,* 369–386.

71. Buryachenko, V. A.; Roy, A.; Effective elastic moduli of nanocomposites with prescribed random orientation of nanofibers, Comp: Part B **2005,** *36(5),* 405–416.

72. Buryachenko, V. A.; Roy, A.; Lafdi, K.; Andeson, K. L.; Chellapilla, S.; Multi-scale mechanics of nanocomposites including interface: experimental and numerical investigation, Comp Sci and Tech **2005,** *65,* 2435–246.

73. Li C, Chou, T. W.; Multiscale modeling of carbon nanotube reinforced polymer composites, J of Nanosci Nanotechnol **2003,** *3,* 423–430.

74. Li C, Chou, T. W.; Multiscale modeling of compressive behavior of carbon nanotube/polymer composites, Comp Sci and Tech **2006,** *66,* 2409–2414.

75. Anumandla, V.; Gibson, R. F.; A comprehensive closed form micromechanics model for estimating the elastic modulus of nanotube-reinforced composites, Com: Part A **2006,** *37,* 2178–2185.

76. Seidel, G. D.; Lagoudas, D. C.; Micromechanical analysis of the effective elastic properties of carbon nanotube reinforced composites, Mech of Mater **2006,** *38,* 884–907.

77. Hashin Z.; Rosen, B.; The elastic moduli of fiber-reinforced materials. Journal of Applied Mechanics **1964,** *31,* 223–232

78. R.; Christensen and, K.; Lo. Solutions for effective shear properties in three phase sphere and cylinder models. Journal of the Mechanics and Physics of Solids, **1979,** *27,* 315–330.

79. Selmi, A.; Friebel, C.; Doghri, I.; Hassis, H.; Prediction of the elastic properties of single walled carbon nanotube reinforced polymers: A comparative study of several micromechanical models, Comp Sci and Tech **2007**, *67*, 2071–2084

80. Luo, D.; Wang, W. X.; Takao, Y.; Effects of the distribution and geometry of carbon nanotubes on the macroscopic stiffness and microscopic stresses of nanocomposites, Comp Sci and Tech **2007**, *67*, 2947–2958.

81. Fu SY, Yue, C. Y.; Hu X, Mai, Y. W.; On the elastic transfer and longitudinal modulus of unidirectional multishort-fiber composites, Compos Sci Technol **2000**, *60*, 3001–3013.

82. Lauke, B.; Theoretical considerations on deformation and toughness of short-fiber reinforced polymers, J Polym Eng **1992**, *11*, 103–154.

83. Tserpes, K. I.; Panikos, P.; Labeas, G.; Panterlakis SpG.; Multi-scale modeling of tensile behavior of carbon nanotube-reinforced composites, Theoret and Appl Fract Mech, **2008**, *49*, 51–60.

84. Spanos, P. D.; Kontsos, A.; A multiscale monte carlo finite element method for determining mechanical properties of polymer nanocomposites, Prob Eng Mech **2008**, doi: 10.1016/j.probengmech.2007.09.002

85. Paiva, M. C.; Zhou, B.; Fernando, K. A. S.; Lin, Y.; Kennedy, J. M.; Sun, Y-P.; Mechanical and morphological characterization of polymer-carbon nanocomposites from functionalized carbon nanotubes. Carbon **2004**, *42*, 2849–54.

86. Zhu, J.; Peng, H.; Rodriguez-Macias, F.; Margrave, J.; Khabashesku, V.; Imam, A.; Lozano, K.; Barrera, E.; "Reinforcing epoxy polymer composites through covalent integration of functionalized nanotubes", Advanced Functional Materials **2004**, *14 (7)*, 643–648.

87. Bhuiyan Md. A.; Raghuram, V.; Pucha, A.; Johnny Worthy, B.; Mehdi Karevan, A.; Kyriaki Kalaitzidou, Defining the lower and upper limit of the effective modulus of CNT/polypropylene composites through integration of modeling and experiments. Composite Structures **2013**, *95*, 80–87.

88. Papanikos, P.; Nikolopoulos, D. D.; Tserpes, K. I.; Equivalent beams for carbon nanotubes. Comput Mater Sci **2008**, *43(2)*, 345–52.

89. Affdl, J. C. H.; Kardos, J. L.; The Halpin–Tsai equations: a review. Polym Eng Sci **1976**, *16(5)*, 344–52.

90. Tandon, G. P.; Weng, G. J.; The effect of aspect ratio of inclusions on the elastic properties of unidirectionally aligned composites. Polym Composite **1984**, *5(4)*, 327–33.

91. Nielsen, L. E.; Mechanical properties of polymers and composites, 2. New York: Marcel Dekker; 1974.

92. Tucker III CL, Liang, E.; Stiffness predictions for unidirectional short-fiber composites: review and evaluation. Compos Sci Technol **1999**, *59(5)*, 655–71.

93. Shokrieh, M. M.; Rafiee, R.; Investigation of nanotube length effect on the reinforcement efficiency in carbon nanotube based composites Composite Structures 92 **2010**, 2415–2420.

94. Shokrieh, M. M.; Rafiee, R.; On the tensile behavior of an embedded carbon nanotube in polymer matrix with nonbonded interphase region. J Compos Struct **2009**, *22*, 23–5.

95. Press Release. US Consulate. World-record-length carbon nanotube grown at US Laboratory. Mumbai-India; September 15, 2004.

96. Evans, J.; Length matters for carbon nanotubes: long carbon nanotubes hold promise for new composite materials. Chemistry World News **2004**, <http://www/rsc.org/chemistryworld/news>.

97. Pan, Z.; Xie, S. S.; Chang, B.; Wang, C.; Very long carbon nanotubes. Nature **1998**, *394*, 631–2.

98. http: //www.carbonsolution.com.

99. http: //www.fibermax.eu/shop/.

100. http: //www.nanoamor.com.

101. www.thomas-swan.co.uk.

102. Rice University's chemical 'Scissors' yield short carbon nanotubes. New process yields nanotubes small enough to migrate through cells. Science Daily [July 2003. <http: //www.sciencedaily.com/releases/2003/07/030723083644.htm>.

103. Chowdhury, S. C.; B. Z. (Gama) Haque, T.; Okabe, J. W.; Gillespie Jr.; Modeling the effect of statistical variations in length and diameter of randomly oriented CNTs on the properties of CNT reinforced nanocomposites, Composites: Part B **2012**, *43*, 1756–1762

104. Tserpes, K. I. Chanteli, A.; Parametric numerical evaluation of the effective elastic properties of carbon nanotube-reinforced polymers, Composite Structures **2013**, *99*, 366–374.

105. Gou, J.; Minaie, B.; Wang, B.; Liang, Z.; Zhang, C.; Computational and experimental study of interfacial bonding of single-walled nanotube reinforced composites. Comput Mater Sci **2004**, *31*, 225–36.

106. Frankland, S. J. V.; Harik, V. M.; Analysis of carbon nanotube pull-out from a polymer matrix. Surf Sci **2003**, *525*, 103–8.

107. Natarajan, U.; Misra, S.; Mattice, W. L.; Atomistic simulation of a polymer-polymer interface: interfacial energy and work of adhesion. Comput Theor Polym Sci **1998**, *8*, 323–9.

108. Lordi V, Yao, N.; Molecular mechanics of binding in carbon-nanotube polymer composites. J Mater Res **2000**, *15*, 2770–9.

109. Wong, M.; Paramsothy, M.; Xu XJ, Ren, Y.; Li S, Liao, K.; Physical interactions at carbon nanotube-polymer interface. Polymer **2003**, *44*, 7757–64.

110. Qian, D.; Liu, W. K.; Ruoff, R. S.; Load transfer mechanism in carbon nanotube ropes. Compos Sci Technol **2003**, *63*, 1561–9.

111. Liu, Y. J.; Chen, X. L.; Continuum models of carbon nanotube-based composites using the boundary element method. J Boundary Elem **2003**, *1*, 316–35.

112. Chen, X. L.; Liu, Y. J.; Square representative volume elements for evaluating the effective material properties of carbon nanotube-based composites. Comput Mater Sci **2004**, *29*, 1–11.

113. Chen, X. L.; Liu, Y. J.; Evaluations of the effective material properties of carbon nanotube-based composites using a nanoscale representative volume element. Mech Mater **2003**, *35*, 69–81.

114. Qian, D.; Dickey, E. C.; Andrews, R.; Rantell, T.; Load transfer and deformation mechanisms in carbon nanotubepolystirene composites. Appl Phys Lett **2000**, *76*, 2868.

115. Thostenson, E. T.; Chou T-W.; Aligned multiwalled carbon nanotube-reinforced composites: processing and mechanical characterization. J Phys D Appl Phys **2002**, *35*, 77–80.

116. Bower, C.; Rosen, R.; Jin, L.; Han, J.; Zhou, O.; Deformation of carbon nanotubes in nanotube-polymer composites. Appl Phys Lett **1999**, *74*, 3317–9.

117. Cooper, C. A.; Cohen, S. R.; Barber, A. H.; Wagner, H. D.; Detachment of nanotubes from a polymer matrix. Appl Phys Lett **2002**, *81*, 3873–5.

118. Qian, D.; Dickey, E. C.; In-situ transmission electron microscopy studies of polymer-carbon nanotube composite deformation. J Microsc **2001**, *204*, 39–45.

119. Schadler, L. S.; Giannaris, S. C.; Ajayan, P. M.; Load transfer in carbon nanotube epoxy composites. Appl Phys Lett **1998**, *73*, 3842.

120. Wagner, H. D.; Nanotube-polymer adhesion: a mechanics approach. Chem Phys Lett **2002**, *361*, 57–61.

121. Ajayan, P. M.; Schadler, L. S.; Giannaris, C.; Rubio, A.; Single-walled carbon nanotube-polymer composites: strength and weakness. Adv Mater **2000**, *12*, 750–3.

122. Cooper, C. A.; Young, R. J.; Halsall, M.; Investigation into the deformation of carbon nanotubes and their composites through the use of Raman spectroscopy. Compos Part A Appl Sci Manuf **2001**, *32*, 401–11.

123. Hadjiev, V. G.; Iliev, M. N.; Arepalli, S.; Nikolaev, P.; Files, B. S.; Raman scattering test of single-wall carbon nanotube composites. Appl Phys Lett **2001**, *78*, 3193.

124. Paipetis, A.; Galiotis, C.; Liu, Y. C.; Nairn, J. A.; Stress transfer from the matrix to the fiber in a fragmentation test: Raman experiments and analytical modeling. J Compos Mater **1999**, *33*, 377–99.

125. Valentini, L.; Biagiotti, J.; Kenny, J. M.; Lopez Manchado, M. A.; Physical and mechanical behavior of singlewalled carbon nanotube/polypropylene/ethylene-propylene-diene rubber nanocomposites. J Appl Polym Sci **2003**, *89*, 2657–63.

126. Qian, D.; Wagner, G. J.; Liu, W. K.; Yu M-F, Ruoff, R. S.; Mechanics of carbon nanotubes. Appl Mech Rev **2002**, *55*, 495–532.

127. Yu M-F, Yakobson, B. I.; Ruo, R. S.; Controlled sliding and pulout of nested shells in individual multiwalled nanotubes. J Phys Chem B **2000**, *104*, 8764–7.

128. Qian, D.; Liu, W. K.; Ruoff, R. S.; Load transfer mechanism in carbon nanotube ropes. Compos Sci Technol **2003**, *63*, 1561–9.

129. Liao, K.; Li, S.; Interfacial characteristics of a carbon nanotubepolystirene composite system. Appl Phys Lett **2001**, *79*, 4225–7.

130. Andrews, R.; Weisenberger, M. C.; Carbon nanotube polymer composites. Curr Opin Solid State Mater Sci **2004**, *8*, 31–7.

131. Lordi, V.; Yao, N.; Molecular mechanics of binding in carbonnanotube- polystirene composite system. J Mater Res **2000**, *5*, 2770–9.

132. Wagner, H. D.; Vaia, R. A.; Nanocomposites: issue at the interface. Mater Today **2004**, *7*, 38–42.

133. Wernik, J. M.; Cornwell-Mott, B. J.; Meguid, S. A.; Determination of the interfacial properties of carbon nanotube reinforced polymer composites using atomistic-based continuum model, Inte. Jour. of Sol. and Struct. **2012**, *49*, 1852–1863.

134. Wernik, J. M.; Meguid, S. A.; Multiscale modeling of the nonlinear response of nanoreinforced polymers. Acta Mech. **2011**, *217*, 1–16.

135. Yang, S.; Yu, S.; Kyoung, W.; Han, D. S.; Cho, M.; Multiscale modeling of size-dependent elastic properties of carbon nanotube/polymer nanocomposites with interfacial imperfections Polymer **2012**, *53*, 623–633.

136. Ayatollahi, M. R. Shadlou, S.; Shokrieh, M. M.; Multiscale modeling for mechanical properties of carbon nanotube reinforced nanocomposites subjected to different types of loadingComposite Structures **2011**, *93*, 2250–2259

137. Jiang, L. Y.; Huang, Y.; Jiang, H.; Ravichandran, G.; Gao, H.; Hwang, K. C.; et al. A cohesive law for carbon nanotube/polymer interfaces based on the van der Waals force. J Mech Phys Solids **2006**, *54*, 2436–52.

138. Tan, H.; Jiang, L. Y.; Huang, Y.; Liu, B.; Hwang, K. C.; The effect of van der Waals-based interface cohesive law on carbon nanotube-reinforced composite materials, Composites Science and Technology **2007**, *67*, 2941–2946.

139. Zalamea, L.; Kim, H.; Pipes, R. B.; Stress transfer in multiwalled carbon nanotubes. Compos Sci Technol **2007**, *67(15–16)*, 3425–33.

140. Shen, G. A.; Namilae, S.; Chandra, N.; Load transfer issues in the tensile and compressive behavior of multiwall carbon nanotubes. Mater Sci Eng A **2006**, *429(1–2)*, 66–73.

141. Gao, X. L.; Li, K.; A shear-lag model for carbon nanotube-reinforced polymer composites. Int J Solids Struct **2005**, *42(5–6)*, 1649–67.

142. Li C.; Chou, T. W.; A structural mechanics approach for the analysis of carbon nanotubes. Int J Solids Struct **2003**, *40(10)*, 2487–99.

143. Tsai, J. L.; Lu, T. C.; Investigating the load transfer efficiency in carbon nanotubes reinforced nanocomposites. Compos Struct **2009**, *90(2)*, 172–9.

144. Lu, T. C.; Tsai, J. L.; Characterizing load transfer efficiency in double-walled carbon nanotubes using multiscale finite element modeling, Composites: Part B **2013**, *44*, 394–402

A DETAILED REVIEW ON FABRICATION AND CHARACTERIZATION OF THE METAL NANO-SIZED BRANCHED STRUCTURES

GUOQIANG XIE, MINGHUI SONG, KAZUO FURUYA, and AKIHISA INOUE

CONTENTS

5.1 INTRODUCTION

Using an electron-beam-induced deposition (EBID) process in a transmission electron microscope, we fabricated self-standing metal nano-sized branched structures including nanowire arrays, nanodendrites, and nanofractal-like trees, as well as their composite nanostructures with controlled size and position on insulator (SiO_2, Al_2O_3) substrates. The fabricated nanostructures were characterized with high-resolution transmission electron microscopy and X-ray energy dispersive spectroscopy. The growth mechanism was discussed. Effect of the electron beam accelerating voltage on crystallization of the nanostructures was investigated. The nanostructures of the different morphologies were obtained by controlling the intensity of the electron beam during the EBID process. A mechanism for the growth and morphology of the nanostructures was proposed involving charge-up produced on the surface of the substrate, and the movement of the charges to and charges accumulation at the convex surface of the substrate and the tips of the deposits. High-energy electron irradiation enhanced diffusion of the metallic atoms in the nanostructures and hence promoted crystallization. More crystallized metal nano-branched structures were achieved by the EBID process using high-energy electron beams.

Metal nanostructures are of great importance in nanotechnology due to their potential applications as building blocks in optoelectronic devices, catalysis, chemical sensors, and other areas [1–3]. The formation of nanostructures with controlled size and morphology has been the focus of intensive research in recent years. Such nanostructures are important in the development of nanoscale devices and in the exploitation of the properties of nanomaterials [4–7]. Many fabrication methods including chemical vapor decomposition [8], the arc-discharge method [9], evaporation [10], hydrothermal reactions [11], and so on, have been developed for the production of nanomaterials with controlled sizes and shapes. However, up to date, fabrication of position controllable nanostructures at selected positions on a substrate is still a challenge.

Among the methods, electron-beam-induced deposition (EBID) process is one of the most promising techniques to fabricate small-sized structures on substrates. An advantage of this process is that the deposited position can be controlled. In this approach, an electron beam in a high vacuum chamber is focused on a substrate surface on which precursor molecules, containing the element to be deposited (e.g., organometallic compound or hydrocarbon), are adsorbed. As a result of complex beam-induced surface reactions, the precursor molecules absorbed in or near to the irradiated area, are dissociated into nonvolatile and volatile parts by energetic electrons. The nonvolatile materials are deposited on surface of the substrate, while the volatile components are pumped away. Due to the controllability of electron beam, fabrication of position controllable nanostructures can be realized. Using the technique, a variety of nanometer-sized structures, such as nanodots, nanowires, nanotubes, nanopatterns, two or three dimensional nano-objects, and so on, have been fabricated [12–22]. Due to easy to receive a stable fabrication condition, conductive substrates are generally used in these fabrications [16]. As a result, compact structures are usually fabricated with this process.

In the case using nonconductive substrates, the nanofabrication with the EBID technique is also very important for the technique to be applied in technology. The accumulation of charges readily occurs on an insulator substrate during the EBID process. The deposition of novel structures may occur under this condition. The growth of carbon fractal-like structures has already been observed on insulator substrates under electron-beam irradiation due to the existence of residual pump oil remaining in the atmosphere of the vacuum chamber [23–25], but the growth of metallic nanostructures on insulator substrates using EBID process was first reported recently by Song et al. [26]. By using insulator substrates such as SiO_2 [27–30] and Al_2O_3 [26, 31–35], characteristic morphologies of nanostructures, such as arrays of nanowhiskers (or nanowires), arrays of nanodendrites, and fractal treelike structures, have been fabricated in transmission electron microscopes (TEMs) by the EBID process [26–35]. The typical size of the diameter of a nanowhisker, the tip of a nanodendrite, and the tip of a nanotree is about 3 nm, and are almost completely independent of the size of the electron beam. In the present review we reported the fabrication and characterization of the metal nano-sized branched structures and the composite nanostructures grown on insulator substrates by the EBID process.

5.2 FABRICATIONS AND CHARACTERIZATION OF NANO-SIZED BRANCHED STRUCTURES

Thin films of SiO_2 and Al_2O_3 were irradiated by an electron beam in TEM with organometallic precursor gasses at room temperature. The EBID process was carried out using JEM-2010 or JEM-ARM1000 TEMs made by JEOL Co., Ltd. The energy of the electron beam used was from 200 to 1000 keV. The pressure in the specimen chambers of the TEMs was on the order of 10^{-5} to 10^{-6} Pa. Insulator SiO_2 and Al_2O_3 substrates suitable for TEM observation were used. An organometallic precursor gas was introduced near the substrate using a special designed system comprizing a nozzle with a diameter smaller than 0.1 mm and a reservoir containing the precursor [26]. Tungsten hexacarbonyl ($W(CO)_6$) or (Methylcyclopentadienyl)trimethylplatinum (Me_3 MeCpPt) powders were used as the precursors for fabricating tungsten (W)-containing, or platinum (Pt)-containing nanostructures, respectively. The vapor pressures of these precursors are several Pa at room temperature. These precursors have been previously typically used to fabricate small objects on conductive substrates by EBID process [16]. The intensity of the electron beam for the EBID process was from 3.2 to 111.7×10^{18} e $cm^{-2} s^{-1}$ (current density: 0.51 to 17.9 A cm^{-2}), which was estimated by measuring the total intensity and the size of the electron beam under operating conditions. The fabricated structure was characterized cm or after fabrication in a TEM. The experiments were performed at room temperature.

Figure 1 shows a series of micrographs of growth process of the W-nanodendrite structures using the TEM at an accelerating voltage of 1000 kV during the electron-beam irradiation with an electron beam current density of 1.6 A cm^{-2} [29]. Figure 1(a) presents a micrograph before the electron-beam irradiation. No deposits are observed at the SiO_2 substrate. Figure 1(b) shows a micrograph taken 3 s after the beginning of the irradiation at a fluence of 3.0×10^{19} e cm^{-2}. The electron beam irradiated the specimen from a direction perpendicular to the plane of the micrograph. Whisker-like deposits begin to nucleate and grow on the surface of the substrate. The deposits are about 3 nm in diameter and about 5 nm in length. The structures grow self-standing at positions separated from each other at a distance of several nanometers, and in parallel and nearly perpendicular to the surface of the substrate within the irradiated area. The whisker-like deposits grow longer and also denser under further electron beam irradiation, and new nucleation and growth deposits are not observed. Figure 1(c) shows a micrograph taken 3 min after the beginning of irradiation at a fluence of 1.8×10^{21} e cm^{-2}. Similarity in length and an almost even thickness of all deposits are observed. Branching is observed at the tips of the deposits. The fabricated deposits have a nanodendrite structure. With further electron beam irradiation, new branches grew from the tips of the grown branches. This process continued as long as the irradiation continued. Figure 1(d) shows a micrograph taken 10 min after the beginning of irradiation at a fluence of 6.0×10^{21} e cm^{-2}. The nanodendrite structures grow at both the tip and trunk. The diameter of the nanodendrite structures increased with the decrease in distance from the substrate, but is about 3 nm at the tips. The average length of the nanodendrite structures increased to about 51.4 nm. Figure 2 shows the length of nanodendrite structures grown at various electron-beam irradiation fluences (namely, electron-beam irradiation time). The length of nanodendrite structures increases approximated linearly with electron-beam irradiation fluence [29].

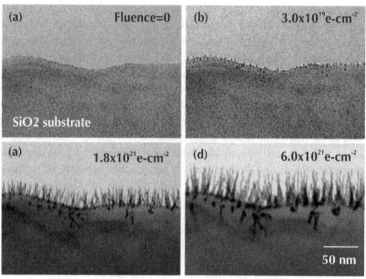

FIGURE 1   Bright-field TEM images of W-nanodendrite structures grown on surface of SiO_2 substrate by the EBID process at an electron beam accelerating voltage of 1000 kV with a current density of $1.6\,A\,cm^{-2}$ after beginning of electron beam irradiation. (a) 0 s; (b) 3 s; (c) 3 min; and (d) 10 min [29].

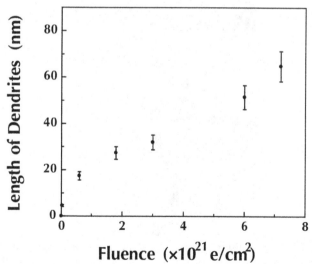

FIGURE 2   Relationship between length of the W-nanodendrite structures grown on surface of SiO_2 substrate by the EBID process at an electron beam accelerating voltage of 1000 kV with a current density of $1.6\,A\,cm^{-2}$ and electron beam irradiating fluence [29].

Furthermore, the morphology of the nanostructures grown can also be controlled by the intensity of the electron beam [26, 36]. Figure 3(a) shows a micrograph of an array of W nanowhiskers (or nanowires) fabricated on an Al_2O_3 substrate irradiated with a 200 keV electron beam at an intensity of 4.7×10^{18} e cm^{-2} s^{-1} (0.75 A cm^{-2}) to about 120 s [26]. The growth speed was very low. By increasing the current density to a higher value such as 1.6 A cm^{-2}, the growth speed was increased. The nanowhiskers were longer and also denser than those grown at the lower current density with an almost even thickness. Branching is observed to take place at the tips of the nanowhiskers, and the morphology becomes dendritic structures [27, 29]. A further increase in the current density to 17.9 A cm^{-2} resulted in extensive branching and the formation of complicated nanotree structures (Figure 3(c)) [26].

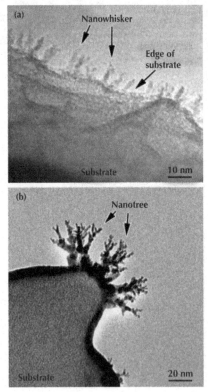

FIGURE 3   (a) W nanowhisker arrays grown on the surface of an Al_2O_3 substrate by the EBID process at an accelerating voltage of 200 kV. The current density of the electron beam was 0.75 A cm^{-2} and the irradiation time was 120 s. (b) Fractal-like W nanotrees grown on the surface of an Al_2O_3 substrate by the EBID process. The current density of the electron beam was 17.9 A cm^{-2} and the irradiation time was 10 s [26].

The composition of the fabricated nanodendrites is important, since they are related to the physical and chemical properties of deposits. The chemical composition of the nanodendrites was examined using an X-ray energy dispersive spectroscopy (EDS). Figure 4 shows the spectra taken using EDS from an as-deposited nanodendrite fabricated at an electron-beam irradiating fluence of 6.0×10^{21} e cm^{-2} after detaching the precursor source [29]. Figure 4(a) presents a spectrum taken from the tips of a nanodendrite structure, and Figure 4(b) shows that taken from its trunk. The size of the electron beam for the analysis using EDS is about 10 nm. The peaks of tungsten dominate these spectra. No obvious differences in the composition between the tip and the trunk of the nanodendrite structure were observed. On the basis of the analyzes using EDS, a relative content of 89.4% W compared with 10.6% C for the spectrum from the tips is obtained. The relative content of W is higher than the reported values for W-deposits fabricated using the same precursor but at a lower voltage of the electron beam (25 kV) by EBID process, which was 75% [17].

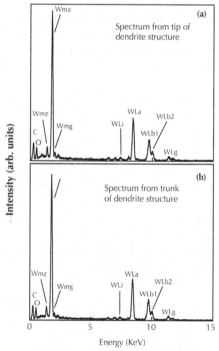

FIGURE 4 EDS spectra taken from the nanodendrite grown on surface of SiO$_2$ substrate by the EBID process at an accelerating voltage of 1000 kV with a current density of 1.6 A cm^{-2} to an electron-beam irradiating fluence of 6.0×10^{21} e cm^{-2} (a) at tip; and (b) at trunk [29].

Using (Methylcyclopentadienyl)trimethylplatinum ($Me_3MeCpPt$) powder as a precursor, one can obtain Pt nanodendritic structures by the EBID process. Figure 5 shows TEM micrographs of Pt nanodendrite-like structures grown on an Al_2O_3 substrate in a 200 kV TEM [34]. The electron beam is defocused to a size of about 600 nm, corresponding to a current density of 0.52 A cm^{-2}. The irradiated time is 1 min. The irradiating fluence of the electron beam is 2.0×10^{20} e cm^{-2}. The edge of the electron beam is indicated by arrows in Fig. 5(a). The contrast inside the irradiated area is obviously dark. The size of dendrite structures is much smaller than the diameter of the electron beam. The dendrites show a tendency to grow at the edge of the substrate. The dendrites have branches at the tip, as observed in Fig. 5(b). The diameter of the nanodendrites become thicker near the substrate, which implies that the deposition takes place at both tip and trunk part. The typical thickness of the tips is less than 3 nm, as observed in Fig. 5(b). TEM observation, diffraction pattern analyzes, and EDS analyzes indicated that a nanodendrite structure with a high Pt content was formed. Figure 6 shows a high resolution TEM (HRTEM) micrograph of the fabricated Pt nanodendrite [34]. It is confirmed that fcc Pt nanoparticles and an amorphous part are contained in the structure. The morphology of the Pt nanodendrite is considerably different from that of the W nanodendrite fabricated on an Al_2O_3 substrate, suggesting that the growth of a nanodendrite depends largely on the properties of the precursor.

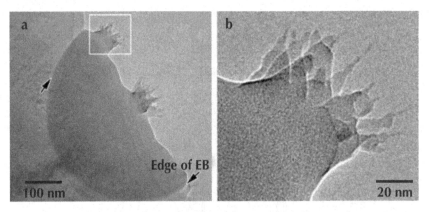

FIGURE 5   (a) Bright-field TEM micrographs of Pt-nanodendrite structures grown on the surface of an Al_2O_3 substrate in a 200 kV TEM with an EBID process 1 min after the start of the electron beam irradiation to a fluence of 2.0×10^{20} e cm^{-2}. (b) Enlargement of the square area in (a), showing the nanodendrites in more detail [34].

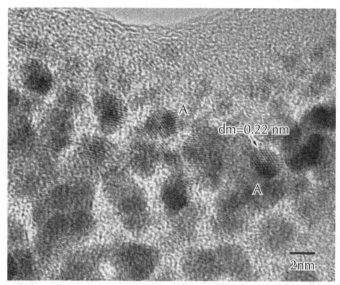

FIGURE 6 An HRTEM micrograph of a nanodendrite structure fabricated on an electron beam irradiating fluence of 2.6×10^{21} e cm^{-2} [34].

5.3 GROWTH MECHANISM OF NANO-SIZED BRANCHED STRUCTURES

It is well known that the EBID process is caused by the dissociation of molecules adsorbed to a surface by energetic electron beam. In this approach, a molecule of a precursor is first adsorbed on surface of a substrate and then decomposed into volatile and nonvolatile parts by further irradiation of the energetic beam. The nonvolatile fraction accumulates to form a deposit, while the volatile component is pumped away by the vacuum system. The dissociation mechanism is complex and not fully understood until the present time because of the huge number of excitation channels available even for small molecules. The details involving the decomposition have been argued to relate to secondary electrons on surface of a specimen produced by incident electron beam and backscattered electrons in an EBID process [17, 19].

If an EBID process is carried out on the surface of an electric grounded conductive substrate, the molecules absorbed may move or not move, but distribute randomly and be decomposed on the surface. Therefore a compact deposit is usually formed. On the other hand, in a case that the deposition is conducted on an insulator substrate, charge-up may take place on the surface due to emission of secondary electrons under energetic beam irradiation. When specimens are exposed to electron bombardments, the molecules absorbed to a surface of a substrate or near the surface in the irradiated area may be polarized by irradiation of an incident electron beam. The irradiated area on the insulator is easily charged, forms local electric potential. It is reasonable to consider that the distribution of charges due to charge-up on the surface may be not even in a nanometer scale. This unevenness may be resulted from the nanoscaled unflatness or atomic steps on surface. Charges may accumulate at some places to some extent. Figure 7(a) shows the growth mechanism of a nanowhisker array [26]. The intensity of the electron beam is weak, but the accumulation of charges occurs. The distribution of the charges on the surface is assumed to be uneven at a nanometer scale. Charges may accumulate at some places (charge centers), where an electric field is generated. The dark dots represent the nonvolatile fraction of the precursor molecules of which the deposits are composed, while the brighter dots represent the volatile fraction. Because the intensity of the electron beam is weak, the adsorbed molecules can move around considerably before being decomposed. The precursor molecules adsorbed on the surface may be attracted to the charge centers since the molecules are easily polarized due to the weak bonding between the atoms of the molecules. The precursor molecules then decompose and form a deposit. After a deposit is formed, the charges on the surface of the substrate or the deposit tend to move and accumulate at the tip of the deposit, since W deposits have been reported to be conductive [17, 19]. Thus, molecules are attracted to the tip, and the deposit increases in length upon further electron beam irradiation. The formation of W nanodot deposits on the surface of a SiO_2 substrate and the growth of W nanowhiskers from these nanodots [30] are consistent with the above mechanism.

Figures 7(b) and (c) schematically show the growth conditions of nanodendrites and fractal-like nanotrees, respectively [26]. When the current density of the electron beam is increased to a definite extent, an adsorbed molecule on the surface may be decomposed in a very short period before it can move. Hence, a compact layer is deposited on the surface. This results in the entire irradiated area becoming conductive. After this stage, the charges on the surface can move a long distance to the substrate edge, particularly when the surface is convex. A strong electric field is thus generated near the convex surface. Once a deposit grows away from the surface at a point, charges then accumulate there. This deposit thus grows preferentially because the charges attract precursor molecules. When the current density is moderate, the electric field generated near a tip may not strongly affect the trajectory of a molecule. Therefore, the deposit grows at both its tip and its main body because the molecules arrive at the tip and the main body at the same time. Therefore, the main body of the deposit increases in thickness with time and develops a dendritic morphology. On the other hand, if the current density of the electron beam is very strong, the generated electric field will become sufficiently strong to affect the trajectory of a molecule near the tip of a deposit. The molecules are thus likely to be attracted to the tip and are not easily adsorbed on the main body of the deposit. Therefore, a fractal-like tree morphology is formed.

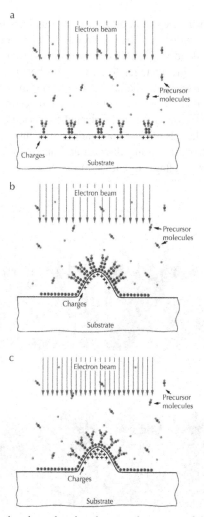

FIGURE 7 Schematic drawings showing the growth process of the nanostructures with different morphologies on insulator substrates by the EBID process. The dark dots represent the nonvolatile fraction of the precursor and the deposit, while the brighter dots represent the volatile fraction of the precursor. (a) The growth of nanowhiskers (or nanowires) inside the area of irradiation with a weak electron beam; (b) the growth of nanodendrites on a convex surface under irradiation with a moderate electron beam; and (c) the growth of fractal-like nanotrees on a convex surface under irradiation with a strong electron beam [26].

5.4 EFFECT OF ACCELERATING VOLTAGE ON CRYSTALLIZATION OF THE NANOSTRUCTURES

The fabricated nanostructure morphologies at various electron beam-accelerating voltages have been investigated with a conventional TEM. Figure 8 shows a set of TEM micrographs of dendrite-like structures grown on SiO_2 substrate at different electron beam accelerating voltages to an electron beam fluence of 6.0×10^{21} e cm^{-2} [27]. The electron beam irradiated the specimen from a direction perpendicular to the plane of the micrograph. The obvious difference in dendrite morphology and their composites for the deposits fabricated by various accelerating voltages is not observed. The length of the nanodendrite structures increases with an increase in electron beam irradiating fluence, and has not obvious effect with the increase of accelerating voltage, as shown in Fig. 9.

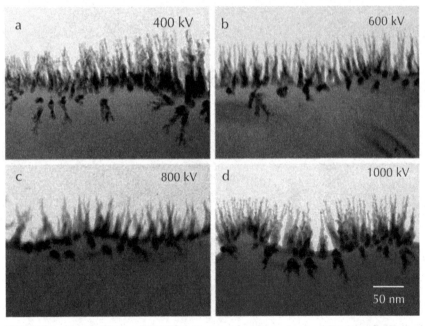

FIGURE 8   A set of bright-field TEM micrographs of the nanodendrite structures grown on SiO_2 substrate irradiated to an electron beam fluence of 6.0×10^{21} e cm^{-2} at different electron beam accelerating voltages. (a) 400 kV; (b) 600 kV; (c) 800 kV; (d) 1000 kV. The current density of electron beam for irradiating the specimens was 1.1 A cm^{-2} for 400 kV, and 1.6 A cm^{-2} for 600, 800 and 1000 kV, respectively [27].

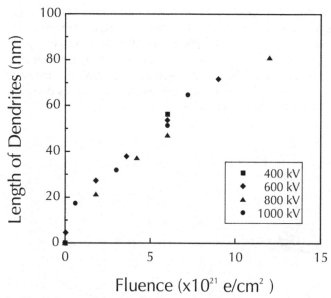

$$\text{Fluence } (\times 10^{21} \text{ e/cm}^2)$$

FIGURE 9 Dependence of the length of nanodendrites on electron beam irradiating fluence at different electron beam accelerating voltages [27].

The effect of electron beam accelerating voltage on crystallization of the nanodendrite structures has been investigated with an HRTEM. Figure 10 shows a series of HRTEM micrographs of some branches of nanodendrite structures fabricated with various accelerating voltages to a fluence of 6.0×10^{21} e cm^{-2} [28]. Figure 10(a) is an HRTEM micrograph of some branches of nanodendrites grown with an accelerating voltage of 400 kV. Lattice fringes are observed at the most places. This indicates that they are crystal in several nanometers. The largest and also the most observed lattice spacing measured from micrographs is 0.22 nm. It is close to the lattice spacing, 0.224 nm, of {110} of bcc W crystals with a deviation smaller than 5%. Moreover, lattice fringes with spacing d = 0.22 nm have interfringe angle of 60 degrees (grain A), as well as that of 90 degrees (grain B), as indicated in Fig. 10. They agree to zone axis of [111] and [001] of bcc W structure, respectively. Combined with nanometer-sized area diffraction pattern analyzes and EDS analyzes, it is clarified that these crystals are W grains in bcc structure. Furthermore, lattice fringes cannot be clearly observed in some places, indicating that a large part of them is in amorphous state, as shown by arrows in Fig. 10. Figure 10(b) shows a micrograph of some branches of nanodendrites grown with an accelerating voltage of 600 kV. The fraction of amorphous state in the as-fabricated nanodendrites decreases obviously. For the dendrites fabricated with an accelerating voltage of 800 kV, as shown in Figure 10(c), the amorphous state is observed only in a few places. Figure 10(d) shows an HRTEM micrograph of some branches of nanodendrites grown with an accelerating voltage of 1000 kV. Lattice fringes are observed clearly in almost all of the grains except in thick region, where small crystals in random orientations overlapped each other, so that their lattice fringes cannot be clearly observed. Therefore, it is indicated that the fraction of amorphous state in the as-fabricated nanodendrites decreases with an increase in electron beam accelerating voltage.

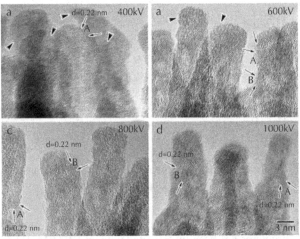

FIGURE 10   A series of HRTEM micrographs of some branches of nanodendrites grown on SiO_2 substrate irradiated to a fluence of 6.0×10^{21} e cm^{-2} with different electron beam accelerating voltages. (a) 400 kV; (b) 600 kV; (c) 800 kV; (d) 1000 kV. Arrows indicate lattice fringes in grains A and B, which have interfringe angles of 60 (A) and 90 (B) degrees, respectively. The current density of electron beam for irradiating the specimens was 1.1 A cm^{-2} for 400 kV, and 1.6 A cm^{-2} for 600, 800 and 1000 kV, respectively [28].

It is known that the growth of a W nanodendrite during the EBID process is controlled by a random accumulation of nonvolatile elements with thermal energy. Therefore the element does not have enough energy to move and form crystalline grains. The electron beam during the EBID process may transfer some energy to the element, but it is not enough for total deposit to transform into crystalline state with a 400 keV electron beam. Therefore, some amorphous structures are remained in deposits (refer to Fig. 10(a)). This also may explain the reason that an amorphous state has often been obtained in the W-deposits fabricated by a lower energy EBID [19, 26]. With an increase in electron beam accelerating voltage, the fraction transformed into crystalline state in the deposits increases, and the fraction of amorphous structure decreases. The results clearly indicate that high-energy electron irradiation enhances crystallization of an amorphous structure.

Song et al. [37] has reported that the effect of 1 MeV electron beam irradiation on crystallization of nanometer-sized W-dendritic structure fabricated on Al_2O_3 substrates with the EBID process at an accelerating voltage of 200 kV. After irradiation at room temperature for 100 min at a current density of 6.4 A cm^{-2}, almost all the grains crystallized. The nanodendrite structure also changes morphology and shrinks its size. In general, there are two important factors in the occurrence of electron irradiation induced crystallization of an amorphous state: (1) the promotion of atomic diffusion by electron irradiation in an amorphous state; and (2) the high stability of crystalline phase against electron irradiation [38]. When the two factors are satisfied simultaneously, electron irradiation induced crystallization of the amorphous structure. The change in morphology and shrinkage in size may be resulted from crystallization and sputtering during the irradiation.

5.5 COMPOSITE NANOSTRUCTURES OF NANOPARTICLES/ NANODENDRITES

Complex-shape nanostructures have attracted great interest because advanced functional materials might be emerged if they can be formed with well-defined three-dimensional (3D) architectures [39–41]. Furthermore, the impact is greater for multielement systems, as in the case of advanced nanostructured systems like Au/Cu, Pt/C composite materials [42, 43] and so on, which are used in catalysis, sensors, energy sources [44, 45] and in many other applications. It has been demonstrated that bimetallic bonding can induce significant changes in the properties of the surface, producing in many cases catalysts that have superior activity and/ or selectivity [42, 43, 46, 47].

The nanostructures fabricated on insulator substrates by the EBID process are good base materials for the fabrication of complex-shape nanostructures because of their superior features. Pt nanoparticles or Au nanoparticles were deposited on W nanodendrites fabricated on insulator Al_2O_3 or SiO_2 substrates, and a composite nanostructure consisting of Pt nanoparticles and W nanodendrites (Pt-nanoparticle-decorated W nanodendrites, or Pt nanoparticle/W nanodendrite), or Au nanoparticle and W nanodendrites were fabricated by ion-sputtering [33]. Figure 11 shows the Pt nanoparticle/W nanodendrite composite structures. Figure 11(a) is a bright-field TEM image of the as-fabricated composite structure produced with an electron beam irradiation fluence of 5.0×10^{21} cm^{-2} and an ion sputtering time of 40 s. The Pt-nanoparticles are nearly uniformly distributed on the W-nanodendrites. Figure 11(b) shows an HRTEM image of a tip of the composite nanostructure. Lattice fringes are observed in the image. By measuring the lattice spacing and the interfringe angle from this image and other images, one can demonstrate that the as-fabricated composite nanostructures consist of nanocrystals of Pt and W. By using selected-area diffraction (SAD) pattern (Fig. 11(c)), these nanocrystals are identified to be equilibrium phases of fcc Pt and bcc W. Figure 11(d) is an EDS spectrum obtained at the tip corresponding to Fig. 11(b). Pt and W peaks dominate in the spectrum although traces of C and O are also observed. Thus, it is confirmed that Pt has been effectively grown on the nanodendritic W structures. From the HRTEM micrographs, the average nanoparticle size at the present ion sputtering conditions was 2.3 nm. The particle size can be easily controlled by variation of the ion sputtering time.

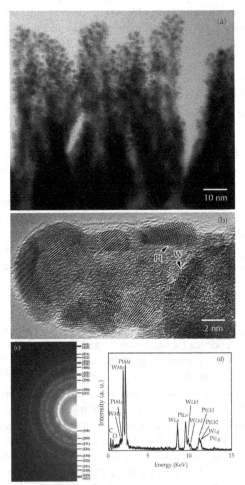

FIGURE 11 (a) Bright-field TEM image of the as-fabricated Pt-nanoparticle/W-nanodendrite composite structures on an Al_2O_3 substrate (b) an HRTEM image, (c) SAD pattern and (d) EDS spectrum taken from a tip of the composite nanostructures [33].

Using this process, various other metal composite nanoparticle/nanodendrite structures, such as Au/W, Mo/W, Au/Pt, and so on, can also be fabricated. Figure 12 shows an example of Au-nanoparticle/W-nanodendrite compound nanostructures grown on an insulator SiO_2 substrate. Therefore, the technique may be easily employed to fabricate metal nanoparticle/nanodendrite composite nanostructures of a wide range of materials. Because the nanodendrite structure possesses a comparatively large specific surface area, it has potential applications in catalysts, sensors, gas storages and so on.

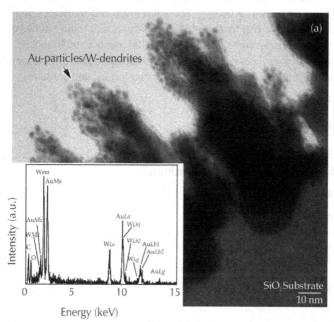

FIGURE 12 (a) Bright-field TEM image of the as-fabricated Au-nanoparticle/W-nanodendrite composite structures on a SiO_2 substrate and (b) an EDS spectrum taken from a tip of the composite nanostructures. The W nanodendrites were fabricated with an electron beam irradiation fluence of 4.4×10^{21} cm^{-2}. Au nanoparticles were deposited by a quick auto coater system (JEOL JFC-1500). The anodic voltage used during sputtering was 1 kV, and anodic current was 7 mA. The ion sputtering time was 7 s. The average Au nanoparticle size was measured to be 2.1 nm [33].

5.6 CONCLUSION

Self-standing metal nano-sized branched structures including nanowire arrays, nanodendrites, and nanofractal-like trees with controlled size and position were grown on insulator (SiO_2, Al_2O_3) substrates by the EBID process. The nanostructures of the different morphologies can be obtained by controlling the intensity of the electron beam during the EBID process. The nucleation and growth of the nano-sized branched structures are proposed to be related to a mechanism involving charge-up produced on the surface of the substrate, and the movement of the charges to and charges accumulation at the convex surface of the substrate and the tips of the deposits. High-energy electron irradiation enhances diffusion of the metallic atoms in the nanostructures and hence promotes crystallization. More crystallized metal nano-branched structures are achieved by the EBID process using high-energy electron beams. The nano-branched structures can be easily decorated by metallic nanoparticles to form composite nano-structures such as the Pt nanoparticles/nanodendrites. Therefore the nano-sized branched structures and their fabrication process may be applied in technology to realize various functional nanomaterials such as catalysts, sensor materials, and emitters.

KEYWORDS

- electron-beam-induced deposition
- nanodendrites
- nanofractal-like trees
- nanowire arrays

REFERENCES

1. Shi, J.; Gider, S.; Babcock, K.; Awschalom, D. D.; *Science* **1996**, *271*, 937.
2. Favier, F.; Walter, E.; Zach, M.; Benter, T.; Penner, R. M.; *Science* **2001**, *293*, 2227.
3. Xia, Y. N.; Yang, P. D.; Sun, Y. G.; Wu, Y. Y.; Mayer, B.; Gates, B.; Yin, Y. D.; Kim, F.; Yan, H. Q.; *Adv. Mater.* **2003**, *15*, 353.
4. Lao, Y. L.; Wen, J. G.; Ren, Z. F.; *Nano Lett.* **2002**, *2*, 1287.
5. Manna, L.; Milliron, D. J.; Meisel, A.; Scher, E. C.; Alivisatos, A. P.; *Nature Mater.* **2003**, *2*, 382.
6. Yan, H.; He, R.; Pham, J.; Yang, P.; *Adv. Mater.* **2003**, *15*, 402.
7. Dick, K. A.; Deppert, K.; Larsson, M. W.; Martensson, T.; Seifert, W.; Wallenberg, L. R.; Samuelson, L.; *Nature Mater.* **2004**, *3*, 380.
8. Zhang, H. F.; Wang, C. M.; Buck, E. C.; Wang, L. S.; *Nano Lett.* **2003**, *3*, 577.
9. Shi, Z. J.; Lian, Y. F.; Liao, F. H.; Zhou, X. H.; Gu, Z. N.; Zhang, T.; Iijima, S.; Li, H. D.; Yue, K. T.; Zhang, S. L.; *J. Phys. Chem. Solids* **2000**, *61*, 1031.
10. Dai, Z. R.; Pan, Z. W.; Wang, Z. L.; *Adv. Funct. Mater.* **2003**, *13*, 9.

11. Jin, Y.; Tang, K. B.; Huang, C. H.; An, L. Y.; *J. Cryst. Growth* **2003**, *253,* 429.
12. Mitsuishi, K.; Shimojo, M.; Han, M.; Furuya, K.; *Appl. Phys. Lett.* **2003,** *83,* 2064.
13. Dong, L. X.; Arai, F.; Fukuda, T.; *Appl. Phys. Lett.* **2002,** *81,* 1919.
14. Utke, I.; Hoffmann, P.; Dwir, B.; Leifer, K.; Kapon, E.; Doppelt, P.; *J. Vac. Sci. Technol. B* **2000,** *18,* 3168.
15. Brückl, H.; Kretz, J.; Koops, H. W.; Reiss, G.; *J. Vac. Sci. Technol. B* **1999,** *17,* 1350.
16. H. Koops, W. P.; J.; Kretz, M.; Rudolph, M.; Weber, G.; Dahm, Lee, K. L.; *Jpn. J. Appl. Phys. Part* **1994,** *1–33,* 7099.
17. H. Koops, W. P.; R.; Weiel, Kern, D. P.; Baum, T. H.; *J.; Vac. Sci. Technol. B* **1988,** *6,* 477.
18. Hiroshima, H.; Suzuki, N.; Ogawa, N.; Komuro, M.; *Jpn. J. Appl. Phys.* **1999,** *1–38,* 7135.
19. Hoyle, P. C.; Cleaver, J. R. A.; Ahmed, H.; *J. Vac. Sci. Technol. B* **1996,** *14,* 662.
20. Kohlmann-von Platen, K. T.; Chlebek, J.; Weiss, M.; Reimer, K.; Oertel, H.; Brünger, W. H.; *J. Vac. Sci. Technol. B* **1993,** *11,* 2219.
21. Matsui, S.; Kaito, T.; Fujita, J.; Komura, M.; Kanda, K.; Haruyama, Y.; *J. Vac. Sci. Technol. B* **2000,** *18,* 3181.
22. Liu, Z. Q.; Mitsuishi, K.; Furuya, K.; *Nanotechnology* **2004,** *15,* S414.
23. Banhart, F.; *Phys. Rev. E* **1995,** *52,* 5156.
24. Zhang, J. Z.; X. Y. Ye, Yang, X. J.; Liu, D.; *Phys. Rev. E* **1997,** *55,* 5796.
25. Wang, H. Z.; Liu, X. H.; Yang, X. J.; Wang, X.; *Mat. Sci. Eng. A* **2001,** *311,* 180.
26. Song, M.; Mitsuishi, K.; Tanaka, M.; Takeguchi, M.; Shimojo, K.; Furuya, *Appl. Phys. A* **2004,** *80,* 1431.
27. Xie, G. Q.; Song, M.; Mitsuishi, K.; Furuya, K.; *J.; Nanosci. Nanotechnol.* **2005,** *5,* 615.
28. Xie, G. Q.; Song, M.; Mitsuishi, K.; Furuya, K.; *Physica E* **2005,** *29,* 564.
29. Xie, G. Q.; Song, M.; Mitsuishi, K.; Furuya, K.; *Jpn. J.; Appl. Phys.* **2005,** *44,* 5654.
30. Song, M.; Mitsuishi, K.; Furuya, K.; *Mater. Trans.* **2007,** *48,* 2551.
31. Song, K.; Mitsuishi, M.; Takeguchi, K.; Furuya, *Appl. Surf. Sci.* **2005,** *241,* 107.
32. Song, M.; Mitsuishi, K.; Furuya, K.; *Physica E* **2005,** *29,* 575.
33. Xie, G. Q.; Song, M.; Furuya, K.; Louzguine, D. V.; Inoue, A.; *Appl. Phys. Lett.* **2006,**88, 263120.
34. Xie, G. Q.; Song, M.; Mitsuishi, K.; Furuya, K.; *J. Mater. Sci.* **2006,** *41,* 2567.
35. Xie, G. Q.; Song, M.; Furuya, K.; *J. Mater. Sci.* **2006,** *41,* 4537.
36. Furuya, K.; Takeguchi, M.; Song, M.; Mitsuishi, K.; Tanaka, M.; *J. Phys. Conf. Ser.* **2008,** *126,* 012024.
37. Song, M.; Mitsuishi, K.; Furuya, K.; *Mater. Sci. Forum* **2005,** *475–479,* 4035.
38. Nagase, T.; Umakoshi, Y.; *Mater. Sci. Eng. A* **2003,** *352,* 251.
39. Gao, P. X.; Ding, Y.; Mai, W.; Hughes, W. L.; Lao, C.; Wang, Z. L.; *Science* **2005,** *309,* 1700.
40. Dick, K. A.; Deppert, K.; Larsson, M. W.; Martensson, T.; Seifert, W.; Wallenberg, L. R.; Samuelson, L. *Nature Mater.* **2004,** *3,* 380.
41. Li, M.; Schnablegger, H.; Mann, S.; *Nature* **1999,** *402,* 393.
42. Pal, U. J.; Ramirez, F. S.; Liu, H. B.; Medina, A.; Ascencio, J. A.; *Appl. Phys. A* **2004,** *79,* 79.

43. Joo, S. H.; Choi, S. J.; Oh, I.; Kwak, J.; Liu, Z.; Terasaki, O.; Ryoo, R.; *Nature* **2001,** *412,* 169.
44. A.; Ruiz, J.; Arbiol, A.; Cirera, A.; Cornet, Morante, J. R.; *Mater. Sci. Eng. C* **2005,** *19,* 105.
45. De Meijer, R. J.; Stapel, C.; Jones, D. G.; Roberts, P. D.; Rozendaal, A.; Macdonald, W. G.; Chen, K. Z.; Zhang, Z. K.; Cui, Z. L.; Zuo, D. H.; Yang, D. Z.; *Nanostruct. Mater.* **1997,** *8,* 205.
46. Wang, A.; Liu, J.; Lin, S.; Lin, T.; Mou, C.; *J.; Catal.* **2005,** *233,* 1486.
47. Liu, P.; Rodriguez, J. A.; Muckerman, J. T.; Hrbek, J.; *Surf. Sci.* **2003,** *530,* L313.

CHAPTER 6

NEW HORIZONS IN NANOTECHNOLOGY

ANAMIKA SINGH, RAJEEV SINGH, KOZLOV GEORGII
VLADIMIROVICH, and YANOVSKII YURII GRIGOR'EVICH

CONTENTS

6.1 NANOTHERAPEUTICS

6.1.1 INTRODUCTION

Nanotechnology expands itself as one of the major area of science [1, 2]. An important and exciting aspect of nanomedicine is the use of nanoparticle for target specific drug delivery systems and it allows a new innovative therapeutic approaches. Due to their small size, these drug delivery systems are promising tools in therapeutic approaches such as selective or targeted drug delivery towards a specific tissue or organ. It also enhances drug transport across biological barriers and intracellular drug delivery. Nanotherapeutic agents work successfully against traditional drugs, which exhibits very less bioavailability and are often associated with gastro-intestinal side effects. Nanotherapeutics improves the delivery of drugs that cannot normally be taken orally and it also improves the safety and efficacy of low molecular weight drugs. It also improves the stability and absorption of proteins that normally cannot be taken orally.

6.1.2 PROPERTIES OF NANOTHERAPEUTICS

A good therapeutic agent or nanoparticle should have ability to: (1) cross one or various biological membranes (e.g., mucosa, epithelium, endothelium) before (2) diffusing through the plasma membrane to (3) finally gain access to the appropriate organelle where the biological target is located.

Different methods of drug delivery are listed below:
1. Oral Drug Delivery
2. Injection Based Drug Delivery
3. Transdermal Drug Delivery
4. Bone Marrow Infusion
5. Control Release Systems
6. Targeted Drug Delivery:
   a. Therapeutical Monoclonal
   b. Antibodies
   c. Liposomes
   d. Microparticles
   e. Modified Blood Cells
   f. Nanoparticles
7. Implant Drug Delivery System

In this chapter, we will discuss about Nanoparticles and its mode of action.

6.1.3 NANOPARTICLES TYPES

6.1.3.1 METAL-BASED NANOPARTICLES

Metallic nanoparticles are either spherical metal or semiconductor particles with nanometer-sized diameters. Similar to nanoparticles "nanocrystal" (NC) or "nanocrystallite" term is used for nanoparticles made up of semiconductors. Generally semiconductor materials from the groups II-VI (CdS, CdSe, CdTe), III-V (InP, InAs) and IV-VI (PbS, PbSe, PbTe) are of particular interest. Metal Based nanoparticle are used for nanoscale transistors, biological sensors, and next generation photovoltaics [3].

6.1.3.2 LIPID-BASED NANOPARTICLES

Lipid based nanoparticles are widely used for drug targeting and drug delivery. Solid lipid nanoparticles (SLN) introduced in 1991 represent an alternative carrier system for drugs [4]. Nanoparticles made from solid lipids are attracting major attention as novel colloidal drug carrier for intravenous applications as they have been proposed as an alternative particulate carrier system. SLN are submicron colloidal carriers ranging from 50 to 1000 nm, which are composed of physiological lipid [5–7].
Lipid based system are many used because:
1. Lipids enhance oral bioavailability and reduce plasma profile variability.
2. Better characterization of lipoid excipients.
3. An improved ability to address the key issues of technology transfers and manufactures scale-up.

6.1.3.3 POLYMER-BASED NANOPARTICLES

Polymeric nanoparticles are nanoparticles, which are prepared from polymers. Polymeric nanoparticles forms (1) the micronization of a material into nanoparticles and (2) the stabilization of the resultant nanoparticles [8]. As for the micronization, one can start with either small monomers or a bulk polymer. The drug is dissolved, entrapped, encapsulated or attached to a nanoparticles and one can obtain different nanoparticles, nanospheres or nanocapsules according to methods of preparation [9]. Gums, Gelatin Sodium alginate Albumin are used for polymer based drug delivery. Polymeric nanoparticles are prepared by Cellulosics, Poly(2-

hydroxy ethyl methacrylate), Poly(N-vinyl pyrrolidone), Poly(vinyl alcohol), Poly(methyl methacrylate), Poly(acrylic acid), Polyacrylamide, Poly(ethylene-covinyl acetate) like polymeric materials. Polymer used in drug delivery must have following qualities like it should be chemically inert, nontoxic and free of leachable impurities [10].

6.1.3.4 BIOLOGICAL NANOPARTICLES

For proper interaction with molecular targets, a biological or molecular coating or layer is required which act as a bioinorganic interface, and it should be attached to the nanoparticle for proper interaction. These biological coating may be anti-bodies, biopolymers like collagen [11]. A layer of small molecules that make the nanoparticles biocompatible are also used as bioinorganic interface [12]. For detection and analysis these nanoparicles should be optically active and they should either fluoresces or change their color in different intensities of light.

6.1.3.5 NANOCARRIERS IN DRUG DELIVERY

Nanoparticles and Liposomes get absorbed by blood streams for its mode of action. For particular function theses nanoparticles must have specific structure and composition. These particles rapidly get cleared from blood stream by macrophages of reticuloendothelial system (RES) [13, 14].

Nanoparticles can be delivered to targets by Phagocytosis pathway and Non-phagocytosis pathway are described below.

6.1.3.5.1 PHAGOCYTOSIS PATHWAY

Phagocytosis occurs by profession phagocytes and by nonprofessional phagocytes. A professional phagocyte includes macrophages, monocytes, neutrophils and dendritic cell, while nonprofessional phagocytes are fibroblast, epithelial and endothelial cell [15, 16].

Phagocytosis takes place by:
1. Recognition of the opsonized particles in the blood stream.
2. Adhesion of particles to macrophages
3. Ingestion of the particle

Opsonization of the nanoparticles is a major step and it takes place before phagocytosis. In this process nanoparticles get tagged by a protein known as *opsonins*. Due to this tagging nanoparticles make a visible complex for macrophages, this whole process takes place in blood streams. Opsonins are immunoglobins (Ig) G, M and complement components like C3, C4 and C5 [17, 18].

These activated particles then get attached to macrophages and this interaction is similar to receptor-ligand interaction [15, 19]. The most important receptor is Fc receptor (FcR) and its complement receptor (CR). This receptor attachment is mediated by Rho family GTPase and surface extension occurs, which leads to formation of pseudopodia around nanoparticles. Small actin fibers get assembled in it and engulfment of nanoparticles takes place. [20, 21].

As actin is depolymerized from the phagosome, the newly denuded vacuole membrane becomes accessible to early endosomes [15, 22]. Through a series of fusion and fission events, the vacuolar membrane and its contents will mature, fusing with late endosomes and ultimately lysosomes to form a phagolysosome (Fig. 1 of Chapter 3). The rate of these events depends on the surface properties of the ingested particle, typically from half to several hours [15]. The phagolysosomes become acidified due to the vacuolar proton pump ATPase located in the membrane and acquire many enzymes, including esterases and cathepsins [16, 23].

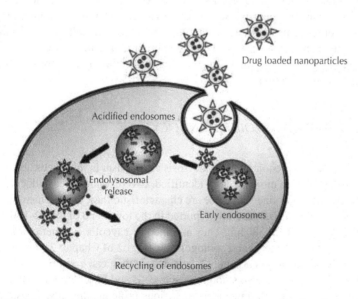

Drug loaded nanoparticles

Acidified endosomes

Endolysosomal release

Early endosomes

Recycling of endosomes

FIGURE 1 Nanoparticles enter the cell through receptor mediated endocytosis and get localized in the endolysosomal compartment.

6.1.3.5.2 NON-PHAGOCYTIC PATHWAY

This is also known as pinocytosis or cell drinking method. It is basically taking of fluids and solutes. Very small nanoparticles get entered in well and absorbed by pinocytosis as phagocytosis is restricted to size of particle. The process of Pinocytosis occurs in specialized cell. It may be clathrin mediated endocytosis, caveolae-mediated endocytosis, macropinocytosis and caveolae-independent endocytosis

6.1.3.5.2.1 CLATHRIN-MEDIATED ENDOCYTOSIS

Endocytosis via clathrin-coated pits, or clathrin-mediated endocytosis (CME), occurs constitutively in all mammalian cells, and fulfills crucial physiological roles, including nutrient uptake and intracellular communication. For most cell types, CME serves as the main mechanism of internalization for macromolecules and plasma membrane constituents. CME via specific receptor-ligand interaction is the best described mechanism, to the extent that it was previously referred to as "receptor-mediated endocytosis" (RME). However, it is now clear that alternative nonspecific endocytosis via clathrin-coated pits also exists (as well as receptor-mediated but clathrin-independent endocytosis). Notably, the CME, either receptor-dependent or independent, causes the endocytosed material to end up in degradative lysosomes. This has an important impact in the drug delivery field since the drug-loaded nanocarriers may be tailored in order to become metabolized into the lysosomes, thus releasing their drug content intracellularly as a consequence of lysosomal biodegradation.

6.1.3.5.2.2 CAVEOLAE-MEDIATED ENDOCYTOSIS

Although CME is the predominant endocytosis mechanism in most cells, alternative pathways have been more recently identified, caveolae-mediated endocytosis (CvME) being the major one. Caveolae are characteristic flask-shaped membrane invaginations, having a size generally reported in the lower end of the 50–100 nm range [24–27], typically 50–80 nm. They are lined by caveolin, a dimeric protein, and enriched with cholesterol and sphingolipids (Fig. 2 of Chapter 3). Caveolae are particularly abundant in endothelial cells, where they can constitute 10–20% of the cell surface [26], but also smooth muscle cells and fibroblasts. CvMEs are involved in endocytosis and trancytosis of various proteins; they also constitute a port of entry for viruses (typically the SV40 virus) [28] and receive increasing attention for drug delivery applications using nanocarriers.

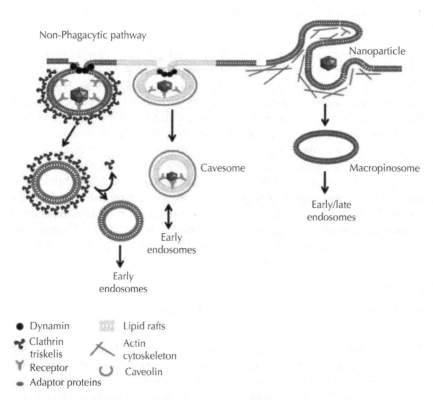

FIGURE 2 Diagrammamtic presentation of (A) Clathrin-mediated endocytosis (B) Caveolae-mediated endocytosis (C) Macropinocytosis

Unlike CME, CvME is a highly regulated process involving complex signaling, which may be driven by the cargo itself [25, 26]. After binding to the cell surface, particles move along the plasma membrane to caveolae invaginations, where they may be maintained through receptor-ligand interactions [25]. Fission of the caveolae from the membrane, mediated by the GTPase dynamin, then generates the cytosolic caveolar vesicle, which does not contain any enzymatic cocktail. Even this pathway is employed by many pathogens to escape degradation by lysosomal enzymes. The use of nanocarriers exploiting CvME may therefore be advantageous to by-pass the lysosomal degradation pathway when the carried drug (e.g., peptides, proteins, nucleic acids, etc.) is highly sensitive to enzymes. On the whole, the uptake kinetics of CvME is known to occur at a much slower rate than that of CME. Ligands known to be internalized by CvME include folic acid, albumin and cholesterol [25].

6.1.3.5.2.3 MACROPINOCYTOSIS

Macropinocytosis is another type of clathrin-independent endocytosis pathway [29], occurring in many cells, including macrophages [24]. It occurs via formation of actin-driven membrane protusions, similarly to phagocytosis. However, in this case, the protusions do not zipper up along the ligand-coated particle; instead, they collapse onto and fuse with the plasma membrane [26] (Fig. 2 of Chapter 3). This generates large endocytic vesicles, called macropinosomes, which sample the extracellular milieu and have a size generally bigger than 1 lm [26] (and sometimes as large as5 lm [24]). The intracellular fate of macropinosomes varies depending on the cell type, but in most cases, they acidify and shrink. They may eventually fuse with lysosomal compartments or recycle their content to the surface [24] (Fig. 2 of Chapter 3). Macropinosomes have not been reported to contain any specific coating, nor do they concentrate receptors. This endocytic pathway does not seem to display any selectivity, but is involved, among others, in the uptake of drug nanocarriers [29].

6.1.3.5.2.4 MAGNETIC NANOPARTICLES (MNPS)

Magnetic nanoparticles (MNPs) have many applications in different areas of biology and medicine. MNPs are used for hyperthermia, magnetic resonance imaging, immunoassay, purification of biologic fluids, cell and molecular separation, tissue engineering. The design of magnetically targeted nanosystems (MNSs) for a smart delivery of drugs to target cells is a promising direction of nanobiotechnology. They traditionally consist on one or more magnetic cores and biological or synthetic molecules, which serve as a basis for polyfunctional coatings on MNPs surface. The coatings of MNSs should meet several important requirements. They should be biocompatible, protect magnetic cores from influence of biological liquids, prevent MNSs agglomeration in dispersion, provide MNSs localization in biological targets and homogeneity of MNSs sizes. The coatings must be fixed on MNPs surface and contain therapeutic products (drugs or genes) and biovectors for recognition by biological systems (Fig. 3 of Chapter 3).

Proteins are promising materials for creation of coatings on MNPs for biology and medicine. When proteins are used as components of coatings it is of the first importance that they keep their functional activity. Protein binding on MNPs surface is a difficult scientific task. Traditionally bifunctional linkers (glutaraldehyde, carbodiimide) are used for protein cross-linking on the surface of MNPs and modification of coatings by therapeutic products and biovectors. Researchers modified MNPs surface with aminosilanes and performed protein molecules attachment using glutaraldehyde. In the issue bovine serum albumin (BSA) was adsorbed on MNPs surface in the presence of carbodiimide. These works revealed several disadvantages of this way of protein fixing which make it unpromizing. Some of them are clusters formation as a result of linking of protein molecules adsorbed on different MNPs, desorption of proteins from MNSs surface as a result of incomplete linking, uncontrollable linking of proteins in solution (Fig. 4 of Chapter 3). The creation of stable protein coatings with retention of native properties of molecules still is an important biomedical problem.

It is known that proteins can be chemically modified in the presence of free radicals with formation of cross-links. The goals of the work were to create stable protein coating on the surface of individual MNPs using a fundamentally novel approach based on the ability of proteins to form interchain covalent bonds under the action of free radicals and estimate activity of proteins in the coating.

6.1.4 CONCLUSION

Now-a-days n-number of drugs are available in market, so nanotherapeutics is particularly very important to minimize health hazards caused by drugs. The most important aspect of nanotherapeutics is that it should be noninvasive and target oriented and it should have very less or no side effects. In-spite of using nonmaterial one can create nanodevices for better functioning. As most of the peptides and proteins have short half-lives so frequent dosing or injections are required, which is not applicable for many situations by using nanotherapeutics one can solve these problems. Thus, nanotherapeutics is expanding day by day in drug delivery system.

6.2 NANOCOMPOSITES STRUCTURES

6.2.1 INTRODUCTION

The modern methods of experimental and theoretical analysis of polymer materials structure and properties allow not only to confirm earlier propounded hypotheses, but also to obtain principally new results. Let us consider some important problems of particulate-filled polymer nanocomposites, the solution of which allows advancing substantially in these materials properties understanding and prediction. Polymer nanocomposites multicomponentness (multiphaseness) requires their structural components quantitative characteristics determination. In this aspect interfacial regions play a particular role, since it has been shown earlier, that they are the same reinforcing element in elastomeric nanocomposites as nanofiller actually [30]. Therefore the knowledge of interfacial layer dimensional character-istics is necessary for quantitative determination of one of the most important parameters of polymer composites in general—their reinforcement degree [31, 32].

The aggregation of the initial nanofiller powder particles in more or less large particles aggregates always occurs in the course of technological process of preparation particulate-filled polymer composites in general [33] and elastomeric nanocomposites in particular [34]. The aggregation process tells on composites (nanocomposites) macroscopic properties [31–33]. For nanocomposites nano-filler aggregation process gains special significance, since its intensity can be the one, that nanofiller particles aggregates size exceeds 100 nm—the value, which is assumed (though conditionally enough [35]) as an upper dimensional limit for nanoparticle. In other words, the aggregation process can result to the situation when primordially supposed nanocomposite ceases to be one. Therefore at present several methods exist, which allow to suppress nanoparticles aggregation process [34, 36]. This also assumes the necessity of the nanoparticles aggregation process quantitative analysis.

It is well known [37, 38] that in particulate-filled elastomeric nanocomposites (rubbers) nanofiller particles form linear spatial structures ("chains"). At the same time in polymer composites, filled with disperse microparticles (microcomposites) particles (aggregates of particles) of filler form a fractal network, which defines polymer matrix structure (analog of fractal lattice in computer simulation) [33]. This results to different mechanisms of polymer matrix structure formation in micro and nanocomposites. If in the first filler particles (aggregates of particles) fractal network availability results to "disturbance" of polymer matrix structure, that is expressed in the increase of its fractal dimension d_f [33], then in case of polymer nanocomposites at nanofiller contents change the value d_f is not changed and equal to matrix polymer structure fractal dimension [39]. As it has been expected, the change of the composites of the indicated classes structure formation mechanism change defines their properties, in particular, reinforcement degree [40, 41]. Therefore, nanofiller structure factuality strict proof and its dimension determination are necessary.

As it is known [42, 43], the scale effects in general are often found at different materials mechanical properties study. The dependence of failure stress on grain size for metals (Holl-Petsch formula) [44] or of effective filling degree on filler particles size in case of polymer composites [45] are examples of such effect. The strong dependence of elasticity modulus on nanofiller particles diameter is observed for particulate-filled elastomeric nanocomposites [5]. Therefore, it is necessary to elucidate the physical grounds of nano- and micromechanical behavior scale effect for polymer nanocomposites.

At present a wide list of disperse material is known, which is able to strengthen elastomeric polymer materials [34]. These materials are very diverse on their surface chemical constitution, but particles small size is a common feature for them. On the basis of this observation the hypothesis was offered that any solid material would strengthen the rubber at the condition that it is in a very dispersed state and it could be dispersed in polymer matrix. Edwards [34] points out, that filler particles small size is necessary and, probably, the main requirement for reinforcement effect realization in rubbers. Using modern terminology, one can say, that for rubbers reinforcement the nanofiller particles, for which their aggregation process is suppressed as far as possible, would be the most effective ones [32, 41]. Therefore, the theoretical analysis of a nanofiller particles size influence on polymer nanocomposites reinforcement is necessary.

This chapter's purpose is to find the solution of the above-considered paramount problems with the help of modern experimental and theoretical techniques on the example of particulate-filled butadiene-stirene rubber.

6.2.2 EXPERIMENTAL PART

The made industrially butadiene-stirene rubber of mark SKS-30, which contains 7.0–12.3% *cis*- and 71.8–72.0% transbonds, with density of 920–930 kg/m³ was used as matrix polymer. This rubber is fully amorphous one.

Fullerene-containing mineral shungite of Zazhoginsk's deposit consists of ~30% globular amorphous metastable carbon and ~70% high-disperse silicate particles. Besides, industrially made technical carbon of mark № 220 was used as nanofiller. The technical carbon, nano- and microshugite particles average size makes up 20, 40 and 200 nm, respectively. The indicated filler content is equal to 37 mass %. Nano- and microdimensional disperse shungite particles were prepared from industrially output material by the original technology processing. The size and polydispersity analysis of the received in milling process shungite particles was monitored with the aid of analytical disk centrifuge (CPS Instruments, Inc., USA), allowing to determine with high precision size and distribution by the sizes within the range from 2 nm up to 50 mcm.

Nanostructure was studied on atomic-forced microscopes Nano-DST (Pacific Nanotechnology, USA) and Easy Scan DFM (Nanosurf, Switzerland) by semicontact method in the force modulation regime. Atomic-force microscopy results were processed with the help of specialized software package SPIP (Scanning Probe Image Processor, Denmark). SPIP is a powerful programs package for processing of images, obtained on SPM, AFM, STM, scanning electron microscopes, transmission electron microscopes, interferometers, confocal microscopes, profilometers, optical microscopes and so on. The given package possesses the whole functions number, which are necessary at images precise analysis, in a number of which the following ones are included:

- the possibility of three-dimensional reflecting objects obtaining, distortions automatized leveling, including Z-error mistakes removal for examination of separate elements and so on;
- quantitative analysis of particles or grains, more than 40 parameters can be calculated for each found particle or pore: area, perimeter, mean diameter, the ratio of linear sizes of grain width to its height, distance between grains, coordinates of grain center of mass a.a. can be presented in a diagram form or in a histogram form.

The tests on elastomeric nanocomposites nanomechanical properties were carried out by a nanointentation method [46] on apparatus Nano Test 600 (Micro Materials, Great Britain) in loads wide range from 0.01 mN up to 2.0 mN. Sample indentation was conducted in 10 points with interval of 30 mcm. The load was increased with constant rate up to the greatest given load reaching (for the rate 0.05 mN/s-1 mN). The indentation rate was changed in conformity with the greatest load value counting, that loading cycle should take 20 s. The unloading was conducted with the same rate as loading. In the given experiment the "Berkovich indentor" was used with the angle at the top of 65.3° and rounding radius of 200 nm. Indentations were carried out in the checked load regime with preload of 0.001 mN.

For elasticity modulus calculation the obtained in the experiment by nanoindentation course dependences of load on indentation depth (strain) in ten points for each sample at loads of 0.01, 0.02, 0.03, 0.05, 0.10, 0.50, 1.0 and 2.0 mN were processed according to Oliver-Pharr method [47].

6.2.3 RESULTS AND DISCUSSION

In Fig. 1, the obtained according to the original methodics results of elasticity moduli calculation for nanocomposite butadiene-styrene rubber/nanoshungite components (matrix, nanofiller particle and interfacial layers), received in interpolation process of nanoindentation data, are presented. The processed in polymer nanocomposite SPIP image with shungite nanoparticles allows experimental determination of interfacial layer thickness l_{if}, which is presented in Fig. 1 as steps on elastomeric matrix-nanofiller boundary. The measurements of 34 such steps (interfacial layers) width on the processed in SPIP images of interfacial layer various section gave the mean experimental value l_{if}=8.7 nm. Besides, nanoindentation results (Fig. 1, figures on the right) showed, that interfacial layers elasticity modulus was only by 23–45% lower than nanofiller elasticity modulus, but it was higher than the corresponding parameter of polymer matrix in 6.0–8.5 times. These experimental data confirm, that for the studied nanocomposite interfacial layer is a reinforcing element to the same extent, as nanofiller actually [30, 32, 41].

Let us fulfill further the value l_{if} theoretical estimation according to the two methods and compare these results with the ones obtained experimentally. The first method simulates interfacial layer in polymer composites as a result of interaction of two fractals—polymer matrix and nanofiller surface [48, 49]. In this case there is a sole linear scale l, which defines these fractals interpenetration distance [50]. Since nanofiller elasticity modulus is essentially higher, than the corresponding parameter for rubber (in the considered case—in 11 times, see Fig. 1), then the indicated interaction reduces to nanofiller indentation in polymer matrix and then $l=l_{if}$. In this case it can be written [50]:

$$ l_{if} \approx a\left(\frac{R_p}{a}\right)^{2(d-d_{surf})/d}, \tag{1} $$

where a is a lower linear scale of fractal behavior, which is accepted for polymers as equal to statistical segment length l_{st} [51], R_p is a nanofiller particle (more precisely, particles aggregates) radius, which for nanoshungite is equal to ~84 nm [52], d is dimension of Euclidean space, in which fractal is considered (it is obvious, that in our case $d=3$), d_{surf} is fractal dimension of nanofiller particles aggregate surface.

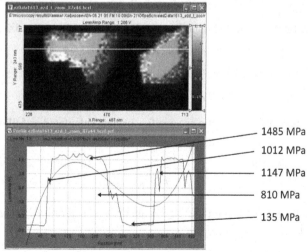

FIGURE 1 The processed in SPIP image of nanocomposite butadiene-styrene rubber/nanoshungite, obtained by force modulation method, and mechanical characteristics of structural components according to the data of nanoindentation (strain 150 nm).

The value l_{st} is determined as follows [53]:

$$l_{st} = l_0 C_\infty,$$ (2)

where l_0 is the main chain skeletal bond length, which is equal to 0.154 nm for both blocks of butadiene-stirene rubber [54], C_∞ is characteristic ratio, which is a polymer chain statistical flexibility indicator [55], and is determined with the help of the equation [51]:

$$T_g = 129 \left(\frac{S}{C_\infty} \right)^{1/2},$$ (3)

where T_g is glass transition temperature, equal to 217 K for butadiene-stirene rubber [32], S is macromolecule cross-sectional area, determined for the mentioned rubber according to the additivity rule from the following considerations. As it is known [56], the macromolecule diameter quadrate values are equal: for polybutadiene—20.7 Å^2 and for polystirene—69.8 Å^2. Having calculated cross-sectional area of macromolecule, simulated as a cylinder, for the indicated polymers according to the known geometrical formulas, let us obtain 16.2 and 54.8 Å^2, respectively. Further, accepting as S the average value of the adduced above areas, let us obtain for butadiene-stirene rubber S=35.5 Å^2. Then according to the Eq. (3) at the indicated values T_g and S let us obtain C_∞=12.5 and according to the Eq. (2)—l_{st}=1.932 nm.

The fractal dimension of nanofiller surface d_{surf} was determined with the help of the equation [3]:

$$S_u = 410 R_p^{d_{surf}-d},$$ (4)

where S_u is nanoshungite particles specific surface, calculated as follows [28]:

$$S_u = \frac{3}{\rho_n R_p},$$ (5)

where ρ_n is the nanofiller particles aggregate density, determined according to the formula [3]:

$$\rho_n = 0.188 \left(R_p \right)^{1/3}$$ (6)

The calculation according to the Eqs. (4)–(6) gives d_{surf}=2.44. Further, using the calculated by the indicated mode parameters, let us obtain from the Eq. (1) the theoretical value of interfacial layer thickness l_{if}^T=7.8 nm. This value is close enough to the obtained one experimentally (their discrepancy makes up ~ 10%).

The second method of value l_{if}^T estimation consists in using of the two following equations [32, 58]:

$$\varphi_{if} = \varphi_n \left(d_{surf} - 2 \right) \tag{7}$$

and

$$\varphi_{if} = \varphi_n \left[\left(\frac{R_p + l_{if}^T}{R_p} \right)^3 - 1 \right], \tag{8}$$

where φ_{if} and φ_n are relative volume fractions of interfacial regions and nanofiller, accordingly.

The combination of the indicated equations allows to receive the following formula for l_{if}^T calculation:

$$l_{if}^T = R_p \left[\left(d_{surf} - 1 \right)^{1/3} - 1 \right] \tag{9}$$

The calculation according to the Eq. (9) gives for the considered nanocomposite l_{if}^T=10.8 nm, that also corresponds well enough to the experiment (in this case discrepancy between l_{if} and l_{if}^T makes up ~19%).

Let us note in conclusion the important experimental observation, which follows from the processed by program SPIP results of the studied nanocomposite surface scan (Fig. 1). As one can see, at one nanoshungite particle surface from one to three (in average—two) steps can be observed, structurally identified as interfacial layers. It is significant that these steps width (or l_{if}) is approximately equal to the first (the closest to nanoparticle surface) step width. Therefore, the indicated observation supposes, that in elastomeric nanocomposites at average two interfacial layers are formed: the first—at the expense of nanofiller particle surface with elastomeric matrix interaction, as a result of which molecular mobility in this layer is frozen and its state is glassy-like one, and the second—at the expense of glassy interfacial layer with elastomeric polymer matrix interaction. The most important question from the practical point of view, whether one interfacial layer or both serve as nanocomposite reinforcing element. Let us fulfill the following quantitative estimation for this question solution. The reinforcement degree (E_n/E_m) of polymer nanocomposites is given by the equation [32]:

$$\frac{E_n}{E_m} = 1 + 11\left(\varphi_n + \varphi_{if}\right)^{1.7},\qquad(10)$$

where E_n and E_m are elasticity moduli of nanocomposite and matrix polymer, accordingly (E_m=1.82 MPa [32]).

According to the Eq. (7) the sum ($\varphi_n + \varphi_{if}$) is equal to:

$$\varphi_n + \varphi_{if} = \varphi_n\left(d_{surf} - 1\right),\qquad(11)$$

if one interfacial layer (the closest to nanoshungite surface) is a reinforcing element and

$$\varphi_n + 2\varphi_{if} = \varphi_n\left(2d_{surf} - 3\right),\qquad(12)$$

if both interfacial layers are a reinforcing element.

In its turn, the value φ_n is determined according to the equation [59]:

$$\varphi_n = \frac{W_n}{\rho_n},\qquad(13)$$

where W_n is nanofiller mass content, ρ_n is its density, determined according to the Eq. (6).

The calculation according to the Eqs. (11) and (12) gave the following E_n/E_m values: 4.60 and 6.65, respectively. Since the experimental value E_n/E_m=6.10 is closer to the value, calculated according to the Eq. (12), then this means that both interfacial layers are a reinforcing element for the studied nanocomposites. Therefore the coefficient 2 should be introduced in the equations for value l_{if} determination (for example, in the Eq. (1)) in case of nanocomposites with elastomeric matrix. Let us remind, that the Eq. (1) in its initial form was obtained as a relationship with proportionality sign, that is, without fixed proportionality coefficient [50].

Thus, the used above nanoscopic methodics allow to estimate both interfacial layer structural special features in polymer nanocomposites and its sizes and properties. For the first time it has been shown, that in elastomeric particulate-filled nanocomposites two consecutive interfacial layers are formed, which are a reinforcing element for the indicated nanocomposites. The proposed theoretical methodics of interfacial layer thickness estimation, elaborated within the frameworks of fractal analysis, give well enough correspondence to the experiment.

For theoretical treatment of nanofiller particles aggregate growth processes and final sizes traditional irreversible aggregation models are inapplicable, since it is obvious, that in nanocomposites aggregates a large number of simultaneous growth takes place. Therefore the model of multiple growth, offered in Ref. [35], was used for nanofiller aggregation description.

In Fig. 2, the images of the studied nanocomposites, obtained in the force modulation regime, and corresponding to them nanoparticles aggregates fractal dimension d_f distributions are adduced. As it follows from the adduced values d_f^{ag} (d_f^{ag} =2.40–2.48), nanofiller particles aggregates in the studied nanocomposites are formed by a mechanism particle-cluster (P-Cl), that is, they are Witten-Sander clusters [61]. The variant A, was chosen which according to mobile particles are added to the lattice, consisting of a large number of "seeds" with density of c_0 at simulation beginning [60]. Such model generates the structures, which have fractal geometry on length short scales with value $d_f \approx 2.5$ (see Fig. 2) and homogeneous structure on length large scales. A relatively high particles concentration c is required in the model for uninterrupted network formation [60].

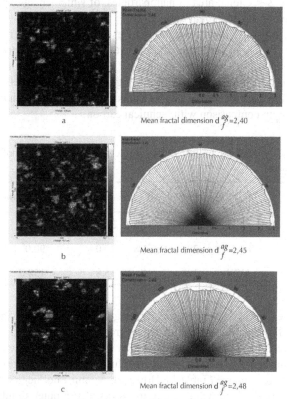

a

Mean fractal dimension d $_f^{ag}$ =2,40

b

Mean fractal dimension d $_f^{ag}$ =2,45

c

Mean fractal dimension d $_f^{ag}$ =2,48

FIGURE 2   The images, obtained on atomic-force microscope in the force modulation regime, for nanocomposites, filled with technical carbon (a), nanoshungite (b), microshungite (c) and corresponding to them fractal dimensions d_f^{ag}.

In case of "seeds" high concentration c_0 for the variant A the following relationship was obtained [60]:

$$R_{max}^{d_f^{ag}} = N = c / c_0 ,$$ (14)

where R_{max} is nanoparticles cluster (aggregate) greatest radius, N is nanoparticles number per one aggregate, c is nanoparticles concentration, c_0 is "seeds" number, which is equal to nanoparticles clusters (aggregates) number.

The value N can be estimated according to the following equation [8]:

$$2R_{max} = \left(\frac{S_n N}{\pi \eta} \right)^{1/2}$$ (15)

where S_n is cross-sectional area of nanoparticles, of which an aggregate consists, η is a packing coefficient, equal to 0.74 [57].

The experimentally obtained nanoparticles aggregate diameter $2R_{ag}$ was accepted as $2R_{max}$ (Table 1) and the value S_n was also calculated according to the experimental values of nanoparticles radius r_n (Table 1). In Table 1 the values N for the studied nanofillers, obtained according to the indicated method, were adduced. It is significant that the value N is a maximum one for nanoshungite despite larger values r_n in comparison with technical carbon.

TABLE 1 The parameters of irreversible aggregation model of nanofiller particles aggregates growth.

Nanofiller	R_{ag}, nm	r_n, nm	N	R_{max}^T, nm	R_{ag}^T, nm	R_c, nm
Technical carbon	34.6	10	35.4	34.7	34.7	33.9
Nanoshungite	83.6	20	51.8	45.0	90.0	71.0
Microshungite	117.1	100	4.1	15.8	158.0	255.0

Further the Eq. (14) allows estimating the greatest radius R_{max}^T of nanoparticles aggregate within the frameworks of the aggregation model [60]. These values R_{max}^T are adduced in Table 1, from which their reduction in a sequence of technical carbon-nanoshungite-microshungite, that fully contradicts to the experimental data, that is, to R_{ag} change (Table 1). However, we must not neglect the fact that the Eq. (14) was obtained within the frameworks of computer simulation, where the initial aggregating particles sizes are the same in all cases [60]. For real nanocomposites the values r_n can be distinguished essentially (Table 1). It is expected, that the value R_{ag} or R_{max}^T will be the higher, the larger is the radius of nanoparticles, forming aggregate, is that is, r_n. Then theoretical value of nanofiller particles cluster (aggregate) radius R_{ag}^T can be determined as follows:

$$R_{ag}^T = k_n r_n N^{1/d_f^{ag}},$$ (16)

where k_n is proportionality coefficient, in the present work accepted empirically equal to 0.9.

The comparison of experimental R_{ag} and calculated according to the Eq. (16) R_{ag}^T values of the studied nanofillers particles aggregates radius shows their good correspondence (the average discrepancy of R_{ag} and R_{ag}^T makes up 11.4%). Therefore, the theoretical model [60] gives a good correspondence to the experiment only in case of consideration of aggregating particles real characteristics and, in the first place, their size.

Let us consider two more important aspects of nanofiller particles aggregation within the frameworks of the model [60]. Some features of the indicated process are defined by nanoparticles diffusion at nanocomposites processing. Specifically, length scale, connected with diffusible nanoparticle, is correlation length ξ of diffusion. By definition, the growth phenomena in sites, remote more than ξ, are statistically independent. Such definition allows connecting the value ξ with the mean distance between nanofiller particles aggregates L_n. The value ξ can be calculated according to the equation [60]:

$$\xi^2 \approx \tilde{n}^{-1} R_{ag}^{d_f^{ag}-d+2},$$ (17)

where c is nanoparticles concentration, which should be accepted equal to nanofiller volume contents φ_n, which is calculated according to the Eqs. (6) and (13).

The values r_n and R_{ag} were obtained experimentally (see histogram of Fig. 3). In Fig. 4 the relation between L_n and ξ is adduced, which, as it is expected, proves to be linear and passing through coordinates origin. This means, that the distance between nanofiller particles aggregates is limited by mean displacement of statistical walks, by which nanoparticles are simulated. The relationship between L_n and ξ can be expressed analytically as follows:

$$L_n \approx 9.6\xi, \text{ nm.}$$ (18)

The second important aspect of the model [60] in reference to nanofiller particles aggregation simulation is a finite nonzero initial particles concentration c or φ_n effect, which takes place in any real systems. This effect is realized at the condition $\xi \approx R_{ag}$, that occurs at the critical value $R_{ag}(R_c)$, determined according to the relationship [60]:

$$c \sim R_c^{d_f^{ag}-d} \tag{19}$$

The Eq. (19) right side represents cluster (particles aggregate) mean density. This equation establishes that fractal growth continues only, until cluster density reduces up to medium density, in which it grows. The calculated according to the Eq. (19) values R_c for the considered nanoparticles are adduced in Table 1, from which follows, that they give reasonable correspondence with this parameter experimental values R_{ag} (the average discrepancy of R_c and R_{ag} makes up 24%).

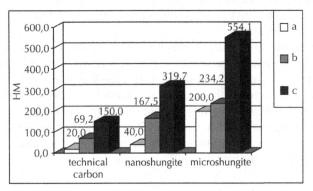

FIGURE 3   The initial particles diameter (a), their aggregates size in nanocomposite (b) and distance between nanoparticles aggregates (c) for nanocomposites, filled with technical carbon, nano- and microshungite.

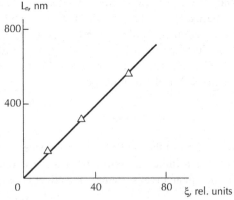

FIGURE 4   The relation between diffusion correlation length ξ and distance between nanoparticles aggregates L_n for considered nanocomposites.

Since the treatment [60] was obtained within the frameworks of a more general model of diffusion-limited aggregation, then its correspondence to the experimental data indicated unequivocally, that aggregation processes in these systems were controlled by diffusion. Therefore let us consider briefly nanofiller particles diffusion. Statistical walkers diffusion constant ζ can be determined with the aid of the relationship [60]:

$$\xi \approx \left(\zeta t\right)^{1/2},$$ (20)

where t is walk duration.

The Eq. (20) supposes (at t=const) ζ increase in a number technical carbon-nanoshungite-microshungite as 196–1069–3434 relative units, that is, diffusion intensification at diffusible particles size growth. At the same time diffusivity D for these particles can be described by the well-known Einstein's relationship [62]:

$$D = \frac{kT}{6\pi\eta r_n \alpha},$$ (21)

where k is Boltzmann constant, T is temperature, η is medium viscosity, α is numerical coefficient, which further is accepted equal to 1.

In its turn, the value η can be estimated according to the equation [63]:

$$\frac{\eta}{\eta_0} = 1 + \frac{2.5\varphi_n}{1-\varphi_n},$$ (22)

where η_0 and η are initial polymer and its mixture with nanofiller viscosity, accordingly.

The calculation according to the Eqs. (21) and (22) shows, that within the indicated above nanofillers number the value D changes as 1.32–1.14–0.44 relative units, that is, reduces in three times, that was expected. This apparent contradiction is due to the choice of the condition t=const (where t is nanocomposite production duration) in the Eq. (20). In real conditions the value t is restricted by nanoparticle contact with growing aggregate and then instead of t the value t/c_0 should be used, where c_0 is the seeds concentration, determined according to the Eq. (14). In this case the value ζ for the indicated nanofillers changes as 0.288–0.118–0.086, that is, it reduces in 3.3 times, that corresponds fully to the calculation according to the Einstein's relationship (the Eq. (21)). This means, that nanoparticles diffusion in polymer matrix obeys classical laws of Newtonian rheology [62].

Thus, the disperse nanofiller particles aggregation in elastomeric matrix can be described theoretically within the frameworks of a modified model of irreversible aggregation particle-cluster. The obligatory consideration of nanofiller initial particles size is a feature of the indicated model application to real systems description. The indicated particles diffusion in polymer matrix obeys classical laws of Newtonian liquids hydrodynamics. The offered approach allows predicting nanoparticles aggregates final parameters as a function of the initial particles size, their contents and other factors number.

At present there are several methods of filler structure (distribution) determination in polymer matrix, both experimental [39, 64] and theoretical [33]. All the indicated methods describe this distribution by fractal dimension D_n of filler particles network. However, correct determination of any object fractal (Hausdorff) dimension includes three obligatory conditions. The first from them is the indicated above determination of fractal dimension numerical magnitude, which should not be equal to object topological dimension. As it is known [65], any real (physical) fractal possesses fractal properties within a certain scales range. Therefore the second condition is the evidence of object self-similarity in this scales range [66]. And at last, the third condition is the correct choice of measurement scales range itself. As it has been shown in Refs. [67, 68], the minimum range should exceed at any rate one self-similarity iteration.

The first method of dimension D_n experimental determination uses the following fractal relationship [69, 70]:

$$D_n = \frac{\ln N}{\ln \rho},$$ (23)

where N is a number of particles with size ρ.

Particles sizes were established on the basis of atomic-power microscopy data (see Fig. 2). For each from the three studied nanocomposites no less than 200 particles were measured, the sizes of which were united into 10 groups and mean values N and ρ were obtained. The dependences $N(\rho)$ in double logarithmic coordinates was plotted, which proved to be linear and the values D_n were calculated according to their slope (see Fig. 5). It is obvious, that at such approach fractal dimension D_n is determined in two-dimensional Euclidean space, whereas real nanocomposite should be considered in three-dimensional Euclidean space. The following relationship can be used for D_n recalculation for the case of three-dimensional space [71]:

$$D3 = \frac{d + D2 \pm \left[(d - D2)^2 - 2 \right]^{1/2}}{2},$$ (24)

where $D3$ and $D2$ are corresponding fractal dimensions in three- and two-dimensional Euclidean spaces, $d=3$.

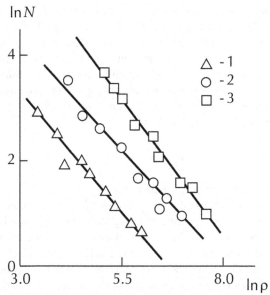

FIGURE 5 The dependences of nanofiller particles number N on their size ρ for nanocomposites BSR/TC (1), BSR/nanoshungite (2) and BSR/microshungite (3).

The calculated according to the indicated method dimensions D_n are adduced in Table 2. As it follows from the data of Table 2, the values D_n for the studied nanocomposites are varied within the range of 1.10–1.36, that is, they characterize more or less branched linear formations ("chains") of nanofiller particles (aggregates of particles) in elastomeric nanocomposite structure. Let us remind that for particulate-filled composites polyhydroxiether/graphite the value D_n changes within the range of ~2.30–2.80 [33, 39], that is, for these materials filler particles network is a bulk object, but not a linear one [65].

TABLE 2 The dimensions of nanofiller particles (aggregates of particles) structure in elastomeric nanocomposites.

Nanocomposite	D_n, the Eq. (23)	D_n, the Eq. (25)	d_0	d_{surf}	φ_n	D_n, the Eq. (29)
BSR/TC	1.19	1.17	2.86	2.64	0.48	1.11
BSR/nanoshungite	1.10	1.10	2.81	2.56	0.36	0.78
BSR/microshungite	1.36	1.39	2.41	2.39	0.32	1.47

Another method of D_n experimental determination uses the so-called "quadrates method" [72]. Its essence consists in the following. On the enlarged nanocomposite microphotograph (see Fig. 2) a net of quadrates with quadrate side size α_i, changing from 4.5 up to 24 mm with constant ratio $\alpha_{i+1}/\alpha_i=1.5$, is applied and then quadrates number N_i, in to which nanofiller particles hit (fully or partly), is counted up. Five arbitrary net positions concerning microphotograph were chosen for each measurement. If nanofiller particles network is a fractal, then the following relationship should be fulfilled [72]:

$$ N_i \sim S_i^{-D_n/2} , \tag{25} $$

where S_i is quadrate area, which is equal to α_i^2.

In Fig. 6 the dependences of N_i on S_i in double logarithmic coordinates for the three studied nanocomposites, corresponding to the Eq. (25), is adduced. As one can see, these dependences are linear, that allows determining the value D_n from their slope. The determined according to the Eq. (25) values D_n are also adduced in Table 2, from which a good correspondence of dimensions D_n, obtained by the two described above methods, follows (their average discrepancy makes up 2.1% after these dimensions recalculation for three-dimensional space according to the Eq. (24)).

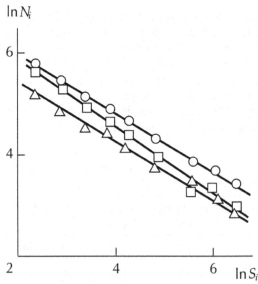

FIGURE 6 The dependences of covering quadrates number N_i on their area S_i, corresponding to the relationship (25), in double logarithmic coordinates for nanocomposites on the basis of BSR. The designations are the same, that in Fig. 5.

As it has been shown in Ref. [73], the usage for self-similar fractal objects at the Eq. (25) the condition should be fulfilled:

$$N_i - N_{i-1} \sim S_i^{-D_n} \qquad (26)$$

In Fig. 7, the dependence, corresponding to the Eq. (26), for the three studied elastomeric nanocomposites is adduced. As one can see, this dependence is linear, passes through coordinates origin, that according to the Eq. (26) is confirmed by nanofiller particles (aggregates of particles) "chains" self-similarity within the selected α_i range. It is obvious, that this self-similarity will be a statistical one [73]. Let us note, that the points, corresponding to α_i=16 mm for nanocomposites butadiene-stirene rubber/technical carbon (BSR/TC) and butadiene-stirene rubber/microshungite (BSR/microshungite), do not correspond to a common straight line. Accounting for electron microphotographs of Fig. 2 enlargement this gives the self-similarity range for nanofiller "chains" of 464–1472 nm. For nanocomposite butadiene-stirene rubber/nanoshungite (BSR/nanoshungite), which has no points deviating from a straight line of Fig. 7, α_i range makes up 311–1510 nm, that corresponds well enough to the indicated above self-similarity range.

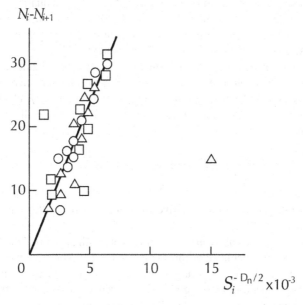

FIGURE 7 The dependences of $(N_i$-$N_{i+1})$ on the value $S_i^{-D_n/2}$, corresponding to the relationship (26), for nanocomposites on the basis of BSR. The designations are the same, that in Fig. 5.

In Refs. [67, 68] it has been shown, that measurement scales S_i minimum range should contain at least one self-similarity iteration. In this case the condition for ratio of maximum S_{max} and minimum S_{min} areas of covering quadrates should be fulfilled [68]:

$$\frac{S_{max}}{S_{min}} > 2^{2/D_n} \qquad (27)$$

Hence, accounting for the defined above restriction let us obtain $S_{max}/S_{min} = 121/20.25 = 5.975$, that is larger than values $2^{2/D_n}$ for the studied nanocomposites, which are equal to 2.71–3.52. This means, that measurement scales range is chosen correctly.

The self-similarity iterations number μ can be estimated from the inequality [68, 69]:

$$\left(\frac{S_{max}}{S_{min}}\right)^{D_n/2} > 2^{\mu} \qquad (28)$$

Using the above-indicated values of the included in the inequality Eq. (28) parameters, $\mu = 1.42–1.75$ is obtained for the studied nanocomposites, that is, in our experiment conditions self-similarity iterations number is larger than unity, that again confirms correctness of the value D_n estimation [64].

And let us consider in conclusion the physical grounds of smaller values D_n for elastomeric nanocomposites in comparison with polymer microcomposites, that is, the causes of nanofiller particles (aggregates of particles) "chains" formation in the first ones. The value D_n can be determined theoretically according to the equation [33]:

$$\varphi_{if} = \frac{D_n + 2.55d_0 - 7.10}{4.18}, \qquad (29)$$

where φ_{if} is interfacial regions relative fraction, d_0 is nanofiller initial particles surface dimension.

The dimension d_0 estimation can be carried out with the help of the Eq. (4) and the value φ_{if} can be calculated according to the Eq. (7). The results of dimension D_n theoretical calculation according to the Eq. (29) are adduced in table 2, from which a theory and experiment good correspondence follows. The Eq. (29) indicates unequivocally to the cause of a filler in nano- and microcomposites different behavior. The high (close to 3, see table 2) values d_0 for nanoparticles and relatively small ($d_0 = 2.17$ for graphite [33]) values d_0 for microparticles at comparable values φ_{if} is such cause for composites of the indicated classes [33, 34].

Hence, the stated above results have shown, that nanofiller particles (aggregates of particles) "chains" in elastomeric nanocomposites are physical fractal within self-similarity (and, hence, factuality [70–73]) range of ~500–1450 nm. In this range their dimension D_n can be estimated according to the Eqs. (23), (25) and (29). The cited examples demonstrate the necessity of the measurement scales range correct choice. As it has been noted earlier [74], the linearity of the plots, corresponding to the Eqs. (23) and (25), and D_n nonintegral value do not guarantee object self-similarity (and, hence, factuality). The nanofiller particles (aggregates of particles) structure low dimensions are due to the initial nanofiller particles surface high fractal dimension.

In Fig. 8, the histogram is adduced, which shows elasticity modulus E change, obtained in nanoindentation tests, as a function of load on indenter P or nanoindentation depth h. Since for all the three considered nanocomposites the dependences $E(P)$ or $E(h)$ are identical qualitatively, then further the dependence $E(h)$ for nanocomposite BSR/TC was chosen, which reflects the indicated scale effect quantitative aspect in the most clearest way.

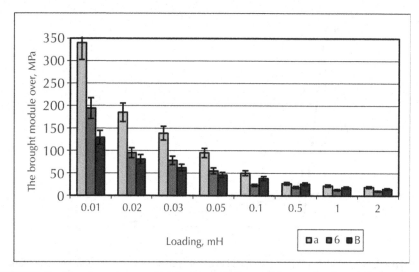

FIGURE 8 The dependences of reduced elasticity modulus on load on indentor for nanocomposites on the basis of butadiene-styrene rubber, filled with technical carbon (a), micro- (b) and nanoshungite (c).

In Fig. 9 the dependence of E on h_{pl} (see Fig. 10) is adduced, which breaks down into two linear parts. Such dependences elasticity modulus—strain is typical for polymer materials in general and is due to intermolecular bonds anharmonicity

[75]. In Ref. [76], it has been shown that the dependence $E(h_{pl})$ first part at $h_{pl} \leq 500$ nm is not connected with relaxation processes and has a purely elastic origin. The elasticity modulus E on this part changes in proportion to h_{pl} as:

$$E = E_0 + B_0 h_{pl},\qquad(30)$$

where E_0 is "initial" modulus, that is, modulus, extrapolated to $h_{pl}=0$, and the coefficient B_0 is a combination of the first and second kind elastic constants. In the considered case $B_0 < 0$. Further Grüneisen parameter γ_L, characterizing intermolecular bonds anharmonicity level, can be determined [76]:

$$\gamma_L \approx -\frac{1}{6} - \frac{1}{2}\frac{B_0}{E_0}\frac{1}{(1-2\nu)},\qquad(31)$$

where ν is Poisson ratio, accepted for elastomeric materials equal to ~0.475 [65].

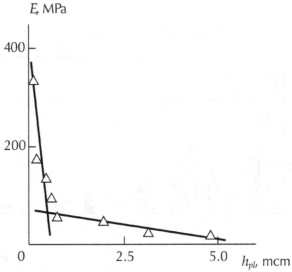

FIGURE 9 The dependence of reduced elasticity modulus E, obtained in nanoindentation experiment, on plastic strain h_{pl} for nanocomposites BSR/TC.

Calculation according to the Eq. (31) has given the following values γ_L: 13.6 for the first part and 1.50—for the second one. Let us note the first from γ_L adduced values is typical for intermolecular bonds, whereas the second value γ_L is much closer to the corresponding value of Grüneisen parameter G for intrachain modes [75].

Poisson's ratio v can be estimated by γ_L (or G) known values according to the formula [75]:

$$\gamma_L = 0.7\left(\frac{1+v}{1-2v}\right) \tag{32}$$

The estimations according to the Eq. (32) gave: for the dependence $E(h_{pl})$ first part $v=0.462$, for the second one—$v=0.216$. If for the first part the value v is close to Poisson's ratio magnitude for nonfilled rubber [65], then in the second part case the additional estimation is required. As it is known [76, 77], a polymer composites (nanocomposites) Poisson's ratio value v_n can be estimated according to the equation:

$$\frac{1}{v_n} = \frac{\varphi_n}{v_{TC}} + \frac{1-\varphi_n}{v_m}, \tag{33}$$

where φ_n is nanofiller volume fraction, v_{TC} and v_m are nanofiller (technical carbon) and polymer matrix Poisson's ratio, respectively.

The value v_m is accepted equal to 0.475 [65] and the magnitude v_{TC} is estimated as follows [78]. As it is known [79], the nanoparticles TC aggregates fractal dimension d_f^{ag} value is equal to 2.40 and then the value v_{TC} can be determined according to the equation [79]:

$$d_f^{ag} = (d-1)(1+v_{TC}) \tag{34}$$

According to the Eq. (34) $v_{TC}=0.20$ and calculation v_n according to the Eq. (33) gives the value 0.283, that is close enough to the value $v=0.216$ according to the Eq. (32) estimation. The obtained by the indicated methods values v and v_n comparison demonstrates, that in the dependence $E(h_{pl})$ ($h_{pl}<0.5$ mcm) the first part in nanoindentation tests only rubber-like polymer matrix ($v=v_m \approx 0.475$) is included and in this dependence the second part—the entire nanocomposite as homogeneous system [80]—$v=v_n \approx 0.22$.

Let us consider further E reduction at h_{pl} growth (Fig. 9) within the frameworks of density fluctuation theory, which value ψ can be estimated as follows [51]:

$$\psi = \frac{\rho_n kT}{K_T}, \tag{35}$$

where ρ_n is nanocomposite density, k is Boltzmann constant, T is testing temperature, K_T is isothermal modulus of dilatation, connected with Young's modulus E by the relationship [75]:

$$K_T = \frac{E}{3(1-\nu)} \tag{36}$$

In Fig. 10, the scheme of volume of the deformed at nanoindentation material V_{def} calculation in case of Berkovich indentor using is adduced and in Fig. 11 the dependence $\psi(V_{def})$ in logarithmic coordinates was shown. As it follows from the data of this Figure, the density fluctuation growth is observed at the deformed material volume increase. The plot $\psi(\ln V_{def})$ extrapolation to $\psi=0$ gives $\ln V_{def} \approx 13$ or $V_{def}(V_{def}^{cr})=4.42\times10^5$ nm^3. Having determined the linear scale l_{cr} of transition to $\psi=0$ as $(V_{def}^{cr})^{1/3}$, let us obtain $l_{cr}=75.9$ nm, that is close to nanosystems dimensional range upper boundary (as it was noted above, conditional enough [35]), which is equal to 100 nm. Thus, the stated above results suppose, that nanosystems are such systems, in which density fluctuations are absent, always taking place in microsystems.

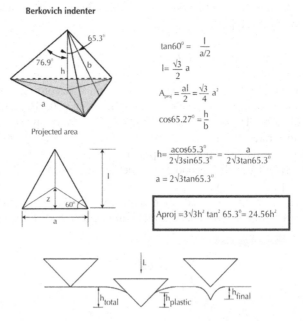

FIGURE 10   The schematic image of Berkovich indentor and nanoindentation process.

As it follows from the data of Fig. 9, the transition from nano- to microsystems occurs within the range h_{pl}=408–726 nm. Both the indicated above values h_{pl} and the corresponding to them values $(V_{def})^{1/3}$≈814–1440 nm can be chosen as the linear length scale l_n, corresponding to this transition. From the comparison of these values l_n with the distance between nanofiller particles aggregates L_n (L_n=219.2–788.3 nm for the considered nanocomposites, see Fig. 3) it follows, that for transition from nano- to microsystems l_n should include at least two nanofiller particles aggregates and surrounding them layers of polymer matrix, that is the lowest linear scale of nanocomposite simulation as a homogeneous system. It is easy to see, that nanocomposite structure homogeneity condition is harder than the obtained above from the criterion ψ=0. Let us note, that such method, namely, a nanofiller particle and surrounding it polymer matrix layers separation, is widespread at a relationships derivation in microcomposite models.

It is obvious, that the Eq. (35) is inapplicable to nanosystems, since $\psi\rightarrow0$ assumes $K_T\rightarrow\infty$, which is physically incorrect. Therefore the value E_0, obtained by the dependence $E(h_{pl})$ extrapolation (see Fig. 9) to h_{pl}=0, should be accepted as E for nanosystems [78].

Hence, the stated above results have shown, that elasticity modulus change at nanoindentation for particulate-filled elastomeric nanocomposites is due to a number of causes, which can be elucidated within the frameworks of anharmonicity conception and density fluctuation theory. Application of the first from the indicated conceptions assumes, that in nanocomposites during nanoindentation process local strain is realized, affecting polymer matrix only, and the transition to macrosystems means nanocomposite deformation as homogeneous system. The second from the mentioned conceptions has shown, that nano- and microsystems differ by density fluctuation absence in the first and availability of ones in the second. The last circumstance assumes, that for the considered nanocomposites density fluctuations take into account nanofiller and polymer matrix density difference. The transition from nano- to microsystems is realized in the case, when the deformed material volume exceeds nanofiller particles aggregate and surrounding it layers of polymer matrix combined volume [78].

In Ref. [32] the following formula was offered for elastomeric nanocomposites reinforcement degree E_n/E_m description:

$$\frac{E_n}{E_m} = 15.2\left[1-\left(d-d_{surf}\right)^{1/t}\right],\qquad(37)$$

where t is index percolation, equal to 1.7 [57].

From the Eq. (37) it follows, that nanofiller particles (aggregates of particles) surface dimension d_{surf} is the parameter, controlling nanocomposites reinforcement degree [81, 82]. This postulate corresponds to the known principle about numerous division surfaces decisive role in nanomaterials as the basis of their properties change [83]. From the Eqs. (4)–(6) it follows unequivocally, that the value d_{surf} is defined by nanofiller particles (aggregates of particles) size R_p only. In its turn, from the Eq. (37) it follows, that elastomeric nanocomposites reinforcement degree E_n/E_m is defined by the dimension d_{surf} only, or, accounting for the said above, by the size R_p only. This means, that the reinforcement effect is controlled by nanofiller particles (aggregates of particles) sizes only and in virtue of this is the true nanoeffect.

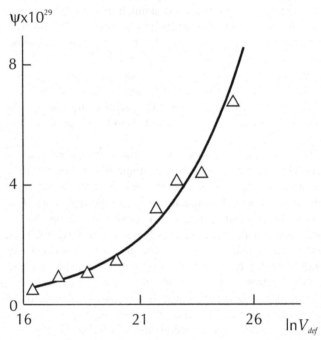

FIGURE 11 The dependence of density fluctuation ψ on volume of deformed in nanoindentation process material V_{def} in logarithmic coordinates for nanocomposites BSR/TC.

In Fig. 12, the dependence of E_n/E_m on $(d-d_{surf})^{1/1.7}$ is adduced, corresponding to the Eq. (37), for nanocomposites with different elastomeric matrices (natural and butadiene-stirene rubbers, NR and BSR, accordingly) and different nanofillers (technical carbon of different marks, nano- and microshungite). Despite the indicated distinctions in composition, all adduced data are described well by the Eq. (37).

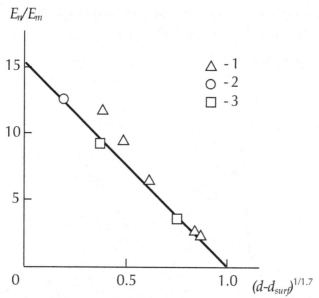

FIGURE 12 The dependence of reinforcement degree E_n/E_m on parameter $(d-d_{surf})^{1/1.7}$ value for nanocomposites NR/TC (1), BSR/TC (2) and BSR/shungite (3).

In Fig. 13, two theoretical dependences of E_n/E_m on nanofiller particles size (diameter D_p), calculated according to the Eqs. (4)–(6) and (37), are adduced. However, at the curve 1 calculation the value D_p for the initial nanofiller particles was used and at the curve 2 calculation—nanofiller particles aggregates size D_p^{ag} (see Fig. 3). As it was expected [34], the growth E_n/E_m at D_p or D_p^{ag} reduction, in addition the calculation with D_p (nonaggregated nanofiller) using gives higher E_n/E_m values in comparison with the aggregated one (D_p^{ag} using). At $D_p \leq 50$ nm faster growth E_n/E_m at D_p reduction is observed than at $D_p > 50$ nm, that was also expected. In Fig. 13 the critical theoretical value D_p^{cr} for this transition, calculated according to the indicated above general principles [83], is pointed out by a vertical shaded line. In conformity with these principles the nanoparticles size in nanocomposite is determined according to the condition, when division surface fraction in the entire nanomaterial volume makes up about 50% and more. This fraction is estimated approximately by the ratio $3l_{if}/D_p$, where l_{if} is interfacial layer thickness. As it was noted above, the data of Fig. 1 gave the average experimental value $l_{if} \approx 8.7$ nm. Further from the condition $3l_{if}/D_p \approx 0.5$ let us obtain $D_p \approx 52$ nm that is shown in Fig. 13 by a vertical shaded line. As it was expected, the value $D_p \approx 52$ nm is a boundary one for regions of slow ($D_p > 52$ nm) and fast ($D_p \leq 52$ nm) E_n/E_m growth at D_p reduction. In other words, the materials with nanofiller particles size $D_p \leq 52$ nm ("superreinforcing" filler according to the terminology of Ref. [34]) should be considered true nanocomposites.

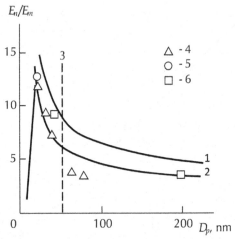

FIGURE 13 The theoretical dependences of reinforcement degree E_n/E_m on nanofiller particles size D_p, calculated according to the equations (4)-(6) and (37), at initial nanoparticles (1) and nanoparticles aggregates (2) size using. 3 – the boundary value D_p, corresponding to true nanocomposite. 4-6 – the experimental data for nanocomposites NR/TC (4), BSR/TC (5) and BSR/shungite (6).

Let us note in conclusion, that although the curves 1 and 2 of Fig. 13 are similar ones, nanofiller particles aggregation, which the curve 2 accounts for, reduces essentially enough nanocomposites reinforcement degree. At the same time the experimental data correspond exactly to the curve 2 that was to be expected in virtue of aggregation processes, which always took place in real composites [33] (nanocomposites [84]). The values d_{surf} obtained according to the Eqs. (4)–(6), correspond well to the determined experimentally ones. So, for nanoshungite and two marks of technical carbon the calculation by the indicated method gives the following d_{surf} values: 2.81, 2.78 and 2.73, whereas experimental values of this parameter are equal to: 2.81, 2.77 and 2.73, that is, practically a full correspondence of theory and experiment was obtained.

6.2.4 CONCLUSIONS

Hence, the stated above results have shown, that the elastomeric reinforcement effect is the true nanoeffect, which is defined by the initial nanofiller particles size only. The indicated particles aggregation, always taking place in real materials, changes reinforcement degree quantitatively only, namely, reduces it. This effect theoretical treatment can be received within the frameworks of fractal analysis. For the considered nanocomposites the nanoparticle size upper limiting value makes up ~52 nm.

KEYWORDS

- **Berkovich indentor**
- **caveolae-mediated endocytosis**
- **clathrin-mediated endocytosis**
- **linear spatial structures**
- **nanocrystallite**
- **opsonins**

REFERENCES

1. Ozin, G. A.; Arsenault, A. C.; *Nanochemistry*; RSC Publishing: Cambridge, **2005**.
2. McFarland, A. D.; Haynes, C. L.; Mirkin, C. A.; "Color My Nanoworld," *J. Chem.* **2004**, *81*, 544A.
3. Mukherjee, S.; Ray, S.; Thakur, R. S.; *Ind. J. Pharm. Sci.* **2009**, 349–358.
4. Abdul Hasan A.; Sathali, M.; Priyanka, K.; Solid lipid nanoparticles: A Review, *Sci. Revs. Chem. Commun.* **2012**, *2(1)*, 80–102.
5. Mozafari, M. R.; *Nano Carrier Technologies: Frontiers of Nano Therapy;* **2006**, 41–50.
6. Houli Li; Xiaobin Zhao; Yukun Ma; Guangxi Zhai; Ling Bing Li; Hong Xiang; Lou. J.; Cont. Release, **2009**, *133*, 238–244.
7. Melike Uner; Gulgun Yener; *Int. J. Nanomed.* **2007**, *2(3)*, 289–300.
8. Kushwaha Anjali; Nanotherapeutics in Drug delivery: A Review, Pharmatutor ART 1291.
9. Mua, L.; Fenga, S. S.; A novel controlled release formulation for the anticancer drug paclitaxl (Taxol®): PLGA nanoparticles containing vitamin E TPGS; *J. Controlled Release*, **2003**, *86*, 33–48.
10. Lu, X. Y. ; Wu, D. C.; Li, Z. J.; Chen, G. Q.; *Polymer nanoparticles. Prog. Mol. Biol. Transl. Sci.* **2011**, *104*, 299–323.
11. Sinani, V. A.; Koktysh, D. S.; Yun, B. G.; Matts, R. L.; Pappas, T. C.; Motamedi, M.; Thomas, S. N.; Kotov, N. A.; Collagen coating promotes biocompatibility of semiconductor nanoparticles in stratified LBL films. *Nano Lett.*, **2003**, *3*, 1177–1182.

12. Zhang, Y.; Kohler, N.; Zhang, M.; Surface modification of superparamagnetic magnetite nanoparticles and their intracellular uptake. *Biomaterials,* **2002,** *23,* 1553–1561.
13. Gregoriadis, G.; Liposomes in the therapy of lysosomal storage diseases. *Nature* **1978,** *275,* 695–696.
14. Grislain, L.; Couvreur, P.; Lenaerts V, Roland, M.; Deprezdecampeneere, D.; Speiser, P.; Pharmacokinetics and distribution of a biodegradable drug-carrier. *Int. J. Pharm.* **1983,** *15,* 335–345.
15. Aderem, A.; Underhill, D.; Mechanisms of phagocytosis in macrophages. *Annu. Rev. Immunol.* **1999,** *17,* 593–623.
16. Rabinovitch, M.; Professional and nonprofessional phagocytes—an introduction. *Trends Cell Biol.* **1995,** *5,* 85–87.
17. Vonarborg, A.; Passirani, C.; Saulnier, P.; Benoit, J.; Parameters influencing the stealthiness of colloidal drug delivery systems. *Biomaterials* **2006,** *27,* 4356–4373.
18. Owens, D.; Peppas, N.; Opsonization, biodistribution, and pharmacokinetics of polymeric nanoparticles. *Int. J. Pharm.* **2006,** *307,* 93–102.
19. Groves, E.; Dart, A.; Covarelli, V.; Caron, E.; Molecular mechanisms of phagocytic uptake in mammalian cells. *Cell. Mol. Life. Sci.* **2008,** *65,* 1957–1976.
20. Vachon, E.; Martin, R.; Plumb, J.; Kwok, V.; Vandivier, R.; Glogauer, M.; Kapus, A.; Wang, X.; Chow, C.; Grinstein, S.; Downey, G.; CD44 is a phagocytic receptor. *Blood* **2006,** *107,* 4149–4158.
21. Caron, E.; Hall, A.; Identification of two distinct mechanisms of phagocytosis controlled by different Rho GTPases. *Science* **1998,** *282,* 1717–1721.
22. Swanson, J. A.; Baer, S. C.; Phagocytosis by zippers and triggers. *Trends Cell Biol.* **1995,** *5,* 89–93.
23. Claus, V.; Jahraus, A.; Tjelle, T.; Berg, T.; Kirschke, H.; Faulstich, H.; Griffiths, G.; Lysosomal enzyme trafficking between phagosomes, endosomes, and lysosomes in J774 macrophages. Enrichment of cathepsin H in early endosomes. *J. Biol. Chem.* **1998,** *273,* 9842–9851.
24. Mukherjee, S.; Ghosh, R. N.; Maxfield, F. R.; Endocytosis. *Physiol. Re*1997, *77,* 759–803.
25. Bareford, L. M.; Swaan, P. W.; Endocytic mechanisms for targeted drug delivery. *Adv. Drug Deliv. Re*2007, *59,* 748–758.
26. Conner, S. D.; Schmid, S. L.; Regulated portals of entry into the cell. *Nature 422,* 37–44. **2003,**
27. Mayor, S.; Pagano, R. E.; Pathways of clathrin-independent endocytosis. *Nat. Rev. Mol. Cell Biol.* **2007,** *8,* 603–612.
28. Swanson, J. A.; Watts, C.; Macropinocytosis. *Trends Cell Biol.* **1995,** *5,* 424–428.
29. Racoosin, E. L.; Swanson, J. A.; M-CSF-induced macropinocytosis increases solute endocytosis but not receptor-mediated endocytosis in mouse macrophages. *J. Cell Sci.* **1992,** *102,* 867–880.
30. Yanovskii, Yu. G.; Kozlov, G. V.; Karnet, Yu. N.; Mekhanika Kompozitsionnykh Materialov i Konstruktsii, **2011,** *17(2),* 203–208.
31. Malamatov, A. Kh.; Kozlov, G. V.; Mikitaev, M. A.; Reinforcement Mechanisms of Polymer Nanocomposites. Moscow, Publishers of the D. I.; Mendeleev RKhTU, **2006,** 240 p.

32. Mikitaev, A. K.; Kozlov, G. V.; Zaikov, G. E.; Polymer Nanocomposites: Variety of Structural Forms and Applications. Moscow, Nauka, **2009**, 278 p.
33. Kozlov, G. V.; Yanovskii, Yu. G.; Karnet, Yu. N.; Structure and Properties of Particulate-Filled Polymer Composites: the Fractal Analysis. Moscow, Al'yanstransatom, **2008**, 363 p.
34. Edwards, D. Mater, Sci.; C. J.; **1990**, *25(12)*, 4175–4185.
35. Buchachenko, A. L.; Uspekhi Khimii, **2003**, *72(5)*, 419–437.
36. Kozlov, G. V.; Yanovskii Yu.G.; Burya, A. I.; Aphashagova Z. Kh. Mekhanika Kompozitsionnykh Materialov i Konstruktsii, **2007**, *13(4)*, 479–492.
37. Lipatov, Yu. S.; The Physical Chemistry of Filled Polymers. Moscow, Khimiya, **1977**, 304 p.
38. Bartenev, G. M.; Zelenev, Yu. V.; Physics and Mechanics of Polymers. Moscow, Vysshaya Shkola, **1983**, 391 p.
39. Kozlov, G. V.; Mikitaev, A. K.; Mekhanika Kompozitsionnykh Materialov i Konstruktsii, **1996**, *2(3–4)*, 144–157.
40. Kozlov, G. V.; Yanovskii, Yu. G.; Zaikov, G. E.; Structure and Properties of Particulate-Filled Polymer Composites: the Fractal Analysis. New York, Nova Science Publishers, Inc.; **2010**, 282 p.
41. Mikitaev, A. K.; Kozlov, G. V.; Zaikov, G. E.; Polymer Nanocomposites: Variety of Structural Forms and Applications. New York, Nova Science Publishers, Inc.; **2008**, 319 p.
42. McClintok, F. A.; Argon, A. S.; Mechanical Behavior of Materials. Reading, Addison-Wesley Publishing Company, Inc.; **1966**, 440 p.
43. Kozlov, G. V.; Mikitaev, A. K.; Doklady AN SSSR, **1987**, *294(5)*, 1129–1131.
44. Honeycombe, R. W. K.; The Plastic Deformation of Metals. Boston, Edward Arnold (Publishers), Ltd.; **1968**, 398 p.
45. Dickie, R. A. In: Polymer Blends. V. 1. Paul D. R.; Newman S. Ed.; New York, San-Francisco, London, Academic Press, **1980**, 386–431.
46. Kornev, Yu. V.; Yumashev, O. B.; Zhogin, V. A.; Karnet Yu.N.; Yanovskii Yu.G.; Kautschuk i Rezina, **2008**, *6*, 18–23.
47. Oliver, W. C.; Pharr, G. M. J.; *Mater. Res.* **1992**, *7(6)*, 1564–1583.
48. Kozlov, G. V.; Yanovskii, Yu. G.; Lipatov, Yu. S.; Mekhanika Kompozitsionnykh Materialov i Konstruktsii, **2002**, *8(1)*, 111–149.
49. Kozlov, G. V.; Burya, A. I.; Lipatov, Yu. S.; Mekhanika Kompozitnykh Materialov, **2006**, *42(6)*, 797–802.
50. Hentschel, H. G. E.; Deutch, J. M.; *Phys. Rev. A*, **1984**, *29(3)*, 1609–1611.
51. Kozlov, G. V.; Ovcharenko, E. N.; Mikitaev, A. K.; Structure of Polymers Amorphous State. Moscow, Publishers of the D. I.; Mendeleev RKhTU, **2009**, 392 p.
52. Yanovskii, Yu. G.; Kozlov Mater, G. V.; VII Intern. Sci.-Pract. Conf. "New Polymer Composite Materials." Nal'chik, KBSU, **2011**, 189–194.
53. Wu, S. J.; *Polymer Sci.: Part B: Polymer Phys.* **1989**, *27(4)*, 723–741.
54. Aharoni, S. M.; *Macromolecules*, **1983**, *16(9)*, 1722–1728.
55. Budtov V. P.; The Physical Chemistry of Polymer Solutions. Sankt-Peterburg, Khimiya, **1992**, 384 p.
56. Aharoni Macromolecules, S. M.; **1985**, *18(12)*, 2624–2630.

57. Bobryshev, A. N.; Kozomazov, V. N.; Babin, L. O.; Solomatov, V. I.; Synergetics of Composite Materials. Lipetsk, NPO ORIUS, **1994**, 154 p.
58. Kozlov, G. V.; Yanovskii, Yu. G.; Karnet, Yu. N.; Mekhanika Kompozitsionnykh Materialov i Konstruktsii, **2005**, *11(3),* 446–456.
59. Sheng, N.; Boyce, M. C.; Parks, D. M.; Rutledge, G. C.; Abes, J. I.; Cohen, R. E.; *Polymer*, **2004**, *45(2),* 487–506.
60. Witten, T. A.; Meakin, P.; *Phys. Rev. B*, **1983**, *28(10),* 5632–5642.
61. Witten, T. A.; Sander, L. M.; *Phys. Rev. B*, **1983**, *27(9),* 5686–5697.
62. Happel J.; Brenner G.; Hydrodynamics at Small Reynolds Numbers. Moscow, Mir, **1976**, 418 p.
63. Mills N. J.; *J. Appl. Polymer Sci.* **1971**, *15(11),* 2791–2805.
64. Kozlov, G. V.; Yanovskii, Yu. G.; Mikitaev, A. K.; Mekhanika Kompozitnykh Materialov, **1998**, *34(4),* 539–544.
65. Balankin, A. S.; Synergetics of Deformable Body. Moscow, Publishers of Ministry Defence SSSR, **1991**, 404 p.
66. Hornbogen, E.; *Intern. Mater. Res.* **1989**, *34(6),* 277–296.
67. Pfeifer, P.; *Appl. Surf. Sci.* **1984**, *18(1),* 146–164.
68. Avnir D.; Farin D.; Pfeifer P.; *J. Colloid Interface Sci.* **1985**, *103(1),* 112–123.
69. Ishikawa Mater, K.; *J. Sci. Lett.* **1990**, *9(4),* 400–402.
70. Ivanova, V. S.; Balankin, A. S.; Bunin, I. Zh.; Oksogoev, A. A.; Synergetics and Fractals in Material Science. Moscow, Nauka, **1994**, 383 p.
71. Vstovskii, G. V.; Kolmakov, L. G.; Terent'ev, V. E.; *Metally*, **1993**, *4,* 164–178.
72. Hansen, J. P.; Skjeitorp, A. T.; *Phys. Rev. B*, **1988**, *38(4),* 2635–2638.
73. Pfeifer, P.; Avnir, D.; Farin D. J.; *Stat. Phys.* **1984**, *36(5/6),* 699–716.
74. Farin, D.; Peleg, S.; Yavin, D.; Avnir, D.; *Langmuir*, **1985**, *1(4),* 399–407.
75. Kozlov, G. V.; Sanditov, D. S.; Anharmonical Effects and Physical-Mechanical Properties of Polymers. Novosibirsk, Nauka, **1994**, 261 p.
76. Bessonov, M. I.; Rudakov Vysokomolek, A. P.; *Soed. B*, **1971**, *13(7),* 509–511.
77. Kubat J.; Rigdahl M.; Welander, M. J.; *Appl. Polymer Sci.* **1990**, *39(5),* 1527–1539.
78. Yanovskii, Yu. G.; Kozlov, G. V.; Kornev, Yu. V.; Boiko, O. V.; Karnet, Yu. N.; Mekhanika Kompozitsionnykh Materialov i Konstruktsii, **2010**, *16(3),* 445–453.
79. Yanovskii, Yu. G.; Kozlov, G. V.; Aloev Mater, V. Z.; Intern. Sci.-Pract. Conf. "Modern Problems of APK Innovation Development Theory and Practice." Nal'chik, KBSSKhA, **2011**, 434–437.
80. Chow Polymer, T. S.; **1991**, *32(1),* 29–33.
81. Ahmed, S.; Jones, F. Mater, Sci.; R. J.; **1990**, *25(12),* 4933–4942.
82. Kozlov, G. V.; Yanovskii Yu.G.; Aloev Mater, V. Z.; Intern. Sci.-Pract. Conf.; dedicated to FMEP 50-th Anniversary. Nal'chik, KBSSKhA, **2011**, 83–89.
83. Andrievskii, R. A.; Rossiiskii Khimicheskii Zhurnal, **2002**, *46(5),* 50–56.
84. Kozlov, G. V.; Sultonov, N. Zh.; Shoranova, L. O.; Mikitaev, A. K.; Naukoemkie Tekhnologii, **2011**, *12(3),* 17–22.

CHAPTER 7

SOME NEW ASPECTS OF POLYMERIC NANOCOMPOSITES

A. K. MIKITAEV, A. Y. BEDANOKOV, and M. A. MIKITAEV

CONTENTS

7.1 INTRODUCTION

The polymeric nanocomposites are the polymers filled with nanoparticles, which interact with the polymeric matrix on the molecular level in contrary to the macrointeraction in composite materials. Mentioned nanointeraction results in high adhesion hardness of the polymeric matrix to the nanoparticles [1, 52].

Usual nanoparticle is less than 100 nm in any dimension, 1 nm being the billionth part of a meter [1, 2].

The analysis of the reported studies tells that the investigations in the field of the polymeric nanocomposite materials are very promizing.

The first notion of the polymeric nanocomposites was given in patent in 1950 [3]. Blumstain pointed in 1961 [4] that polymeric clay – based nanocomposites had increased thermal stability. It was demonstrated using the data of the thermogravimetric analysis that the polymethylmetacrylate intercalated into the Na^+ – methylmetacrylate possessed the temperature of destruction 40–50 °C higher than the initial sample.

This branch of the polymeric chemistry did not attract much attention until 1990 when the group of scientists from the Toyota Concern working on the polyamide – based nanocomposites [5–9] found two – times increase in the elasticity modulus using only 4.7 wt% of the inorganic compound and 100 °C increase in the temperature of destruction, both discoveries widely extending the area of application of the polyamide. The polymeric nanocomposites based on the layered silicates began being intensively studied in state, academic and industrial laboratories all over the world only after that.

7.2 STRUCTURE OF THE LAYERED SILICATES

The study of the polymeric nanocomposites on the basis of the modified layered silicates (broadly distributed and well – known as various clays) is of much interest. The natural layered inorganic structures used in producing the polymeric nanocomposites are the montmorillonite [10–12], hectorite [13], vermiculite [14], kaolin, saponine [15] and others. The sizes of inorganic layers are about 220 and 1 nm in length and width, respectively [16, 17].

The perspectives ones are the bentonite breeds of clays, which include at least 70% of the minerals from the montmorillonite group.

Montmorillonite $(Na, K, Ca)(Al, Fe, Mg)[(Si, Al)_4O_{10}](OH)_2 \cdot nH_2O$, named after the province Montmorillion in France, is the high – dispersed layered aluminous silicate of white or gray color in which appears the excess negative charge due to the nonstoichiometric replacements of the cations of the crystal lattice, charge being balanced by the exchange cations from the interlayer space. The main feature of the montmorillonite is its ability to adsorb ions, generally cations, and to exchange them. It produces plastic masses with water and may enlarge itself 10 times. Montmorillonite enters the bentonite clays (the term "bentonite" is given after the place Benton in USA).

The inorganic layers of clays arrange the complexes with the gaps called layers or galleries. The isomorphic replacement within the layers (such as Mg^{2+} replacing Al^{3+} in octahedral structure or Al^{3+} replacing Si^{4+} in tetrahedral one) generates the negative charges, which electrostatically are compensated by the cations of the alkali, or alkali-earth metals located in the galleries (Fig. 1) [18]. The bentonite is very hydrophilic because of this. The water penetrates the interlayer space of the montmorillonite, hydrates its surface and exchanges cations what results in the swelling of the bentonite. The further dilution of the bentonite in water results in the viscous suspension with bold tixotropic properties.

The more pronounced cation – exchange and adsorption properties are observed in the bentonites montmorillonite of which contains predominantly exchange cations of sodium.

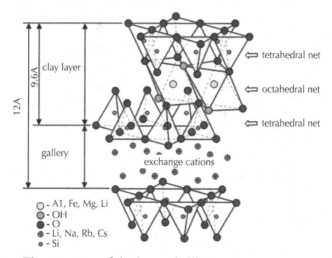

FIGURE 1   The structure of the layered silicate.

7.3   MODIFICATION OF THE LAYERED SILICATES

Layered silicates possess quite interesting properties – sharp drop of hardness at wetting, swelling at watering, dilution at dynamical influences and shrinking at drying.

The hydrophility of aluminous silicates is the reason of their incompatibility with the organic polymeric matrix and is the first hurdle needs to be overridden at producing the polymeric nanocomposites.

One way to solve this problem is to modify the clay by the organic substance. The modified clay (organoclay) has at least two advantages: (1) it can be well dispersed in polymeric matrix [19] and (2) it interacts with the polymeric chain [13].

The modification of the aluminous silicates can be done with the replacement of the inorganic cations inside the galleries by the organic ones. The replacing by the cationic surface – active agents like bulk ammonium and phosphonium ions increases the room between the layers, decreases the surface energy of clay and makes the surface of the clay hydrophobic. The clays modified such a way are more compatible with the polymers and form the layered polymeric nanocomposites [52]. One can use the nonionic modifiers besides the organic ones, which link themselves to the clay surface through the hydrogen bond. Organoclays produced with help of non – ionic modifiers in some cases become more chemically stable than the organoclays produced with help of cationic modifiers (Fig. 2a) [20].

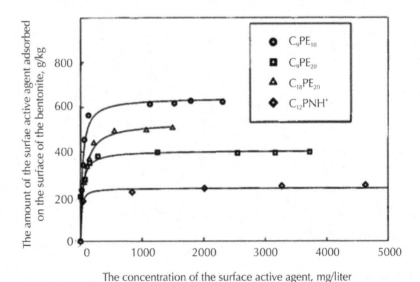

FIGURE 2A   The adsorption of different modifiers on the clay surface.

The least degree of desorption is observed for non – ionic interaction between the clay surface and organic modifier (Fig. 2b). The hydrogen bonds between the ethylenoxide grouping and the surface of the clay apparently make these organoclays more chemically stable than organoclays produced with nonionic mechanism.

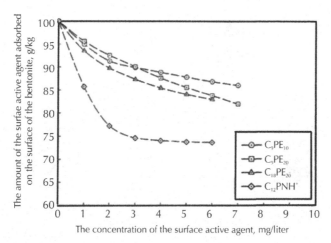

FIGURE 2B The desorption of different modifiers from the clay surface:
$C_9PE_{10} - C_9H_{19}C_6H_4(CH_2\ CH_2O)_{10}OH$;
$C_9PE_{20} - C_9H_{19}C_6H_4(CH_2\ CH_2O)_{20}OH$;
$C_{18}E_{20} - C_{18}H_{37}(CH_2\ CH_2O)_{20}OH$;
$C_{12}PNH^+ - C_{12}H_{25}C_6H_4NH^+Cl^-$.

7.4 STRUCTURE OF THE POLYMERIC NANOCOMPOSITES ON THE BASIS OF THE MONTMORILLONITE

The study of the distribution of the organoclay in the polymeric matrix is of great importance because the properties of composites obtained are in the direct relation from the degree of the distribution.

According Giannelis [21], the process of the formation of the nanocomposite goes in several intermediate stages (Fig. 3). The formation of the tactoid happens on the first stage – the polymer surrounds the agglomerations of the organoclay. The polymer penetrates the interlayer space of the organoclay on the second stage. Here the gap between the layers may reach 2–3 nm [22]. The further separation of the layers, third stage, results in partial dissolution and disorientation of the layers. Exfoliation is observed when polymer shifts the clay layers on more than 8–10 nm.

All mentioned structures may be present in real polymeric nanocomposites in dependence from the degree of distribution of the organoclay in the polymeric matrix. Exfoliated structure is the result of the extreme distribution of the organoclay. The excess of the organoclay or bad dispersing may born the agglomerates of the organoclays in the polymeric matrix what finds experimental confirmation in the X-ray analysis [11, 12, 21, 23].

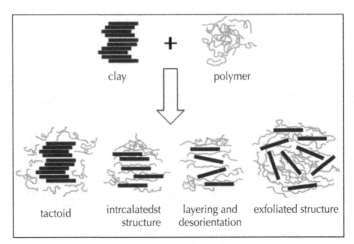

FIGURE 3   The schematic formation of the polymeric nanocomposite [24].

In the following subsections, we describe a number of specific methods used at studying the structure of the polymeric nanocomposites.

7.4.1 DETERMINATION OF THE INTERLAYER SPACE

The X-ray determination of the interlayer distance in the initial and modified layered silicates as well as in final polymeric nanocomposite is one of the main methods of studying the structure of the nanocomposite on the basis of the layered silicate. The peak in the small – angle diapason ($2q = 6$–8 °C) is characteristic for pure clays and responds to the order of the structure of the silicate. This peak drifts to the smaller values of the angle $2q$ in organomodified clays. If clay particles are uniformly distributed in the bulk of the polymeric matrix then this peak disappears, what witnesses on the disordering in the structure of the layered silicate. If the amount of the clay exceeds the certain limit of its distribution in the polymeric matrix, then the peak reappears again. This regularity was demonstrated on the instance of the polybutylenterephtalate (Fig. 4) [11].

The knowledge of the angle $2q$ helps to define the size of the pack of the aluminous silicate consisting of the clay layer and interlayer space. The size of such pack increases in a row from initial silicate to polymeric nanocomposite according to the increase in the interlayer space. The average size of that pack for montmorillonite is 1.2–1.5 nm but for organomodified one varies in the range of 1.8–3.5 nm.

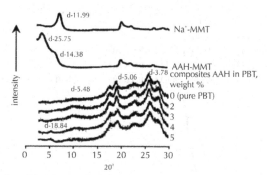

FIGURE 4   The data of the X-ray analysis for clay, organoclay and nanocomposite PBT/organoclay.

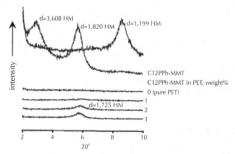

FIGURE 5   The data of the X-ray analysis for clay, organoclay and nanocomposite PET/organoclay.

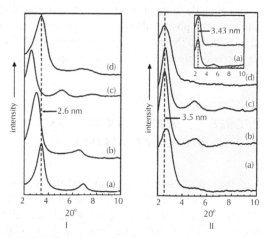

FIGURE 6   The data of the X-ray analysis for:
I. (a) – dimetyldioctadecylammonium (DMDODA) – hectorite;
(b) – 50% polystirene (PS)/50% DMDODA – hectorite;
(c) – 75% polyethylmetacrylate (PEM)/25% DMDODA;

(d) – 50% PS/ 50 % DMDODA – hectorite after 24 h of etching in cyclohexane.
II. mix of PS, PEM and ofganoclay:
(a) – 23.8% PS/71.2% PEM/5% DMDODA – hectorite;
(b) – 21.2% PS/63.8% PEM/15% DMDODA – hectorite;
(c) – 18.2% PS/54.8% PEM/27% DMDODA – hectorite;
(d) – 21.2% PS/63.8% PEM/15% DMDODA – hectorite after 24 h of etching in cyclohexane.

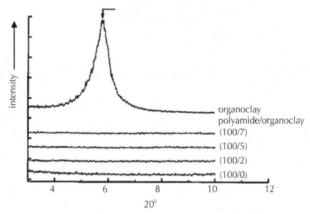

FIGURE 7   The data of the X-ray analysis for organoclay and nanocomposite polyamide acid/ organoclay.

Summing up we conclude that comparing the data of the X-ray analysis for the organoclay and nanocomposite allows for the determination of the optimal clay amount need be added to the composite. The data from the scanning tunneling (STM) and transmission electron (TEM) microscopes [27, 28] can be used as well.

7.4.2   THE DEGREE OF THE DISTRIBUTION OF THE CLAY PARTICLES IN THE POLYMERIC MATRIX

The two structures, namely the intercalated and exfoliated ones, could be distinguished with the respect to the degree of the distribution of the clay particles, Fig.8. One should note that clay layers are quite flexible though they are shown straight in the figure. The formation of the intercalated or exfoliated structures depends on many factors, for example the method of the production of the nanocomposite or the nature of the clay, and so forth [29].

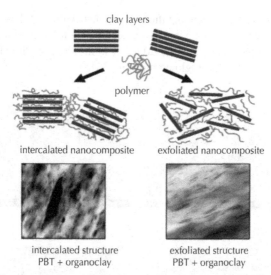

FIGURE 8 The formation of the intercalated and exfoliated structures of the nanocomposite.

The TEM images of the surface of the nanocomposites can help to find out the degree of the distribution of the nanosized clay particles, see plots (a) to (d) in the Fig. 9.

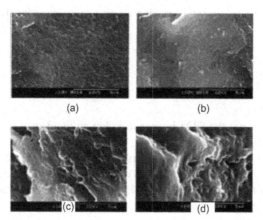

FIGURE 9 The images from scanning electron microscope for the nanocomposite surfaces:

(a) – pure PBT;

(b) – 3 wt% of organoclay in PBT;

(c) – 4 wt% of organoclay in PBT;

(d) – 5 wt% of organoclay in PBT.

The smooth surface tells about the uniform distribution of the organoclay particles. The surface of the nanocomposite becomes deformed with the increasing
amount of the organoclay, see plots (a) to (d) in the Fig. 10. Probably, this is due
to the influence of the clay agglomerates [30, 31].

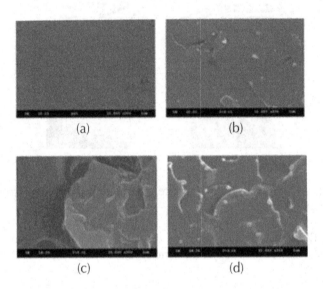

FIGURE 10   The images from scanning electron microscope for the nanocomposite
surfaces:
(a) – pure PET;
(b) – 3 wt% of organoclay in PET;
(c) – 4 wt% of organoclay in PET;
(d) – 5 wt% of organoclay in PET.

Also one can use the STM images to judge on the degree of the distribution
of the organoclay in the nanocomposite, Figs. 11 and 12. If the content of the organoclay is 2–3 wt% then the clay layers are separated by the polymeric layer of
4 to 10 nm width, Fig. 11. If the content of the organoclay reaches 4–5 wt % then
the majority of the clay becomes well distributed, however, the agglomerates of
4 to 8 nm may appear.

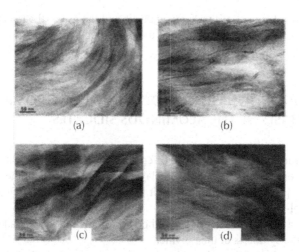

FIGURE 11 The images from tunneling electron microscope for the nanocomposite surfaces:
(a) – 2 wt% of organoclay in PBT;
(b) – 3 wt% of organoclay in PBT;
(c) – 4 wt% of organoclay in PBT;
(d) – 5 wt% of organoclay in PBT.

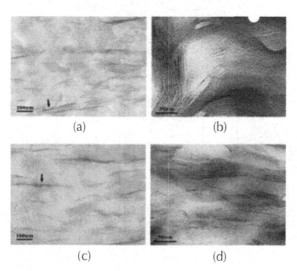

FIGURE 12 The images from tunneling electron microscope for the nanocomposite surfaces:
(a) – 1 wt% of organoclay in PET;
(b) – 2 wt% of organoclay in PET;
(c) – 3 wt% of organoclay in PET;
(d) – 4 wt% of organoclay in PET.

So the involvement of the X-ray analysis and the use of the microscopy data tell that the nanocomposite consists of the exfoliated clay at the low content (below 3 wt %) of the organoclay.

7.5 PRODUCTION OF THE POLYMERIC NANOCOMPOSITES ON THE BASIS OF THE ALUMINOUS SILICATES

Different groups of authors [32–35] offer following methods for obtaining nanocomposites on the basis of the organoclays: (1) in the process of the synthesis of the polymer [33, 36, 37], (2) in the melt [38, 39], (3) in the solution [40–46] and (4) in the sol-gel process [47–50].

The most popular ones are the methods of producing in melt and during the process of the synthesis of the polymer.

The producing of the polymeric nanocomposite in situ is the intercalation of the monomer into the clay layers. The monomer migrates through the organoclay galleries and the polymerization happens inside the layers [19, 51], Fig. 13.

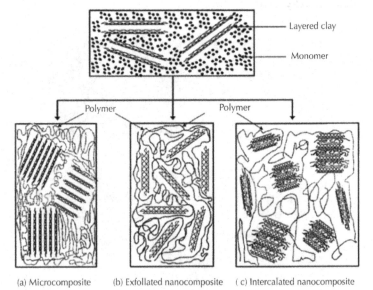

(a) Microcomposite   (b) Exfollated nanocomposite   (c) Intercalated nanocomposite

FIGURE 13   The production of the nanocomposite in situ:

(a) – microcomposite;

(b) – exfoliated nanocomposite;

(c) – intercalated nanocomposite [51].

The polymerization may be initiated by the heat, irradiation or other source. Obviously, the best results on the degree of the distribution of the clay particles in the polymeric matrix must emerge if using given method. This is associated with the fact that the separation of the clay layers happens in the very process of the inclusion of the monomer in the interlayer space. In other words, the force responsible for the separation of the clay layers is the growth of the polymeric chain whereas the main factor for reaching the necessary degree of the clay distribution in solution or melt is just satisfactory mixing. The most favorable condition for synthesizing the nanocomposites is the vacuuming or the flow of the inert gas. Besides, one has to use the fast speeds of mixing for satisfactory dispersing of the organoclay in the polymeric matrix.

The method of obtaining the polymeric nanocomposites in melt (or the method of extrusion) is the mixing of the polymer melted with the organoclay. The polymeric chains lose the considerable amount of the conformational entropy during the intercalation. The probable motive force for this process is the important contribution of the enthalpy of the interaction between the polymer and organoclay at mixing. One should add that polymeric nanocomposite on the basis of the organoclays could be successively produced by the extrusion [22]. The advantage of the extrusion method is the absence of any solvents what excludes the unnecessary leaks. Moreover the speed of the process is several times more and the technical side is simpler. The extrusion method is the best one in the industrial scales of production of the polymeric nanocomposites what acquires the lesser source expenses and easier technological scheme.

If one produces the polymer – silicate nanocomposite in solution then the organosilicate swells in the polar solvent such as toluene or N, N-dimethylformamide. Then the added is the solution of the polymer, which enters the interlayer space of the silicate. The removing of the solvent by means of evaporation in vacuum happens after that. The main advantage of the given method is that "polymer-layered silicate" might be produced from the polymer of low polarity or even nonpolar one. However, this method is not widely used in industry because of much solvent consumption [52].

The sol-gel technologies find application at producing nanocomposites on the basis of the various ceramics and polymers. The initial compounds in these technologies are the alcoholates of specific elements and organic oligomers.

The alcoholates are firstly hydrolyzed and obtained hydroxides being polycondensated then. The ceramics from the inorganic 3D net is formed as a result. Also the method of synthesis exists in which the polymerization and the formation of the inorganic glass happen simultaneously. The application of the nanocomposites on the basis of ceramics and polymers as special hard defensive coverages and like optic fibers [53] is possible.

7.6   PROPERTIES OF THE POLYMERIC NANOCOMPOSITES

Many investigations in physics, chemistry and biology have shown that the jump from macroobjects to the particles of 1–10 nm results in the qualitative transformations in both separate phases and systems from them [54].

One can improve the thermal stability and mechanical properties of the polymers by inserting the organoclay particles into the polymeric matrix. It can be done by means of joining the complexes of properties of both the organic and inorganic substances, that is combining the light weight, flexibility and plasticity of former and durability, heat stability and chemical resistance of latter.

Nanocomposites demonstrate essential change in properties if compared to the nonfilled polymers. So, if one introduces modified layered silicates in the range of 2 to 10 wt% into the polymeric matrix then he observes the change in mechanical (tensile, compression, bending and overall strength), barrier (penetrability and stability to the solvent impact), optical and other properties. The increased heat and flame resistance even at low filler content is among the interesting properties too. The formation of the thermal isolation and negligible penetrability of the charred polymer to the flame provide for the advantages of using these materials.

The organoclay as a nanoaddition to the polymers may change the temperature of the destruction, refractoriness, rigidity and rupture strength. The nanocomposites also possess the increased rigidity modulus, decreased coefficient of the heat expansion, low gas-penetrability, increased stability to the solvent impact and offer broad range of the barrier properties [54]. In Table 1, we gather the characteristics of the nylon-6 and its derivative containing 4.7 wt% of the organomodified montmorillonite.

TABLE 1   The properties of the nylon-6 and composite based on it [54].

	Rigidity modulus, GPa	Tensile strength, MPa	Temperature of the deformation, °C	Impact viscosity, kJ/m^2	Water consumption, weight %	Coefficient of the thermal expansion (x, y)
Nylon-6	1.11	68.6	65	6.21	0.87	13×10^{-5}
Nanocomposite	1.87	97.2	152	6.06	0.51	6.3×10^{-5}

It is important that the temperature of the deformation of the nanocomposite increases on 87 °C.

The thermal properties of the polymeric nanocomposites with the varying organoclay content are collected in Table 2.

TABLE 2 The main properties of the polymeric nanocomposites.

Property	Composition								
	Polybutyleneterephtalate + AAX-montmorillonite					Polyethyleneterephtalate + C_{12}PPh-montmorillonite			
	Organoclay content, %								
	0	2	3	4	5	0	1	2	3
Viscosity, dliter/g	0.84	1.16	0.77	0.88	0.86	1.02	1.26	0.98	1.23
T_g, °C	27	33	34	33	33	—	—	—	—
T_m, °C	222	230	230	229	231	245	247	245	246
T_d, °C	371	390	388	390	389	370	375	384	386
$W_{tR}^{600\,c}$, %	1	6	7	7	9	1	8	15	21
Strength limit, MPa	41	50	60	53	49	46	58	68	71
Rigidity modulus, GPa	1.37	1.66	1.76	1.80	1.86	2.21	2.88	3.31	4.10
Relative enlargement, %	5	7	6	7	7	3	3	3	3

The inclusion of the organoclay into the polybutyleneterephtalate leads to the increase in the glass transition temperature (T_g) from 27 to 33 degrees centigrade if the amount of the clay raises from zero to 2 wt%. That temperature does not change with the further increase of the organoclay content. The increase in T_g may be the result of two reasons [56–59]. The first is the dispersion of the small amount of the organoclay in the free volume of the polymer, and the second is the limiting of the mobility of the segments of the polymeric chain due to its interlocking between the layers of the organoclay.

The same as the T_g, the melting temperature T_m increases from the 222 to 230 °C if the organoclay content raises from 0 to 2 wt% and stays constant up to 5 wt%, see Table 2. This increase might be the consequence of both complex multilayer structure of the nanocomposite and interaction between the organoclay and polymeric chain [60, 61]. Similar regularities have been observed in other polymeric nanocomposites also.

The thermal stability of the nanocomposites polybutyleneterephtalate, briefly PBT, (or polyethyleleterephtalate)/organoclay determined by the thermogravimetric analysis is presented in Table 3 and in Figs. 14 and 15 [11, 12].

TABLE 3   The thermal properties of the fibers from PET with varying organoclay content.

Organoclay content, weight %	η_{inh} [a]	T_m (°C)	H_m [b] (J/g)	T_D^{i} [c] (°C)	Wt_R^{600} [d] (%)
0 (pure PET)	1.02	245	32	370	1
1	1.26	247	32	375	8
2	0.98	245	33	384	15
3	1.23	246	32	386	21

[a]viscosities were measured at 30 °C using 0.1 g of polymer on 100 mL of solution in mix phenol/tetrachlorineethane (50/50);
[b]change in enthalpy of melting;
[c]initial temperature of decomposition;
[d]weight percentage of the coke remnant at 600 °C.

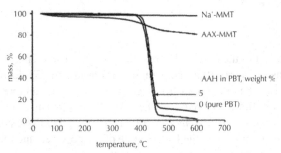

FIGURE 14   The thermogravimetrical curves for the montmorillonites, PBT and nanocomposites PBT/organoclay.

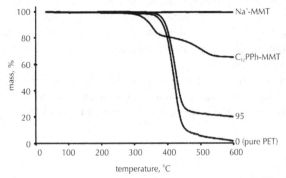

FIGURE 15   The thermogravimetrical curves for the montmorillonites, PET and nanocomposites PET/organoclay.

The temperature of the destruction, T_D, increases with the organoclay content up to 350 °C in case of the composite PBT/organoclay. The thermogravimetrical curves for pure and composite PBTs have similar shapes below 350 °C. The values of temperature T_D depends on the amount of organoclay above 350 °C. The organoclay added becomes a barrier for volatile products being formed during the destruction [61, 62]. Such example of the improvement of the thermal stability was studied in papers [63, 64]. The mass of the remnant at 600 °C increases with organoclay content.

Following obtained data authors draw the conclusion that the optimal results for thermal properties are being obtained if 2 wt% of the organoclay is added [11, 12, 19].

The great number of studies on the polymeric composite organoclay – based materials show [11, 12, 13, 19] that the inclusion of the inorganic component into the organic polymer improves the thermal stability of the latter, see Tables 3 and 4.

TABLE 4 The basic properties of the nanocomposite based on PBT with varying organoclay content.

Organoclay content, weight %	I.V.[a]	T_g	T_m (°C)	$T_D^{i\ [b]}$ (°C)	Wt_R^{60} (°C) (%)
0 (pure PBT)	0.84	27	222	371	1
2	1.16	33	230	390	6
3	0.77	34	230	388	7
4	0.88	33	229	390	7
5	0.86	33	231	389	9

[a]viscosities were measured at 30 °C using 0.1 g of polymer on 100 mL of solution in mix phenol/tetrachlorineethane (50/50);
[b]initial temperature of the weight loss;
[c]weight percentage of the coke remnant at 600 °C.

The values of the melting temperature increase from 222 to 230 °C if the amount of the organoclay added reaches 2 wt% and then stay constant. This effect can be explained by both thermal isolation of the clay and interaction between the polymeric chain and organoclay [43, 64]. The increase in the glass transition temperature also occurs what can be a consequence of several reasons [55, 56, 58]. One of the main among them is the limited motion of the segments of the polymeric chain in the galleries within the organoclay.

If the organoclay content in the polymeric matrix of the PBT reaches 2 wt% then both the temperature of the destruction increases and the amount of the coke remnant increase at 660 °C and then both stay practically unchanged with the further increase of the organoclay content up to 5 wt%. The loss of the weight due to the destruction of the polymer in pure PBT and its composites looks familiar in all cases below 350 °C. The amount of the organoclay added becomes important above that temperature because the very clays possess good thermal stability and make thermal protection by their layers and form a barrier preventing the volatile products of the decomposition to fly off [43, 60]. Such instance of the improvement in thermal properties was observed in many polymeric composites [64–68]. The weight of the coke remnant increases with the rizing organoclay content up to 2 wt% and stays constant after that. The increase of the remnant may be linked with the high thermal stability of the organoclay itself. Also it is worth noticing that the polymeric chain closed in interlayer space of the organoclay has fewer degrees of oscillatory motion at heating due to the limited interlayer space and the formation of the abundant intermolecular bonds between the polymeric chain and the clay surface. And the best result is obtained at 2–3 wt% content of the organoclay added to the polymer.

If one considers the influence of the organoclay added to the polyethelenetere-phtalate, briefly PET, [69] then the temperature of the destruction increases on 16 °C at optimal amount of organoclay of 3 wt%. The coke remnant at 600 °C again increases with the rizing organoclay content, see Table 3.

Regarding the change in the temperature of the destruction in the cases of PET and PBT versus the organoclay content one can note that both trends look similar. However the coke remnant considerably increases would the tripheyldodecy-lphosphonium cation be present within the clay. The melting temperature does not increase in case of the organoclay added into the PET in contrary to the case of PBT. Apparently this may be explained by the more crystallinity of the PBT and the growth of the degree of crystallinity with the organoclay content.

It becomes obvious after analyzing the above results that the introduction of the organoclay into the polymer increases the thermal stability of the latter according the (1) thermal isolating effect from the clay layers and (2) barrier effect in relation to the volatile products of destruction.

The studying of the mechanical properties of the nanocomposites, see Table 2, have shown that the limit of the tensile strength increases with the organoclay added up to 3 wt% for the majority of the composites. Further addition of the organoclay, up to 5 wt%, results in the decreasing limit of the tensile strength. We explain this by the fact that agglomerates appear in the nanocomposite when the organoclay content exceeds the 3 wt% value [61, 70, 71]. The proof for the formation of the agglomerates have been obtained from the X-ray study and using the data from electron microscopes.

Nevertheless the rigidity modulus increases with the amount of the organoclay added into the polymeric matrix, the resistance of the clay itself being the explanation for that. The oriented polymeric chains in the clay layers also participate in the increase of the rigidity modulus [72]. The percentage of enlargement at breaking became 6–7 wt% for all mixes.

Using data of the Table 2 we explain the improvement in the mechanical properties of the nanocomposites with added organoclay up to 3 wt% by the good degree of distribution of the organoclay within the polymeric matrix. The degree of the improvement also depends on the interaction between the polymeric chain and clay layers.

The study of the influence of the degree of the extract of fibers on the mechanical properties has shown that the limit of strength and the rigidity modulus both increase in PBT whereas they decrease in nanocomposites, Table 5. This can be explained by the breaking of the bonds between the organoclay and PBT at greater degree of extract. Such phenomena have been observed in numerous polymeric composites [73–75].

TABLE 5   The ability to stretch of the nanocomposites PBT/organoclay at varying degrees of extract.

Organoclay content, weight %	Limit of strength, MPa			Rigidity modulus, GPa		
	DR=1	DR=3	DR=6	DR=1	DR=3	DR=6
0 (pure PBT)	41	50	52	1.37	1.49	1.52
3	60	35	29	1.76	1.46	1.39

The first notions on the lowered flammability of the polymeric nanocomposites on the organoclay basis appeared in 1976 in the patent on the composite based on the nylon-6 [5]. The serious papers in the field were absent till the 1995 [76].

The use of the calorimeter is very effective for studying the refractoriness of the polymers. It can help at measuring the heat release, the carbon monoxide depletion and others. The speed of the heat release is one of the most important parameters defining the refractoriness [77]. The data on the flame resistance in various polymer/organoclay systems such as layered nanocomposite nylon-6/organoclay, intercalated nanocomposites polystyrene (or polypropylene)/organoclay were given in paper [78] in where the lowered flammability was reported, see Table 6. And the lowered flammability have been observed in systems with low organoclay content, namely in range from 2 to 5 wt%.

TABLE 6   Calorimetric data.

Sample	Remnant (%)±0.5	Peak of the HRR (Δ%) (kW/m^2)	Middle of the HRR (Δ%) (kW/m^2)	Average value H_c (MJ/kg)	Average value SEA (m^2/kg)	Average CO left (kg/kg)
Nylon-6	1	1010	603	27	197	0.01

TABLE 6   *(Continued)*

Nylon-6/organoclay, 2%, delaminated	3	686 (32%)	390 (35%)	27	271	0.01
Nylon-6/organoclay, 5%, delaminated	6	378 (63%)	304 (50%)	27	296	0.02
Polystirene	0	1120	703	29	1460	0.09
PS/organoclay, 3%, bad mixing	3	1080	715	29	1840	0.09
PS/organoclay, 3%, intercalated/delaminated	4	567 (48%)	444 (38%)	27	1730	0.08
PS w/DBDPO/Sb$_2$O$_3$, 30%	3	491 (56%)	318 (54%)	11	2580	0.14
Polypropylene	0	1525	536	39	704	0.02
PP/organoclay, 2%, intercalated	5	450 (70%)	322 (40%)	44	1028	0.02

H$_c$ – heat of combustion;
SEA – specific extinguishing area;
DBDPO – dekabrominediphenyloxide;
HRR – speed of the heat release.

The curve of the heat release for the polypropylene and the nanocomposite on its basis (organoclay content varying from 2 to 4 wt%) is given in the Fig.16 from which one can see that the speed of the heat release for the nanocomposite enriched with the 4 wt% organoclay (the interlayer distance 3.5 nm) is 75% less than for pure polypropylene.

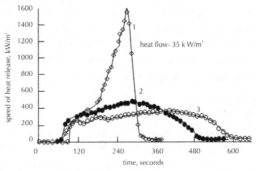

FIGURE 16   The speed of the heat release for:
1 – pure polypropylene;
2 – nanocomposite with 2 wt% of organoclay;
3 – nanocomposite with 4 wt% of organoclay.

The comparison of the experimental data for the nanocomposites on the basis of the nylon-6, polypropylene and polystirene gathered in Table 7 show that the heat of combustion, the smoke release and the amount of the carbon monoxide are almost constant at varying organoclay content. So we conclude that the source for the increased refractoriness of these materials is the stability of the solid phase and not the influence of the vapor phase. The data for the polystirene with the 30% of the dekabrominediphenyloxide and Sb_2O_3 are given in Table 6 as the proof of the influence of the vapor phase of bromine. The incomplete combustion of the polymeric material in the latter case results in low value of the heat of the combustion and high quantity of the carbon monoxide released [79].

One should note that the mechanism for the increased fire resistance of the polymeric nanocomposites on the basis of the organoclays is not, in fact, clear at all. The formation of the barrier from the clay layers during the combustion at their collapse is supposed to be the main mechanism. That barrier slows down the combustion [80]. In our paper we study the influence of the nanocomposite structure on the refractoriness. The layered structure of the nanocomposite expresses higher refractoriness comparing to that in intercalated nanocomposite, see Fig. 17.

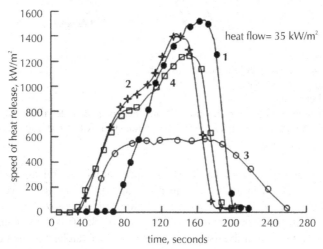

FIGURE 17   The speed of the heat release for:
1 – pure polystirene (PS);
2 – PS mixed with 3 wt% of Na$^+$ MMT;
3 – intercalated/delaminated PS (3 wt% 2C18-MMT) extruded at 170 °C;
4 – intercalated PS (3 wt% C14-FH) extruded at 170 °C.

The data on the polymeric polystirene – based nanocomposites presented in Fig.17 are for (1) initial ammoniumfluorine hectorite and (2) quaternary ammonium montmorillonite. The intercalated nanocomposite was produced in first case whereas the layered-intercalated nanocomposite was produced in the second one. But because the chemical nature and the morphology of the organoclay used was quite different it is very difficult to draw a unique conclusion about the flame resistance in polymeric nanocomposites produced. Nonetheless, one should point out that good results of the same quality were obtained for both layered and intercalated structures when studying the aliphatic groupings of the polyimide nanocomposites based on these clays. The better refractoriness is observed in case of polystirene embedded in layered nanocomposite while intercalated polystirene – based nanocomposite (with MMT) also exhibits increased refractoriness.

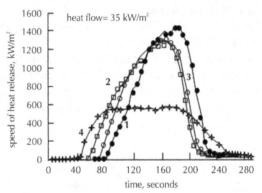

FIGURE 18   The speed of the heat release for:

1 – pure polysterene (PS);

2 – polystirene with Na-MMT;

3 – intercalated PS with organomontmorillonite obtained in extruder at 185 °C;

4 – intercalated/layered PS with organoclay obtained in extruder at 170 °C in nitrogen atmosphere or in vacuum.

As one can see the from the Fig.18, the speed of the heat release for the nanocomposite produced in nitrogen atmosphere at 170 °C is much lower than for other samples. Probably, the reason for the low refractoriness of the nanocomposite produced in extruder without the vacuuming at 180 °C is the influence of the high temperature and of the oxygen from the air what can lead to the destruction of the polymer in such conditions of the synthesis.

It is impossible to give an exact answer on the question about how the refractoriness of organoclay – based nanocomposites increases basing on only the upper experimental data but the obvious fact is that the increased thermal stability and refractoriness are due to the presence of the clays existing in the polymeric matrix as nanoparticles and playing the role of the heat isolators and elements preventing the flammable products of the decomposition to fly off.

There are still many problems unresolved in the field but indisputably polymeric nanocomposites will take the leading position in the chemistry of the advanced materials with high heat and flame resistance. Such materials can be used either as itself or in combination with other agents reducing the flammability of the substances.

The processes of the combustion are studied for the number of a polymeric nanocomposites based on the layered silicates such as nylon-6.6 with 5 wt% of Cloisite 15A – montmorillonite being modified with the dimethyldialkylammonium (alkyls studied C_{18}, C_{16}, C_{14}), maleinated polypropylene and polyethylene, both (1.5%) with 10 wt% Cloisite 15A. The general trend is two times reduction of the speed of the heat release. The decrease in the period of the flame induction is reported for all nanocomposites in comparison with the initial polymers [54].

The influence of the nanocomposite structure on its flammability is reflected in the Table 7. One can see that the least flammability is observed in delaminated nanocomposite based on the polystyrene whereas the flammability of the intercalated composite is much higher [54].

TABLE 7   Flammability of several polymers and composites

Sample	Coke remnant, wt%	Max speed of heat release, kW/m^2	Average value of heat release, kW/m^2	Average heat of combustion, MJ/kg	Specific smoke release, m^2/kg	CO release, kg/kg
Nylon-6	1	1010	603	27	197	0.01
Nylon-6 + 2% of silicate (delaminated)	3	686	390	27	271	0.01
Nylon-6 + 5% of silicate (delaminated)	6	378	304	27	296	0.02
Nylon-12	0	1710	846	40	387	0.02
Nylon-12 + 2% of silicate (delaminated)	2	1060	719	40	435	0.02
Polystyrene	0	1562	803	29	1460	0.09
PS + 3% of silicate Na-MMT	3	1404	765	29	1840	0.09
PS + 3% of silicate C14-FH (intercalated)	4	1186	705	28	1790	0.09
PS + 3% of silicate 2C18-MMT (delaminated)	4	567	444	28	1730	0.08
Polypropylene	0	1525	536	39	704	0.02
PP + 2% of silicate (intercalated)	3	450	322	40	1028	0.02

The optical properties of the nanocomposites are of much interest too. The same materials could be either transparent or opaque depending on certain conditions. For example in Fig.19 we see transparency, plot (a), and turbidity, plot (c), of the material in dependence of the frequency of the current applied.

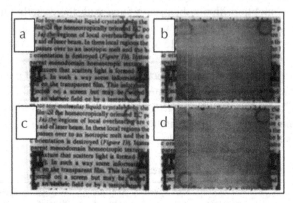

FIGURE 19   The optical properties of the clay – based nanocomposites in dependence of the applied electric current:
(a) – low frequency, switched on;
(b) – low frequency, switched off;
(c) – high frequency, switched on;
(d) – high frequency, switched off.

The effect in Fig. 19 is reversible and can be innumerately repeated. The transparent and opaque states exhibit the memory effect after the applied current switched off, plots (b) and (d) in Fig. 19. The study of the intercalated nanocomposites based on the smectite clays reveals that the optical and elecrooptical properties depend on the degree of intercalation [81].

7.7   CONCLUSIONS

The quantity of the papers in the field of the nanocomposite polymeric materials has grown multiple times in recent years. The possibility to use almost all polymeric and polycondensated materials as a matrix is shown. The nanocomposites from various organoclays and polymers have been synthesized. Here is just a small part of the compounds for being the matrix referenced in literature: polyacrylate [83], polyamides [82, 84, 85], polybenzoxasene [86], polybutyleneterephtalate [11, 82, 87], polyimides [88], polycarbonate [89], polymethylmetacrylate [90], polypropylene [91, 92], polystirene [90], polysulphones [93], polyurethane

[94], polybuthyleneterephtalate and polyethyleneterephtalate [10, 65, 68, 79, 95, 99–107], polyethylene [96], epoxies [97].

The organomodified montmorillonite is of the special interest because it can be an element of the nanotechnology and it can also be a carrier of the nanostructure and of asymmetry of length and width in layered structures. The organic modification is being usually performed using the ion-inducing surface-active agents. The nonionic hydrophobization of the surface of the layered structures have been reported either. The general knowledge about the methods of the study is being formed and the understanding of the structure of the nanocomposite polymeric materials is becoming clear. Also scientists come closer to the realizing of the relations between the deformational and strength properties and the specifics of the nanocomposite structure. The growth of researches and their direction into the nanoarea forecasts the fast broadening of the industrial involvement to the novel and attractive branch of the materials science.

KEYWORDS

- bentonite
- methylmetacrylate
- organoclays
- polybutylenterephtalate
- polymethylmetacrylate

REFERENCES

1. Romanovsky, B. V.; Makshina, E. V.; Sorosovskii obrazovatelniu zhurnal, **2004**, *8(2)*, 50–55 (in Russian)
2. Golovin, Yu. I.; Priroda, 1–2004 (in Russian)
3. Carter, L. W.; Hendrics, J. G.; Bolley, D. S.; United States Patent **1950**, *2*, 531,396.
4. Blumstain A.; *Bull Chem. Soc.* **1961**, 899–905.
5. Fujiwara S.; Sakamoto T.; Japanese Application *(109)*, 998; 1976.
6. Usuki A.; Kojima Y.; Kawasumi M.; Okada A.; Fukushima Y.; Kurauchi T.; Kamigatio O.; Journal of Appl polym science. **1995**, *55*, 119.
7. Usuki A.; Koiwai A.; Kojima Y.; Kawasumi M.; Okada A.; Kurauchi T.; Kamigaito O.; Journal of Appl polym science. **1995**, *55*, 119
8. Okada A.; Usuki A.; Mater Sci Engng. **1995**, *3*, 109.
9. Okada A.; Fukushima Y.; Kawasumi M.; Inagaki S.; Usuki A.; Sugiyama S.; Kurauchi T.; Kamigaito O.; United States Patent **1988**, *4*, 739,007.

10. Mikitaev, M. A.; Lednev, O. B.; Kaladjian, A. A.; Beshtoev, B. Z.; Bedanokov, A. Yu.; Mikitaev, A. K.; Second International Conference (Nalchik 2005).

11. Chang, J. -H.; An, Y. U.; Kim, S. J.; S. Im. Polymer, **2003**, *44,* 5655–5661.

12. Mikitaev, A. K.; Bedanokov, A. Y.; Lednev, O. B.; Mikitaev, M. A. Polymer/silicate nanocomposites based on organomodified clays/Polymers, Polymer Blends, Polymer Composites and Filled Polymers. Synthesis, Properties, Application. Nova Science Publishers: New York **2006**.

13. Delozier, D. M.; Orwoll, R. A.; Cahoon, J. F.; Johnston, N. J.; Smith, J. G.; Connell, J. W.; *Polymer,* **2002**, *43,* 813–822.

14. Kelly, P.; Akelah, A.; Moet, A. J.; *Mater. Sci.* **1994**, *29,* 2274–2280.

15. Chang, J. -H.; An, Y. U.; Cho, D.; Giannelis, E. P.; *Polymer,* **2003**, *44,* 3715–3720.

16. Yano, K.; Usuki, A.; Okada A. *J. Polym. Sci. Part A: Polym. Chem.* **1997**, *35,* 2289.

17. Garcia-Martinez, J. M.; Laguna, O.; Areso, S.; Collar EP. *J. Polym. Sci. Part B: Polym. Phys.* **2000**, *38,* 1564.

18. Giannelis, E. P.; Krishnamoorti R.; Manias E.; Advances in Polymer Science, 138 Springer-Verlag: Berlin Heidelberg, **1999**.

19. Delozier, D. M.; Orwoll, R. A.; Cahoon, J. F.; Ladislaw, J. S.; Smith, J. G.; Connell, Polymer, J. W.; **2003**, *44,* 2231–2241.

20. Shen, Y. -H.; *Chemosphere,* **2001**, *44,* 989–995.

21. Giannelis, E. P.; *Adv. Mater.* **1996**, *8,* 29–35.

22. Dennis, H. R.; Hunter, D. L.; Chang, D.; Kim, S.; White, J. L.; Cho, J. W.; Paul, Polymer, D. R.; **2001**, *42,* 9513–9522.

23. Kornmann, X.; Lindberg, H.; Berglund, L. A.; *Polymer* **2001**, *42,* 1303–1310.

24. Fornes, T. D.; Paul, D. R.; Formation and properties of nylon 6 nanocomposites. Polímeros 13 no.4 São Carlos Oct./Dec. 2003.

25. Voulgaris, D.; Petridis, D.; *Polymer,* **2002**, *43,* 2213–2218.

26. Tyan, H. -L.; Liu, Y. -C.; Wie, K. -H.; *Polymer,* **1999**, *40,* 4877–4886.

27. Davis, C. H.; Mathias, L. J.; Gilman, J. W.; Schiraldi, D. A.; Shields, J. R.; Trulove, P.; Sutto, T. E.; Delong, H. C. *J. Polym. Sci. Part B: Polym. Phys.* **2002**, *40,* 2661.

28. Morgan, A. B.; Gilman, J. W. *J. Appl. Polym. Sci.* **2003**, *87,* 1329.

29. John N.; Hay and Steve J.; Shaw. Organic-inorganic hybridssthe best of both worlds? Europhysics News **2003**, *34(3).*

30. Chang, J. H.; An YU, Sur GS. *J. Polym. Sci. Part B: Polym. Phys.* **2003**, *41,* 94.

31. Chang, J. H.; Park, D. K.; Ihn, K. J. *J. Appl. Polym. Sci.* **2002**, *84,* 2294.

32. Pinnavaia, T. J.; Science **1983**, *220,* 365.

33. Messersmith, P. B.; Giannelis, E. P.; *Chem. Mater.* **1993**, *5,* 1064.

34. Vaia, R. A.; Ishii, H.; Giannelis, E. P.; *Adv. Mater.* **1996**, *8,* 29.

35. Gilman, J. W.; *Appl. Clay. Sci.* **1999**, *15,* 31.

36. Fukushima, Y.; Okada, A.; Kawasumi, M.; Kurauchi, T.; Kamigaito O.; *Clay Miner.* **1988**, *23,* 27.

37. Akelah, A.; Moet A. J Mater Sci **1996**, *31,* 3589.

38. Vaia, R. A.; Ishii, H.; Giannelis, E. P.; *Adv. Mater.* **1996**, *8,* 29.

39. Vaia, R. A.; Jandt, K. D.; Kramer, E. J.; Giannelis, E. P.; Macromolecules **1995**, *28,* 8080.

40. Greenland DG. J Colloid Sci **1963**, *18,* 647.

41. Chang, J. H.; Park, K. M.; *Polym. Eng. Sci.* **2001**, *41,* 2226.

42. Greenland DG. J Colloid Sci **1963**, *18,* 647.
43. Chang, J. H.; Seo, B. S.; Hwang, D. H.; Polymer **2002,** *43,* 2969.
44. Vaia, R. A.; Jandt, K. D.; Kramer, E. J.; Giannelis, E. P.; Macromolecules **1995,** *28,* 8080.
45. Fukushima, Y.; Okada, A.; Kawasumi, M.; Kurauchi, T.; Kamigaito O.; *Clay Miner.* **1988,** *23,* 27.
46. Chvalun S. N.; Priroda **2000,** 7 (in Russian)
47. Brinker, C. J.; Scherer G. W.; Sol-Gel Science. Boston, 1990.
48. Mascia, L.; Tang, T.; Polymer **1998,** *39,* 3045.
49. Tamaki, R.; Chujo, Y.; *Chem. Mater.* **1999,** *11,* 1719.
50. Serge Bourbigot, E. A. Investigation of Nanodispersion in Polystirene–Montmorillonite Nanocomposites by Solid-State NMR.; Journal of Polymer Science: Part B: Polymer Physics, *41,*3188–3213 **2003,**
51. Lednev, O. B.; Kaladjian, A. A.; Mikitaev, M. A.; Tlenkopatchev, M. A.; New polybutylene terephtalate and organoclay nanocomposite materials. Abstracts of the International Conference on Polymer materials (México, 2005)
52. Tretiakov A. O.; Oborudovanie I instrument dlia professionalov (02(37) **2003,** (in Russian)
53. Sergeev Ros, G. B.; Chem. J. (Journal of D. I. Mendeleev Russian Chemical Society), **2002,** XLVI, 5 (in Russian)
54. Lomakin, S. M.; Zaikov Visokomol, G. E.; Soed. B. **2005,** *47(1),* 104–120 (in Russian)
55. Xu H, Kuo, S. W.; Lee, J. S.; Chang FC.; Macromolecules **2002,** *35,* 8788.
56. Haddad, T. S.; Lichtenhan, J. D.; Macromolecules **1996,** *29,* 7302.
57. Mather, P. T.; Jeon, H. G.; Romo-Uribe, A.; Haddad, T. S.; Lichtenhan, J. D.; Macromolecules **1996,** *29,* 7302.
58. Hsu, S. L. C.; Chang KC.; Polymer **2002,** *43,* 4097.
59. Chang, J. H.; Seo, B. S.; Hwang, D. H.; Polymer **2002,** *43,* 2969.
60. Fornes, T. D.; Yoon, P. J.; Hunter, D. L.; Keskkula, H.; Paul DR.; Polymer **2002,** *43,* 5915.
61. Chang, J. H.; Seo, B. S.; Hwang, D. H.; Polymer **2002,** *43,* 2969.
62. Fornes, T. D.; Yoon, P. J.; Hunter, D. L.; Keskkula, H.; Paul DR.; Polymer **2002,** *43,* 5915.
63. Wen, J.; Wikes, G. L.; *Chem. Mater.* **1996,** *8,* 1667.
64. Zhu, Z. K.; Yang, Y.; Yin, J.; Wang, X.; Ke Y, Qi Z. *J. Appl. Polym. Sci.* **1999,** *3,* 2063.
65. Mikitaev, M. A.; Lednev, O. B.; BeshtoevB. Z.; Bedanokov, A. Yu.; Mikitaev A. K.; Second International conference "Polymeric composite materials and covers" (Yaroslavl **2005,** may) (in Russian)
66. Fischer, H. R.; Gielgens, L. H.; Koster TPM.; Acta Polym **1999,** *50,* 122.
67. Petrovic, X. S.; Javni, L.; Waddong, A.; Banhegyi GJ. *J. Appl. Polym. Sci.* **2000,** *76,* 133.
68. Lednev, O. B.; Beshtoev, B. Z.; Bedanokov, A. Yu.; Alarhanova, Z. Z.; Mikitaev A. K.; Second International Conference (Nalchik 2005) (in Russian)
69. Chang, J. -H.; Kim, S. J.; Joo, Y. L.; S. Im. Polymer, **2004,** *45,* 919–926.
70. Lan, T.; Pinnavaia, T. J.; *Chem. Mater.* **1994,** *6,* 2216.

71. Masenelli-Varlot, K.; Reynaud, E.; Vigier, G.; Varlet, J. *J. Polym. Sci. Part B: Polym. Phys.* **2002,** *40,* 272.

72. Yano, K.; Usuki, A.; Okada A. J Polym Sci Part A: Polym Chem **1997,** *35,* 2289.

73. Shia, D.; Hui, Y.; Burnside, S. D.; Giannelis, E. P.; *Polym. Eng. Sci.* **1987,** *27,* 887.

74. Curtin WA. J Am Ceram Soc **1991,** *74,* 2837.

75. Chawla KK.; Composite materials science and engineering. New York: Springer; 1987.

76. Burnside, S.; D.; Giannelis, E. P.; *Chem. Mater.* **1995,** *7,* 4597.

77. Babrauskas, V.; Peacock, R. D.; *Fire Safety J.* **1992,** *18,* 225.

78. Gilman, J.; Kashiwagi, T.; Lomakin, S.; Giannelis, E.; Manias, E.; Lichtenhan, J.; Jones, P.; In: *Fire Retardancy of Polymers: the Use of Intumescence.* The Royal Society of Chemistry, Cambridge, **1998,** 203–221.

79. Mikitaev, A. K.; Kaladjian, A. A.; Lednev, O. B.; Mikitaev M. A.; Plastic masses **2004,** *12,* 45–50 (in Russian)

80. Gilman, J.; Morgan A. 10th Annual BCC Conference, May 24–26, 1999.

81. John N.; Hay and Steve, J.; Shaw. Organic-inorganic hybrids: the best of both worlds? Europhysics News **2003,** *34(3).*

82. Delozier, D. M.; Orwoll, R. A.; Cahoon, J. F.; Johnston, N. J.; Smith, J. G.; Connell, Polymer, J. W.; **2002,** *43,* 813–822.

83. Chen, Z.; Huang, C.; Liu, S.; Zhang, Y.; Gong K. J Apply Polym Sci **2000,** *75,* 796–801.

84. Okado, A.; Kawasumi, M.; Kojima, Y.; Kurauchi, T.; Kamigato O.; *Mater. Res. Soc. Symp. Proc.* **1990,** *171,* 45.

85. Leszek A.; Utracki, Jorgen Lyngaae-Jorgensen. Rheologica Acta, **2002,** *41,* 394–407.

86. Wagener, T. R.; Reizinger, J. G.; Polymer, **2003,** *44,* 7513–7518.

87. Li, X.; Kang, T.; Cho, W. J.; Lee, J. K.; Ha, C. S.; Macromol Rapid Commun.

88. Tyan, H. -L.; Liu, Y. -C.; Wie, K. -H.; *Polymer,* **1999,** *40,* 4877–4886.

89. Vaia, R.; Huang, X.; Lewis, S.; Brittain W.; Macromolecules **2000,** *33,* 2000–4.

90. Okamoto, M.; Morita, S.; Taguchi, H.; Kim, Y.; Kotaka, T.; Tateyama H.; Polymer **2000,** *41,* 3887–90.

91. Chow, W. S.; Mohd Ishak, Z. A.; Karger-Kocsis, J.; Apostolov, A. A.; Ishiaku, Polymer, U. S.; **2003,** *44,* 7427–7440.

92. Antipov, E. M.; Guseva, M. A.; Gerasin, V. A.; Korolev, Yu.M.; Rebrov, A. V.; Fisher, H. R.; Razumovskaya I. V.; Visokomol Soed. A. **2003,** *45(11),* 1885–1899 (in Russian).

93. Sur, G.; Sun, H.; Lyu, S.; Mark, J.; *Polymer* **2001,** *42,* 9783–9.

94. Wang, Z.; Pinnavaia T.; *Chem. Mater.* **1998,** *10,* 3769–71.

95. Bedanokov A. Yu.; Beshtoev B. Z. /Malij polimernij congress (Moscow 2005) (in Russian)

96. Antipov, E. M.; Guseva, M. A.; Gerasin, V. A.; Korolev, Yu.M.; Rebrov, A. V.; Fisher, H. R.; Razumovskaya Visokomol, I. V.; Soed, A. **2003,** *45(11),* 1874–1884 (in Russian).

97. Lan T.; Kaviartna P.; Pinnavaia T.; Proceedings of the ACS PMSE **1994,** *71,* 527–8.

98. Kawasumi, C., et al. Nematic liquid crystal/clay mineral composites. Science and Engineering, *6,* 135–143, 1998.

99. Lednev, O. B.; Kaladjian, A. A.; Mikitaev M. A.; Second International Conference (Nalchik 2005) (in Russian)

100. Mikitaev, A. K.; Kaladjian, A. A.; Lednev, O. B.; Mikitaev, M. A.; Davidov, E. M.; Plastic masses **2005**, *4*, 26–31(in Russian)

101. Eid, A.; Mikitaev, M. A.; Bedanokov, A. Y.; Mikitaev, A. K. Recycled Polyethylene Terephthalate/Organo-Montmorillanite Nanocomposites, Formation And Properties The first Afro-Asian Conference on Advanced Materials Science and Technology (AMSAT 06), Egypt, 2006.

102. Mikitaev, A. K.; Bedanokov, A. Y.; Lednev, O. B.; Mikitaev, M. A.; Polymer/silicate nanocomposites based on organomodified clays/ Polymers, Polymer Blends, Polymer Composites and Filled Polymers. Synthesis, Properties, Application. Nova Science Publishers. New York 2006.

103. Malamatov, A. H.; Kozlov, G. V.; Mikitaev M. A.; Mechanismi uprochnenenia polimernih nanokompozitov, (Moscow, RUChT 2006) 240 p. (in Russian)

104. Eid ,A.; Doctor Thesis, (Moscow, RUChT 2006) 121 p. (in Russian)

105. Lednev, O. B.; Doctor Thesis (Moscow, RUChT 2006) 128 p. (in Russian)

106. Malamatov, A. H.; Professor Thesis (Nalchik, KBSU 2006) 296 p. (in Russian)

107. Borisov, V. A.; Bedanokov, A. Yu.; Karmokov, A. M.; Mikitaev, A. K.; Mikitaev, M. A.; Turaev, E. R.; *Plastic Mass.* 2007, 5 (in Russian).

CHAPTER 8

QUANTUM – CHEMICAL MODELING

M. A. CHASHKIN, V. I. KODOLOV, A. I. ZAKHAROV,
YU. M. VASILCHENKO, M. A. VAKHRUSHINA, V. V. TRINEEVA,
and G. E. ZAIKOV

CONTENTS

8.1 INTRODUCTION

This chapter is dedicated to the investigation of modification processes of cold hardened epoxy compositions with metal/carbon nanocomposites to improve their operational characteristics (adhesive strength and thermal stability). The work is theoretically substantiated with quantum-chemical modeling.

The chapter contains the investigation results of viscous properties of metal/carbon nanocomposite fine suspensions and thermal and physical properties of metal/carbon nanocomposite/epoxy compositions obtained.

The quantum-chemical modeling of the systems imitating the behavior of copper/carbon nanocomposite fine suspensions and epoxy resins cold hardened with them is given. IR spectra of polyethylene polyamine and metal/carbon nanocomposite fine suspensions based on it with concentrations 0.001–0.03% from polyethylene polyamine mass are analyzed and their comparative analysis is presented. The possible processes flowing in fine suspensions during the interaction of copper/carbon nanocomposite and polyethylene polyamine are described, as well as the processes influencing the increase in adhesive strength and thermal stability of metal/carbon nanocomposite/epoxy compositions.

This decade has been heralded by a large-scale replacement of conventional metal structures with structures from polymeric composite materials (PCM). Currently, we are facing the tendency of production growth of PCM with improved operational characteristics. In practice, PCM characteristics can be improved when applying modern manufacturing technologies, for example, the application of "binary" technologies of prepreg production [1], as well as the synthesis of new polymeric PCM matrixes or modification of the existing polymeric matrixes with different fillers.

The most cost-efficient way to improve operational characteristics is to modify the existing polymeric matrixes; therefore, currently the group of polymeric materials modified with nanostructures (NS) is of special interest. NS are able to influence the supermolecular structure, stimulate self-organization processes in polymeric matrixes in supersmall quantities, thus contributing to the efficient formation of a new phase "medium modified-nanocomposite" and qualitative improvement of the characteristics of final product – PCM. This effect is especially visible when NS activity increases which directly depends on the size of specific surface, shape of the particle and its ultimate composition [2]. Metal ions in the NS used in this chapter also contribute to the activity increase as they stimulate the formation of new bonds.

The increase in the attraction force of two particles is directly proportional to the growth of their elongated surfaces; therefore, the possibility of NS coagulation and decrease in their efficiency as modifiers increases together with their activity growth. This fact and the fact that the effective concentrations to modify polymeric matrixes are usually in the range below 0.01 mass % impose specific methods for introducing NS into the material modified. The most justified and widely applicable are the methods for introducing NS with the help of fine suspensions (FS). This introduction method allows most uniformly distributing particles in the volume of the medium modified, decreasing the possibility of their coagulation and preserving their activity during storage.

8.2 QUANTUM – CHEMICAL INVESTIGATION OF FRAGMENTS INTERACTION IN EPOXY POLYMERS MODIFIED WITH METAL/ CARBONIC NANOCOMPOSITES

In the process of quantum-chemical modeling the fragments imitating the initial reagents are optimized: epoxy diane resin (EDR), polyethylene polyamine (PEPA), cobalt/carbon nanocomposite (Co/C NC), nickel/carbon nanocomposite (Ni/C NC), copper/carbon nanocomposite (Cu/C NC) with the inclusion of CO_2^+, Ni^{2+}, Cu^{2+} ions (Fig. 1). For each of these fragments the absolute value of binding energy is defined E_{EDR}, E_{PEPA}, $E_{NC(Me)}$, where Me – cobalt, nickel or copper ion.

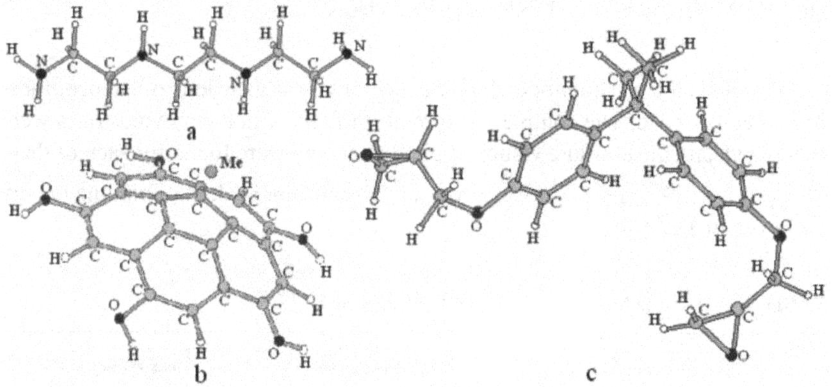

FIGURE 1   Fragments of initial substances: (a) PEPA fragment; (b) Me/C NC fragment; (c) EDR fragment.

As the modification process initially assumed the production of fine suspensions (FS) of NC on PEPA basis, the fragments imitating the behavior of the corresponding suspensions of Co/C, Ni/C, Cu/C nanocomposites were optimized (Fig. 2) and their absolute values of binding energy $E_{PEPA\text{-}NC(Co)}$, $E_{PEPA\text{-}NC(Ni)}$, $E_{PEPA\text{-}NC(Cu)}$ were defined.

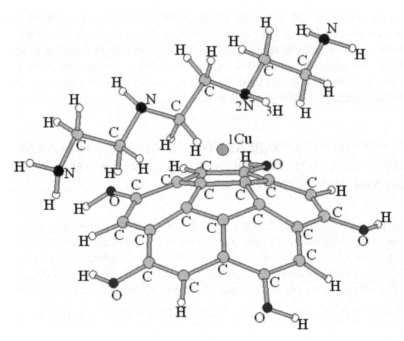

FIGURE 2   Fragment of Cu/C NC FS on PEPA basis.

The next step was to model the influence of fine suspensions of nanocomposites on epoxy resin. The complexes formed similarly with the previous ones were optimized, and the absolute values of binding energy were found for each of them $E_{PEPA-NC(Co)}$, $E_{PEPA-NC(Ni)}$, $E_{PEPA-NC(Cu)}$. The absolute values of binding energy are given in Table 1.

TABLE 1   Absolute binding energies of the fragments.

	$E_{NC(Me)}$, KJ/mol	$E_{PEPA-NC(Me)}$, KJ/mol	$E_{EDR-PEPA-NC(Me)}$, KJ/mol
CO_2^+	−19116.50	−29992.05	−51486.96
Ni^{2+}	−18562.38	−29098.04	−50621.15
Cu^{2+}	−18340.32	−28764.72	−50315.94
E_{EDR}, KJ/mol	−21424.25		
E_{PEPA}, KJ/mol	−10131.36		

Using the data from Table 1 by the following formula:

$$E_1 = E_{EDR-PEPA-NC(Me)} - E_2 - E_{EDR}$$

$$E_2 = E_{PEPA-NC(Me)} - E_{PEPA}$$

the relative interaction energies of molecular complexes E_1 are calculated and the diagram is arranged (Fig. 3).

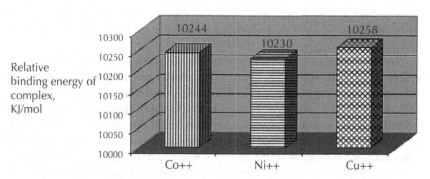

FIGURE 3   Diagram of relative interaction energies of molecular complexes.

From Fig. 3, it is seen that the relative interaction energy of molecular complexes with Cu/C NC is higher in comparison with the complexes with Co/C NC and Ni/C NC content. As the polymer forms the strongest complexes with Cu/C NC, therefore after the modification it will be the most effective. The detailed analysis of the lengths of the bonds formed and effective charges before and after the optimization of fragments imitating PEPA interaction with Cu/C NC (Tables 2 and 3) indicates that stable coordination bonds were formed between NC fragment and PEPA (between copper ion and nitrogen atom of amine group NH of PEPA). It was found that after the interaction of two fragments studied a part of electron density of N atom participating in the bond shifted to Cu atom, thus resulting in NH bond weakening.

TABLE 2   Bond lengths.

Bond designation	Bond length before optimization, Å	Bond length after optimization, Å
Cu-N	2.82	1.95
Cu-C	2.65	2.25

TABLE 3   Effective charges.

Atom number (see Fig. 2)	Atom designation	Effective charge before optimization	Effective charge after optimization
1	Cu (copper)	0.138	−0.250
2	N (nitrogen)	−0.066	0.420
3	H (hydrogen)	0.045	0.027

The bond weakening is indirectly confirmed by the increase in the effective charge of H atom and slight change in the wave number in oscillatory spectra calculated. The wave number of NH bond before the optimization was 3360 cm⁻¹, and after 3354 cm⁻¹, which correlates with the data of IR spectra obtained with the help of IR Fourier spectrometer. For instance, in the spectrum of PEPA and NC suspension on PEPA basis with nanocomposite concentration 0.03% the shift of wave numbers of peaks of amine groups is observed from 3280 cm⁻¹ to 3276 cm⁻¹.

The modeling of hardening process with the participation of epoxy resin, polyethylene polyamine and Cu/C nanocomposite is given in Figs. 4a and 4b, respectively. From the geometry of optimized molecular systems it is seen that the introduction of Cu/C NC into the system leads to its self-organization (Fig. 4b) and formation of presumably coordination polymer. The formation of coordination polymer due to the introduction of nanocomposite active particles can result in increasing the adhesive strength and thermal stability as the total number of bonds in polymer grid grows and more energy will be required for its thermal destruction. At the same time, the formation of nanocomposite metal coordination bond with PEPA nitrogen can increase the stability of fine suspension formed. Such coordination results in polyethylene polyamine activity growth during the hardening of epoxy diane resin.

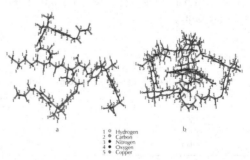

a   1 ○ Hydrogen   b
                     2 ● Carbon
                     3 ● Nitrogen
                     4 ● Oxygen
                     5 ● Copper

FIGURE 4   Influence of Cu/C NC on epoxy resin (ER) hardening: (a) ER hardening without Cu/C NC; (b) ER hardening in the presence of Cu/C NC.

Thus, quantum-chemical modeling allows predicting the interaction processes of components when hardening the epoxy resin with polyethylene polyamine and active participation of copper/carbon nanocomposite.

8.2 EXPERIMENTAL PART

8.2.1 MATERIALS

Currently there is a huge need in modern epoxy systems with improved operational characteristics [3] that can be reached, as previously mentioned, when modifying them with nanostructures. Therefore in this chapter the modification processes of cold-hardened model epoxy composition containing epoxy diane resin ED-20 State Standard (GOST) 10587-84 in the amount of 100 weight fractions and polyethylene polyamine (PEPA) grade A Technical Condition (TU) 2413-357-00203447-99 in the amount of 10 weight fractions were considered as the research object. The modification was carried out when introducing Metal/C nanocomposite into the epoxy resin.

PEPA represents the mixture of linear branched ethylene polyamines with average molecular mass 200–250 and very wide molecular-mass distribution. The general structural formula of PEPA is as follows:

$$NH_2[-CH_2CH_2NH-]_nH, \text{ where } n = 2-8.$$

Synthesized Co/C, Ni/C, Cu/C nanocomposites were studied with the help of transmission electron microscopy (TEM) and electron microdiffraction (EMD) in the shared centers of research institutions in Ekaterinburg and Moscow.

The investigations of Cu/C and Ni/C nanocomposites revealed that their average sizes (r) differ approximately in two times: r of Cu/C NC is 25 nm, r of Ni/C NC is about 11 nm. The average size of Co/C nanocomposite practically equals the average size of Ni/C NC. Such correlation of average sizes of nanocomposites is apparently connected with the ability to form metal containing clusters. Atom magnetic moments of nanocomposites, as demonstrated in [], are greater than the corresponding moments of microparticles of the same metals. The availability of magnetic properties of nanoproducts widens their application possibilities.

8.2.2 PROCESSING OF FINE SUSPENSIONS

The fine suspension was prepared in a number of stages:
  1. Preliminary grinding of Cu/C NC in mechanical mortar.

2. Mechanical and chemical activation of suspension components when combining PEPA and Cu/C nanocomposite.
3. Ultrasound processing for complete and uniform dispergation of Cu/C NC particles in PEPA volume.

8.2.3 PROCESSING OF EPOXY RESINS MODIFIED WITH CU/C NANOCOMPOSITE

The samples of epoxy polymer modified with Cu/C NC were produced mixing the epoxy resin heated up to 60 °C and fine suspension of Cu/C NC on PEPA basis in proportion 10:1. The mixing took 5–10 min. A part of the polymer modified with NC was poured into the mold with copper wire to test the adhesive strength (Fig. 5) and was further hardened in the mold. Another part was hardened in the form of plates to grind the samples for thermogravimetric investigation.

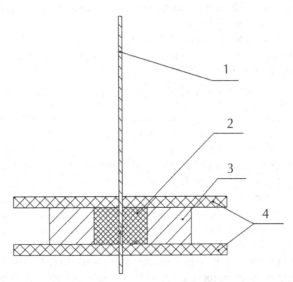

FIGURE 5 Sample for determining the adhesive strength of cold-hardened epoxy composition (CHEC): (1) copper wire; (2) hardened composition; (3) metal mold; (4) antiadhesive liner.

Two sets of samples were produced to determine the adhesive strength, including the reference and modified FS of Cu/C NC with concentrations 0.001%,

0.01% and 0.03%. For oxide film removal and degreasing the copper wire was treated with 0.1 M solution of hydrochloric acid and acetone.

For thermogravimetric investigation also two sets of samples were produced, including the reference and modified FS of Cu/C NC with concentrations 0.001%, 0.01%, 0.03%, 0.05%.

8.2.4   INVESTIGATION TECHNIQUE

1. *Quantum-chemical modeling.* In the frameworks of this method the fragments imitating the behavior of Co/C, Ni/C, Cu/C nanocomposites, fine suspensions of nanocomposites and systems being epoxy resins modified with nanocomposites were constructed and optimized with the help of software HyperChem v. 6.03 and semiempirical methods. At the same time, the absolute binding energies of certain components of the molecular systems formed were found and their relative interaction energies were calculated. The oscillatory spectra of fine suspensions of Cu/C nanocomposite were calculated with the help of semiempirical method PM3.

2. *IR spectroscopy.* For IR spectra IR Fourier-spectrometer FSM 1201 was used. The IR spectra of liquid films of PEPA, NC FS with concentrations 0.001%, 0.01% and 0.03% were taken. The spectra were taken on KBr glasses in wave number range 399–4500 cm^{-1}.

3. *Optical spectroscopy.* Spectral photometer KFK-03-01 was used to define the optimal time period of NC FS ultrasound processing. The optimal time period value was the interval when the optical density (D) was at the maximum, that is, corresponded to the maximum saturation of FS with NC. The optical density of NC FS samples with the concentration 0.01% processed in ultrasound bath Sapfir UZV 28 within 0, 3, 7, 10, 15, 20, 30 min, respectively, at US power 0.5 kW and frequency 35 kHz was measured in 5-mL quartz cuvettes. The work wavelength was found when defining the optical density in the range λ=320–920 nm.

4. *Determination of relative viscosity.* The relative viscosity of PEPA and FS on its basis was found with the help of viscometer VZ-246 in accordance with State Standard (GOST) 8420-74. The relative viscosity values were translated into the kinematic following GOST 8420-74. The viscosity was measured for PEPA processed and not processed with ultrasound, and also for fine suspensions of NC with the concentrations 0.001%, 0.01% and 0.03% processed and not processed with ultrasound.

5. *Thermogravimetric technique.* Thermal stability was found by the destruction temperatures of modified epoxy resins. The temperatures of destruction beginning were determined by thermogravimetric (TG) curves. Derivatographer DIAMOND TG/DTA was applied to obtain TG curves. TG curves of modified

epoxy resins with NC concentrations 0.001%, 0.03%, 0.05% from PEPA weight was taken. The sample heating rate was 5 °C/min.

6. *Determination of adhesive strength.* The adhesive strength of NC/EC was found with the technique described (Fig. 5). The tests were carried out on the tensile testing machine. The strength was found comparing breaking stresses of CHEC and NC/EC samples.

8.3 RESULTS AND DISCUSSION

8.3.1 OPTIMAL TIME PERIOD OF ULTRASOUND PROCESSING OF FINE SUSPENSIONS OF COPPER/CARBON NANOCOMPOSITE

In the process of selecting the work wavelength (λ) on spectral photometer KFK-03-01 to define the optimal time period of ultrasound processing of FS on PEPA basis the optimal wavelength 413 nm was found. This wavelength was selected as the optical density of FS with the concentration of Cu/C NC 0.01% not processed with ultrasound was 0.800 that corresponded to the middle of the operational range of this instrument (operational range D = 0.001–1.5). At the wavelength λ = 413 nm the optical densities of all FS investigated were defined. This is reflected in the graphic dependence of optical density D upon the time period of ultrasound processing τ_{us} (Fig. 6).

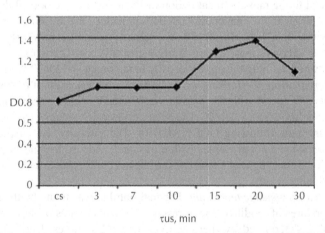

FIGURE 6 Dependence of the change in optical density (D) of Cu/C NC FS upon the time period of ultrasound processing τ_{us}.

The analysis of optical density dependence upon the time period of ultrasound processing (Fig. 6) demonstrates that the optimal processing time of Cu/C NC fine suspension on PEPA basis is 20 min. Further, ultrasound processing is useless as in 20 min the maximum optical density 1.37 is reached.

Taking into account the above data for defining the optimal time period of ultrasound processing, the preparation of fine suspension for IR investigations, for finding the viscosity and for producing the modified epoxy resin was carried out with preliminary ultrasound processing within 20 min.

8.3.2 VISCOUS PROPERTIES OF FINE SUSPENSIONS OF COPPER/CARBON NANOCOMPOSITE

The following diagram was prepared based on the results of viscosity measurements of PEPA and fine suspension (FS) with different Cu/C NC content before and after US processing (Fig. 7).

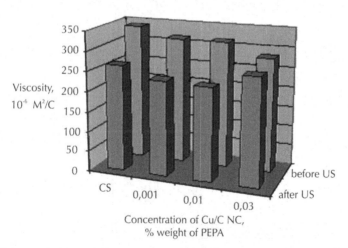

FIGURE 7 Diagram of the dependence of FS kinematic viscosity on Cu/C NC concentration.

The decrease in FS viscosity with the increase in Cu/C NC concentration on the diagram (Fig. 7) is explained by the fact that the system consisting of two phases, in our case (PEPA – dispersion medium and Cu/C NC – disperse phase) tends to surface energy decrease. This tendency is expressed by self-decrease in interface surface due to sorption. PEPA molecules start sorbing on Cu/C NC surface, probably producing the interface between Cu/C NC particles being the obstacle for coagulation. Due to such localization the interaction between medium particles diminishes, resulting in intermolecular friction decrease, and, consequently, decrease in system viscosity (Fig. 8a). When FS are processed with ultrasound, the size of disperse phase decreases, thus leading to the growth of surface energy and increase in sorption ability, the action region on disperse medium increases (Fig. 8b). The action regions of Cu/C nanocomposite in fine suspension with concentrations 0.001% and 0.01% do not overlap that is confirmed by viscosity decrease, and for FS with concentration 0.03% the action regions overlap and, consequently, the increase in intermolecular friction and viscosity are observed.

Taking into account that the action degree on the medium mainly depends on the action time of Cu/C NC on the medium modified, FS were studied for 48 h. After this time interval FS layered into floccular structures with no sediment on the bottom observed (Fig. 8c).

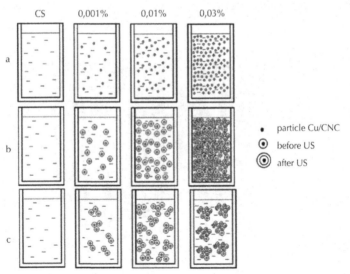

FIGURE 8 Distribution of Cu/C NC in PEPA: (a) before ultrasound processing; (b) after ultrasound processing; (c) after 48 h.

Mechanical action resulted in FS recovery, which is indirectly confirmed by the availability of PEPA stable layer between Cu/C nanocomposite particles preventing the coagulation and possibility of suspension production with the ability to recover the distribution of nanocomposite particle distribution.

8.3.3 IR SPECTROSCOPIC INVESTIGATION

The comparative IR spectroscopic investigation of *bis*(2-aminoethyl)amine – one of PEPA olygomers demonstrates the consistency of wave numbers appropriate for these compounds.

In both spectra the wave numbers appropriate for the oscillations of amine groups $v_s(NH_2)$ 3352 cm⁻¹ and asymmetric $v_{as}(NH_2)$ 3280 cm⁻¹ are available, there are wave numbers that refer to symmetric $v_s(CH_2)$ 2933 cm⁻¹ and asymmetric valence $v_{as}(CH_2)$ 2803 cm⁻¹, deformation wagging oscillations $v_d(CH_2)$ 1349 cm⁻¹ of methylene groups, deformation oscillations v_d (NH) 1596 cm⁻¹ and $v_d(NH_2)$ 1456 cm⁻¹ of amine groups, and also the oscillations of skeleton bonds are vivid $v(CN)$ 1059–1273 cm⁻¹ and $v(CC)$ 837 cm⁻¹.

IR spectra of PEPA and Cu/C NC FS were taken (Fig. 9).

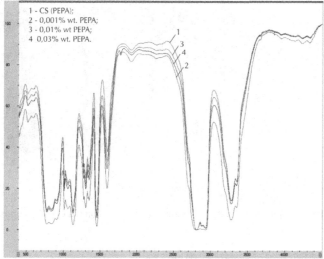

FIGURE 9   IR spectra of PEPA and FS of Cu/C nanocomposite.

The comparison of IR spectra of polyethylene polyamine and fine suspension of Cu/C NC on PEPA basis (Fig. 9) indicates that practically all changes of wave numbers in the spectra are within the error ± 2 cm⁻¹. However, in FS spectra the vivid increase in peak intensity corresponding to deformation oscillations of NH bonds is observed. These changes can spread onto the vast areas arranging a certain supermolecular structure, apparently involving the adjoining amine groups into the process, which is demonstrated by the intensity change of these peaks.

8.3.4 ADHESION

The tests for defining the adhesion of modified epoxy resin to copper wire were carried out on tensile testing machine, the values of destruction load were found. The adhesive strength was calculated by the following formula:

$$\sigma = F/A \text{ [MPa]}$$

where F – average load values at which the breaking-off took place, (kgs); A – area of wire interaction with the hardened composition, (cm²).

$$A = 2 \cdot \pi \cdot r \cdot h = 2 \cdot 3,\ 14 \cdot 0,\ 05 \cdot 1 = 0.314 \text{ (cm}^2\text{)},$$

where r – wire radius 0.05 (cm); h – height of metal mold 1.00 (cm).

The following diagram was prepared based on tests (Fig. 10)

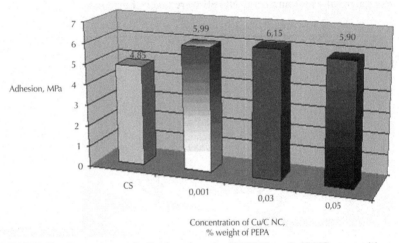

FIGURE 10   Dependence of adhesive strength of NC/EC on Cu/C NC composition.

8.3.5 THERMOGRAVIMETRIC INVESTIGATIONS

To define the influence of nanocomposite on thermal stability of epoxy composition, a number of thermogravimetric investigations were carried out on reference and modified samples. The concentrations 0.001%, 0.03% and 0.05% from PEPA weight were used. Based on the results of thermogravimetric investigations the following diagram was prepared (Fig. 11).

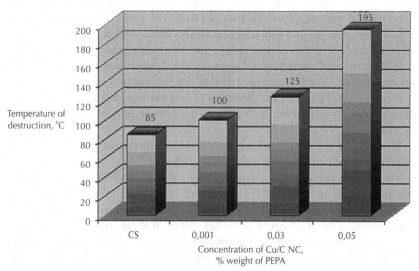

FIGURE 11   Dependence of thermal stability of cold-hardened epoxy composition on the concentration of copper/carbon nanocomposite

The data from the diagram of adhesive strength (Fig. 10) indicate that the strength maximum of the modified epoxy resin is reached when Cu/C nanocomposite concentration is 0.03%. Probably the maximum strength of the modified epoxy resin at this concentration is conditioned by the optimal number of a new phase growth centers. The adhesion decrease with the concentration increase indicates that the number of Cu/C nanocomposite particles exceeded the critical value, which depends on their activity [2]. Therefore probably the number of cross-links in the polymer grid increased – the material became brittle. At the same time, when the concentration of nanocomposite elevates, the growth of polymer thermal stability is observed (Fig. 11). The thermal stability growth is apparently connected with the increase in the number of coordination bonds in epoxy polymer.

8.4   CONCLUSION

Quantum-chemical modeling methods allow quite precisely defining the typical reaction of interaction between the system components, predict the properties of molecular systems, decrease the number of experiments due to the imitation of technological processes. The computational results are completely comparable with the experimental modeling results.

The optimal composition of nanocomposite was found with the help of software HyperChem v. 6.03. Cu/C nanocomposite is the most effective for modifying the epoxy resin. Nanosystems formed with this NC have higher interaction energy in comparison with nanosystems produced with Ni/C and Co/C nanocomposites. The effective charges and geometries of nanosystems were found with semiempirical methods. The fact of producing stable fine suspensions of nanocomposite on PEPA basis and increasing the operational characteristics of epoxy polymer was ascertained. It was demonstrated that the introduction of Cu/C nanocomposite into PEPA facilitates the formation of coordination bonds with nitrogen of amine groups, thus resulting in PEPA activity increase in epoxy resin hardening reactions.

It was found out that the optimal time period of ultrasound processing of copper/carbon nanocomposite fine suspension is 20 min.

The dependence of Cu/C nanocomposite influence on PEPA viscosity in the concentration range 0.001–0.03% was found. The growth of specific surface of NC particles contributes to partial decrease in PEPA kinematic viscosity at concentrations 0.001%, 0.01% with its further elevation at the concentration 0.03%.

IR investigation of Cu/C nanocomposite fine suspension confirms the quantum-chemical computational experiment regarding the availability of NC interactions with PEPA amine groups. The intensity of these groups increased in several times when Cu/C nanocomposite was introduced.

The test for defining the adhesive strength and thermal stability correlate with the data of quantum-chemical calculations and indicate the formation of a new phase facilitating the growth of cross-links number in polymer grid when the concentration of Cu/C nanocomposite goes up. The optimal concentration for elevating the modified epoxy resin (ER) adhesion equals 0.003% from ER weight. At this concentration the strength growth is 26.8%. From the concentration range studied, the concentration 0.05% from ER weight is optimal to reach a high thermal stability. At this concentration the temperature of thermal destruction beginning increases up to 195 °C.

Thus, in this chapter the stable fine suspensions of Cu/C nanocomposite were obtained. The modified polymers with increased adhesive strength (by 26.8%) and thermal stability (by 110 °C) were produced based on epoxy resins and fine suspensions.

KEYWORDS

- epoxy compositions
- metal/carbon nanocomposites
- polymeric composite materials
- quantum-chemical modeling

REFERENCES

1. Panfilov, B. F. Composite materials: production, application, market tendencies. *Polymeric Materials*, **2010**, *2–3,* 40–43.
2. Kodolov, V. I.; Khokhriakov, N. V.; Trineeva, V. V.; Blagodatskikh, I. I. Activity of nanostructures and its expression in nanoreactors of polymeric matrixes and active media. *Chemical Physics and Mesoscopy*, **2008**, *10(4),* 448–460.
3. Bobylev, V. A.; Ivanov, A. V. New epoxy systems for glues and sealers produced by CJSC "Chimex Ltd." *Glues. Sealers. Paints*, **2008**, *2*, 162–166.
4. Kodolov, V. I.; Kovyazina, O. A.; Trineeva, V. V.; Vasilchenko, Yu. M.; Vakhrushina, M. A.; Chmutin I. A. On the production of metal/carbon nanocomposites, water and organic suspensions on their basis. VII International Scientific-Technical Conference "Nanotechnologies to the production, 2010." *Proceedings Fryazino*, **2010**, 52–53.

CHAPTER 9

NOVEL PHASE OF ELEMENTAL SILVER NANO-PARTICLES FORMATION

N. I. NAUMKINA, O. V. MIKHAILOV, and T. Z. LYGINA

CONTENTS

9.1   INTRODUCTION

The process of "reprecipitation" of elemental silver according to scheme $Ag \rightarrow Ag_4$ $[Fe(CN)_6] \rightarrow Ag$ into gelatin-immobilized matrices, on the first stage of which these matrices are processed with water-alkaline solutions containing potassium hexacyanoferrate(II) and potassium hexacyanoferrate(III), on the second stage, with water-alkaline solutions containing tin(II) dichloride and organic or inorganic substance forming rather stable soluble coordination compounds with Ag(I). $D = f(D^{Ag})$ and $D = f(C_{Ag}^{V})$ dependences where D is an optical density of "reprecipitated" silver in the matrix corresponding to initial density D^{Ag} and volume concentrations of elemental silver C_{Ag}^{V}. It has been found that a degree of influence of complex-formed substance (D/D^{Ag}) is determined with its nature as well as its concentration in reducing solution; besides, the most considerable this influence is in the case of ethanediamine-1,2, the least considerable one, in the case of ammonia. It has been found that silver formed in gelatin matrix, contains two various phases distinguished between themselves by optical and XRD parameters; in addition, one of these phases consists of nano-particles of elemental silver.

Even more 40 years ago in Ref. [1] it was mentioned about colloid element silver with rather small size of the particles, formed in gelatin layers at development of silver halide photographic emulsions. A number of works, in which there are indications on existence of a separate phase of the element silver consisting from nano-particles and received as a result of photochemical reduction of Ag(I) salts, was appeared lately in the literature, in particular Refs. [2–11]. Before in Refs. [10, 11], it has been noted that at development of gelatin layers of silver halide photographic emulsions by water-alkaline solutions containing tin(II) dichloride and some inorganic or organic substance forming rather stable coordination compounds with Ag(I), formation only of element silver occurs, too, however, besides the gelatin layer is tinged brown or red but not black color as it takes place at standard development by using hydroquinone developers. It is significant that with increase of optical densities of the gelatin layer containing such elemental silver, red tone in coloring of gelatin layer becomes more and more clearly expressed. The similar phenomenon takes place, too, when instead of silver halide AgHal in a gelatin matrix there is such silver(I) compound as silver(I) hexacyanoferrate(II) $Ag_4[Fe(CN)_6]$. Whether is this totality of particles a novel phase of element silver? Or it is only a variety of known phases of the given simple substance? This question remains till now opened and deserves special consideration.

As it is known through Refs. [12–15], physical-chemical processes in the so-called gelatin-immobilized matrix systems, in particular the reaction of nucleophilic substitution, ion exchange, and "self-assembly" of metal chelates are often accompanied by the formation of such compounds that can not be obtained by carrying out similar processes in solution or solid phase. The reason for this is, on the one hand, the fact that these processes occur in the intermolecular cavities of gelatin; on the other hand, the fact that in this particular reaction system, there is a preliminary reduction of entropy, which are made possible through such processes that normally thermodynamic forbidden. In this connection, it may be expected that in the gelatin-immobilized matrix systems, specific redox processes, in particular, with the formation of particles of elemental metals, can occur. In this connection, Chapter 9 is devoted to consideration of redox-reactions connected with obtaining element silver.

9.2 EXPERIMENTAL PART

As initial material to obtain silver-containing gelatin-immobilized matrix implants (hereafter **GIM**), X-ray film *Structurix D-10 (Agfa-Gevaert, Belges)* was. Samples of the given film (which actually is nothing but AgHal-**GIM**) having format 20×30 cm^2 were exposed to X-ray radiation with an irradiation dose at a range 0.05–0.50 Röntgen. These exposed sample were further subjected to processing according to the following technology [10, 11]:

- Development in D-19 standard developer as it was indicated in Ref. [10, 11], for 6 min at 20–25 °C;
- Washing with running water for 2 min at 20–25 °C;
- Fixing in 25% water solution of sodium trioxosulphidosulfate(VI) (Na-$_2$S$_2$O$_3$) for 10 min at 20–25 °C;
- Washing with running water for 15 min at 18–25 °C.

Three first stages of standard processing indicated (development, washing and fixing) were carried out at nonactinic green-yellow light, final washing – at natural light. The samples of **GIM** obtained which contained elemental silver (Ag-**GIM**), then were processed according to next technology:

1. Oxidation in water solution containing (g×L^{-1})

Potassium hexacyanoferrate(III)	50.0
Potassium hexacyanoferrate(II)	20.0
Potassium hydroxide	10.0
Sodium trioxocarbonate(IV)	5.0
(Na$_2$CO$_3$) Water	up to 1000 ml for 6 min at 20–25 °C;

2. Washing with running water for 2 min at 20–25 °C;
3. Reduction in water solution containing (g×L^{-1})

Tin(II) chloride	50.0
Sodium N, N'-ethylenediaminetetraacetate	35.0
Potassium hydroxide	50.0
Reagent formed water-soluble complex with Ag(I)	1.0–100.0
	up to 1000 ml
Water	for 1 min at 20–25 °C;

4. Washing with running water for 15 min at 18–25 °C;
5. Drying for 2–3 hours at 20–25 °C.

As complex-forming reagents that form water-soluble complexes with Ag(I), ammonia NH_3, potassium thiocyanate KSCN, sodium trioxosulphidosulfate(VI) $Na_2S_2O_3$, ethanediamine-1,2 $H_2N-CH_2-CH_2-NH_2$, 2-aminoethanol $H_2N-CH_2-CH_2-OH$ and 3-(2-hydroxyethyl)-3-azapenthanediol-1,5 $N(CH_2-CH_2-OH)_3$, were used. At the first stage of given processing of Ag-**GIM** obtained, conversion of Ag-**GIM** into $Ag_4[Fe(CN)_6]$-**GIM** occurred, on the second stage, reduction of $Ag_4[Fe(CN)_6]$-**GIM** with Sn(II) to elemental silver took place. And so, peculiar "reprecipitation" of elemental silver into gelatin matrix occurred incidentally.

An isolation substances from Ag-**GIM** was carried out by means of influence on them of water solutions of some proteolytic enzymes (e.g., trypsin or *Bacillus mesentericus*) destroying the polymeric carrier of a **GIM** (gelatin), and the subsequent separation of a solid phase from mother solution according to a technique described in Ref. [16]. The substances isolated thus from **GIM** further analyzed by X-ray diffraction method with using of spectrometer D8 Advance (Bruker, Germany). A scanning was carried out in an interval from 3 up to 65° 2θ, a step was 0.05–2θ.

Calculation of intensities of reflexes (*I*) and interplane distances (*d*) carried out with application of standard software package EVA. Theoretical XRD spectra (X-ray patterns) were calculated under *PowderCell* program described in Refs. [17, 18]. Optical density Ag-**GIM** was measured by means of Macbeth TD504 photometer (Kodak, USA) in a range 0.1–5.0 units with accuracy of ±2 % (rel.).

9.3   RESULTS AND DISCUSSION

Already at visual observation over a course of process of transformation $Ag_4[Fe(CN)_6]$-**GIM** into Ag-**GIM,** following circumstance attracts its attention. Ag-**GIM** received as a result of standard processing of exposed AgHal-**GIM**, at rather small optical density (D^{Ag}) have gray color, at big D^{Ag}, black color. Coloring

Ag-**GIM** containing the "reprecipitated" element silver, varies from black-brown to red depending on the nature and quantity of complex-forming reagent contained in solution.

It is significant, however, that absorption spectra of both initial and the "reprecipitated" element silver in visible area do not contain any accurately expressed maxima. Besides, optical density Ag-**GIM** with the "reprecipitated" silver (D^{Ag}), at the same volume concentration of element silver (C_{Ag}^V) in **GIM**, as a rule, is essentially more than D^{Ag} values and, also, depends on nature and quantity of complex-forming reagent in solution contacting with **GIM**. Examples of $D = f(D^{Ag})$ и $D = f(C_{Ag}^V)$ dependences for inorganic and organic reagents are presented in Figs. 1–6. It is significant that (D/D^{Ag}) value, as a rule, greater than 1.0, and in some cases, it reaches very high values (as in the case of potassium thiocyanate – nearly 5.0). Attention is drawn to the fact that the stronger the color of the gelatin layer with the "reprecipitated" elemental silver is different from the gray-black tones of the gelatinous layer initially Ag-gelatin-immobilized matrix, the greater is the (D/D^{Ag}) value. The maximal degree of amplification (D/D^{Ag})$_{max}$ is also very much depends on the nature of the complexing agent (Table 1); the most profound effect on this parameter has etandiamin-1,2 [(D/D^{Ag})$_{max}$ = 5.80], the least severe – ammonia, the degree of possibility of which, nevertheless, is also quite high [(D/D^{Ag})$_{max}$ = 3.40]. For the ammonia, the growth (D/D^{Ag}) values is typical with increasing concentration of NH_3 in reducing solution to a relatively small its value (~0.30 mol×L^{-1}), after which the optical density **D** begin to fall (Fig. 1). It is noteworthy that red-brown color of gelatin layer attained at the indicated concentration, with a further increase in the concentration of ammonia does not change. Analogous situation occurs in the case of the other two we studied inorganic complexing agents – trioxosulfidosulfate(VI) and the thiocyanate anion (Figs. 2–3), with the only difference being that in the case of $S_2O_3^{2-}$ maximum degree of amplification is achieved with less high in comparison with NH_3 concentration (0.15 mol×L^{-1}), in the case of SCN$^-$ – at a higher concentration (~ 0.70 mol×L^{-1}). In this regard, it was quite natural to try the difference marked with different stability of 1:2 complexes formed by Ag(I) with NH_3, $S_2O_3^{2-}$ and SCN$^-$ (pK = 7.25, 13.32 and 8.39, respectively). However, in the presence of such a correlation, the value of concentration indicated for SCN$^-$ should be lower than for NH_3, which in fact is not observed. In reality, for the three organic complex-forming substances studied by us, molar concentrations at which the maximum value (D/D^{Ag}) is reached, significantly greater than those for inorganic complex-forming substances (Table 1). Stability of the complexes of silver(I) with each of these ligands is lower than with NH_3, and the correlation function between these concentrations and the stability of coordination compounds of Ag(I) with given ligands in varying degrees, still visible. But any explosion-term relationship that very stability of the (D/D^{Ag}) values do not see: how it may be easily noticed when comparing the data of Table 1, the maximum degree of amplification decreases in the direction of ethanediamine-1,2 > 2-aminoethanol > SCN⁻ > $S_2O_3^{2-}$ > NH_3 > 3-(2-hydroxyethyl)-3-azapenthane-diol-1,5, while the resistance formed by these ligands complexes with silver(I) – in the direction $S_2O_3^{2-}$ > SCN⁻ > ethanediamine-1,2 > NH_3 > 2-aminoethanol > 3-(2-hydroxyethyl)-3-azapenthanediol-1,5. Thus, the complexing is though important, but not the sole determinant of the degree of influence of complex-forming agents on the redox process considered.

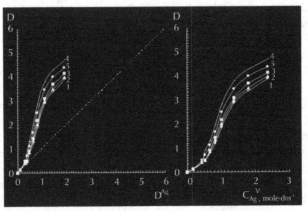

FIGURE 1 Dependences of **D** = f(D^{Ag}) and **D** = f(C_{Ag}^V) of reduction process $Ag_4[Fe(CN)_6] \rightarrow Ag$ with using of NH_3 at concentration 1.5 g×L⁻¹ (curve 1), 3.0 g×L⁻¹ (2), 4.5 g×L⁻¹ (3), 6.0 g×L⁻¹ (4) and 7.5 g×L⁻¹ (5). Optical densities D^{Ag} and **D** were measured with blue light-filter with a transmission maximum at 450 nm.

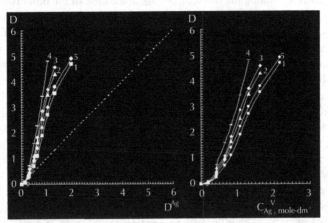

FIGURE 2 Dependences of **D** = f(D^{Ag}) and **D** = f(C_{Ag}^V) of reduction process $Ag_4[Fe(CN)_6] \rightarrow Ag$ with using of $Na_2S_2O_3$ at concentration 2.0 g×L⁻¹ (curve 1), 4.0 g×L⁻¹ (2), 8.0 g×L⁻¹ (3), 24.0 g×L⁻¹ (4) and 40.0 g×L⁻¹ (5). Optical densities D^{Ag} and **D** were measured with blue light-filter with a transmission maximum at 450 nm.

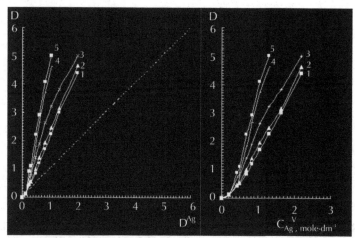

FIGURE 3 Dependences of $D = f(D^{Ag})$ and $D = f(C_{Ag}^V)$ of reduction process $Ag_4[Fe(CN)_6] \rightarrow Ag$ with using of KSCN at concentration 2.0 g×L^{-1} (curve 1), 4.0 g×L^{-1} (2), 8.0 g×L^{-1} (3), 24.0 g×L^{-1} (4) and 60.0 g×L^{-1} (5). Optical densities D^{Ag} and D were measured with blue light-filter with a transmission maximum at 450 nm.

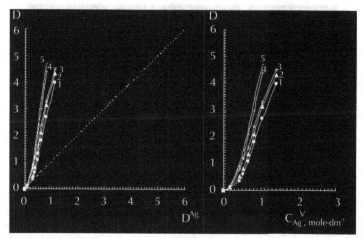

FIGURE 4 Dependences of $D = f(D^{Ag})$ and $D = f(C_{Ag}^V)$ of reduction process $Ag_4[Fe(CN)_6] \rightarrow Ag$ with using of 2-aminoethanol $H_2N–(CH_2)_2–OH$ at concentration 7.5 g×L^{-1} (curve 1), 15.0 g×L^{-1} (2), 55.0 g×L^{-1} (3), 110.0 g×L^{-1} (4) and 150.0 g×L^{-1} (5). Optical densities D^{Ag} and D were measured with blue light-filter with a transmission maximum at 450 nm.

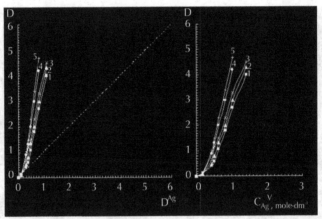

FIGURE 5 Dependences of $\mathbf{D} = f(D^{Ag})$ and $\mathbf{D} = f(C_{Ag}^{V})$ of reduction process $Ag_4[Fe(CN)_6] \rightarrow Ag$ with using of ethanediamine-1,2 $H_2N-(CH_2)_2-NH_2$ at concentration 5.0 g×L⁻¹ (curve 1), 10.0 g×L⁻¹ (2), 20.0 g×L⁻¹ (3), 40.0 g×L⁻¹ (4) and 80.0 g×L⁻¹ (5). The optical densities D^{Ag} and $\mathbf{D}$ were measured with blue light-filter with a transmission maximum at 450 nm.

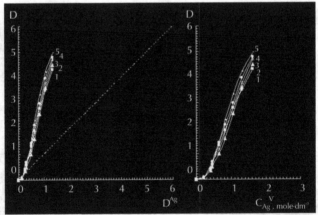

FIGURE 6 Dependences of $\mathbf{D} = f(D^{Ag})$ and $\mathbf{D} = f(C_{Ag}^{V})$ of reduction process $Ag_4[Fe(CN)_6] \rightarrow Ag$ with using of 3-(2-hydroxyethyl)-3-azapenthanediol-1,5 $N(CH_2-CH_2-NH_2)_3$ at concentration 10.0 g×L⁻¹ (curve 1), 20.0 g×L⁻¹ (2), 35.0 g×L⁻¹ (3), 50.0 g×L⁻¹ (4) and 100.0 g×L⁻¹ (5). The optical densities D^{Ag} and $\mathbf{D}$ were measured with blue light-filter with a transmission maximum at 450 nm.

TABLE 1 The maximal (D/D^{Ag}) and pK_s values of Ag(I) complexes for various complex-forming reagents.

Complex-forming reagent	$(D/D^{Ag})_{max}$	Concentration of complex-forming reagent in solution at which reaches $(D/D^{Ag})_{max}$, $g \times L^{-1}$ $(mol \times L^{-1})$	pK_s of Ag(I) complex having 1:2 composition
NH_3	3.40	4.5 (0.27)	7.25
$Na_2S_2O_3$	4.13	23.7 (0.15)	13.32
KSCN	4.92	69.7 (0.70)	8.39
$HO-(CH_2)_2-NH_2$	5.40	109.8 (1.80)	6.62
$H_2N-(CH_2)_2-NH_2$	5.80	77.0 (1.28)	7.84
$N(CH_2-CH_2-OH)_3$	3.93	99.8 (0.67)	3.64

At the first stage of the given process, reaction, which may be described by Eq. (1), takes place (in the braces {....}, formulas of gelatin-immobilized chemical compounds have been indicated):

$$4\{Ag\} + 4[Fe(CN)_6]^{3-} \rightarrow \{Ag_4[Fe(CN)_6]\} + 3[Fe(CN)_6]^{4-} \qquad (1)$$

Each of complex-forming reagents under examination forms with Ag(I) *soluble* complex having a metal ion: ligand ratio of 1:2. That is why, formation of silver(I) complex with corresponding *CR* will occur to some extent when $Ag_4[Fe(CN)_6]$-**GIM** is at the contact with the solution containing any of complexation reagent indicated. Gelatin-immobilized silver(I) hexacyanoferrate(II) as well as any of these soluble complexes, can participate in process of reduction with Sn(II). In this connection, proceeding two parallel processes Ag(I)→Ag(0) will take place at contact of $Ag_4[Fe(CN)_6]$-**GIM** with the above indicated solution, containing Sn(II) and complexation reagent:

- gelatin-immobilized silver(I) hexacyanoferrate(II) reduction proceeding in *a polymer layer*;
- Ag(I) complex with complexation reagent reduction proceeding *on interface of phases* a **GIM**/solution.

In water solutions at pH=12–13, Sn(II) is mainly in a form of hydroxo-complex $[Sn(OH)_3]^-$. In this connection, Eq. (2) may be offered for the first of these processes in the case of negative charged "acid"-ligands may be ascribed (**L** – symbol of ligand, z – its charge). The particles of element silver formed as a result of Eqs. (3) and (4), theoretically should have smaller sizes than the particles of element silver arising in polymer layer of **GIM**. To be a part of substance immobilized in **GIM**, these particles should place freely in intermolecular cavities of a gelatin layer. Only in this case, they may diffuse in **GIM** and may be immobilized in gelatin mass.

$$\{Ag_4[Fe(CN)_6]\}+2[Sn(OH)_3]^- + 6OH^- \rightarrow 4\{Ag\} + 2[Sn(OH)_6]^{2-} + [Fe(CN)_6]^{4-} \quad (2)$$

For the second of these processes, Eq. (3)

$$2[AgL_2]^+ + [Sn(OH)_3]^- + 3OH^- \rightarrow 2Ag + 4L + [Sn(OH)_6]^{2-} \qquad (3)$$

in the case of noncharged ligands and Eq. (4)

$$2[AgL_2]^{(2z-1)-} + [Sn(OH)_3]^- + 3OH^- \rightarrow 2Ag + 4L^{z-} + [Sn(OH)_6]^{2-} \qquad (4)$$

Gelatin has an extremely high surface area and an extensive system of micropores. The fragment of its structure has been shown in Fig. 7; as may be seen, it contains many intermolecular cavities. It may be valued the average size of intermolecular cavity in the gelatin structure [12–15].

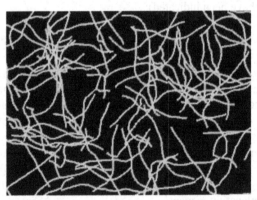

FIGURE 7   The fragment of gelatin structure containing intermolecular cavities.

For example, the volume of polymer layer of **GIM** (V_{gl}) having area 1 cm^2 and thickness 20 μm is $(1.0 \times 1.0 \times 20 \times 10^{-4})$ cm^3 = 2.0×10^{-3} cm^3, so that the mass of gelatin containing in such a layer, at average value of its density 0.5 g cm^{-3}, is $(0.5 \times 2.0 \times 10^{-3})$ g = 1.0×10^{-3} g. Molecular mass of gelatin (M_{Gel}) is known to be $\sim(2.0–3.0) \times 10^5$ c.u. [12, 13], the number of its molecules in given mass will be $(1.0 \times 10^{-3}/M_{Gel}) \times (6.02 \times 10^{23}) = (2.0–3.0) \times 10^{15}$. As discussed above, gelatin molecule in average has a length ~285,000 pm and diameter ~1400 pm, and if it may be considered as narrow cylinder, total volume of gelatin molecules V_M will be equal to $(1/4)\pi D^2 h = (1/4) \times 3.14 \times (285,000 \times 10^{-10}$ cm$) \times (1400 \times 10^{-10}$ cm$)^2$ $= 4.38 \times 10^{-19}$ cm^3. In the case of maximal compact arrangement, these molecules occupy total volume equal to 4.38×10^{-19} cm^3 $(2.0-3.0) \times 10^{15} = (8.76-13.15) \times 10^{-4}$ cm^3. It may be postulated that the volume of cavities indicated, is equal to total volume of polymer massif minus the volume occupied by gelatin molecules, namely $(2.0 \times 10^{-3} - (8.76-13.15) \times 10^{-4})$ cm^3 that will be in the end $(0.69-1.12) \times 10^{-3}$ cm^3. Then, the average volume of one intermolecular cavity may be found as a quotient from division of their total volume into number of gelatin molecules and, as it may be easily noted, will be $(3.4-5.6) \times 10^{-19}$ cm$^3 = (3.4-5.6) \times 10^{11}$ pm^3. The linear size of such an "average" cavity in the case when it has *spherical* form, will be equal to $d = (6 \ V/\pi)^{1/3} = [6 \times (3.7-5.6) \times 10^{11}$ pm$^3/3.14]^{1/3} = (89.1-102.2) \times 10^2$ pm; when it has *cubic* form, equal to $a = V^{1/3} = [(3.7-5.6) \times 10^{11}$ pm$^3)]^{1/3} = (71.8-82.4) \times 10^2$ pm. As one can see from these values, these cavities are nano-sized. Therefore, only *nano-particles* of substance can entry into these cavities. By entering into such cavities, nano-particles of element silver are isolated from each other. In consequence of thereof, their aggregation with each other becomes rather difficult.

With growth of concentration of any of complexation reagent mentioned above, concentration of coordination compounds formed given complexation reagent with Ag(I) must increase. Correspondingly, the quantity of nano-particles of the element silver formed as a result of reduction of these coordination compounds by $[Sn(OH)_3]^-$ complex, should increase, too. In this connection, it may be expected that when concentration of these complexation reagent in a solution increases, the share of nano-particles contained in the "reprecipitated" elemental silver, should accrue gradually. Thus, at the same concentration nano-particles of element silver owing to their higher dispersion degree in comparison with microparticles should provide higher degree of absorption of visible light (and, accordingly, higher optical density) polymeric layer **GIM**. The experimental data presented in Figs. 1–6, are in full conformity with the given prediction.

The particles of element silver formed as a result of Eqs. (3) and (4), are one- or two-nuclear. While it is not enough of them (it occurs, when concentration of Ag(I) complexes on interface of phases a **GIM**/solution is low enough), these particles owing to their remoteness from each other have not time to be aggregated. They diffuses in polymeric layer of **GIM**, where are immobilized without change of their sizes. With increase of complex-forming reagent concentration (and, accordingly, of concentration of Ag(I) complex with given reagent), quantity of nano-particles indicated on interface of phases the **GIM**/solution accrues. It leads to increase of a number of such particles, diffused into **GIM**. However, at some rather high concentration complex-forming reagent in a solution, the effect of aggregation of nano-particles of element silver starts to affect. One- and two-nuclear particles of element silver formed at reduction of corresponding Ag(I) complex, begin to unite to some extent with each other in larger particles. Polynuclear particles of element silver resulting such an association are not so mobile and, consequently, will be not diffuse into polymeric layer of **GIM**. They will be precipitated in it near to interface **GIM**/solution (or even to escape as a solid phase in the solution contacting with **GIM**). As a result, rates of an increment of number of one-and two-nuclear particles of elemental silver with further growth of concentration complex-forming reagent begin to be slowed down. Thus, inevitably there should come the moment when the number of similar particles will reach some limiting value. That is why, since certain "threshold" concentration complex-forming reagent in a solution, growth of D^{Ag} values must stop. Moreover, at excess of this "threshold" concentration, certain decrease D^{Ag} should begin. The point is that an alignment between number of the aggregated particles and number one- and two-nuclear ones with growth of concentration of Ag(I) complex continuously grows and has no restrictions. These polynuclear particles are precipitated in frontier zone **GIM** on rather small depth and form, as a matter of fact, the same kind of elemental silver, as the microparticles of elemental silver formed as a result of reduction of gelatin-immobilized $Ag_4[Fe(CN)_6]$ according to Eq. (2). That is why, D^{Ag} values with increase of complex-forming reagent concentration at first increase, reach a maximum and then decrease.

By taking into consideration the foregoing, it may be assumed that "reprecipitated" gelatin-immobilized silver should contain, as a minimum, two phases of the silver particles, one of which is formed by nano-particles, and another, by microparticles. In order to corroborate the given conclusion, we carried out the analysis of elemental silver isolated from initial Ag-**GIM** and elemental silver isolated from Ag-**GIM** after end of "reprecipitation" process by X-ray powder diffraction method. X-ray powder diffraction patterns (XRD-patterns) of samples obtained are presented in Figs. 8–10. As may be seen from them, XRD-pattern of initial elemental silver with gray-black color of gelatin layer (Fig. 8) and XRD-patterns of "reprecipitated" elemental silver (Figs. 9 and 10), rather essentially differ from each other. So, in XRD-patterns of "reprecipitated" elemental silver obtained at an availability of any of studied *CR* in a solution contacting with **GIM**, there are accurate reflexes having $d = 333.6, 288.5, 166.7$ and 129.1 pm that are absent in XRD-pattern of initial elemental silver. At the same time, reflexes with $d = 235.7, 204.1, 144.4, 123.1$ and 117.9 pm are observed on them. These reflexes are characteristic for the known phase of elemental silver isolated from initial Ag-**GIM**. In this connection, there are all reasons to believe that the "reprecipitated" elemental silver obtained with using of a solution containing any of complex-forming reagent indicated above, contains at least two structural modifications of elemental silver.

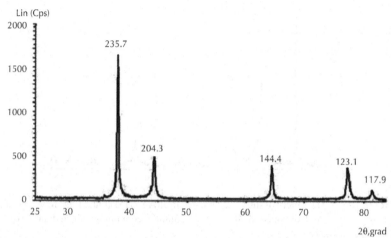

FIGURE 8   XRD-pattern of elemental silver isolated from initial Ag-**GIM**.

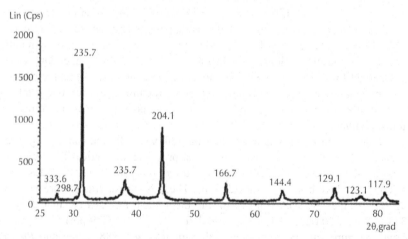

FIGURE 9   XRD-pattern of substance isolated from Ag-**GIM** containing "reprecipitated" elemental silver and obtained with using of solution containing $Na_2S_2O_3$ in concentration 20.0 gxL^{-1}.

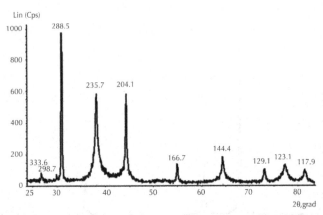

FIGURE 10   XRD-pattern of substance isolated from Ag-**GIM** containing "reprecipitated" elemental silver and obtained with using of solution containing ethanediamine-1,2 in concentration 20.0 g×L^{-1}.

The next curious circumstance attracts its attention: reflexes with d = 333.6, 288.5, 204.2, 166.7 and 129.1 pm are rather close to d values of reflexes of silver(I) bromide AgBr (number of card PDF 06–0438, parameter of an elementary cell a_0 = 577.45 pm, face-centered lattice, cubic syngonia, *Fm3 m* group of symmetry according to the international classification [14, 15]). In this connection, it may be assumed that the structure of the novel phase contained in "reprecipitated" elemental silver, at least in outline, resembles structure AgBr and its crystal lattice is similar to a lattice of silver(I) bromide where positions of atoms Br occupy atoms of silver.

To answer on the question, whether the reflexes indicated can belong to elemental silver with such a space structure in principle, theoretical XRD-patterns of assumed structure of element silver with use of program *PowderCell* described in Ref. [18], have been constructed by us. These XRD-patterns are presented in Fig. 11. As may be seen from them, the theoretical d values calculated by us (333.6, 288.7, 204.2, 174.1, 166.7, 144.4, 132.5, 129.1, 117.9 pm) for specified above structure with an elementary cell parameter a = 288.72 pm and *Pm3 m* symmetry group, practically coincide with d values experimentally observed in XRD-pattern of the "reprecipitated" elemental silver (d = 333.6, 288.5, 166.7 and 129.1 pm). It should be noted in this connection that d values calculated theoretically for the elemental silver isolated from initial Ag-**GIM** (235.4, 204.3, 144.5, 123.2 and 118.0 pm), correspond to compact-packed crystal structure having an elementary cell parameter a = 408.62 pm and *Fm3 m* symmetry group, interplane distances in which are 235.7, 204.1, 144.4, 123.1 and 117.9 pm.

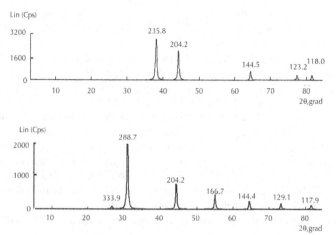

FIGURE 11   Theoretical XRD-patterns of elemental silver contained in initial Ag-**GIM** (*a*) and of elemental silver formed in **GIM** as a result of "reprecipitation" process (*b*).

Thus, formation of a novel phase of elemental silver, which probably was not described in the literature up to now, takes place here indeed.

ACKNOWLEDGEMENT

The Russian Foundation of Basic Researches (RFBR) is acknowledged for the financial support (Grant No. 09-03-97001).

KEYWORDS

- *Bacillus mesentericus*
- gelatin matrix
- *Powder Cell* program
- reprecipitation

REFERENCES

1. Skillman, D. G.; Berry, C. R. Effect of particle shape on the spectral absorption of colloid silver in gelatin. *J. Chem. Phys.* **1968**, *48(7)*, 3297–3304.

2. Linnert, T.; Mulvaney, P.; Henglein, A.; Weller, H. Long-lived nonmetallic silver clusters in aqueous solution: preparation and photolysis. *J. Am. Chem. Soc.* **1990**, *112(12)*, 4657–4664.

3. Fedrigo, S.; Harbich, W.; Butter, J. Collective dipole oscillations in small silver clusters embedded in raregas matrices. *Phys. Rev.* **1993**, *47(23)*, 10706–10715.

4. Satoh, N.; Hasegawa, H.; Tsujii, K.; Kimura, K. Photoinduced coagulation of Ag nanocolloides. *J. Phys. Chem.* **1994**, *98(7)*, 2143–2147.

5. Sato, T.; Ishikawa, T.; Ito, T.; Yonazawa, Y.; Kodono, K.; Sakaguchi, T.; Miya, M.; *Chem. Phys. Lett.* **1995**, *242(3)*, 310–314.

6. Al-Obaidi, A. H. R.; Rigbi, S. J.; McGarvey, J. J.; Wamsley, D. G.; Smith, K. W.; Hellemans, I.; Snauwaert, J. Microstructural and spectroscopies studies of metal liquidlike films of silver and gold. *J. Phys. Chem.* **1994**, *98(24)*, 11163–11168.

7. Ershov, B. G.; Henglein, A. Reduction of Ag$^+$ on polyacrilate chains in aqueous solutions. *J. Phys. Chem. B* **1998**, *102(24)*, 10663–10666.

8. Kapoor, S. Surface modification of silver particles. *Langmuir*, **1998**, *14(5)*, 1021–1025.

9. Sergeev, B. M.; Kiryukhin, M. V.; Prusov, A. N. Effect of light on the disperse composition of silver hydrosols stabilized by partially decarboxylated polyacrylate. *Mendeleev Commun.* **2001**, *11(2)*, 68–69.

10. Mikhailov, O. V.; Guseva, M. V.; Krikunenko, R. I. An amplification of silver photographic images by using of processes changing disperse of image carrier. *Zh. Nauchn. Prikl. Foto-Kinematogr.* **2003**, *48(4)*, 52–57 (in rus.)

11. Mikhailov, O. V.; Kondakov, A. V.; Krikunenko, R. I. Image intensification in silver halide photographic materials for detection of high-energy radiation by repreciptation of elemental silver. *High Energy Chem.*; **2005**, *39(5)*, 324–329.

12. Mikhailov, O. V. Reactions of nucleophilic, electrophilic substitution and template synthesis in the metalhexacyanoferrate(II) gelatin-immobilized matrix. *Rev. Inorg. Chem.*; **2003**, *23(1)*, 31–74.

13. Mikhailov, O. V. Gelatin-Immobilized Metalcomplexes: Synthesis and Employment. *J. Coord. Chem.* **2008**, *61(7)*, 1333–1384.

14. Mikhailov, O. V. Self-Assembly of Molecules of Metal Macrocyclic Compounds in Nanoreactors on the Basis of Biopolymer-Immobilized Matrix Systems. *Nanotechnol. Russ.* **2010**, *5(1–2)*, 18–25.

15. Mikhailov, O. V. Soft template synthesis of Fe(II, III), Co(II, III), Ni(II) and Cu(II) metalmacrocyclic compounds into gelatin-immobilized matrix implants. *Rev. Inorg. Chem.*; **2010**, *30(4)*, 199–273.

16. Mikhailov, O. V. Enzyme-assisted matrix isolation of novel dithiooxamide complexes of nickel(II). *Indian J. Chem.* **1991**, *30A(2)*, 252–254.

17. Kraus, W.; Nolze, G. Powder cell – a program for the representation and manipulation of crystal structures and calculation of the resulting X-ray powder patterns. *J. Appl. Cryst.* **1996**, *29(3)*, 301–303.

18. Powder Diffract File. Search Manual Fink Method. Inorganic. USA, Pennsylvania: JCPDS – International Centre for Diffraction Data, **1995** (release 2000).

CHAPTER 10

A STUDY ON POLYCARBONATE MODIFIED WITH CU/C NANOCOMPOSITE

YU. V. PERSHIN and V. I. KODOLOV

CONTENTS

10.1   INTRODUCTION

Modification of polycarbonate with copper/carbon nanocomposite was investigated. The changes in spectral, optical and thermal-physical characteristics of compositions based on polycarbonate modified with copper/carbon nanocomposites were observed.

Recently, the materials based on polycarbonate have been modified to improve their thermal-physical and optical characteristics and to apply new properties to use them for special purposes. The introduction of nanostructures into materials facilitates self-organizing processes in them. These processes depend on surface energy of nanostructures, which is connected with energy of their interaction with the surroundings. It is known [1] that the surface energy and activity of nanoparticles increase when their sizes decrease.

For nanoparticles, the surface and volume are defined by the defectiveness and form of conformation changes of film nanostructures depending on their crystallinity degree. However, the possibilities of changes in nanofilm shapes with the changes in medium activity are greater in comparison with nanostructures already formed. At the same time, sizes of nanofilms formed and their defectiveness, i.e., tears and cracks on the surface of nanofilms, play an important role [2].

When studying the influence of super-small quantities of substances introduced into polymers and considerably changing their properties, apparently we should consider the role, which these substances play in polymers possessing highly organized super-molecular regularity, both in crystalline and amorphous states. It can be assumed that the mechanism of this phenomenon is in the nanostructure energy transfer to polymer structural formations through the interface resulting in the changes in their surface energy and mobility of structural elements of the polymeric body. Such mechanism is quite realistic as polymers are structural-heterogenic (highly dispersed) systems.

Polycarbonate modification with super-small quantities of Cu/C nanocomposite is possible using fine suspensions of this nanocomposite, which contributes to uniform distribution of nanoparticles in polycarbonate solution.

10.2   EXPERIMENTAL PART

Polycarbonate "Actual" was used as the modified polycarbonate.

Fine suspensions of copper/carbon nanocomposite were prepared combining 1.0; 0.1; 0.01; 0.001% of nanocomposite in polycarbonate solution in ethylene dichloride. The suspensions underwent ultrasonic processing.

To compare the optical density of nanocomposite suspension in polycarbonate solution in ethylene dichloride, as well as polycarbonate and polycarbonate samples modified with nanocomposites, spectrophotometer KFK-3-01a was used.

Samples in the form of modified and nonmodified films for studying IR spectra were prepared precipitating them from suspension or solution under vacuum. The obtained films about 100 mcm thick were examined on Fourier-spectrometer FSM 1201.

To investigate the crystallization and structures formed the high-resolution microscope (up to 10 mcm) was used.

To examine thermal-physical characteristics the lamellar material on polycarbonate and polymethyl methacrylate basis about 10 mm high was prepared. Three layers of polymethyl methacrylate and two layers of polycarbonate were used. Thermal-physical characteristics (specific thermal capacity and thermal conductivity) were investigated on calorimeters IT-c-400 and IT-λ-400.

10.3   RESULTS AND DISCUSSION

During the investigation the nanocomposite fine suspension in polycarbonate solution in ethylene dichloride was studied, polycarbonate films modified with different concentrations of nanocomposite were compared with the help of optical spectroscopy, microscopy, IR spectroscopy and thermal-physical methods of investigation.

The results of investigation of optical density of Cu/C nanocomposite fine suspension (0.001%) based on polycarbonate solution in ethylene dichloride are given in Fig. 1. As seen in Fig. 1, the introduction of nanocomposite and polycarbonate into ethylene dichloride resulted in transmission increase in the range 640–690 nm (approximately in three times). At the same time, at 790 nm and 890 nm the significant increase in optical density was observed. The comparison of optical densities of films of polycarbonate and modified materials after the introduction of different quantities of nanocomposite (1.0; 0.1; 0.01%) into polycarbonate is interesting (Fig. 2).

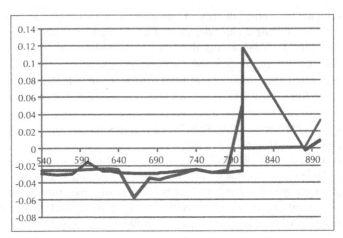

FIGURE 1   Curves of optical density of suspension based on ethylene dichloride diluted with polycarbonate and Cu/C nanocomposite in the concentration to polycarbonate 0.001% (1) and ethylene dichloride (2).

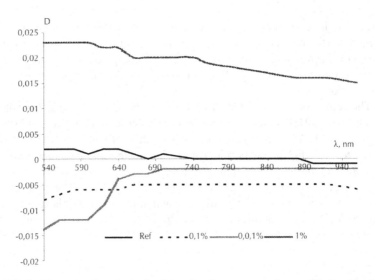

FIGURE 2   Curves of optical density of reference sample modified with Cu/C nanocomposites in concentration 1%, 0.1%, and 0.01%.

Comparison of optical density of suspension containing 0.001% of nanocomposite and optical density polycarbonate sample modified with 0.01% of nanocomposite indicates the proximity of curves character. Thus the correlation of optical properties of suspensions of nanocomposites and film materials modified with the same nanocomposites is quite possible.

Examination of curves of samples optical density demonstrated that when the nanostructure concentration was 1% from polycarbonate mass, the visible-light spectrum was absorbed by about 4.2% more if compared with the reference sample. When the nanocomposite concentration was 0.1%, the absorption decreased by 0.7%. When the concentration was 0.01%, the absorption decreased in the region 540–600 nm by 2.3%, and in the region 640–960 nm by 0.5%.

During the microscopic investigation of the samples the schematic picture of the structures formed was obtained at 20-mcm magnification. The results are given in Fig. 3. From the schematic pictures it is seen that volumetric structures of regular shape surrounded by micellae were formed in polycarbonate modified with 0.01% Cu/C nanocomposite. When 0.1% of nanocomposite was introduced, the linear structures distorted in space and surrounded by micellae were formed. When Cu/C nanocomposite concentration was 1%, large aggregates were not observed in polycarbonate.

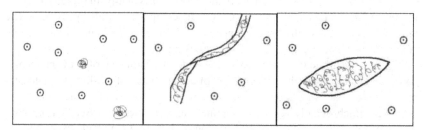

FIGURE 3   Schematic picture of structure formation in polycarbonate modified with Cu/C nanocomposites in concentrations 1%, 0.1%, and 0.01% (from left to the right), 20-mcm magnification

Apparently, the decrease in nanocomposite concentration in polycarbonate can result in the formation of self-organizing structures of bigger size. In Ref. [3], there is the hypothesis on the transfer of nanocomposite oscillations onto the molecules of polymeric composition, the intensity of bands in IR spectra of which sharply increases even after the introduction of super-small quantities of nanocomposites. In our case, this hypothesis was checked on modified and non-modified samples of polycarbonate films.

In Fig. 4, you can see the IR spectra of polycarbonate and polycarbonate modified with 0.001% of Cu/C nanocomposite.

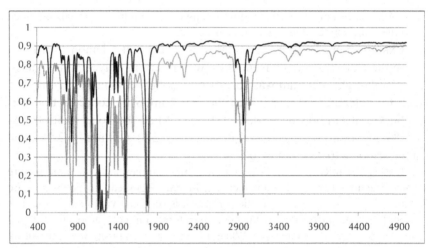

FIGURE 4   IR spectra of reference sample (upper) and polycarbonate modified with Cu/C nanocomposites in concentration 0.001% (lower).

As seen in IR spectra, the intensity increases in practically all regions, indicating the influence of oscillations of Cu/C nanocomposite on all the system. The most vivid changes in the band intensity are observed at 557 cm^{-1}, in the region 760–930 cm^{-1}, at 1400 cm^{-1}, 1600 cm^{-1}, 2970 cm^{-1}.

Thus, polycarbonate self-organization under the influence of Cu/C nanocomposite takes place with the participation of certain bonds, for which the intensity of absorption bands goes up. At the same time, the formation of new phases is possible, which usually results in thermal capacity increase. But thermal conductivity can decrease due to the formation of defective regions between the aggregates formed.

The Table 1 contains the results of thermal-physical characteristics studied.

TABLE 1   Thermal-physical characteristics of polycarbonate and its modified analogs.

	Nanocomposite content in polycarbonate, %			
	0%, ref.	1%	0.1%	0.01%
m 10^{-3}, kg	1.9	1.955	1.982	1.913
h 10^{-3}, m	9.285	9.563	9.76	9.432
Csp, J/kg K	1440	1028	1400	1510
Λ	0.517	0.503	0.487	0.448

It is demonstrated that when the nanostructure concentration in the material decreases, thermal capacity goes up that is confirmed by the results of previous investigations. Thermal conductivity decline, when the nanostructure concentration decreases, is apparently caused by the material defectiveness.

10.4 CONCLUSION

When Cu/C nanocomposites are introduced into the material modified, the nanostructures can be considered as the generator of molecules excitation, which results in wave process in the material.

It is found that polycarbonate modification with metal/carbon containing nanocomposites results in the changes in polycarbonate structure influencing its optical and thermal-physical properties.

KEYWORDS

- copper/carbon nanocomposite
- ethylene dichloride
- IR spectra
- polycarbonate

REFERENCES

1. Kodolov, V. I.; Khokhriakov, N. V. Chemical physics of the processes of formation and transformation of nanostructures and nanosystems. Izhevsk, Publishing office of Izhevsk State Agricultural Academy, 2009, *1–2*, 728 p.
2. Kodolov, V. I.; Khokhriakov, V. N. V.; Trineeva, V.; Blagodatskikh, I. I. Activity of nanostructures and its demonstration in nanoreactors of polymeric matrixes and active media. *Chem. Phys. Mesoscopy*, 2008, *10(4)*, p. 448.
3. Kodolov, V, V. I.; Trineeva, V. Perspectives of idea development about nanosystems self-organization in polymeric matrixes; Publishing office of Izhevsk State Agricultural Academy, 2012, *1(4)*, 650 p.

CHAPTER 11

NANOFIBERS AND SOLAR CELLS

V. MOTTAGHITALAB, M. SAJEDI, M. S. MOTLAGH, M. ABBASI,
and A. K. HAGHI

CONTENTS

11.1 INTRODUCTION

Nowadays, the wearable wireless electronic device and system extensively needs a sustainable energy supply as potential alternative in order to overcome the urgent needs to regular battery with finite power. The basic concept of photovoltaic effect reviewed and the structure of a range of energy conversion devices including inorganic either homojunction or multijunction, DSSC and polymer based solar cells briefly introduced to consolidate the idea regarding to the development of textile solar cells. This chapter attempts to presents an overview on photovoltaic devices by focusing on flexible solar cells composed of nanofibers as key element. A detailed review on cell material selection and their effect on energy conversion are considered to elucidate the role of nanofiber in energy conversion. The chapter also presents promizing potentials of electrospun nanofibers web for being a key element in photovoltaic devices. As a new alternative, the transparent ITO glass can be replaced by the transparent conductive fabric. The sprayed anatase TiO_2 layer introduced on transparent conductive fabric as blocking layer, electrospun anatase TiO_2 nanofibers web laminated on TiO_2 coated conductive fabric as the photoanode, and a transparent PET film employed as the seal film to protect the gel electrolyte. The preparation of core–shell nanofibers by coaxial electrospinning process of two components such as poly(3-hexyl thiophene) (P3HT) (a conducting polymer) or P3HT/PCBM as the core and poly(vinyl pyrrolidone) (PVP) as the shell material open a new horizon for integration of nanofibers in textile solar cells. Latest experience regarding to polymer solar cell and low band gap material needs to be considered for ambitious plans with nanofiber morphology.

In recent years, renewable energies attract considerable attention due to the inevitable end of fossil fuels and due to global warming and other environmental problems. Photovoltaic solar energy is being widely studied as one of the renewable energy sources with key significance potentials and a real alternate to fossil fuels. Solar cells are in general packed between w80, brittle and rigid glass plates. Therefore, increasing attention is being paid to the construction of lighter, portable, robust, multipurpose and flexible substrates for solar cells. Textiles substrates are fabricated by a wide variety of processes, such as weaving, knitting, braiding and felting. These fabrication techniques offer enormous versatility for allowing a fabric to conform to even complex shapes. Textile fabrics not only can be rolled up for storage and then unrolled on site but also they can also be readily installed into structures with complex geometries.

Textiles are engaging as flexible substrates in that they have an enormous variety of uses, ranging from clothing and household articles to highly sophisticated technical applications. Last innovations on photovoltaic technology have allowed obtaining flexible solar cells, which offer a wide range of possibilities, mainly in wearable applications that need independent systems. Nowadays, entertainment, voice and data communication, health monitoring, emergency, and surveillance functions, all of which rely on wireless protocols and services and sustainable energy supply in order to overcome the urgent needs to regular battery with finite power. Because of their steadily decreasing power demand, many portable devices can harvest enough energy from clothing-integrated solar modules with a maximum installed power of 1–5 W. [1]

Increasingly textile architecture is becoming progressively of a feature as permanent or semipermanent constructions. Tents, such as those used by the military and campers, are the best known textile constructions, as are sun shelter, but currently big textile constructions are used extensively for exhibition halls, sports complexes and leisure and recreation centers. Although all these structures provide protection from the weather, including exposure to the sun, but solar concept offers an additional precious use for providing power. Many of these large textile architectural constructions cover huge areas, sufficient to supply several kilowatts of power. Even the fabric used to construct a small tent is enough to provide a few hundred watts. In addition to textile architecture, panels made from robust solar textile fabrics could be positioned on the roofs of existing buildings. Compared to conventional and improper solar panels for roof structures lightweight and flexible solar textile panels is able to tolerate load-bearing weight without shattering.

Moreover, natural disaster extensively introduces the huge potential needs the formulation of unusual energy package based on natural source. Over the past five years, more than 13 million *people have lost their home* and possessions because of earthquake, bush fire, flooding or other natural disaster. The victims of these disasters are commonly housed in tents until they are able to rebuild their homes. Whether they stay in tented accommodation for a short or long time, tents constructed from solar textile fabrics could provide a source of much needed power. This power could be stored in daytime and used at night, when the outdoor temperature can often fall. There are also a number of other important potential applications. The military would benefit from tents and field hospitals, especially those in remote areas, where electricity could be generated as soon as the structure is assembled.

11.2 THE BASIC CONCEPT OF SOLAR CELL

In 1839 French scientist, Edmond Becquerel found out photovoltaic effect when he observed increasing of electricity generation while light exposure to the two metal electrodes immersed in electrolytic solution [2]. Light is composed of energy packages known as photons. Typically, when a matter exposed to the light, electrons are excited to a higher level within material, but they return to their initial state quickly. When electrons take sufficient energy more than a certain threshold (band gap), move from the valance band to the conduction band holes with positive charge will be created. In the photovoltaic effect electron-hole pairs are separated and excited electrons are pulled and fed to an external circuit to buildup electricity [3] (Fig. 1).

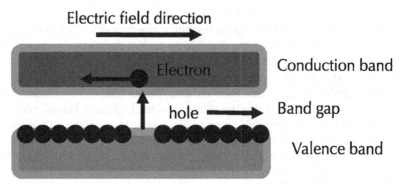

FIGURE 1   Electron excitation from valence band to conduction band.

An effective solar cell generally comprises an opaque material that absorbs the incoming light, an electric field that arises from the difference in composition between the semiconducting layers comprising the absorber, and two electrodes to carry the positive and negative charges to the electrical load. Designs of solar cells differ in detail but all must include the above features (Fig. 2).

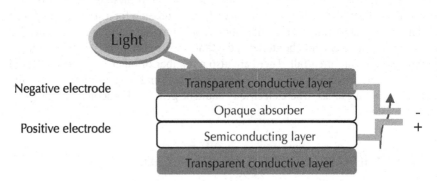

FIGURE 2   The general structure of a Solar cell.

Generally, Solar cells categorized into three main groups consisting of inorganic, organic and hybrid solar cells (Dye sensitized solar cells). Inorganic solar cells cover more than 95% of commercial products in solar cells industry.

11.2.1   ELECTRICAL MEASUREMENT BACKGROUND

All current–voltage characteristics of the photovoltaic devices were measured with a source measure unit in the dark and under simulated solar simulator source was calibrated using a standard crystalline silicon diode. The current-voltage characteristics of Photovoltaic devices are generally characterized by the short-circuit current (Isc), the open-circuit voltage (Voc), and the fill factor (FF). The photovoltaic power conversion efficiency (η) of a solar cell is defined as the ratio between the maximum electrical power (Pmax) and the incident optical power and is determined by Eq. (1). [4]

$$\eta = \frac{I_{sc} \times V_{oc} \times FF}{P_{in}} \tag{1}$$

In Eq. (1), The short circuit current (I_{sc}) is the maximum current that can run through the cell. The open circuit voltage (V_{oc}) depends on the highest occupied molecular orbital (homo) level of the donor (p-type semiconductor quasi Fermi level) and the lowest unoccupied molecular orbital (lumo) level of the acceptor (n-type semiconductor quasi Fermi level), linearly. Pin is the incident light power density. FF, the fill-factor, is calculated by dividing Pmax by the multiplication of Isc and Voc and this can be explained by the following equation (2):

$$FF = \frac{I_{mpp} \times V_{mpp}}{I_{sc} \times V_{oc}} \tag{2}$$

In the Eq. (2), Vmpp and Impp represent, respectively, the voltage and the current at the maximum power point (MPP), where the product of the voltage and current is maximized [4].

11.2.2   INORGANIC SOLAR CELL

Inorganic solar cells based on semiconducting layer architecture can be divided into four main categories including P–N homo junction, hetrojunction either P–I–N or N–I–P and multi junction.

The P–N homojunction is the basis of inorganic solar cells in which two different doped semiconductors (n-type and p-type) are in contact to make solar cells (Fig. 3a). P-type semiconductors are atoms and compounds with fewer electrons in their outer shell, which could create holes for the electrons within the lattice of p-type semiconductor. Unlike p-types semiconductors, the n-type have more electrons in their outer shell and sometimes there are exceed amount of electron on n-type lattice result lots of negative charges [5].

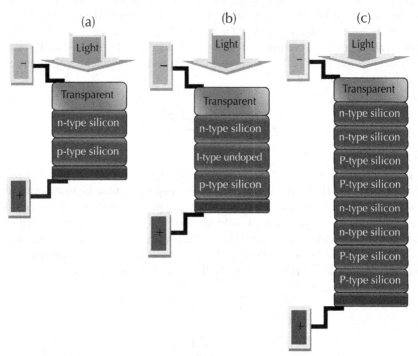

FIGURE 3   A general scheme of inorganic solar cell (a) P–N homojunction (b) P–I–N hetrojunction (c) multijunction

Compared to homojunction structure amorphous silicon thin-film cells use a P–I–N hetrojunction structure, whereas cadmium telluride (CdTe) cells use a N–I–P arrangement. The overall picture embraces a three-layer sandwich with a middle intrinsic (i-type or undoped) layer between an N-type layer and a P-type layer (Fig. 3b). Multiple junction cells have several different semiconductor layers stacked together to absorb different wavebands in a range of spectrum, producing a greater voltage per cell than from a single junction cell which most of the solar spectrum to electricity lies in the red (Fig. 3c).

Variety of semiconducting material such as single and poly crystal silicon, amorphous silicon, Cadmium-Telluride (CdTe), Copper Indium/Gallium Di Selenide (CIGS) have been employed to form inorganic solar cell based on layers configuration to enhance absorption efficiency, conversion efficiency, production and maintenance cost.

11.2.3 ORGANIC SOLAR CELLS (OSCS)

Photoconversion mechanism in organic or excitonic solar cells is differing from conventional inorganic solar cells in which exited mobile state are made by light absorption in electron donor. While, light absorption creates free electron-hole pairs in inorganic solar cells [5]. It is due to law dielectric constant of organic materials and weak noncovalent interaction between organic molecules. Consequently, exciton dissociation of electron-hole pairs occurs at the interface between electron donor and electron acceptor components [6] . Electron donor and acceptor act as semiconductor p-n junction in inorganic solar cells and should be blended together to prevent electron-hole recombination (Fig. 4).

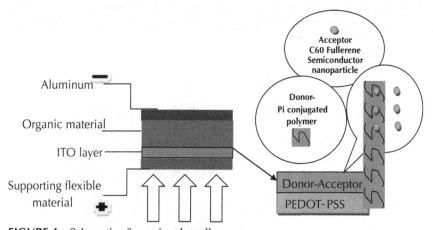

FIGURE 4 Schematic of organic solar cell.

There are two main types of PSCs including: bilayer heterojunction and bulk-heterojunction [7]. Bulk-heterojunction PSCs are more attractive due to their high surface area junction that increases conversion efficiency. This type of polymer solar cell consists of Glass, ITO, PEDOT-PSS, active layer, calcium and aluminum in which conjugated polymer are used as active layer [8]. The organic solar cells with maximum conversion efficiency about 6% still are at the beginning of development and have a long way to go to compete with inorganic solar cells. Indeed, the advantages of polymers including low-cost deposition in large areas, low weight on flexible substrates and sufficient efficiency are promizing advent of new type of solar cells [9]. Conjugated small molecule attracted as an alternative approach of organic solar cells. Development of small molecule for OSCs interested because of their properties such as well-defined molecular structure, definite molecular weight, high purity, easy purification, easy mass-scale production, and good batch-to-batch reproducibility [10–12].

11.2.4   DYE-SENSITIZED SOLAR CELLS

Dye-sensitized solar cells (DSSC) use a variety of photosensitive dyes and common, flexible materials that can be incorporated into architectural elements such as windowpanes, building paints, or textiles. DSSC technology mimic photosynthesis process whereby the leaf structure is replaced by a porous titania nanostructure, and the chlorophyll is replaced by a long-life dye. The general scheme of DSSC process is shown in Fig. 5. Although traditional silicon-based photovoltaic solar cells currently have higher solar energy conversion ratios, dye-sensitive solar cells have higher overall power collection potential due to low-cost operability under a wider range of light and temperature conditions, and flexible application [13].

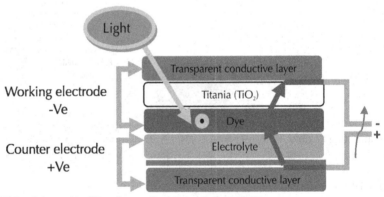

FIGURE 5   Schematic of Dye Senetesized Solar Cell (DSSC).

Oxide semiconductors materials such as TiO_2, ZnO_2 and SnO_2 have a relatively wide band gap and cannot absorb sunlight in visible region and create electron. Nevertheless, in sensitization process, visible light could be absorbed by photosensitizer organic dye results creation of electron. Consequently, excited electrons are penetrated into the semiconductor conduction band. Generally, DSSC structures consist of a photoelectrode, photosensitizer dye, a redox electrolyte, and a counter electrode. Photoelectrodes could be made of materials such as metal oxide semiconductors. Indeed, oxide semiconductor materials, particularly TiO_2, are choosing due to their good chemical stability under visible irradiation, nontoxicity and cheapness. Typically, TiO_2 thin film photoelectrode prepared via coating the colloidal solution or paste of TiO_2 and then sintering at 450 °C to 500 °C on the surface of substrate, which led to increase of dye absorption drastically by TiO_2 [14]. The substrate must have high transparency and low ohmic resistance to high performance of cell could be achieved. Recently, many researches focused on the both organic and inorganic dyes as sensitizer regarding to their extinction coefficients and performance. Among them, B4 (N3): RuL2(NCS)2:L=(2,2'-bipyridyl-4,4'-dicarboxylic acid) and B2(N719): {cis–bis (thiocyanato)-bis(2,20-bipyridyl-4,40-dicarboxylato)-ruthenium(II) bis-tetrabutylammonium} due to its outstanding performance was interested (Scheme 1).

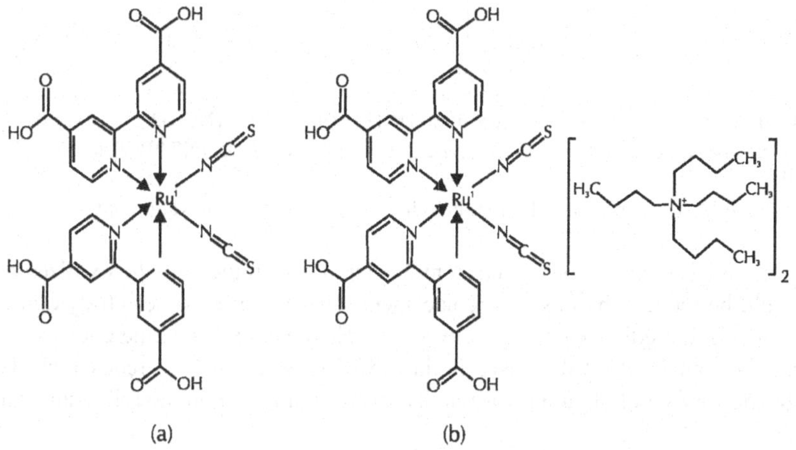

(a)   (b)

SCHEME 1 The chemical structure of (a) B4 (N3): RuL2(NCS)2 L(2,2'-bipyridyl-4,4'-dicarboxylic acid), (b) B2(N719) {cis–bis (thiocyanato)-bis(2,20-bipyridyl-4,40-dicarboxylato)-ruthenium(II) bis-tetrabutylammonium

B2 is the most common high performance dye and a modified form of B4 to increase cell voltage. Up to now different methods have been performed to develop of new dyes with high molar extinction coefficients in the visible and near-IR in order to outperform N719 as sensitizers in a DSSC [15]. DNH2 is a hydrophobic dye, which very efficiently sensitizes wide band-gap oxide semiconductors, like titanium dioxide. DBL (otherwise known as "black dye") is designed for the widest range spectral sensitization of wide band-gap oxide semiconductors, like titanium dioxide up to wavelengths beyond 800 nm (Scheme 2).

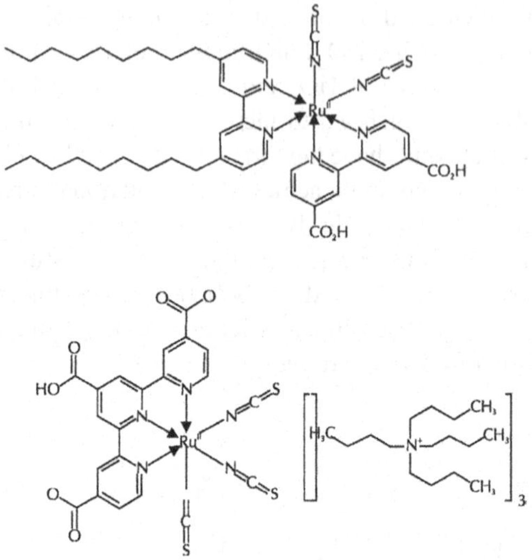

SCHEME 2　The chemical structure of (a) DNH2 (Z907) RuLL'(NCS)2, L=2,2'-bipyridyl-4,4'-dicarboxylic acid, L'= 4,4'-dinonyl-2,2'-bipyridine (b) DBL (N749) [RuL(NCS)3]: 3 TBA
L= 2,2':6,'2''-terpyridyl-4,4,'4''-tricarboxylic acid TBA=tetra-n-butylammonium.

In order to continuous electron movement through the cell, the oxidized dye should be reduced by electron replacement. The role of redox electrolyte in the DSSCs is to mediate electrons between the photoelectrode and the counter electrode. Common electrolyte used in the DSSC is based on I^-/I_3^- redox ions [16]. The mechanism of photon to current has been summarized in the following equations:

$$\text{Dye} + \text{light} \rightarrow \text{Dye*} + e^- \tag{a}$$

$$\text{Dye*} + e^- + \text{TiO}_2 \rightarrow e^-(\text{TiO2}) + \text{oxidized Dye} \tag{b}$$

$$\text{Oxidized Dye} + 3/2 I^- \rightarrow \text{Dye} + 1/2 I^{-3} \tag{c}$$

$$1/2 I_3^- + e^-(\text{counter electrode}) \rightarrow 3/2 I^- \tag{d}$$

Despite of many advantages of DSSCs, still lower efficiency compared to commercialized inorganic solar cells is a challenging area. Recently, one-dimensional nanomaterials, such as nanorods, nanotubes and nanofibers, have been proposed to replace the nanoparticles in DSSCs because of their ability to improve the electron transport leading to enhanced electron collection efficiencies in DSSCs.

Subsequent sections attempts to provide fundamental knowledge to general concept of textile solar cells and their recent progress based on. Of particular interest are electrospun TiO_2 nanofibers playing the role as a key material in DSSCs and other organic solar cell, which have been shown to improve the electron transport efficiency and to enhance the light harvesting efficiency by scattering more light in the red part of the solar spectrum. A detailed review on cell material selection and their effect on energy conversion are considered to elucidate the potential role of nanofiber in energy conversion for textile solar cell applications.

11.3   TEXTILE SOLAR CELLS

Clothing materials either for general or specific use are passive and the ability to integrate electronics into textiles provides great opportunity as smart textiles to achieve revolutionary improvements in performance and the realization of capabilities never before imagined on daily life or special circumstances such as battlefield. In general, smart textiles address diverse function to withstand an interactive wearable system. Development, incorporation and interconnection of flexible electronic devices including sensor, actuator, data processing, communication, internal network and energy supply beside basic garment specifications sketch the road map toward smart textile architecture. Regardless of the subsystem functions, energy supply and storage play a critical role to propel the individual functions in overall smart textile systems.

The integration of photovoltaic (PVs) into garments emerges new prospect of having a strictly mobile and versatile source of energy in communications equipment, monitoring, sensing and actuating systems. Despite of extremely good power efficiency, most conventional crystalline silicon based semiconductor PVs is intrinsically stiff and incompatible with the function of textiles where flexibility is essential. Extensive research has been conducted to introduce the novel potential candidates for shaping the textile solar cell (TSC) puzzle. In particular, polymer-based organic solar cell materials have the advantages of low price and ease of operation in comparison with silicon-based solar cells. Organic semiconductors, such as conductive polymers, dyes, pigments, and liquid crystals, can be manufactured cheaply and used in organic solar cell constructions easily. In the manufacturing process of organic solar cells, thin films are prepared using specific techniques, such as vacuum evaporation, solution processing, printing [17, 18], or nanofiber formation [19] and electrospinning [20] at room temperatures. Dipping, spin coating, doctor blading, and printing techniques are mostly used for manufacturing organic solar cells based on conjugated polymers [17].Recent TSC studies revealed two distinctive strategies for developing flexible textile solar cell and its sophisticated integration.

1.   The first strategy involved the simple incorporation of a polymer PV on a flexible substrate Such as poly thyleneterphthalate (PET) directly into the clothing as a structural power source element.

2. The second strategy was more complicated and involved the lamination of a thin anti reflective layer onto a suitably transparent textile material followed by plasma, thermal or chemical treatment. The next successive step focuses on application of a photoanode electrode onto the textile material. Subsequent procedure led to the deposition of the active material and finally evaporation of the cathode electrode complete the device as a textile PV composed of organic, inorganic and also their composites

Regardless of many gaps need to be bridged before large-scale application of this technology, the TSC fabrication based on second strategy may be envisaged through two routes to solve pertinent issues of efficiency and stability. The solar cell architecture in first approach is founded based on knitted or woven textile substrate, however second alternative follows a roadmap to develop a wholly PV fiber for further knitting or weaving process that may form energy-harvesting textile structures in any shape and structure.

Irrespective to fabric or fiber shaped of the photovoltaic unit, the light penetration and scattering in photo anode layer needs a waveguide layer. This basic requirement naturally mimicked by polar bear hair. The optical functions of the polar bear hair are scattering of incident light into the hair, luminescence wave shift and wave-guide properties due to total reflection. The hair has an opaque, rough-surfaced core, called the medulla, which scatters incident light. The simulated synthetic coreshell fiber can be manufactured through spinning of a core fiber or with sufficient wave-guiding properties followed by finishing with an optically active, that is, fluorescing, coating as a shell to achieve a polar bear hair' effect [21]. As described in Tributsch's original work [22], a high-energy conversion can be expected from high frequency shifts as difference between the frequencies of absorbed and emitted light. According to Tributsch et al. [22] for the polar bear hair, the frequency shift is of the order of 2×10^{14} Hz.

In principle, the refractive index varies over the fiber diameter and this would provide a certain wave-guiding property of the fiber. In charge of wave-guiding property most specifically With regard to manufacturing on a larger scale, fiber morphology, crystalinity, alignment, diameter and geometry can be altered for preferred optical performance as solar energy transducer. The focus on both approached of second strategy and illuminates their opportunities and challenges are main area of interest in next parts.

11.3.1   RECENT PROGRESS OF PV FIBERS

Fiber shaped organic solar cells has been subject of a few patent, project and research papers. Polymers, small molecules and their combinations were used as light-absorbing layers in previous studies. A range of synthetic substrate with various level of flexibility including optical [23], polyimide [24] and poly propylene (PP) [25] subjected to processes in which functional electrodes and light absorbing layer continuously forms on fiber scaffold. In recent PV fiber studies, conducting polymers such as P3HT in combination to small molecules nanostructure materials such as branched fullerene plays main role as photoactive material. For instance, Ref. [23] introduces a light-absorbing layer on optical fiber composed of poly (3-hexylthiophene) (P3HT): phenyl-C61-butyric acid methyl ester (P3HT:PCBM) (Fig. 6). While the light was traveling through the optical fiber and generating hole–electron pairs, the 100-nm top metal electrode (which does not let the light transmit from outside) was used to collect the electrons [23].

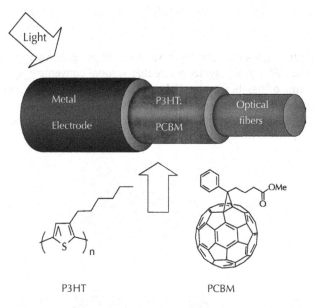

P3HT   PCBM

FIGURE 6 Simplified pattern of hetrojunction photoactive layer in fiber solar cell.

One of the important challenges of flexible solar cell concentrated in hole collecting electrode, which is most widely used ITO as a transparent conducting material. However, the inclusion of ITO layer in flexible solar cell could not be applicable. The restrictions are mostly due to the low availability and expense of indium, employment of expensive vacuum deposition techniques and providing high temperatures to guarantee highly conductive transparent layers. Accordingly, there are some ITO-free alternative approaches, such as using carbon nanotube (CNT) layers or different kinds of poly(3,4-ethylenedioxythiophene):poly(stirene sulfonate) (PEDOT:PSS) and its mixtures [26–28], or using a metallic layer [29] to perform as a hole-collecting electrode (Scheme 3).

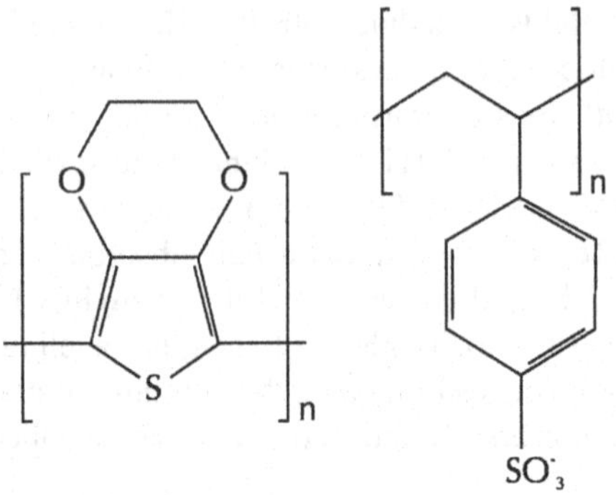

SCHEME 3   The chemical structure of poly (3,4-ethylenedioxythiophene):poly(stirenes ulfonate).

The ITO free hole-collecting layer was realized using highly conductive solution of PEDOT:PSS as a polymer anode that is more convenient for textile substrates in terms of flexibility, material cost, and fabrication processes compared with ITO material. Based on procedure described in Ref. [25] a sophisticated and simple design was presented to show how thin and flexible could be a solar cell panel (Fig. 7).

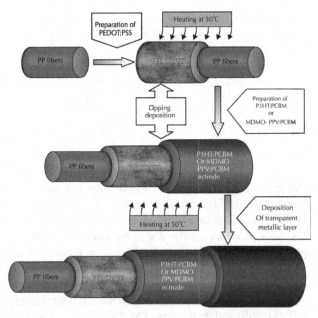

FIGURE 7   The Schematic of procedure for preparation of PV fiber.

TABLE 1 Photoelectrical characteristics of photovoltaic fibers having different having different photoactive layers (P3HT: PCBM and MDMO-PPV:PCBM) [25].

Solar cell pattern	Voc (mV)	Isc (mA/cm^2)	FF (%)	η (%)
PP|PEDOT:PSS| P3HT:PCBM|LiF/Al	360	0.11	24.5	0.010
PP|PEDOT:PSS| MDMO-PPV:PCBM| LiF/Al	300	0.27	26	0.021
ITO|PEDOT:PSS| MDMO-PPV:PCBM| LiF/Al	740	4.56	43.4	1.46

Based on implemented pattern, the sunbeams entered into the photoactive layer with 4–10 mm^2 active area by passing through a 10 nm of lithium fluoride/aluminum (LiF/Al) layer as semitransparent cathode outer electrode.

Table 1 gives the current density data versus voltage characteristics of the photovoltaic fibers consisting of P3HT: PCBM and MDMO-PPV: PCBM blends. Based on given results of open-circuit voltage, short-circuit current density and fill factor for two types of photoactive material, and also Eqs. (1) and (2) in Section 11.2.1, the power conversion efficiency of the MDMO-PPV: PCBM based photovoltaic fiber was higher than the P3HT: PCBM based photovoltaic fiber.

Comparing the solar cell characteristics of second and third pattern shows greater performance of ITO\|PEDOT:PSS\| MDMO-PPV:PCBM\| LiF/Al rigid organic solar cell compared to PP\|PEDOT:PSS\| MDMO-PPV:PCBM\| LiF/Al fiber solar cell. Since a same cathode, anode and photoactive material used in both pattern, the higher power conversion efficiency of ITO solar cell can be attributed to different wave-guide property and transparency of cell pattern in sun's ray entrance angle (i.e., LiF/AL versus ITO glass).

Since using ITO is strictly restricted for PV fiber, enhancing the power conversion efficiency of needs to improving existing materials and techniques. In particular, the optical band gap of the polymers used as the active layer in organic solar cells is very important. Generally, the best bulk heterojunction devices based on widely studied P3HT: PCBM materials are active for wavelengths between 350 and 650 nm. Polymers with narrow band gaps can absorb more light at longer wavelengths, such as infrared or near-infra-red, and consequently enhance the device efficiency. Low band gap polymers (<1.8 eV) can be an alternative for better power efficiency in the future, if they are sufficiently flexible and efficient for textile applications [30, 31].The variety of factors influence on polymer band gaps which can be categorized as intrachain charge transfer, substituent effect, π-conjugation length.

Systematically the fused ring low band gap copolymer composes of a low energy level electron acceptor unit coupled with a high energy level electron donor unit. The band gap of the donor/acceptor copolymer is determined by the HOMO of the donor and LUMO of the acceptor, and therefore a high energy level of the HOMO of the donor and a low energy level of the LUMO of the acceptor results in a low band gap [32].

The substituent on the donor and acceptor units can affect the band gap. The energy level of the HOMO of the donor can be enhanced by attaching electron-donating groups (EDG), such as thiophene and pyrrole. Similarly, the energy level of the LUMO of the acceptor is lowered, when electron-withdrawing groups (EWG), such as nitrile, thiadiazole and pyrazine, are attached. This will result in improved donor and acceptor units, and hence, the band gap of the polymer is decreased [33].

11.3.2 PV INTEGRATION IN TEXTILE

Having a complete functional textile solar cell motivates the researchers to attempt an approach for direct incorporation of photovoltaic cell elements onto the textile. The textile substrates inherently scatter most part of the incident light outward. Therefore, it was found necessary to apply a layer of the very flexible polymer PE onto the textile substrate to have a surface compatible with a layered device. The textile-PE substrate was plasma treated before application of the transparent PEDOT electrode in order to obtain good adhesion of the PEDOT layer to the PE carrier. Then screen-printing was employed for the application of the active polymer poly[2-methoxy-5-(2'-ethylhexyloxy)-p-phenylene vinylene] (MEH-PPV) [33].

The traditional solar cell geometry was reinvented in fractal forms that allow the building of structured modules by sewing the 25–40 cm cells realized. Figure 8 shows an step-by-step approach for fabrication textile solar cell pattern based on polymer photo absorbing layer.

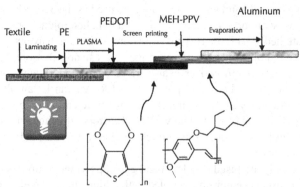

FIGURE 8   A typical fabrication procedure and key elements of textile solar cell.

The pattern designed was particularly challenging for application in solar cells and reduced the active area to 190 cm^2 (19% of the real area). The best module output power was found to be 0.27 mW with a Isc = 3:8 µA, Voc = 275 mV and a FF% of 25.7%. The pattern designed allows connections in different site of the cloth cell with reproducible performances within 5–10%.

11.4 NANOFIBERS AS A POTENTIAL KEY ELEMENT IN TEXTILE SOLAR CELLS

Previous sections present variety of solar cell structure and their corresponding elements and power conversion performance to indicate opportunities and challenges of producing of solar energy harvesting module based on a wholly flexible textile based photovoltaic unit. Current state of Textile Solar Cells is extremely far from commercial inorganic hetrojunction solar cells that showing around 45% conversion efficiency. Current section addresses promizing potential of nanofiber 1D morphology to be used as solar cell elements. Of particular, enhancement of photovoltaic unit demanding properties is a great of importance. Two different strategies can be presumed including integration of functional photoanode, photo cathode, scattering layer, photoactive or acceptor—donor materials in the form of nanofiber on to textile substrate or developing fully integrated multilayer nonwoven solar cloth.

11.4.1 ELECTROSPUN NANOFIBER

Fibers with a diameter of around 100 nm are generally classified as nanofibers. What makes nanofibers of great interest is their extremely small size. Nanofibers compared to conventional fibers, with higher surface area to volume ratios and smaller pore size, offer an opportunity for use in a wide variety of applications. To date, the most successful method of producing nanofibers is through the process of electrospinning. The electrospinning process uses high voltage to create an electric field between a droplet of polymer solution at the tip of a needle and a collector plate. When the electrostatic force overcomes the surface tension of the drop, a charged, continuous jet of polymer solution is ejected. As the solution moves away from the needle and toward the collector, the solvent evaporates and jet rapidly thins and dries. On the surface of the collector, a nonwoven web of randomly oriented solid nanofibers is deposited. Material properties such as melting temperature and glass transition temperature as well as structural characteristics of nanofiber webs such as fiber diameter distribution, pore size distribution and fiber orientation distribution determine the physical and mechanical properties of the webs. The surface of electrospun fibers is important when considering end-use applications. For example, the ability to introduce porous surface features of a known size is required if nanoparticles need to be deposited on the surface of the fiber.

The conventional setup for producing a nonwoven layer can be manipulated to fabricate diverse profile and morphology including oriented [34], Core shell [35] and hollow [36] nanofiber. Figure 9 shows the latest nanofiber profiles and its corresponding electrospinning production instrument. The variety and propagation of nanofiber products opens new horizon for development of functional profile respect to demanding application. Amongst developed techniques, coaxial electrospinning forms coreshell and/or hollow nanofiber through combination of different materials in the core or shell side, novel properties and functionalities for nanoscale devices can be found.

Increasing demands for the manufacturing of bi-component structures, in which one is surrounded by the other or the particles of one are encapsulated in the matrix of the other, at the micro or nano level, show potential for a wide range of uses.

Application includes minimizing chances of decomposition of an unstable material, control releasing a substance to a particular receptor and improving mechanical properties of a core polymer by its reinforcing with another material. The electro-spinneret consists of concentric inner and outer syringe by witch two fluids are introduced to the spinneret, one in the core of the inner syringe and the other in the space between in the inner syringe and outer syringe.

(a)

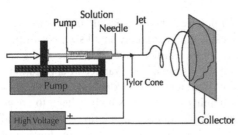

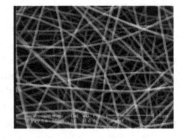

FIGURE 9   *(Continued)*

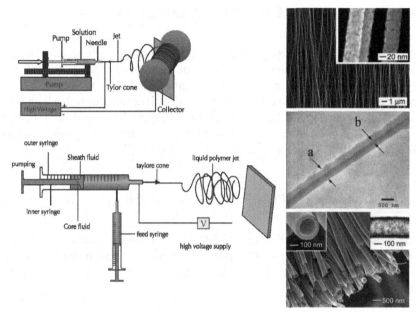

FIGURE 9   Electrospinning setup and its corresponding nanofiber profile (a) conventional nanofiber] (b) Oriented nanofiber [34] (c) Core shell nanofiber [35], (d) Hollow nanofiber [36].

The droplet of the sheath solution elongates and stretches due to the repulsing between charges and form a conical shape. When the applied voltage increases, the charge accumulation reaches a certain value so a thin jet extends from the cone. The stresses are generated in the sheath solution cause to the core liquid to deform into the conical shape and a compound coaxial jet develops at the tip of the cones (Fig. 10).

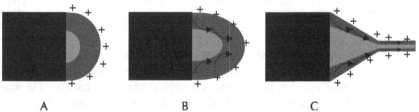

FIGURE 10   Schematic illustration of compound Taylor cone formation (a) Surface charges on the sheath solution, (b) viscous drag exerted on the core by the deformed sheath droplet, © Sheath-core compound Taylor cone formed due to continuous viscous drag).

11.4.2 NANOFIBERS IN DSSC SOLAR CELLS

It can be presumed that the electrospun nanofiber offers high specific surface areas (ranging from hundreds to thousands of square meters per gram) and bigger pore sizes than nanoparticle or film. Meanwhile, referring to Section 2.3, the particle-based titanium dioxide layers have low efficiencies due to the high density of grain boundaries, which exist between nanoparticles. The 1D morphology of metal oxide fibers attracts more interest because of the lower density of grain boundaries compared to those of sintered nanoparticles [37].

11.4.2.1 ELECTROSPUN SCATTERING LAYER

Reference [38] compared the effect of TiO_2 nanofiber and nanoparticle as scattering layer and indicated the significant enhancement of all photovoltaic specifications. In another attempt ZNO nanofiber used instead of TiO2 nanofiber to form photo anode [39].

TABLE 2 The effect of TiO_2 nanofiber on cell performance for DSSC solar cell [38–40].

Solar cell pattern	Voc (mV)	Isc (mA/cm²)	FF (%)	η (%)	Ref.
FTO|TiO_2 nanoparticle|N719|LiI/ I2/TBT|Pt|FTO	630	11.3	54	3.85	[38]
FTO|TiO_2 nanofiber|N719|LiI/I2/ TBT|Pt|FTO	660	14.3	53	4.9	[38]
FTO|ZNO nanofiber|N719|LiI/I2/ TBT|Pt|FTO	690	2.87	44	0.88	[39]
FTO|TiO_2 nanofiber: Ag nano P|N719|LiI/I2/TBT|Pt|FTO	800	7.57	55	3.3	[40]

However, the measured photocurrent density–voltage shows poorer results compared to TiO_2. The influence of Ag nanoparticle was also studied showed nearly same fill factor but lower power conversion efficiency compared to neat TiO_2 nanofiber scattering layer [40]. Recently, a composite anatase TiO_2 nanofibers/nanoparticle electrode was fabricated through electrospinning [41]. This method avoided the mechanical grinding process, and offered a higher surface area, so conversion efficiencies of 8.14% and 10.3% for areas of 0.25 and 0.052 cm², respectively, were reported Hybrid TiO_2 nanofibers with moderate multi walled carbon nanotubes (MWCNTs) content also can prolong electron recombination lifetimes [41]. Since MWCNTs can quickly transport charges generated during photocatalysis, the opportunity for charge recombination is reduced. Furthermore, MWCNTs decrease the agglomeration of TiO_2 nanoparticles and increase the surface area of TiO_2. These advantages make this hybrid electrode a promizing candidate for DSSCs.

11.4.2.2   NANOFIBER ENCAPSULATED ELECTROLYTE

Nanofiber can be considered as promizing candidate for preparation of solid or semi solid electrolyte. This is mostly because of the inherent long-term instability of electrolyte used in DSSCs usually consists of triiodide/iodide redox coupled in organic solvents [42]. Many solid or semisolid viscous electrolytes with low level of penetration to TiO_2 layer such as ionic liquids [43], and gel electrolytes [44] used to triumph over these problems. However, nanofiber with may increase the penetration of viscous polymer gel electrolytes through large and controllable pore sizes.

A few research conducted to fabricate the Electrospun PVDF-HFP membrane by electrospinning process from a solution of poly(vinylidenefluoride-cohexafluoropropylene) in a mixture of acetone/N, N-dimethylacetamide to encapsulate electrolyte solution [45, 46]. Although the solar energy-to electricity conversion efficiency of the quasi-solid-state solar cells with the electrospun PVDF-HFP membrane was slightly lower than the value obtained from the conventional liquid electrolyte solar cells, this cell exhibited better long-term durability because of the prevention of electrolyte solution leakage.

11.4.2.3   FLEXIBLE NANOFIBER AS COUNTER ELECTRODE

Efficient charge transfer from a counter electrode to an electrolyte is a key process during the operation of dye-sensitized solar cells. One of the greatest flexible counter electrode could be polyaniline (PAni) nanofibers on graphitized polyimide (GPi) carbon films for use in a tri-iodide reduction. These results are due to the high electrocatalytic activity of the PAni nanofibers and the high conductivity of the flexible GPi film. In combination with a dye-sensitized TiO_2 photoelectrode and electrolyte, the photovoltaic device with the PAni counter electrode shown in Table 3 energy conversion efficiency of 6.85%. Short-term stability tests indicate that the photovoltaic device with the PAni counter electrode approximately preserves its initial performance [47].

TABLE 3 Current–voltage characteristics of the dye-synthesized solar cells with various electrodes [47].

Solar cell pattern	Voc (mV)	Isc (mA/cm^2)	FF (%)	η (%)
FTO|TiO$_2$ nanoparticle|N719|PMII/I2/ TBP |Pt|FTO	820	12.61	62.3	6.44
FTO|TiO$_2$ nanoparticle|N719|PMII/I2/ TBP |PAni|FTO	831	12.22	62.1	6.31
FTO|TiO$_2$ nanoparticle|N719|PMII/I2/ TBP |PAni	856	11.59	58.7	5.82
FTO|TiO$_2$ nanoparticle|N719|PMII/I2/ TBP |PAni|GPi	901	9.68	28.3	2.49

The major concern for the application of alternative counter electrodes to conventional platinized TCOs in DSSCs is long-term stability. Many publications indicate that during prolonged exposure in corrosive electrolyte, catalysts will detach from the substrate and deposit onto the surface of the semiconductor photoelectrode.

11.4.2.4   A PROPOSED MODEL FOR DSSC TEXTILE SOLAR CELL USING NANOFIBERS

Choosing proper material and structure for DSSC textile solar cell using previously mentioned nanofiber propose potential candidates for designing of an integrated photovoltaic unit. As can be seen in Fig. 11 a multilayer textile DSSC solar cell composed of a complicated pattern while nanofiber is dominant in step-by-step fabrication process.

The major concern regarding the DSSC textile solar is TiO$_2$ nanofiber that needs to subjected to high temperature for being scattering layer. This strategy is not compatible with other textile element and it is believed that the usage of Anatase TiO$_2$ spinning solution provide the possibility to avoid high temperature treatment. The proposed strategy is under intensive investigation in our laboratory and future results probably mostly illuminate the opportunities and challenges.

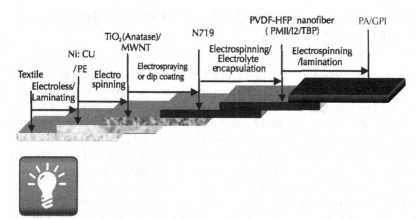

FIGURE 11 A proposed model for DSSC textile solar cell using nanofibers in successive layer.

11.4.3 ELECTROSPUN NONWOVEN ORGANIC SOLAR CELL

The idea of generating nonwoven photovoltaic (PV) cloths using organic conducting polymers by electrospinning is quite new and has not been intensively investigated. Based on previously reported PV fiber composed of conjugated hole conducting polymer (see Section 11.3.3), a novel methodology was reported to generate a nonwoven organic solar cloth. The fabrication of core–shell nanofibers has been achieved by coelectrospinning of two components such as poly(3-hexyl thiophene) (P3HT) (a conducting polymer) or P3HT/PCBM as the core and poly(vinyl pyrrolidone) (PVP) as the shell using a coaxial electrospinning set up [see section 4.1].[48]

Initial measurements of the current density vs. voltage of the P3HT/PCBM solar cloth were carried out and showed current density (Isc), open circuit voltage (Voc) and fill factor (FF) of the fiber cloth around 3.2×10^{-6} mA/cm^2, -0.12 V and 22.1%, respectively. In addition, a six order of magnitude lower photo conversion efficiency of the fiber cloth around 8.7×10^{-8} was observed that might sound disappointing. The low photovoltaic (PV) parameters of the fiber cloth could be attributed to the following factors:

(a) The fiber cloth processing steps including electrospinning as well as ethanol washing were carried out under ambient conditions.

(b) The thickness of the fiber cloth was ~5 μm compared to the diffusion length of the charge carriers in organic solar cells, which is only several nm.

Therefore, most of the charge carriers were lost in the fiber matrix itself. Drop casted films of the same thickness also showed similar PV parameters.

11.4.4 FUTURE PROSPECTS OF ORGANIC NANOFIBER TEXTILE SOLAR

Nanofiber revolutionizes the future trend of material selection to enhance the characteristics of textile solar cell. Latest experience regarding to polymer solar cell and low band gap material needs to be considered for ambitious plans with nanofiber morphology. Figure 12 shows a schematic for layer-by-layer hetrojunction textile solar cell according to previously mentioned concerns and promizing potential solar absorbing and photoactive material. The usage of anti reflective and protective materials is extremely crucial for industrial scale production. A protective layer will save the organic material from moisture and oxygen. The antireflective layer in solar cells can obstruct the reflection of light and also contribute to the device performance. This is an important point that should be overcome in the case of large-scale production of textile solar cell. The PEDOT:PSS combination with CNTs forms first P–N junction in the form of bi-layer nanofiber. The second layer composed of core shell or general bi-layer nanofiber of MDMO/PPV:PCBM, which absorb visible spectrum. The proposed plan can be realized in large-scale production through a needles electrospinning setup. The continuity of process also is not beyond expectation and a range of materials as nanofiber has been already provided on given substrates.

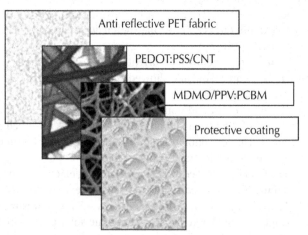

FIGURE 12   A general scheme of hetrojunction Organic nanofiber textile solar.

11.5 CONCLUSION

The multilayer solar cell energy conversion unit although obey simple theory of light scattering, absorption, electron excitation, charge transfer and its compensation but each layer for specific prescribed functions needs to be intensively investigated for being applicable in diverse circumstances. The commercial solar cell products including silicon either homo or heterojunctions, dye synthesized solar cell and organic solar cell subjected to demanding research and reached to high level of maturity. The huge experience in silicon solar cell and other photovoltaic system should be reviewed for developing new generation of flexible solar cell. Therefore, current work in first part has a quick glance on variety of solar cells including inorganic, organic and hybrid structures. In overall, regardless of type of solar cell and its corresponding elements, following points needs to be addressed to find fascinating performance:
- The wider absorption wavebands in a range of spectrum;
- The thinner the solar cell;
- The lowering the band gap;
- The higher surface area per unit mass;
- The lowering the cathode thicknesses;
- Using antireflective coating;
- Using protective coating;
- The multijunction cell to cover range of spectrum;
- Using semitransparent wave guide material;
- The lifetime enhancement.

Incorporation of organic solar cells into textiles has been realized reaching encouraging performances. Stability issues need to be solved before future commercialization can be envisaged. The mechanical stability of the devices was not limiting the function of the devices prepared. It would seem that low power conversion efficiency much more pertinent than the mechanical stability on the timescale of commercial.

KEYWORDS

- **antireflective coating**
- **photovoltaic system**
- **protective coating**
- **silicon solar cell**

REFERENCES

1. Schubert, M. B.; In: Conf. Rec. 31st IEEE Photovolt. Specialists Conf.; IEEE, New York, **2005**, 1488.
2. Nelson, J.; The Physics Of Solar Cells: Imperial College Press, **2003**, 1–16.
3. Miles, R. W.; Forbes. H.; I.; Photovoltaic solar cells: An overview of state-of-the-art cell development and environmental issues. Progress in Crystal Growth and Characterization of Materials, **2005**, *51*, 1–42.
4. Günes, S.; Beugebauer, H.; and Sariciftci, N. S.; Conjugated Polymer-based Organic Solar Cells, *Chem. Rev.;* **2007**, *107*, 1324–1338.
5. Castafier, T. M.; Solar Cells. In: Castafier, T. M. L. (Ed.), Practical Handbook of Photovoltaics: Fundamentals and Applications: Elsevier, **2003**, 71–95.
6. Thompson, B. C.; Polymer–Fullerene Composite Solar Cells. Angew. *Chem. Int. Ed.* **2008**, *47*, 58–77.
7. Shrotriya, V.; Yao, Y, Moriarty, T, Emery, K.; Yang.Y.; Accurate Measurement and Characterization of Organic Solar Cells. *Adv. Funct. Mater.* **2006**, *16*, 2016–2023.
8. Krebs, F. C. In F. C.; Krebs (Ed.), Polymer Photovoltaics A Practical Approach: SPIE, 1–10. **2008**,
9. Cai, W.; Cao, Y.; Polymersolarcells: Recent development and possible routes for improvement in the performance. *Solar Energy Materials and Solar Cells* **2010**, *94*, 114–127.
10. Dutta, P.; Eom, S. H.; Lee, S. H.; Synthesis and characterization of triphenylamine flanked thiazole-based small molecules for high performance solution processed organic solar cells. Organic Electronics, **2012**, *13(2)*, 273–282.
11. Soa, S.; Koa, H. M.; Kima, C.; Paeka, S.; Choa, N.; Songb, K.; Leec, J. K.; Koa, J.; Novel unsymmetrical push–pull squaraine chromophores for solution processed small molecule bulk heterojunction solar cells. *Solar Energy Materials and Solar Cells*, **2012**, *98*, 224–232.
12. Lina, Y.; Liua, Y.; Shia, Q.; Hua, W.; Y.; Lia, Zhan, X.; Small molecules based on bithiazole for solution-processed organic solar cells. Organic Electronics, **2012**, *13(4)*, 673–680.
13. Gratzel, M.; *Nature* **2001**, *414*, 338–344.
14. Ginger, D. S.; N. C. G. **2004**, Electrical Properties of Semiconductor Nanocrystals. In V. I.; Klimov (Ed.), Semiconductor and Metal Nanocrystals (pp. 236–285): Marcel Dekker, Inc. Ferrazza, F. **2002**, Large size multicrystalline silicon ingots. Proc. E-MRS **2001**, Spring Meeting, Symposium E on Crystalline Silicon Solar Cells. *Sol. Energy Mater. Sol. Cells, 72*, 77–81.
15. Kisserwan, H.; Enhancement of photovoltaic performance of a novel dye, "T18," with ketene thioacetal groups as electron donors for high efficiency dye-sensitized solar cells. *Inorganica Chimica Acta*, **2010**, *363*, 2409–2415.
16. Wei, D. **2010**, Dye Sensitized Solar Cells, *Int. J. Mol. Sci. 11,*1103–1113.
17. Günes, S.; Beugebauer, H.; and Sariciftci, N. S.; Conjugated Polymer-based Organic Solar Cells, *Chem. Rev.; 107*, 1324–1338 **2007**,
18. Brabec, C. J.; Dyakonov, V.; Parisi, J.; and Sariciftci, N. S.; "Organic Photovoltaics Concepts and Realization," 1st Ed.; Springer: New York, **2003**.

19. Berson, S.; de Bettignies, R.; Bailly, S.; and Guillerez S.; Poly(3-hexylthiophene) Fibers for Photovoltaic Applications, *Adv. Funct. Mater.* **2007**, *17*, 1377–1384.

20. Gonzalez, R.; and Pinto, N. J.; Electrospun poly(3-hexylthiophene-2,5-diyl) Fiber Field Effect Transistor, *Synthetic Metals*, **2005**, *151*, 275–278.

21. Bahners, T.; Schlosser, U.; Gutmann, R.; Schollmeyer, E.; Textile solar light collectors based on models for polar bear hair, *Solar Energy Materials and Solar Cells* **2008**, *92*, 1661–1667.

22. Tributsch, H.; Goslowski, H.; Ku, U.; Wetzel, H.; Light collection and solar sensing through the polar bear pelt, *Sol. Energy Mater.* **1990**, *21*, 219–236.

23. Liu, J.; Namboothiry, M. A. G.; Carroll, D. L.; Fiber based Architectures for Organic Photovoltaics, *Appl. Phys. Lett.* **2007**, *90*, 063501.

24. O'Connor, B.; Pipe, K. P.; Shtein, M.; Fiber Based Organic Photovoltaic Devices, *Appl. Phys. Lett.* **2008**, *92*, 193306.

25. Bedeloglu, A.; Demir, A.; Bozkurt, Y.; Sariciftci, N. S.; A Photovoltaic Fiber Design for Smart Textiles, *Textile Res. J.* **2007**, *80(11)*, 1065–1074.

26. Ouyang, J.; Chu, C. W.; Chen, F. -C.; Xu, Q.; Yang, Y.; High-conductivity Poly 3, 4-ethylenedioxythiophene): Poly(stirene sulfonate) Film and its Application in Polymer Optoelectronic Devices, *Adv. Funct. Mater.* **2005**, *15*, 203–208.

27. Kushto, G. P.; Kim, W.; Kafafi, Z. H.; Flexible Organic Photovoltaics using Conducting Polymer Electrodes, *Appl. Phys. Lett.* **2005**, *86*, 093502.

28. Huang, J.; Wang, X.; Kim, Y.; deMello, A. J.; Bradley, D. D. C.; deMello, J. C.; High Efficiency Flexible ITO-free Polymer/fullerene Photodiodes, *Phys. Chem. Chem. Phys.* **2006**, *8*, 3904–3908

29. Tvingstedt, K.; Inganäs, O.; Electrode Grids for ITO-free Organic Photovoltaic Devices, *Adv. Mater.* **2007**, *19*, 2893–2897.

30. Perzon, E.; Wang, X.; Admassie, S.; Inganäs, O.; Andersson, M. R.; An Alternating Low Band-gap Polyfluorene for Optoelectronic Devices, *Polymer*, **2006**, *47*, 4261–4268.

31. Campos, L. M.; Tontcheva, A.; Günes, S.; Sonmez, G.; Neugebauer, H.; Sariciftci, N. S.; Wudl, F.; Extended Photocurrent Spectrum of a Low Band Gap Polymer in a Bulk eterojunction Solar Cell, *Chem. Mater.* **2005**, *17*, 4031–4033.

32. Shaheen, S. E.; Brabec, C. J.; Sariciftci, N.; Padinger, F.; Fromherz, T, Hummelen, J. C.; *Appl. Phys. Lett.* **2001**, *78*, 841–843.

33. Scharber, M. S.; Mühlbacher, D.; Koppe, M, Denk, P, C.; Waldauf, Heeger, A. J.; Brabec, C. J.; *Adv. Mater.* **2006**, *18*, 789–794.

33. Krebs, F. C.; Biancardo, M.; Jensen, B. W.; Spanggard, H.; Alstrup, J, Strategies for incorporation of polymer photovoltaics into garments and textiles, *Solar Energy Materials and Solar Cells*, **2006**, *90*, 1058–1067.

34. Li, D.; Wang, Y.; Y.; Xia, Electrospinning Nanofibers as Uniaxially Aligned Arrays and Layer-by-Layer Stacked Films, *Adv. Mat.*, **2004**, *16(14)*, 361–366.

35. Yu, J. H.; Fridrikh, S. V.; Rutledge, G. C.; Production of Submicrometer Diameter Fibers by Two-Fluid Electrospinning, *Adv. Mat.*, **2004**, *16(17)*, 1562–1566.

36. Li, D.; Xia, Y.; Direct Fabrication of Composite and Ceramic Hollow Nanofibers by Electrospinning, *Nano Lett.*, **2004**, *4(5)*, 933–938.

37. Chuangchote, S.; Sagawa, T.; Yoshikawa, S.; *Appl. Phys. Lett.* **2008**, *93*, 033310.

38. Zhao, X.; Lin, H.; Li, X.; Li, J.; The application of freestanding titanate nanofiber paper for scattering layers in dye-sensitized solar cells, *Materials Lett.* **2011,** *65*, 1157–1160.

39. Li, S.; Zhang, X, Jiao, X.; Lin, H.; One-step large-scale synthesis of porous ZnO nanofibers and their application in dye-sensitized solar cells, *Materials Lett.* **2011,** *65*, 2975–2978.

40. Li, J.; Chen, X, Ai, N.; Hao, J.; Chen, Q.; Strauf, S.; Shi, Y, Silver nanoparticle doped TiO2 nanofiber dye sensitized solar cells, *Chem. Phys. Lett.* **2011,** *514*, 141–145.

41. Hu, G. J.; Meng, X. F.; Feng, X. Y.; Ding, Y. F.; Zhang, S. M.; Yang, M. S.; *J. Mat. Sci.* **2007,** *42*, 7162–7170.

42. Kubo, W.; Kitamura, T.; Hanabusa, K.; Wada, Y.; Yanagida, S.; Chemical Communication. **2002,** 374–375.

43. Wang, P.; Zakeeruddin, S. M.; Comte, P.; Exnar, I.; Gratzel, M.; *J. Am. Chem. Soc.,* **2003,** *125*, 1166–1167.

44. Wang, P.; Zakeeruddin, S. M.; Moser, J. E.; Nazeeruddin, M. E.; Sekiguchi, T.; Gratzel, M.; *Natural Material.* **2003,** *2,* 402–407.

45. Park, S. H.; Kim, J. U.; Lee, S. Y.; Lee, W. K.; Lee, J. K.; Kim, M. R.; *J. Nanosci. Nanotech.* **2008,** *8*, 4889–4894.

46. Kim, J. U.; Park, S. H.; Choi, H. J.; Lee, W. K.; Lee, J. K.; Kim, M. R.; Sol. Energy Mater. Sol. Cells **2009,** *93*, 803–807.

47. Chen, J.; Lia, B.; Zheng, J.; Zhao, J.; Jing, Zhu, Z.; Polyaniline nanofiber/carbon film as flexible counter electrodes in platinum-free dye-sensitized solar cells, Electrochimica Acta **2011,** *56*, 4624–4630.

48. Sundarrajan, S.; Murugan, R.; Nair, A. S.; Ramakrishna, S.; Fabrication of P3HT/PCBM solar cloth by electrospinning technique, *Materials Lett.* **2010,** *64*, 2369–2372.

CHAPTER 12

NEW GENERATION OF NANOMOLECULAR STRUCTURES

A. L. IORDANSKII, S. Z. ROGOVINA, I. AFANASOV,
and A. A. BERLIN

CONTENTS

12.1 INTRODUCTION

Challenges and development perspectives on nanopatterned implants loaded with drugs intended to replace or improve human organs and tissues are analyzed. An innovative approach to polymer and composite design combines the surface modification on the molecular and nanosized levels, formation of the implant matrix as hybrid nanocomposite, and drug encapsulation aimed at ensuring their targeted and programed delivery. The economic and scientometric situations in the world development of composite and hybrid implant systems are briefly described. The major fields in nanoimplantology for the nearest decade are represented, including cardiology, ophthalmology, genitourology and orthopedy. The prospects for biodegradable polymers, such as poly(α-hydroxy acids) (PLA PGA, PLGA) and poly(β-hydroxyalkanoates) (PHB, PHVB), are considered, as well as nanoscale biochips and sensors, miniature electromechanical systems (MEMS and NEMS), and neurological conduits.

When designing the new generation of medical implants, it should be considered many factors and processes that guarantee biomedical efficiency, the retention of the necessary properties during the service life, patient's comfort, as well as other parameters characterizing implant behavior in biological media [1–5]. In addition to the above requirements, the implants should also sustain drug delivery [6–10] and be arranged with nanoscale structures [11, 12]. The latter requirement is determined by implant—cell interactions at the receptor and immune levels, which are most effective if composite implants, have a nanostructured surface [13, 14].

The implanted medical device (IMD) is an intracorporeal system (device or material) intended for (a) the replacement of lost organs, tissues, or cell assembly; (b) an enhance in the efficiency of the body's biological function; and (c) a patient postsurgery rehabilitation and mechanical support at arrangement in vivo. Implants differ from transplants in their artificial origin, that is, implants are manufactured under plant, pilot, or laboratory conditions. Traditionally, implants are produced from metals, inorganic compounds, and polymers. Recently, biocomposites, multilayered and hybrid constructs, and nanoscale modifications of implants have been elaborated. IMD specific features are their biocompatibility and, when necessary, controlled biodegradability. State-of-the-art implants include electronic or computer devices (pacemakers and auricular prostheses), trigger regulators, and feedback systems, preferably wireless. A separate group of implants is the combined systems for dosed and targeted drug delivery in the body. Unlike traditional drug formulation, the active substance in this case is encapsulated directly into the matrix or reservoir of an implanted device as, for example, in the combined cardiac stents with an active polymer coating. This group of implants is appropriate for therapeutic procedures, surgery (injectors), and diagnosing (in vivo electronic chips).

Note that, if both passive and active transports may be implemented in the case of the nanoparticles assisted drug delivery, the drug encapsulated in the implant enters the adjacent tissues exclusively via a combination of the hydrolytic (enzymatic) process [15] and diffusion [16]. This suggests that, in order to provide the appropriate drug flux from the implant's matrix that is necessary for therapy, the optimally high sorption capacity of implant is required, that is, the critical concentration of the functional groups in polymer for drug immobilization.

To ensure the efficient function and safety of the implants and coordination of the efforts on their design and commercial promotion, European Union has issued a set of directives (for example, [17] that strictly defines what is an IMD. Over the last three years, Russian metrology has also formulated a number of requirements to IMDs. Analogous directives have also been elaborated by the corresponding authorities in the United States [18] and Canada [19].

In 2010, the manufacture of biomedical devices and functional implants worldwide reached $272.2 billion at market price. According to the expert forecasts, by the year 2015, this value will increase by 65%, despite the economic crisis and will almost double in Eastern European countries (from $4.1 billion in 2010 to $7.8 billion in 2015). The United States, which occupies the widest sector on the market of medical products and devices (over 40%), is first in the list of manufacturers of implanted systems, followed by EU countries (Western European countries, 27%) [20].

This development of the market reflects a high demand for medical devices and implants in clinics and diagnostic centers worldwide. Only in the United States, the number of patients using various IMDs exceeds 25 million [20]. Over 2 million stents are annually implanted in cardiac surgery and urology. An even higher number of patients use temporary medical devices, such as catheters (400 million) and kidney dialyzers (25 million) [21]. First on the list of the manufactured IMDs are cardiovascular implants (stents, defibrillators, and pacemakers), followed by orthopedic implants. The next group includes neurological stimulators and signal transducers, as well as gastrointestinal bands and laparoscopy systems. Auditory prostheses are the most dynamically developing group of polymer implants with the production prospect of $27.9 billion by the year of 2016. Similar to most high technologies and taking into account the regulations developed by state control institutions, the period from laboratory research to commercial production for the nanomedical innovations is 12–15 years. The first successful results in the field of nanomedicine were obtained in the early 2000 s; correspondingly, a considerable increase in the number of nanoscale drugs and devices can be expected as soon as the end of the current decade. By the end of 2010, FDA approved 22 ready-to-use nanopharmaceutical implanted systems, and the number of systems that undergo different phase of clinical trials is even higher. Concurrently, about 25 items of IMDs in Europe are in the final stage of clinical trials.

Note that the commercialization of nanotechnology products is accompanied by exponential growth in the corresponding research and development activities; moreover, as is evident from Fig. 1, the rate of increase in the number of publications (patents and research papers) in the field of nanobiomedicine exceeds that rate of the corresponding publications in the remaining fields of nanotechnology. Note also that the improvement of regulations and legislation for design and use of implants and prostheses, which took place in 1989, had a significant effect on the intensity of the corresponding research, as is reflected by the breakpoint in curve 2, which is denoted for clarity by the corresponding symbol. According to the location within or outside the body, all the currently known implants fall into intracorporeal and intercorporeal. In addition, the devices intended for analysis, diagnosis, and drug delivery are traditionally used under in vivo, ex vivo, and in vitro conditions. Excluding all intercorporeal implants, such as ocular contact lenses, dental prostheses and fillings, periodontal membranes, transdermal systems, wound coverings, and so on from consideration, this review will consider intracorporeal miniature systems except for the traditional orthopedic devices, cartilage implants, surgical meshes, pericardium, and other macroscale medical constructs.

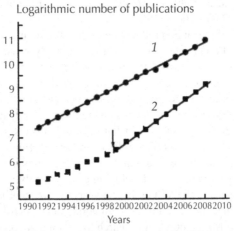

FIGURE 1   Exponential growth in the number of publications (research papers and patents) in (1) all nanotechnology areas and (2) nanobiomedicine area. Arrow denotes approval datum of the first regulation documents for implants in the United States and one year later in Europe.

12.2 MODIFIED CARDIAVASCULAR STENTS AS IMPLANTS PROVIDING DRUG DELIVERY FUNCTION

Cardiovascular diseases are first on the list of mortality and morbidity, exceeding the cancer diseases and casualties. Cardiovascular pathologies are among the basic causes of short lifespan and deteriorating demographic situation. In 2008, the World Health Organization (WHO) acknowledged Romania as the most unfavorable country, with cardiovascular diseases accounting for 52% of fatal cardiac cases. In 2005, Russian Federation headed this list in Europe, with 57% of fatal cases, which is twofold higher than the most trouble-free developed countries [22]. According to the World Health Organization (WHO), 17.3 million people died from cardiovascular related diseases in 2008, and this number is projected to rise to an estimated 23.6 million by 2030. Therefore, it is quite natural that the prevention and therapy of cardiovascular diseases is considered to be a most important priority in the field of demography, health promotion, and increase in the lifespan of population [22, 23].

The latest advance in the field of nanotechnology most pronouncedly appeared in construction of cardiovascular stents. First and foremost, this concerns the main concept in design of the stents that provide a prolonged and local drug delivery. In addition, the obligatory requirement to cardiovascular stents is effective hemocompatibility. The combination of controlled drug delivery and biocompatibility may be attained by nanostructuring the surfaces of cardiac implants, for example, constructing meso- and nanoporous layers responsible for controllable release, protein adsorption, and adhesion of blood cells (platelets, erythrocytes, leukocytes, etc.).

A typical complication when implanting a vascular stent is restenosis as a result of neointimal hyperplasia, on the one hand, and thrombosis activated by platelets, on the other hand. After the active clinical and technical search for efficient macro and micro constructs (the search for an optimal geometry of bare metal spiral stents) have been depleted, it has become clear that the classic paradigm for introduction of metal devices should be replaced with elaboration of surface modified devices able to dose drugs and provide their prolonged elution. The pioneers in modern technologies for producing next generation cardiac implants with a modified surface (drug eluting stents) are several companies, that is, Boston Scientific, Johnson & Johnson, Medtronic, Guidant and the others. Currently, designs of the newest stents with biodegradable or inert coatings provide for the targeted delivery of various cardiac drugs, such as Paclitaxl, Sirolimus, Zotrolimus, and so forth, which are responsible for the inhibition of cell proliferation and, thereby, decrease the incidence rate of restenosis compared to helical metal stents of the first generation.

However, the antiproliferation effect is most pronounced only soon after implanting (the first six months); however, as recently shown [24, 25], these constructs can initiate thrombosis because of a decrease in cell proliferation and partial degeneration of the neointima [26]. Construction of the nanoporous surface structures with aluminum oxide or titanium oxide for release of Tacrolimus

[27], carbon nanoparticles with chromium additions that form the surface layer of the implants for dispensing Paclitaxl [28], and gold [29] is aimed at solving a polyfunctional problem, namely, concurrent provision of (a) biocompatibility of implant and (b) the necessary kinetic profile of drug release. When improving biocompatibility, it is necessary to take into account the topography characteristic of the stent surface (e.g., roughness), which should simulate the natural vascular surface in a nanoscale range. Achieving this effect should lead to improved cell adhesion and stent endothelization and, as a consequence, decrease the probability of thrombosis due to the tight contact between the stent surface and endothelium [30]. The results of constructing the surface structures with nickel titanate [31] or hydroxyapatite [32], which are analogous to the specific features of natural vascular surface, suggest with a certain degree of reserve that one of the promizing directions in nanotechnology stent constructs is designing of their nanopatterned surfaces. These surfaces should concurrently provide the following two functions: targeted drug delivery and regulated cell adhesion as the factor that limits restenosis in short-terms and the factor that prevents long-term thrombosis as well as promote endotheliazation [33] see the scheme in Fig. 2. In the case of specific therapy of vascular tissues and concurrent diagnostics, it is proposed to use magnetized metals as a stent core structure [34]; these metals attract superparamagnetic nanoparticles (e.g., iron oxides) from the bloodstream, thereby allowing not only targeted drug delivery in the absence of external magnetic field, but also to contrast abnormal regions on the vascular surface [35].

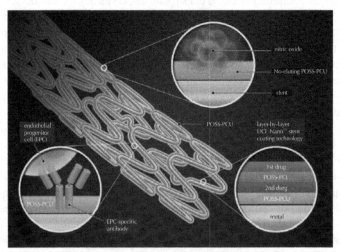

FIGURE 2   The novel generation cardiovascular stents coating with special nanocomposite polymers like POSS-PCU (polyhedral oligomeric silsesquioxane and copolymer carbonate-urea urethane: trade named UCL-NanoTM) has been developed specifically to capture oxide nitrogen and EPC specific antibodies (factor of endothelialization) simultaneously.

Another promising direction in the development of implanted stents is a complete rejecting of metal alloys as basic coiled structures. It is proposed to use for these purposes biodegradable polymeric materials, namely, polylactide-coglycolides, PLGAs, or poly(3-hydroxybutyrate-co3-hydroxyvalerate), PHBV. Controlled biodegradation allows, first, for a prolonged drug release to local regions of the cardiovascular system and, second, for a decrease in the risk of repeated thrombosis and hyperplasia with biodegradation of stent [36]. When designing this type of implants, two important conditions should be met, that is, (1) the rate of stent biodegradation should be approximately equal to the cell proliferation rate; and (2) the biodegradation products should not be toxic or induce undesirable immune responses. The mentioned biopolymers almost completely satisfy these criteria, since their intermediate degradation products are members of the physiological Krebs cycle and the final products are molecules of CO_2 and H_2O [37].

12.3   MODERN ORTHOPEDIC IMPLANTS AND THEIR FURTHER DEVELOPMENT

The modern orthopedic implants (OIs) belong to the class of nanocomposites with rather intricate structural organization. They should not only effectively withstand the stresses, but also should be biocompatible and provide a prolonged targeted drug delivery of biologically active substances. In the general case, OI biocompatibility is determined by the cell response of bone tissues to the implanted material, which characterizes the degree of OI integration into the implanted region. The innovative approach to OI construction as nanobiocomposites is in that the material used to replace bone tissue is represented by a composite with a successive hierarchical structure of spatial elements of organic and inorganic origins. In the nanoscale range, the used structure forming elements are collagen fibers, hydroapatite, and proteoglycans, which together provide for the structural stiffness, stability, and concurrent plasticity of a natural bone.

The effective use of a bone implant is based on its biocompatibility at the cell level. Most researchers who work in this field believe that it is possible to influence the cell behavior via biochemical mediators (with involvement of biologically active substances), surface topography of implant (creation of modifying surface layer or coating), and induction of excitation signals [38]. All of these processes provide a stable cell adhesion and, consequently, form a perfect contact between nonphysiological material and bone tissue. In this situation, the implants maximally biologically mimic the natural bone material. In the absence of such correspondence, the cells in contact with the implant surface change at both the morphological and functional levels, as well as alter their behavior during adhesion, proliferation, endocytosis, and gene regulation. The OI surface is modified using various state-of-the-art technologies that form a nanoscale structure, including nanolithography, phase separation, chemical etching, molecular self-assembly, and electrospinning of micro and nanofibers.

Nanostructuring of implants creates highly developed surfaces, while the implants acquire fundamentally new mechanical, electric, optical, superparamagnetic, and other characteristics. The set of these properties will allow for the successful application of OI in targeted delivery of low molecular weight drugs, active macromolecules, and cells in the nearest future, as well as their use as diagnostic and biosensor elements. The current concepts of the interactions between the bone tissue and composite materials include adsorption of specific proteins, such as fibronectin and vitronectin, and subsequent adhesion of osteoblasts and other bone cells [39, 40]. In this case, both the composition and structure of the protein layer determine the quality and intensity of cell adhesion and, consequently, the integration ability of bone tissue which determines the differentiation, growth, and death of bone cells on the surface of implant [39, 41].

Currently, intensive studies in this field aim to clarify how the surface energy, chemical composition, charge, morphology, and nanoscale topography of the phase boundary influence the adsorption of specific proteins [42, 43]. In particular, fibronectin predominantly adsorbs on the calcium–phosphate surface of composite [44]. Then, the adhesion intensity of osteoblasts increases in the presence of the amino acid motif of the absorbed protein on the surface, such as arginine–glycine–asparagine (RGD sequence in Fig. 3), as well as when osteoprosthesis is heparinized [45]. In an ideal situation, the protein adsorption layer should stimulate adhesion of the cells that positively influence biocompatibility of the implant, for example, osteoblasts, and prevent adhesion of fibroblasts, which negatively influence the biocompatibility of implants [39]. As we already mentioned, one of the most important factors in OI biocompatibility, along with adsorption processes, is the surface topography of the implant. This particular factor has a considerable effect on the proliferation and differentiation of osteoblasts, the young cells responsible for the formation of new bone tissue [46]. It is not completely clear how the nanoscale surface topological elements act on the behavior of these cells [47]. However, there are serious reasons to believe that surface structuring in a nanoscale range modulates the cell membrane receptors, thereby modifying the communication between cells [48]. In addition, a nanoscale landscape of the implant surface influences the adsorption and conformational changes of integrin-binding proteins (see Fig. 3), which also activates the cell signaling receptors [49].

CURRENT STATE AND DEVELOPMENTAL PROSPECTS

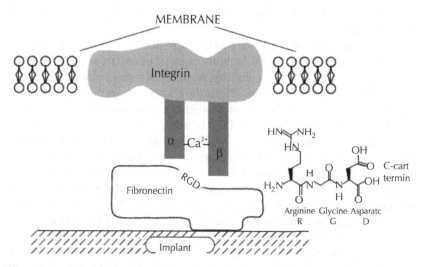

FIGURE 3 Protein adsorption on ostheoimplant surface and following interactions adsorbed proteins with cells of bone tissue.

Nanoscale implant surface topography comprises structural elements, such as pores, corners, slot-shaped hollows, nanofiber structures, nodular intertangling, and their combinations. The specific topographic features of the surface considerably influence its interaction with the cells of surrounding tissue, and this effect on cell adhesion and proliferation is frequently more significant compared to the effect of implant's chemical structure [50]. The situation is somewhat complicated by that individual cell types respond differently to the surface topography. For example, the adhesion of osteoblasts increases with the surface energy of carbon nanofibers, whereas the adhesion of fibroblasts and smooth muscle cells decreases [51]. This is why the detailed mechanism underlying the cell interactions with implant's surface elements is still vague and, in most cases, designers form the topographic landscape in a rather empirical manner.

The surface chemistry of implants plays an important role in their service. During implantation, proteins and other biomolecules adsorb on the implant's surface. The changes in native conformations of biomolecules on a non-physiological surface initiate a trigger cascade of inflammatory (immune) responses, which interfere with the integration of the implant into the living tissue or enhance the formation of fiber capsule. A certain success in the prevention of the negative processes is achieved via a controlled administration of anti-inflammatory drugs. The formation of the surface layers, for example, of gold modified with thiol groups and carrying spatially structured functional groups (–CH3, –OH, –COOH, and – NH2) able to bind integrin or regulating cell adhesion, is no less promizing [52]. Hydrophilic–hydrophobic balance of the implant surface influences the behavior of an adhered cell.

In addition to modern methods (atomic force microscopy, transmission electron microscopy, etc.), surface wettability (method of contact angle) is used as a simplified rapid assay for hydrophilicity. Amazing as it may seem, a macroscopic characteristic, such as wettability, reflects the OI structure on the nanoscale level. In particular, the surface wetting of aluminum can be increased by decreasing the particle size from 167 to 24 nm [53]. In another work [54], ultrathin titanium crystals displayed a high wettability, which stimulates osteoblasts to proliferation and adhesion and, as a consequence, accelerates the implant integration with the bone tissue. Regarding the bone tissue as a nanocomposite, the current strategy of bone engineering is based on creating a combination of inorganic compounds (calcium triphosphate, hydroxyapatite, organic silicon glass, as well as carbon tubes [55, 56] and organic polymers, such as copolymer of lactic-coglycolic acid, PLGA; poly(3-hydroxybutyrate), PHB; polycaprolactone; polyphosphazene; chitin; and chitosan. Polymeric matrices form 3D structures for the repair or regeneration of the bone tissue. An original method for OI construction consists of the incorporation of a nanoceramic-reinforcing agent that ensures an optimal surface topography and increases the strength of material [55]. The presence of mineral nanoparticles on the OI surface stabilizes its contacts with the cell tissue (integration into tissue) and strengthens the signal transduction of growing cells, thereby enhancing osteogenic activity.

The matrices of the nanofibers impregnated with drugs display the characteristics necessary for growing organs and tissues in cell engineering. Characteristic of the polymer nanofibers produced by electrospinning are high surface-to-volume ratios, large specific porosity, and adjustable size of nanopores. This set of characteristics makes the nanofiber matrices the most promizing material for growing bone and other biological tissues [57, 58]. The ultrathin fiber sizes and their branched surface create optimal conditions for immobilization of growing cells, while the system of pores enhances the metabolic processes associated with cell growth and vital activities [59]. Stem cells are of considerable interest in this connection [60]. Another nanotechnology approach to this problem consists of improving the biocompatibility of the surface. This is attained by coating the implant surface with nanofibers and increasing the flow of drug encapsulated in them [61].

Biocompatibility can be improved by introducing biologically active substances into the modified surface layers of a bone nanoimplant. Controlled release of antiseptics, peptides, extracellular proteins, osteoinductive growth factor, and osteogenic cells is used to stimulate osteogenesis (formation of new bone tissue) and the replacement of biodegradable regions of the implant with bone tissue [62, 63].

Thus, state-of-the-art bone implants should simultaneously contain nano-structured biocompatible surface and the active compounds in their surface layer, which creates an antiseptic effect, as well as controls the state and growth of bone tissue at the boundary with the implant. The adhesion and subsequent colonization of bacteria on the OI surface lead to subsequent infectious complications after implantation. Despite preliminary aseptic procedures, bacterial infections present a serious problem during post-surgery period, when IO is in the body. In particular, the rate of complications in hip arthroplasty reaches 3% and the rate of reinfection, even 14%. Antibiotics, the immobilization of antimicrobial agents within the matrix or on its surface, the addition of antiseptic metals (such as silver, copper, or titanium), and the encapsulation of nitrogen oxide are used to prevent the propagation of pathogenic microorganisms [62,64]. Pioletti et al. [65] have reviewed the latest data on the controlled drug release from OIs. A considerable segment of the innovative research and developments includes nanotechnology modifications of traditional implants, first and foremost, involving titanium and its alloys, in particular with nickel (TiNi). The specific feature of new generation titanium implants is their nanostructured surface. For this purpose, titanium nanotubes vertically oriented relative to the OI surface are currently synthesized to provide the encapsulation of the drug with its subsequent controlled release into the bone tissue. The rate of drug elution from titanium nanotubes can be regulated by the thickness and composition of biodegradable polymer layer used to coat the tube surface and represent an additional barrier for drug diffusion (see Fig. 4) [66].

 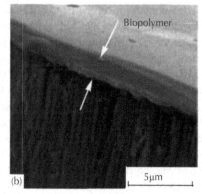

FIGURE 4 Nanostructured surface of a titanium osteoimplant (a) formed as tubes and (b) covered with a layer of poly(lactide-coglycolide).

Researchers from Brown University (United States) have demonstrated that the quality of bone titanium implants can be considerably improved by coating them with a film of polypyrrole (a conductive polymer). This makes it possible to control and program the elution rate of antibiotics and anti-inflammatory drugs by exciting an electric pulse in this electroconductive polymer [67]. Bisphosphonates, correctors of metabolism in the bone and cartilage tissues, prevent the lysis of the bone tissue near the implant and, consequently, enhance its long-term fixation at the implantation site. Several representatives of this class of drugs, for example, Zoledronate® (both salt and acid variants), pamidronate, and ibandronate, encapsulated in hydroxyapatite nanoparticles display a local prolongation effect for inhibiting the activity of osteoclasts (giant cells) and block calcium desorption from bones as a manifestation of osteoporosis. According to the definition by the FDA [66], in this case, OIs are used as a combined system that comprises a support function and the role of the matrix for controllable drug release.

A factor that complicates the use of nanostructured composite implants and scaffolds is the destruction of their elements in the body. Because of the potential cytotoxicity of the products of implant enzymatic hydrolysis, it is necessary to use polymeric materials that are degradable to nontoxic products, in particular CO_2 and H_2O, as is the situation with the biodegradable polymers of poly(3-hydroxyalkanoates) (PHAs and their derivatives) or poly(α-hydroxyacids) (PGA, PLA, and their copolymers PLGA). The use of carbon in implants carries a certain toxicological risk; however, clearance may be increased via hydrophilization of their walls with the functional groups, such as –COOH, –OH, and –NH2 [69]. Application of biodegradable materials in the current OIs or scaffolds requires serious studies aimed at regulating the rate of biodegradation [70], since, in an ideal situation, the rate of implant resorption should very precisely coincide with the growth rate of bone tissue in (or its penetration into) OI pores.

In summary, it can be expected that new commercial OIs will appear in clinics in the next years that will differ from the currently available implants in (a) nanopatterned surface and/or matrix; (b) combination of orthopedic and therapeutic functions; and (c) high biocompatibility and antibacterial resistance. Currently, the well-known manufacturers like Johnson and Johnson, Biomet, Apex Surgical LLC, Encore Orthopedics Inc., Optetrack Inc., Osteoimplant Technology Inc., Smith and Nephew Inc., Stryker Howmedica Osteonics, Zimmer Inc., Sulzer Orthopedics, and Depuy Co., have brought to the market the composite osteoimplants for replacing bone tissue in knees and hips, in spinal cord surgery, maxillofacial surgery, and so on. In 2009, Medtronic Inc. (Minneapolis, United States) announced the production of Mastergraft ceramic scaffolds for growing spinal implants that elute the corresponding drugs. SurModics Inc. (Minnesota, United States) has reached certain success in surface modification and subsequent drug immobilization in the surface layer. However, OIs that meet all three above characteristics are practically still absent and will be the focus of innovative developments in this decade.

12.4 MODERN GENITOURINARY IMPLANTS

Gynitourinary implants (GIs) are widely used for contraception and prolonged therapy (1 year and more) of female pelvic organs, for example, in endometriosis. Specific construction features of such implants are their elasticity, compatibility with adjacent tissues, and, which is most important, drug encapsulation in their matrix. However, in some cases, specialized laparoscopically introduced micro-implants are used to induce a cell rejection response and subsequent growth of fibrous tissue for blocking fallopian tubes, thereby preventing pregnancy.

A polymer-metal device, Essure (Conceptus Inc., San Carlo, California, United States), registered and approved by the FDA, is an example [71]. In 2006, the contraceptive implants Implanon (Organon USA Inc.) and Nexplanon (Merck and Co. Inc., USA) were registered by the FDA and appeared on the market; these microimplants are multilayered copolymer constructs carrying up to 70 mg of etonogestrel. Unlike the widely known contraceptive rings and their analogs (Norplant and Jadelle), these microimplants provide a daily dose of ~30 μg and more of long-term drug elution (for 3 years) without hurting patient tissues [72, 73]. To visualize their location, these implants are provided with contrasting (X-ray or NMR) characteristics, for example, by copper doping [74]. Among the latest devices of this type, one should note the Mirena (Bayer, Germany) polysi-loxane implant, produced in 2007, which is capable of the controlled release of hormonal compounds for 5 years. The disadvantages of all the currently available commercial GIs are, on one hand, a limited contribution of innovative technologies to their production and, on the other, the lack of the possibility to modify the rate of drug elution according to circadian and monthly body cycles. The current chronobiology recommends synchronizing a targeted drug delivery to the activity of the corresponding organ. A programed pattern of targeted drug delivery can soon be achieved by implanting microchips into the problem areas of the body. These feedback devices are currently produced by innovative methods (nano-lithography, etching, electron micrography, imprinting, etc.). Depending on the changes in biochemical composition of the ambience, these miniature-combined chips provide for regulated drug release, similar to the transdermal and insulin-containing electromechanical systems [75].

Nanotechnology is applied to designing new methods of treating urological diseases in two main directions. When dealing with functional abnormalities, such as bladder dyskinesia, prostatitis, and hyperplasia, various types of nanopar-ticles are used as diagnostic and therapeutic agents. The second direction of appli-cation of nanobiomedicine in urology is the reconstruction of lost or congenitally abnormal elements of the genitourinary system using cell-engineering methods. This involves the restoration of ureters, bladder repair, cosmetic surgery of geni-tals, and so on. In this review, the latter aspect associated with the implantation of reconstituted urological elements and tissues is of interest. In this case, a key role of cell engineering is evident, namely, the growth of cells, tissues, and elements of organs under special conditions and with the help of specialized 2D and 3D matrix systems (scaffolds). Simplified sketch of tissue growth on matrices is displayed in Fig. 5.

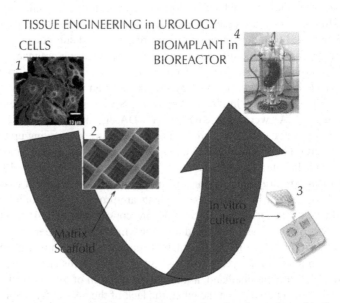

FIGURE 5   (a) The principal stages of tissue engineering for urology. (b) Compilation elements adopted from Ref. [76]. Perspectives on the role of nanotechnology in bone tissue Engineering. (c) Ref. [77] (d) Cit. David Mack/Science Photo Library [78]. Recent advances in the field of tissue bioengineering have taken scientists closer to growing organs in the laboratory.

Cell technology today has started to compensate for the deficiencies in tissues acquired due to traumas or surgery. In particular, Pattison et al. [79] pioneered the construction of a nanoscale biodegradable 3D matrix for the successful growth of bladder smooth muscle cells. The fibrinogen nanofibers produced by electrospinning were used to design a new type of porous matrices to produce cell material in urology [80]. Bioartificial (hybrid) implants as analogs of the kidney [81] are developed at the University of Michigan (United States). Their working element is a nanoporous silicon membrane that concurrently works as a filter and a matrix for immobilization of renal tubular cells. These devices are intended for the cases of acute kidney failure. As is mentioned above, a drug with antiseptic and cell stimulating functions can be immobilized in a polymeric material, for example, biodegradable poly(lactide-coglycolide).

Trauma, toxic or inflammatory circumstances as well as congenital diseases of a genitourinary tract may result in serious damages and even the loss of function. In this situation novel relevant biomaterials are crucially demanded to substitute damaged organs. For example the search for an alternative reservoir to replace the native bladder has proved to be crucially intricate. Synthetic, nonbiodegradable materials, such as polyolefines, silicone, rubber, polytetrafluoroethylene and others were initially used in urological surgery. However, these genitourinary implants quickly covered with a crust, were prone to infection and subject to host–foreign body reactions [82]. Currently the tissue engineering is playing an increasing role in the management of genitourinary implant variety. Scaffolds for cell cultivation may use not only biopolymers but also artificial tissues or combination of the two. In particular, a combined scaffold of collagen and polyglycolic acid even enhances even tissue regeneration [83]. Sharma et al. [84] performed bladder augmentation in rats with a special polymeric scaffold (elastomeric poly(1,8-octanediol-cocitrate)). The polymer matrix was seeded on opposing surfaces with human mesenchymal stem cells (MSC) and urothelial cells (UC). At 10 weeks morphometry of MSC-polymer-UC sandwiches revealed muscle/collagen ratios approximately 2 times more than controls. Additionally the data of work [83] displayed that MSC colony support partial regeneration of bladder tissue in vivo, A new approach in the area of biomaterials also for the lower urinary tract could be the use of electrospun nanofibers which reveal a superior viability in comparison to other materials, most probably due to their high surface/volume ratio imitating an extracellular matrix structure. Growth of stem cells on the nanofiber material and differentiation into the different cell types has been confirmed in histological analysis [85].

To make the picture complete, the innovative works in the field of urological surgery should be mentioned. In particular, an EnSeal™ implant (SurgRx, Palo Alto, California, United States) used in laparoscopic resection of the prostate and comprising millions of nanoparticles, which enhances healing injured small veins, thereby minimizing the tissue damage in the involved organ [86]. Submicron tweezers, a potential urological surgery tool, is now used in surgery of spermatic ducts or, for example, in vasectomies [87]. High-molecular-weight implants, such as hydrogels or elastic polymeric constructs, are used when treating and replacing genitourinary elements. Hydrogel membranes are formed for a number of implanted devices to separate the ambience and drug container. The therapy of prostate tumors, as well as the regulation of adolescent sexual maturation, involve the administration of hormones; for this purpose, the Hydron, Supprelin, and Vantas gel implants with about 1-year-long drug elution have been developed [88]. The reservoir systems that use an osmotic pump principle, that is, ALZET (Alza) [89] and DUROS [90], have been designed; their separation membranes are formed of cellulose ester and polyurethane, respectively, and they elute at a constant rate a protein, LHRH agonist, administered to treat prostate cancer.

Biodegradable polymers represent a separate group of polymer implants in urology. By analogy with bioresorbable suture materials and nanoparticles, the following groups of natural polymers have been used for their production: poly(α-hydroxyacids) (polylactides and their copolymers) and poly(β-hydroxyalkanoates) (PHAs) and the copolymers with hydroxyvalerate, hydrooxyhexanoate, and so forth. An example is the multifunctional implanted system Zoladex™ (Astra Zeneca, Canada), which was designed over 20 years ago and is currently widely used in clinical practice for the long-term therapy of endometriosis, breast tumors, and prostate tumors [91]. Note that polyhydroxyacids (PLA and PGLA) and poly(ortho esters) were certified by the FDA US in 2009. Catheters of various types and diameters are now widely used as transient implants. They are temporarily introduced into various body parts, such as the bladder, to replace ureters, to enhance the therapy of renal tissues, and to prevent gall bladder dyskinesia. The main requirements to polymeric catheters are nontoxicity; biocompatibility; and, in particular, their antiseptic effect. For this purpose, similar to vascular catheters, their surfaces are modified and enriched for encapsulated drugs, most frequently, a bactericidal agent. The controlled delivery of the drug desorbed from the walls of an implant in this situation allows for the avoidance of numerous complications caused by urine-containing, bacterially active media.

12.5 NANOIMPLANTS WITH ENCAPSULATED DRUGS FOR OPHTHALMOLOGY

The eye diseases present a serious problem, first and foremost, for the population of industrially developed countries. These diseases include age related retinal changes, diabetic retinopathy, structural alterations in the lens, and so on. A systemic drug delivery to the anatomically intricate organ, such as the eye, and especially to the intraocular elements is not a simple task, which interferes with the therapy of the uvea, vitreous body, and optic nerve. Because of multiple tissue barriers and lacrimal drainage, only 1–3% of the active substance of traditionally administered drugs (e.g., as eye drops) reaches the problem areas within the eye [92=83]. Therefore, it is planned in the nearest future to widely use nanoparticles and polymer implants with encapsulated drugs for prolonged and targeted drug delivery. The eye lens spatially separates the eyeball into the anterior segment, which comprises a cornea, sclera, and conjunctiva, and the posterior segment, with all of its elements localized beyond the lens deep in the eye (retina, vascular membrane, vitreous body, and optic nerve). This is why the ocular implants are classified according to the eye anatomical structure into anterior ocular and (e.g., contact lenses) and intraocular devices.

The installation of ocular micro and macroscale implants is associated with a number of issues that should be briefly mentioned in this section. First, the eye is a paired organ; correspondingly, this provides for an objective comparison of the outcome of innovative treatment, be it a surgery or therapy, of one eye with the intact state of the other eye, using it as a control [94]. Second, the eyeball structure is rather transparent, which allows for an efficient use of optical and photosensitive methods in diagnosis and control. Third, the anatomical structure of the eye includes miniature elements in both the anterior (in front of the lens) and posterior (beyond the lens) chambers. Correspondingly, the ocular implants are limited by small sizes, including their nanoscale geometry. Fourth, the loss of sight is not necessarily associated with the loss of life, but considerably complicates it; therefore, the efforts for expanding the market of ocular preparations are very active, and the ophthalmological market by 2011 had reached several billions of dollars (Table 1).

TABLE 1   Dynamics and forecast for the world market of implanted medical devices and prostheses (billions of $$) [18].

Years	2004	2006	2008	2010	2015 (forecast)
United States and Canada	83.14	93.97	105.88	118.23	164.63
Western Europe	52.50	59.04	65.65	73.11	103.00
Japan	24.14	26.64	29.31	32.65	44.76
Asian countries (without Japan)	15.72	18.36	23.13	30.20	54.97
South America	3.48	3.94	4.83	6.11	11.18
Eastern Europe	3.98	4.50	5.96	7.73	14.90
Africa	2.49	2.82	3.41	4.13	7.79
Total production	185.45	209.27	238.17	272.16	401.23

The current state and nearest future of implants in ophthalmology are to a certain degree reflected by Tables 2 and 3, which lists the IMDs for local drug delivery both commercially produced and at various stages of clinical trials. All the tools are partitioned according to the target of their prolonged delivery, namely, the anterior eye chamber (cornea, conjunctiva, sclera, etc.) and the posterior chamber (retina, optic nerve, vitreous body, macula, etc.). According to the current trends, the analogous drug forms already tested in ophthalmology but designed for gene therapy will be further developed. In particular, it has been shown at the initial stage of clinical trials in the United States that some RNAs (such as low molecular weight interfering RNAs) implanted into biodegradable polymer complexes display a high efficiency in treatment of the congenital retinal degeneration and macular degeneration [95]. In addition, it is purposeful to focus on the design and production of the drainage systems for treatment of glaucoma [95], one of the most widespread diseases. In an ideal situation, these systems (ANDIs) should contain the polymer lines for drainage of lacrimal fluid and exudate and provide for a targeted delivery of the drugs decreasing intraocular pressure and preventing destruction of the optic nerve. Advance in designing the new generation implants for the anterior eye chamber will allow the miniature sensors for biochemical eye surface monitoring [96] and concurrent drug elution [97] to be installed directly into contact lenses or into posterior compartment (see Fig. 6).

TABLE 2   Promising implanted devices for local drug delivery to the anterior eye chamber [65].

Active agent	Trademark	Type	Matrix	For treating of	Stage of development
Betaxolol	Betoptic S®	Eye drops	Amberlite IRP-69	Glaucoma	In market
Azithromycin	AzaSite®	Eye drops	Polycarbophil	Bacterial conjunctivitis	In market
Ketotifen		Soft contact lenses	–	Allergic conjunctivitis	Phase 3
Latanoprost		Tampon cylinder	PVA and cellulose	Glaucoma	Phase 2
Latanoprost		Implant of eye connective tissue (subconjuctival)	PVA–PLGA composite	Glaucoma	Phase 2
				Keratoconjunctivitis	Phase 3
Cyclosporine (LKh-201)		Implant in the eye epis- clera	Silicon	Uveitis	Phase 2
Dexamethasone phosphate EGP-437	EyeGate II®	Device for microiono- phoresis	–	Difficulties in lacrima- tion (dry eye symptom)	Phase 3

TABLE 3   Promising implanted devices for local drug delivery to the posterior eye chamber [65].

Active agent	Trademark	Type	Matrix	For treating of	Stage of development
Ganciclovir	Vitrasert®	Intravitreal implant	Poly(ethylene-co-vinyl alcohol)	Viral retinitis (inflamma- tion of retina)	In market
Fluocinolone acetonide	Retisert®	Intravitreal implant	PVA–silicone	Uveitis	In market
Fluocinolone acetonide	Iluvien®	Intravitreal implant	Polyimide–PVA	Diabetic macular edema	Phase 3
Dexamethasone	Ozurdex®	Intravitreal implant	Biodegradable PLGA	Renal venous congestion	In production
Brimonidine	–	Intravitreal implant	Biodegradable PLGA	Age-related macular de- generation	Phase 2
Triamcinolone acetonide	I-vation™ TA	Intravitreal implant	PMMA/ethylvinyl ace- tate	Diabetic macular edema	Phase 2
Triamcinolone acetonide	IBI-20089	Intravitreal implant	Oil medium	Renal venous congestion	Phase 1
Triamcinolone acetonide	RETAAC	Intravitreal implant	Biodegradable PLGA	Diabetic macular edema	Phase 1

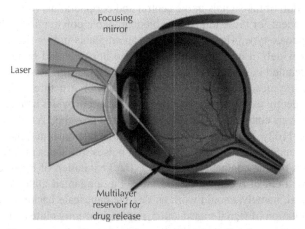

FIGURE 6 Photodynamically activated system for controlled drug delivery within the intraocular region of eye.

The current nanomedicine provides the possibilities for replacement of several eye elements with artificial implants. The replacement therapy in ophthalmology includes the repair of the retinal nerve (formation of polymer conduit) and optical nerve [95]. A separate field in cell engineering uses 2D and 3D biodegradable scaffolds that enhance cell proliferation and function for the growth of eye tissues and, in the nearest future, also elements of the eye, for example, the cornea [98]. In this case, the surface topography of the polymer should meet special requirements, since the nanostructured surface of the scaffold provides controllable cell adhesion and migration. Thus, we again meet the problem of controllable drug elusion from polymer matrix. For example, the delivery of the peptides capable of spatial self-organization and self-assembly will be used in the nearest future for the regeneration of the retinal axon [99].

12.6 PROSPECTS OF NEUROLOGICAL IMPLANTS

Neurology is an interesting and most promizing field for implant application. Typical implants used for this purpose are polymer conduits intended for the restoration of the integrity and conductivity of nervous conductors. It is known that the nerve cells have only limited ability to repair and the nerve tissue has a limited ability to regenerate after injuries, for example, craniocerebral traumas. Moreover, an increase in the rate of diseases, such as Alzheimer's or Parkinson's disease, as well as of cerebrovascular impairments and subsequent dystrophy of nerve fibers, requires an ever-increasing number of micro and nanoimplants providing for transduction of the signals in the central and peripheral nervous systems. The major component for these implants is biodegradable polymers of polylactide or polyoxyalkanoate type. The prospects for nanoimplants in this field, third in the list of most widely used devices (after cardiac implants and bone prostheses), have been detailed in several reviews [100, 101]. Here, organic silicon (polymer) systems are the most promizing. Useful properties of structured silicon dioxide have been recently discovered; its nanoporous surface provides a more stable contact with neurons compared to smooth surfaces and prevents tissue overproliferation (gliosis) [102]. The pioneer experiments have demonstrated that the activation of semiconductor nanoparticles with light pulses stimulates rat brain neurons, which allows for the contactless (electrode-free) excitation of a neuronal signal [103]. In this field of regeneration medicine, the implanted materials should be polyfunctional, namely, should work as flexible nanoscale backbone constructs, enhance excitation of signal transduction, and concurrently elute a drug that activate development of a nerve impulse.

12.7 MINIATURE ELECTROMECHANICAL IMPLANTS FOR PROGRAMMED DELIVERY OF BIOLOGICALLY ACTIVE SUBSTANCES

Innovative developments in the chemistry of polymers and supramolecular chemistry, composite engineering, and design of micro and nanoscale electromechanical systems (MEMS and NEMS) are directed towards obtaining of fundamentally new therapeutic systems for controllable release of low-molecular-weight drugs, macromolecules, and cells [104, 105]. Polymer and inorganic implants equipped with electronic chips create kinetic profiles with different complexities to release the biologically active component. Several implanted systems additionally contain mechanical manipulators, drives, and nanocantilevers. The combination of computer, electronic, and mechanical elements in an implant determines that it is a principally new electromechanical device (MEMS/NEMS) [106–108]. The miniaturization of these devices, which use state-of-the-art technologies (nanolithography and imprinting), allows these implants to be considered an independent universal class of the systems for targeted drug delivery [109–111]. The introduction of MEMS/NEMS as drug-delivery-programed systems into clinical practice is determined to a considerable degree by advances in a new field of biomedicine, personalized medicine, which is a custom iced therapy of each patient that takes into account his/her individual genetic features and specificity of biochemical processes [112, 113]. The information about symptomatology of disease and data on the metabolites assayed with biomarkers are input into a chip to provide a programed release of a set of drugs from a system of containers according to these integrate data. Here, the most impressive results are attained when using an integrated treatment of complex pathologies, such as tumor diseases, liver cirrhosis, and blood disorders.

Using a feedback principle, an MEMS is used to design a vitally important organ, the artificial pancreas, which monitors the sugar level in the blood and releases the necessary dose of insulin according to the sugar concentration [114–116]. Currently, there are three hindrances to the commercialization of this system for diabetes control, namely, (a) a reliable programed device; (b) glucose sensors stable in their characteristics and able to stably function for a long time in the specific aggressive medium, such as blood; and (c) still high price of the final product [117]. Nonetheless, taking into account the high rate of diabetic diseases, in the next 10–12 years, these implanted systems should become a mass therapeutic tool.

Once the first two problems are associated with advances in nanotechnology, the third issue has a rather moral and ethical aspect. No matter how large the expenditures for treating a patient using precision and expensive electromechanical platforms, an improvement in his/her quality of life cannot be limited by material and financial parameters. The last statement may be illustrated by one rare disease with a very limited number of cases. It is known that metabolic disorders require constant therapy to sustain a patient's comfort. In this situation, implanted NEMS devices for the controllable release of deficient enzymes find their own niche in the world pharmaceutical market. For example, Genzyme Therapeutics (United States) [117] specializes in the production of a Cerezyme implant, which elutes recombinant proteins for the long-term therapy of the Gaucher disease. Despite that this disease is very rare (1 per 200,000 in the population), in 2008, this company acquired $1.2 billion from selling this device, which accounts for 27% of its annual profit. As in the case with the NEMS implants, large expenditures for implantation and therapy ($200000 USD annually per patient) [112] are completely justified, since any alternative method that alleviates suffering and prevents a preliminary fatal outcome is absent.

The specific construction features of the MEMS-and NEMS-type implants include not only their miniature size and stable operation, but also the use of bioinert and biocompatible materials. In their operation principle, these devices are miniature reservoirs that contain a drug, which is eluted in response to electrical signal or movement of piston/micropump walls. The common characteristics for the main types of electromechanical implants are listed in Table 4.

At the micro and nanoscales, several electromechanical devices for controllable drug release act as pumps with either a drive or by the effect of osmotic forces [118]. In the latter case, there is no need for any electromechanical drive, and strictly speaking, such devices cannot be regarded as MEMS/NEMS. Only a very limited number of implanted systems that use a micropump principle are manufactured in batches. The examples include a MiniMed Medtronic Pump for controlled insulin release and SynchroMed II Programmable Pump (commercially available since 1988) for critical therapy (a high level of pain and neurological indications). A series of the Debiotech pumps, which use the piezoelectric effect in their design, allows the drug solution to be eluted at a rate of 100 μL/h [119]. Another operating principle of a micropump is to use materials with a shape memory effect [119]; they allow the therapeutic rate of drug delivery to be reduced to 5 μL/h.

TABLE 4 Comparative characteristics of nanoimplants utilizing different drug release mechanisms

Principle of drug release	Main characteristics	Therapeutic potential
Implanted electronic chip	Limitations on volume of device Only for highly active compounds High probability of tissue capsule formation Stricter requirements to bioinert and nontoxic materials in long-term operation Invasive implantation procedure High price	Local and more economical drug delivery Precise and time-controllable drug delivery More preferable than parenteral drug administration Possibility of pulse and more complex kinetic profile of drug release Has advantages as compared with aerosols and oral administration Combined therapy and diagnosing (monitoring)
Implanted micropump	Limitations on volume of device Only for highly active compounds Exclusively for release in liquid form (difficulties in releasing polypeptides due to their low stability in aqueous medium) Possible structural changes in active component because of shear stress Precipitation and subsequent plugging of pump piston with the drug precipitated from solution Invasive implantation procedure Considerable expenditures for implantation and production	Local and more economical drug delivery More preferable than parenteral drug administration Possibility of pulse and more complex kinetic profile of drug release; Has advantages as compared with aerosol, oral, and parenteral drug administration routes
Implanted polymer chip	Low invasive Limitations on volume of device Only for highly active compounds Stricter requirements to drug stability (absence of hydrolysis) Broad therapeutic window Difficulties in precise dosing	Local delivery provides advantages compared to traditional drug injection and aerosol administrations Perfect alternative to parenteral drug administration Tolerance to delivery via systemic or local administration routes Biodegradable elements do not require repeated invasion for extracting worked out system
Implanted selective membranes	Limitations on volume of device Only for highly active compounds Requires surgical intervention Risk for stoppage of membrane pores with loss of production capacity and selectivity Appropriate only for drug solutions Favorable conditions for cell material	Due to local delivery, has advantages as compared with traditional drug injection and aerosol administration Perfect alternative to parenteral drug administration Depending on implantation site, allows for controlled delivery via bloodstream or to local body area Does not require repeated invasion for extracting worked out system

12.8 CONCLUSIONS

The transition from a macroscopic description of implanted systems to molecular and submolecular consideration is a necessary condition for the development of new generation bioimplants. Nowadays, there are good grounds to believe that the structuring of the implant surface in a nanoscale range modulates cell membrane receptors that modify the communication between cells and has a positive effect on adsorption and conformational changes in the proteins responsible for immune and toxic responses. The successive consideration of the cardiac implants based on the example of new generation stents and bone implants, as well as ophthalmological, urological, and others, demonstrates that the combination of the following three factors determines the nearest prospects in implantology: (1) topography and chemical composition of the surface corresponding to the natural organ; (2) the ability to perform mechanical, optical, ion conducting, and other functions characteristic of the replaced element; and (3) the ability to encapsulate drugs with the possibility of their controllable (programed) and prolonged targeted delivery to a local area of the body.

KEYWORDS

- biodegradable polymers
- Gynitourinary implants
- hybrid nanocomposite
- implanted medical device
- modern orthopedic implants
- neurological implants

REFERENCES

1. Logothetidis, S.; *NanoScience and Technology.* **2012,** *61,* 1–26.
2. Huebsch, N.; Mooney, D. J.; *Nature,* **2009,** *462(7272),* 426–432.
3. Hasirci, V.; Vrana, E.; Zorlutuna, P.; Ndreu, A.; Yilgor, P.; Basmanav, F. B.; Aydin, E.; *Journal of Biomaterials Science, Polymer* **2006,** *17 (11),* 1241–1268.
4. Bazhenov, S. L.; Berlin, A. A.; Kul'kov, A. A.; and Oshmyan, V. G.; *Polimernye kompozitsionnye materialy.Prochnost'i tekhnologiya* (Polymer Composite Materials. Durability and Technology), Rus. Edition. Dolgoprudnyi: Intellekt, 2010.
5. Jorfi, M.; Roberts, M. N.; Foster, E. J.; Weder, C. *ACS Applied Materials and Interfaces.* **2013,** *5 (4),* 1517–1526.
6. Frima, H. J.; Gabellieri, C.; Nilsson, M. -I.; *Journal of Controlled Release.* **2012,** *161 (2),* 409–415. Drug delivery research in the European Union's Seventh Framework Program for Research

7. Drug Delivery Systems US Industry Study with Forecasts for **2012,** and **2017,** Study #2294 | March **2008,** 338 The Freedonia Group. Cleveland, OH. USA. www. freedoniagroup.com

8. Staples, M.; *Wiley Interdisciplinary Reviews: Nanomedicine and Nanobiotechnology* **2010,** *2 (4),* 400–417.

9. Leucuta, S. E.; *Current Clinical Pharmacology* **2012,** *7 (4),* 282–317.

10. *Stevenson, C. L.; Santini Jr.; J. T.; Langer R.;* Advanced Drug Delivery Reviews.2012. In press doi: 10.1016/j.addr.2012.02.005. Reservoir-Based Drug Delivery Systems Utilizing Microtechnology.

11. Sirivisoot, S.; Pareta, R. A.; Webster, T. J.; *Recent Patents on Biomedical Engineering* **2012,** *5 (1),* 63–73.

12. Global Industry Analysts Inc.; Nanomedicine: a global strategic business report, 2009.

13. Murday, J. S., et al. Nanomedicine: Nanotechnology, Biology, and Medicine, **2009,** *5,* 251–273.

14. Ekdahl, K. N.; Lambris, J. D.; H.; Elwing, D.; Ricklin, Nilsson, P. H.; Y.; Teramura, I. A.; Nicholls Bo Nilsson. *Advanced Drug Delivery Reviews* **2011,** *63,* 1042–1050.

15. Bonartsev, A. P.; Boskhomedgiev, A. P.; Iordanski, A. L. et al. In: Kinetics, Catalysis and Mechanism of Chemical Reactions. From pure to applied science. V. 2 -Tomorrow and perspectives. (eds. Islamova, R. M.; Kolesov, S. V.; Zaikov, G. E.) Nova Science Publishers: New York **2012,** Ch.27. 335–350. Degradation of poly(3-hydroxybytyrare) and its derivatives: Characterization and kinetic behavior.

16. Iordanskii, A. L.; Rudakova, T. E.; Zaikov, G. E.; *Interaction of Polymers with Bioactive and Corrosive Media. Ser. New Concepts in Polymer Science.* VSP Science Press. Utrecht –Tokyo Japan. **1994,** 298p.

17. Directive 2007–47-EC of the European Parliament and of the council.pdf

18. http://www.fda.gov/MedicalDevices/DeviceRegulationandGuidance/Overview/ ClassifyYourDevice/ucm051512.htm

19. http://laws-lois.justice.gc.ca/eng/regulations/CRC,c.870/index.html

20. Market Report: World Medical Devices Market September **2007,** Acmite Market Intellegence. http://www.acmite.com/market-reports/medicals/world-medical-devicesmarket.html.

21. Report of Ministry of Healthcare and social development RF Review [online]. http:// www.mzsrrf.ru/press_smi/388.html.

22. (a) The World Health Organization Report, **2006,** [online]. http://www.who.int/entity/ whr/2006/whr06_en.pdf. (b) The World Health Organization Report, **2011,** Cardiovascular disease. http://www.who.int/cardiovascular diseases/en/

23. Document: Conception of demography development in Russia until **2015,** http:// www.demoscope.ru/weekly/knigi/koncepciya/koncepciya.html

24. Kastrati, A. et al. *N. Engl. J. Med.* **2007,** *356,* 1030.

25. Stone, G. W. et al. *N. Engl. J. Med.* **2007,** *356,* 998.

26. Lagerqvist, B. et al. *N. Engl. J. Med.* **2007,** *356,* 1009.

27. Wieneke, H. et al. *Catheter Cardiovasc. Interv.* **2003,** *60,* 399.

28. Bhargava, B. et al. *Catheter Cardiovasc. Interv.* **2006,** *67,* 698.

29. Erlebacher, J. et al. *Nature* **2001,** *410,* 450.

30. Caves, J. M.; Chaikof, E. Vasc, L. J.; Surg. **2006,** *44,* 1363.

31. Samaroo, H. D. et al. Int. J.; Nanomedicineто **2008**, *3*, 75.
32. Tana, Y.; Farhatnia, Achala de Mel, J.; Rajadas, Alavijeh, M. S.; Seifalian, A. M.; Journal of Biotechnology **2013**, *164*, 151–170. Inception to actualization: Next generation coronary stent coatings incorporating nanotechnology.
33. Godin, B.; Sakamoto, J. H.; Serda ,R. E.; Grattoni, A.; Bouamrani ,A.; Ferrari, M. Trends in Pharmacological Sciences. **2010**, *31 (5)*, 199.
34. Rosengart, A. J.; Kaminski, M. D.; Chen H.; Caviness, P. L.; Ebner, A. D.; Ritter J. Magn, A. *J. Magn. Mater.* **2005**, *293*, 633.
35. Cregg, P. J.; Murphy K.; Mardinoglu A.; Prina-Mello A.; Journal of Magnetism and Magnetic Materials. **2010**, 322, 2087.
36. Zilberman M.; Nelson, K. D.; Eberhart R. C.; Mechanical Properties and In Vitro Degradation of Bioresorbable Fibers and Expandable Fiber-Based Stents. Published online 30 June **2005**, in Wiley InterScience (www.interscience.wiley.com). DOI: 10.1002/jbm.b.30319
37. Yun-Xuan Weng, Xiu-Li Wang, Yu-Zhong Wang. Polymer Testing. **2011**, *30*, 372.
38. Kumbar, S. G.; Kofron, M. D.; Nair, L. S.; Laurencin CT.; Cell behavior toward nanostructured surfaces. In: Gon-salves, K. E.; Laurencin, C. T.; Halberstadt, C.; Nair L. S., eds. Biomedical Nanostructures. New York: John Wiley & Sons; **2008**, 261–295.
39. Balasundarama G.; Webster T. Mater, J. J.; Chem. **2006**, 16, 3737.
40. Schakenraad J. M.; Biomaterial Science / Eds. Ratner, B. D.; Hoffman, A. S.; Schoen, F. J.; Lemons J. E. (San Diego: Academic Press, Inc.; 1996), 140.
41. Kennedy, S. B.; Washburn N. R, Simon, C. G.; Amis Biomaterials, *E. J.;* **2006**, 20, 3817.
42. Hersel U.; Dahmen C.; Kessler H.; Biomaterials. **2003**, *24*, 4385.
43. Feng Y.; Mrksich M.; Biochemistry, **2004**, *43*, 15811.
44. El-Ghannam A.; Ducheyne P.; Shapiro I. *M. J.;* Orthop. Res. **1999**, *17*, 340.
45. Sagnella S.; Anderson E.; Sanabria N.; Marchant, R. E.; Kottke-Marchant K.; Tissue Eng. **2005**, *11*, 226.
46. Anselme K.; Bigerell M. *J. Mater. Sci. Mater. Med.* **2006**, *17*, 471.
47. Curtis A. *IEEE Trans. Nanobiosci.* **2004**, *3*,P. 293.
48. Price, R. L.; Ellison K.; Haberstroh, K. M.; Webster T. *J. Biomed, J. Mater. Res.* **2004**, V. *70*, 129.
49. Webster, T. J.; Schadler, L. S.; Siegel, R. W.; Bizios R.; *Tissue Eng.* **2001**, *7*, 291.
50. Curtis, A.; Britland S.; Surface modification of biomaterials by topographic and chemical patterning. In: Advanced Biomaterials in Biomedical Engineering and Drug Delivery Systems. Ogata, N.; Kim, S. W.; Feijen, J.; Okano, T.; eds. Tokyo: Springer-Verlag, **1996**, 158–167.
51. Woo, K. M.; Chen V. J.; Ma P. X. *J. Biomed. Mater. Res..* **2003**, *67A (2)*, 531.
52. Keselowsky, B. G.; Collard, D. M.; Garcia A. J. J Biomed. Mater. Res. **2003**, *66A (2)*, 247.
53. Webster, T. J.; Ergun C.; Doremus, R. H.; Siegel, R. W.; Bizios R. *J. Biomed. Mater. Res..* **2000**, 51. 3, 475.
54. Faghihi, S.; Azari, F.; Zhilyaev, A. P.; et al. Biomaterials **2007**, V. 28. (27 P. 3887.
55. Khan, Y.; El-Amin S. F, Laurencin C. T. Conference Proceedings of the IEEE Engineering in Medicine and Biology Society.; New York, **2006**, *1*, 529.
56. Harrison B. S, Atala A.; Biomaterials **2007**, *28(2)*, 344.

57. Li, W. J.; Laurencin, C. T.; Caterson, E. J.; Tuan, R. S.; Ko F. Biomed, K. J.; Mater. Res. **2002,** 60, 613.
58. Nukavarapu, S. P.; Kumbar, S. G.; Nair, L. S.; Laurencin C. T.; Biomedical Nanostructures / Eds. Gonsalves, K. E.; Laurencin, C. T.; Halberstadt C.; Nair L. S. (NewYork: Wiley, 2008) P. 377.
59. Kumbar, S. G.; Nukavarapu, S. P.; Roshan R.; Nair, L. S.; Laurencin C. T. . Biomed Mater **2008,** 3, 1.
60. Pelled G.; Tai K.; Sheyn D.; Zilberman Y.; Kumbar, S. G.; et al. J.; Biomech **2007,** 40, 399.
61. Kumbar, S. G.; Nair, L. S.; Bhattacharyya S.; Laurencin C. T. J Nanosci. Nanotechnol. **2006,** 6, 2591–2607.
62. Nablo, B. J.; Rothrock, A. R.; Schoenfisch M. H.; Biomaterials **2005,** 26, 917.
63. Simchi A.; Tamji E.; Pishbin F.; Boccaccini Nanomedicine, Nanotechnology, A. R.; Biology, and Medicine. **2011,** 7, 22.
64. Popat K. C, Eltgroth, M.; LaTempa T. J, Grimes C. A, Desai T. A.; Small **2007,** 3, 1878.
65. Pioletti, D. P.; Gauthier O.; Stadelmann V. A. et al. Orthopedic implant used as drug delivery system: Clinical situation and state of the research. Current Drug Delivery, **2008,** 5, 59–63.
66. Gulati K.; Ramakrishnan S.; Aw, M. S.; Atkins, G. J.; Findlay, D. M.; Losic D.; Acta Biomaterialia **2012,** 8, 449.
67. Sirivisoot S.; Pareta R.; and Webster, T. J.; Electrically controlled drug release from nanostructured polypyrrole coated on titanium. **2011,** *Nanotechnology* 22–085101 doi: 10.1088/0957–4484/22/8/085101
68. http: //www.fda.gov/oc/combination/
69. Laurencin, C. T.; Kumbar, S. G.; Nukavarapu Interdiscipl, S. P.; Rev. Nanomed. Nanobiotechnol. **2009,** 1, 6.
70. Ivantsova, E. L.; Kosenko R. Yu.; Iordanskii, A. L.; et al. Polymer Science, Ser. A, **2012,** 54,No. 2,pp. 87–93
71. Shavell V. I.; Abdallah, M. E.; Shade Jr, G. H.; Diamond, M. P.; Berman J. M. J Min. Invasive Gynecol. **2000,** *16(1),* 22.
72. Shulman L. P, Gabriel H.; Contraception **2006,** 73, 325.
73. http: //www.merck.com/newsroom/news-release-archive/prescription-medicine-news/2011_1109.html
74. Correia L.; Ramos A. B, Machado, A. I.; Rosa D.; Marques C.; Contraception **2011,** in press doi: 10.1016/j.contraception.2011.10.011
75. Staples M.; Microchips and controlled release drug reservoirs. WIREs Nanomed Nanobiotechnol 2010–2 400–417.
76. E.; Saiz, Zimmermann, E. A.; Lee, J. S.; U. Wegst, G. K.; Tomsi, A. P.; Dental Materials **2013,** 29, 103–115.
77. Fennema E.; Rivron N.; Rouwkema J.; van Blitterswijk C.; de Boer J.Trends in Biotechnology, **2013,** *31(2),* 271–282. http: //dx.doi.org/10.1016/j.tibtech.2012.12.003.
78. Chopa N.; Kayes O. Trends in Urology and men's health. **2013,** No 1–2 14–16. Recent advances in the field of tissue bioengineering have taken scientists closer to growing organs in the laboratory. Cit. David Mack/Science Photo Library.
79. Pattison M. A, Wurster, S.; Webster, T. J.; Haberstroh K. M.; Biomaterials **2005,** 26, 2491.

80. McManus, M.; Boland, E.; Bowlin, G.; Simpson, D.; Espy, P.; Koo H.; American Urological Association Annual Meeting, May 8–13, **2004,** San Francisco, CA, USA.
81. Humes H. D.; *Semin. Nephrol.* **2000,** *20,* 71.
82. Lee, D.; Lee, J. T.; Sheperd, D.; Abrahams H.; Lee, D.; Preliminary use of the Enseal™ system for sealing of the dorsal venous complex during robotic assisted laparoscopic prostatectomy. American Urological Association Annual Meeting, May 21–26, **2005,** San Antonio, TX, USA.
83. Eberli D.; Yoo, J. J.; Atala A.; Urologe. **2007,** A*46,* 32.
84. Sharma, A. K.; Hota, P. V.; Matoka, D. J., et al. Biomaterials. **2010,** *31 (24),* 6207–62178.
85. Feil G.; Daum L.; Amend B.; et.al. Advanced Drug Delivery Reviews. **2011,** 63 375–378. From tissue engineering to regenerative medicine in urology — The potential and the pitfalls.
86. Murphy, D. G.; Costello Nanotechnology, A. J.; In: New Technology in Urology. (P.; Dasgupta et al., eds) Springer Verlag. London 201 268.
87. Cruz C. G. M. PHD Thesis's. pH-Triggered Dynamic Molecular Tweezers for Drug Delivery Applications. Queen's University. Kingston, Ontario, Canada September **2011,** 180.
88. Wright J. C.; Hoffman, A. S.; Historical Overview of Long Acting Injections and Implants in Long Acting Injections and Implants (J. C.; Wright and Burgess D. J.; (eds.)), Advances in Delivery Science and Technology, Chapter 2. DOI 10.1007/978–1-4614–0554–2_2, © Controlled Release Society, Springer. 2012.
89. Theeuwes, F.; Yum S. I. Ann Biomed. Eng. **1976,** 4343–353. Principles of the design and operation of generic osmotic pumps for the delivery of semisolid or liquid drug formulations
90. Wright, J. C.; Leonard, S. T.; Stevenson, C. L.; Beck, J. C.; Chen, G.; Jao, R. M.; Johnson, P. A.; Leonard, J.; Skowronski, R. An in vivo/in vitro comparison with a leuprolide osmotic implant for the treatment of prostate cancer. J Control Release. **2001,** 75, 1–10.
91. Duncan R.; Nature Review. **2006,** *6,* 688–701. Polymer conjugates as anticancer nanomedicines.
92. Maurice, D. M.; Mishima, S.; Ocular pharmacokinetics, in Handbook of Experimental Pharmacology, Sears, M. L.; Ed.; Berlin Heidelberg: Springer, **1984,** 16.
93. Kuno, N.; Fujii, S.; Polymers, **2011,** *3,* 193.
94. Thomson, H.; Lotery, A.; Nanomedicine, **2009,** *4,* 599.
95. Pan, T.; Brown, J. D.; and Ziaie, B.; IEEE Eng. Med. Biol. Soc.; **2006,** *1,* 3174.
96. Parviz, B. A.; Shen, T. T.; and Ho, H.; Invest. Ophtalmol. Vis. Sci.; **2008,** *49,* Abstr. 4783, Functional contact lens with integrated inorganic microstructures. www.arvo.org
97. Kapoor, Y.; Thomas, J. C.; Tan, G.; John, V. T.; Chauhan, A.; Biomaterials, **2009,** *30,* p. 867.
98. Liu, B. S.; Yao, C. H.; Hsu, S. H.; et al.; *J. Biomater. Appl.* **2004,** *19,* 21.
99. Elder J. B, Liu, C. Y.; Apuzzo MLJ.; Neurosurgery in the realm of 10–9, Part 2, Applications of nanotechnology to neurosurgery—present and future. Neurosurgery. **2008,** 62 269–84.
100. Ellis-Behnke R. Med Clin of North America. **2007,** 91, 937–962.

101. Lebedev, M. A.; Nicolelis MAL. Brain-machine interfaces past, present, future. Trends Neurosci **2006**, 29, 536–46.

102. Moxon, K. A.; Hallman, S.; Aslani, A.; Kalkhoran, N. M.; Lelkes P. I.; Bioactive properties of nano-structured porous silicon for enhancing electrode to neuron interfaces. *J Biomater Sci Polym.* **2007**,18, 1263–1281.

103. Zhao, Y.; Larimer, P.; Pressler, R. T.; Strowbridge, B. W.; Burda C.; Wireless activation of neurons in brain slices using nanostructured semiconductor photoelectrodes. Angew Chem Int Edn. **2009**, 48, 2407–2410.

104. Madou M. J.; Fundamentals of Microfabrication: The Science of Miniaturization, 2nd ed.; CRC Press: Boca Raton, **2002**, 752.

105. Borzenko A. G.; Scientifical instrument engineering. **2005**, *15(3)*, 8–24. Analytical devices: Mili- micro and nanohorizons. (In Rus: Борзенко А. Г. Аналитические приборы: мили—, микро, и нано— горизонты // Научное приборостроение. 15 **2005**, *3*, 8–24.)

106. Kim, K.; Lee J-B. MEMS for drug delivery, Chapter 12. In: WangW, Soper, S. A.; eds. Bio-MEMS: Technologies and Applications. Boca Raton, FL: CRC Press; **2006**, 325–348.

107. Santini J. T. Jr., Richards A. C, Scheidt R. A, Cima M. J, Langer R. S.; Microchip technology in drug delivery. Ann Med, **2000**, *32*, 377–379.

108. Santini Jr, J. T.; Cima, M. J. Langer, R. A controlled-release microchip. *Nature*, **1999**, *397,* 335–338.

109. 99. Gardner, D. P.; Microfabricated nanochannel implantable drug delivery devices: trends, limitations and possibilities. Exp Opin Drug Deliv, **2006**, *3*, 479–487.

110. Shrivistava, S. D.; Dash Applying nanotechnology to human health. J Nanotech **2009**, 1–14.

111. US Food and Drug Administration. Nanotechnology page. Available at: http://www. fda. gov/nanotechnology/. (Accessed February 1, 2010).

112. The Personalized Medicine Coalition. Home page. Available at: http://www. personalized medicinecoalition.org/sciencepolicy/personalmed-101 overview.php. (Accessed January 2, 2010).

113. Amgen J. Future of biotechnology in healthcare, Chapter 9. In: An Introduction to Biotechnology. **2009**, 31–35. Available at: http://www.amgenscholars.eu/web/guest/ futureof-biotechnology-in-healthcare (accessed January 2, 2010).

114. Heller. Integrated medical feedback systems for drug delivery. Am Inst Chem Eng, J.; **2005**, 51, 1054–1066.

115. Brunetti, P.; Federici, M. O.; Benedetti, M. M.; The artificial pancreas. Artif Cells Blood Substit Immobil Biotechnol. **2003**, *31*, 127–138.

116. M.; Staples. Microchips and controlled release drug reservoirs. WIREs Nanomed Nanobiotechnol**2010**, 2, 400–417.

117. Genzyme website. The cost of enzyme replacement therapy. Available at: http: // www.genzyme.com/commitment/patients/costof treatment.asp. (Accessed January 2, 2010).

118. Shoji, S.; Esashi M.; Microflow devices and systems. J.; Micromech Microeng. **1994**, *4,* 157–171.

119. Amirouche, F.; Zhou, Y.; Johnson T.; Current micropump technologies and their biomedical applications. Microsys Tech, **2009**, *15,* 647–667.

CHAPTER 13

INFLUENCE OF VARIOUS METAL/ CARBON NANOCOMPOSITES ON CHANGES IN THE PROPERTIES OF COMPOSITIONS ON LIQUID GLASS BASIS

L. F. AKHMETSHINA, V. I. KODOLOV, and G. E. ZAIKOV

CONTENTS

13.1   INTRODUCTION

This chapter presents the investigation results of the influence of metal/carbon nanocomposites on heat-physical and viscous properties of silicate compositions. The introduction of nanocomposite facilitates the decrease in material temperature conductivity by nearly 50%. The article discusses the investigation results of liquid glass suspensions with IR spectroscopy. The changes in liquid glass properties depend on nanocomposite. It is found that relative viscosity of iron/carbon nanocomposites decreases, if compared with nickel/carbon.

Nanotechnology is based on self-organization of nanosize particles resulting in the improvement of material properties and obtaining of products and items with unique characteristics. The modification of silicates and liquid glass with nanocomposites containing metal clusters is perspective. The interest to liquid glass is conditioned, first of all, by its ecological friendliness and production and application simplicity, inflammability and nontoxicity, biological stability, and raw material availability. Nanostructures can be used to improve such characteristics of paints as elasticity, adhesion to the base, hydrophobic behavior and also helps solving the problems connected with coating flaking-off from the base, discoloration, limited color range, and so forth. Due to the unique properties of nanostructures we can apply new properties to silicate paints, for example, to produce the coating protecting from electromagnetic action.

The application of nanostructures in silicate materials to improve their stability to external action is known [1].

The introduction of nanosize structures into the liquid glass allows qualitatively changing the material behavior in a positive way. This possibly occurs in the process of binder supermolecular structure change.

Based on the aforesaid, it is important to improve operational properties of compositions on liquid glass basis modifying them with metal/carbon nanocomposites, providing a wider field of their application.

The aim of this chapter is the investigation and analysis of the influence of various metal/carbon nanocomposites on changes in the properties of compositions on liquid glass basis.

13.2   EXPERIMENTAL PART

13.2.1   NANOCOMPOSITES STRUCTURE

To modify the materials nickel and iron containing nanocomposites were obtained with low-temperature synthesis in nanoreactors of polymeric matrixes of polyvinyl chloride [2] and polyvinyl alcohol [3]. Iron oxide (3) was selected as metal containing phase, which in the process of nanostructures obtaining is reduced to magnetite, and nickel oxide, which is also reduced.

To define the sizes, shape and structure the nanoproduct obtained was investigated with transmission electron microscopy (TEM) and electron diffraction (ED). Iron/carbon nanocomposite (Fe/C NC) is mainly represented as film nanostructures with metal inclusions (Fig. 1).

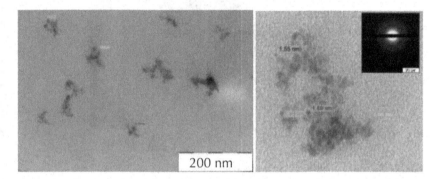

FIGURE 1   TEM picture of Fe/C nanocomposite.

The particles of metal containing phase have a globular shape and are located between the layers of carbon films connected with them. There are also nanostructures close to a spherical shape. In TEM pictures of Fe/C NC dark spots characterize metal, light ones—carbon films. Electron-diffraction patterns are presented as ring reflexes. This indicates the availability of amorphous phase in the sample. The average size of particles constituting the aggregates is 17 nm.

Fig. 2 demonstrates large particles and aggregations of nickel clusters. A significant number of particles have the shape close to spherical, however, there are elongated particles as well. Globular film nanostructures with the inclusion of nickel nanoparticles and its compounds are present in the samples. The electron-diffraction pattern is presented as ring reflexes, however dot reflexes are also present. This indicates the availability of both crystalline and amorphous phases in the sample. The average size of particles is 11 nm.

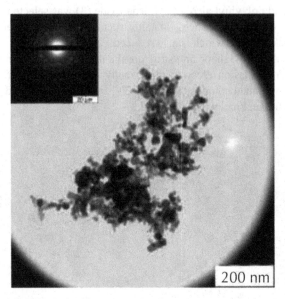

FIGURE 2   TEM picture of Ni/C nanocomposite.

13.2.2   EQUIPMENT AND INVESTIGATION TECHNIQUE

13.2.2.1   SPECTROPHOTOMETRY

To define the interaction between liquid glass and nanocomposite the optical density of liquid glass sample films with nanostructures was investigated. The films were produced applying a thin layer of silicate modified with nanostructures onto transparent film with further drying to remove the moisture. Films without nanocomposites were used as the reference. The investigations were carried out with spectrophotometer KFK-3-01 in the wave range 950–350 nm.

13.2.2.2   ULTRASOUND PROCESSING OF SUSPENSIONS

Spectrophotometer KFK-3-01 was also used to define the optimal action time interval on nanocomposite fine suspension processed with ultrasound. In the process of selecting the work wavelength (λ) on spectrophotometer KFK-3-01 to define the optimal time interval of ultrasound processing of fine suspension on liquid glass basis, the optimal wavelength of 430 nm was found, as with this wavelength the optical density of liquid glass with NC before the US processing was 0.818, thus corresponding to the midpoint of the work range of this instrument (work range: optical density (D) = 0.001–1.5). At λ = 430 nm the optical densities of all fine suspensions investigated were found. The optical density was measured for the samples of nanocomposite fine suspension and liquid glass with the concentration 0.003% processed in ultrasound bath Sapfir UZV 28 with US power 0.5 kW and frequency 35 kHz. Liquid glass not modified with nanostructures was used as a reference sample. The processing time with maximum D was selected as optimal.

13.2.2.3   INVESTIGATION OF HEAT-PHYSICAL CHARACTERISTICS

Heat capacity of the samples was investigated with dynamic technique measuring the specific heat capacity of solids with c-calorimeter IT-c-400 [4]. The measuring device was based on comparative technique of c-calorimeter with heat meter and adiabatic shell. Temperature delay time on heat meter was measured with stopwatch at room temperature and the specific heat capacity value was found by the following formula:

$$C = K_T/m_0(\tau_T - \tau_T^{\circ}),$$

where C is the specific heat capacity, J/kg*K; m_0 is the sample mass, g; K_T is the instrument constant which depends on the temperature and is found by the instrument calibration; τ_T° is the temperature delay time with empty ampoule, s; τ_T is the temperature delay time on heat meter, s.

λ-calorimeter IT-λ-400 based on monotonous heating mode was used to measure heat conductivity. In the experiment, the readings of n_0 is the temperature difference by plate thickness was measured with a stopwatch by the following formula to define the heat resistance:

$$R_s = ((1 + \sigma_c) \, S_{n0}/K_T \times n_T) - R_K,$$

where R_s is the sample heat resistance, $m^2 \times$ K/W; S is the sample area, m^2; σ_c = $C_0/2(C_0 + C_c)$ is the allowance taking into account the sample heat capacity C_0 (C_0 is the total heat capacity of the sample tested, J/(kg*K); found before with the instrument IT-s-400; C_c, K_T, and R_K are the constants of the measuring instrument. The time was measured with a stopwatch with the accuracy 0.01 s.

13.2.2.4   DETERMINATION OF RELATIVE AND ABSOLUTE VISCOSITY OF SUSPENSIONS

Relative viscosity of the modified liquid glass was found with viscometer VZ-246 [5]. The viscometer was placed in the rack and fixed horizontally with the balance. The vessel was placed under the viscometer nozzle. The nozzle hole was closed with a finger, the material to be tested was poured into the viscometer to excess in order to have convex meniscus above the upper edge of the viscometer. The viscometer was filled slowly to avoid air bubbles. Then the nozzle hole was opened and as soon as the material being tested is appeared, the stopwatch was started, then stopped and the discharge time was calculated. The values of relative viscosity were translated into the kinematic based on Russian Standard 8420-74.

The absolute (dynamic) viscosity of liquid glass was measured with a falling sphere viscometer (Geppler viscometer) intended for precise measurement of transparent Newton liquids and gasses. In accordance with Geppler principle, the liquid viscosity is proportional to the falling time being measured. The falling time of a sphere rolling down inside the inclined pipe filled with the liquid being tested was measured. The time required for the sphere to cover the distance between the upper and lower ring marks on the pipe with the sample was measured with a stopwatch with the resolution 0.01 s. Further the values of dynamic viscosity were calculated.

13.2.2.5   IR SPECTROMETRY TECHNIQUE

IR Fourier-spectrometer FSM 1201 was used to obtain IR spectra of suspensions based on liquid glass and nanocomposites. The spectra were taken in the range of wave numbers 399–4500 cm^{-1}.

13.3   RESULTS AND DISCUSSION

13.3.1   PREPARATION OF FINE SUSPENSION

According to reference sources, the most advantageous and urgent technique to introduce nanostructures into material is the application of fine suspensions of nanoproducts. They provide the fine modifier uniform distribution through the volume of the material modified.

The necessary concentration of nanostructures in suspension was chosen based on the basic binder mass and varied from 0.003 up to 0.3%. The suspensions were prepared in mechanical mortar mixing nanocomposites with liquid glass following the preselected mode with further processing in ultrasound bath

(Fig. 3) to prevent the nanostructure coagulation. The metal in nanostructures will possibly interact with liquid glass forming insoluble silicates. Thus, the hydrosilicic acid is isolated from the liquid glass as an insoluble gel that results in compressing the liquid glass structure and increasing its moisture stability.

13.3.2 ANALYSIS OF COMPOSITIONS ON LIQUID GLASS BASIS

13.3.2.1 ULTRASOUND PROCESSING

According to reference sources, the ultrasound technique is an effective dispergation method. Such techniques are widely applied today. However, the lengthy soaking in ultrasound baths can result in partial or complete decomposition of nanostructures and their recoagulation. The decomposition of nanostructures due to lengthy ultrasound action is conditioned by high temperature and pressure during the cavitation. The coagulation can be caused with the decomposition of solvate shell on disperse phase particles. It is necessary to define the processing time interval during which the optical density will be at the maximum, i.e., it will correspond to the maximal saturation of nanocomposite suspension.

To define the time interval of ultrasound processing four suspensions were prepared for each soaking period—3, 5, 10, 15 min, respectively, and also the reference solution without US processing—for comparison. The concentration in all solutions was 0.003%.

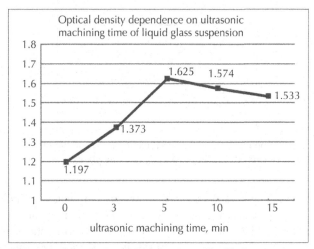

FIGURE 3   Diagram of suspension optical density on liquid glass.

Thus the optimal time interval for ultrasound processing is 5 min. Further processing of suspensions for investigation will be carried out within this interval.

13.3.2.2   SPECTROPHOTOMETRIC INVESTIGATIONS

In accordance with the results of spectrophotometric investigations of modified films in the glue sample with Fe and glue with Ni the shift of optical density is observed at some wavelengths, indicating the changes taking place when nanostructures are introduced (Fig. 4).

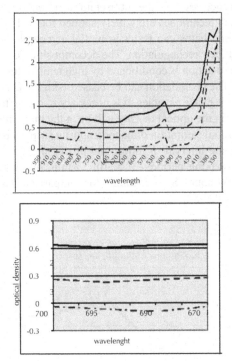

FIGURE 4   Shift of optical density for Fe in relation to Ni and liquid glass (1—Fe/C NC and liquid glass, 2—Ni/C NC and liquid glass, 3—liquid glass).

At other wavelength values the optical density only increases indicating the possibility to apply nanostructures as coloring pigments in paints. At the same time, the increase in the optical density of films with nanostructures in comparison with liquid glass films possibly indicates the increase in the density of compositions and formation of new structural elements in them.

13.3.2.3   INVESTIGATION OF HEAT-PHYSICAL PROPERTIES

To obtain the details of changes in heat-physical characteristics, the heat capacity and thermal conductivity of the samples based on cardboard and modified liquid glass were investigated. The samples were prepared gluing several layers of cardboard with liquid glass modified with nanostructures. The sample dimensions were found by the technique for measuring heat capacity and thermal conductivity. The sample being tested was 15 mm in diameter and 10 mm in height. The sample dimensions were measured with micrometer with 0.01 mm accuracy. The sample mass was measured with the allowance not exceeding 0.001 g. Heat capacity of the samples was investigated on c-calorimeter IT-s-400. To find thermal conductivity the calorimeter IT-λ-400 was used. Further the specific thermal conductivity of the sample was calculated as follows:

$$\lambda = h/R_s,$$

where λ is the specific thermal conductivity, W/m*K; h is the sample thickness, m.

The results of heat-physical investigations are given in Table 1.

TABLE 1   Heat-physical characteristics of the samples.

Samples	Cardboard / Glue	Cardboard / Glue with Fe (change in %)	Cardboard / Glue with Ni (change in %)
Density, kg/m^3	624.5	744 (↑19%)	669 (↑7%)
Heat capacity C_{spec}, J/kg*K	1790	2156 (↑20%)	2972 (↑66%)
Thermal conductivity λ, W/m*K	0.083	0.061 (↓27%)	0.064 (↓23%)

Heat capacity of the sample with Fe increased by 20% in comparison with the reference, sample with Ni by 66%. At the same time, thermal conductivity decreased by 27% and 23% for the samples with Fe and Ni, respectively. So when nanostructures are introduced, in the average the characteristics change as follows: density increases by 13%, heat capacity by 40%, thermal conductivity decreases by 25%. Further, using the experimental results, the temperature conductivity is calculated by the following formula:

$$a = \lambda/c\rho,$$

where a is the temperature conductivity coefficient, λ is the heat conductivity coefficient, c is the heat capacity, ρ is the density.

Inserting the experimental data into the formula, we can define the temperature conductivity values (in percent):

$$a = \lambda/c\rho = 0.75\lambda_0/1.4c_0 \times 1.13\rho_0 = 0.47a_0$$

Or calculate separately:

$$a_1/a_0 = \lambda_1\rho_0C_0/\lambda_0\rho_1C_1 = 0.51$$
$$a_2/a_0 = \lambda_2\rho_0C_0/\lambda_0\rho_2C_2 = 0.43$$

where a_1, λ_1, ρ_1, C_1 is the characteristics of the sample with Fe/C NC; a_2, λ_2, ρ_2, C_2 is the characteristics of the sample with Ni/C NC; a_0, λ_0, ρ_0, C_0 is the characteristics of nonmodified sample.

Thus, the temperature conductivity decreased by nearly 50% in comparison with the initial values (a_0).

When nanostructures are introduced, self-organization takes place. Nanoparticles structure the silicate matrix leading to the formation of new elements in the structure, thus increasing the material density and influencing its heat-physical characteristics. When additional structural elements and new bonds are formed, the system internal energy increases leading to heat capacity elevation and, consequently, temperature conductivity decrease. Thermal conductivity decrease of silicate paints when applied as a coating allows improving heat-physical characteristics of the whole protective structure of a building. In turn, temperature conductivity decrease results in decreasing the amount of heat passing through the coating, thus preserving adhesive characteristics of the coating for a long time.

13.3.2.4   IR SPECTROSCOPY

To determine the interactions between nanostructures and liquid glass the IR spectroscopy investigations were carried out as well. The spectra were taken in relation to water as the suspensions contained nanocomposite solution with liquid glass and water (Fig. 5). The spectra were read with the reference tables. Water spectra demonstrate the region 2750–3750 cm^{-1} connected with O–H bond oscillations. They were observed in other regions as indicated by the values of wave numbers: 2380, 3850, and 3840 cm^{-1}. At the same time, the values 3200, and 3500 cm^{-1} appropriate for valence bound OH appeared in the spectra of nanostructures on Fe basis. In the region 1600–1650 cm^{-1} the bands of deformation oscillations of OH-groups were observed. Bands appropriate to the oscillations of Si-O-Me and Si-O-bonds (1100–400 cm^{-1}) [5] were seen in low frequency region.

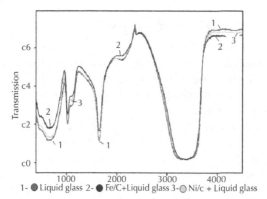

1- ● Liquid glass 2- ● Fe/C+Liquid glass 3-○ Ni/c + Liquid glass

FIGURE 5   IR spectrum of suspensions on the basis of liquid glass and various types of nanocomposites (1—liquid glass, 2—Fe/C NC and liquid glass, 3—Ni/C NC and liquid glass).

IR spectra contained wave numbers reflecting the interaction between metal/ carbon nanocomposites and liquid glass, for example, peaks with wave numbers 407–420 cm^{-1} (Si-Me). The spectra of suspensions vividly demonstrated peaks at 1020 and 1100 cm^{-1}, appropriate to the oscillations of Si-O-Si and Si-O-C. However, in the spectrum of liquid glass containing Fe/C nanocomposite those values were shifted from 1104 to 1122 cm^{-1}, there was also the shift from 600 to 660 cm^{-1} caused by the change in oscillations of some bonds under the nanocomposite action. The spectrum obtained resembles the liquid glass spectrum from reference sources [5].

The spectra demonstrate the absence of strong interactions, mostly the intensity changes thus indicating the formation of a large number of definite bonds, for example, Si-O showing the structuring of liquid medium.

For suspensions on liquid glass basis the relative and absolute viscosity were found. Taking into account the aforementioned data for defining the optimal ultrasound processing time, the suspensions for viscosity test were prepared with ultrasound preprocessing within 5 min.

13.3.2.5   DETERMINATION OF RELATIVE AND ABSOLUTE VISCOSITY OF SUSPENSIONS

Dynamic viscosity is the ration between force unit required to shift the liquid layer for distance unit and layer area unit. In metric system it is given as dyne-second per square centimeter called "poise."

$$\eta = \tau \cdot \left(\rho_1 - \rho_2 \right) \cdot \kappa \cdot F ,$$

where η is the dynamic viscosity [mPa·s], τ is the sphere movement time [s], ρ_1 is the sphere density according to the test certificate [g/cm³], ρ_2 is the sample density, [g/cm³], κ is the sphere constant according to the test certificate [mPa·cm³/g], F is the work angle constant.

The experiment was carried out with pipe inclination for the sample 80° => F = 1.0

ρ_1 = 15.2 g/cm³
ρ_2 = 1.45 g/cm³
κ = 0.7
Liquid glass viscosity: τ_{cp} = 9.60 s

$$\eta = 9.6(15.2–1.45) \times 0.7 \times 1 = 92.4 \text{ Pa·s}$$

Viscosity of liquid glass modified with iron containing nanocomposite:

$$\eta = 6.97(15.2–1.45) \times 0.7 \times 1 = 67.09 \text{ Pa·s}$$

Viscosity of liquid glass modified with nickel containing nanocomposite:

$$\eta = 8.76(15.2–1.45) \times 0.7 \times 1 = 84.32 \text{ Pa·s}$$

Kinematic viscosity is the ratio between dynamic viscosity and liquid density.

Based on the data obtained when measuring the viscosity of liquid glass and fine suspension with Fe/C and Ni/C NC, the diagrams were constructed (Figs. 6 and 7).

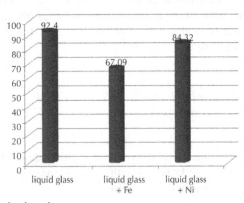

FIGURE 6 Dynamic viscosity.

Having analyzed this diagram, we can conclude that when a nanocomposite is introduced the viscosity decreases. Such phenomenon will have a positive effect in the process of silicate material production and application. For example, better application—in silicate paint, much easier foaming process—in foam glass.

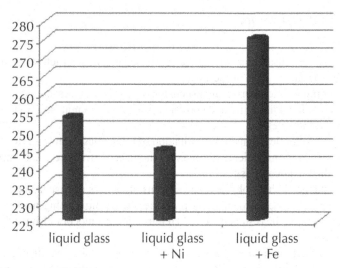

FIGURE 7   Kinematic viscosity

In Fig. 7, we can see that nanostructures work differently. The viscosity of fine suspension with Ni/C NC is less than kinematic viscosity with Fe/C NC. This can be caused by coagulation, aggregation of Fe/C NC (Table 2).

TABLE 2   Characteristics of kinematic and dynamic viscosity of liquid glass and its analogs modified with Ni/C and Fe/C nanocomposites.

Sample	Kinematic viscosity, $mm^2 \cdot s^{-1}$	Dynamic viscosity, Pa·s (change in %)
Liquid glass	253.3	92.4
Liquid glass with Ni/C nanocomposite	244.3 (↓4%)	84.32 (↓9%)
Liquid glass with Fe/C nanocomposite	275.1 (↑9%)	67.09 (↓27%)

The viscosity of compositions on liquid glass basis depends on nanocomposite composition: in some cases the viscosity increase is observed, in other cases—the decrease. The viscosity increase is connected with the elevation of intermolecular friction, which can be caused by the influence of nanoparticles on liquid glass structuring and density increase. The decrease in suspension viscosity can be explained by the fact that NC is surrounded by liquid glass molecules resulting in the decrease in their interactions and intermolecular friction.

13.4 CONCLUSION

Based on the results obtained it can be concluded that nanostructures have a positive effect on liquid glass properties.

The optimal time interval of suspension ultrasound processing was found experimentally. It equals 5 min and corresponds to the maximum solution saturation with nanocomposites.

The data of optical density investigation of liquid glass films and IR spectra of liquid glass solutions indicate the changes in material structure when introducing metal/carbon nanocomposites.

The changes in heat-physical and viscous properties of suspensions on liquid glass indicate the self-organization processes when introducing nanostructures. At the same time, the results differ when introducing nanocomposites of different compositions.

The application of nanocomposites will allow increasing the life of paints and, consequently, their storage and covering ability. Also not only potassium but sodium liquid glass can be used as well, since potassium liquid glass is significantly more expensive than sodium one and its manufacturing is limited. To improve the characteristics only small concentrations of nanoproduct are required, which will positively affect the material production cost.

KEYWORDS

- **Geppler principle**
- **IR spectroscopy**
- **metal/carbon nanocomposites**
- **poise**
- **silicate compositions**

REFERENCES

1. Kodolov, V. I.; Khokhriakov, N. V.; In: Chemical physics of formation and transformation processes of nanostrcutures and nanosystems. Izhevsk: Izhevsk State Agricultural Academy, **2009**, *1*, 365 p, *2*, 415 p.
2. Patent 2393110 Russia Technique of obtaining carbon metal containing nanostructures; Kodolov, V. I.; Vasilchenko, Yu. M.; Akhmetshina, L. F.; Shklyaeva, D. A.; Trineeva, V. V.; Sharipova, A. G.; Volkova, E. G.; Ulyanov, A. L.; Kovyazina, O. A.; declared on 17.10.2008, published on 27.06.2010.
3. Patent 2337062 Russia Technique of obtaining carbon nanostructures from organic compounds and metal containing substances; Kodolov, V. I.; Kodolova, V. V. (Trineeva), Semakina, N. V.; Yakovlev, G. I.; Volkova, E. G., et al.; declared on 28.08.2006, published on 27.10.2008.
4. Platunov, E. S.; Buravoy, S. E.; Kurepin, V. V.; Petrov, G. S.; In: Heat-physical measurements and instruments. Platunov E. S., ed.; L.: Mashinostroenie, 1986.
5. Plyusnina I. I. In: IR spectra of silicates. M.: Publishing House of Moscow State University, **1967,** 190 p.

INDEX

Printed in the United States
by Baker & Taylor Publisher Services

One-Dimensional Finite Elements

Andreas Öchsner · Markus Merkel

One-Dimensional Finite Elements

An Introduction to the FE Method

Second Edition

 Springer

Andreas Öchsner
Faculty of Mechanical Engineering
Esslingen University of Applied Sciences
Esslingen
Germany

Markus Merkel
Institute for Virtual Product Development
Aalen University of Applied Sciences
Aalen
Germany

ISBN 978-3-030-09157-6   ISBN 978-3-319-75145-0   (eBook)
https://doi.org/10.1007/978-3-319-75145-0

Printed on acid-free paper

This Springer imprint is published by the registered company Springer International Publishing AG part of Springer Nature
The registered company address is: Gewerbestrasse 11, 6330 Cham, Switzerland

This book is dedicated to our families.

Preface

The title of this book 'One-Dimensional Finite Elements: An Introduction to the FE Method' stands for content and focus. Nowadays, much literature regarding the topic of the finite element method exists. The different works reflect the multifaceted perceptions and application possibilities. The basic idea of this introduction into the finite element method relies on the concept of explaining the complex method with the help of *one-dimensional* elements. It is the goal to introduce the manifold aspects of the finite element method and to enable the reader to get a methodical understanding of important subject areas. The reader learns to understand the assumptions and derivations at different physical problems in structural mechanics. Furthermore, he/she learns to critically evaluate possibilities and limitations of the finite element method. Additional comprehensive mathematical descriptions, which solely result from advanced illustrations for two- or three-dimensional problems, are omitted. Hence, the mathematical description largely remains simple and clear. The focus on one-dimensional elements, however, is not just a pure limitation on a more simple and clearer formal illustration of the necessary equations. Within structural engineering, there are various structures—for example bridges or high transmission towers —which are usually modeled via one-dimensional elements. Therefore, this work also contains a 'set of tools', which can also be applied in practice.

The concentration on one-dimensional elements is new for a textbook and allows the treatment of various basic and demanding physical questions of structural mechanics within one single textbook. This new concept therefore allows a methodical understanding of important subject areas (e.g., plasticity or composite materials), which occur to a prospective engineer during professional work, which, however, are seldomly treated in this way at universities. Consequently, simple access is possible, also in supplementary areas of application of the finite element method.

This book originates from a collection of lecture notes which were developed as written material for lectures and training documents for specialized courses on the finite element method. Especially, the calculated examples and the supplementary problems refer to typical questions which are raised by students and course participants.

A prerequisite for a good understanding is the basics in linear algebra, physics, materials science and strength of materials, the way they are typically communicated in the basic studies of a technical subject in the field of mechanical engineering.

Within the first chapters, the one-dimensional elements will be introduced, due to which the basic load cases of tension/compression, torsion, and bending can be illustrated. In each case, the differential equation as well as the basic equations from the strength of materials (this is the kinematic relationship, the constitutive relationship, and the equilibrium equation) are being derived. Subsequently, the finite elements with the usual definitions for force and displacement parameters are introduced. With the help of examples, the general procedure is illustrated. Short solutions for supplementary problems are attached in the appendix.

Chapter 6 deals with questions which are independent of the loading type and the therewith connected element formulation. A general one-dimensional finite element, which can be constructed from the combination of basic elements, the transformation of elements in the general three-dimensional space and the numerical integration as an important tool for the implementation of the finite element method is dealt with.

The complete analysis of an entire structure is introduced in Chap. 7. The total stiffness relation results from the single stiffness relation of the basic elements under consideration of the relations to each other. A reduced system results due to the boundary conditions. Unknown parameters are derived from the reduced system. The procedure will be introduced as examples on plane and general three-dimensional structures.

Chapters 8–12 deal with topics which are usually not part of a basic textbook. The beam element with shear consideration is introduced in Chap. 8. The Timoshenko beam theory serves as a basis for this.

Within Chap. 9, a special class of material—composite materials—is introduced into a finite element formulation. First, various ways of description for direction-dependent material behavior are introduced. Fiber composites are addressed briefly. A composite element is demonstrated by examples of a composite bar and a composite beam.

Chapters 10, 11, and 12 deal with nonlinearities. In Chap. 10, the different types of nonlinearities are introduced. The case of the nonlinear elasticities is dealt with more closely. The problem is illustrated for bar elements. First, the principal finite element equation is derived under consideration of the strain dependency. The direct iteration as well as the complete and modified Newton–Raphson iterations are derived for the solution of a nonlinear system of equations. In addition, many examples serve as a demonstration of this issue.

Chapter 11 considers elasto-plastic behavior, one of the most common forms of material nonlinearities. First, the continuum-mechanical basics for plasticity in the case of the one-dimensional continuum bar are composed. The yield condition, the flow rule, the hardening law, and the elasto-plastic modulus are introduced for uniaxial, monotonic load cases. Within the hardening behavior, the description is limited to isotropic hardening. For the integration of the elasto-plastic constitutive equation, the incremental predictor–corrector method is generally introduced and derived for the case of the fully implicit and the semi-implicit backward-Euler algorithm. On crucial points, the difference between one- and three-dimensional descriptions will be pointed out, to guarantee a simple transfer of the derived methods to general problems.

Chapter 12 deals with stability, which is an issue that is especially relevant for the design and dimensioning of lightweight components. Finite elements developed for this type of nonlinearities are used for the solving of the Euler's buckling loads.

Chapter 13 serves to introduce a finite element formulation for dynamic problems. Stiffness matrices as well as mass matrices will be established. Different assumptions for the distribution of the masses, whether continuously or concentrated, lead to different formulations. The issue is discussed by example of axial vibrations of the bar.

As an illustration, each chapter is recessed both with precisely calculated and commented examples as well as with supplementary problems—including short solutions. Each chapter concludes with an extensive bibliography.

Preface to the 2nd Edition

The basic concept for the treatment of the finite element method with one-dimensional questions has been preserved in the 2nd edition. Additionally included is the stationary heat conduction, and the principle of virtual work became another method for the derivation of finite element formulations. Added is Chap. 14 with special elements for the modeling of elastic foundations, singularities, and infinite extension of elements.

Bexbach, Hüttlingen   Andreas Öchsner
December 2017   Markus Merkel

Acknowledgements

Critical questions and comments of students and course participants are gratefully acknowledged. Their input contributed to the actual form of this book.

Finally, we like to thank our families for the understanding and patience during the preparation of this book.

Contents

Symbols and Abbreviations

Latin Symbols (Capital Letters)

A	Surface, cross-sectional area
$\boldsymbol{B}$	Matrix with derivatives of the shape functions
$\boldsymbol{C}$	Elasticity matrix
	Damping matrix
C^{elpl}	Elasto-plastic stiffness matrix
D	Diameter
$\boldsymbol{D}$	Elasticity matrix
E	Modulus of elasticity
E^{elpl}	Elasto-plastic modulus
E^{pl}	Plastic modulus
$\tilde{E}$	Average modulus
F	Yield condition
	Force
$\boldsymbol{F}^{\text{ext}}$	Column matrix of the external loads
G	Shear modulus
I	Second moment of area
K	Bulk modulus
$\boldsymbol{K}$	Total stiffness matrix
$\boldsymbol{K}_{\text{T}}$	Tangent stiffness matrix
L	Element length
$\mathcal{L}_1$	Differential operator of first order
$\mathcal{L}_i^n$	Lagrange polynomial
L_{crit}	Buckling length
M	Moment
$\boldsymbol{M}$	Mass matrix
N	Shape function
$\boldsymbol{N}$	Row matrix of the shape functions, $\boldsymbol{N} = [N_1 \; N_2 \; \dots \; N_n]$
Q	Plastic potential
$\boldsymbol{Q}$	Elasticity matrix (plane case)

	Shear force
R	Radius
S	Bar force
$\boldsymbol{S}$	Compliance matrix
T	Torsional moment
$\boldsymbol{T}$	Transformation matrix
V	Volume
W	Weight function
X	Global spatial coordinate
Y	Global spatial coordinate
Z	Global spatial coordinate

Latin Symbols (Small Letters)

a	Geometric dimension
b	Geometric dimension, width
c	Integration constant
d	Geometric dimension
h	Geometric dimension, height
$\boldsymbol{e}$	Unit vector
f	Function
g	Function
	Gravitational acceleration
$\boldsymbol{h}$	Function of the stabilization modification
i	Increment number
	Variable
j	Iteration index
	Variable
k	Spring stiffness
	Yield stress
k_{s}	Shear correction factor
$\boldsymbol{k}^{\mathrm{e}}$	Element stiffness matrix
m	Number of elements
	Slope
	Polynomial degree
	Mass
m_{t}	Continuously distributed torsional moment per length
$\boldsymbol{m}$	Column matrix of residual functions
n	Number of nodes
	Variable
	Condition
q	Load
	Integration order
	Modal coordinates
$\boldsymbol{q}$	Matrix of the inner variables

r	Function of the flow direction
	Radius
	Residual
$\boldsymbol{r}$	Vector of the flow direction
t	Time
	Geometric dimension
t_{ij}	Component of the transformation matrix
u_x	Displacement in x-direction
u_y	Displacement in y-direction
u_z	Displacement in z-direction
$\boldsymbol{u}$	Column matrix of the nodal displacements
υ	Argument vector (NEWTON procedure)
x	Spatial coordinate
y	Spatial coordinate
z	Spatial coordinate

Greek Symbols (Capital Letters)

Γ	Boundary
Λ	Parameter (TIMOSHENKO beam)
Π	Energy
$\bar{\Pi}$	Complementary energy
Π_{ext}	Potential of the external loads
Π_{int}	Elastic strain energy
$\boldsymbol{\Phi}$	Modal matrix
Ω	Domain, volume

Greek Symbols (Small Letters)

α	Thermal expansion coefficient
	Constant
	Angle
β	Angle
	Constant
γ	Shear strain
δ	Virtual
ε	Strain
ε_{ij}	Strain tensor
ε	Column matrix of the strain components
$\varepsilon_{\text{eff}}^{\text{pl}}$	Equivalent plastic strain
κ	Inner variable (plasticity), curvature (beam bending)
λ	Eigenvalue
$\text{d}\lambda$	Consistency parameter
ν	Lateral contraction ratio (POISSON's ratio)

ζ	Unit coordinate $(-1 \leq \xi \leq 1)$
π	Volume specific work
	Volume specific energy
σ	Stress, normal stress
ρ	Density
σ_{n+1}^{trial}	Test stress state
σ_{ij}	Stress tensor
$\boldsymbol{\sigma}$	Column matrix of the stress components
τ	Shear stress
η	Unit coordinate $(-1 \leq \eta \leq 1)$
ζ	Unit coordinate $(-1 \leq \zeta \leq 1)$
ψ	Phase angle
ϕ	Rotation angle
φ	Rotation angle
ω	Eigenfrequency

Indices, Superscript

$\ldots^{V}$	Composite
$\ldots^{e}$	Element
$\ldots^{el}$	Elastic
$\ldots^{ext}$	External parameter
$\ldots^{geo}$	Geometric
$\ldots^{glo}$	Global
$\ldots^{init}$	Initial
$\ldots^{lo}$	Local
$\ldots^{pl}$	Plastic
$\ldots^{red}$	Reduced
$\ldots^{trial}$	Test condition (back projection)

Indices, Subscript

$\cdots_{Im}$	Imaginary part of an imaginary number
$\cdots_{Re}$	Real part of an imaginary number
$\cdots_{b}$	Bending
$\cdots_{c}$	Pressure ('compression')
	Damping
$\cdots_{eff}$	Effective value
$\cdots_{f}$	Fiber in the composite
$\cdots_{krit}$	Critical
$\cdots_{l}$	Lamina
$\cdots_{m}$	Matrix in the composite
	Inertia
$\cdots_{p}$	Nodal value

$\cdots$ s	Shear
$\cdots$ t	Torsion
	Tension
$\cdots$ w	Wall

Mathematical Symbols

$(\ldots)^{\mathrm{T}}$	Transpose
$\lvert\cdots\rvert$	Absolute value
$\lVert\ldots\rVert$	Norm
$\otimes$	Dyadic product
sgn	Sign function
$\mathbb{R}$	Set of real numbers

Abbreviations

1D	One-dimensional
2D	Two-dimensional
CAD	Computer-aided design
FCM	Fiber composite material
FE	Finite element
FEM	Finite element method
inc	Increment number

Chapter 1
Introduction

Abstract In this first chapter the content as well as the focus will be classified in various aspects. First, the development of the finite element method will be explained and considered from different perspectives.

1.1 The Finite Element Method at a Glance

Seen chronologically, the roots of the finite element method lie in the middle of the last century. Therefore, this method is a relatively young tool in comparison with other tools and aids for the design and dimensioning of components. The development of the finite element method started in the 1950s. Scientists and users have brought in ideas from quite different fields and have therefore turned the method into a universal tool which has nowadays become an indispensable part in research and development of engineering applications. Initially rather basic questions were covered, for example on questions regarding the principle solvability. Regarding the software implementation, only rudimental resources were available — from today's point of view. The preprocessing consisted of punching of cards, which were fed in batches to a calculating machine. Mistakes during the programming were directly displayed with blinking lights. With progressive computer development, the programming environment has become more comfortable and algorithms could be tested and optimized on more challenging examples. From the point of view of engineering application, the problems, which were analyzed via the finite element method, were limited to simple examples. The computer capacities only allowed a quite rough modeling.

Nowadays, the basic questions have been clarified, the central issue of the problem rather lies on the user side. Finite element program packages are available in a large variety and are used in the most different forms. On the one hand there are program packages, which are primarily used in teaching. It is the goal to illustrate the systematic approach. Source codes are available for such programs. On the other hand, there are commercial program packages which are used to their full capac-

© Springer International Publishing AG, part of Springer Nature 2018

A. Öchsner and M. Merkel, *One-Dimensional Finite Elements*,

https://doi.org/10.1007/978-3-319-75145-0_1

1

ity regarding program technology as well as the content. Especially the program modules, which have been adapted to a computer platform or computer architecture (parallel computing) are quite efficient and make the processing of very comprehensive problems possible. Regarding the content, the authors venture to say that there is no physical discipline for which no finite element program exists.

In regard to the development of the finite element method, the focus is nowadays on the cooperation and integration with other development tools, as for example the point of intersection with engineering design. Both classical disciplines, i.e. calculation and design, become more and more connected and are partly already fused underneath by a common graphical interface. Besides single finite element software packages there are also in a computer aided design (CAD) system integrated solutions available on the market. From the view of the user an appropriate finite element preprocessing and postprocessing of *his/her* special problem is in the foreground. The time intensive process steps of the geometry preparation should not involve a considerable extra effort for the application of the finite element method. Calculation results are supposed to be integrated seamlessly in the according process chain.

Regarding the application areas, there are no limits for the application of the finite element method. The dimensioning and configuration of elements, subsystems or complete machines surely is the focus in mechanical and plant engineering.

The application of the finite element method or in general of simulation tools in the product development is often seen as a competing tool to the real or lab-scaled experiment or test. The authors rather see an ideal complement at this point. Therefore, a single test stand or complete test scenarios can be optimized ex ante via finite element simulation. In return, experimental results help to create more precise simulation models.

1.2   Foundations of Modeling

A model of a physical or technical problem represents the initial situation for the application of the finite element method. A complete description of the problem comprises

- the geometry for the description of the domain,
- the field equations in the domain,
- the boundary conditions and
- the initial conditions for time dependent problems.

Within this book solely *one-dimensional* elements will be regarded. The general procedure for two- and three-dimensional problems is similar. The mathematical demand however is much more complex.

Usually the problems can be described via the differential equation. Here, differential equations of second order are focused on. As an example, the differential equations of a certain class of physical problems can in general be described as follows:

Table 1.1 Physical problems in the context of the differential equation Eq. (1.1). Adapted from [1]

Problem	Field parameter	Coefficient	
	$u(x)$	a	f
Heat conduction	Temperature Θ	Heat conduction λ	Heat sources q
Pipe flow	Pressure p	Pipe resistance $1/R$	
Viscous flow	Velocity v_x	Viscosity ν	Pressure gradient $\frac{dp}{dx}$
Elastic bars	Displacement u	Stiffness EA	Distributed loads f
Elastic torsion	Rotation φ	Stiffness GI_p	Torsional moments m_t
Electrostatics	Electrical potential Φ	Dielectricity ϵ	Charge density ρ

$$-\frac{d}{dx}\left[a\,\frac{du(x)}{dx}\right] - f = 0. \tag{1.1}$$

Depending on the physical problem a different meaning is assigned to the variable $u(x)$ and the parameters a and f. Table 1.1 lists the meaning of the parameters for a few physical problems [1].

To describe a problem completely, the statement about the boundary conditions is necessary besides the differential equation. The local boundary conditions can generally be divided into three groups:

• Boundary condition of the 1st kind or DIRICHLET boundary condition (also referred to as essential, fundamental, geometric or kinematic boundary condition):
A boundary condition of the 1st kind exists, if the boundary condition is being expressed in parameters in which the differential equation is being formulated.
• Boundary condition of the 2nd kind or NEUMANN boundary condition (also referred to as natural or static boundary condition):
A boundary condition of the 2nd kind exists, if the boundary condition purports the derivative in the direction of the normal of the boundary Γ.
• Boundary condition of the 3rd kind or CAUCHY boundary condition (also referred to as mixed or ROBIN boundary condition):
Defines a weighted sum of DIRICHLET and NEUMANN boundary conditions on the boundary.

These three types of boundary conditions are summarized in Table 1.2, along with their formulas.

Table 1.2 Different boundary conditions of a differential equation

Differential equation	DIRICHLET	NEUMANN	CAUCHY
$\mathcal{L}\{u(x)\} = b$	u	$\frac{du}{dx}$	$\alpha u + \beta\frac{du}{dx}$

It needs to be considered at this point that one talks about *homogeneous boundary conditions* if the corresponding variables are zero on the boundary.

Within this book, the finite element method will be highlighted from the view of mathematics, physics or the engineering application. From a mathematical view, the finite element method is an appropriate tool to solve partial differential equations. From a physical view a multitude of physical problems can be worked on via the finite element method. The areas go from electrostatics via diffusion problem all the way to elasticity theory. Engineers make use of the finite element method for the configuration and the dimensioning of products. Regarding the physical problems, at this point solely elastomechanical problems will be discussed. Within statics

- the tension bar,
- the torsion bar and
- the bending beam with and without shear contribution

will be covered. Vibrations of bars and beams will be covered as dynamic problems. This book is already the second edition and a German translation is as well available [2, 3].

References

1. Reddy JN (2006) An introduction to the finite element method. McGraw Hill, Singapore
2. Merkel M, Öchsner A (2014) Eindimensionale Finite Elemente: Ein Einstieg in die Methode. Springer Vieweg, Berlin
3. Öchsner A, Merkel M (2013) One-dimensional finite elements: an introduction to the FE method. Springer, Berlin

Chapter 2
Motivation for the Finite Element Method

Abstract The approach to the finite element method can be derived from different motivations. Essentially there are three ways:

- a rather descriptive way, which has its roots in the engineering working method,
- a physical or
- mathematically motivated approach.

Depending on the perspective, different initial formulations result in the same principal finite element equation. The different formulations will be elaborated in detail based on the following descriptions:

- matrix methods,
- energy methods and
- weighted residual method.

The finite element method is used to solve different physical problems. Here solely finite element formulations related to structural mechanics are considered.

2.1 From the Engineering Perspective Derived Methods

Matrix methods can be regarded in elastostatics as the initial point for the application of the finite element method to analyze complex structures. As an example, consider the plane structure which is shown in Fig. 2.1. This example is adapted from [1].

The structure consists of various substructures I, II, III and IV. The substructures are referred to as elements. The elements are coupled at the nodes 2, 3, 4 and 5. The entire structure is supported on nodes 1 and 6, an external load is applied at node 4.

Unknown are

- the displacements and reaction forces on every single inner node and
- the support reactions

in consequence of the acting load.

© Springer International Publishing AG, part of Springer Nature 2018
A. Öchsner and M. Merkel, *One-Dimensional Finite Elements*,
https://doi.org/10.1007/978-3-319-75145-0_2

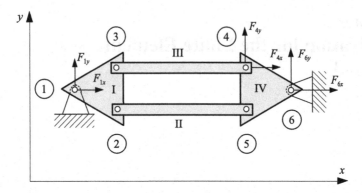

Fig. 2.1 Plane structure, adapted from [1]

To solve the problem, matrix methods can be used. In the context of matrix methods one distinguishes between the force method (static method), which is based on a direct determination of the statically indeterminate forces, and the displacement method (kinematic method), which considers the displacements as unknown parameters.

Both methods allow the determination of the unknown parameters. The decisive advantage of the displacement method is that during the application it is not necessary to distinguish between statically determinate and statically indeterminate structures. Due to the generality this method is applied in the following.

2.1.1 The Matrix Stiffness Method

It is the primary subgoal to establish the stiffness relation for the entire structure from Fig. 2.1. The following stiffness relation serves as the basis for the matrix displacement method:

$$F = K\,u\,. \tag{2.1}$$

F and u are column matrices, K is a square matrix. F summarizes all nodal forces and u summarizes all nodal displacements. The matrix K represents the stiffness matrix of the entire structure. One single element is identified as the basic unit for the problem and is characterized by the fact that it is coupled with other elements via nodes. Displacements and forces are introduced at every single node.

To solve the entire problem

- the compatibility and
- the equilibrium

have to be fulfilled.

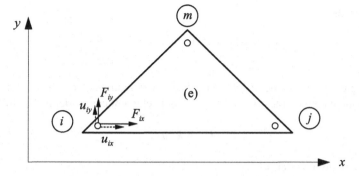

Fig. 2.2 Single elements with displacements and forces

In the scope of the matrix displacement method, one introduces the nodal displacements as essential unknowns. The displacement vector at a node is defined to be valid for all elements connected at this node. Therewith the compatibility of the entire structure is a priori fulfilled.

A Single Element

Forces and displacements are introduced for each node of the single element (see Fig. 2.2).

For a unique representation the *nodal* forces and the *nodal* displacements are provided with an index 'p' to highlight that these are parameters, which are defined at nodes.

The column matrices of the nodal displacements $\boldsymbol{u}_p$ or alternatively nodal forces $\boldsymbol{F}_p$ in general consist of various components for the respective coordinates. An additional index 'e' indicates to which element the parameters relate.[1] Therewith the nodal forces result, according to Fig. 2.2, in

$$\boldsymbol{F}_i^e = \begin{bmatrix} F_{ix} \\ F_{iy} \end{bmatrix}, \quad \boldsymbol{F}_j^e = \begin{bmatrix} F_{jx} \\ F_{jy} \end{bmatrix}, \quad \boldsymbol{F}_m^e = \begin{bmatrix} F_{mx} \\ F_{my} \end{bmatrix}, \tag{2.2}$$

and the nodal displacements in

$$\boldsymbol{u}_i = \begin{bmatrix} u_{ix} \\ u_{iy} \end{bmatrix}, \quad \boldsymbol{u}_j = \begin{bmatrix} u_{jx} \\ u_{jy} \end{bmatrix}, \quad \boldsymbol{u}_m = \begin{bmatrix} u_{mx} \\ u_{my} \end{bmatrix}. \tag{2.3}$$

If one summarizes all nodal forces and nodal displacements to column matrices each at one element, the

$$\text{entire nodal forces } \boldsymbol{F}_p^e = \begin{bmatrix} \boldsymbol{F}_i \\ \boldsymbol{F}_j \\ \boldsymbol{F}_m \end{bmatrix} \tag{2.4}$$

---

[1] The additional index 'e' is to be dropped at displacements since the nodal displacement is identical for each linked element in the displacement method.

as well as the

$$\text{entire nodal displacements } \boldsymbol{u}_p = \begin{bmatrix} \boldsymbol{u}_i \\ \boldsymbol{u}_j \\ \boldsymbol{u}_m \end{bmatrix} \tag{2.5}$$

for a single element are described. With the column matrices for the nodal forces and displacements the stiffness relation for a single element can be defined as follows:

$$\boldsymbol{F}_p^e = \boldsymbol{k}^e \, \boldsymbol{u}_p , \tag{2.6}$$

or alternatively for each node:

$$\boldsymbol{F}_r^e = \boldsymbol{k}_{rs}^e \boldsymbol{u}_s \qquad (r, s = i, j, m) . \tag{2.7}$$

The single stiffness matrix $\boldsymbol{k}^e$ connects the nodal forces with the nodal displacements. In the present example the single stiffness relation is formally defined as

$$\begin{bmatrix} F_{ix} \\ F_{iy} \\ \hline F_{jx} \\ F_{jy} \\ \hline F_{mx} \\ F_{my} \end{bmatrix} = \begin{bmatrix} k_{ii}^e & k_{ij}^e & k_{im}^e \\ \hline k_{ji}^e & k_{jj}^e & k_{jm}^e \\ \hline k_{mi}^e & k_{mj}^e & k_{mm}^e \end{bmatrix} \begin{bmatrix} u_{ix} \\ u_{iy} \\ \hline u_{jx} \\ u_{jy} \\ \hline u_{mx} \\ u_{my} \end{bmatrix} . \tag{2.8}$$

For further progression it needs to be assumed that the single stiffness matrices of the elements I, II, III and IV are known. The single stiffness relations of one-dimensional elements will explicitly be derived in the following chapters for different deformation types.

The Overall Stiffness

The equilibrium condition of each single element is fulfilled via the single stiffness relation in Eq. (2.6). The overall equilibrium is satisfied by the fact that each node is set into equilibrium. As an example the equilibrium will be set for node 4 in Fig. 2.3:
    With

$$\boldsymbol{F}_4^{\text{ext}} = \begin{bmatrix} F_{4x} \\ F_{4y} \end{bmatrix} \tag{2.9}$$

the following is valid:

$$\boldsymbol{F}_4^{\text{ext}} = \sum_e \boldsymbol{F}_4^e = \boldsymbol{F}_4^{\text{III}} + \boldsymbol{F}_4^{\text{IV}} . \tag{2.10}$$

If one substitutes the nodal forces by the single stiffness relations times the nodal displacements, this yields

Fig. 2.3 Equilibrium on
node 4 for the problem of
Fig. 2.1

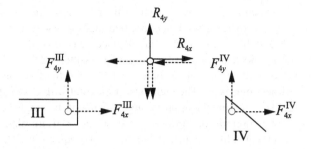

$$F_4^{\text{ext}} = k_{43}^{\text{III}} u_3 + \left(k_{44}^{\text{III}} + k_{44}^{\text{IV}} \right) u_4 + k_{45}^{\text{IV}} u_5 + k_{46}^{\text{IV}} u_6. \tag{2.11}$$

If one sets up the equilibrium on each node accordingly and arranges all relations in
the form of a matrix equation, the *overall* stiffness relation results

$$\mathbf{F} = \mathbf{K}\, \mathbf{u} \tag{2.12}$$

with

$$\mathbf{K} = \sum_e k_{ij}^{\text{e}}, \tag{2.13}$$

or alternatively in detail

$$
\begin{bmatrix}
R_1 \\
\hline
0 \\
\hline
0 \\
\hline
F_4^{\text{ext}} \\
\hline
0 \\
\hline
R_6
\end{bmatrix}
=
\begin{bmatrix}
k_{11}^{\text{I}} & k_{12}^{\text{I}} & k_{13}^{\text{I}} & 0 & 0 & 0 \\
\hline
k_{21}^{\text{I}} & k_{22}^{\text{I}} + k_{22}^{\text{II}} & k_{23}^{\text{I}} & 0 & k_{25}^{\text{II}} & 0 \\
\hline
k_{31}^{\text{I}} & k_{32}^{\text{I}} & k_{33}^{\text{I}} + k_{33}^{\text{III}} & k_{34}^{\text{III}} & 0 & 0 \\
\hline
0 & 0 & k_{43}^{\text{III}} & k_{44}^{\text{III}} + k_{44}^{\text{IV}} & k_{45}^{\text{IV}} & k_{46}^{\text{IV}} \\
\hline
0 & k_{52}^{\text{II}} & 0 & k_{54}^{\text{IV}} & k_{55}^{\text{II}} + k_{55}^{\text{IV}} & k_{56}^{\text{IV}} \\
\hline
0 & 0 & 0 & k_{64}^{\text{IV}} & k_{65}^{\text{IV}} & k_{66}^{\text{IV}}
\end{bmatrix}
\begin{bmatrix}
0 \\
\hline
u_2 \\
\hline
u_3 \\
\hline
u_4 \\
\hline
u_5 \\
\hline
0
\end{bmatrix}
. \tag{2.14}
$$

This equation is also referred to as the *principal finite element equation*. The column matrix of the external loads (applied loads or support reactions) is on the left-hand side and the vector of all nodal displacements is on the right-hand side. The external forces and reaction forces are assigned with an F^{ext} and R in order to distinguish between element related forces and external forces. Both are coupled via the total stiffness matrix K. The elements of the total stiffness matrix result according to Eq. (2.13) by adding the appropriate elements of the single stiffness matrices.

The support conditions $u_1 = 0$ and $u_6 = 0$ are already considered in the displacement column matrix. From the matrix equations (2.2)–(2.5) the unknown nodal displacements u_2, u_3, u_4 and u_5 can be derived. If these are known, one receives, through insertion into the matrix equations (2.1) and (2.6), the unknown support reactions R_1 and R_6.

The matrix displacement method is precise as long as the single stiffness matrices can be defined and as long as elements are coupled in well defined nodes. This is the case for example in truss and frame structures within the heretofore valid theories.

With the so far introduced method the nodal displacements and forces in dependency on the external loads can be determined. For the analysis of the strength of a single element the strain and stress state inside each element is of relevance. Usually the displacement field is described via the nodal displacements u_{p} and shape functions. The strain field can be defined via the kinematics relation and the stress field via the constitutive equation.

2.1.2   Transition to the Continuum

In the previous section, the matrix displacement method was discussed for a joint supporting structure. In contrast to this, in the continuum (Fig. 2.4), the *virtual* discretized finite elements are connected at infinitely many nodal points. However, in a real application of the matrix displacement method, only a finite number of nodes can be considered. Therewith it is not possible to exactly fulfill both demanded conditions for compatibility and for equilibrium at the same time. Either the compatibility or the equilibrium will be fulfilled *on average*.

Fig. 2.4 Continuum with load and boundary conditions

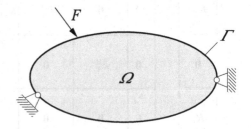

In principle, the procedure can be demonstrated with the force method or the displacement method can be illustrated. In the following, only the *displacement method* will be considered. Here

- the compatibility is *exactly* fulfilled and
- the equilibrium *on average*.

The following approach results:

1. The continuum is discretized, meaning for two-dimensional problems it is divided by virtual lines and for three-dimensional problems through virtual surfaces in subregions, so-called finite elements.
2. The flux of force from element to element occurs in discrete nodes. The displacements of these nodes are introduced as principal unknowns (*displacement method*!).
3. The displacement state within an element is formulated as a function of the nodal displacements. The displacement formulations are compatible with the adjacent neighboring elements.
4. Through the kinematics equation the strain state within the element and through the constitutive equation the stress state are known as function of the nodal displacement.
5. Via the *principle of virtual work*, the stresses along the virtual element boundaries are assigned to statically equivalent resulting nodal forces *on average*.
6. To maintain the overall equilibrium all nodal equilibria have to be fulfilled. Via this condition one gets to the total stiffness relation, from which the unknown nodal displacements can be calculated after considering the kinematic boundary conditions.
7. If the nodal displacements are known, one knows the displacement field and strain field and therefore also the stress state of each single element.

Comments to the Single Steps

Discretization

Through discretization the entire continuum is divided into elements. An element is in contact with one or various neighboring elements. In the two-dimensional case lines result as contact regions, in the three-dimensional case surfaces occur. Figure 2.5 illustrates a discretization for a plane case.

The discretization can be interpreted as follows: Single points do not change their geometric position within the continuum. The relation to the neighboring points however does change. While each point within the continuum is in interaction with its neighboring point, in the virtual discretized continuum this is only valid within one element. If two points lie within two different elements they are not directly linked.

Nodes and Displacements

The information flow between single elements only occurs via the nodes. In the displacement method, displacements are introduced at the nodes as principal unknowns (see Fig. 2.6).

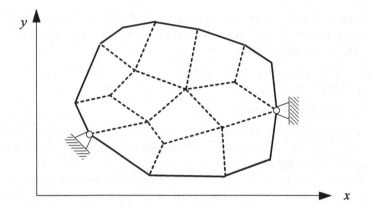

Fig. 2.5 Discretization of a plane area

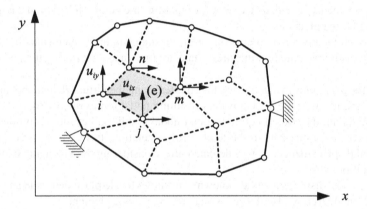

Fig. 2.6 Nodes with displacements

The displacements are identical for each on the node neighboring elements. Forces are only transferred via the nodes, no forces are transferred via the element boundaries even though the element boundaries are geometrically identical.

Approximation of the Displacement Field

A typical way to describe the displacement field $u^e(x)$ inside an element is to approximate the field through the displacements at the nodes and so-called shape functions (see Fig. 2.7):

$$u^e(x) = N(x)\, u_p \,. \tag{2.15}$$

The discretization must not lead to gaps in the continuum. To ensure the compatibility between single elements a suitable description of the displacement field has to be

Fig. 2.7 Approximation of the displacement field in the element

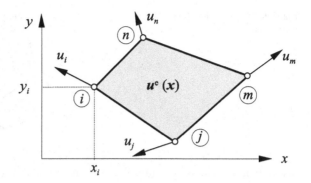

Fig. 2.8 Displacement, strain and stress in the element

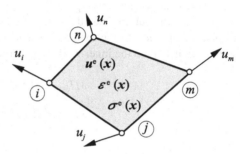

chosen. The choice of the shape functions has a significant influence on the quality of the approximation and will be discussed in detail in Sect. 6.4.

Strain Field and Stress Field

From the displacement field $u^e(x)$ one can get to the strain field

$$\varepsilon^e(x) = \mathcal{L}_1 u^e(x) \tag{2.16}$$

via the above kinematics relation. Thereby $\mathcal{L}_1$ is a differential operator (Fig. 2.8).[2] The stress within an element can be determined via the constitutive equation:

$$\sigma^e(x) = D\varepsilon^e(x) = D\mathcal{L}_1 N(x) u_p = DB(x) u_p. \tag{2.17}$$

The expression $\mathcal{L}_1 N(x)$ contains the derivatives of the shape functions. Usually a new matrix termed B is introduced.

Principle of Virtual Work, Single Stiffness Matrices

While any point can interact with a neighboring point within the continuum, this is only possible within an element in the discretized structure. A direct exchange beyond the element boundaries is not foreseen. The principle of virtual work represents an

---

[2]In the one-dimensional case, the differential operator simplifies to the first-order derivative $\frac{d}{dx}$.

Fig. 2.9 Principle of virtual work at one element

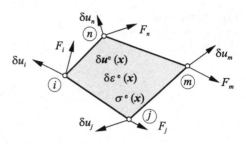

appropriate tool to assign statically equivalent nodal forces to the stress along the virtual element boundaries (Fig. 2.9).

For this, one summarizes the nodal forces to a column matrix F_p^e. The virtual displacements δu_p do the external virtual work $\delta \Pi_{ext}$ with the nodal forces, the virtual strains $\delta \varepsilon$ do the inner work $\delta \Pi_{int}$ with the stresses σ^e inside:

$$\delta \Pi_{ext} = (F_p^e)^T \, \delta u_p \,,$$

$$\delta \Pi_{int} = \int_\Omega (\sigma^e)^T \, \delta \varepsilon^e \mathrm{d}\Omega. \tag{2.18}$$

According to the principle of virtual work the following is valid:

$$\delta \Pi_{ext} = \delta \Pi_{int} \,. \tag{2.19}$$

If one transposes the equation to

$$(\delta u_p)^T \, F_p^e = \int_\Omega (\delta \varepsilon^e)^T \, \sigma^e \, \mathrm{d}\Omega$$

and if one inserts (2.16) and (2.17) accordingly, this yields

$$(\delta u_p)^T \, F_p^e = (\delta u_p)^T \int_\Omega B^T \, D \, B \, \mathrm{d}\Omega \, u_p. \tag{2.20}$$

From this one receives the single stiffness relation

$$F_p^e = k^e \, u_p \tag{2.21}$$

with the element stiffness matrix

$$k^e = \int_\Omega B^T \, D \, B \mathrm{d}\Omega \,. \tag{2.22}$$

Total Stiffness Relation

One receives the total stiffness relation

$$F = K u \tag{2.23}$$

from the overall equilibrium. This can be achieved by setting up the equilibrium on every single node. The unknown displacements cannot be gained from the total stiffness relation yet. In the context of equation solving the system matrix is not regular. Only after taking at least the rigid-body motion (translation and rotation) from the overall system, a reduced system results

$$F^{\text{red}} = K^{\text{red}} u_{\text{p}}^{\text{red}} , \tag{2.24}$$

which can be solved. A description of the equation solution can be found in Sect. 7.2 and in the Appendix A 1.5.

Determination of Element Specific Field Parameters

After the equation's solution the nodal displacements are known. Therewith the displacement, strain and stress field inside of every single element can be defined. In addition the support reactions can be determined.

2.2 Integral Principles

The derivation of the finite element method is often based on so-called energy principles. Therefore this chapter serves as a short summary about a few important principles. The overall potential or the total potential energy of a system can generally be written as

$$\Pi = \Pi_{\text{int}} + \Pi_{\text{ext}}, \tag{2.25}$$

whereupon Π_{int} represents the elastic strain energy and Π_{ext} represents the potential of the external loads. The elastic strain energy — or work of the internal forces — results in general for linear elastic material behavior via the column matrix of the stresses and strains into:

$$\Pi_{\text{int}} = \frac{1}{2} \int_{\Omega} \sigma^{\text{T}} \varepsilon \, d\Omega . \tag{2.26}$$

The potential of the external loads — which corresponds with the negative work of the external loads — can be written as follows for the column matrix of the external loads F and the displacements u:

$$\Pi_{\text{ext}} = -\boldsymbol{F}^{\text{T}}\boldsymbol{u}. \tag{2.27}$$

• **Principle of Virtual Work**:

The principle of virtual work comprises the principle of virtual displacements and the principle of virtual forces. The principle of virtual displacements states that if an element is in equilibrium, the entire internal virtual work equals the entire external virtual work for arbitrary, compatible, small, virtual displacements, which fulfill the geometric boundary conditions:

$$\int_{\Omega} \boldsymbol{\sigma}^{\text{T}} \delta \boldsymbol{\varepsilon} \, d\Omega = \boldsymbol{F}^{\text{T}} \delta \boldsymbol{u} . \tag{2.28}$$

Accordingly the principle of virtual forces results in:

$$\int_{\Omega} \delta \boldsymbol{\sigma}^{\text{T}} \boldsymbol{\varepsilon} \, d\Omega = \delta \boldsymbol{F}^{\text{T}} \boldsymbol{u} . \tag{2.29}$$

• **Principle of Minimum of Potential Energy**:

According to this principle the overall potential energy takes an extreme value in the equilibrium state:

$$\Pi = \Pi_{\text{int}} + \Pi_{\text{ext}} = \text{minimum} . \tag{2.30}$$

• **Castigliano's Theorem**:

CASTIGLIANO's first theorem states that the partial derivative of the complementary strain energy, see Fig. 2.10a with respect to an external force F_i leads to the displacement of the force application point in the direction of this force. Accordingly it results that the partial derivative of the complementary strain energy with respect to an external moment M_i leads to the rotation of the moment application point in the direction of this moment:

$$\frac{\partial \bar{\Pi}_{\text{int}}}{\partial F_i} = u_i , \tag{2.31}$$

$$\frac{\partial \bar{\Pi}_{\text{int}}}{\partial M_i} = \varphi_i . \tag{2.32}$$

CASTIGLIANO's second theorem states that the partial derivative of the strain energy (see Fig. 2.10a) with respect to the displacements u_i leads to the force F_i in direction to the considered displacement u_i. An analogous connection is valid for the rotation and the moment:

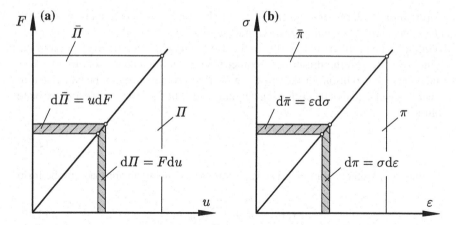

Fig. 2.10 Definition of the strain energy and the complementary strain energy: **a** absolute; **b** volume specific

$$\frac{\partial \Pi_{\text{int}}}{\partial u_i} = F_i , \tag{2.33}$$

$$\frac{\partial \Pi_{\text{int}}}{\partial \varphi_i} = M_i . \tag{2.34}$$

2.3 Weighted Residual Method

The initial point of the weighted residual method is the differential equation, which describes the physical problem. In the one-dimensional case such a physical problem within the domain Ω can in general be described via the differential equation

$$\mathcal{L}\{u^0(x)\} = b \quad (x \in \Omega) \tag{2.35}$$

as well as via the boundary conditions, which are prescribed on the boundary Γ. The differential equation is also referred to as the *strong form* of the problem since the problem is exactly described in every point x of the domain. In Eq. (2.35) $\mathcal{L}\{\ldots\}$ represents an arbitrary differential operator, which can, for example, take the following forms:

$$\mathcal{L}\{\ldots\} = \frac{d^2}{dx^2}\{\ldots\}, \tag{2.36}$$

$$\mathcal{L}\{\ldots\} = \frac{d^4}{dx^4}\{\ldots\}, \tag{2.37}$$

$$\mathcal{L}\{\ldots\} = \frac{d^4}{dx^4}\{\ldots\} + \frac{d}{dx}\{\ldots\} + \{\ldots\}. \tag{2.38}$$

Furthermore b represents a given function in Eq. (2.35), whereupon one talks about a *homogeneous differential equation* in the case of $b = 0$: $\mathcal{L}\{u^0(x)\} = 0$. The exact solution of the problem, $u^0(x)$, fulfills the differential equation in every point of the domain $x \in \Omega$ and the prescribed geometric and static boundary conditions on Γ. Since the exact solution for the most engineering problems cannot be calculated in general, it is the goal of the following derivation to define a best possible approximate solution

$$u(x) \approx u^0(x). \tag{2.39}$$

For the approximate solution in Eq. (2.39) in the following an approach in the form

$$u(x) = \alpha_0 + \sum_{k=1}^{n} \alpha_k \varphi_k(x) \tag{2.40}$$

is chosen, whereupon α_0 needs to fulfill the non-homogeneous boundary conditions, $\varphi_k(x)$ represents a set of linear independent basis functions and α_k are the free parameters of the approximation approach, which are defined via the approximation procedure in a way so that the exact solution u^0 of the approximate solution u is approximated in the best way.

2.3.1 Procedure on Basis of the Inner Product

If one incorporates the approximate formulation for u^0 into the differential equation (2.35), one receives a local error, the so-called *residual r*:

$$r = \mathcal{L}\{u(x)\} - b \neq 0. \tag{2.41}$$

Within the weighted residual method this error is weighted with a weight function $W(x)$ and is integrated over the entire domain Ω, so that the error disappears on average:

$$\int_{\Omega} W r \, d\Omega = \int_{\Omega} W(\mathcal{L}\{u(x)\} - b) \, d\Omega \stackrel{!}{=} 0. \tag{2.42}$$

This formulation is also referred to as the *inner product*. One notes that the weight or test function $W(x)$ allows to weigh the error differently within the domain Ω. However, the overall error must on average, meaning integrated over the domain, become zero. The structure of the weight function is most of the time set in a similar way as with the approximate function $u(x)$

$$W(x) = \sum_{k=1}^{n} \beta_k \psi_k(x),\tag{2.43}$$

whereupon β_k represent *arbitrary* coefficients and $\psi_k(x)$ linear independent shape functions. The approach (2.43) includes — depending on the choice of the amount of the summands k and the functions $\psi_k(x)$ — the class of the procedures with equal shape functions for the approximate solution and the weight function ($\varphi_k(k) = \psi_k(x)$) and the class of the procedures, at which the shape functions are chosen differently ($\varphi_k(k) \neq \psi_k(x)$). Depending on the choice of the weight function the following classic methods can be distinguished [2, 3]:

• **Point-Collocation Method**: $\psi_k(x) = \delta(x - x_k)$

The point-collocation method takes advantage of the properties of the delta function. The error r disappears exactly on the n freely selectable points $x_1, x_2, \ldots, x_n$, with $x_k \in \Omega$, the so-called collocation points. Therefore, the approximate solution fulfills the differential equation exactly in the collocation points. The weight function can therefore be set as

$$W(x) = \beta_1 \underbrace{\delta(x - x_1)}_{\psi_1} + \cdots + \beta_n \underbrace{\delta(x - x_n)}_{\psi_n} = \sum_{k=1}^{n} \beta_k \delta(x - x_k),\tag{2.44}$$

whereupon the delta function is defined as follows:

$$\delta(x - x_k) = \begin{cases} 0 & \text{for } x \neq x_k \\ \infty & \text{for } x = x_k \end{cases}.\tag{2.45}$$

If one incorporates this approach into the inner product according to Eq. (2.42) and considers the properties of the delta function,

$$\int_{-\infty}^{\infty} \delta(x - x_k)\,dx = \int_{x_k-\varepsilon}^{x_k+\varepsilon} \delta(x - x_k)\,dx = 1,\tag{2.46}$$

$$\int_{-\infty}^{\infty} f(x)\delta(x - x_k)\,dx = \int_{x_k-\varepsilon}^{x_k+\varepsilon} f(x)\delta(x - x_k)\,dx = f(x_k),\tag{2.47}$$

n linear independent equations result for the calculation of the free parameters α_k:

$$r(x_1) = \mathcal{L}\{u(x_1)\} - b = 0,\tag{2.48}$$
$$r(x_2) = \mathcal{L}\{u(x_2)\} - b = 0,\tag{2.49}$$

$$\vdots$$

$$r(x_n) = \mathcal{L}\{u(x_n)\} - b = 0. \tag{2.50}$$

One considers that the approximate approach has to fulfill all boundary conditions, meaning the essential and natural boundary conditions. Due to the property of the delta function, $\int_\Omega r\, W(\delta)\mathrm{d}\Omega = r = 0$, no integral has to be calculated within the point-collocation procedure, meaning no integration via the inner product. One therefore does not need to do an integration and receives the approximate solution faster — compared to the example of the GALERKIN procedure. A disadvantage is however that the collocation points can be chosen freely. These can therefore also be chosen unfavorable.

- **Subdomain-Collocation Procedure**: $\psi_k(x) = 1$ in Ω_k and otherwise zero

This procedure is as well a collocation method. However, besides the demand that the error has to disappear on certain points, here it is demanded that the integral of the error becomes zero over the different domains, the subdomains:

$$\int_{\Omega_i} r\, \mathrm{d}\Omega_i = 0 \quad \text{for a subregion } \Omega_i\,. \tag{2.51}$$

With this procedure the finite difference method can, for example, be derived.

- **Method of Least Squares**: $\psi_k(x) = \dfrac{\partial r}{\partial \alpha_k}$

The average quadratic error is optimized by the method of least squares

$$\int_\Omega (\mathcal{L}\{u(x)\} - b)^2 \mathrm{d}\Omega = \text{minimum}\,, \tag{2.52}$$

or alternatively

$$\frac{\mathrm{d}}{\mathrm{d}\alpha_k} \int_\Omega (\mathcal{L}\{u(x)\} - b)^2 \mathrm{d}\Omega = 0\,, \tag{2.53}$$

$$\int_\Omega \frac{\mathrm{d}(\mathcal{L}\{u(x)\} - b)}{\mathrm{d}\alpha_k} (\mathcal{L}\{u(x)\} - b)\mathrm{d}\Omega = 0\,. \tag{2.54}$$

- **Petrov–Galerkin Procedure**: $\psi_k(x) \neq \varphi_k(x)$

This term summarizes all procedures, at which the shape functions of the weight function and the approximate solution are different. Therefore for example the subdomain-collocation method can be allocated to this group.

- **Galerkin Procedure**: $\psi_k(x) = \varphi_k(x)$

The basic idea of the GALERKIN or BUBNOV–GALERKIN method is to choose the *same* shape function for the approximate approach and the weight function approach. Therefore, the weight function results in the following for this method:

$$W(x) = \sum_{k=1}^{n} \beta_k \varphi_k(x) \,. \tag{2.55}$$

Since the same shape functions $\varphi_k(x)$ were chosen for $u(x)$ and $W(x)$ and the coefficients β_k are arbitrary, the function $W(x)$ can be written as a variation of $u(x)$ (with $\delta\alpha_0 = 0$):

$$W(x) = \delta u(x) = \delta\alpha_1 \varphi_1(x) + \cdots + \delta\alpha_n \varphi_n(x) = \sum_{k=1}^{n} \delta\alpha_k \varphi_k(x) \,. \tag{2.56}$$

The variations can be virtual parameters, as for example virtual displacements or velocities. The incorporation of this approach into the inner product according to Eq. (2.42) yields a set of n linear independent equations for a linear operator for the definition of n unknown free parameters α_k:

$$\int_{\Omega} \left(\mathcal{L}\{u(x)\} - b \right) \cdot \varphi_1(x) \, \mathrm{d}\Omega = 0 \,, \tag{2.57}$$

$$\int_{\Omega} \left(\mathcal{L}\{u(x)\} - b \right) \cdot \varphi_2(x) \, \mathrm{d}\Omega = 0 \,, \tag{2.58}$$

$$\vdots$$

$$\int_{\Omega} \left(\mathcal{L}\{u(x)\} - b \right) \cdot \varphi_n(x) \, \mathrm{d}\Omega = 0 \,. \tag{2.59}$$

Conclusion regarding the procedure based on the inner products:

These formulations demand that the shape functions — which have been assumed to be defined over the entire domain Ω — fulfill all boundary conditions, meaning the essential and natural boundary conditions. This demand, as well as the demanded differentiability of the shape functions ($\mathcal{L}$ operator) often lead to a difficulty in finding appropriate functions in the practical application. Furthermore, in general, unsymmetric coefficient matrices occur (if the $\mathcal{L}$ operator is symmetric the coefficient matrix of the GALERKIN method is also symmetric).

2.3.2 *Procedure on Basis of the Weak Formulation*

For the derivation of another class of approximate procedures the inner product is partially integrated again and again until the derivative of $u(x)$ and $W(x)$ has the same order and one reaches the so-called *weak formulation*. Within this formulations the demand regarding the differentiability for the approximate function is diminished, the demand regarding the weight function however increased. If one uses the idea of the GALERKIN method, meaning equal shape functions for the approximate approach and the weight function, the demand regarding the differentiability of the shape functions is reduced in total.

For a differential operator of second or fourth order, meaning

$$\int_\Omega \mathcal{L}_2\{u(x)\} W(x)\, d\Omega\,, \tag{2.60}$$

$$\int_\Omega \mathcal{L}_4\{u(x)\} W(x)\, d\Omega\,, \tag{2.61}$$

a one-time partial integration of Eq. (2.60) yields the weak form

$$\int_\Omega \mathcal{L}_1\{W(x)\}\mathcal{L}_1\{u(x)\} d\Omega = [W(x)\mathcal{L}_1\{u(x)\}]_\Gamma\,, \tag{2.62}$$

or alternatively two-times partial integration the weak form of Eq. (2.61):

$$\int_\Omega \mathcal{L}_2\{W(x)\}\mathcal{L}_2\{u(x)\} d\Omega = [\mathcal{L}_1\{W(x)\}\mathcal{L}_2\{u(x)\} - W(x)\mathcal{L}_3\{u(x)\}]_\Gamma\,. \tag{2.63}$$

For the derivation of the finite element method one switches to domain-wise defined shape functions. For such a domain, meaning a finite element with $\Omega^e < \Omega$ and a local element coordinate x^e the weak formulation of (2.62) for example results in:

$$\int_{\Omega^e} \mathcal{L}_2\{W(x^e)\}\mathcal{L}_2\{u(x^e)\} d\Omega^e = \left[\mathcal{L}_1\{W(x^e)\}\mathcal{L}_2\{u(x^e)\} - W(x^e)\mathcal{L}_3\{u(x^e)\}\right]_{\Gamma^e}\,. \tag{2.64}$$

Since the weak formulation contains the natural boundary conditions — for this also see sample Problem 2.2 —, it can be demanded in the following that the approach[3] for $u(x)$ only has to fulfill the essential boundary conditions. According to the GALERKIN

---

[3] The index 'e' of the element coordinate is neglected in the following — in the case it does not affect the understanding.

method it is demanded for the derivation of the principal finite element equation that
the same shape functions for the approximate and weight function are chosen. Within
the framework of the finite element method the nodal values u_k are chosen for the free
values α_k and the shape functions $\varphi_k(x)$ are referred to as form or shape functions
$N_k(x)$. Therefore, the following illustrations result for the approximate solution and
the weight function:

$$u(x) = N_1(x)u_1 + N_2(x)u_2 + \cdots N_n(x)u_n = \sum_{k=1}^{n} N_k(x)u_k \,, \qquad (2.65)$$

$$W(x) = \delta u_1 N_1(x) + \delta u_2 N_2(x) + \cdots \delta u_n N_n(x) = \sum_{k=1}^{n} \delta u_k N_k(x) \,, \qquad (2.66)$$

whereupon n represents the number of nodes per element. It is important for this
procedure that the error on the nodes, whose position has to be defined by the user,
is minimized. This is a significant difference to the classic GALERKIN method on
the basis of the inner product, which has found the points with $r = 0$ itself. For the
further derivation of the principal finite element equation the approaches (2.65) and
(2.66) have to be written in matrix form and inserted into the weak form. For further
details of the derivation refer to the explanations in Chaps. 3 and 5 at this point.

Within the framework of the finite element method the so-called RITZ method is
often mentioned. The classic procedure takes into account the overall potential Π
of a system. Within this overall potential an approximate approach in the form of
Eq. (2.40) is used, which is however defined for the entire domain Ω in the RITZ
method. The shape functions φ_k have to fulfill the geometric, however not the static
boundary conditions[4]. Via the derivative of the potential with respect to the unknown
free parameters α_k, meaning determination of the extremum of Π, a system of equations results for the definition of k free parameters, the so-called RITZ coefficients.
In general however it is difficult to find shape functions with unknown free values,
which fulfill *all* geometric boundary conditions of the problem. However, if one
modifies the classic RITZ method in a way so that only the domain Ω^e of a finite
element is considered and one makes use of an approximate approach according to
Eq. (2.65) one also achieves the finite element method at this point.

2.3.3 Procedure on Basis of the Inverse Formulation

Finally it needs to be remarked that the inner product can be partially integrated
again and again for the derivation of another class of approximate procedures until

---

[4]Since the static boundary conditions are implicitly integrated in the overall potential, the shape
functions do not have to fulfill those. However, if the shape functions fulfill the static boundary
conditions additionally, an even more precise approximation can be achieved.

the derivative of $u(x)$ can be completely shifted onto $W(x)$. Therewith one achieves the so-called *inverse formulation*. Depending on the choice of the weight function one receives the following methods:

- Choice of W so that $\mathcal{L}(W) = 0$ or $\mathcal{L}(u) \neq 0$.
  Procedure: *Boundary element method (Boundary integral equation of the first kind)*.

- Use of a so-called fundamental solution $W = W^*$, meaning a solution, which fulfills the equation $\mathcal{L}(W^*) = (-)\delta(\xi)$.
  Procedure: *Boundary element method (Boundary integral equation of the second kind)*.
  The coefficient matrix of the corresponding system of equations is fully occupied and not symmetric. What is decisive for the application of the method is the knowledge about a fundamental solution for the $\mathcal{L}$ operator (in elasticity theory such an analytical solution is known through the KELVIN solution — concentrated load at a point of an infinite elastic medium).

- Equal shape functions for approximation approach and weight function approach.
  Procedure: TREFFTZ *method*.

- Equal shape functions for approximate approach and weight function and $\mathcal{L}(u) = \mathcal{L}(W) = 0$ is valid.
  Procedure: *Variation of the* TREFFTZ *method*.

2.4 Sample Problems

2.1 Example: Galerkin Method on Basis of the Inner Product

Since the term GALERKIN method is an often used term within the finite element method, the original GALERKIN method needs to be explained in the following within the framework of this example. For this the differential equation, which is defined in the domain $0 < x < 1$ is considered

$$\mathcal{L}\{u(x)\} - b = \frac{d^2 u^0}{dx^2} + x^2 = 0 \quad (0 < x < 1) \tag{2.67}$$

with the homogeneous boundary conditions $u^0(0) = u^0(1) = 0$. For this problem the exact solution

$$u^0(x) = \frac{x}{12}(-x^3 + 1) \tag{2.68}$$

can be defined via integration and subsequent consideration of the boundary conditions. Define the approximate solution for an approach with two free values.

2.1 Solution

For the construction of the approximate solution $u(x)$ according to the GALERKIN method the following approach with two free parameters can be made use of:

$$u^0(x) \approx u(x) = \alpha_1 \varphi_1(x) + \alpha_2 \varphi_2(x), \tag{2.69}$$

$$= \alpha_1 x(1-x) + \alpha_2 x^2(1-x), \tag{2.70}$$

$$= \alpha_1 x + (\alpha_2 - \alpha_1)x^2 - \alpha_2 x^3. \tag{2.71}$$

One needs to consider that the functions $\varphi_1(x)$ and $\varphi_2(x)$ are chosen in a way so that the boundary conditions, meaning $u(0) = u(1) = 0$, are fulfilled. Therefore, polynomials of first order are eliminated since a linear slope could only connect the two zero points as a horizontal line. Furthermore, both functions are chosen in a way so that they are linearly independent. The first derivatives of the approximation approach result in

$$\frac{du(x)}{dx} = \alpha_1 + 2(\alpha_2 - \alpha_1)x - 3\alpha_2 x^2, \tag{2.72}$$

$$\frac{d^2 u(x)}{dx^2} = 2(\alpha_2 - \alpha_1) - 6\alpha_2 x, \tag{2.73}$$

and the residual function results in the following via the second derivative from Eq. (2.41):

$$r(x) = \frac{d^2 u}{dx^2} + x^2 = 2(\alpha_2 - \alpha_1) - 6\alpha_2 x + x^2. \tag{2.74}$$

The insertion of the weight function, meaning

$$W(x) = \delta u(x) = \delta\alpha_1 x(1-x) + \delta\alpha_2 x^2(1-x), \tag{2.75}$$

into the residual equation yields

$$\int_0^1 \underbrace{\left(2(\alpha_2 - \alpha_1) - 6\alpha_2 x + x^2\right)}_{r(x)} \times \underbrace{\left(\delta\alpha_1 x(1-x) + \delta\alpha_2 x^2(1-x)\right)}_{W(x)} dx = 0 \tag{2.76}$$

or generally split into two integrals:

$$\delta\alpha_1 \int_0^1 r(x)\varphi_1(x)\,dx + \delta\alpha_2 \int_0^1 r(x)\varphi_2(x)\,dx = 0. \tag{2.77}$$

Since the $\delta\alpha_i$ are arbitrary coefficients and the shape functions $\varphi_i(x)$ are linearly independent, the following system of equations results herefrom:

$$\delta\alpha_1 \int_0^1 \left(2(\alpha_2 - \alpha_1) - 6\alpha_2 x + x^2\right) \times (x(1-x))\,dx = 0, \tag{2.78}$$

$$\delta\alpha_2 \int_0^1 \left(2(\alpha_2 - \alpha_1) - 6\alpha_2 x + x^2\right) \times \left(x^2(1-x)\right)\,dx = 0. \tag{2.79}$$

After the integration, a system of equations results for the definition of the two unknown free parameters α_1 and α_2

$$\frac{1}{20} - \frac{1}{6}\alpha_2 - \frac{1}{3}\alpha_1 = 0, \tag{2.80}$$

$$\frac{1}{30} - \frac{2}{15}\alpha_2 - \frac{1}{6}\alpha_1 = 0, \tag{2.81}$$

or alternatively in matrix notation:

$$\begin{bmatrix} \frac{1}{3} & \frac{1}{6} \\ \frac{1}{6} & \frac{2}{15} \end{bmatrix} \begin{bmatrix} \alpha_1 \\ \alpha_2 \end{bmatrix} = \begin{bmatrix} \frac{1}{20} \\ \frac{1}{30} \end{bmatrix}. \tag{2.82}$$

From this system of equations the free parameters result in $\alpha_1 = \frac{1}{15}$ and $\alpha_2 = \frac{1}{6}$. Therefore, the approximate solution and the residual function finally result in:

$$u(x) = x\left(-\frac{1}{6}x^2 + \frac{1}{10}x + \frac{1}{15}\right), \tag{2.83}$$

$$r(x) = x^2 - x + \frac{1}{5}. \tag{2.84}$$

The comparison between the approximate solution and exact solution is illustrated in Fig. 2.11a. One can see that the two solutions coincide on the boundaries — one needs to consider that the approximate approach has to fulfill the boundary conditions — as well as on two other locations.

Fig. 2.11 Approximate
solution according to the
GALERKIN method, **a** exact
solution and **b** residual as a
function of the coordinate

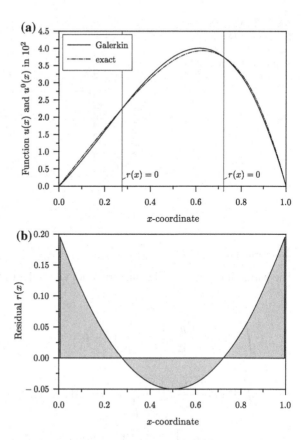

It needs to be remarked at this point that the residual function — see Fig. 2.11b — does not illustrate the difference between the exact solution and the approximate solution. Rather it is about the error, which results from inserting the approximate solution into the differential equation. To illustrate this, Fig. 2.12 shows the absolute difference between the exact solution and approximate solution.

Finally it can be summarized that the advantage of the GALERKIN method is that the procedure itself is in search of the points with $r = 0$. This is quite an advantage in comparison to the collocation method. However within the GALERKIN method the integration needs to be performed and therefore this method is in comparison to the collocation more complex and slower.

2.2 Example: Finite Element Method

For the differential equation (2.67) and the given boundary conditions one needs to calculate, based on the weak formulation, a finite element solution, based on two equidistant elements with linear shape functions.

Fig. 2.12 Absolute
difference between exact
solution and approximate
solution as function of the
coordinate

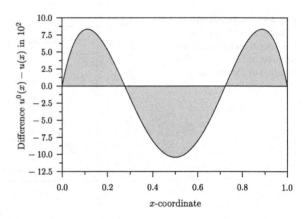

2.2 Solution

The partial integration of the inner product yields the following formulation:

$$\int_0^1 W(x) \left(\frac{d^2 u(x)}{dx^2} + x^2 \right) dx = 0, \tag{2.85}$$

$$\int_0^1 W(x) \frac{d^2 u(x)}{dx^2} dx + \int_0^1 W(x) x^2 dx = 0, \tag{2.86}$$

$$\left[W(x) \frac{du(x)}{dx} \right]_0^1 - \int_0^1 \frac{dW(x)}{dx} \frac{du(x)}{dx} dx + \int_0^1 W(x) x^2 dx = 0, \tag{2.87}$$

or alternatively the weak form in its final form:

$$\int_0^1 \frac{dW(x)}{dx} \frac{du(x)}{dx} dx = \left[W(x) \frac{du(x)}{dx} \right]_0^1 + \int_0^1 W(x) x^2 dx. \tag{2.88}$$

For the derivation of the finite element method one merges into domain-wise defined shape functions. For such a domain $\Omega^e < \Omega$, namely a finite element[5] of the length L^e, the weak formulation results in:

---

[5]Usually a separate *local* coordinate system $0 \le x^e \le L^e$ is introduced for each element 'e'. The coordinate in Eq. (2.88) is then referred to as global coordinate and receives the symbol X.

Fig. 2.13 Global coordinate system X and local coordinate system x_i for every element

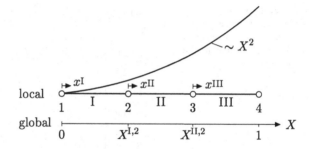

$$\int\limits_0^{L^e} \frac{dW(x^e)}{dx^e}\frac{du(x^e)}{dx^e}\,dx^e = \left[W(x^e)\frac{du(x^e)}{dx^e}\right]_0^{L^e} + \int\limits_0^{L^e} W(x^e)(x^e+c^e)^2\,dx^e. \quad (2.89)$$

In the transition from Eqs. (2.88) to (2.89), meaning from the global formulation to the consideration on the element level, in particular the quadratic expression on the right-hand side of Eq. (2.88) needs to be considered. To ensure that the in the global coordinate system defined expression X^2 is considered appropriately in the description on the element level, a coordinate transformation has to be performed for every element 'e' via a term c^e. From Fig. 2.13 it can be seen that the term c turns zero for the first element (I) since global and local coordinate system coincide. For the second element (II) $c^{II} = X^{1,2} = \frac{1}{3}$ results with an equidistant division and $c^{III} = X^{II,2} = \frac{2}{3}$ results accordingly for the third element (III).

Since the weak formulation contains the natural boundary conditions — for this see the boundary expression in Eq. (2.89) — it can be demanded in the following that the approach for $u(x)$ has to fulfill the essential boundary conditions only. According to the GALERKIN method it is demanded for the derivation of the principal finite element equation that the same shape functions for the approximate and weight function are chosen. Within the framework of the finite element method the nodal values u_k are chosen for the free parameters α_k and the shape functions $\varphi_k(x)$ are referred to as form and shape functions $N_k(x)$. For linear shape functions the following illustrations result for the approximate solution and the weight function:

$$u(x) = N_1(x)u_1 + N_2(x)u_2, \quad (2.90)$$
$$W(x) = \delta u_1 N_1(x) + \delta u_2 N_2(x). \quad (2.91)$$

For the chosen linear shape functions, an element-wise linear course of the approximate function and a difference between the exact solution and the approximate approach as shown in Fig. 2.14 are obtained. It is obvious that the error is minimal on the nodes, at the best identical with the exact solution.

Fig. 2.14 Absolute
difference between exact
solution and finite element
solution as a function of the
coordinate

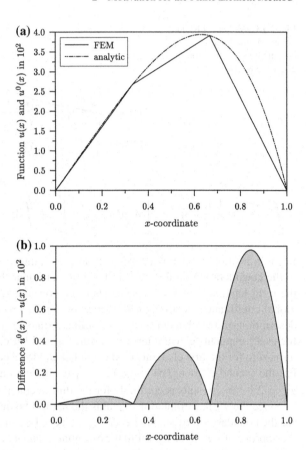

References

1. Kuhn G, Winter W(1993) Skriptum Festigkeitslehre Universität Erlangen-Nürnberg
2. Brebbia CA, Telles JCF, Wrobel LC (1984) Boundary element techniques: theory and applications. Springer, Berlin
3. Zienkiewicz OC, Taylor RL (2000) The finite element method volume 1: the basis. Butterworth-Heinemann, Oxford

Chapter 3
Bar Element

Abstract The bar element describes the basic load cases tension and compression. First, the basic equations known from the strength of materials will be introduced. Subsequently the bar element will be introduced, according to the common definitions for load and deformation quantities, which are used in the handling of the FE method. The derivation of the stiffness matrix will be described in detail. Apart from the simple prismatic bar with constant cross-section and material properties also more general bars, where the size varies along the body axis will be analyzed in examples and exercises.

3.1 Basic Description of the Bar Element

In the simplest case, the bar element can be defined as a prismatic body with constant cross-sectional area A and constant modulus of elasticity E, which is loaded with a concentrated force F in the direction of the body axis (see Fig. 3.1).

The unknown quantities are

- the extension ΔL and
- the strain ε and stress σ inside the bar

depending on the external load.

The following three basic equations are known from the strength of materials: By

$$\varepsilon(x) = \frac{\mathrm{d}u(x)}{\mathrm{d}x} = \frac{\Delta L}{L} \tag{3.1}$$

the kinematics equation describes the relation between the strains $\varepsilon(x)$ and the deformations $u(x)$. By

$$\sigma(x) = E\,\varepsilon(x) \tag{3.2}$$

© Springer International Publishing AG, part of Springer Nature 2018
A. Öchsner and M. Merkel, *One-Dimensional Finite Elements*,
https://doi.org/10.1007/978-3-319-75145-0_3

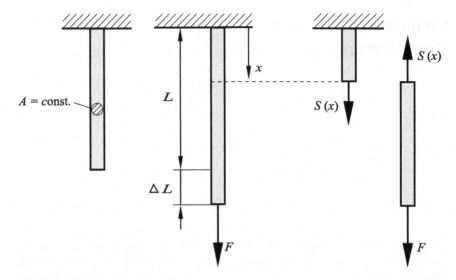

Fig. 3.1 Tensile bar loaded by a single force

the constitutive equation describes the relation between the stress $\sigma(x)$ and the strain $\varepsilon(x)$ and the equilibrium condition results in

$$\sigma(x) = \frac{S(x)}{A(x)} = \frac{S(x)}{A} = \frac{F}{A}. \tag{3.3}$$

The connection between the force F and the length variation ΔL of the bar can easily be described with these three equations:

$$\frac{F}{A} = \sigma = E\varepsilon = E\frac{\Delta L}{L} \tag{3.4}$$

or with

$$F = \frac{EA}{L}\Delta L. \tag{3.5}$$

The relation between force and length variation is described as axial stiffness. Hence, the following occurs for the bar under tensile loading[1]:

$$\frac{F}{\Delta L} = \frac{EA}{L}. \tag{3.6}$$

For the derivation of the differential equation the force equilibrium at an infinitesimal small bar element has to be regarded (see Fig. 3.2). A continuously distributed line load $q(x)$ acts as the load in the unit force per unit length.

---

[1]The parlance tension bar includes the load case compression.

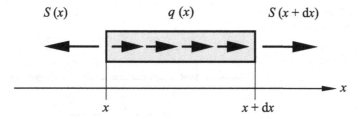

Fig. 3.2 Force equilibrium at an infinitesimal small bar element

The force equilibrium in the direction of the body axis delivers:

$$- S(x) + q(x)dx + S(x + dx) = 0. \tag{3.7}$$

After a series expansion of $S(x + dx) = S(x) + dS(x)$ the following occurs

$$- S(x) + q(x)dx + S(x) + dS(x) = 0 \tag{3.8}$$

or in short:

$$\frac{dS(x)}{dx} = -q(x). \tag{3.9}$$

The Eqs. (3.1)–(3.3) for the kinematics, the constitutive and the equilibrium relations continue to apply. If Eqs. (3.1) and (3.3) are inserted in Eq. (3.2), one obtains

$$EA(x)\frac{du(x)}{dx} = S(x). \tag{3.10}$$

After the differentiation and insertion of Eq. (3.9) one obtains

$$\frac{d}{dx}\left[EA(x)\frac{du(x)}{dx}\right] + q(x) = 0 \tag{3.11}$$

as the differential equation for a bar with continuously distributed load. This is a differential equation of 2nd order within the displacements. Under constant cross-section A and constant modulus of elasticity E the term simplifies to

$$EA\frac{d^2u(x)}{dx^2} + q(x) = 0. \tag{3.12}$$

Fig. 3.3 Definition of the
finite element tension bar

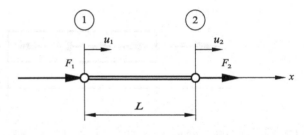

Fig. 3.4 Tension bar
modeled as a linear spring

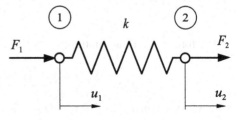

3.2   The Finite Element Tension Bar

The tension bar is defined as a prismatic body with a single body axis. Nodes are introduced at both ends of the tension bar, where forces and displacements, as sketched in Fig. 3.3 are positively defined.

The main objective is to achieve a stiffness relation for this element in the form

$$\boldsymbol{F}^{\mathrm{e}} = \boldsymbol{k}^{\mathrm{e}} \boldsymbol{u}_{\mathrm{p}}$$

or

$$\begin{bmatrix} F_1 \\ F_2 \end{bmatrix} = \begin{bmatrix} \cdot & \cdot \\ \cdot & \cdot \end{bmatrix} \begin{bmatrix} u_1 \\ u_2 \end{bmatrix}. \tag{3.13}$$

With this stiffness relation the bar element can be integrated in a structure. Furthermore the displacements, the strains and the stresses *inside* the element are unknown.

At first, an easy approach is introduced, in which the bar is modeled as a linear spring, see Fig. 3.4.

This is possible when

- the cross-sectional area A and
- the modulus of elasticity E

are constant along the body axis. The previously derived axial stiffness (see Eq. (3.4)) of the tension bar can then be interpreted as a spring constant or spring stiffness of a linear spring through

$$\frac{F}{\Delta L} = \frac{EA}{L} = k. \tag{3.14}$$

For the derivation of the stiffness relation, which is required for the finite element method, a thought experiment is conducted. If, within the spring model at first only the spring force F_2 is in effect and the spring force F_1 is being faded out, the equation

$$F_2 = k\Delta u = k(u_2 - u_1) \tag{3.15}$$

then describes the relation between the spring force and the length variation of the spring. If subsequently only the spring force F_1 is in effect and the spring force F_2 is being faded out, the equation

$$F_1 = k\Delta u = k(u_1 - u_2) \tag{3.16}$$

then describes the relation between the spring force and the length variation of the spring. Both situations can be superimposed and summarized compactly in matrix-form as

$$\begin{bmatrix} F_1 \\ F_2 \end{bmatrix} = \begin{bmatrix} k & -k \\ -k & k \end{bmatrix} \begin{bmatrix} u_1 \\ u_2 \end{bmatrix}. \tag{3.17}$$

With that the desired stiffness relation between the forces and deformations on the nodal points is derived.

The efficiency of this simple model however is limited. Thus no statements regarding the displacement, strain and stress distribution on the inside can be made. Therefore, a more elaborated model is necessary. This will be introduced in the following.

At first the displacement distribution $u^e(x)$ inside a bar will be described through shape functions $N(x)$ and the displacements u_p at the nodes:

$$u^e(x) = N(x)\, u_p. \tag{3.18}$$

In the simplest case, the displacement distribution is approximated linearly for the tension bar (see Fig. 3.5). With the following approach

$$u^e(x) = \alpha_1 + \alpha_2 x \tag{3.19}$$

the displacements at the nodes

Fig. 3.5 Linear approximation of the displacement distribution in the tension bar

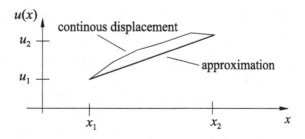

$$\begin{bmatrix} u_1 \\ u_2 \end{bmatrix} = \begin{bmatrix} 1 & x_1 \\ 1 & x_2 \end{bmatrix} \begin{bmatrix} \alpha_1 \\ \alpha_2 \end{bmatrix} \tag{3.20}$$

can be described. After the elimination of α_i the following results for the displacement distribution:

$$u^e(x) = \frac{x_2 - x}{x_2 - x_1} u_1 + \frac{x - x_1}{x_2 - x_1} u_2 \tag{3.21}$$

or summarized

$$u^e(x) = \frac{1}{L}(x_2 - x) u_1 + \frac{1}{L}(x - x_1) u_2 . \tag{3.22}$$

By this the shape functions $N_1(x)$ and $N_2(x)$ can be described with

$$N_1(x) = \frac{1}{L}(x_2 - x) \quad \text{and} \quad N_2(x) = \frac{1}{L}(x - x_1). \tag{3.23}$$

The displacement distribution results in a compact form in:

$$u^e(x) = N_1(x) u_1 + N_2(x) u_2 = [N_1 \, N_2] \begin{bmatrix} u_1 \\ u_2 \end{bmatrix} = N(x) \, u_p . \tag{3.24}$$

Through the kinematics relation the strain distribution results

$$\varepsilon^e(x) = \frac{d}{dx} u^e(x) = \frac{d}{dx} N(x) \, u_p = B \, u_p \tag{3.25}$$

and because of the constitutive equation the stress distribution results in

$$\sigma^e(x) = E\varepsilon^e(x) = E B u_p , \tag{3.26}$$

where the matrix B for the derivatives of the shape functions is introduced. For the linear approximation of the displacement distribution the derivatives of the shape functions result in:

$$\frac{d}{dx} N_1(x) = -\frac{1}{L} \quad , \quad \frac{d}{dx} N_2(x) = \frac{1}{L} \tag{3.27}$$

and therefore the matrix B results in

$$B = \frac{1}{L}[-1 \quad 1]. \tag{3.28}$$

For the derivation of the element stiffness matrix the following integral has to be evaluated

$$k^e = \int_{\Omega} B^{\mathsf{T}} D B \, d\Omega . \tag{3.29}$$

The elasticity matrix D is represented only through the modulus of elasticity E. Assuming $A = \text{const.}$:

$$\int_{\Omega} B^{\mathrm{T}} D B \, \mathrm{d}\Omega = A \int_{L} B^{\mathrm{T}} D B \, \mathrm{d}x . \tag{3.30}$$

For the tension bar, the stiffness matrix therefore results in:

$$k^{\mathrm{e}} = A E \int_{L} \frac{1}{L} \begin{bmatrix} -1 \\ 1 \end{bmatrix} \frac{1}{L} [-1 \quad 1] \, \mathrm{d}x = \frac{E A}{L^2} L \begin{bmatrix} 1 & -1 \\ -1 & 1 \end{bmatrix} . \tag{3.31}$$

In a compact form, the element stiffness matrix is evaluated as:

$$k^{\mathrm{e}} = \frac{E A}{L} \begin{bmatrix} 1 & -1 \\ -1 & 1 \end{bmatrix} . \tag{3.32}$$

There are also other ways to derive the stiffness matrix, which are introduced in the following sections.

3.2.1   Derivation Through Potential

The elastic potential energy[2] of a one-dimensional problem according to Fig. 3.1 with linear-elastic material behaviour results in:

$$\Pi_{\text{int}} = \frac{1}{2} \int_{\Omega} \varepsilon_x \sigma_x \mathrm{d}\Omega . \tag{3.33}$$

If stress and strain are substituted by use of the formulations according to Eqs. (3.25) and (3.26) and if $\mathrm{d}\Omega = A \mathrm{d}x$ is taken into consideration, the following applies:

$$\Pi_{\text{int}} = \frac{1}{2} \int_{0}^{L} E A \left(B u_{\mathrm{p}} \right)^{\mathrm{T}} B u_{\mathrm{p}} \mathrm{d}x . \tag{3.34}$$

If the relation for the transpose of a product of two matrices[3] is taken into account the following results

---

[2] The form $\Pi_{\text{int}} = \frac{1}{2} \int_{\Omega} \varepsilon^{\mathrm{T}} \sigma \mathrm{d}\Omega$ can be used in the general three-dimensional case, where σ and ε represent the column matrices with the stress and strain components.
[3] $(A B)^{\mathrm{T}} = B^{\mathrm{T}} A^{\mathrm{T}}$.

$$\Pi_{\text{int}} = \frac{1}{2} \int\limits_0^L E A u_{\text{p}}^{\text{T}} B^{\text{T}} B u_{\text{p}} \text{d}x . \tag{3.35}$$

Since the nodal values u_{p} do not represent a function of x, both column matrices can be eliminated from the integral:

$$\Pi_{\text{int}} = \frac{1}{2} u_{\text{p}}^{\text{T}} \left[\int\limits_0^L E A B^{\text{T}} B \text{d}x \right] u_{\text{p}} . \tag{3.36}$$

Under consideration of the B matrix definition according to Eq. (3.28) the following results for constant axial stiffness $E A$:

$$\Pi_{\text{int}} = \frac{1}{2} u_{\text{p}}^{\text{T}} \underbrace{\left[\frac{E A}{L^2} \int\limits_0^L \begin{bmatrix} 1 & -1 \\ -1 & 1 \end{bmatrix} \text{d}x \right]}_{k^{\text{e}}} u_{\text{p}} . \tag{3.37}$$

The last equation is equivalent to the general formulation of the potential energy of a finite element

$$\Pi_{\text{int}} = \frac{1}{2} u_{\text{p}}^{\text{T}} k^{\text{e}} u_{\text{p}} \tag{3.38}$$

and allows the identification of the element stiffness matrix k^{e} as

$$k^{\text{e}} = \frac{E A}{L} \begin{bmatrix} 1 & -1 \\ -1 & 1 \end{bmatrix} . \tag{3.39}$$

3.2.2 Derivation Through Castigliano's Theorem

If the stress in the formulation for the elastic potential energy according to Eq. (3.33) is substituted by use of HOOKE's law according to Eq. (3.2) and if $\text{d}\Omega = A\text{d}x$ is taken into consideration, the following results:

$$\Pi_{\text{int}} = \frac{1}{2} \int\limits_L E A \varepsilon_x^2 \text{d}x . \tag{3.40}$$

If now the strain is substituted using the kinematics relation according to Eq. (3.1) and introduces the approach for the displacement distribution according to Eq. (3.24), the elastic potential energy for constant axial stiffness $E A$ finally results in:

$$\Pi_{\text{int}} = \frac{EA}{2} \int_0^L \left(\frac{dN_1(x)}{dx} u_1 + \frac{dN_2(x)}{dx} u_2 \right)^2 dx . \qquad (3.41)$$

The application of CASTIGLIANO's theorem on the potential energy with reference to the nodal displacement u_1 leads to the external force F_1 on the node 1:

$$\frac{d\Pi_{\text{int}}}{du_1} = F_1 = EA \int_0^L \left(\frac{dN_1(x)}{dx} u_1 + \frac{dN_2(x)}{dx} u_2 \right) \frac{dN_1(x)}{dx} dx . \qquad (3.42)$$

From the differentiation regarding the other deformation parameter u_2 the following arises accordingly:

$$\frac{d\Pi_{\text{int}}}{du_2} = F_2 = EA \int_0^L \left(\frac{dN_1(x)}{dx} u_1 + \frac{dN_2(x)}{dx} u_2 \right) \frac{dN_2(x)}{dx} dx . \qquad (3.43)$$

Equations (3.42) and (3.43) can be summarized as the following matrix formulation:

$$EA \int_0^L \begin{bmatrix} \dfrac{dN_1(x)}{dx}\dfrac{dN_1(x)}{dx} & \dfrac{dN_2(x)}{dx}\dfrac{dN_1(x)}{dx} \\ \dfrac{dN_1(x)}{dx}\dfrac{dN_2(x)}{dx} & \dfrac{dN_2(x)}{dx}\dfrac{dN_2(x)}{dx} \end{bmatrix} dx \begin{bmatrix} u_1 \\ u_2 \end{bmatrix} = \begin{bmatrix} F_1 \\ F_2 \end{bmatrix} . \qquad (3.44)$$

After introducing the shape functions according to Eq. (3.23) and executing the integration the element stiffness matrix, which is given in Eq. (3.32), results.

3.2.3 Derivation Through the Weighted Residual Method

In the following, the differential equation for the displacement field according to Eq. (3.12) is being considered. This formulation assumes that the axial stiffness EA is constant and it results in

$$EA \frac{d^2 u^0(x)}{dx^2} + q(x) = 0 , \qquad (3.45)$$

where $u^0(x)$ represents the exact solution of the problem. The last equation with the exact solution is exactly fulfilled at every position x on the bar and is also referred to as the *strong formulation* of the problem. If the exact solution in Eq. (3.45) is substituted through an approximate solution $u(x)$, a residual or remainder r results:

$$r = EA\frac{d^2u(x)}{dx^2} + q(x) \neq 0. \tag{3.46}$$

Due to the introduction of the approximate solution $u(x)$ it is in general not possible to fulfill the differential equation at every position x of the bar. As an alternative, it is demanded in the following that the differential equation is fulfilled over a certain length (and not at every position x) and therefore ends up with the following integral demand[4]

$$\int_0^L W^T(x)\left(EA\frac{d^2u(x)}{dx^2} + q(x)\right) dx \overset{!}{=} 0, \tag{3.47}$$

which is also referred to as the *inner product*. $W(x)$ as part of Eq. (3.47) represents the so-called weight function, which distributes the error or the residual over the regarded length.

The following results through partial integration[5] of the first expression in the parentheses of Eq. (3.47)

$$\int_0^L \underbrace{W^T}_{f} EA \underbrace{\frac{d^2u(x)}{dx^2}}_{g'} dx = EA\left[W^T(x)\frac{du(x)}{dx}\right]_0^L - EA\int_0^L \frac{dW^T(x)}{dx}\frac{du(x)}{dx}dx. \tag{3.48}$$

Under consideration of Eq. (3.47) the so-called *weak formulation* of the problem results in:

$$EA\int_0^L \frac{dW^T(x)}{dx}\frac{du(x)}{dx}dx = EA\left[W^T\frac{du(x)}{dx}\right]_0^L + \int_0^L W^T(x)q(x)dx. \tag{3.49}$$

When considering the weak form it becomes obvious that one derivative of the approximate solution was shifted to the weight function through the partial integration and that now with reference to the derivatives a symmetric form arose. This symmetry with reference to the derivative of the approximate solution and the weight function will subsequently guarantee that a symmetric element stiffness matrix for the bar element results.

In the following, first the left-hand side of the Eq. (3.49) needs to be considered to derive the element stiffness matrix for a linear bar element.

The basic idea of the finite element method now is to no longer approximate the unknown displacement distribution $u(x)$ in the total domain, but to approximately describe the displacement distribution through

---

[4]The use of the transposed 'T' for the scalar weight function W is not obvious at the first glance. However, the following matrix operations will clarify this approach.

[5]A usual representation of the partial integration of two functions $f(x)$ and $g(x)$ is: $\int fg'dx = fg - \int f'gdx$.

$$u^e(x) = N(x)u_p = [N_1 \ N_2] \times \begin{bmatrix} u_1 \\ u_2 \end{bmatrix} \qquad (3.50)$$

for a subdomain, the so-called finite element. Within the context of the finite element method the same approach as for the displacement is chosen for the weight function:

$$W(x) = N(x) \, \delta u_p = [N_1 \ N_2] \times \begin{bmatrix} \delta u_1 \\ \delta u_2 \end{bmatrix}, \qquad (3.51)$$

whereupon δu_i represent the so-called arbitrary or virtual displacements. The derivative of the weight function results in

$$\frac{\mathrm{d}W(x)}{\mathrm{d}x} = \frac{\mathrm{d}}{\mathrm{d}x}\left(N \delta u_p\right) = B \delta u_p. \qquad (3.52)$$

In the following it remains to be seen that the virtual displacements can be eliminated with an identical expression on the right-hand side of Eq. (3.49) and no further consideration will be necessary at this point. When considering the approaches for the displacement and the weight function on the left-hand side of Eq. (3.49), the following results for constant axial stiffness EA:

$$EA \int_0^L \left(\delta u_p^T B^T\right)\left(B u_p\right) \mathrm{d}x \qquad (3.53)$$

or under consideration that the vector of the nodal displacement can be regarded as constant:

$$\delta u_p^T \, EA \underbrace{\int_0^L B^T B \, \mathrm{d}x}_{k^e} u_p. \qquad (3.54)$$

The expression δu_p^T can be eliminated with an identical expression on the right-hand side of the Eq. (3.49) and u_p represents the column matrix of the unknown nodal displacements. Therefore, the stiffness matrix can be calculated due to the derivative of the shape function according to Eq. (3.28) and finally the formulation according to Eq. (3.32) for the element stiffness matrix results.

In the following, the right-hand side of Eq. (3.49) is considered to derive the column matrix of the total load for a linear bar element. The first part of the right-hand half is

$$EA \left[W^T \frac{\mathrm{d}u(x)}{\mathrm{d}x} \right]_0^L \qquad (3.55)$$

with the definition of the weight function according to Eq. (3.51)

$$EA \left[\left(N\, \delta u_{\mathrm{p}}\right)^{\mathrm{T}} \frac{\mathrm{d}u(x)}{\mathrm{d}x} \right]_0^L = EA \left[\delta u_{\mathrm{p}}^{\mathrm{T}} N^{\mathrm{T}} \frac{\mathrm{d}u(x)}{\mathrm{d}x} \right]_0^L \tag{3.56}$$

results, or in components

$$\delta u_{\mathrm{p}}^{\mathrm{T}} EA \left[\begin{bmatrix} N_1 \\ N_2 \end{bmatrix} \frac{\mathrm{d}u(x)}{\mathrm{d}x} \right]_0^L . \tag{3.57}$$

The virtual displacements $\delta u_{\mathrm{p}}^{\mathrm{T}}$ in the last equation can be eliminated with the corresponding expression in Eq. (3.54). Furthermore the last equation represents a system of two equations, which have to be evaluated on the boundary of integration at $x = 0$ and $x = L$. The first row results in:

$$\left(N_1 EA \frac{\mathrm{d}u}{\mathrm{d}x} \right)_{x=L} - \left(N_1 EA \frac{\mathrm{d}u}{\mathrm{d}x} \right)_{x=0} . \tag{3.58}$$

Under consideration of the shape functions boundary values, meaning $N_1(L) = 0$ and $N_1(0) = 1$, the following results:

$$-EA \frac{\mathrm{d}u}{\mathrm{d}x} \bigg|_{x=0} \overset{(3.10)}{=} -S(x = 0) . \tag{3.59}$$

The value of the second row can be calculated accordingly:

$$EA \frac{\mathrm{d}u}{\mathrm{d}x} \bigg|_{x=L} \overset{(3.10)}{=} S(x = L) . \tag{3.60}$$

It must be noted that the forces S represent the internal reactions according to Fig. 3.2, hence the external loads with the positive direction according to Fig. 3.3 result from the internal reactions by reversing the positive direction on the left-hand section and by maintaining the positive direction of the internal reaction on the right-hand section.

The second part of the right-hand side of Eq (3.49), meaning after eliminating of δu^{T}

$$\int_0^L N(x)^{\mathrm{T}} q(x) \mathrm{d}x \tag{3.61}$$

represents the general calculation rule for the definition of the equivalent nodal loads in the case of arbitrarily distributed loads. It should be noted at this point that the evaluation of Eq. (3.61) for a constant distributed load q results in the following load matrix:

$$F_q = \frac{qL}{2} \begin{bmatrix} 1 \\ 1 \end{bmatrix}.$$ (3.62)

3.2.4   Derivation Through Virtual Work

The principle of virtual work was already briefly mentioned in Sect. 2.2. In the following, we are going to use a sub-principle, i.e. the principle of virtual displacements, to derive the week form which contains as primary unknowns the displacements [1]. Let us consider a tension bar as shown in Fig. 3.3 ($x_1 = 0$), where single forces act at both ends and which can freely deform. It should be noted at this point that the virtual displacements must fulfill the geometrical constraints.

The equilibrium equation for the domain $\Omega =]0, L[$ results from Eq. (3.9) for an infinitesimal small bar element to:

$$\frac{dS(x)}{dx} + q(x) = 0.$$ (3.63)

In a similar way, one can state the equilibrium conditions between the inner bar forces (S_i) and the external loads (F_i) at the boundaries, i.e. at $x = 0$ and $x = L$, as:

$$F_1 + S(0) = 0,$$ (3.64)
$$F_2 - S(L) = 0.$$ (3.65)

Under the influence of the so-called virtual displacements δu, the structure virtually moves away from its state of equilibrium. The resulting forces according to Eqs. (3.63)–(3.65) do some virtual work along the virtual displacements:

$$\int_0^L \left(\frac{dS(x)}{dx} + q(x) \right) \delta u dx + (F_2 - S(L)) \delta u|_L + (F_1 + S(0)) \delta u|_0 = 0.$$ (3.66)

The round brackets in the last equation can be expanded to obtain the following expression:

$$\int_0^L \frac{dS(x)}{dx} \delta u dx + \int_0^L q(x) \delta u dx + F_2 \delta u|_L - S(L) \delta u|_L + F_1 \delta u|_0 + S(0) \delta u|_0 = 0.$$
                                                                                         (3.67)

Partial integration of the left-hand expression for the internal bar force S, i.e.

$$\int\limits_0^L \frac{\mathrm{d}S(x)}{\mathrm{d}x}\delta u \mathrm{d}x = \underbrace{S(L)\delta u|_L - S(0)\delta u|_0}_{[S(x)\,\delta u]_0^L} - \int\limits_0^L S(x)\frac{\mathrm{d}\delta u}{\mathrm{d}x}\mathrm{d}x\,, \tag{3.68}$$

allows the simplification of Eq. (3.67):

$$-\int\limits_0^L S(x)\frac{\mathrm{d}\delta u}{\mathrm{d}x}\mathrm{d}x + \int\limits_0^L q(x)\delta u\mathrm{d}x + F_2\delta u|_L + F_1\delta u|_0 = 0\,. \tag{3.69}$$

Replacing in the last equation—under the assumption of constant tensile stiffness EA— the internal bar force via Eq. (3.3), i.e. $S(x) = \sigma A$, and continuing to use the constitutive equation (3.2) and the kinematics equation (3.1), one obtains the following relation:

$$EA\int\limits_0^L \frac{\mathrm{d}\delta u}{\mathrm{d}x}\frac{\mathrm{d}u}{\mathrm{d}x}\mathrm{d}x = \int\limits_0^L q(x)\delta u\mathrm{d}x + F_2\delta u|_L + F_1\delta u|_0\,. \tag{3.70}$$

Now we replace in the last equation the displacement distribution by the approximation according to Eq. (3.24) and the virtual displacements according to the approach for the weight function in Eq. (3.51), i.e. $\delta u = \delta u_\mathrm{p}^\mathrm{T} N^\mathrm{T}(x)$. Elimination of $\delta u_\mathrm{p}^\mathrm{T}$ allows to express the weak form of the problem as:

$$EA\int\limits_0^L \frac{\mathrm{d}N^\mathrm{T}}{\mathrm{d}x}\frac{\mathrm{d}N}{\mathrm{d}x}\mathrm{d}x\, u_\mathrm{p} = \int\limits_0^L N^\mathrm{T} q(x)\mathrm{d}x + F_1 N^\mathrm{T}\big|_0 + F_2 N^\mathrm{T}\big|_L\,. \tag{3.71}$$

Let us show at the end of this section that the formulation according to Eq. (3.66) can be split in the inner and external virtual work. For this, consider the intermediate step according to Eq. (3.69) and replace the bar force by $S = \sigma A = \sigma \int_A \mathrm{d}A$ to obtain the following expression for the entire virtual work:

$$\underbrace{-\int\limits_\Omega \frac{\mathrm{d}\delta u}{\mathrm{d}x}\sigma\,\mathrm{d}\Omega}_{\text{inner virtual work}} + \underbrace{\int\limits_0^L q(x)\delta u\mathrm{d}x + F_2\delta u|_L + F_1\delta u|_0}_{\text{external virtual work}} = 0\,. \tag{3.72}$$

3.3 Sample Problems and Supplementary Problems

3.3.1 Sample Problems

3.1 Tension Bar with Variable Cross-Section

So far the cross-section $A(x)$ was assumed to be constant along the body axis. As an enhancement to that the cross-section needs to be variable. The cross-section $A(x)$ should change linearly along the body axis. The modulus of elasticity is still regarded to be constant. Unknown is the stiffness matrix (Fig. 3.6).

Solution

The integral

$$k^e = \int_\Omega B^T D B \, d\Omega \tag{3.73}$$

has to be evaluated to derive the element stiffness matrix. The displacement distribution should be approximated linearly, as in the derivation above. Nothing changes for the shape functions or their derivatives. The following results for matrix B

$$B = \frac{1}{L}[-1 \quad 1]. \tag{3.74}$$

In contrast to the prismatic bar with constant cross-section, the area $A(x)$ remains under the integral. The constant modulus of elasticity E in

$$k^e = \int_L \frac{1}{L}\begin{bmatrix} -1 \\ 1 \end{bmatrix} E \frac{1}{L}[-1 \quad 1] A(x) \, dx \tag{3.75}$$

can be drawn in front of the integral. It remains:

$$k^e = \frac{E}{L^2}\begin{bmatrix} 1 & -1 \\ -1 & 1 \end{bmatrix}\int_L A(x) \, dx. \tag{3.76}$$

Fig. 3.6 Tension bar with variable cross-section

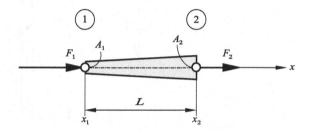

The linear course of the cross-section can be described through the following:

$$A(x) = A_1 + \frac{A_2 - A_1}{L} x. \tag{3.77}$$

After the execution of the integration

$$\int_L A(x)\, dx = \int_L \left[A_1 + \frac{A_2 - A_1}{L} x \right] dx = \frac{1}{2}(A_1 + A_2)\, L \tag{3.78}$$

the stiffness matrix

$$k^e = \frac{E}{L} \frac{A_1 + A_2}{2} \begin{bmatrix} 1 & -1 \\ -1 & 1 \end{bmatrix} \tag{3.79}$$

for a tension bar with linear changeable cross-section results.

3.2 Tension Bar under Dead Weight

Given is a bar with length L with constant cross-section A, constant modulus of elasticity E and constant density ρ along the bar axis. The bar is now loaded through its dead weight (see Fig. 3.7).
   Unknown are:

1. The analytical solution and
2. the finite element solution for a single bar element with linear approximation of the displacement distribution.

Analytical Solution for the Tension Bar under Dead Weight

Equation (3.12) is the basis for the solution. The dead weight force needs to be interpreted as a continuously distributed load $q(x)$, which is constant throughout the length of the bar:

$$q(x) = q_0 = \rho g A. \tag{3.80}$$

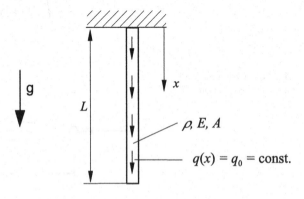

Fig. 3.7 Tension bar under dead weight

Starting from the differential equation of 2nd order

$$EA\frac{d^2}{dx^2}u(x) = EAu''(x) = -q_0 , \tag{3.81}$$

one receives the first derivative of the displacement through a one-time integration

$$EA\frac{d}{dx}u(x) = EAu'(x) = -q_0 x + c_1 , \tag{3.82}$$

and due to a further integration one receives the function of the displacement:

$$EAu(x) = -\frac{1}{2}q_0 x^2 + c_1 x + c_2 . \tag{3.83}$$

The constants of integration c_1 and c_2 are adjusted through the boundary conditions. The displacement is zero at the fixed support and the following applies:

$$u(x = 0) = 0 \implies c_2 = 0 . \tag{3.84}$$

The end of the bar is without force and the following results from Eq. (3.82):

$$EAu'(x = L) = 0 \implies -q_0 L + c_1 = 0 \implies c_1 = q_0 L . \tag{3.85}$$

If the constants of integration c_1 and c_2 are inserted with the term for the distributed load, the following results for the displacement field along the bar axis

$$u(x) = \frac{1}{EA}\left[-\frac{1}{2}q_0 x^2 + q_0 L x\right] = \frac{\rho g L^2}{E}\left[-\frac{1}{2}\left(\frac{x}{L}\right)^2 + \left(\frac{x}{L}\right)\right] . \tag{3.86}$$

The strain field is obtained from the kinematics relation

$$\varepsilon(x) = \frac{du(x)}{dx} = \frac{\rho g L}{E}\left[1 - \frac{x}{L}\right] , \tag{3.87}$$

while the stress field results from the constitutive relation:

$$\sigma(x) = E\varepsilon(x) = \rho g L\left[1 - \frac{x}{L}\right] . \tag{3.88}$$

FE Solution for the Tension Bar under Dead Weight

The basis for the finite element solution is the stiffness relation

$$\begin{bmatrix} k & -k \\ -k & k \end{bmatrix}\begin{bmatrix} u_1 \\ u_2 \end{bmatrix} = \frac{1}{2}q_0 L\begin{bmatrix} 1 \\ 1 \end{bmatrix} \tag{3.89}$$

with a linear approximation of the displacement distribution. If the formulations

$$k = \frac{EA}{L} \quad , \quad q_0 = \rho g A \tag{3.90}$$

are inserted for the stiffness k and the distributed load q_0, the following compact form results

$$\begin{bmatrix} 1 & -1 \\ -1 & 1 \end{bmatrix} \begin{bmatrix} u_1 \\ u_2 \end{bmatrix} = \frac{1}{2} \frac{\rho g L^2}{E} \begin{bmatrix} 1 \\ 1 \end{bmatrix} , \tag{3.91}$$

from which, the displacement at the lower end of the bar

$$u_2 = \frac{1}{2} \frac{\rho g L^2}{E} \tag{3.92}$$

can be read off, after introducing the boundary condition ($u_1 = 0$). The displacement at the lower end of the bar matches with the analytical solution. The displacement is assigned to be linearly distributed on the inside of the bar. The error towards the analytical solution with a quadratic distribution can be minimized or eliminated through the use of more elements or elements with quadratic shape functions.

3.3   Tension Bar under Dead Weight, Two Elements

Given is the tension bar with length L under dead weight, as in Exercise 3.2. For the determination of the solution on the basis of the FE method, *two* elements with linear shape functions should be used.

Solution

The basis for the solution is the single stiffness relation for the bar under consideration of a distributed load. One receives the total stiffness relation with two elements through the development of two single stiffness relations.[6]
    With the formulations for the stiffness k and the distributed load q_0

$$k = \frac{EA}{L} \quad , \quad q_0 = \rho g A \tag{3.93}$$

compact form results

$$\frac{EA}{\frac{1}{2}L} \begin{bmatrix} 1 & -1 & 0 \\ -1 & 1+1 & -1 \\ 0 & -1 & 1 \end{bmatrix} \begin{bmatrix} u_1 \\ u_2 \\ u_3 \end{bmatrix} = \frac{1}{2} \rho g A \frac{1}{2} L \begin{bmatrix} 1 \\ 1+1 \\ 1 \end{bmatrix} . \tag{3.94}$$

The first row and column can be eliminated due to the boundary condition ($u_1 = 0$). It remains a system of equations with two unknowns

---

[6]Here the FE solution is shown in brief. A detailed derivation for the development of a total stiffness matrix, for the introduction of boundary conditions and for the identification of the unknown is introduced in Chap. 7.

$$\begin{bmatrix} 2 & -1 \\ -1 & 1 \end{bmatrix} \begin{bmatrix} u_2 \\ u_3 \end{bmatrix} = \frac{1}{8} \frac{\rho g L^2}{E} \begin{bmatrix} 2 \\ 1 \end{bmatrix}. \tag{3.95}$$

After a short transformation

$$\begin{bmatrix} 2 & -1 \\ 0 & 1 \end{bmatrix} \begin{bmatrix} u_2 \\ u_3 \end{bmatrix} = \frac{1}{8} \frac{\rho g L^2}{E} \begin{bmatrix} 2 \\ 4 \end{bmatrix} \tag{3.96}$$

the displacement at the end node

$$u_3 = \frac{1}{2} \frac{\rho g L^2}{E} \tag{3.97}$$

and through insertion into Eq. (3.96) the displacement at the mid-node

$$u_2 = \frac{1}{2} \left[\frac{1}{2} + \frac{1}{8} \right] \frac{\rho g L^2}{E} = \frac{3}{8} \frac{\rho g L^2}{E} \tag{3.98}$$

can be identified.

3.3.2 Supplementary Problems

3.4 Tension Bar with Quadratic Approximation

Given is a prismatic tension bar with length L with constant cross-section A and modulus of elasticity E. In contrast to the derivation above, the displacement distribution on the inside of the bar element needs to be approximated through a quadratic shape function. Unknown is the stiffness matrix.

3.5 Tension Bar with Variable Cross-Section and Quadratic Approximation

The cross-section $A(x)$ changes linearly along the body axis. The modulus of elasticity is still constant. The displacement distribution on the inside of the bar element needs to be approximated through quadratic shape functions. Unknown is the stiffness matrix.

Reference

1. Argyris JH, Mlejnek H-P (1986) Die Methode der finiten Elemente in der elementaren Struktmechanik, vol 1. Verschiebungsmethode in der Statik. Friedrich Vieweg & Sohn, Braunschweig

Chapter 4
Equivalences to Tension Bar

Abstract The procedure applied to the one-dimensional tension bar can be used to describe other physical field problems. As examples, we present the one-dimensional torsion bar and the case of one-dimensional heat conduction in a bar. The first part of the chapter treats the torsion bar. First, the basic equations known from the strength of materials will be introduced. Subsequently, the torsion bar will be introduced, according to the common definitions for the torque and angle variables, which are used in the handling of the FE method. The explanations are limited to torsion bars with circular cross-section. The stiffness matrix will be derived according to the procedure for the tension bar. The second part follows a similar approach to elaborate on the hear flux bar.

4.1 Basic Description of the Torsion Bar

In the simplest case, the torsion bar can be defined as a prismatic body with constant circular cross-section (outside radius R) and constant shear modulus G, which is loaded with a torsional moment M in the direction of the body axis. Figure 4.1a illustrates the torsion bar with applied load and Fig. 4.1b shows the free body diagram.

The unknown quantities are

- the rotation $\Delta\varphi$ of the end cross-sections,
- the rotation $\varphi(x)$, the shear strain $\gamma(x)$ and the shear stress $\tau(x)$ distribution along the principal axes of the bar in dependence of the external load.

The following three basic equations are known from the strength of materials. The interrelationships of the kinematic state variables are shown in Fig. 4.2 under consideration of a cylindrical coordinate system (x, r, φ).[1]

---

[1]Besides the shear strain $\gamma_{x\varphi}(r, x)$ and the deformation $u_\varphi(x, r)$ no further deformation parameters occur during the torsion of circular cross-sections. For clarity reasons the indexing for clear dimensions is omitted.

© Springer International Publishing AG, part of Springer Nature 2018   51
A. Öchsner and M. Merkel, *One-Dimensional Finite Elements*,
https://doi.org/10.1007/978-3-319-75145-0_4

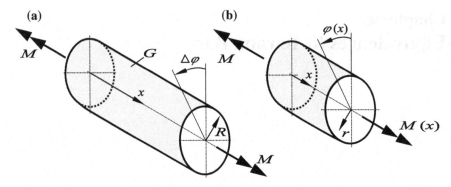

(a)   **(b)**

Fig. 4.1 Torsion bar **a** with applied load and **b** free body diagram

Fig. 4.2 Torsion bar with
state variables

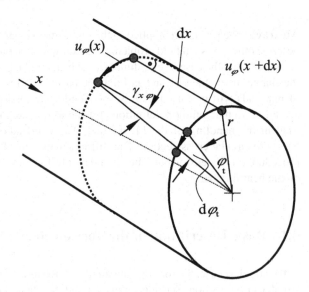

Kinematics describes the relation between the shear strain and the change of angle:

$$\gamma(x) = \frac{\mathrm{d}u_\varphi}{\mathrm{d}x} = r\,\frac{\mathrm{d}\varphi(x)}{\mathrm{d}x}. \tag{4.1}$$

The constitutive equation describes the relation between the shear stress and the shear strain with

$$\tau(x) = G\gamma(x). \tag{4.2}$$

The internal moment $M(x)$ is calculated through

$$M(x) = \int_A r\,\tau(x)\mathrm{d}A, \tag{4.3}$$

and with the kinematics relation from Eq. (4.1) and the constitutive equation from Eq. (4.2) the following results

$$M(x) = G\frac{\mathrm{d}\varphi}{\mathrm{d}x} \int_A r^2 \mathrm{d}A = G I_\mathrm{p}\frac{\mathrm{d}\varphi}{\mathrm{d}x}.$$ (4.4)

Hereby the elastic behavior regarding torsion can be described through

$$\frac{\mathrm{d}\varphi(x)}{\mathrm{d}x} = \frac{M(x)}{G I_\mathrm{p}}.$$ (4.5)

On the basis of this equation the interrelation between the rotation $\Delta\varphi$ of the two end cross-sections and the torsional moment M can be described easily:

$$\Delta\varphi = \frac{M}{G I_\mathrm{p}} L.$$ (4.6)

The expression $G I_\mathrm{p}$ is called the torsional stiffness. The stiffness for the torsion bar results from the relation between the moment and the rotation of the end cross-section:

$$\frac{M}{\Delta\varphi} = \frac{G I_\mathrm{p}}{L}.$$ (4.7)

For the derivation of the differential equation the equilibrium at the infinitesimal small torsion bar element has to be regarded (see Fig. 4.3). A continuously distributed load $m_\mathrm{t}(x)$ in the unit moment per unit length serves as the external load.

The moment equilibrium in the direction of the body axis provides the following:

$$- M(x) + m_\mathrm{t}(x)\mathrm{d}x + M(x + \mathrm{d}x) = 0.$$ (4.8)

After a series expansion of $M(x + \mathrm{d}x) = M(x) + \mathrm{d}M(x)$ the following results

$$- M(x) + m_\mathrm{t}(x)\mathrm{d}x + M(x) + \mathrm{d}M(x) = 0$$ (4.9)

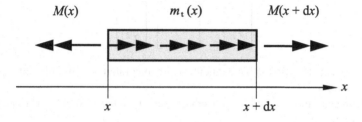

Fig. 4.3 Equilibrium at the infinitesimal small torsion bar element

or in short:

$$\frac{\mathrm{d}M(x)}{\mathrm{d}x} + m_\mathrm{t}(x) = 0. \tag{4.10}$$

Equations (4.1)–(4.3) for the kinematics, the constitutive and the equilibrium equation furthermore apply. If Eqs. (4.1) and (4.2) are inserted in Eq. (4.3), one receives

$$GI_\mathrm{p}(x) \frac{\mathrm{d}\varphi(x)}{\mathrm{d}x} = M(x). \tag{4.11}$$

After differentiating and inserting of Eq. (4.10) one obtains

$$\frac{\mathrm{d}}{\mathrm{d}x}\left[GI_\mathrm{p}(x)\frac{\mathrm{d}\varphi(x)}{\mathrm{d}x}\right] + m_\mathrm{t}(x) = 0 \tag{4.12}$$

as the differential equation for a torsion bar with continuously distributed load. This is a differential equation of 2nd order. At constant torsional stiffness GI_p the term simplifies to

$$GI_\mathrm{p}\frac{\mathrm{d}^2\varphi(x)}{\mathrm{d}x^2} + m_\mathrm{t}(x) = 0. \tag{4.13}$$

4.2   The Finite Element Torsion Bar

The handling of the torsion bar occurs analogous to the handling of the tension bar. The procedure is identical. The matrices, occurring within the frame of the FE method are similar.

The torsion bar is defined as a prismatic body with constant circular cross-section (outside radius R) along the body axis. Nodes are introduced at both ends of the torsion bar, at which moments and angles, as drafted in Fig. 4.4 are positively defined.

It is the objective to achieve a stiffness relation in the form

$$\boldsymbol{T}^\mathrm{e} = \boldsymbol{k}^\mathrm{e}\,\boldsymbol{\varphi}_\mathrm{p} \tag{4.14}$$

or

$$\begin{bmatrix} T_1 \\ T_2 \end{bmatrix} = \begin{bmatrix} \cdot & \cdot \\ \cdot & \cdot \end{bmatrix} \begin{bmatrix} \varphi_1 \\ \varphi_2 \end{bmatrix} \tag{4.15}$$

for this element. The torsion bar element can be integrated in a structure through this stiffness relation.

First, an easy approach will be introduced, at which the torsion bar is modeled as a linear torsion spring.

This is possible, when the torsional stiffness GI_p is constant along the body axis. The previously derived stiffness of the torsion bar can then be interpreted with

Fig. 4.4 Definition for the finite element torsion bar

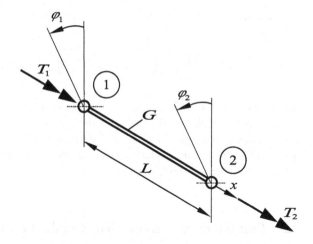

$$\frac{G\,I_p}{L} = k_t \tag{4.16}$$

as spring constant or spring stiffness of a linear torsion spring. To avoid confusion with the stiffness of the tension bar, the torsional stiffness is exposed with the index 't'.

For the derivation of the stiffness relation, which is requested for the finite element method, a thought experiment is conducted. If, within the spring model at first only the torsional moment T_2 is in effect and the moment T_1 is faded out, the equation

$$T_2 = k_t \Delta\varphi = k_t(\varphi_2 - \varphi_1) \tag{4.17}$$

then describes the relation between the spring moment and the torsion angle of the end cross-sections. If subsequently only the torsional moment T_1 is in effect and the moment T_2 is faded out, the equation

$$T_1 = k_t \Delta\varphi = k_t(\varphi_1 - \varphi_2) \tag{4.18}$$

then describes the relation between the spring moment and the torsion angle of the end cross-sections. Both situations can be superimposed and summarized compactly in matrix form as

$$\begin{bmatrix} T_1 \\ T_2 \end{bmatrix} = \begin{bmatrix} k_t & -k_t \\ -k_t & k_t \end{bmatrix} \begin{bmatrix} \varphi_1 \\ \varphi_2 \end{bmatrix}. \tag{4.19}$$

With that the desired stiffness relation between the torsional moments and the rotations on the nodal points is derived.

The element stiffness matrix for the finite element torsion bar is called

$$k^e = k_t \begin{bmatrix} 1 & -1 \\ -1 & 1 \end{bmatrix} = \frac{GI_p}{L} \begin{bmatrix} 1 & -1 \\ -1 & 1 \end{bmatrix} \qquad (4.20)$$

and is similar to the stiffness matrix of the tension bar.

The field variables on the inside of the elements are approximated through the nodal values and shape functions. The derivation of this description as well as the derivation of the stiffness relation through other ways are omitted. The proceeding is identical to that for the tension bar.

4.3 Fundamentals on the Heat Conduction Bar

A simple application of steady state heat transfer is one-dimensional heat flux. The heat flux $\dot{Q}$ flows through a prismatic body with cross section A and different temperature levels at both ends. In Fig. 4.5a a bar is defined by temperature levels and heat flux at both sides. Figure 4.5b shows the free body diagram.

In the most simplest case of an ideal isolation at the skin surfaces, it is assumed that no convection occurs. There is input of heat at one end and output of heat at the other end.

Unknowns are

- the temperatures Θ_1, Θ_2 or
- the heatfluxes $\dot{Q}$ at the ends of the bar and
- the temperature distribution $\Theta(x)$ inside the bar along the axes

in dependence on defined boundary conditions for temperature and heat flux at the ends of the bar. The unit for temperature Θ is Kelvin [K], for heat flux $\dot{Q}$ is Watt [W], for heat flux density $\dot{q}$ is [W/m²] and for heat conductivity λ is [W/m K].

The Fourier law for steady state heat transfer

$$\dot{q}(x) = -\lambda \operatorname{grad} \Theta \qquad (4.21)$$

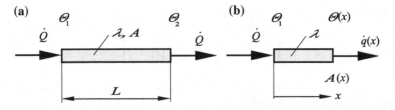

Fig. 4.5 Steady state heat flux bar **a** state variables **b** free body diagram

describes the relation between heat flux density and temperature gradient. In the one-dimensional case the equation is simplified as

$$-\lambda \frac{d\Theta(x)}{dx} = \frac{\dot{Q}(x)}{A(x)} \ . \tag{4.22}$$

The relation between the difference of temperatures $\Delta\Theta$ at both ends and the heat flux $\dot{Q}$ at the ends of the bar is formulated as

$$\dot{Q} = -\frac{\lambda A}{L}\Delta\Theta. \tag{4.23}$$

The ratio of heat flux and difference of temperatures at the end is evaluated as:

$$\frac{\dot{Q}}{\Delta\Theta} = -\frac{\lambda A}{L} \ . \tag{4.24}$$

While Eq. (4.21) represents a vector equation the other equations are scalar. In combination with the direction of the heat flux and the normal of the surface at the ends a physical correct interpretation is achieved.

In order to derive the differential equation the equilibrium is evaluated at an infinitesimal small sized bar element (see Fig. 4.6). A continuously distributed heat flux density $\dot{q}_w(x)$ acts as load, the unit is heat flux density per unit length [W/m^3].

The equilibrium in direction of the body axis provides:

$$-\dot{q}(x) + \dot{q}_w(x)\, dx + \dot{q}(x + dx) = 0 \ . \tag{4.25}$$

A series expansion of $\dot{q}(x + dx) = \dot{q}(x) + d\dot{q}(x)$ delivers

$$-\dot{q}(x) + \dot{q}_w(x)\, dx + \dot{q}(x) + d\dot{q}(x) = 0 \tag{4.26}$$

or in short:

$$-\frac{d\dot{q}(x)}{dx} + \dot{q}_w(x) = 0. \tag{4.27}$$

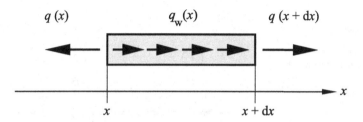

Fig. 4.6 Equilibrium at an infinitesimal small sized bar element

Inserting Eq. (4.21) in Eq. (4.27) the differential equation

$$\frac{d}{dx}\left[\lambda \frac{d\Theta(x)}{dx}\right] + \dot{q}_w(x) = 0 \tag{4.28}$$

is obtained for one-dimensional heat transfer including continuously distributed heat sources. This is an differential equation of 2nd order in the temperature. For constant heat conductivity λ the term simplifies to

$$\lambda \frac{d^2\Theta(x)}{dx^2} + \dot{q}_w(x) = 0. \tag{4.29}$$

4.4 The Finite Element Heat Flux Bar

The evolution of the one-dimensional heat flux bar appears similar to the one-dimensional tension or torsion bar. The procedure is nearly identical. Vectors and matrices have similar meanings in the frame of the FE method.

A heat flux bar is defined as prismatic body with constant cross-section A along the body axis. Nodes are introduced at both ends of the bar, at which temperatures and heat fluxes are defined by a positive orientation as drafted in Fig. 4.7.

The objective is to achieve a relation in the form

$$\dot{Q} = k^e \, \Theta \,, \tag{4.30}$$

or

$$\begin{bmatrix} \dot{Q}_1 \\ \dot{Q}_2 \end{bmatrix} = \begin{bmatrix} \cdot & \cdot \\ \cdot & \cdot \end{bmatrix} \begin{bmatrix} \Theta_1 \\ \Theta_2 \end{bmatrix}. \tag{4.31}$$

Through this relation a heat flux bar can be integrated in a FE structure.

Firstly, a simple approach without heat sources will be discussed. The heat conductivity λ and the cross-section A are assumed to be constant along the bar axis. The above evaluated equation

$$-\frac{\lambda A}{L} = k_\Theta \tag{4.32}$$

Fig. 4.7 Definition of the Finite Element heat flux bar

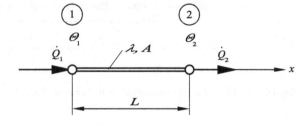

can be interpreted as a temperature stiffness. In order to avoid any misunderstanding with other stiffnesses, the index Θ is added.

A thought experiment helps to evaluate a stiffness relation required in the sense of the FE method. Firstly, the heat flux $\dot{Q}_2$ acts exclusively and the heatflux $\dot{Q}_1$ is suppressed. Then, the equation

$$\dot{Q}_2 = k_\Theta \Delta\Theta = k_\Theta (\Theta_2 - \Theta_1) \tag{4.33}$$

describes the relation between heat flux and the differences of temperatures at the bar ends. In a next step the heat flux $\dot{Q}_1$ acts and the heat flux $\dot{Q}_2$ is suppressed. Then, the equation

$$\dot{Q}_1 = k_\Theta \Delta\Theta = k_\Theta (\Theta_1 - \Theta_2) \tag{4.34}$$

describes the relation between heat flux and the differences of temperatures at the bar ends. Both situations can be superimposed. Reordering the equations leads to a compact matrix form

$$\begin{bmatrix} \dot{Q}_1 \\ \dot{Q}_2 \end{bmatrix}^e = \begin{bmatrix} k_\Theta & -k_\Theta \\ -k_\Theta & k_\Theta \end{bmatrix} \begin{bmatrix} \Theta_1 \\ \Theta_2 \end{bmatrix}. \tag{4.35}$$

This equation represents the relation between heat flux and temperatures at the nodes of a bar.

The matrix for the finite element heat flux bar

$$k^e = k_\Theta \begin{bmatrix} 1 & -1 \\ -1 & 1 \end{bmatrix} = -\frac{\lambda A}{L} \begin{bmatrix} 1 & -1 \\ -1 & 1 \end{bmatrix} \tag{4.36}$$

is similar to the stiffness matrices for the tension and torsion bar.

Temperatures inside a bar are approximated by nodal values and shape functions.

4.1 Sample Problem

A prismatic bar with length $2L$ and constant cross-section A is divided in two regions with different material properties.

For the left-hand region I the heat conductivity λ^I is assigned, for the right-hand region II the heat conductivity λ^{II}. At the left-hand end (position B) the temperature is 500 K, at the right-hand (position D) a heat flux of 0.1 W is defined (Figs. 4.8 and 4.9).

Assuming one-dimensional heat transfer, the temperatures at the coupling point between both regions (position C) and at the right-hand end are unknowns as well as the heat flux at the left-hand end.

Defined: $L = 1\,\mathrm{m}$, $A = 10^{-4}\,\mathrm{m}^2$, $\Theta_B = 500\,\mathrm{K}$, $\dot{Q}_D = 1\,\mathrm{W}$,
$\lambda^I = 100\,\mathrm{W/mK}$, $\lambda^{II} = 200\,\mathrm{W/mK}$.

Solution

The bar is discretized by two finite elements.

For the bar element I the general relation is adapted to

$$k^{\mathrm{I}} \begin{bmatrix} 1 & -1 \\ -1 & 1 \end{bmatrix} \begin{bmatrix} \Theta_1 \\ \Theta_2 \end{bmatrix}^{\mathrm{I}} = \begin{bmatrix} \dot{Q}_1 \\ \dot{Q}_2 \end{bmatrix}^{\mathrm{I}} \qquad (4.37)$$

with

$$k^{\mathrm{I}} = -\frac{\lambda^{\mathrm{I}} A}{L} . \qquad (4.38)$$

For the bar element II the general relation is adapted to

$$k^{\mathrm{II}} \begin{bmatrix} 1 & -1 \\ -1 & 1 \end{bmatrix}^{\mathrm{II}} \begin{bmatrix} \Theta_2 \\ \Theta_3 \end{bmatrix}^{\mathrm{II}} = \begin{bmatrix} \dot{Q}_2 \\ \dot{Q}_3 \end{bmatrix}^{\mathrm{II}} \qquad (4.39)$$

with

$$k^{\mathrm{II}} = -\frac{\lambda^{\mathrm{II}} A}{L} . \qquad (4.40)$$

The overall system is summarized as

$$\begin{bmatrix} k^{\mathrm{I}} & -k^{\mathrm{I}} & 0 \\ -k^{\mathrm{I}} & k^{\mathrm{I}} + k^{\mathrm{II}} & -k^{\mathrm{II}} \\ 0 & -k^{\mathrm{II}} & k^{\mathrm{II}} \end{bmatrix} \begin{bmatrix} \Theta_1 \\ \Theta_2 \\ \Theta_3 \end{bmatrix} = \begin{bmatrix} \dot{Q}_1 \\ \dot{Q}_2 \\ \dot{Q}_3 \end{bmatrix} . \qquad (4.41)$$

Including the numbers, the relation is evaluated as

Fig. 4.8 Sample for one-dimensional heat transfer

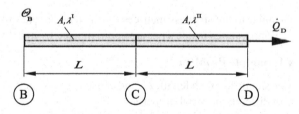

Fig. 4.9 Discretization of sample for one-dimensional heat transfer

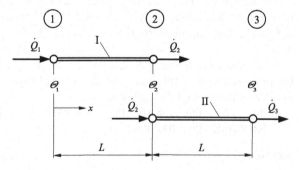

$$- 10^{-2}\,\text{W/K} \begin{bmatrix} 1 & -1 & 0 \\ -1 & 3 & -2 \\ 0 & -2 & 2 \end{bmatrix} \begin{bmatrix} 500\,\text{K} \\ \Theta_2 \\ \Theta_3 \end{bmatrix} = \begin{bmatrix} \dot{Q}_1 \\ 0 \\ 1\,\text{W} \end{bmatrix}. \tag{4.42}$$

Applying the boundary conditions the system of equations is reduced to

$$\begin{bmatrix} 3 & -2 \\ -2 & 2 \end{bmatrix} \begin{bmatrix} \Theta_2 \\ \Theta_3 \end{bmatrix} = \begin{bmatrix} 500\,\text{K} \\ -100\,\text{K} \end{bmatrix}. \tag{4.43}$$

This leads to $\Theta_2 = 400\,\text{K}$ and $\Theta_3 = 350\,\text{K}$. The heat flux $\dot{Q}_1$ is evaluated by back substitution to $-1\,\text{W}$.

4.5   Supplementary Problems

4.2   Torsion bar under torsional load

In this example a bar with rotational and constant cross-section is restraint one-sided. The other end is loaded with a torsional moment T_2. Unknown is the torsional moment T_1 at the fixed end, the torsion angle φ_2 and the shear stress.

4.3   Heat conduction bar

A bar with constant cross-section A has a temperature of 400 K on the left-hand end and a temperature of 300 K on the right-hand end. Assuming one-dimensional heat flux and ideal isolation, calculate the heat flux $\dot{Q}_1$ and $\dot{Q}_2$ at the ends of the bar. The bar is 500 mm long and has a 100 mm^2 cross-section. The heat conductivity is assumed to be 50 $\frac{W}{mK}$.

References

1. Gross D, Hauger W, Schröder J, Werner EA (2008) Hydromechanik, Elemente der Höheren Mechanik, Numerische Methoden. Springer, Berlin
3. Kwon YW, Bang H (2000) The finite element method using MATLAB. CRC Press, Boca Raton

Chapter 5
Bending Element

Abstract This element describes the basic deformation mode of bending. First, several elementary assumptions for modeling will be introduced and the beam element used in this chapter will be differentiated from other element formulations. The basic equations from the strength of materials, meaning kinematics, equilibrium and constitutive equation will be introduced and used for the derivation of the differential equation of the bending line. Analytical solutions will conclude the section of the basic principles. Subsequently, the bending element will be introduced, according to the common definitions for load and deformation parameters, which are used in the handling of the FE method. The derivation of the stiffness matrix is carried out through various methods and will be described in detail. Besides the simple, prismatic beam with constant cross-section also variable cross-sections, generalized loads between the nodes and orientation in the plane and the space will be analyzed.

5.1 Introductory Remarks

In the following, a prismatic body will be examined, at which the load occurs perpendicular to the center line and therefore bends. Perpendicular to the center line means that either the line of action of a force or the direction of a momentum vector are oriented orthogonally to the center line of the element. Consequently a different type of deformation can be modeled with this prismatic body compared to a bar (see Chaps. 3 and 4), see Table 5.1. A general element, which includes all these deformation mechanisms will be introduced in Chap. 6.

Basically, one distinguishes in beam statics between shear rigid and shear flexible models. The classic, shear rigid beam, also called the BERNOULLI beam, disregards the shear deformation from the shear force. With this modeling approach, one assumes that a cross-section, which was at the right angle to the beam axis before the deformation is also at right angles to the beam axis after the deformation, see Fig. 5.1a. Furthermore, it is assumed that a plane cross-section stays plane and unwarped. These two assumptions are also called the BERNOULLI's hypothesis. Altogether one

© Springer International Publishing AG, part of Springer Nature 2018   63
A. Öchsner and M. Merkel, *One-Dimensional Finite Elements*,
https://doi.org/10.1007/978-3-319-75145-0_5

Table 5.1 Differentiation between the bar and beam element; center line parallel to the x-axis

	Bar	Beam
Force	Along the bar axis	Perpendicular to the beam axis
Unknown	Displacement in or rotation around bar axis	Displacement perpendicular to and rotation perpendicular to the beam axis

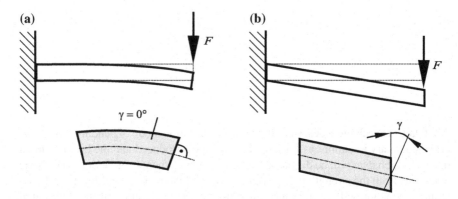

Fig. 5.1 Different deformation of a bending beam: **a** shear rigid and **b** shear flexible. Adapted from [1]

imagines the cross-sections fixed on the center line of the beam,[1] so that a change of the center line affects the entire deformation. Consequently, it is also assumed that the geometric dimensions of the cross-section[2] do not change. Regarding a shear flexible beam, also referred to as the TIMOSHENKO beam besides the bending deformation also the shear deformation is considered, and the cross-sections will be rotated by an angle γ compared to the perpendicular position, see Fig. 5.1b. In general the shear part for beams, which length is 10–20 times larger than a characteristic dimension of the cross-section[3] is disregarded in the first approximation.

The different load types, meaning a pure bending moment or a shear force, lead to different stress fractions within a beam in bending. For a BERNOULLI beam solely loading occurs through normal stresses, which rise linearly over the cross-section. Hence, a tension — alternatively compression maximum on the upper — alternatively lower side of the beam occurs, see Fig. 5.2a.

The zero crossing[4] occurs for symmetric cross-sections in the middle of the cross-section. The shear stress exhibits, for example, for a rectangular cross-section a parabolic shape and features zero at the free boundaries.

---

[1]More precisely this is the neutral fiber or the bending line.

[2]Consequently the width b and the height h of a, for example, rectangular cross-section remain the same.

[3]For this see the explanations in Chap. 8.

[4]The sum of all points with $\sigma = 0$ along the beam axis is called the neutral fiber.

(a)   **(b)**

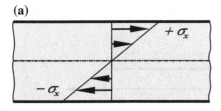
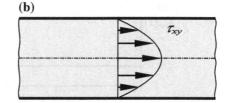

Fig. 5.2 Different stress distributions for a beam in bending using the example of a rectangular cross-section for linear-elastic material behavior: **a** normal stress; **b** shear stress

Table 5.2 Analogy between the beam and plate theories

	Beam theory	Plate theory
Dimensionality	1D	2D
Shear rigid	BERNOULLI beam	KIRCHHOFF plate
Shear flexible	TIMOSHENKO beam	REISSNER–MINDLIN plate

Finally, it needs to be noted that the one-dimensional beam theory has a counterpart in the two-dimensional space, see Table 5.2. In plate theories the BERNOULLI beam is equal to the shear rigid KIRCHHOFF plate and the TIMOSHENKO beam is equal to the shear flexible REISSNER–MINDLIN plate, see [2–4].

Further details regarding the beam theory and the corresponding basic definitions and assumptions can be found in Refs. [5–8]. In the following part of the chapter solely the BERNOULLI beam is considered. The consideration of the shear part on the deformation takes place in Chap. 8.

5.2 Basic Description of the Beam

5.2.1 Kinematics

A beam with length L under constant moment loading $M_z(x) =$ constant, meaning under *pure* bending, is considered for the derivation of the kinematics relation, see Fig. 5.3. One can see that both external single moments in the left- and right-hand beam border lead to a positive bending moment distribution $M_z(x)$ within the beam. The vertical position of a point in regards to the center line of the beam *without action* of an external load is described through the y-coordinate. The vertical *displacement* of a point on the center line of the beam, meaning for a point with $y = 0$, under action of the external load is indicated with u_y. The deformed center line is represented by the sum of these points with $y = 0$ and is referred to as the bending line $u_y(x)$.

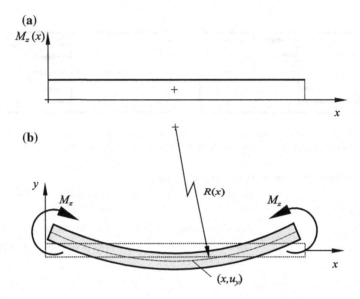

Fig. 5.3 Beam under pure bending: **a** internal bending moment distribution; **b** deformed beam. Note that the deformation is an overdrawn illustration. For the deformations considered in this chapter the following applies: $R \gg L$

Only the center line ($y = 0$) of the deformed beam is considered in the following. Through the relation for an arbitrary point (x, u_y) on a circle with radius R around the central point (x_0, y_0), meaning

$$(x - x_0)^2 + (u_y(x) - y_0)^2 = R^2 , \tag{5.1}$$

through differentiation regarding the x-coordinate, one obtains

$$2(x - x_0) + 2(u_y(x) - y_0)\frac{du_y}{dx} = 0 \tag{5.2}$$

alternatively after another differentiation:

$$2 + 2\frac{du_y}{dx}\frac{du_y}{dx} + 2(u_y(x) - y_0)\frac{d^2 u_y}{dx^2} = 0. \tag{5.3}$$

Equation (5.3) provides the vertical distance of the regarded point on the center line of the beam in regards to the center of the circle to

$$(u_y - y_0) = -\frac{1 + \left(\frac{du_y}{dx}\right)^2}{\frac{d^2 u_y}{dx^2}} , \tag{5.4}$$

while the difference of the x-coordinate results from Eq. (5.2):

$$(x - x_0) = -(u_y - y_0)\frac{du_y}{dx}.$$ (5.5)

If the expression according to Eq. (5.4) is used in Eq. (5.5) the following results:

$$(x - x_0) = \frac{du_y}{dx}\frac{1 + \left(\frac{du_y}{dx}\right)^2}{\frac{d^2u_y}{dx^2}}.$$ (5.6)

Introducing both expressions for the coordinate differences according to Eqs. (5.6) and (5.4) in the circle relation according to Eq. (5.1) leads to:

$$R^2 = (x - x_0)^2 + (u_y - y_0)^2$$ (5.7)

$$= \left(\frac{du_y}{dx}\right)^2 \frac{\left(1 + \left(\frac{du_y}{dx}\right)^2\right)^2}{\left(\frac{d^2u_y}{dx^2}\right)^2} + \frac{\left(1 + \left(\frac{du_y}{dx}\right)^2\right)^2}{\left(\frac{d^2u_y}{dx^2}\right)^2}$$

$$= \left(\left(\frac{d^2u_y}{dx^2}\right)^2 + 1\right)\frac{\left(1 + \left(\frac{du_y}{dx}\right)^2\right)^2}{\left(\frac{d^2u_y}{dx^2}\right)^2}$$

$$= \frac{\left(1 + \left(\frac{du_y}{dx}\right)^2\right)^3}{\left(\frac{d^2u_y}{dx^2}\right)^2}.$$ (5.8)

Since the circle configuration, which is shown in Fig. 5.3, is a 'left-handed curve' $\left(\frac{du_y^2}{dx^2} > 0\right)$, the radius of curvature R results in:

$$R = \frac{+}{(-)}\frac{\left(1 + \left(\frac{du_y}{dx}\right)^2\right)^{3/2}}{\left(\frac{d^2u_y}{dx^2}\right)}.$$ (5.9)

Note that the expression curvature, which results as the reciprocal value from the radius of curvature, $\kappa = \frac{1}{R}$, is used here as well. For small bending deflections, meaning $u_y \ll L$, it results that $\frac{du_y}{dx} \ll 1$ and Eq. (5.9) simplifies to:

Fig. 5.4 Segment of a beam
under pure bending. Note
that the deformation is
overdrawn for better
illustration

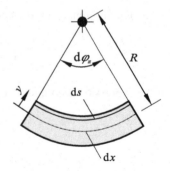

$$R = \frac{1}{\frac{d^2 u_y}{dx^2}} \quad \text{or} \quad \kappa = \frac{d^2 u_y}{dx^2}. \tag{5.10}$$

For the determination of the strain one goes back to the basic definition, meaning extension referred to the initial length. With the expression from Fig. 5.4 the longitudinal extension for a fiber with the distance y to the neutral fiber results in:

$$\varepsilon_x = \frac{ds - dx}{dx}. \tag{5.11}$$

The lengths of the circular arcs ds and dx result from the corresponding radii and the central angle in radian measure for both sectors to:

$$dx = R d\varphi_z, \tag{5.12}$$
$$ds = (R - y)d\varphi_z. \tag{5.13}$$

If these relations are used for the circular arcs in Eq. (5.11), the following results:

$$\varepsilon_x = \frac{(R - y)d\varphi_z - R d\varphi_z}{dx} = -y\frac{d\varphi_z}{dx}. \tag{5.14}$$

It results from Eq. (5.12) that $\frac{d\varphi_z}{dx} = \frac{1}{R}$ and together with relation (5.10) the strain can finally be expressed as follows:

$$\varepsilon_x = -y\frac{d^2 u_y(x)}{dx^2} \overset{(5.10)}{=} -y\kappa. \tag{5.15}$$

An alternative derivation of the kinematics relation results from consideration of Fig. 5.5.

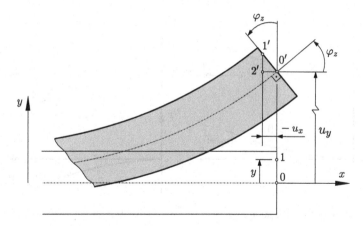

Fig. 5.5 Alternative consideration for the derivation of the kinematics relation. Note that the deformation is overdrawn for better illustration

From the relation for the rectangular triangle 0'1'2', meaning $\sin \varphi_z = \frac{-u_x}{y}$, the following[5] results for small angles ($\sin \varphi_z \approx \varphi_z$):

$$u_x = -y\varphi_z. \tag{5.16}$$

Furthermore, it continues to apply that the rotation angle of the slope equals the center line for small angles:

$$\tan \varphi_z = \frac{\mathrm{d}u_y(x)}{\mathrm{d}x} \approx \varphi_z. \tag{5.17}$$

The definition of the positive and negative rotation angle is illustrated in Fig. 5.6.
If Eqs. (5.17) and (5.16) are summarized, the following results

$$u_x = -y\frac{\mathrm{d}u_y(x)}{\mathrm{d}x}. \tag{5.18}$$

The last relation equals $(\mathrm{d}s - \mathrm{d}x)$ in Eq. (5.11) and differentiation with respect to the x-coordinate leads directly to Eq. (5.15).

5.2.2 Equilibrium

The equilibrium conditions are derived from an infinitesimal beam element of length $\mathrm{d}x$, which is loaded by a constant distributed load q_y, see Fig. 5.7. The internal

---

[5]Note that according to the assumption for the BERNOULLI beam the lengths 01 and 0'1' remain unchanged.

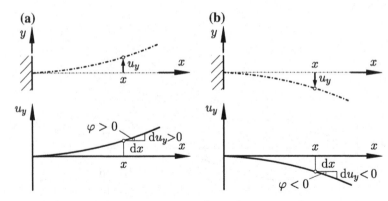

Fig. 5.6 Definition of the rotation angle: **a** $\varphi_z = \frac{\mathrm{d}u_y}{\mathrm{d}x}$ positive; **b** $\varphi_z = \frac{\mathrm{d}u_y}{\mathrm{d}x}$ negative

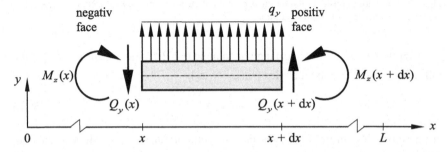

Fig. 5.7 Infinitesimal beam element with internal reactions and load through constant distributed load at deformation in the x–y plane

reactions are drawn on both cut faces, meaning the locations x and $x + \mathrm{d}x$. One can see that the internal shear force at the positive[6] cut face points in positive y-direction and that the internal bending moment features the same rotational direction as the positive z-axis (right-hand grip rule[7]). The positive orientation of the shear force and the bending moment are reversed at the negative cut face to neutralize the effect of the internal reactions in sum. This convention for the direction of the internal reactions is maintained in the following. Furthermore, it can be concluded from Fig. 5.7 that an upwards directed *external* force or alternatively a, in a mathematically sense positively rotating *external* moment on the right-hand boundary of a beam lead to a positive shear force or alternatively a positive internal moment. Accordingly it results that on the left-hand boundary of a beam a downwards directed *external* force

---

[6]The *positive* cut face is defined by the surface normal on the cutting plane which features the same orientation as the positive x-axis. It should be regarded that the surface normal is always directed outwardly. Regarding the *negative* cut face the surface normal and the positive x-axis are oriented antiparallel.

[7]If the axis is grasped with the right-hand in a way so that the spread out thumb points in the direction of the positive axis, the bent fingers then show the direction of the positive rotational direction.

or alternatively a, in a mathematically sense negatively rotating *external* moment, lead to a positive shear force or alternatively positive internal moment.

Let us state in the following the vertical force equilibrium. Assuming that forces in the direction of the positive y-axis are applied positively, the following results:

$$- Q(x) + Q(x + dx) + q_y dx = 0. \tag{5.19}$$

If the shear force on the right-hand face is expanded in a TAYLOR's series of first order, meaning

$$Q(x + dx) \approx Q(x) + \frac{dQ(x)}{dx} dx, \tag{5.20}$$

Equation (5.19) results in

$$- Q(x) + Q(x) + \frac{dQ(x)}{dx} dx + q_y dx = 0, \tag{5.21}$$

or alternatively after simplification finally to:

$$\frac{dQ(x)}{dx} = -q_y. \tag{5.22}$$

For the special case that no distributed load occurs ($q_y = 0$), Eq. (5.22) simplifies to:

$$\frac{dQ(x)}{dx} = 0. \tag{5.23}$$

The equilibrium of moments around the reference point at $x + dx$ delivers:

$$M_z(x + dx) - M_z(x) + Q_y(x)dx - \frac{1}{2} q_y dx^2 = 0. \tag{5.24}$$

If the bending moment on the right-hand face is expanded into a TAYLOR's series of first order according to Eq. (5.20) and consideration that the term $\frac{1}{2} q_y dx^2$ as infinitesimal small size of higher order can be disregarded, finally the following results:

$$\frac{dM_z(x)}{dx} = -Q_y(x). \tag{5.25}$$

The combination of Eqs. (5.22) and (5.26) leads to the relation between the bending moment and the distributed load to:

$$\frac{d^2 M_z(x)}{dx^2} = -\frac{dQ_y(x)}{dx} = q_y. \tag{5.26}$$

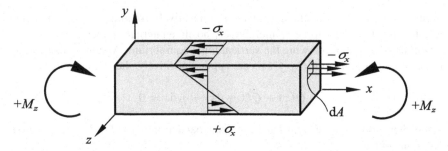

Fig. 5.8 Schematic representation of the normal stress distribution $\sigma_x = \sigma_x(y)$ of a beam. In addition, definition and position of an infinitesimal surface element for the derivation of the resulting effect of moments of the normal stress distribution

5.2.3 Constitutive Equation

The one-dimensional HOOKE's law according to Eq. (3.2) can also be used in the case of the beam in bending, since according to the assumptions only normal stresses are regarded in this chapter:

$$\sigma_x(x, y) = E\varepsilon_x . \tag{5.27}$$

Through the kinematics relation according to Eq. (5.15), the stress results as a function of the deflection to:

$$\sigma_x(x, y) = -Ey\frac{d^2u_y(x)}{dx^2} . \tag{5.28}$$

The stress distribution as shown in Fig. 5.8 generates the internal moment, which acts at this point. To calculate the effect of this moment, the stress is multiplied with a surface area, so that the resulting force is obtained. Multiplication with the corresponding lever arm then delivers the internal moment. Since this is a matter of a variable stress the consideration takes place on an infinitesimal small surface element:

$$dM_z = (+y)(-\sigma_x)dA = -y\sigma_x dA . \tag{5.29}$$

Therefore, the entire moment results via integration over the entire surface A in:

$$M_z = -\int_A y\sigma_x dA \overset{(5.28)}{=} +\int_A yEy\frac{d^2u_y}{dx^2}dA . \tag{5.30}$$

Assuming that the modulus of elasticity is constant and under the consideration of Eq. (5.10) the internal moment around the z-axis results in:

$$M_z = E \frac{\mathrm{d}^2 u_y}{\mathrm{d}x^2} \underbrace{\int_A y^2 \mathrm{d}A}_{I_z} \,. \tag{5.31}$$

The integral in Eq. (5.31) is the so-called axial second moment of area or axial surface moment of 2nd order in the SI unit m^4. This factor is only dependent on the geometry of the cross-section and is also a measure for the stiffness of a plane cross-section against bending. The values of the axial second moment of area for simple geometric cross-sections are collected in Table A.7. Consequently the internal moment can also be shown as

$$M_z = E I_z \frac{\mathrm{d}^2 u_y}{\mathrm{d}x^2} \overset{(5.10)}{=} \frac{E I_z}{R} = E I_z \kappa. \tag{5.32}$$

Equation (5.32) describes the bending line $u_y(x)$ as a function of the bending moment and is therefore also referred to as the bending line-moment relation. The product $E I_z$ in Eq. (5.32) is also called the bending stiffness. If the result from Eq. (5.32) is used in the relation for the bending stress according to Eq. (5.28), the distribution of stress across the cross-section results in:

$$\sigma_x(y) = -\frac{M_z}{I_z} y \,. \tag{5.33}$$

The minus sign causes, that after the introduced sign convention for deformations in the x–y plane, a positive bending moment (see Fig. 5.3) in the upper beam half, meaning for $y > 0$, leads to a compressive stress. The corresponding equations for a deformation in the x–z plane are collected at the end of Sect. 5.2.5. In the case of plane bending with $M_z(x) \neq$ const. the bending line can be approximated in each case locally through a circle of curvature $R(x)$, see Fig. 5.9. Therefore, the result for pure bending according to Eq. (5.32) can be transferred to the case of plane bending:

$$E I_z \frac{\mathrm{d}^2 u_y(x)}{\mathrm{d}x^2} = M_z(x) \,. \tag{5.34}$$

Finally, the three elementary basic equations for a beam under arbitrary moment loading $M_z(x)$ at bending in the x–y plane are summarized in Table 5.3.

5.2.4 Differential Equation of the Bending Line

Two times differentiation of Eq. (5.32) and consideration of the relation between the bending moment and distributed load according to Eq. (5.26) lead to the classical formulation of differential equation of the bending line,

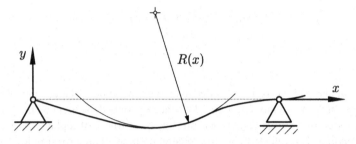

Fig. 5.9 Deformation of a BERNOULLI beam at plane bending, meaning for $M_z(x) \neq$ const.

Table 5.3 Elementary basic equations for the BERNOULLI beam at deformation in the x–y plane

Relation	Equation
Kinematics	$\varepsilon_x(x, y) = -y\dfrac{\mathrm{d}^2 u_y(x)}{\mathrm{d}x^2}$
Equilibrium	$\dfrac{\mathrm{d}Q_y(x)}{\mathrm{d}x} = -q_y(x);\ \dfrac{\mathrm{d}M_z(x)}{\mathrm{d}x} = -Q_y(x)$
Constitution	$\sigma_x(x, y) = E\varepsilon_x(x, y)$
Stress	$\sigma_x(x, y) = -\dfrac{M_z(x)}{I_z}y(x)$
Diff. equation	$EI_z\dfrac{\mathrm{d}^2 u_y(x)}{\mathrm{d}x^2} = M_z(x)$
	$EI_z\dfrac{\mathrm{d}^3 u_y(x)}{\mathrm{d}x^3} = -Q_y(x)$
	$EI_z\dfrac{\mathrm{d}^4 u_y(x)}{\mathrm{d}x^4} = q_y(x)$

$$\frac{\mathrm{d}^2}{\mathrm{d}x^2}\left(EI_z\frac{\mathrm{d}^2 u_y}{\mathrm{d}x^2}\right) = q_y,\tag{5.35}$$

which is also referred to as the bending line-distributed load relation. For a beam with constant bending stiffness EI_z along the beam axis, the following results:

$$EI_z\frac{\mathrm{d}^4 u_y}{\mathrm{d}x^4} = q_y.\tag{5.36}$$

The differential equation of the bending line can be also expressed in terms the bending moment or the shear force as

$$EI_z\frac{\mathrm{d}^2 u_y}{\mathrm{d}x^2} = M_z\quad\text{or}\tag{5.37}$$

$$EI_z\frac{\mathrm{d}^3 u_y}{\mathrm{d}x^3} = -Q_y.\tag{5.38}$$

Finally, the three different formulations for the differential equation for the BERNOULLI beam at bending in the x–y plane are summarized in Table 5.3.

5.2.5 Analytical Solutions

The analytical calculation of the bending line for simple statically determinate support cases will be considered in the following. The differential equation of the bending line has to be integrated analytically according to Eqs. (5.36), (5.37) or (5.38). The constants of integration occurring in this integration can be determined with the help of the boundary conditions, see Table 5.4.

If the distributed load (or moment or shear force distribution) cannot be represented in a closed form for the entire beam because supports, pin-joints, effects of jumps or kinks in the load function occur, the integration has to be done in sections. The additional constants of integration then have to be defined through the transition conditions. The following transition conditions (conditions of continuity) for the illustrated beam divisions in Fig. 5.10 can for example be named:

$$u_y^{\mathrm{I}}(a) = u_y^{\mathrm{II}}(a), \tag{5.39}$$

$$\frac{\mathrm{d}u_y^{\mathrm{I}}(a)}{\mathrm{d}x} = \frac{\mathrm{d}u_y^{\mathrm{II}}(a)}{\mathrm{d}x}. \tag{5.40}$$

Table 5.4 Boundary conditions at bending in the x–y plane

Symbol	Type of bearing	u_y	$\dfrac{\mathrm{d}u_y}{\mathrm{d}x}$	M_z	Q_y
	Simply supported	0	–	0	–
	Roller support	0	–	0	–
	Free end	–	–	0	0
	Fixed support	0	0	–	–
	Support with shear force link	–	0	–	0
	Spring support	$\dfrac{F}{c}$	–	0	–

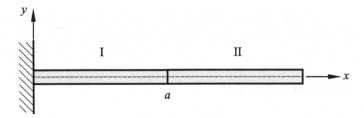

Fig. 5.10 For the definition of the transition condition between different sections of a beam

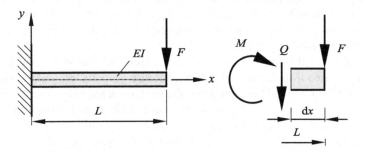

Fig. 5.11 Calculation of the bending line for the BERNOULLI beam under single force

The analytical calculation of the bending line is shown in the following for the example of a cantilever beam subjected to a single force, see Fig. 5.11. The differential equation of the bending line in the form with the fourth-order derivative according to Eq. (5.36) is chosen as an initial point. Four times integration gradually leads to the following equations:

$$EI_z \frac{d^3 u_y}{dx^3} = c_1 \, (= -Q_y) , \tag{5.41}$$

$$EI_z \frac{d^2 u_y}{dx^2} = c_1 x + c_2 \, (= M_z) , \tag{5.42}$$

$$EI_z \frac{d u_y}{dx} = \frac{1}{2} c_1 x^2 + c_2 x + c_3 , \tag{5.43}$$

$$EI_z u_y = \frac{1}{6} c_1 x^3 + \frac{1}{2} c_2 x^2 + c_3 x + c_4 , \tag{5.44}$$

where the four constants of integration $c_1, \ldots, c_4$ in the general solution have to be adjusted to the particular boundary conditions according to Fig. 5.11a.

The conditions $u_y(0) = 0$ and $\frac{d u_y(0)}{dx} = 0$ apply for the fixed support on the left-hand boundary ($x = 0$), see Table 5.4. These boundary conditions immediately result together with Eqs. (5.43) and (5.44) in $c_3 = c_4 = 0$. For the determination of the remaining constants of integration one cannot use Table 5.4. In fact the external load has to be put in relation to the internal reactions. To do so, consider the infinitesimal

Fig. 5.12 Calculation of the bending line for a beam under single force on the basis of the moment distribution

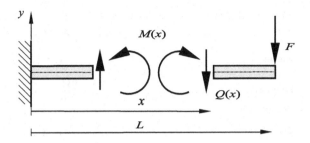

element shown in Fig. 5.11b, at which the external force F acts. The equilibrium between the external load and the internal reactions has to be formulated at position $x = L$, thus at the load application point of the external force. Therefore, the external force does not create a moment action since the case $dx \rightarrow 0$ or in other words the position $x = L$ is considered. The equilibrium of moments,[8] meaning $M_z(x = L) = 0$, together with Eq. (5.42) leads to the relation $c_2 = -c_1 L$. The vertical equilibrium of forces according to Fig. 5.11b leads to $Q_y(x = L) = -F$. Through Eq. (5.41) it results therefrom that $c_1 = F$. Consequently, the equation of the bending line can be formulated as

$$u_y(x) = \frac{1}{EI_z} \left(\frac{1}{6} F x^3 - \frac{1}{2} F L x^2 \right). \tag{5.45}$$

The maximum deflection on the right-hand boundary results for $x = L$ as:

$$u_y(L) = -\frac{FL^3}{3EI_z}. \tag{5.46}$$

The calculation of the bending line can alternatively also start, for example, from the moment distribution $M_z(x)$. To do so, the beam has to be 'cut' in two parts at an arbitrary position x, see Fig. 5.12. Subsequently it is enough to consider only one of the two parts for the equilibrium condition.

The equilibrium of moments on the right-hand part around the reference point at position x delivers $+M_z(x) + (L - x)F = 0$, or alternatively solved for the moment distribution:

$$M_z(x) = (x - L)F. \tag{5.47}$$

The differential equation of the bending line in the form with the 2nd order derivative according to Eq. (5.37) is chosen as the initial point. Two times integration gradually leads to the following equations:

---

[8] Just for the case that an external moment M^{ext} would act at position $x = L$, the internal moment would result in: $M_z(x = L) = M^{\text{ext}}$. Hereby it was assumed that the external moment M^{ext} would be positive in a mathematical sense.

Table 5.5 Elementary basic equations for the BERNOULLI beam at deformation in the x–z plane

Name	Equation
Kinematics	$\varepsilon_x(x, z) = -z \dfrac{d^2 u_z(x)}{dx^2}$
Equilibrium	$\dfrac{dQ_z(x)}{dx} = -q_z(x); \quad \dfrac{dM_y(x)}{dx} = Q_z(x)$
Constitution	$\sigma_x(x, z) = E\varepsilon_x(x, z)$
Stress	$\sigma_x(x, z) = \dfrac{M_y(x)}{I_y} z(x)$
Diff. equation	$EI_y \dfrac{d^2 u_z(x)}{dx^2} = -M_y(x)$
	$EI_y \dfrac{d^3 u_z(x)}{dx^3} = -Q_z(x)$
	$EI_y \dfrac{d^4 u_z(x)}{dx^4} = q_z(x)$

$$EI_z \frac{d^2 u_y}{dx^2} = M_z(x) = (x - L)F \,, \tag{5.48}$$

$$EI_z \frac{du_y}{dx} = \left(\frac{1}{2}x^2 - Lx \right) F + c_1 \,, \tag{5.49}$$

$$EI_z u_y(x) = \frac{1}{6}x^3 F - \frac{1}{2}Lx^2 F + c_1 x + c_2 \,. \tag{5.50}$$

The consideration of the boundary conditions on the fixed support, meaning $u_y(0) = 0$ and $\frac{du_y(0)}{dx} = 0$, finally leads to Eq. (5.45) and the maximum deflection according to Eq. (5.46).

For the case of bending in the x–z plane, the basic equations have to be slightly modified at some points since the positive orientation of the angles or rather moments are defined around the positive y-axis. The corresponding basic equations are summarized in Table 5.5 and apply irrespective of the orientation — either positive upwards or positive downwards — of the vertical z-axis.

5.3 The Finite Element Method of Plane Bending Beams

The bending element is defined as a prismatic body with the x-axis along the center line and the y-axis orthogonally to the center line. Nodes are introduced at both ends of the bending element, at which displacements and rotations or alternatively forces and moments are defined, see Fig. 5.13. The deformation and loading parameters are assumed to be positive in the drafted direction.

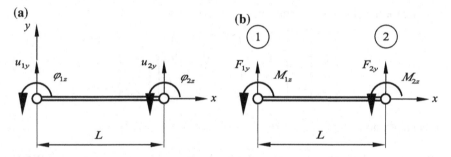

Fig. 5.13 Definition of the positive direction for the bending element at deformation in the x–y plane: **a** deformation parameters; **b** load parameters

Since deformation parameters are present at both nodes, meaning u_y and $\varphi_z = \frac{du_y}{dx}$, a polynomial with four unknown parameters will be assessed in the following for the displacement field:

$$u_y(x) = \alpha_0 + \alpha_1 x + \alpha_2 x^2 + \alpha_3 x^3 = \begin{bmatrix} 1 & x & x^2 & x^3 \end{bmatrix} \begin{bmatrix} \alpha_0 \\ \alpha_1 \\ \alpha_2 \\ \alpha_3 \end{bmatrix} = \chi^{\mathrm{T}} \alpha. \qquad (5.51)$$

The rotational field is obtained by differentiation with respect to the x-coordinate:

$$\varphi_z(x) = \frac{du_y(x)}{dx} = \alpha_1 + 2\alpha_2 x + 3\alpha_3 x^2. \qquad (5.52)$$

Evaluation of the deformation distributions $u_y(x)$ and $\varphi_z(x)$ at both nodes, meaning for $x = 0$ and $x = L$, delivers:

$$\text{Node 1:} \quad u_{1y}(0) = \alpha_0, \qquad (5.53)$$
$$\varphi_{1z}(0) = \alpha_1, \qquad (5.54)$$
$$\text{Node 2:} \quad u_{2y}(L) = \alpha_0 + \alpha_1 L + \alpha_2 L^2 + \alpha_3 L^3, \qquad (5.55)$$
$$\varphi_{2z}(L) = \alpha_1 + 2\alpha_2 L + 3\alpha_3 L^2, \qquad (5.56)$$

and in matrix notation:

$$\begin{bmatrix} u_{1y} \\ \varphi_{1z} \\ u_{2y} \\ \varphi_{2z} \end{bmatrix} = \underbrace{\begin{bmatrix} 1 & 0 & 0 & 0 \\ 0 & 1 & 0 & 0 \\ 1 & L & L^2 & L^3 \\ 0 & 1 & 2L & 3L^2 \end{bmatrix}}_{X} \begin{bmatrix} \alpha_0 \\ \alpha_1 \\ \alpha_2 \\ \alpha_3 \end{bmatrix}. \qquad (5.57)$$

Solving for the unknown coefficients $\alpha_1, \ldots, \alpha_4$ yields:

$$\begin{bmatrix} \alpha_0 \\ \alpha_1 \\ \alpha_2 \\ \alpha_3 \end{bmatrix} = \begin{bmatrix} 1 & 0 & 0 & 0 \\ 0 & 1 & 0 & 0 \\ -\frac{3}{L^2} & -\frac{2}{L} & \frac{3}{L^2} & -\frac{1}{L} \\ \frac{2}{L^3} & \frac{1}{L^2} & -\frac{2}{L^3} & \frac{1}{L^2} \end{bmatrix} \begin{bmatrix} u_{1y} \\ \varphi_{1z} \\ u_{2y} \\ \varphi_{2z} \end{bmatrix} \tag{5.58}$$

or in matrix notation:

$$\boldsymbol{\alpha} = \boldsymbol{A}\boldsymbol{u}_{\mathrm{p}} = \boldsymbol{X}^{-1}\boldsymbol{u}_{\mathrm{p}} . \tag{5.59}$$

The row matrix of the shape functions[9] results from $\boldsymbol{N} = \boldsymbol{\chi}^{\mathrm{T}}\boldsymbol{A}$ and includes the following components:

$$N_{1u}(x) = 1 - 3\left(\frac{x}{L}\right)^2 + 2\left(\frac{x}{L}\right)^3 , \tag{5.60}$$

$$N_{1\varphi}(x) = x - 2\frac{x^2}{L} + \frac{x^3}{L^2} , \tag{5.61}$$

$$N_{2u}(x) = 3\left(\frac{x}{L}\right)^2 - 2\left(\frac{x}{L}\right)^3 , \tag{5.62}$$

$$N_{2\varphi}(x) = -\frac{x^2}{L} + \frac{x^3}{L^2} . \tag{5.63}$$

A graphical illustration of the shape functions is given in Fig. 5.14.

In compact form the displacement distribution herewith results in:

$$u_y^{\mathrm{e}}(x) = N_{1u}u_{1y} + N_{1\varphi}\varphi_{1z} + N_{2u}u_{2y} + N_{2\varphi}\varphi_{2z} \tag{5.64}$$

$$= \begin{bmatrix} N_{1u} & N_{1\varphi} & N_{2u} & N_{2\varphi} \end{bmatrix} \begin{bmatrix} u_{1y} \\ \varphi_{1z} \\ u_{2y} \\ \varphi_{2z} \end{bmatrix} = \boldsymbol{N}(x)\boldsymbol{u}_{\mathrm{p}} . \tag{5.65}$$

The strain distribution results from the the kinematics relation according to Eq. (5.15) in:

$$\varepsilon_x^{\mathrm{e}}(x, y) = -y\frac{\mathrm{d}^2 u_y^{\mathrm{e}}(x)}{\mathrm{d}x^2} = -y\frac{\mathrm{d}^2}{\mathrm{d}x^2}\left(\boldsymbol{N}(x)\boldsymbol{u}_{\mathrm{p}}\right) = -y\frac{\mathrm{d}^2\boldsymbol{N}(x)}{\mathrm{d}x^2}\boldsymbol{u}_{\mathrm{p}} . \tag{5.66}$$

According to the procedure for the bar element in Chap. 3, a generalized $\boldsymbol{B}$-matrix can be introduced at this point for the bending element. Thus, one obtains an equivalent formulation as in Eq. (3.25), meaning $\varepsilon_x^{\mathrm{e}} = \boldsymbol{B}\boldsymbol{u}_{\mathrm{p}}$, with

$$\boldsymbol{B} = -y\frac{\mathrm{d}^2\boldsymbol{N}(x)}{\mathrm{d}x^2} . \tag{5.67}$$

---

[9]Alternatively the expression interpolation or form function is used.

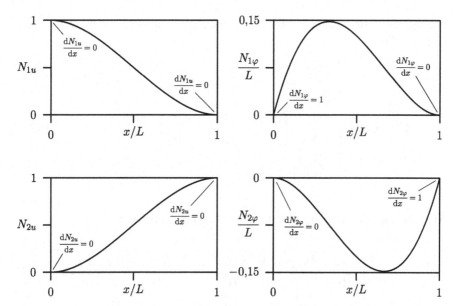

Fig. 5.14 Shape functions for the BERNOULLI element at bending in the $x-y$ plane

The stress distribution results from the constitutive law according to Eq. (5.27) in:

$$\sigma_x^e(x, y) = E\varepsilon_x^e(x, y) = E\mathbf{B}\mathbf{u}_p. \tag{5.68}$$

The general approach for the derivation of the element stiffness matrix, meaning

$$k^e = \int_\Omega \mathbf{B}^T \mathbf{D} \mathbf{B} \mathrm{d}\Omega, \tag{5.69}$$

can be further simplified. The elasticity matrix $\mathbf{D}$ reduces to the modulus of elasticity E for this one-dimensional case. Consequently the following results:

$$k^e = \int_\Omega \left(-y\frac{\mathrm{d}^2 \mathbf{N}^T(x)}{\mathrm{d}x^2}\right) E \left(-y\frac{\mathrm{d}^2 \mathbf{N}(x)}{\mathrm{d}x^2}\right) \mathrm{d}\Omega. \tag{5.70}$$

If the cross-section of the beam along the x-axis is constant, the following results:

$$k^e = E \int_L \left(\int_A y^2 \mathrm{d}A\right) \frac{\mathrm{d}^2 \mathbf{N}^T(x)}{\mathrm{d}x^2} \frac{\mathrm{d}^2 \mathbf{N}(x)}{\mathrm{d}x^2} \mathrm{d}x = E I_z \int_L \frac{\mathrm{d}^2 \mathbf{N}^T(x)}{\mathrm{d}x^2} \frac{\mathrm{d}^2 \mathbf{N}(x)}{\mathrm{d}x^2} \mathrm{d}x. \tag{5.71}$$

The stiffness matrix can be written based on matrix components for the shape functions as follows

$$
k^{e} = E I_{z} \int_{0}^{L}
\begin{bmatrix}
\dfrac{d^{2}N_{1u}}{dx^{2}} \\[2ex]
\dfrac{d^{2}N_{1\varphi}}{dx^{2}} \\[2ex]
\dfrac{d^{2}N_{2u}}{dx^{2}} \\[2ex]
\dfrac{d^{2}N_{2\varphi}}{dx^{2}}
\end{bmatrix}
\begin{bmatrix}
\dfrac{d^{2}N_{1u}}{dx^{2}} & \dfrac{d^{2}N_{1\varphi}}{dx^{2}} & \dfrac{d^{2}N_{2u}}{dx^{2}} & \dfrac{d^{2}N_{2\varphi}}{dx^{2}}
\end{bmatrix}
dx .
\tag{5.72}
$$

After all multiplications are executed, the following expression results:

$$
k^{e} = E I_{z} \int_{0}^{L}
\begin{bmatrix}
\dfrac{d^{2}N_{1u}}{dx^{2}}\dfrac{d^{2}N_{1u}}{dx^{2}} & \dfrac{d^{2}N_{1u}}{dx^{2}}\dfrac{d^{2}N_{1\varphi}}{dx^{2}} & \dfrac{d^{2}N_{1u}}{dx^{2}}\dfrac{d^{2}N_{2u}}{dx^{2}} & \dfrac{d^{2}N_{1u}}{dx^{2}}\dfrac{d^{2}N_{2\varphi}}{dx^{2}} \\[3ex]
\dfrac{d^{2}N_{1\varphi}}{dx^{2}}\dfrac{d^{2}N_{1u}}{dx^{2}} & \dfrac{d^{2}N_{1\varphi}}{dx^{2}}\dfrac{d^{2}N_{1\varphi}}{dx^{2}} & \dfrac{d^{2}N_{1\varphi}}{dx^{2}}\dfrac{d^{2}N_{2u}}{dx^{2}} & \dfrac{d^{2}N_{1\varphi}}{dx^{2}}\dfrac{d^{2}N_{2\varphi}}{dx^{2}} \\[3ex]
\dfrac{d^{2}N_{2u}}{dx^{2}}\dfrac{d^{2}N_{1u}}{dx^{2}} & \dfrac{d^{2}N_{2u}}{dx^{2}}\dfrac{d^{2}N_{1\varphi}}{dx^{2}} & \dfrac{d^{2}N_{2u}}{dx^{2}}\dfrac{d^{2}N_{2u}}{dx^{2}} & \dfrac{d^{2}N_{2u}}{dx^{2}}\dfrac{d^{2}N_{2\varphi}}{dx^{2}} \\[3ex]
\dfrac{d^{2}N_{2\varphi}}{dx^{2}}\dfrac{d^{2}N_{1u}}{dx^{2}} & \dfrac{d^{2}N_{2\varphi}}{dx^{2}}\dfrac{d^{2}N_{1\varphi}}{dx^{2}} & \dfrac{d^{2}N_{2\varphi}}{dx^{2}}\dfrac{d^{2}N_{2u}}{dx^{2}} & \dfrac{d^{2}N_{2\varphi}}{dx^{2}}\dfrac{d^{2}N_{2\varphi}}{dx^{2}}
\end{bmatrix}
dx .
\tag{5.73}
$$

The derivatives of the single shape functions in Eq. (5.73) result from Eqs. (5.60) up to (5.63):

$$
\frac{dN_{1u}(x)}{dx} = -\frac{6x}{L^{2}} + \frac{6x^{2}}{L^{3}} ,
\tag{5.74}
$$

$$
\frac{dN_{1\varphi}(x)}{dx} = 1 - \frac{4x}{L} + \frac{3x^{2}}{L^{2}} ,
\tag{5.75}
$$

$$
\frac{dN_{2u}(x)}{dx} = \frac{6x}{L^{2}} - \frac{6x^{2}}{L^{3}} ,
\tag{5.76}
$$

$$
\frac{dN_{2\varphi}(x)}{dx} = -\frac{2x}{L} + \frac{3x^{2}}{L^{2}} ,
\tag{5.77}
$$

or alternatively the second-order derivatives:

$$\frac{d^2 N_{1u}(x)}{dx^2} = -\frac{6}{L^2} + \frac{12x}{L^3} , \tag{5.78}$$

$$\frac{d^2 N_{1\varphi}(x)}{dx^2} = -\frac{4}{L} + \frac{6x}{L^2} , \tag{5.79}$$

$$\frac{d^2 N_{2u}(x)}{dx^2} = \frac{6}{L^2} - \frac{12x}{L^3} , \tag{5.80}$$

$$\frac{d^2 N_{2\varphi}(x)}{dx^2} = -\frac{2}{L} + \frac{6x}{L^2} . \tag{5.81}$$

The integration in Eq. (5.73) can be carried out analytically and, after a short calculation, the element stiffness matrix of the BERNOULLI beam in compact form results:

$$k^e = \frac{EI_z}{L^3} \begin{bmatrix} 12 & 6L & -12 & 6L \\ 6L & 4L^2 & -6L & 2L^2 \\ -12 & -6L & 12 & -6L \\ 6L & 2L^2 & -6L & 4L^2 \end{bmatrix} . \tag{5.82}$$

Taking into account the external loads and deformations shown in Fig. 5.13, the principal finite element equation on element level yields:

$$\frac{EI_z}{L^3} \begin{bmatrix} 12 & 6L & -12 & 6L \\ 6L & 4L^2 & -6L & 2L^2 \\ -12 & -6L & 12 & -6L \\ 6L & 2L^2 & -6L & 4L^2 \end{bmatrix} \begin{bmatrix} u_{1y} \\ \varphi_{1z} \\ u_{2y} \\ \varphi_{2z} \end{bmatrix} = \begin{bmatrix} F_{1y} \\ M_{1z} \\ F_{2y} \\ M_{2z} \end{bmatrix} . \tag{5.83}$$

5.3.1   Derivation through Potential

The elastic potential energy of an one-dimensional problem[10] with linear-elastic material can be expressed the following:

$$\Pi_{int} = \frac{1}{2} \int_{\Omega} \sigma_x \varepsilon_x d\Omega . \tag{5.84}$$

If stress and strain are formulated via the shape functions and the node deformations according to Eq. (5.68), the following results:

$$\Pi_{int} = \frac{1}{2} \int_{\Omega} E \left(B u_p \right)^{T} B u_p d\Omega . \tag{5.85}$$

---

[10]In the general three-dimensional case the form $\Pi_{int} = \frac{1}{2} \int_{\Omega} \varepsilon^T \sigma d\Omega$ can be applied, whereat σ and ε represent the column matrix with the stress and strain components.

If the relation for the transpose of the product of two matrices, meaning $(AB)^{\mathrm{T}} = B^{\mathrm{T}} A^{\mathrm{T}}$, is considered, the following results:

$$\Pi_{\mathrm{int}} = \frac{1}{2} \int_{\Omega} E u_{\mathrm{p}}^{\mathrm{T}} B^{\mathrm{T}} B u_{\mathrm{p}} \mathrm{d}\Omega . \tag{5.86}$$

Since the nodal values do not represent a function, the transposed column matrix of the deformations can be taken out from the integral:

$$\Pi_{\mathrm{int}} = \frac{1}{2} u_{\mathrm{p}}^{\mathrm{T}} \left[\int_{\Omega} E B^{\mathrm{T}} B \mathrm{d}\Omega \right] u_{\mathrm{p}} . \tag{5.87}$$

Through the definition for the generalized B-matrix according to Eq. (5.67) the following results here from:

$$\Pi_{\mathrm{int}} = \frac{1}{2} u_{\mathrm{p}}^{\mathrm{T}} \left[\int_{\Omega} E(-y) \frac{\mathrm{d}^2 N^{\mathrm{T}}(x)}{\mathrm{d}x^2} (-y) \frac{\mathrm{d}^2 N(x)}{\mathrm{d}x^2} \mathrm{d}\Omega \right] u_{\mathrm{p}} . \tag{5.88}$$

The axial second moment of area can also be identified at this point, so that the last equation can be written as follows:

$$\Pi_{\mathrm{int}} = \frac{1}{2} u_{\mathrm{p}}^{\mathrm{T}} \left[\int_0^L \left(\int_A y^2 \mathrm{d}A \right) E \frac{\mathrm{d}^2 N^{\mathrm{T}}(x)}{\mathrm{d}x^2} \frac{\mathrm{d}^2 N(x)}{\mathrm{d}x^2} \mathrm{d}x \right] u_{\mathrm{p}} . \tag{5.89}$$

Hence the elastic potential energy for constant material and cross-section values can be written as:

$$\Pi_{\mathrm{int}} = \frac{1}{2} u_{\mathrm{p}}^{\mathrm{T}} \underbrace{\left[E I_z \int_0^L \frac{\mathrm{d}^2 N^{\mathrm{T}}(x)}{\mathrm{d}x^2} \frac{\mathrm{d}^2 N(x)}{\mathrm{d}x^2} \mathrm{d}x \right]}_{k^{\mathrm{e}}} u_{\mathrm{p}} . \tag{5.90}$$

The last equation complies with the general formulation of the potential energy of a finite element

$$\Pi_{\mathrm{int}} = \frac{1}{2} u_{\mathrm{p}}^{\mathrm{T}} k^{\mathrm{e}} u_{\mathrm{p}} \tag{5.91}$$

and therefore allows the identification of the element stiffness matrix.

The derivation of the principal finite element equation including the stiffness matrix often takes place through extremal or variational principles as for example the principle of virtual work[11] or the HELLINGER–REISSNER principle, [10–12]. The principal finite element equation will be derived in the following by means of CASTIGLIANO's theorem,[12] see [13, 14]. The elastic potential energy will be regarded as the starting point for the derivation, see Eq. (5.84). Consideration of the kinematics relation (5.15) and the constitutive law (5.27) allows the following transformation:

$$\Pi_{\text{int}} = \frac{1}{2} \int_\Omega E\varepsilon_x^2 d\Omega = \frac{1}{2} \int_\Omega E\left(-y\frac{d^2 u_y(x)}{dx^2}\right)^2 d\Omega \tag{5.92}$$

$$= \frac{1}{2} \int_L E\left(\int_A y^2 dA\right)\left(\frac{d^2 u_y(x)}{dx^2}\right)^2 dx \tag{5.93}$$

$$= \frac{E I_z}{2} \int_0^L \left(\frac{d^2 u_y(x)}{dx^2}\right)^2 dx . \tag{5.94}$$

Through the approach for the displacement distribution according to Eq. (5.64) the following results herefrom:

$$\Pi_{\text{int}} = \frac{E I_z}{2} \int_0^L \left(\frac{d^2 N_{1u}}{dx^2} u_{1y} + \frac{d^2 N_{1\varphi}}{dx^2} \varphi_{1z} + \frac{d^2 N_{2u}}{dx^2} u_{2y} + \frac{d^2 N_{2\varphi}}{dx^2} \varphi_{2z}\right)^2 dx . \tag{5.95}$$

The application of the second CASTIGLIANO's theorem on the potential energy regarding the nodal distribution u_{1y} leads to the external force F_{1y} at node 1:

$$\frac{d\Pi_{\text{int}}}{d u_{1y}} = F_{1y} = E I_z \int_0^L \left(\frac{d^2 N_{1u}}{dx^2} u_{1y} + \frac{d^2 N_{1\varphi}}{dx^2} \varphi_{1z} + \right.$$

$$\left. + \frac{d^2 N_{2u}}{dx^2} u_{2y} + \frac{d^2 N_{2\varphi}}{dx^2} \varphi_{2z}\right) \frac{d^2 N_{1y}}{dx^2} dx . \tag{5.96}$$

Accordingly, the following results from the differentiation with regards to the other deformation parameters on the nodes:

---

[11] The principle of virtual work encompasses the principle of the virtual displacements and the principle of the virtual forces [9].

[12] CASTIGLIANO's theorems were formulated by the Italian builder, engineer and scientist Carl Alberto CASTIGLIANO (1847–1884). The second theorem signifies: The partial derivative of the stored potential energy in a linear-elastic body with regards to the displacement u_i yields the force F_i in the direction of the displacement at the considered point. An analog coherence also applies for the rotation and the moment.

$$\frac{\mathrm{d}\Pi_{\text{int}}}{\mathrm{d}\varphi_{1z}} = M_{1z} = E I_z \int_0^L \left(\frac{\mathrm{d}^2 N_{1u}}{\mathrm{d}x^2} u_{1y} + \frac{\mathrm{d}^2 N_{1\varphi}}{\mathrm{d}x^2} \varphi_{1z} + \right.$$

$$\left. + \frac{\mathrm{d}^2 N_{2u}}{\mathrm{d}x^2} u_{2y} + \frac{\mathrm{d}^2 N_{2\varphi}}{\mathrm{d}x^2} \varphi_{2z} \right) \frac{\mathrm{d}^2 N_{1\varphi}}{\mathrm{d}x^2} \mathrm{d}x \,, \qquad (5.97)$$

$$\frac{\mathrm{d}\Pi_{\text{int}}}{\mathrm{d}u_{2y}} = F_{2y} = E I_z \int_0^L \left(\frac{\mathrm{d}^2 N_{1u}}{\mathrm{d}x^2} u_{1y} + \frac{\mathrm{d}^2 N_{1\varphi}}{\mathrm{d}x^2} \varphi_{1z} + \right.$$

$$\left. + \frac{\mathrm{d}^2 N_{2u}}{\mathrm{d}x^2} u_{2y} + \frac{\mathrm{d}^2 N_{2\varphi}}{\mathrm{d}x^2} \varphi_{2z} \right) \frac{\mathrm{d}^2 N_{2y}}{\mathrm{d}x^2} \mathrm{d}x \,, \qquad (5.98)$$

$$\frac{\mathrm{d}\Pi_{\text{int}}}{\mathrm{d}u_{2\varphi}} = M_{2y} = E I_z \int_0^L \left(\frac{\mathrm{d}^2 N_{1u}}{\mathrm{d}x^2} u_{1y} + \frac{\mathrm{d}^2 N_{1\varphi}}{\mathrm{d}x^2} \varphi_{1z} + \right.$$

$$\left. + \frac{\mathrm{d}^2 N_{2u}}{\mathrm{d}x^2} u_{2y} + \frac{\mathrm{d}^2 N_{2\varphi}}{\mathrm{d}x^2} \varphi_{2z} \right) \frac{\mathrm{d}^2 N_{2\varphi}}{\mathrm{d}x^2} \mathrm{d}x \,. \qquad (5.99)$$

After carrying out the integration, Eqs. (5.96)–(5.99) can be summarized in the principal finite element equation in matrix form, see Eq. (5.83).

The total potential is also often used for the derivation of the principal finite element equation. The total potential or the entire potential energy of a beam in bending can be generally written as

$$\Pi = \Pi_{\text{int}} + \Pi_{\text{ext}} \,, \qquad (5.100)$$

where Π_{int} represents the elastic strain energy (energy of elastic deformation) and Π_{ext} the potential of the external load. The entire potential energy under influence of the external loads can be stated as follows

$$\Pi = \frac{1}{2} \int_\Omega \sigma_x \varepsilon_x \mathrm{d}\Omega - \sum_{i=1}^m F_{iy} u_{iy} - \sum_{i=1}^{m'} M_{iz} \varphi_{iz} \,, \qquad (5.101)$$

where F_{iy} and M_{iz} represent the external forces and moments acting on the nodes.

5.3.2 Derivation through Weighted Residual Method

The partial differential equation of the displacement field $u_y(x)$ according to Eq. (5.35) will be considered in the following. Let us assume for simplicity that the bending stiffness EI_z is constant and that no distributed load ($q_y = 0$) occurs. Thus, the partial differential equation of the displacement field is simplified to:

$$EI_z \frac{d^4 u_y^0(x)}{dx^4} = 0, \qquad (5.102)$$

where $u_y^0(x)$ represents the exact solution of the problem. Equation (5.102) is exactly fulfilled at every position x on the beam and is also referred to as the *strong form* of the problem. If the exact solution in Eq. (5.102) is substituted through an approximate solution $u_y(x)$ a residual or remainder r results in:

$$r(x) = EI_z \frac{d^4 u_y(x)}{dx^4} \neq 0. \qquad (5.103)$$

Due to the introduction of the approximate solution $u_y(x)$ it is in general not possible to fulfill the partial differential equation at every position x of the beam. Alternatively, it is demanded in the following that the differential equation is fulfilled throughout a certain domain (and not at every position x) and one receives the following integral[13] requirement

$$\int_0^L W^T(x) \underbrace{EI_z \frac{d^4 u_y(x)}{dx^4}}_{r(x)} dx \overset{!}{=} 0, \qquad (5.104)$$

which is also called the *inner product*. $W(x)$ in Eq. (5.104) represents the so-called weight function, which distributes the error or the residual throughout the regarded domain.

Partial integration[14] of Eq. (5.104) yields:

$$\int_0^L \underbrace{W^T}_{f} EI_z \underbrace{\frac{d^4 u_y}{dx^4}}_{g'} dx = EI_z \left[W^T \frac{d^3 u_y}{dx^3} \right]_0^L - \int_0^L EI_z \frac{dW^T}{dx} \frac{d^3 u_y}{dx^3} dx = 0. \quad (5.105)$$

Partial integration of the integral on the right-hand side of Eq. (5.105) results in:

---

[13]The use of the transposed 'T' for the scalar weight function is not obvious at the first glance. However, the following matrix operations will clarify this approach.
[14]A common representation of the partial integration of two functions $f(x)$ and $g(x)$ is: $\int fg' \, dx = fg - \int f'g dx$.

$$\int_0^L EI_z \underbrace{\frac{\mathrm{d}W^{\mathrm{T}}}{\mathrm{d}x}}_{f} \underbrace{\frac{\mathrm{d}^3 u_y}{\mathrm{d}x^3}}_{g'} \,\mathrm{d}x = EI_z \left[\frac{\mathrm{d}W^{\mathrm{T}}}{\mathrm{d}x}\frac{\mathrm{d}^2 u_y}{\mathrm{d}x^2}\right]_0^L - \int_0^L EI_z \frac{\mathrm{d}^2 W^{\mathrm{T}}}{\mathrm{d}x^2}\frac{\mathrm{d}^2 u_y}{\mathrm{d}x^2}\,\mathrm{d}x . \quad (5.106)$$

The combination of Eqs. (5.105) and (5.106) yields the *weak form* of the problem to:

$$\int_0^L EI_z \frac{\mathrm{d}^2 W^{\mathrm{T}}}{\mathrm{d}x^2}\frac{\mathrm{d}^2 u_y}{\mathrm{d}x^2}\,\mathrm{d}x = EI_z \left[-W^{\mathrm{T}}\frac{\mathrm{d}^3 u_y}{\mathrm{d}x^3} + \frac{\mathrm{d}W^{\mathrm{T}}}{\mathrm{d}x}\frac{\mathrm{d}^2 u_y}{\mathrm{d}x^2}\right]_0^L . \quad (5.107)$$

Regarding the weak form it becomes obvious that due to the partial integration, two derivatives (differential operators) were shifted from the approximate solution to the weight function and regarding the derivatives, a symmetric form results. This symmetry concerning the derivatives of the weight function and the approximate solution will guarantee in the following that a symmetric stiffness matrix results for the bending element.

First the left-hand side of Eq. (5.107) will be considered to derive the stiffness matrix for a bending element with two nodes.

The basic idea of the finite element method is not to approximate the unknown displacement u_y for the entire domain but to approximately describe the displacement distribution for a subsection, the so-called finite element through

$$u_y^{\mathrm{e}}(x) = N(x)u_{\mathrm{p}} = \begin{bmatrix} N_{1u} & N_{1\varphi} & N_{2u} & N_{2\varphi} \end{bmatrix} \times \begin{bmatrix} u_{1y} \\ \varphi_{1z} \\ u_{2y} \\ \varphi_{2z} \end{bmatrix} . \quad (5.108)$$

Within the framework of the finite element method, the same approach is chosen for the weight function as for the displacement:

$$W(x) = N(x)\delta u_{\mathrm{p}} = \begin{bmatrix} N_{1u} & N_{1\varphi} & N_{2u} & N_{2\varphi} \end{bmatrix} \times \begin{bmatrix} \delta u_{1y} \\ \delta \varphi_{1z} \\ \delta u_{2y} \\ \delta \varphi_{2z} \end{bmatrix} , \quad (5.109)$$

in which δu_i represent arbitrary displacements or rotations. The following will show that these arbitrary or so-called virtual values can be eliminated with an identical expression on the right-hand side of Eq. (5.107) and no further considerations are required.

If Eqs. (5.108) and (5.109) on the left-hand side of Eq. (5.107) are considered the following results for constant bending stiffness:

$$EI_z \int_0^L \frac{\mathrm{d}^2}{\mathrm{d}x^2}\left(N(x)\delta u_{\mathrm{p}}\right)^{\mathrm{T}} \frac{\mathrm{d}^2}{\mathrm{d}x^2}\left(N(x)u_{\mathrm{p}}\right)\mathrm{d}x \quad (5.110)$$

or

$$\delta u_{\mathrm{p}}^{\mathrm{T}} \, E I_z \underbrace{\int_0^L \frac{\mathrm{d}^2}{\mathrm{d}x^2}\left(\boldsymbol{N}^{\mathrm{T}}(x)\right) \frac{\mathrm{d}^2}{\mathrm{d}x^2}\left(\boldsymbol{N}(x)\right) \mathrm{d}x}_{k^{\mathrm{e}}} \, \boldsymbol{u}_{\mathrm{p}} . \tag{5.111}$$

The expression $\delta u_{\mathrm{p}}^{\mathrm{T}}$ can be eliminated with a corresponding expression on the right-hand side of Eq. (5.107) and u_{p} represents the column matrix of the unknown nodal deformation. Consequently the stiffness matrix can be illustrated through the single shape functions according to Eq. (5.72).

In the following, the right-hand side of Eq. (5.107) is considered in order to derive the column matrix of the external loads for a bending element with two nodes. Considering in

$$E I_z \left[-W^{\mathrm{T}} \frac{\mathrm{d}^3 u_y}{\mathrm{d}x^3} + \frac{\mathrm{d}W^{\mathrm{T}}}{\mathrm{d}x} \frac{\mathrm{d}^2 u_y}{\mathrm{d}x^2} \right]_0^L \tag{5.112}$$

the definition of the weight function according to Eq. (5.109), the following results

$$E I_z \left[-\delta u_{\mathrm{p}}^{\mathrm{T}} \boldsymbol{N}^{\mathrm{T}}(x) \frac{\mathrm{d}^3 u_y}{\mathrm{d}x^3} + \frac{\mathrm{d}}{\mathrm{d}x}\left(\delta u_{\mathrm{p}}^{\mathrm{T}} \boldsymbol{N}^{\mathrm{T}}(x)\right) \frac{\mathrm{d}^2 u_y}{\mathrm{d}x^2} \right]_0^L , \tag{5.113}$$

or in components:

$$\delta u_{\mathrm{p}}^{\mathrm{T}} E I_z \left[- \begin{bmatrix} N_{1u} \\ N_{1\varphi} \\ N_{2u} \\ N_{2\varphi} \end{bmatrix} \frac{\mathrm{d}^3 u_y}{\mathrm{d}x^3} + \frac{\mathrm{d}}{\mathrm{d}x} \begin{bmatrix} N_{1u} \\ N_{1\varphi} \\ N_{2u} \\ N_{2\varphi} \end{bmatrix} \frac{\mathrm{d}^2 u_y}{\mathrm{d}x^2} \right]_0^L . \tag{5.114}$$

$\delta u_{\mathrm{p}}^{\mathrm{T}}$ from the last equation can be eliminated with the corresponding expression in Eq. (5.111). Furthermore, Eq. (5.114) represents a system of four equations, which needs to be evaluated on the integration boundaries, meaning at $x = 0$ and $x = L$. The first row of Eq. (5.114) results in:

$$\left(-N_{1u} E I_z \frac{\mathrm{d}^3 u_y}{\mathrm{d}x^3} + \frac{\mathrm{d}N_{1u}}{\mathrm{d}x} \frac{\mathrm{d}^2 u_y}{\mathrm{d}x^2} \right)_{x=L} - \left(-N_{1u} E I_z \frac{\mathrm{d}^3 u_y}{\mathrm{d}x^3} + \frac{\mathrm{d}N_{1u}}{\mathrm{d}x} \frac{\mathrm{d}^2 u_y}{\mathrm{d}x^2} \right)_{x=0} . \tag{5.115}$$

Under consideration of the boundary values of the shape functions or alternatively their derivatives according to Fig. 5.14, meaning $N_{1u}(L) = 0$, $\frac{\mathrm{d}N_{1u}}{\mathrm{d}x}(L) = \frac{\mathrm{d}N_{1u}}{\mathrm{d}x}(0) = 0$ and $N_{1u}(0) = 1$, the following results:

$$+ EI_z \left. \frac{d^3 u_y}{dx^3} \right|_{x=0} \overset{(5.38)}{=} -Q_y(0) \,. \tag{5.116}$$

Accordingly the values of the three other rows in Eq. (5.114) can be calculated:

$$\text{Line 2:} \quad - EI_z \left. \frac{d^2 u_y}{dx^2} \right|_{x=0} \overset{(5.37)}{=} -M_z(0) \,, \tag{5.117}$$

$$\text{Line 3:} \quad - EI_z \left. \frac{d^3 u_y}{dx^3} \right|_{x=L} \overset{(5.38)}{=} +Q_y(L) \,, \tag{5.118}$$

$$\text{Line 4:} \quad + EI_z \left. \frac{d^2 u_y}{dx^2} \right|_{x=L} \overset{(5.37)}{=} +M_z(L) \,. \tag{5.119}$$

It needs to be considered that the results in Eqs. (5.116) up to (5.119) are the internal reactions according to Fig. 5.7. The external loads with the positive direction according to Fig. 5.13b therefore result from the internal reactions[15] through reversing the positive direction on the left-hand border and through maintaining the positive direction of the internal reaction on the right-hand border.

5.3.3 Derivation through Virtual Work

The derivation follows the course of reasoning as introduced for the bar element in Sect. 3.2.4. Let us consider a beam in bending as shown in Fig. 5.13 which is loaded by single forces and moments and which can freely deform. The equilibrium equations for the domain $\Omega =]0, L[$ are obtained according to Eqs. (5.22) and (5.26) as:

$$\frac{dQ_y(x)}{dx} + q_y(x) = 0 \,, \tag{5.120}$$

$$\frac{dM_z(x)}{dx} + Q_y(x) = 0 \,. \tag{5.121}$$

In a similar way, one can state the equilibrium conditions between the inner reactions and the external loads (F_{iy}, M_{iz}) at the boundaries, i.e. at $x = 0$ and $x = L$, as:

---

[15] See Sect. 5.2.2 with the executions for the internal reactions and external loads.

$$F_{1y} + Q_y(0) = 0, \quad M_{1z} + M_z(0) = 0, \tag{5.122}$$
$$F_{2y} - Q_y(L) = 0, \quad M_{2z} - M_z(L) = 0. \tag{5.123}$$

Combining the equilibrium conditions according to Eqs. (5.120) and (5.121) and following the procedure as outlined in Sect. 5.2.4, gives for constant bending stiffness the equilibrium condition for the domain as follows[16]:

$$- EI_z \frac{\mathrm{d}^4 u_y(x)}{\mathrm{d}x^4} + q_y(x) = 0. \tag{5.124}$$

Under the influence of the virtual displacements δu_y and the virtual rotations $\delta\varphi_z$, the beam virtually moves away from its state of equilibrium. The resulting forces and moments according to Eqs. (5.122)–(5.124) do some virtual work along the virtual displacements and rotations:

$$\int_0^L \left(-EI_z \frac{\mathrm{d}^4 u_y(x)}{\mathrm{d}x^4} + q_y(x)\right) \delta u_y \mathrm{d}x + \left(F_{1y} + Q_y(0)\right)\big|_0 \delta u_y + \left(F_{2y} - Q_y(L)\right)\big|_L \delta u_y + \left(M_{1z} + M_z(0)\right)\big|_0 \delta\varphi_z + \left(M_{2z} - M_z(L)\right)\big|_L \delta\varphi_z = 0. \tag{5.125}$$

The round brackets in the last equation can be expanded to obtain the following expression:

$$- \int_0^L EI_z \frac{\mathrm{d}^4 u_y(x)}{\mathrm{d}x^4} \delta u_y \mathrm{d}x + \int_0^L q_y(x)\delta u_y \mathrm{d}x + F_{1y}\big|_0 \delta u_y + Q_y(0)\big|_0 \delta u_y$$
$$+ F_{2y}\big|_L \delta u_y - Q_y(L)\big|_L \delta u_y + M_{1z}\big|_0 \delta\varphi_z + M_z(0)\big|_0 \delta\varphi_z + M_{2z}\big|_L \delta\varphi_z$$
$$- M_z(L)\big|_L \delta\varphi_z = 0. \tag{5.126}$$

Two times partial integration of the left-hand expression, i.e.

$$\int_0^L EI_z \underbrace{\frac{\mathrm{d}^4 u_y(x)}{\mathrm{d}x^4}}_{f'} \underbrace{\delta u_y}_{g} \mathrm{d}x = EI_z \left[\frac{\mathrm{d}^3 u_y(x)}{\mathrm{d}x^3} \delta u_y\right]_0^L - \int_0^L EI_z \underbrace{\frac{\mathrm{d}^3 u_y(x)}{\mathrm{d}x^3}}_{f'} \underbrace{\frac{\mathrm{d}\delta u_y}{\mathrm{d}x}}_{g} \mathrm{d}x$$
$$= EI_z \left[\frac{\mathrm{d}^3 u_y(x)}{\mathrm{d}x^3} \delta u_y - \frac{\mathrm{d}^2 u_y(x)}{\mathrm{d}x^2}\frac{\mathrm{d}\delta u_y}{\mathrm{d}x}\right]_0^L + \int_0^L EI_z \frac{\mathrm{d}^2 u_y(x)}{\mathrm{d}x^2}\frac{\mathrm{d}^2\delta u_y}{\mathrm{d}x^2}\mathrm{d}x, \tag{5.127}$$

---

[16]Pay attention to the sign: the virtual displacements have the same direction as the external loads.

allows, under consideration of $EI_z \frac{d^3 u_y(x)}{dx^3} = -Q_y(x)$, $EI_z \frac{d^2 u_y(x)}{dx^2} = M_z(x)$ and $\frac{d\delta u_y}{dx} = \delta\varphi_z$, the simplification of Eq. (5.126):

$$\int_0^L EI_z \frac{d^2 u_y(x)}{dx^2} \frac{d^2 \delta u_y}{dx^2} dx = \int_0^L q_y(x)\delta u_y dx + F_{1y}\big|_0 \delta u_y + F_{2y}\big|_L \delta u_y$$

$$+ M_{1z}\big|_0 \delta\varphi_z + M_{2z}\big|_L \delta\varphi_z. \qquad (5.128)$$

Now we replace in the last equation the displacement distribution by the approximation according to Eq. (5.65) and the virtual displacements and rotations according to the approach for the weight function in Eq. (5.109). Elimination of δu_p^T and evaluation of the boundary expressions allow to express the weak form of the problem as:

$$EI_z \int_0^L \frac{d^2 N^T(x)}{dx^2} \frac{d^2 N(x)}{dx^2} dx\, u_p = \int_0^L q_y(x) N^T(x)\, dx + \begin{bmatrix} F_{1y} \\ M_{1z} \\ F_{2y} \\ M_{2z} \end{bmatrix}. \qquad (5.129)$$

5.3.4 Comments on the Derivation of the Shape Functions

The shape functions were derived in Sect. 5.3 via a polynomial with four unknown parameters, see Eq. (5.51). The derivation of the shape function can also be achieved through a clearer method. The general feature of a shape function N_i, i.e. adopting the value 1 at its node i and turning zero on all other nodes has to be considered herefore. Furthermore, it should be considered in the case of the beam in bending that the displacement and rotation field should be decoupled at the nodes. Consequently, a shape function for the displacement field has to adopt the value 1 on 'its' node as well as the slope zero. On all other nodes j, the value of the function as well as the slope turn zero:

$$N_{iu}(x_i) = 1, \qquad (5.130)$$

$$N_{iu}(x_j) = 0, \qquad (5.131)$$

$$\frac{d N_{iu}(x_i)}{dx} = 0, \qquad (5.132)$$

$$\frac{d N_{iu}(x_j)}{dx} = 0. \qquad (5.133)$$

Accordingly it results that a shape function for the rotation field reaches the slope 1 at 'its' node but the function value zero. At all other nodes the function value and the

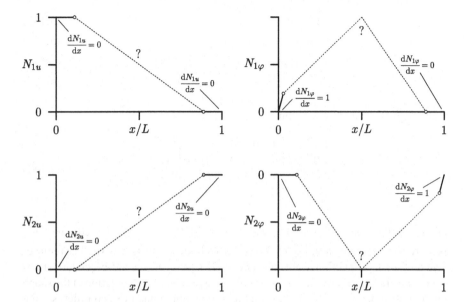

Fig. 5.15 Boundary conditions for the shape functions for the bending element at bending in the x–y plane. Note that the sections for the given slopes are overdrawn for better illustration

slope are equally zero. Therefore, the boundary conditions shown in Fig. 5.15 result for the four shape functions.

If the course of the shape functions should be without discontinuities, meaning kinks, then every shape function has to change its curvature. Therefore, at least a polynomial of 3rd order has to be applied, so that a linear function results for the curvature, meaning the second-order derivative:

$$N(x) = \alpha_0 + \alpha_1 x + \alpha_2 x^2 + \alpha_3 x^3 \,. \tag{5.134}$$

Since a polynomial of 3rd order usually exhibits four unknowns, $\alpha_0, \ldots, \alpha_3$, all unknowns can be determined due to this approach via the four boundary conditions – two for the function values and two for the slopes.

In the following, the first shape function is regarded as an example. The boundary conditions in this case result in:

$$N_{1u}(0) = 1 \,, \tag{5.135}$$

$$\frac{\mathrm{d}N_{1u}}{\mathrm{d}x}(0) = \frac{\mathrm{d}N_{1u}}{\mathrm{d}x}(L) = 0 \,, \tag{5.136}$$

$$N_{1u}(L) = 0 \,. \tag{5.137}$$

If the boundary conditions are evaluated according to the approach given in Eq. (5.134), the following results:

$$1 = \alpha_0 , \tag{5.138}$$

$$0 = \alpha_1 , \tag{5.139}$$

$$0 = \alpha_0 + \alpha_1 L + \alpha_2 L^2 + \alpha_3 L^3 , \tag{5.140}$$

$$0 = \alpha_1 + 2\alpha_2 L + 3\alpha_3 L^2 , \tag{5.141}$$

or alternatively in matrix notation:

$$\begin{bmatrix} 1 \\ 0 \\ 0 \\ 0 \end{bmatrix} = \begin{bmatrix} 1 & 0 & 0 & 0 \\ 0 & 1 & 0 & 0 \\ 1 & L & L^2 & L^3 \\ 0 & 1 & 2L & 3L^2 \end{bmatrix} \begin{bmatrix} \alpha_0 \\ \alpha_1 \\ \alpha_2 \\ \alpha_3 \end{bmatrix} . \tag{5.142}$$

Solving for the unknown constants leads to $\alpha = \begin{bmatrix} 1 & 0 & -\frac{3}{L^2} & \frac{2}{L^3} \end{bmatrix}^{\mathrm{T}}$. Exactly the same shape functions as given in Eq. (5.60) result with these constants. Another requirement for the shape function results from Eq. (5.73). Here, the second-order derivatives of the shape functions are contained. Therefore, a reasonable formulation of the shape function for a bending element has to at least be a polynomial of 2nd order, so that derivatives different from zero result. To conclude it needs to be remarked that the shape functions for the bending beam are the so-called HERMITE's polynomials. A continuous displacement and rotation occur at the nodes for this type of interpolation since the nodal values as well as the slopes are considered at the nodes.

5.4 The Finite Element BERNOULLI Beam with Two Deformation Planes

Let us consider in the following a BERNOULLI beam, which can deform in two mutually orthogonal planes. The stiffness matrix for bending in the x–y plane is given according to Eq. (5.82) as follows:

$$k_{xy}^{\mathrm{e}} = \frac{EI_z}{L^3} \begin{bmatrix} 12 & 6L & -12 & 6L \\ 6L & 4L^2 & -6L & 2L^2 \\ -12 & -6L & 12 & -6L \\ 6L & 2L^2 & -6L & 4L^2 \end{bmatrix} . \tag{5.143}$$

In the orthogonal plane, meaning for bending in the x–z plane, a slightly modified stiffness matrix results, since the positive orientation of the angles around the y-axis is now in the clockwise direction, see Fig. 5.16. Under consideration of the definition of the positive rotation angle according to $\varphi_y(x) = -\frac{du_z(x)}{dx}$, the stiffness matrix for bending in the x–z-plane results in[17]:

---

[17] Also see Table 5.5 and supplementary problems 5.6.

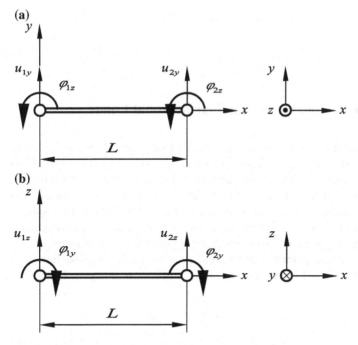

Fig. 5.16 Definition of the positive deformation parameters at bending in the **a** x–y plane and **b** x–z plane

$$k^e_{xz} = \frac{EI_y}{L^3}\begin{bmatrix} 12 & -6L & -12 & -6L \\ -6L & 4L^2 & 6L & 2L^2 \\ -12 & 6L & 12 & 6L \\ -6L & 2L^2 & 6L & 4L^2 \end{bmatrix}. \tag{5.144}$$

Both stiffness matrices for the deformation in the x–y and x–z plane can easily be superposed, so that the following form for an element with two orthogonal deformation planes results

$$k^e = \frac{E}{L^3}\begin{bmatrix} 12I_z & 0 & 0 & 6I_zL & -12I_z & 0 & 0 & 6I_zL \\ 0 & 12I_y & -6I_yL & 0 & 0 & -12I_y & -6I_yL & 0 \\ 0 & -6I_yL & 4I_yL^2 & 0 & 0 & 6I_yL & 2I_yL^2 & 0 \\ 6I_zL & 0 & 0 & 4I_zL^2 & -6I_zL & 0 & 0 & 2I_zL^2 \\ -12I_z & 0 & 0 & -6I_zL & 12I_z & 0 & 0 & -6I_zL \\ 0 & -12I_y & 6I_yL & 0 & 0 & 12I_y & 6I_yL & 0 \\ 0 & -6I_yL & 2I_yL^2 & 0 & 0 & 6I_yL & 4I_yL^2 & 0 \\ 6I_zL & 0 & 0 & 2I_zL^2 & -6I_zL & 0 & 0 & 4I_zL^2 \end{bmatrix}, \tag{5.145}$$

whereupon the deformation and load matrices are represented as follows:

$$\boldsymbol{u}_\mathrm{p} = \begin{bmatrix} u_{1y} & u_{1z} & \varphi_{1y} & \varphi_{1z} & u_{2y} & u_{2z} & \varphi_{2y} & \varphi_{2z} \end{bmatrix}^\mathrm{T}, \tag{5.146}$$

$$\boldsymbol{F}^\mathrm{e} = \begin{bmatrix} F_{1y} & F_{1z} & M_{1y} & M_{1z} & F_{2y} & F_{2z} & M_{2y} & M_{2z} \end{bmatrix}^\mathrm{T}. \tag{5.147}$$

5.5 Determination of Equivalent Nodal Loads

Within the finite element method, external loads can only be applied at the nodes. Distributed loads or point loads[18] between the nodes must be converted into equivalent nodal loads. The procedure can be demonstrated in the following for the example of a beam which is clamped at both ends, see Fig. 5.17. A pragmatic approach is based on an equivalent statical system, see Fig. 5.17b. Support reactions, consisting of vertical forces and moments occur on the fixed supports, whereby our goal is to determine the inner reactions, which are acting at the beam borders.

As a starting point the differential equation of the bending line according to Eq. (5.36) in the form according to our problem, meaning with negative load, is chosen:

$$E I_z \frac{\mathrm{d}^4 u_y}{\mathrm{d}x^4} = -q_y. \tag{5.148}$$

Four times integration yields the general approach for the bending line:

$$E I_z \frac{\mathrm{d}^3 u_y}{\mathrm{d}x^3} = -q_y x + c_1 \overset{!}{=} -Q_y(x), \tag{5.149}$$

$$E I_z \frac{\mathrm{d}^2 u_y}{\mathrm{d}x^2} = -\frac{1}{2} q_y x^2 + c_1 x + c_2 \overset{!}{=} M_z(x), \tag{5.150}$$

$$E I_z \frac{\mathrm{d}^1 u_y}{\mathrm{d}x^1} = -\frac{1}{6} q_y x^3 + \frac{1}{2} c_1 x^2 + c_2 x + c_3, \tag{5.151}$$

$$E I_z u_y(x) = -\frac{1}{24} q_y x^4 + \frac{1}{6} c_1 x^3 + \frac{1}{2} c_2 x^2 + c_3 x + c_4. \tag{5.152}$$

If the boundary conditions are taken into consideration, meaning $u_y(0) = u_y(L) = 0$ and $\varphi_z(0) = \varphi_z(L) = 0$, the four integration constants result in:

$$c_3 = c_4 = 0, \tag{5.153}$$

$$c_2 = -\frac{1}{12} q_y L^2, \tag{5.154}$$

$$c_1 = \frac{1}{2} q_y L. \tag{5.155}$$

---

[18]If point loads appear between nodes, the discretization can of course be further sub-divided, so that a new node is positioned on the location of the loading point. However, this chapter considers only the case that no further subdivision is regarded.

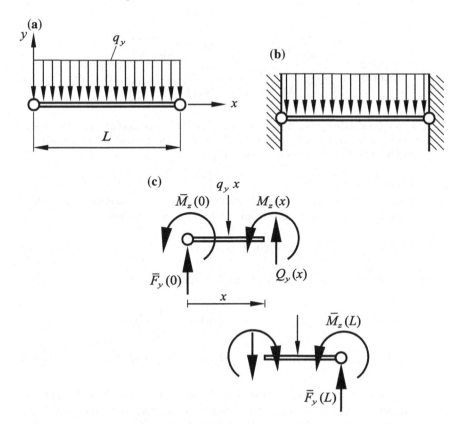

Fig. 5.17 Calculation of equivalent nodal loads: **a** example configuration; **b** equivalent statical system; **c** free-body diagram with support reactions

Through these integration constants and the relations in Eqs. (5.149) and (5.150), one obtains the shear force and bending moment distribution within the element:

$$Q_y(x) = -\frac{1}{2}q_y L + q_y x, \tag{5.156}$$

$$M_z(x) = -\frac{1}{12}q_y L^2 + \frac{1}{2}q_y L x - \frac{1}{2}q_y x^2. \tag{5.157}$$

Evaluation at the boundaries, meaning for $x = 0$ and $x = L$, leads to the following values of the internal reactions:

$$Q_y(0) = -\frac{1}{2}q_y L, \qquad\qquad M_z(0) = -\frac{1}{12}q_y L^2, \tag{5.158}$$

$$Q_y(L) = +\frac{1}{2}q_y L, \qquad\qquad M_z(L) = -\frac{1}{12}q_y L^2. \tag{5.159}$$

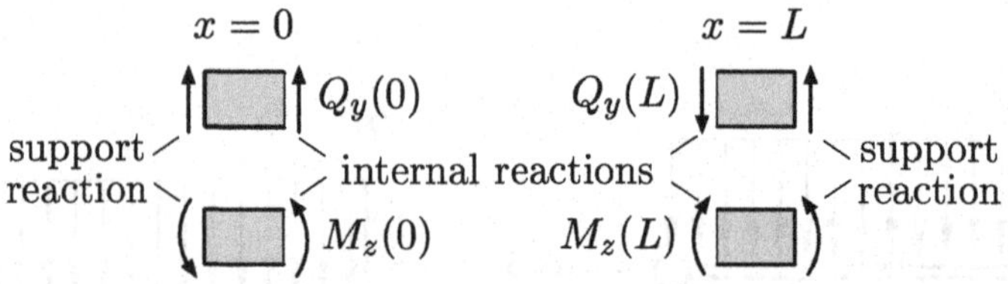

Fig. 5.18 Support and internal reactions at the boundaries of the beam from Fig. 5.17. The support reactions have the direction, which are defined in Fig. 5.17; the internal reactions are according to Fig. 5.7 positive oriented on the corresponding cutting planes

The support reactions can be defined through the load and moment equilibrium according to Fig. 5.18. Thus, the vertical equilibrium of forces yields

$$+ \bar{F}_y(0) + Q_y(0) = 0 \,, \tag{5.160}$$

and accordingly all support reactions $\bar{F}_y$ and $\bar{M}_z$ result:

$$\bar{F}_y(0) = -\frac{1}{2} q_y L \,, \qquad\qquad \bar{M}_z(0) = -\frac{1}{12} q_y L^2 \,, \tag{5.161}$$

$$\bar{F}_y(L) = +\frac{1}{2} q_y L \,, \qquad\qquad \bar{M}_z(L) = -\frac{1}{12} q_y L^2 \,. \tag{5.162}$$

Taking into consideration the definition of the positive direction of the external loads of a beam element according to Fig. 5.13, the equivalent loads on nodes F_{iy} and M_{iz} result through evaluation of the internal reactions Q_y and M_z to:

$$F_{1y} = -\frac{1}{2} q_y L \,, \qquad\qquad M_{1z} = -\frac{1}{12} q_y L^2 \,, \tag{5.163}$$

$$F_{2y} = -\frac{1}{2} q_y L \,, \qquad\qquad M_{2z} = +\frac{1}{12} q_y L^2 \,. \tag{5.164}$$

It should to be remarked at this point that the equivalent nodal loads are *not* the support reactions. The equivalent nodal loads have to cause the support reactions. At the end of this derivation it is appropriate to point out that we distinguished between the following parameters:

- internal reactions $Q_y(x)$, $M_z(x)$,
- support reactions $\bar{F}_y$, $\bar{M}_z$ and
- equivalent nodal loads F_{iy}, M_{iz}.

Alternatively, the derivation of the equivalent nodal loads can also take place through the equivalence of the potential of the external loads, meaning the distributed load and the equivalent nodal loads:

$$\Pi_{\text{ext}} = -\int_0^L q_y(x)u_y(x)\mathrm{d}x \overset{!}{=} -\left(F_{1y}u_{1y} + M_{1z}\varphi_{1z} + F_{2y}u_{2y} + M_{2z}\varphi_{2z}\right).$$

(5.165)

Through application of the approximation for the displacement $u_y(x)$ according to Eq. (5.64) the potential of a distributed load can be written as

$$\Pi_{\text{ext}} = -\int_0^L q_y(x)\left(N_{1u}u_{1y} + N_{1\varphi}\varphi_{1z} + N_{2u}u_{2y} + N_{2\varphi}\varphi_{2z}\right)\mathrm{d}x,$$

(5.166)

or under consideration that the nodal values of the deformation can be considered as constant for the integration, as

$$\Pi_{\text{ext}} = -\left(\int_0^L q_y(x)N_{1u}(x)\mathrm{d}x\, u_{1y} + \int_0^L q_y(x)N_{1\varphi}(x)\mathrm{d}x\, \varphi_{1z}+\right.$$
$$\left. + \int_0^L q_y(x)N_{2u}(x)\mathrm{d}x\, u_{2y} + \int_0^L q_y(x)N_{2\varphi}(x)\mathrm{d}x\, \varphi_{2z}\right).$$

(5.167)

A comparison of the two potentials finally delivers the equivalent nodal loads to

$$F_{1y} = \int_0^L q_y(x)N_{1u}(x)\,\mathrm{d}x,$$

(5.168)

$$M_{1z} = \int_0^L q_y(x)N_{1\varphi}(x)\,\mathrm{d}x,$$

(5.169)

$$F_{2y} = \int_0^L q_y(x)N_{2u}(x)\,\mathrm{d}x,$$

(5.170)

$$M_{2z} = \int_0^L q_y(x)N_{2\varphi}(x)\,\mathrm{d}x,$$

(5.171)

where at the shape functions according to Eqs. (5.60)–(5.63) have to be used. If, for example, at the location $x = a$ an external force F acts on the beam, the external potential results in

$$\Pi_{\text{ext}} = -Fu_y(a).$$

(5.172)

A comparison of the two potentials delivers for this case the equivalent nodal loads to:

$$F_{1y} = FN_{1u}(a), \tag{5.173}$$

$$M_{1z} = FN_{1\varphi}(a), \tag{5.174}$$

$$F_{2y} = FN_{2u}(a), \tag{5.175}$$

$$M_{2z} = FN_{2\varphi}(a). \tag{5.176}$$

Equivalent nodal loads for simple loading cases are summarized in Table A.8. It needs to be remarked at the end of this chapter, that the column matrix of the equivalent nodal loads can be reached in an easier way, if during the application of the weighted residual method the differential equation (5.36) under consideration of the distributed load is used. Under consideration of an arbitrary distributed load the inner product results in:

$$\int_{0}^{L} W^{\mathrm{T}}(x)\left(EI_z\frac{\mathrm{d}^4 u_y(x)}{\mathrm{d}x^4} - q_y(x)\right)\mathrm{d}x \overset{!}{=} 0. \tag{5.177}$$

After the introduction of the approach for the weight function, meaning $W(x) = N(x)\delta u_{\mathrm{p}}$, the expression with the distributed load can be brought on the right-hand side and after elimination of $\delta u_{\mathrm{p}}^{\mathrm{T}}$ the additional load matrix results:

$$\cdots = \cdots + \int_{0}^{L} q_y(x)\begin{bmatrix} N_{1u} \\ N_{1\varphi} \\ N_{2u} \\ N_{2\varphi} \end{bmatrix}\mathrm{d}x. \tag{5.178}$$

This expression equals exactly Eqs. (5.168)–(5.171).

5.6 Sample Problems and Supplementary Problems

5.6.1 Sample Problems

5.1 Bending of beam under point load or moment – Approximation through a single finite element

Determine the displacement and the rotation of the right-hand end of the beam, which is shown in Fig. 5.19, through a single finite element. Furthermore, determine the course of the bending line $u_y = u_y(x)$ and compare the finite element result with the analytical solution.

Solution (a) The finite element equation on element level according to Eq. (5.83) reduces for the illustrated loading case to:

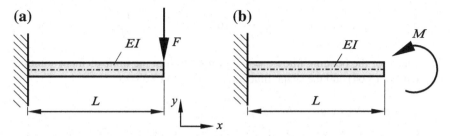

Fig. 5.19 Sample problem BERNOULLI beam: **a** point load; **b** single moment

$$\frac{EI_z}{L^3}\begin{bmatrix} 12 & 6L & -12 & 6L \\ 6L & 4L^2 & -6L & 2L^2 \\ -12 & -6L & 12 & -6L \\ 6L & 2L^2 & -6L & 4L^2 \end{bmatrix}\begin{bmatrix} u_{1y} \\ \varphi_{1z} \\ u_{2y} \\ \varphi_{2z} \end{bmatrix} = \begin{bmatrix} 0 \\ 0 \\ -F \\ 0 \end{bmatrix}. \tag{5.179}$$

Since the displacement and the rotation are zero on the left-hand boundary due to the fixed support, the first two rows and columns of the system of equations can be eliminated:

$$\frac{EI_z}{L^3}\begin{bmatrix} 12 & -6L \\ -6L & 4L^2 \end{bmatrix}\begin{bmatrix} u_{2y} \\ \varphi_{2z} \end{bmatrix} = \begin{bmatrix} -F \\ 0 \end{bmatrix}. \tag{5.180}$$

Solving for the unknown deformations yields:

$$\begin{bmatrix} u_{2y} \\ \varphi_{2z} \end{bmatrix} = \frac{L^3}{EI_z}\begin{bmatrix} 12 & -6L \\ -6L & 4L^2 \end{bmatrix}^{-1}\begin{bmatrix} -F \\ 0 \end{bmatrix} \tag{5.181}$$

$$= \frac{L^3}{EI_z(48L^2 - 36L^2)}\begin{bmatrix} 4L^2 & 6L \\ 6L & 12 \end{bmatrix}\begin{bmatrix} -F \\ 0 \end{bmatrix} = \begin{bmatrix} -\frac{FL^3}{3EI_z} \\ -\frac{FL^2}{2EI_z} \end{bmatrix}. \tag{5.182}$$

According to Table A.10 the analytical displacement results in:

$$u_y(x = L) = -\frac{F}{6EI_z}(3L^3 - L^3) = -\frac{FL^3}{3EI_z}. \tag{5.183}$$

The analytical solution for the rotation results from differentiation of the general displacement distribution according to Table A.10 for $a = L$ to:

$$\varphi_z(x) = \frac{du_y(x)}{dx} = -\frac{F}{6EI_z} \times \left[6Lx - 3x^2\right], \tag{5.184}$$

or alternatively on the right-hand boundary:

$$\varphi_z(x = L) = -\frac{F}{6EI_z} \times \left[6L^2 - 3L^2\right] = -\frac{FL^2}{2EI_z}. \tag{5.185}$$

The course of the bending line $u_y = u_y(x)$ results from the finite element solution through Eq. (5.64) and the shape functions (5.62) and (5.63) to:

$$
\begin{aligned}
u_y(x) &= N_{2u}(x)u_{2y} + N_{2\varphi}(x)\varphi_{2z} \\
&= \left[3\left(\frac{x}{L}\right)^2 - 2\left(\frac{x}{L}\right)^3\right]\left(-\frac{FL^3}{3EI_z}\right) + \left[-\frac{x^2}{L} + \frac{x^3}{L^2}\right]\left(-\frac{FL^2}{2EI_z}\right) \\
&= \frac{F}{6EI_z}\left(x^3 - 3Lx^2\right).
\end{aligned} \tag{5.186}
$$

According to Table A.10 this course matches with the analytical solution.

Conclusion: Finite element solution and analytical solution are identical!

Solution (b) The reduced system of equations in this case results in:

$$\frac{EI_z}{L^3}\begin{bmatrix} 12 & -6L \\ -6L & 4L^2 \end{bmatrix}\begin{bmatrix} u_{2y} \\ \varphi_{2z} \end{bmatrix} = \begin{bmatrix} 0 \\ M \end{bmatrix}. \tag{5.187}$$

Solving for the unknown deformations yields:

$$\begin{bmatrix} u_{2y} \\ \varphi_{2z} \end{bmatrix} = \frac{L^3}{12EI_zL^2}\begin{bmatrix} 4L^2 & 6L \\ 6L & 12 \end{bmatrix}\begin{bmatrix} 0 \\ M \end{bmatrix} = \begin{bmatrix} \frac{ML^2}{2EI_z} \\ \frac{ML}{EI_z} \end{bmatrix}. \tag{5.188}$$

The analytical solution according to Table A.10 delivers

$$u_y(x = L) = -\frac{M}{2EI_z}\left(-L^2\right) = \frac{ML^2}{2EI_z}, \tag{5.189}$$

or alternatively the rotation in general for $a = L$ to:

$$\varphi_z(x) = \frac{du_y(x)}{dx} = -\frac{M}{2EI_z}(-2x) \tag{5.190}$$

or only on the right-hand boundary:

$$\varphi_z(x = L) = -\frac{M}{2EI_z}(-L) = \frac{ML}{EI_z}. \tag{5.191}$$

The course of the bending line $u_y = u_y(x)$ results from the finite element solution through Eq. (5.64) and the shape functions (5.62) and (5.63) to:

$$u_y(x) = N_{2u}(x)u_{2y} + N_{2\varphi}(x)\varphi_{2z}$$

$$= \left[3\left(\frac{x}{L}\right)^2 - 2\left(\frac{x}{L}\right)^3\right]\left(\frac{ML^2}{2EI_z}\right) + \left[-\frac{x^2}{L} + \frac{x^3}{L^2}\right]\left(\frac{ML}{EI_z}\right)$$

$$= \frac{Mx^2}{2EI_z}. \tag{5.192}$$

According to Table A.10 this course matches with the analytical solution.

Conclusion: Finite element solution and analytical solution are identical!

5.2 Bending of beam under constant distributed load – Approximation through a single finite element

Determine through a single finite element the displacement and the rotation (a) of the right-hand boundary and (b) in the middle of a beam under constant distributed load, see Fig. 5.20. Furthermore, determine the course of the bending line $u_y = u_y(x)$ and compare the finite element result with the analytical solution.

Solution

To solve the problem, we must first convert the constant distributed load into equivalent nodal loads. These equivalent nodal loads can be extracted from Table A.8 for the considered case, and the finite element equation results to:

$$\frac{EI_z}{L^3}\begin{bmatrix} 12 & 6L & -12 & 6L \\ 6L & 4L^2 & -6L & 2L^2 \\ -12 & -6L & 12 & -6L \\ 6L & 2L^2 & -6L & 4L^2 \end{bmatrix}\begin{bmatrix} u_{1y} \\ \varphi_{1z} \\ u_{2y} \\ \varphi_{2z} \end{bmatrix} = \begin{bmatrix} -\frac{qL}{2} \\ -\frac{qL^2}{12} \\ -\frac{qL}{2} \\ +\frac{qL^2}{12} \end{bmatrix}. \tag{5.193}$$

(a) Consideration of the support conditions from Fig. 5.20a, meaning the fixed support on the left-hand boundary, and solving for the unknowns yields:

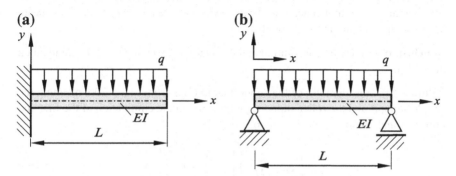

Fig. 5.20 Sample problem BERNOULLI beam under constant distributed load at different supports

$$\begin{bmatrix} u_{2y} \\ \varphi_{2z} \end{bmatrix} = \frac{L}{12EI_z} \begin{bmatrix} 4L^2 & 6L \\ 6L & 12 \end{bmatrix} \begin{bmatrix} -\frac{qL}{2} \\ +\frac{qL^2}{12} \end{bmatrix} = \begin{bmatrix} -\frac{qL^4}{8EI_z} \\ -\frac{qL^3}{6EI_z} \end{bmatrix}. \tag{5.194}$$

The analytical solution according to Table A.10 yields

$$u_y(x = L) = -\frac{q}{24EI_z}(6L^4 - 4L^4 + L^4) = -\frac{qL^4}{8EI_z}, \tag{5.195}$$

or alternatively the rotation in general for $a_1 = 0$ and $a_2 = L$ to:

$$\varphi_z(x) = \frac{du_y(x)}{dx} = -\frac{q}{24EI_z}(12L^2x - 12Lx^2 + 4x^3) \tag{5.196}$$

or only on the right-hand boundary:

$$\varphi_z(x = L) = -\frac{q}{24EI_z}(12L^3 - 12L^3 + 4L^3) = -\frac{qL^3}{6EI_z}. \tag{5.197}$$

The course of the bending line $u_y = u_y(x)$ results from the finite element solution through Eq. (5.64) and the shape functions (5.62) and (5.63) to:

$$\begin{aligned} u_y(x) &= N_{2u}(x)u_{2y} + N_{2\varphi}(x)\varphi_{2z} \\ &= \left[3\left(\frac{x}{L}\right)^2 - 2\left(\frac{x}{L}\right)^3 \right] \left(-\frac{qL^4}{8EI_z} \right) + \left[-\frac{x^2}{L} + \frac{x^3}{L^2} \right] \left(-\frac{qL^3}{6EI_z} \right) \\ &= -\frac{q}{24EI_z}(-2Lx^3 + 5L^2x^2), \end{aligned} \tag{5.198}$$

however the analytical course according to Table A.10 results in $u_y(x) = -\frac{q}{24EI_z}(x^4 - 4Lx^3 + 6L^2x^2)$, meaning the analytical and therefore the exact course is not identical with the numerical solution between the nodes $(0 < x < L)$, see Fig. 5.21. One can see that between the nodes a small difference between the two solutions arises. If a higher accuracy is demanded between those two nodes, the beam has to be divided into more elements.

Conclusion: Finite element solution and the analytical solution are only identical on the nodes!

Solution (b) Consideration of the support conditions from Fig. 5.20b, meaning the simple support and the roller support, yields through the elimination of the first and third row and column of the system of Eq. (5.193):

$$\frac{EI_z}{L^3} \begin{bmatrix} 4L^2 & 2L^2 \\ 2L^2 & 4L^2 \end{bmatrix} \begin{bmatrix} \varphi_{1z} \\ \varphi_{2z} \end{bmatrix} = \begin{bmatrix} -\frac{qL^2}{12} \\ +\frac{qL^2}{12} \end{bmatrix}. \tag{5.199}$$

Fig. 5.21 Comparison of the analytical and the finite element solution for the beam according to Fig. 5.20a

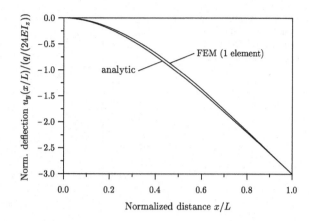

Solving for the unknowns yields:

$$\begin{bmatrix} \varphi_{1z} \\ \varphi_{2z} \end{bmatrix} = \frac{1}{12EI_z L} \begin{bmatrix} 4L^2 & -2L^2 \\ -2L^2 & 4L^2 \end{bmatrix} \begin{bmatrix} -\frac{qL^2}{12} \\ +\frac{qL^2}{12} \end{bmatrix} = \begin{bmatrix} -\frac{qL^3}{24EI_z} \\ +\frac{qL^3}{24EI_z} \end{bmatrix}. \tag{5.200}$$

The course of the bending line $u_y = u_y(x)$ results from the finite element solution through Eq. (5.64) and the shape functions (5.61) and (5.63) to:

$$u_y(x) = N_{1\varphi}(x)\varphi_{1z} + N_{2\varphi}(x)\varphi_{2z}$$

$$= \left[x - 2\frac{x^2}{L} + \frac{x^3}{L^2} \right] \left(-\frac{qL^3}{24EI_z} \right) + \left[-\frac{x^2}{L} + \frac{x^3}{L^2} \right] \left(+\frac{qL^3}{24EI_z} \right)$$

$$= -\frac{q}{24EI_z} \left(-L^2 x^2 + L^3 x \right), \tag{5.201}$$

however the analytical course according to Table A.10 results in $u_y(x) = -\frac{q}{24EI_z} \left(x^4 - 2Lx^3 + L^3 x \right)$, meaning the analytical and therefore exact course is also at this point not identical with the numerical solution between the nodes ($0 < x < L$), see Fig. 5.22.

The numerical solution for the deflection in the middle of the beam yields $u_y(x = \frac{1}{2}L) = \frac{-4qL^4}{384EI_z}$, however the exact solution is $u_y(x = \frac{1}{2}L) = \frac{-5qL^4}{384EI_z}$.

Conclusion: Finite element solution and analytical solution are only identical on the nodes!

5.3 Bending of a beam with variable cross-section

The beam, which is illustrated in Fig. 5.23, has a variable cross-section along the x-axis. Derive for

(a) a circular cross-section, and

Fig. 5.22 Comparison of the analytical and the finite element solution for the beam according to Fig. 5.20b

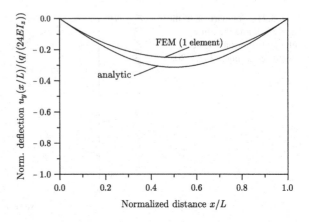

Fig. 5.23 Sample problem BERNOULLI beam with variable cross-section: **a** change along the x-axis; **b** circular cross-section; **c** square cross-section

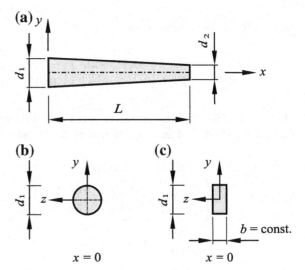

(b) a square cross-section

the element stiffness matrix for the case $d_1 = 2h$ and $d_2 = h$.

Solution (a) Square cross-section:

Equation (5.71) can be used as an initial point for the derivation of the stiffness matrix:

$$k^e = E \int_x \left(\underbrace{\int_A y^2 dA}_{I_z} \right) \frac{d^2 N^T(x)}{dx^2} \frac{d^2 N(x)}{dx^2} dx . \qquad (5.202)$$

Since the axial second moment of area changes along the x-axis, we must first derive the corresponding function. An elegant method is to use the polar second moment of area of a circle and the change along the x-axis. Hereby the relation, that the polar second moment of area is the sum of the two axial second moments of area I_y and I_z, is used:

$$I_p = \int_A r^2 dA = I_y + I_z . \tag{5.203}$$

Since the axial second moments of area of a circle are identical, the following expression can be derived for I_z:

$$I_z(x) = \frac{1}{2} I_p(x) = \frac{1}{2} \int_A r^2 dA = \frac{1}{2} \int_{\alpha=0}^{2\pi} \int_0^{r(x)} \underbrace{\hat{r}^2 \, \hat{r} d\hat{r} d\alpha}_{dA} \tag{5.204}$$

$$= \pi \int_0^{r(x)} \hat{r}^3 d\hat{r} = \pi \left[\frac{1}{4} \hat{r}^4 \right]_0^{r(x)} = \frac{\pi}{4} r(x)^4 . \tag{5.205}$$

The change of the radius along the x-axis can be easily derived from Fig. 5.23a:

$$r(x) = h \left(1 - \frac{x}{2L} \right) = \frac{h}{2} \left(2 - \frac{x}{L} \right) . \tag{5.206}$$

Therefore, the axial second moment of area results in

$$I_z(x) = \frac{\pi h^4}{64} \left(2 - \frac{x}{L} \right)^4 \tag{5.207}$$

and can be used in Eq. (5.202):

$$k^e = E \frac{\pi h^4}{64} \int_L \left(2 - \frac{x}{L} \right)^4 \frac{d^2 N^T(x)}{dx^2} \frac{d^2 N(x)}{dx^2} dx . \tag{5.208}$$

The integration can be carried out through the second-order derivatives of the shape function according to Eqs. (5.78)–(5.81). As an example for the first component of the stiffness matrix

$$k_{11} = E \frac{\pi h^4}{64} \int_L \left(2 - \frac{x}{L} \right)^4 \left(-\frac{6}{L^2} + \frac{12x}{L^3} \right)^2 dx , \tag{5.209}$$

is used and the entire stiffness matrix finally results after a short calculation:

$$
k^e_{\text{circle}} = \frac{E\,\pi h^4}{L^3\,64}
\begin{bmatrix}
\frac{2988}{35} & \frac{1998}{35}L & -\frac{2988}{35} & \frac{198}{7}L \\[4pt]
\frac{1998}{35}L & \frac{1468}{35}L^2 & -\frac{1998}{35}L & \frac{106}{7}L^2 \\[4pt]
-\frac{2988}{35} & -\frac{1998}{35}L & \frac{2988}{35} & -\frac{198}{7}L \\[4pt]
\frac{198}{7}L & \frac{106}{7}L^2 & -\frac{198}{7}L & \frac{92}{7}L^2
\end{bmatrix} .
\tag{5.210}
$$

Solution (b) Square cross-section:

Regarding the square cross-section, Eq. (5.202) serves as a basis as well. However, in this case it seems to be a good idea to go back to the definition of I_z immediately:

$$
I_z(x) = \int_A y^2 \mathrm{d}A = \int_{-y(x)}^{y(x)} \hat{y}^2 \underbrace{b\,\mathrm{d}\hat{y}}_{\mathrm{d}A} = b\left[\frac{1}{3}\hat{y}^3\right]_{-y(x)}^{y(x)} = \frac{2b}{3}y(x)^3 .
\tag{5.211}
$$

The course of the function $y(x)$ of the cross-section is identical with the radius of the task in part (a), meaning $y(x) = h(1 - \frac{x}{2L})$, and the second moment of area in this case results in:

$$
I_z(x) = \frac{2bh^3}{3}\left(1 - \frac{x}{2L}\right)^3 = \frac{bh^3}{12}\left(2 - \frac{x}{L}\right)^3 .
\tag{5.212}
$$

Due to the special form of the second moment of area, the stiffness matrix therefore results in

$$
k^e = E\frac{bh^3}{12}\int_L \left(2 - \frac{x}{L}\right)^3 \frac{\mathrm{d}^2 N^{\mathrm{T}}(x)}{\mathrm{d}x^2}\frac{\mathrm{d}^2 N(x)}{\mathrm{d}x^2}\,\mathrm{d}x
\tag{5.213}
$$

or after the integration finally as:

$$
k^e_{\text{square}} = \frac{E\,bh^3}{L^3\,12}
\begin{bmatrix}
\frac{243}{5} & \frac{156}{5}L & -\frac{243}{5} & \frac{87}{5}L \\[4pt]
\frac{156}{5}L & \frac{114}{5}L^2 & -\frac{156}{5}L & \frac{42}{5}L^2 \\[4pt]
-\frac{243}{5} & -\frac{156}{5}L & \frac{243}{5} & -\frac{87}{5}L \\[4pt]
\frac{87}{5}L & \frac{42}{5}L^2 & -\frac{87}{5}L & 9L^2
\end{bmatrix} .
\tag{5.214}
$$

5.6.2 Supplementary Problems

5.4 Equilibrium relation for infinitesimal beam element with variable load

Determine the vertical balance of forces and the equilibrium of moments for the beam element, which is illustrated in Fig. 5.24.

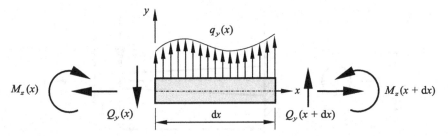

Fig. 5.24 Infinitesimal beam element with internal reactions and loading through a variable distributed load

Fig. 5.25 Quadratic distributed load

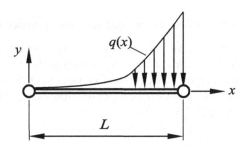

5.5 Weighted residual method with variable distributed load

Derive the finite element equation through the weighted residual method. The initial point therefore should be the bending differential equation *with* an arbitrary distributed load $q_y(x)$. Furthermore it should to be assumed, that the bending stiffness EI_z is constant.

5.6 Stiffness matrix at bending in the x-z plane

Derive the stiffness matrix for a beam element at bending in the x–z plane. For this see Eq. (5.144) and Fig. 5.16b.

5.7 Bending beam with variable cross-section

Solve problem 5.3 for arbitrary values of D_1 and D_2!

5.8 Equivalent nodal loads for quadratic distributed load

Calculate the equivalent nodal loads for the BERNOULLI beam, which is illustrated in Fig. 5.25 in the case of:

(a) $q(x) = q_0 x^2$,

(b) $q(x) = q_0 \left(\dfrac{x}{L}\right)^2$.

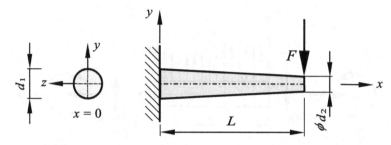

Fig. 5.26 Bending beam with variable cross-section at loading through point load

5.9 Bending beam with variable cross-section under point load

For the beam with variable cross-section, which is illustrated in Fig. 5.26, calculate for $d_1 = 2h$ and $d_2 = h$ the vertical displacement of the right-hand boundary. For this purpose, a single finite element has to be used and the numerical solution has to be compared with the exact solution. Advice: The stiffness matrix can be taken from example 5.3.

References

1. Hartman F, Katz C (2007) Structural analysis with finite elements. Springer, Berlin
2. Timoshenko S, Woinowsky-Krieger S (1959) Theory of plates and shells. McGraw-Hill Book Company, New York
3. Gould PL (1988) Analysis of shells and plates. Springer, New York
4. Altenbach H, Altenbach J, Naumenko K (1998) Ebene Flächentragwerke: Grundlagen der Modellierung und Berechnung von Scheiben und Platten. Springer, Berlin
5. Szabó I (2003) Einführung in die Technische Mechanik: Nach Vorlesungen István Szabó. Springer, Berlin
6. Gross D, Hauger W, Schröder J, Wall WA (2009) Technische Mechanik 2: Elastostatik. Springer, Berlin
7. Budynas RG (1999) Advanced strength and applied stress analysis. McGraw-Hill Book, Singapore
8. Hibbeler RC (2008) Mechanics of materials. Prentice Hall, Singapore
9. Szabó I (1996) Geschichte der mechanischen Prinzipien und ihrer wichtigsten Anwendungen. Birkhäuser, Basel
10. Szabó I (2001) Höhere Technische Mechanik: Nach Vorlesungen István Szabó. Springer, Berlin
11. Betten J (2004) Finite Elemente für Ingenieure 2: Variationsrechnung, Energiemethoden, Näherungsverfahren, Nichtlinearitäten, Numerische Integrationen. Springer, Berlin
12. Oden JT, Reddy JN (1976) Variational methods in theoretical mechanics. Springer, Berlin
13. Betten J (2001) Kontinuumsmechanik: Elastisches und inelastisches Verhalten isotroper und anisotroper Stoffe. Springer, Berlin
14. Hutton DV (2004) Fundamentals of finite element analysis. McGraw-Hill Book, Singapore

Chapter 6
General 1D Element

Abstract The three basic modes of deformation, i.e. tension, torsion and bending, can occur in an arbitrary combination. This chapter serves to introduce how the stiffness relation for a general 1D element can be composed. The stiffness relations of the basic types build the foundation. For 'simple' loadings the three basic types can be regarded separately and can easily be superposed. A mutual dependency is nonexistent. The generality of the 1D element also relates to the arbitrary orientation within space. Transformation rules from local to global coordinates are provided. As an example, structures in the plane as well as in three-dimensional space will be discussed. Furthermore, there will be a short introduction in the subject of numerical integration.

6.1 Superposition to a General 1D Element

A general 1D element can be derived from the basic types of tension, bending and torsion without mutual dependency. For an arbitrary point, the 3 forces and 3 moments can be represented as

- normal force $N(x)$,
- respectively a shear force and a bending moment around an axis of the cross-section: $Q_z(x)$, $M_{yb}(x)$, $Q_y(x)$, $M_{zb}(x)$ and
- torsional moment $M_t(x)$ around the body axis.

The six kinematic parameters are described as follows:

- the three displacements $u_x(x)$, $u_y(x)$ and $u_z(x)$. Usually the displacement in the body axis equals the displacement $u_x(x)$.
- the three rotations $\varphi_x(x)$, $\varphi_y(x)$, $\varphi_z(x)$.

Figure 6.1 shows the kinematic parameters, the forces and the moments.

The arrangement of the single parameters in the column matrices defines the structure of the total stiffness matrix. If the kinematic parameters are arranged in the order that follows

$$u = [u_x, u_y, u_z, \varphi_x, \varphi_y, \varphi_z]^\mathrm{T}, \tag{6.1}$$

© Springer International Publishing AG, part of Springer Nature 2018

A. Öchsner and M. Merkel, *One-Dimensional Finite Elements*,

https://doi.org/10.1007/978-3-319-75145-0_6

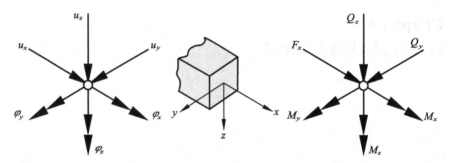

Fig. 6.1 State variables for the general three-dimensional case

the order of entries for the column matrix of *generalized forces* in the stiffness relation results in:

$$F = [N_x, Q_y, Q_z, M_x, M_y, M_z]^\mathrm{T} \ . \tag{6.2}$$

An alternative order results, if the column matrix of *generalized forces* is established in the following order

- normal force (in the direction of the x-axis),
- bending (around the y-axis and around the z-axis) and
- torsion (around the x-axis),

meaning

$$F = [N_x, Q_z, M_y, Q_y, M_z, M_x]^\mathrm{T}. \tag{6.3}$$

For this order, the single stiffness relation in Eq. (6.4) is illustrated. Under the assumption of a two-node element the stiffness matrix consists of the 6 respective entries on both nodes. The dimensions of the stiffness matrix results in 12×12.

$$
\begin{bmatrix} N_{1x} \\ Q_{1z} \\ M_{1y} \\ Q_{1y} \\ M_{1z} \\ M_{1x} \\ \hline N_{2x} \\ Q_{2z} \\ M_{2y} \\ Q_{2y} \\ M_{2z} \\ M_{2x} \end{bmatrix}
=
\left[\begin{array}{cccccc|cccccc}
Z & 0 & 0 & 0 & 0 & 0 & Z & 0 & 0 & 0 & 0 & 0 \\
0 & B_y & B_y & 0 & 0 & 0 & 0 & B_y & B_y & 0 & 0 & 0 \\
0 & B_y & B_y & 0 & 0 & 0 & 0 & B_y & B_y & 0 & 0 & 0 \\
0 & 0 & 0 & B_z & B_z & 0 & 0 & 0 & 0 & B_z & B_z & 0 \\
0 & 0 & 0 & B_z & B_z & 0 & 0 & 0 & 0 & B_z & B_z & 0 \\
0 & 0 & 0 & 0 & 0 & T & 0 & 0 & 0 & 0 & 0 & T \\
\hline
Z & 0 & 0 & 0 & 0 & 0 & Z & 0 & 0 & 0 & 0 & 0 \\
0 & B_y & B_y & 0 & 0 & 0 & 0 & B_y & B_y & 0 & 0 & 0 \\
0 & B_y & B_y & 0 & 0 & 0 & 0 & B_y & B_y & 0 & 0 & 0 \\
0 & 0 & 0 & B_z & B_z & 0 & 0 & 0 & 0 & B_z & B_z & 0 \\
0 & 0 & 0 & B_z & B_z & 0 & 0 & 0 & 0 & B_z & B_z & 0 \\
0 & 0 & 0 & 0 & 0 & T & 0 & 0 & 0 & 0 & 0 & T
\end{array} \right]
\begin{bmatrix} u_{1x} \\ u_{1z} \\ \varphi_{1y} \\ u_{1y} \\ \varphi_{1z} \\ \varphi_{1x} \\ \hline u_{2x} \\ u_{2z} \\ \varphi_{2y} \\ u_{2y} \\ \varphi_{2z} \\ \varphi_{2x} \end{bmatrix}
\tag{6.4}
$$

The stiffness matrix contains entries,

- which are marked with Z, for entries of the single stiffness matrix of the tension bar,
- which are marked with B_y and B_z, for the entries of the single stiffness matrix of beams bending around the y- and z-axis and
- which are marked with T, for the entries of the single stiffness matrix of the torsion bar.

The stiffness matrix contains 0-entries in many cells. This documents the decoupling of the basic types. To analyze a general three-dimensional problem, a user can choose between various ways for the choice of the elements. Generally the general stiffness matrix can be allocated to each 1D element. This however leads to an increased storage effort and lengthens computing times, since for many elements 'unnecessary' ballast is dragged along. Undoubtedly a preselection by the user makes sense. Commercial program packages mostly contain the basic types within their element library as well as several special cases.

6.1.1 Sample 1: Bar Under Tension and Torsion

In principle, a total stiffness matrix can be established through an arbitrary combination of basic types. Within this example the stiffness relation needs to be established through the basic types *tension bar* and *torsion bar*. Figure 6.2 illustrates the state variables, Fig. 6.2a the force parameter and Fig. 6.2b the deformation parameters.

The total stiffness relation for the 1D element

$$\begin{bmatrix} N_{1x} \\ M_{1x} \\ N_{2x} \\ M_{2x} \end{bmatrix} = \begin{bmatrix} Z & 0 & Z & 0 \\ 0 & T & 0 & T \\ Z & 0 & Z & 0 \\ 0 & T & 0 & T \end{bmatrix} \begin{bmatrix} u_{1x} \\ \varphi_{1x} \\ u_{2x} \\ \varphi_{2x} \end{bmatrix} \tag{6.5}$$

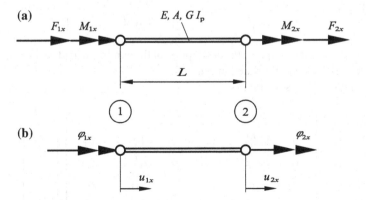

Fig. 6.2 Finite element for tension and torsion: **a** load parameters and **b** deformation parameters

consists of the basic types *tension bar* and *torsion bar*: In the matrix, the positions,

- which are marked with Z, state entries of the single stiffness matrix of the tension bar,
- which are marked with T, state entries of the single stiffness matrix of the torsion bar.

In detail, the total stiffness relation via the geometrical and material parameters is:

$$
\begin{bmatrix} N_{1x} \\ M_{1x} \\ N_{2x} \\ M_{2x} \end{bmatrix} = \begin{bmatrix} \dfrac{EA}{L} & 0 & -\dfrac{EA}{L} & 0 \\ 0 & \dfrac{GI_t}{L} & 0 & -\dfrac{GI_t}{L} \\ -\dfrac{EA}{L} & 0 & \dfrac{EA}{L} & 0 \\ 0 & -\dfrac{GI_t}{L} & 0 & \dfrac{GI_t}{L} \end{bmatrix} \begin{bmatrix} u_{1x} \\ \varphi_{1x} \\ u_{2x} \\ \varphi_{2x} \end{bmatrix}.
\tag{6.6}
$$

6.1.2 Sample 2: Beam in the Plane with Tension Part

For the bending beam with a normal force part the two basic load types bending and tension have to be combined. First the bending in the x-y plane needs to be described. The state variables of the combined load types are illustrated in Fig. 6.3.

The single stiffness relation is:

$$
\begin{bmatrix} N_{1x} \\ Q_{1y} \\ M_{1z} \\ N_{2x} \\ Q_{2y} \\ M_{2z} \end{bmatrix} = \begin{bmatrix} \frac{EA}{L} & 0 & 0 & -\frac{EA}{L} & 0 & 0 \\ 0 & 12\frac{EI_z}{L^3} & 6\frac{EI_z}{L^2} & 0 & -12\frac{EI_z}{L^3} & 6\frac{EI_z}{L^2} \\ 0 & 6\frac{EI_z}{L^2} & 4\frac{EI_z}{L} & 0 & -6\frac{EI_z}{L^2} & 2\frac{EI_z}{L} \\ -\frac{EA}{L} & 0 & 0 & \frac{EA}{L} & 0 & 0 \\ 0 & -12\frac{EI_z}{L^3} & -6\frac{EI_z}{L^2} & 0 & 12\frac{EI_z}{L^3} & -6\frac{EI_z}{L^2} \\ 0 & 6\frac{EI_z}{L^2} & 2\frac{EI_z}{L} & 0 & -6\frac{EI_z}{L^2} & 4\frac{EI_z}{L} \end{bmatrix} \begin{bmatrix} u_{1x} \\ u_{1y} \\ \varphi_{1z} \\ u_{2x} \\ u_{2y} \\ \varphi_{2z} \end{bmatrix}.
\tag{6.7}
$$

For the bending in the x-z plane the description of the combined load occurs similarly. The state variables are illustrated in Fig. 6.4.

The single stiffness relation is:

$$
\begin{bmatrix} N_{1x} \\ Q_{1z} \\ M_{1y} \\ N_{2x} \\ Q_{2z} \\ M_{2y} \end{bmatrix} = \begin{bmatrix} \frac{EA}{L} & 0 & 0 & -\frac{EA}{L} & 0 & 0 \\ 0 & 12\frac{EI_y}{L^3} & -6\frac{EI_y}{L^2} & 0 & -12\frac{EI_y}{L^3} & -6\frac{EI_y}{L^2} \\ 0 & -6\frac{EI_y}{L^2} & 4\frac{EI_y}{L} & 0 & 6\frac{EI_y}{L^2} & 2\frac{EI_y}{L} \\ -\frac{EA}{L} & 0 & 0 & \frac{EA}{L} & 0 & 0 \\ 0 & -12\frac{EI_y}{L^3} & 6\frac{EI_y}{L^2} & 0 & 12\frac{EI_y}{L^3} & 6\frac{EI_y}{L^2} \\ 0 & -6\frac{EI_y}{L^2} & 2\frac{EI_y}{L} & 0 & 6\frac{EI_y}{L^2} & 4\frac{EI_y}{L} \end{bmatrix} \begin{bmatrix} u_{1x} \\ u_{1z} \\ \varphi_{1y} \\ u_{2x} \\ u_{2z} \\ \varphi_{2y} \end{bmatrix}.
\tag{6.8}
$$

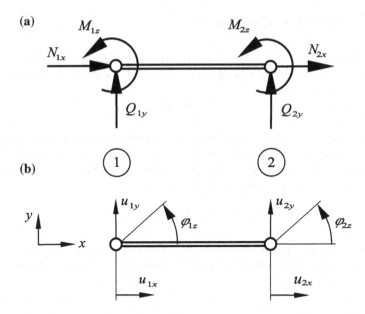

Fig. 6.3 Bending in the x-y plane with normal force: **a** load parameters and **b** deformation parameters

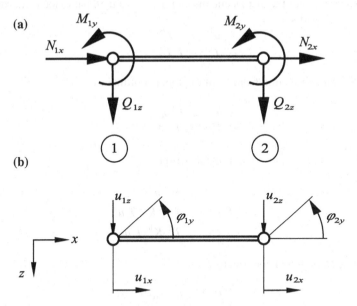

Fig. 6.4 Bending in the x-z plane with normal force: **a** load parameters and **b** deformation parameters

6.2 Coordinate Transformation

So far, the stiffness relation was formulated for a single element. The basis was the local (lo) coordinate system related to one element. For the simple tension bar the stiffness relation is called:

$$F^{\text{lo}} = k^{\text{lo}} \, u^{\text{lo}} \, . \tag{6.9}$$

However within a plane or general three-dimensional structure the single elements can be oriented arbitrarily within space. Usually a fixed, global coordinate system will be defined. The transformation rule for a vector between the local (lo) and global (glo) coordinate system is generally referred to as:

$$[\,\cdot\,]^{\text{lo}} = T \, [\,\cdot\,]^{\text{glo}} \, . \tag{6.10}$$

T is referred to as a transformation matrix. The mathematical characteristics are described in detail in the appendix. For the following derivation the relation

$$T^{-1} = T^{\text{T}} \tag{6.11}$$

is relevant. This transformation matrix is used for the conversion of all parameters. For the transformation of the displacements and forces from global to local coordinates, the following results

$$\begin{aligned} u^{\text{lo}} &= T \, u^{\text{glo}} \,, \\ F^{\text{lo}} &= T \, F^{\text{glo}} \,, \end{aligned} \tag{6.12}$$

and for the transformation from local to global coordinates

$$\begin{aligned} u^{\text{glo}} &= T^{\text{T}} \, u^{\text{lo}} \,, \\ F^{\text{glo}} &= T^{\text{T}} \, F^{\text{lo}} \, . \end{aligned} \tag{6.13}$$

The stiffness relation, after the transformation

$$\begin{aligned} F^{\text{lo}} &= K^{\text{lo}} \, u^{\text{lo}} \\ T^{-1} \, T \, F^{\text{glo}} &= T^{-1} \, K^{\text{lo}} \, T \, u^{\text{glo}} \\ F^{\text{glo}} &= T^{\text{T}} \, K^{\text{lo}} \, T \, u^{\text{glo}} \end{aligned}$$

can be written in global coordinates:

$$F^{\text{glo}} = K^{\text{glo}} \, u^{\text{glo}} \tag{6.14}$$

with

$$K^{\text{glo}} = T^{\text{T}} \, K^{\text{lo}} \, T \quad . \tag{6.15}$$

Within the following sections the transformation will be introduced with the help of examples for the rotation in the plane and in the general three-dimensional space.

6.2.1 Plane Structures

The transformation for plane structures can be shown graphically. The local x-y coordinate system is rotated with the angle α compared to the X-Y global coordinate system. The angle α is defined anticlockwise, in mathematical positive direction of rotation.[1]

The transformation relation between the local and the global coordinate system for a vector is called (Fig. 6.5):

$$\begin{bmatrix} \cdot \\ \cdot \end{bmatrix}^{\text{lo}} = \begin{bmatrix} \cos \alpha & \sin \alpha \\ -\sin \alpha & \cos \alpha \end{bmatrix} \begin{bmatrix} \cdot \\ \cdot \end{bmatrix}^{\text{glo}} . \tag{6.16}$$

First, the transformation for the element tension bar will be shown. To guarantee a better overview Fig. 6.6 only illustrates the forces. For the displacements, an equal procedure applies.

Fig. 6.5 Coordinate transformation in the plane

Fig. 6.6 Coordinate transformation for the tension bar in the plane

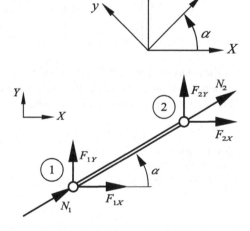
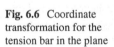

---

[1] Other definition results in a different transformation.

With two nodes the transformation matrix is:

$$T = \begin{bmatrix} \cos\alpha & \sin\alpha & 0 & 0 \\ -\sin\alpha & \cos\alpha & 0 & 0 \\ 0 & 0 & \cos\alpha & \sin\alpha \\ 0 & 0 & -\sin\alpha & \cos\alpha \end{bmatrix}. \tag{6.17}$$

Based on the description in local coordinates, the state vectors are brought into the same dimension (4 components)

$$F^{\text{lo}} = \begin{bmatrix} N_1 \\ 0 \\ N_2 \\ 0 \end{bmatrix} , \quad F^{\text{glo}} = \begin{bmatrix} F_{1X} \\ F_{1Y} \\ F_{2X} \\ F_{2Y} \end{bmatrix} \tag{6.18}$$

and

$$u^{\text{lo}} = \begin{bmatrix} u_1 \\ 0 \\ u_2 \\ 0 \end{bmatrix} , \quad u^{\text{glo}} = \begin{bmatrix} u_{1X} \\ u_{1Y} \\ u_{2X} \\ u_{2Y} \end{bmatrix}. \tag{6.19}$$

For the transformation of the single stiffness matrix, one has to conduct the transformation rule of Eq. (6.15) and finally obtains

$$k = k^{\text{T}} = \frac{EA}{L} \begin{bmatrix} \cos^2\alpha & \cos\alpha\sin\alpha & -\cos^2\alpha & -\cos\alpha\sin\alpha \\ & \sin^2\alpha & -\cos\alpha\sin\alpha & -\sin^2\alpha \\ & & \cos^2\alpha & \cos\alpha\sin\alpha \\ \text{sym.} & & & \sin^2\alpha \end{bmatrix}. \tag{6.20}$$

For the rotation of the bending beam in the plane, normal force, shear force and the momental vector are already considered in the local coordinate system (see Fig. 6.7). This one is at a right angle and also remains so during a rotation.

Fig. 6.7 Coordinate transformation for the bending beam in the X-Y plane

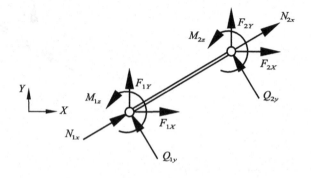

The following transformation matrix results for the bending in the X-Y plane

$$T^{XY} = \begin{bmatrix} \cos\alpha & \sin\alpha & 0 & 0 & 0 & 0 \\ -\sin\alpha & \cos\alpha & 0 & 0 & 0 & 0 \\ 0 & 0 & 1 & 0 & 0 & 0 \\ 0 & 0 & 0 & \cos\alpha & \sin\alpha & 0 \\ 0 & 0 & 0 & -\sin\alpha & \cos\alpha & 0 \\ 0 & 0 & 0 & 0 & 0 & 1 \end{bmatrix} \qquad (6.21)$$

for the beam element with tensile loading. A corresponding transformation rule can be formulated for the bending in the X-Z plane.

6.2.2 General Three-Dimensional Structures

The transformation for general, three-dimensional structures can formally be also described via Eq. (6.10). Graphically however it is not so easy anymore to display the transformation. The local (x, y, z) coordinate system is defined via three coordinate axes. These can be rotated arbitrarily compared to a global (X, Y, Z) coordinate system (see Fig. 6.8).

A one-dimensional finite element 0-1 is represented by the vector

$$L = (X_1 - X_0)e_X + (Y_1 - Y_0)e_Y + (Z_1 - Z_0)e_Z, \qquad (6.22)$$

which is oriented in the direction of the local x-axis.

A vector in the direction of the local y-axis can, according to Fig. 6.8, be illustrated as follows:

$$Y = (X_2 - X_0)e_X + (Y_2 - Y_0)e_Y + (Z_2 - Z_0)e_Z. \qquad (6.23)$$

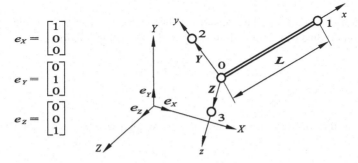

Fig. 6.8 Rotatory transformation of a 1D element in the space. The unit vectors in the direction of the global X-, Y- and Z-axis are labelled with e_i

The direction cosines between the local y-axis and the global coordinate axes result via the global nodal coordinates to

$$t_{yX} = \cos(y, X) = \frac{X_2 - X_0}{|\boldsymbol{Y}|},$$ (6.24)

$$t_{yY} = \cos(y, Y) = \frac{Y_2 - Y_0}{|\boldsymbol{Y}|},$$ (6.25)

$$t_{yZ} = \cos(y, Z) = \frac{Z_2 - Z_0}{|\boldsymbol{Y}|},$$ (6.26)

whereupon the length of the vector $\boldsymbol{Y}$ results in

$$|\boldsymbol{Y}| = \sqrt{(X_2 - X_0)^2 + (Y_2 - Y_0)^2 + (Z_2 - Z_0)^2}.$$ (6.27)

Accordingly a vector in the direction of the local z-axis results in:

$$\boldsymbol{Z} = (X_3 - X_0)\boldsymbol{e}_X + (Y_3 - Y_0)\boldsymbol{e}_Y + (Z_3 - Z_0)\boldsymbol{e}_Z$$ (6.28)

and the direction cosines in

$$t_{zX} = \cos(z, X) = \frac{X_3 - X_0}{|\boldsymbol{Z}|},$$ (6.29)

$$t_{zY} = \cos(z, Y) = \frac{Y_3 - Y_0}{|\boldsymbol{Z}|},$$ (6.30)

$$t_{zZ} = \cos(z, Z) = \frac{Z_3 - Z_0}{|\boldsymbol{Z}|},$$ (6.31)

whereupon the length of the vector $\boldsymbol{Z}$ results in

$$|\boldsymbol{Z}| = \sqrt{(X_3 - X_0)^2 + (Y_3 - Y_0)^2 + (Z_3 - Z_0)^2}.$$ (6.32)

An arbitrary vector v can be transformed between the local (x, y, z) and global (X, Y, Z) coordinate system with the help of the following relation:

$$\boldsymbol{v}_{xyz} = \boldsymbol{T}^{3D}\boldsymbol{v}_{XYZ},$$ (6.33)

$$\boldsymbol{v}_{XYZ} = \boldsymbol{T}^{3D^T}\boldsymbol{v}_{xyz},$$ (6.34)

whereupon the transformation matrix can be written as follows via the direction cosines:

$$\boldsymbol{T}^{3D} = \begin{bmatrix} t_{xX} & t_{xY} & t_{xZ} \\ t_{yX} & t_{yY} & t_{yZ} \\ t_{zX} & t_{zY} & t_{zZ} \end{bmatrix} = \begin{bmatrix} \cos(x, X) & \cos(x, Y) & \cos(x, Z) \\ \cos(y, X) & \cos(y, Y) & \cos(y, Z) \\ \cos(z, X) & \cos(z, Y) & \cos(z, Z) \end{bmatrix}.$$ (6.35)

A general one-dimensional element reveals 6 kinematic state variables and 6 'force parameters' at each node. For a two-node element a transformation matrix with the dimensions 12×12 results for the transformation of a state variable. With Eq. (6.36) the transformation relation from global to local coordinates for the kinematic state variables is represented here as an example.

$$
\begin{bmatrix} u_{1x} \\ u_{1y} \\ u_{1z} \\ \varphi_{1x} \\ \varphi_{1y} \\ \varphi_{1z} \\ u_{2x} \\ u_{2y} \\ u_{2z} \\ \varphi_{2x} \\ \varphi_{2y} \\ \varphi_{2z} \end{bmatrix} = \underbrace{\begin{bmatrix} t_{xX} & t_{xY} & t_{xZ} & 0 & 0 & 0 & 0 & 0 & 0 & 0 & 0 & 0 \\ t_{yX} & t_{yY} & t_{yZ} & 0 & 0 & 0 & 0 & 0 & 0 & 0 & 0 & 0 \\ t_{zX} & t_{zY} & t_{zZ} & 0 & 0 & 0 & 0 & 0 & 0 & 0 & 0 & 0 \\ 0 & 0 & 0 & t_{xX} & t_{xY} & t_{xZ} & 0 & 0 & 0 & 0 & 0 & 0 \\ 0 & 0 & 0 & t_{yX} & t_{yY} & t_{yZ} & 0 & 0 & 0 & 0 & 0 & 0 \\ 0 & 0 & 0 & t_{zX} & t_{zY} & t_{zZ} & 0 & 0 & 0 & 0 & 0 & 0 \\ 0 & 0 & 0 & 0 & 0 & 0 & t_{xX} & t_{xY} & t_{xZ} & 0 & 0 & 0 \\ 0 & 0 & 0 & 0 & 0 & 0 & t_{yX} & t_{yY} & t_{yZ} & 0 & 0 & 0 \\ 0 & 0 & 0 & 0 & 0 & 0 & t_{zX} & t_{zY} & t_{zZ} & 0 & 0 & 0 \\ 0 & 0 & 0 & 0 & 0 & 0 & 0 & 0 & 0 & t_{xX} & t_{xY} & t_{xZ} \\ 0 & 0 & 0 & 0 & 0 & 0 & 0 & 0 & 0 & t_{yX} & t_{yY} & t_{yZ} \\ 0 & 0 & 0 & 0 & 0 & 0 & 0 & 0 & 0 & t_{zX} & t_{zY} & t_{zZ} \end{bmatrix}}_{T} \begin{bmatrix} u_{1X} \\ u_{1Y} \\ u_{1Z} \\ \varphi_{1X} \\ \varphi_{1Y} \\ \varphi_{1Z} \\ u_{2X} \\ u_{2Y} \\ u_{2Z} \\ \varphi_{2X} \\ \varphi_{2Y} \\ \varphi_{2Z} \end{bmatrix}. \quad (6.36)
$$

The transformation of the stiffness matrix results in:

$$
k_{XYZ}^{e} = T^{3D^{T}} k_{xyz}^{e} T . \quad (6.37)
$$

The small letters stand for the axes of the local coordinate system and the capital letters for the axes of the global coordinate system.

6.3 Numerical Integration of a Finite Element

Within this chapter a short introduction into numerical integration is conducted. For a comprehensive overview, the reader is referred to the relevant literature at this point [1]. The subject will be introduced with regard to one-dimensional problems.

For the approximate calculation of certain integrals, a number of numerical algorithms or so-called quadrature rules is available. The basic idea is to fractionize the integral

$$
\int_{a}^{b} f(x)\,\mathrm{d}x \approx \sum_{i=1}^{q} f(x_{i})\,\Delta x_{i} \quad (6.38)
$$

into subintervals and to subsequently sum up. Graphically this is shown in Fig. 6.9.

Formulated more generally, the integral consists of partial contributions, which are each calculated through a function value and a weighting coefficient:

Fig. 6.9 Numerical
integration of a function

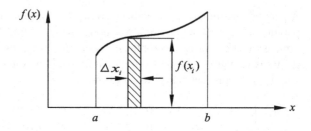

$$\int\limits_{a}^{b} f(x)\,\mathrm{d}x \approx \sum_{i=1}^{q} f(\xi_i)\,W_i\,. \tag{6.39}$$

Within the integration formula, the ξ_i are referred to as the integration points, the $f(\xi_i)$ as the discrete function values on these integration points, and W_i as the weighting coefficient. q stands for the integration order. Numerical integration thus means a multiplication of function values of the integrand on discrete points of support with weights. Subsequently the partial contributions are summed up.

Independent from the boundaries of integration, numerical integration is mostly conducted in the unit domain between -1 and $+1$, which means that the integration interval is transformed on the domain $-1 \leq \xi \leq 1$.

The Gauss Quadrature

Within the framework of the finite element method mostly the quadrature formula according to GAUSS is used for the numerical integration. An essential advantage is that a polynomial of the order

$$m = 2q - 1 \tag{6.40}$$

with q integration points can be exactly integrated. Two integration points consequently allow a cubic polynomial to be exactly integrated and with three integration points a polynomial of 5th order can be exactly integrated. The position of the integration points as well as the corresponding weights can be found in tables. Table 6.1 shows the values up to an integration order $q = 3$.

Table 6.1 Points of support and weights for the numerical integration according to GAUSS–LEGENDRE

q	Points of support ξ_i	Weights W_i
1	0.00000	2.00000
2	$\pm\dfrac{1}{\sqrt{3}} = \pm 0.57735$	1.00000
3	$\pm\sqrt{\dfrac{3}{5}} = \pm 0.77459$	$\pm\dfrac{5}{9} = \pm 0.55555$
	0.00000	$+\dfrac{8}{9} = 0.88888$

The integration illustrated for the one-dimensional case can easily be expanded to integrals with a higher dimension.

Example

The integral

$$\int_{-1}^{+1} (1 + 2x + 3x^2) \, dx \qquad (6.41)$$

is to be evaluated with the help of the quadrature formula according to GAUSS.

Solution

The exact solution results in

$$\int_{-1}^{+1} (1 + 2x + 3x^2) \, dx = \left[x + x^2 + x^3\right]_{-1}^{+1} = \qquad (6.42)$$

$$(1 + 1 + 1) - (-1 + 1 - 1) = 4 \ .$$

The integrand is a polynomial of second order. From $m = 2 = 2q - 1$ the necessary integration order is calculated to $q = 1.5$. Since this has to be integer, the integration order is appointed with $q = 2$. The positions of the integration points ξ_i with the corresponding weights W_i

$$\xi_{1/2} = \pm \frac{1}{\sqrt{3}} \quad , \quad W_{1/2} = 1.0 \qquad (6.43)$$

for the integration order $q = 2$ are taken from Table 6.1. The numerical integration

$$\int_{-1}^{+1} (1 + 2x + 3x^2) \, dx \approx \sum_{i=1}^{q=2} f(\xi_i) \, W_i = f(\xi_1) \, W_1 + f(\xi_2) \, W_2$$

$$= \left[1 + 2\left(-\frac{1}{\sqrt{3}}\right) + 3\left(-\frac{1}{\sqrt{3}}\right)^2\right] \cdot 1.0 \qquad (6.44)$$

$$+ \left[1 + 2\left(+\frac{1}{\sqrt{3}}\right) + 3\left(+\frac{1}{\sqrt{3}}\right)^2\right] \cdot 1.0$$

$$= \left[1 + 2\left(-\frac{1}{\sqrt{3}}\right) + 1\right] + \left[1 + 2\left(+\frac{1}{\sqrt{3}}\right) + 1\right] = 4$$

delivers the exact result.

6.4 Shape Function

Within the framework of the finite element method, functions have to be approximated. In previous chapters shape functions have already been introduced to approximate the displacement distribution within elements. Now this subject will be introduced in detail. It is the goal to describe the distribution of a physical parameter as easily as possible. As an example, the distribution of the displacement, the strain and the stress along the center line of a bar needs to be named. A common approach is to describe the real distribution of a function through a combination of function values on selected positions, the nodes of a element and functions between those points of support. This approach is also referred to as the nodal approach. The displacement distribution within an element is approximated with

$$u^{e}(x) = N^{e}(x) \, u_{p}. \tag{6.45}$$

The parameter $u^{e}(x)$ describes the distribution of the displacement within the element, N stands for the shape function. The index 'e' in the equation stands for the parameter, which is related to an element. The index 'p' labels the node p, at which the nodal point displacement u_{p} was introduced.

In principle, arbitrary functions can be chosen for the interpolation, however the following conditions have to be fulfilled:

- The shape function has to be continuous on the inside of an element.
- The shape function also has to be constant on the boundaries towards neighboring elements.
- It has to be possible to describe a rigid-body motion with the shape function, so that no strains or stresses in the element are caused as a result.

In general, polynomials fulfill these requirements. In the framework of the FEM such special polynomials are called LAGRANGE polynomials. A LAGRANGE polynomial of the order $n - 1$ is defined through n function values on the coordinates $x_1, x_2, x_3, x_i, ..., x_n$:

$$L_{i}^{n}(x) = (x - x_1) \, (x - x_2) \, (x - x_{i-1})(x - x_{i+1}) \cdots (x - x_n) . \tag{6.46}$$

Especially it needs to be remarked that the LAGRANGE polynomial

- $L_{i}^{n}(x_j)$ takes on the function values $L_{i}^{n}(x_j) = 0$ on the locations $x_j = 1, 2, ..., n$, ($j \neq i$) and
- $L_{i}^{n}(x_i)$ takes on the function value $L_{i}^{n}(x_i) \neq 0$ at the position x_i.

If the points of support of the LAGRANGE polynomial are put on the nodes of an element and the non-zero function value is used for the scaling, then as a result appropriate shape functions are constructed:

Fig. 6.10 Coordinates **a** physical and **b** natural

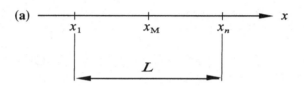

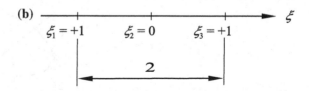

$$N_i(x) = \frac{L_i^n(x)}{L_i^n(x_i)} = \prod_{j=1, j \neq i}^{n} \frac{(x - x_j)}{(x_i - x_j)}. \tag{6.47}$$

For the description of a physical parameter, it can makes sometimes sense to define another coordinate system. Element-own coordinates, so-called *natural coordinates* ξ are introduced. Figure 6.10a shows the local and Fig. 6.10b the natural coordinates.

The transformation can be described through

$$\xi = \frac{x - x_M}{L} \times 2. \tag{6.48}$$

The centre of the element is described with x_M or in natural coordinates with $\xi = 0$. The beginning and the end of the element are described with the natural coordinates $\xi = -1$ and $\xi = +1$. The shape function can also be formulated in natural coordinates

$$N_i(\xi) = L_i(\xi) = \prod_{j=1, j \neq i}^{n} \frac{(\xi - \xi_j)}{(\xi_i - \xi_j)}. \tag{6.49}$$

For a bar element with *linear* shape function, with the two nodes on the coordinates $\xi_1 = -1$ and $\xi_2 = +1$ the two shape functions result in

$$
\begin{aligned}
N_1(\xi) &= \frac{(\xi - \xi_2)}{(\xi_1 - \xi_2)} = \frac{(\xi - 1)}{(-1 - (+1))} = \frac{1}{2}(1 - \xi), \\
N_2(\xi) &= \frac{(\xi - \xi_1)}{(\xi_2 - \xi_1)} = \frac{(\xi - (-)1)}{(+1 - (-1))} = \frac{1}{2}(1 + \xi).
\end{aligned}
\tag{6.50}
$$

Figure 6.11 shows the two shape functions for the linear approach.

For a bar element with *quadratic* shape functions, with the three nodes on the coordinates $\xi_1 = -1$, $\xi_2 = 0$ and $\xi_3 = +1$, the three shape functions result in

Fig. 6.11 Shape functions, linear approach

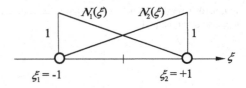

Fig. 6.12 Shape function, quadratic approach

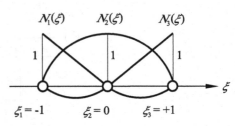

$$N_1(\xi) = \frac{(\xi - \xi_2)(\xi - \xi_3)}{(\xi_1 - \xi_2)(\xi_1 - \xi_3)} = \frac{(\xi - 0)(\xi - 1)}{(-1 - 0)(-1 - (+1))} = \frac{1}{2}\xi(\xi - 1),$$

$$N_2(\xi) = \frac{(\xi - \xi_1)(\xi - \xi_3)}{(\xi_2 - \xi_1)(\xi_2 - \xi_3)} = \frac{(\xi - (-1))(\xi - 1)}{(0 - (-1)(0 - (+1))} = (1 - \xi^2), \qquad (6.51)$$

$$N_3(\xi) = \frac{(\xi - \xi_1)(\xi - \xi_2)}{(\xi_3 - \xi_1)(\xi_3 - \xi_2)} = \frac{(\xi - (-1))(\xi - 0)}{(+1 - (-1)(+1 - 0)} = \frac{1}{2}\xi(1 + \xi).$$

Figure 6.12 shows the three shape functions for a quadratic approach.

The definition of the shape functions for a cubic approach represents the content of exercise 6.1.

6.5   Unit Domain

Within the process of finite element analysis, numerous vectors and matrices are defined via integration through a state variable X. This is formulated as follows:

$$\int_{\Omega} X \, d\Omega . \qquad (6.52)$$

Hereby X mainly is a parameter, which is dependent on the shape functions N or their derivatives. As an example, the stiffness matrix

$$K = \int_{\Omega} B^{\mathrm{T}} D B \, d\Omega \qquad (6.53)$$

needs to be given. Two transformations are necessary for the execution of the integration. In the first step, a transformation from global to local coordinates is conducted.

This step was already discussed in Sect. 6.2. Secondly the integration domain is transformed into a unit domain:

$$\int_\Omega X(x, y, z)\, d\Omega = \int_{-1}^{+1}\int_{-1}^{+1}\int_{-1}^{+1} \overline{X}(\xi, \eta, \zeta)\, J(\xi, \eta, \zeta)\, d\xi\, d\eta\, d\zeta \tag{6.54}$$

with

$$X(x, y, z) = \overline{X}(x(\xi, \eta, \zeta), y(\xi, \eta, \zeta), z(\xi, \eta, \zeta)) = \overline{X}(\xi, \eta, \zeta). \tag{6.55}$$

In Eq. (6.54) the following expression

$$J(\xi, \eta, \zeta) = \frac{\partial(x, y, z)}{\partial(\xi, \eta, \zeta)} \tag{6.56}$$

stands for the JACOBIan matrix. If the integration is only conducted in one dimension, the transformation rule can be simplified to

$$\int_L X(x)\, dx = \int_{-1}^{+1} \overline{X}(\xi)\, J(\xi)\, d\xi = \int_{-1}^{+1} \overline{X}(\xi)\, \frac{\partial x}{\partial \xi}\, d\xi = \tag{6.57}$$

with

$$X(x) = \overline{X}(x(\xi)) = \overline{X}(\xi). \tag{6.58}$$

6.6 Supplementary Problems

6.1 Cubic displacement distribution inside the tension bar

Approximate the displacement distribution for a bar element through LAGRANGE polynomials. Unknown are the four shape functions in natural coordinates ξ for a cubic approximation of the displacement distribution in the bar element.

6.2 Coordination transformation for a tension bar in the plane

In a plane the local coordinate system for a bar is rotated by $\alpha = 30°$ compared to the global X-Y coordinate system. The bar is represented by 2 nodes (Fig. 6.13).
   The following has to be defined

1. the transformation matrix and
2. the single stiffness relation in the global X-Y coordinate system.

Fig. 6.13 Rotated bar in the plane

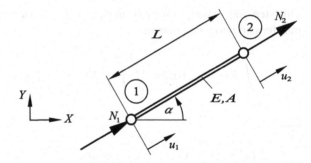

Reference

1. Onate E (2009) Structural analysis with the finite element method. Springer, Berlin

Chapter 7
Plane and Spatial Frame Structures

Abstract The procedure for the analysis of a load-bearing structure will be introduced in this chapter. Structures will be considered, which consist of multiple elements and are connected with each other on coupling points. The structure is supported properly and subjected with loads. Unknown are the deformations of the structure and the reaction forces at the supports. Furthermore, the internal reactions of the single element are of interest. The stiffness relation of the single elements are already known from the previous chapters. A global stiffness relation forms on the basis of these single stiffness relations. From a mathematical point of view the evaluation of the global stiffness relation equals the solving of a linear system of equations. As examples plane and general three-dimensional structures of bars and beams will be introduced.

7.1 Assembly of the Global Stiffness Relation

It is the goal of this section to formulate the stiffness relation for an entire structure. It is assumed that the stiffness relations for each element are known and can be set up. Each element is connected with the neighboring elements through nodes. One obtains the global stiffness relation by setting up the equilibrium of forces on each node. The structure of the global stiffness relation is therefore predetermined:

$$Ku = F.$$ (7.1)

The dimensions of the column matrices F and u equal the sum of the degrees of freedom on all nodes. The assembly of the global matrix K can be illustrated graphically by *sorting* all submatrices k^e in the global stiffness matrix. Formally this can be noted as follows:

$$K = \sum_e k^e.$$ (7.2)

© Springer International Publishing AG, part of Springer Nature 2018
A. Öchsner and M. Merkel, *One-Dimensional Finite Elements*,
https://doi.org/10.1007/978-3-319-75145-0_7

The setup of the global stiffness relation occurs in multiple steps:

1. The single stiffness matrix k^e is known for each element.
2. It is known which nodes are attached to each element. The single stiffness relation can therefore be formulated for each element in local coordinates:

$$F^e = k^e \, u_p.$$

3. The single stiffness relation, formulated in local coordinates has to be formulated in global coordinates.
4. The dimension of the global stiffness matrix is defined via the sum of the global degrees of freedom on all nodes.
5. A numeration of the nodes and the degrees of freedom on each node has to be defined.
6. The entries from the single stiffness matrix have to be sorted in the corresponding positions in the global stiffness matrix.

This can be shown with the help of a simple example.

Given is a bar-similar structure with length $2L$ with constant cross-section A. The structure is divided into two parts with length L with differing material (this means different moduli of elasticity). The structure has a fixed support on one side and is loaded with a point load F on the other side (Fig. 7.1).

For further analysis the structure will be divided into two parts each of length L. The example therefore consists of two finite elements and three nodes, which are numbered in the sequence 1 - 2 - 3 (see Fig. 7.2). The nodes 1 and 2 are attached to element I. The single stiffness relation for the element I is:

Fig. 7.1 Bar-shaped structure with the length $2L$

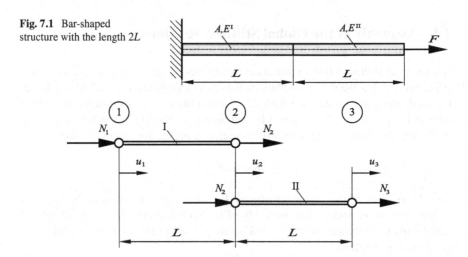

Fig. 7.2 Discretized structure with two finite elements

$$\begin{bmatrix} k & -k \\ -k & k \end{bmatrix}^{\mathrm{I}} \begin{bmatrix} u_1 \\ u_2 \end{bmatrix}^{\mathrm{I}} = \begin{bmatrix} N_1 \\ N_2 \end{bmatrix}^{\mathrm{I}} \tag{7.3}$$

with

$$k^{\mathrm{I}} = \frac{EA}{L}. \tag{7.4}$$

Nodes 2 and 3 are attached to element II. The single stiffness relation for the element II is:

$$\begin{bmatrix} k & -k \\ -k & k \end{bmatrix}^{\mathrm{II}} \begin{bmatrix} N_2 \\ N_3 \end{bmatrix}^{\mathrm{II}} = \begin{bmatrix} u_2 \\ u_3 \end{bmatrix}^{\mathrm{II}} \tag{7.5}$$

with

$$k^{\mathrm{II}} = \frac{EA}{L}. \tag{7.6}$$

Since the local and global coordinate systems are identical for the present problem, a coordinate transformation is omitted. The dimension of the global stiffness relation results in 3×3, since one degree of freedom exists on each node. The numbering of the degrees of freedom is defined in the sequence 1 - 2 - 3. The global stiffness relation results by assembling all submatrices in the global stiffness matrix:

$$\begin{bmatrix} k^{\mathrm{I}} & -k^{\mathrm{I}} & 0 \\ -k^{\mathrm{I}} & k^{\mathrm{I}} + k^{\mathrm{II}} & -k^{\mathrm{II}} \\ 0 & -k^{\mathrm{II}} & k^{\mathrm{II}} \end{bmatrix} \begin{bmatrix} u_1 \\ u_2 \\ u_3 \end{bmatrix} = \begin{bmatrix} N_1 \\ N_2 \\ N_3 \end{bmatrix}. \tag{7.7}$$

The numbering of the nodes has an influence on the structure of the global stiffness matrix. Instead of numbering the nodes with 1 - 2 - 3, the sequence 1 - 3 - 2 can be defined (Fig. 7.3).

Accordingly the total stiffness relation results in:

$$\begin{bmatrix} k^{\mathrm{I}} & -k^{\mathrm{I}} & 0 \\ -k^{\mathrm{I}} & k^{\mathrm{I}} + k^{\mathrm{II}} & -k^{\mathrm{II}} \\ 0 & -k^{\mathrm{II}} & k^{\mathrm{II}} \end{bmatrix} \begin{bmatrix} u_1 \\ u_3 \\ u_2 \end{bmatrix} = \begin{bmatrix} N_1 \\ N_3 \\ N_2 \end{bmatrix}. \tag{7.8}$$

For the case when the numbering is chosen in ascending order (1-2-3), the following results:

$$\begin{bmatrix} k^{\mathrm{I}} & 0 & -k^{\mathrm{I}} \\ 0 & k^{\mathrm{II}} & -k^{\mathrm{II}} \\ -k^{\mathrm{I}} & -k^{\mathrm{II}} & k^{\mathrm{I}} + k^{\mathrm{II}} \end{bmatrix} \begin{bmatrix} u_1 \\ u_2 \\ u_3 \end{bmatrix} = \begin{bmatrix} N_1 \\ N_2 \\ N_3 \end{bmatrix}. \tag{7.9}$$

Fig. 7.3 Alternative numbering of nodes for a structure with two bars

Fig. 7.4 Structure consisting
of four bar elements

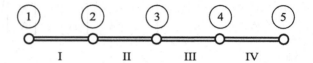

Compared to Eq. (7.7) the zero entries in the system matrix are in different positions. The numbering of the nodes can influence the result. With an exact number
representation and a non-occurring rounding error while conducting mathematical
operations the numbering of the nodes would not have an influence on the final result.
However in practice, exclusively numerical methods are used. Herewith, for example the sequence of the single mathematical operations and the computer internal
number representation have an influence on the subresult as well as the final result.
Especially the structure of the system matrix is of importance for the result of the
equation. Thus, the assembly process influences speed and quality of the result.

The just now considered example can easily be expanded to multiple elements.
Four bar elements are arranged behind each other in Fig. 7.4.

The global stiffness relation is:

$$
\begin{bmatrix}
k^{\mathrm{I}} & -k^{\mathrm{I}} & 0 & 0 & 0 \\
-k^{\mathrm{I}} & k^{\mathrm{I}}+k^{\mathrm{II}} & -k^{\mathrm{II}} & 0 & 0 \\
0 & -k^{\mathrm{II}} & k^{\mathrm{II}}+k^{\mathrm{III}} & -k^{\mathrm{III}} & 0 \\
0 & 0 & -k^{\mathrm{III}} & k^{\mathrm{III}}+k^{\mathrm{IV}} & -k^{\mathrm{IV}} \\
0 & 0 & 0 & -k^{\mathrm{IV}} & k^{\mathrm{IV}}
\end{bmatrix}
\begin{bmatrix}
u_1 \\ u_2 \\ u_3 \\ u_4 \\ u_5
\end{bmatrix}
=
\begin{bmatrix}
N_1 \\ N_2 \\ N_3 \\ N_4 \\ N_5
\end{bmatrix} . \qquad (7.10)
$$

The band structure in the system matrix is clearly visible. Around the main diagonal
respectively one secondary diagonal is occupied. Large domains have zero entries. By
tendency the domains with zero entries become proportionally bigger with increasing
number of finite elements within a structure, the domains with non-zero entries
become smaller. The concentration of the non-zero entries around the main diagonal
cannot be enforced at all times. Structures with coupling points, at which multiple
elements concur, lead to non-zero entries in the zero domains.

The unknown parameters can be found through the global stiffness relation. Therefore at first proper preconditions have to be established. In a mathematical sense the
system matrix is still singular. Degrees of freedom have to be taken from the system.
Graphically this means that at least so many degrees of freedom have to be taken
until the rigid-body motion of the remaining system becomes impossible. A reduced
system results from the total stiffness relation:

$$
\boldsymbol{K}^{\mathrm{red}}\, \boldsymbol{u}^{\mathrm{red}} = \boldsymbol{F}^{\mathrm{red}} . \qquad (7.11)
$$

Herefrom the unknown parameters can be determined. Within the following sequence
the solving of the system equation will be elaborated in detail.

7.2 Solving of the System Equation

The solving of a linear system of equations such as system (7.11) is part of the basic tasks in mathematics. A common illustration yields:

$$A\,x = b.\qquad(7.12)$$

The matrix A is referred to as a system matrix, the vector x contains the unknowns and the vector b represents the right-hand side. The right-hand side represents a load case from the mechanical point of view. Multiple load cases result in a right-side-matrix, whose number of columns equals the number of load cases. If Eqs. (7.11) and (7.12) are compared,

- the system matrix A with the reduced stiffness matrix K^{red},
- the vector of the unknown x with u^{red} and
- the column matrix of the right-hand side b with F^{red}

can be identified. To solve the linear system of equations basically two methods are possible:

- Direct and
- iterative methods.

For an in-depth discussion it can be referred to corresponding literature at this point [1]. From the point of view of the user the following criteria are paramount:

- the reliability of the solvers,
- the accuracy of the solution,
- the time, which is needed for the solving and
- the resources, which are made use of.

The direct methods can be characterized through the following attributes:

- The system is solvable for a well constructed problem.
- The direct solver is implementable as a *black box*.
- Multiple load cases and therefore more right-hand sides can be handled without significant additional expenditure.
- The calculating time is basically defined through the dimensions of the system matrix.
- The accuracy of the solution is basically defined through the computer internal number representation.

For iterative methods interim solutions are determined according to a fixed algorithm, based on an initial solution. The essential attributes are:

- The convergence of an iterative method cannot be guaranteed for every application.
- The accuracy of the solution can be influenced and given by the user.
- Multiple right-hand sides demand multiple computing times (n right-hand sides mean n-times solution of the system of equations).

With iterative methods, solutions for many applications can be found very quickly. The computing times for a load case can be a few percent of the computing times of the direct solver. In commercial program packages mainly direct methods are used for the solution of the equation system. Extended computing times seem to be more reasonable to the user as a possible termination of the iterative solution algorithm.

7.3 Postprocessing

After the solving of the linear system of equations, the following displacements u_p are known on each node for general problems in a global coordinate system. For a further evaluation in the single elements, the displacements are each transformed in the element-own local coordinate system. With the shape functions the displacement field in each element

$$u^\mathrm{e}(x) = N(x)u_\mathrm{p} \qquad (7.13)$$

can be defined. Through the kinematics relation furthermore the strain field

$$\varepsilon^\mathrm{e}(x) = \mathcal{L}_1 u^\mathrm{e}(x) = \mathcal{L}_1 N(x) u_\mathrm{p} \qquad (7.14)$$

and through the constitutive equation the stress field in the element

$$\sigma^\mathrm{e}(x) = D \varepsilon^\mathrm{e}(x) \qquad (7.15)$$

can be defined. Furthermore, the unknown support reactions can be determined with the nodal displacement values via a so-called follow-up calculation.

7.4 Examples in the Plane

This section serves to discuss structures which are located in a plane. The first example deals with a structure, which consists of two bars. The second example deals with a beam and a bar.

7.4.1 Plane Structure with Two Bars

As a first and simple example a structure will be discussed which is made of two bars (see Fig. 7.5). Both bars have the same length $\sqrt{2} \times L$ and the same cross-section A, consist of the same material (same modulus of elasticity E), are each simply supported at one end and are pin-jointed at the position C. A single force F acts on position C.

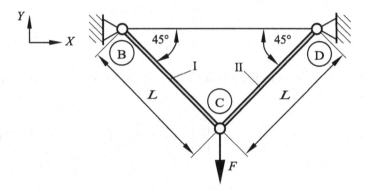

Fig. 7.5 Plane structure made of two bars

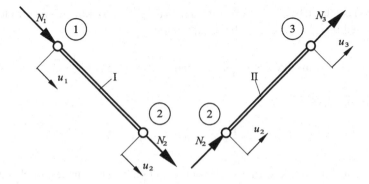

Fig. 7.6 Plane structure with two bar elements

Given: E, A, L and F
Unknown are

- the displacement on position C and
- the strain and stress in the elements.

Solution

The simplest discretization of the structure is obvious. The structure is divided into two elements. The nodes with the numbers 1, 2 and 3 are introduced at the positions B, C and D (see Fig. 7.6).

Nodes 1 and 2 are attached to element I. The single stiffness relation for element I is, in local coordinates:

$$\begin{bmatrix} k & -k \\ -k & k \end{bmatrix}^{\mathrm{I}} \begin{bmatrix} u_1 \\ u_2 \end{bmatrix}^{\mathrm{I}} = \begin{bmatrix} N_1 \\ N_2 \end{bmatrix}^{\mathrm{I}} \tag{7.16}$$

with

$$k^{\mathrm{I}} = \frac{EA}{\sqrt{2}L} . \tag{7.17}$$

Nodes 2 and 3 are attached to element II. The single stiffness relation for element II is, in local coordinates:

$$\begin{bmatrix} k & -k \\ -k & k \end{bmatrix}^{\mathrm{II}} \begin{bmatrix} u_2 \\ u_3 \end{bmatrix}^{\mathrm{II}} = \begin{bmatrix} N_2 \\ N_3 \end{bmatrix}^{\mathrm{II}} \tag{7.18}$$

with

$$k^{\mathrm{II}} = \frac{EA}{\sqrt{2}L} . \tag{7.19}$$

Element I is rotated by $\alpha^{\mathrm{I}} = -45°$ compared to the global coordinate system, element II is rotated by $\alpha^{\mathrm{II}} = +45°$. With the expressions

$$\sin(-45°) = -\frac{1}{2}\sqrt{2} \quad \text{and} \quad \cos(-45°) = \frac{1}{2}\sqrt{2} \tag{7.20}$$

for element I and

$$\sin(+45°) = +\frac{1}{2}\sqrt{2} \quad \text{and} \quad \cos(+45°) = \frac{1}{2}\sqrt{2} \tag{7.21}$$

for element II the single stiffness relations are, in global coordinates for element I:

$$\frac{1}{2}k^{\mathrm{I}} \begin{bmatrix} 1 & -1 & -1 & 1 \\ -1 & 1 & 1 & -1 \\ -1 & 1 & 1 & -1 \\ 1 & -1 & -1 & 1 \end{bmatrix} \begin{bmatrix} u_{1X} \\ u_{1Y} \\ u_{2X} \\ u_{2Y} \end{bmatrix} = \begin{bmatrix} F_{1X} \\ F_{1Y} \\ F_{2X} \\ F_{2Y} \end{bmatrix} \tag{7.22}$$

and for element II:

$$\frac{1}{2}k^{\mathrm{II}} \begin{bmatrix} 1 & 1 & -1 & -1 \\ 1 & 1 & -1 & -1 \\ -1 & -1 & 1 & 1 \\ -1 & -1 & 1 & 1 \end{bmatrix} \begin{bmatrix} u_{2X} \\ u_{2Y} \\ u_{3X} \\ u_{3Y} \end{bmatrix} = \begin{bmatrix} F_{2X} \\ F_{2Y} \\ F_{3X} \\ F_{3Y} \end{bmatrix} . \tag{7.23}$$

One obtains the global stiffness relation by inserting the single stiffness relations in the according positions:

$$\frac{1}{2}\begin{bmatrix} k^{\mathrm{I}} & -k^{\mathrm{I}} & -k^{\mathrm{I}} & k^{\mathrm{I}} & 0 & 0 \\ -k^{\mathrm{I}} & k^{\mathrm{I}} & k^{\mathrm{I}} & -k^{\mathrm{I}} & 0 & 0 \\ -k^{\mathrm{I}} & k^{\mathrm{I}} & k^{\mathrm{I}}+k^{\mathrm{II}} & -k^{\mathrm{I}}+k^{\mathrm{II}} & -k^{\mathrm{II}} & -k^{\mathrm{II}} \\ k^{\mathrm{I}} & -k^{\mathrm{I}} & -k^{\mathrm{I}}+k^{\mathrm{II}} & k^{\mathrm{I}}+k^{\mathrm{II}} & -k^{\mathrm{II}} & -k^{\mathrm{II}} \\ 0 & 0 & -k^{\mathrm{II}} & -k^{\mathrm{II}} & k^{\mathrm{II}} & k^{\mathrm{II}} \\ 0 & 0 & -k^{\mathrm{II}} & -k^{\mathrm{II}} & k^{\mathrm{II}} & k^{\mathrm{II}} \end{bmatrix}\begin{bmatrix} u_{1X} \\ u_{1Y} \\ u_{2X} \\ u_{2Y} \\ u_{3X} \\ u_{3Y} \end{bmatrix} = \begin{bmatrix} F_{1X} \\ F_{1Y} \\ F_{2X} \\ F_{2Y} \\ F_{3X} \\ F_{3Y} \end{bmatrix}.$$

$$(7.24)$$

In the next step, the boundary conditions will be introduced.

- The displacement on node 1 and on node 3 are zero in each case.
- The external force on node 2 acts in the global Y direction.

Therewith the global stiffness relation is:

$$\frac{1}{2}\begin{bmatrix} k^{\mathrm{I}} & -k^{\mathrm{I}} & -k^{\mathrm{I}} & k^{\mathrm{I}} & 0 & 0 \\ -k^{\mathrm{I}} & k^{\mathrm{I}} & k^{\mathrm{I}} & -k^{\mathrm{I}} & 0 & 0 \\ -k^{\mathrm{I}} & k^{\mathrm{I}} & k^{\mathrm{I}}+k^{\mathrm{II}} & -k^{\mathrm{I}}+k^{\mathrm{II}} & -k^{\mathrm{II}} & -k^{\mathrm{II}} \\ k^{\mathrm{I}} & -k^{\mathrm{I}} & -k^{\mathrm{I}}+k^{\mathrm{II}} & k^{\mathrm{I}}+k^{\mathrm{II}} & -k^{\mathrm{II}} & -k^{\mathrm{II}} \\ 0 & 0 & -k^{\mathrm{II}} & -k^{\mathrm{II}} & k^{\mathrm{II}} & k^{\mathrm{II}} \\ 0 & 0 & -k^{\mathrm{II}} & -k^{\mathrm{II}} & k^{\mathrm{II}} & k^{\mathrm{II}} \end{bmatrix}\begin{bmatrix} 0 \\ 0 \\ u_{2X} \\ u_{2Y} \\ 0 \\ 0 \end{bmatrix} = \begin{bmatrix} F_{1X} \\ F_{1Y} \\ 0 \\ F_{2Y} \\ F_{3X} \\ F_{3Y} \end{bmatrix}.$$

$$(7.25)$$

After the crossing out of the rows and columns 1, 2, 5 and 6 a reduced system

$$\frac{1}{2}\begin{bmatrix} k^{\mathrm{I}}+k^{\mathrm{II}} & -k^{\mathrm{I}}+k^{\mathrm{II}} \\ -k^{\mathrm{I}}+k^{\mathrm{II}} & k^{\mathrm{I}}+k^{\mathrm{II}} \end{bmatrix}\begin{bmatrix} u_{2X} \\ u_{2Y} \end{bmatrix} = \begin{bmatrix} 0 \\ -F \end{bmatrix} \tag{7.26}$$

with the rows and columns 3 and 4 remains. So far, the stiffnesses for the elements are labeled with the indexes I and II, even though they are identical. For the further approach the stiffnesses are consistently referred to as $k^{\mathrm{I}} = k^{\mathrm{II}} = k = \frac{EA}{\sqrt{2}L}$. The simplified global stiffness relation is:

$$\begin{bmatrix} k & 0 \\ 0 & k \end{bmatrix}\begin{bmatrix} u_{2X} \\ u_{2Y} \end{bmatrix} = \begin{bmatrix} 0 \\ -F \end{bmatrix}. \tag{7.27}$$

Herefrom the unknown displacements u_{2X} and u_{2Y} can be determined:

$$u_{2X} = 0 \quad , \quad u_{2Y} = -\frac{F}{k} = -\frac{F}{EA}\sqrt{2}L. \tag{7.28}$$

Through transformation of the displacements u_{2X} and u_{2Y} in the element-own local coordinate systems the following results:

$$u_2^{\mathrm{I}} = \frac{1}{2}\sqrt{2}u_{2X} - \frac{1}{2}\sqrt{2}u_{2Y} = \frac{1}{2}\sqrt{2}\left(0 - \left(-\frac{F}{k}\right)\right) = +\frac{1}{2}\sqrt{2}\frac{F}{k} = +\frac{F}{EA}L,$$

$$(7.29)$$

$$u_2^{II} = \frac{1}{2}\sqrt{2}u_{2X} + \frac{1}{2}\sqrt{2}u_{2Y} = \frac{1}{2}\sqrt{2}\left(0 + \left(-\frac{F}{k}\right)\right) = -\frac{1}{2}\sqrt{2}\frac{F}{k} = -\frac{F}{EA}L\,.$$
(7.30)

From the local displacements the strain in element I

$$\varepsilon^I(x) = \frac{1}{\sqrt{2}L}(+u_2^I - u_1^I) = \left(\frac{F}{EA}L - 0\right)\frac{1}{\sqrt{2}L} = +\frac{1}{2}\sqrt{2}\frac{F}{EA}$$
(7.31)

and in element II

$$\varepsilon^{II}(x) = \frac{1}{\sqrt{2}L}(+u_3^{II} - u_2^{II}) = \left(0 - \left(-\frac{F}{EA}L\right)\right)\frac{1}{\sqrt{2}L} = +\frac{1}{2}\sqrt{2}\frac{F}{EA}$$
(7.32)

can be determined.

After the local displacements in the particular elements are known, the local forces can be determined via the single stiffness relation:

Bar I:
$$N_1^I = k\,(+u_1^I - u_2^I) = k(0 - \tfrac{1}{2}\sqrt{2}\tfrac{F}{k}) = -\tfrac{1}{2}\sqrt{2}F\,,$$
$$N_2^I = k\,(-u_1^I + u_2^I) = k(0 + \tfrac{1}{2}\sqrt{2}\tfrac{F}{k}) = +\tfrac{1}{2}\sqrt{2}F\,.$$
(7.33)

Bar II:
$$N_2^{II} = k\,(+u_2^{II} - u_3^{II}) = k(-\tfrac{1}{2}\sqrt{2}\tfrac{F}{k} - 0) = -\tfrac{1}{2}\sqrt{2}F\,,$$
$$N_3^{II} = k\,(-u_2^{II} + u_3^{II}) = k(-(-\tfrac{1}{2}\sqrt{2}\tfrac{F}{k}) + 0) = +\tfrac{1}{2}\sqrt{2}F\,.$$
(7.34)

From the definition of the bar force it becomes obvious that both bar I and also bar II are tension bars. The normal stress in bar I results in:

$$\sigma^I = \frac{1}{2}\sqrt{2}\frac{F}{A}$$
(7.35)

and in bar II:

$$\sigma^{II} = \frac{1}{2}\sqrt{2}\frac{F}{A}\,.$$
(7.36)

Herewith also the loading conditions in the single elements are known.

7.4.2   Plane Structure: Beam and Bar

As a second simple example a structure will be discussed which is made up of a beam and a bar (see Fig. 7.7). The beam has a fixed support on one end (position B). The beam is pin-jointed with the bar at position C, which is simply supported at position D. The entire structure is loaded with a point force F.

Given: E, I, A, L and F

Unknown are

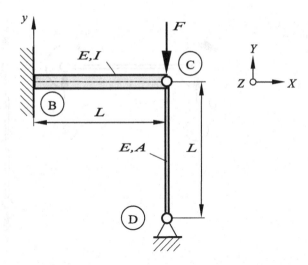

Fig. 7.7 Plane structure composed of a beam and a bar element

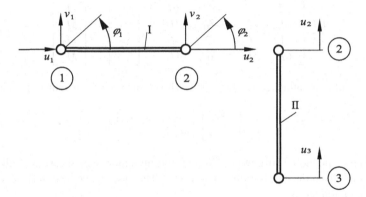

Fig. 7.8 Plane structure with kinematic parameters on the single elements

- the displacements and rotations on position C and
- the reaction forces on the fixed supports.

Two ways for solving will be introduced. They can be distinguished in the use of a global coordinate system. First, the approach with the introduction of a global coordinate system will be explained.

The discretization of the structure is obvious. The beam is element I, the bar is element II. The nodes with the numbers 1, 2 and 3 are introduced at the positions B, C and D. For the single elements the kinematic parameters in Fig. 7.8 and the 'load parameters' in Fig. 7.9 are illustrated.

Nodes 1 and 2 are attached to element I. Herewith the stiffness relation for element I is:

Fig. 7.9 Plane structure
with 'load parameters' on the
single elements

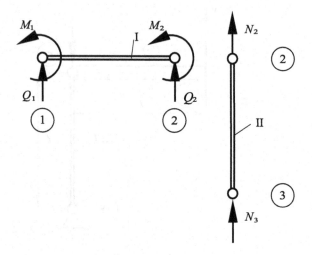

$$\frac{EI}{L^3}\begin{bmatrix} 12 & 6L & -12L & 6L \\ 6L & 4L^2 & -6L & 2L^2 \\ -12 & -6L & 12 & -6L \\ 6L & 2L^2 & -6L & 4L^2 \end{bmatrix}\begin{bmatrix} v_1 \\ \varphi_1 \\ v_2 \\ \varphi_2 \end{bmatrix} = \begin{bmatrix} Q_1 \\ M_1 \\ Q_2 \\ M_2 \end{bmatrix}. \qquad (7.37)$$

Nodes 3 and 2 are attached to element II. The element stiffness matrix for element
II is:

$$\begin{bmatrix} k & -k \\ -k & k \end{bmatrix}\begin{bmatrix} u_3 \\ u_2 \end{bmatrix} = \begin{bmatrix} N_3 \\ N_2 \end{bmatrix}. \qquad (7.38)$$

One obtains the global stiffness relation by setting up the overall force equilibrium.
This results from the equilibrium on all nodes. The system can be described via the
following parameters:

- on position B: displacement v_1 and the rotation φ_1,
- on position C: displacement v_2 and the rotation φ_2,
- on position D: displacement u_3.

On the coupling point C

- the displacement v_2 on the beam equals the displacement u_2 on the bar and
- the shear force Q_2 on the beam equals the normal force N_2 on the bar.

The dimensions of the global stiffness relation is therewith determined: a 5×5
system. For clarity reasons the entries in the following illustration are not conducted
in detail. The abbreviation

- Z stands for the tension and compression bar and
- B stands for the bending beam:

$$\begin{bmatrix} B & B & B & B & 0 \\ B & B & B & B & 0 \\ B & B & B+Z & B & Z \\ B & B & B & B & 0 \\ 0 & 0 & Z & 0 & Z \end{bmatrix} \begin{bmatrix} v_1 \\ \varphi_1 \\ v_2 \\ \varphi_2 \\ u_3 \end{bmatrix} = \begin{bmatrix} Q_1 \\ M_1 \\ Q_2 \\ M_2 \\ N_3 \end{bmatrix}. \tag{7.39}$$

It is already included in the total stiffness relation

- that the displacement v_2 regarding the bending beam on node 2 is identical with the displacement u_2 on the bar and
- that the shear force Q_2 regarding the bending beam on node 2 is identical with the normal force of the tension bar N_2.

In the next step the boundary conditions will be inserted into the total stiffness relation.

- On node 1 the displacement v_1 and the rotation φ_1 are zero.
- On node 2 there is the external force $-F$, the moment M_2 is zero.
- On node 3 the displacement u_3 is zero.

Herewith the rows and columns 1, 2 and 5 can be removed from the total stiffness relation. A reduced system of equations remains:

$$\begin{bmatrix} B+Z & B \\ B & B \end{bmatrix} \begin{bmatrix} v_2 \\ \varphi_2 \end{bmatrix} = \begin{bmatrix} -F \\ 0 \end{bmatrix}. \tag{7.40}$$

The reduced system of equations in detail is:

$$\begin{bmatrix} -F \\ 0 \end{bmatrix} = \begin{bmatrix} 12\frac{EI}{L^3}+\frac{EA}{L} & -6\frac{EI}{L^2} \\ -6\frac{EI}{L^2} & 4\frac{EI}{L} \end{bmatrix} \begin{bmatrix} v_2 \\ \varphi_2 \end{bmatrix}. \tag{7.41}$$

Herefrom the unknown displacement v_2 and the rotation φ_2 can be determined:

$$v_2 = -\frac{F}{3\frac{EI}{L^3}+\frac{EA}{L}}, \quad \varphi_2 = -\frac{3}{2L}\frac{F}{3\frac{EI}{L^3}+\frac{EA}{L}}. \tag{7.42}$$

Since the parameters v_2 and φ_2 are known now, the support reactions can be defined by inserting those in Eq. (7.39).

In this example, the definition of a global coordinate system and therewith the transformation from local to global coordinates was relinquished. Generally this is not possible. In this example, due to the right-angled position to each other, parameters of one element can be identified with those on the other element.

For the sake of completeness the approach with the coordinate transformation will also be introduced. The single stiffness relations in local coordinates are already known. For both elements the kinematic parameters in Fig. 7.10 as well as the 'load parameters' in Fig. 7.11 are illustrated.

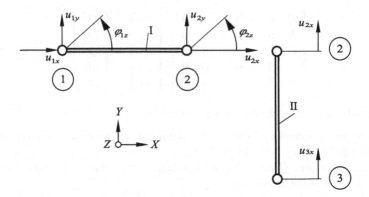

Fig. 7.10 Plane structure with kinematic parameters in local coordinates on the single elements

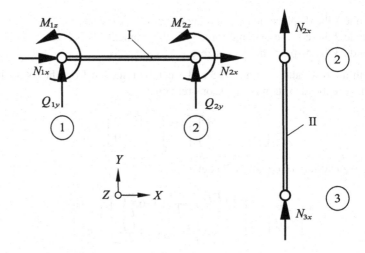

Fig. 7.11 Plane structure with 'load parameters' in local coordinates on the single elements

In the next step, the single stiffness relations will be illustrated in global coordinates. First, a global coordinate system will be defined. The local and global coordinate system are identical for element I. The single stiffness relation for element I is:

$$\frac{EI}{L^3}\begin{bmatrix} 0 & 0 & 0 & 0 & 0 & 0 \\ 0 & 12 & 6L & 0 & -12L & 6L \\ 0 & 6L & 4L^2 & 0 & -6L & 2L^2 \\ 0 & 0 & 0 & 0 & 0 & 0 \\ 0 & -12 & -6L & 0 & 12 & -6L \\ 0 & 6L & 2L^2 & 0 & -6L & 4L^2 \end{bmatrix}\begin{bmatrix} u_{1X} \\ u_{1Y} \\ \varphi_{1Z} \\ u_{2X} \\ u_{2Y} \\ \varphi_{2Z} \end{bmatrix} = \begin{bmatrix} F_{1X} \\ F_{1Y} \\ M_{1Z} \\ F_{2X} \\ F_{2Y} \\ M_{2Z} \end{bmatrix}. \qquad (7.43)$$

The single stiffness relation in local coordinates for element II is:

$$\begin{bmatrix} \frac{EA}{L} & 0 & -\frac{EA}{L} & 0 \\ 0 & 0 & 0 & 0 \\ -\frac{EA}{L} & 0 & \frac{EA}{L} & 0 \\ 0 & 0 & 0 & 0 \end{bmatrix} \begin{bmatrix} u_{3x} \\ u_{3y} \\ u_{2x} \\ u_{2y} \end{bmatrix} = \begin{bmatrix} N_{3x} \\ N_{3y} \\ N_{2x} \\ N_{2y} \end{bmatrix}. \tag{7.44}$$

Element II is rotated by $\alpha = 90°$ compared to the global coordinate system. The transformation matrix for a vector is:

$$\boldsymbol{T} = \begin{bmatrix} 0 & 1 & 0 & 0 \\ -1 & 0 & 0 & 0 \\ 0 & 0 & 0 & 1 \\ 0 & 0 & -1 & 0 \end{bmatrix}. \tag{7.45}$$

The single stiffness relation therefore results for element II in global coordinates:

$$\begin{bmatrix} 0 & 0 & 0 & 0 \\ 0 & \frac{EA}{L} & 0 & -\frac{EA}{L} \\ 0 & 0 & 0 & 0 \\ 0 & -\frac{EA}{L} & 0 & \frac{EA}{L} \end{bmatrix} \begin{bmatrix} u_{3X} \\ u_{3Y} \\ u_{2X} \\ u_{2Y} \end{bmatrix} = \begin{bmatrix} F_{3X} \\ F_{3Y} \\ F_{2X} \\ F_{2Y} \end{bmatrix}. \tag{7.46}$$

The dimensions of the global stiffness relation result in 8×8. The kinematic parameters are

- on node 1: $u_{1X}, u_{1Y}, \varphi_{1Z}$,
- on node 2: $u_{2X}, u_{2Y}, \varphi_{2Z}$ and
- on node 3: u_{3X}, u_{3Y}.

The 'load parameters' are

- on node 1: F_{1X}, F_{1Y}, M_{1Z},
- on node 2: F_{2X}, F_{2Y}, M_{2Z} and
- on node 3: F_{3X}, F_{3Y}.

The global stiffness relation therefore results in:

$$\begin{bmatrix} 0 & 0 & 0 & 0 & 0 & 0 & 0 & 0 \\ 0 & B & B & 0 & B & B & 0 & 0 \\ 0 & B & B & 0 & B & B & 0 & 0 \\ 0 & 0 & 0 & 0 & 0 & 0 & 0 & 0 \\ 0 & B & B & 0 & B+Z & B & 0 & Z \\ 0 & B & B & 0 & B & B & 0 & 0 \\ 0 & 0 & 0 & 0 & 0 & 0 & 0 & 0 \\ 0 & 0 & 0 & 0 & Z & 0 & 0 & Z \end{bmatrix} \begin{bmatrix} u_{1X} \\ u_{1Y} \\ \varphi_{1Z} \\ u_{2X} \\ u_{2Y} \\ \varphi_{2Z} \\ u_{3X} \\ u_{3Y} \end{bmatrix} = \begin{bmatrix} F_{1X} \\ F_{1Y} \\ M_{1Z} \\ F_{2X} \\ F_{2Y} \\ M_{2Z} \\ F_{3X} \\ F_{3Y} \end{bmatrix}. \tag{7.47}$$

The boundary conditions are inserted into the global stiffness relation.

- On node 1, the fixed support of the beam, the displacement u_{1X}, u_{1Y} and the angle φ_{1Z} are zero.
- On node 2 the external force acts against the Y-direction.
- On node 3 the displacements u_{3X} and u_{3Y} are zero.

Therewith the corresponding rows and columns $(1, 2, 3, 4, 7, 8)$ can be removed from the total stiffness relation. A reduced system of the dimensions 2×2 remains:

$$\begin{bmatrix} B+Z & B \\ B & B \end{bmatrix} \begin{bmatrix} u_{2Y} \\ \varphi_{2Z} \end{bmatrix} = \begin{bmatrix} -F \\ 0 \end{bmatrix}. \tag{7.48}$$

This system of equations is similar to the system shown above, which resulted from the description in local coordinates. The displacement v_2 equals u_{2Y} and the rotation φ_2 equals φ_{2Z}. The latter approach is identical.

7.5   Examples in the Three-Dimensional Space

The structure consists of three plane sections, which are oriented differently within space. The sections are each arranged mutually right-angled (see Fig. 7.12). The entire structure is fixed supported at position B and is loaded with a point load F at position G.

Given: E, ν, A, L and F

Unknown are

Fig. 7.12 General structure in space

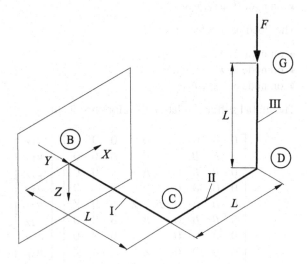

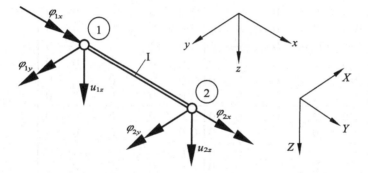

Fig. 7.13 Element I with state variables in local coordinates

- the displacement and rotation on each coupling point (these are the positions B, C, D) and
- the reaction force at the fixed support (position B).

Solution:

Each section is represented as a single 1D finite element. First the elements and nodes are labeled with numbers. For the description of the displacement and rotation a global (X, Y, Z) coordinate system is defined.

In principle, the global stiffness matrix of a general 1D element could be used for any element. This leads to quite extensive descriptions. Alternatively, the corresponding stiffness matrices of the basic load types for each element can be named:

- Element I is loaded by bending and torsion,
- element II by bending and
- element III by compression.

The shear part in element I and II is disregarded.

Figure 7.13 illustrates the state variables for element I in local coordinates. For clarity reasons only the parameters relevant for the description of the bending and compression load are considered.

In local coordinates the column matrices of the state variables are

$$\left[u_{1x}, u_{1y}, u_{1z}, \varphi_{1x}, \varphi_{1y}, \varphi_{1z}, u_{2x}, u_{2y}, u_{2z}, \varphi_{2x}, \varphi_{2y}, \varphi_{2z} \right]^{\mathrm{T}} , \tag{7.49}$$

$$\left[F_{1x}, F_{1y}, F_{1z}, M_{1x}, M_{1y}, M_{1z}, F_{2x}, F_{2y}, F_{2z}, M_{2x}, M_{2y}, M_{2z} \right]^{\mathrm{T}} \tag{7.50}$$

and the stiffness matrix for element I:

$$\begin{bmatrix}
0 & 0 & 0 & 0 & 0 & 0 & 0 & 0 & 0 & 0 & 0 & 0 \\
0 & 0 & 0 & 0 & 0 & 0 & 0 & 0 & 0 & 0 & 0 & 0 \\
0 & 0 & 12\frac{EI_y}{L^3} & 0 & -6\frac{EI_y}{L^2} & 0 & 0 & 0 & -12\frac{EI_y}{L^3} & 0 & -6\frac{EI_y}{L^3} & 0 \\
0 & 0 & 0 & \frac{GI_t}{L} & 0 & 0 & 0 & 0 & 0 & -\frac{GI_t}{L} & 0 & 0 \\
0 & 0 & -6\frac{EI_y}{L^2} & 0 & 4\frac{EI_y}{L} & 0 & 0 & 0 & 6\frac{EI_y}{L^2} & 0 & 2\frac{EI_y}{L} & 0 \\
0 & 0 & 0 & 0 & 0 & 0 & 0 & 0 & 0 & 0 & 0 & 0 \\
0 & 0 & 0 & 0 & 0 & 0 & 0 & 0 & 0 & 0 & 0 & 0 \\
0 & 0 & 0 & 0 & 0 & 0 & 0 & 0 & 0 & 0 & 0 & 0 \\
0 & 0 & -12\frac{EI_y}{L^3} & 0 & 6\frac{EI_y}{L^2} & 0 & 0 & 0 & 12\frac{EI_y}{L^3} & 0 & 6\frac{EI_y}{L^2} & 0 \\
0 & 0 & 0 & -\frac{GI_t}{L} & 0 & 0 & 0 & 0 & 0 & \frac{GI_t}{L} & 0 & 0 \\
0 & 0 & -6\frac{EI_y}{L^2} & 0 & 2\frac{EI_y}{L} & 0 & 0 & 0 & 6\frac{EI_y}{L^2} & 0 & 4\frac{EI_y}{L} & 0 \\
0 & 0 & 0 & 0 & 0 & 0 & 0 & 0 & 0 & 0 & 0 & 0
\end{bmatrix}. \tag{7.51}$$

The transformation rule from local to global coordinates for a vector regarding element I is:

$$\begin{bmatrix}
\cos(\frac{\pi}{2}) & \cos(0) & \cos(\frac{\pi}{2}) \\
\cos(\pi) & \cos(\frac{\pi}{2}) & \cos(\frac{\pi}{2}) \\
\cos(-\frac{\pi}{2}) & \cos(-\frac{\pi}{2}) & \cos(0)
\end{bmatrix}. \tag{7.52}$$

Element I has two nodes with respectively 6 scalar parameters. Therefore, a 12×12 system results:

$$T^{\mathrm{I}} = \begin{bmatrix}
0 & 1 & 0 & 0 & 0 & 0 & 0 & 0 & 0 & 0 & 0 & 0 \\
-1 & 0 & 0 & 0 & 0 & 0 & 0 & 0 & 0 & 0 & 0 & 0 \\
0 & 0 & 1 & 0 & 0 & 0 & 0 & 0 & 0 & 0 & 0 & 0 \\
0 & 0 & 0 & 0 & 1 & 0 & 0 & 0 & 0 & 0 & 0 & 0 \\
0 & 0 & 0 & -1 & 0 & 0 & 0 & 0 & 0 & 0 & 0 & 0 \\
0 & 0 & 0 & 0 & 0 & 1 & 0 & 0 & 0 & 0 & 0 & 0 \\
0 & 0 & 0 & 0 & 0 & 0 & 0 & 1 & 0 & 0 & 0 & 0 \\
0 & 0 & 0 & 0 & 0 & 0 & -1 & 0 & 0 & 0 & 0 & 0 \\
0 & 0 & 0 & 0 & 0 & 0 & 0 & 0 & 1 & 0 & 0 & 0 \\
0 & 0 & 0 & 0 & 0 & 0 & 0 & 0 & 0 & 0 & 1 & 0 \\
0 & 0 & 0 & 0 & 0 & 0 & 0 & 0 & 0 & -1 & 0 & 0 \\
0 & 0 & 0 & 0 & 0 & 0 & 0 & 0 & 0 & 0 & 0 & 1
\end{bmatrix}. \tag{7.53}$$

After the transformation in global coordinates the single stiffness relation for element I is:

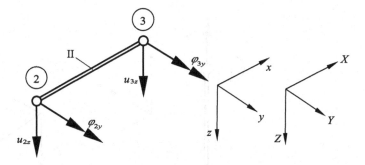

Fig. 7.14 Element II with state variables in local coordinates

$$
\begin{bmatrix}
0 & 0 & 0 & 0 & 0 & 0 & 0 & 0 & 0 & 0 & 0 & 0 \\
0 & 0 & 0 & 0 & 0 & 0 & 0 & 0 & 0 & 0 & 0 & 0 \\
0 & 0 & 12\frac{EI_y}{L^3} & 6\frac{EI_y}{L^2} & 0 & 0 & 0 & 0 & -12\frac{EI_y}{L^3} & 6\frac{EI_y}{L^2} & 0 & 0 \\
0 & 0 & 6\frac{EI_y}{L^2} & 4\frac{EI_y}{L} & 0 & 0 & 0 & 0 & -6\frac{EI_y}{L^2} & 2\frac{EI_y}{L} & 0 & 0 \\
0 & 0 & 0 & 0 & \frac{GI_t}{L} & 0 & 0 & 0 & 0 & 0 & -\frac{GI_t}{L} & 0 \\
0 & 0 & 0 & 0 & 0 & 0 & 0 & 0 & 0 & 0 & 0 & 0 \\
0 & 0 & 0 & 0 & 0 & 0 & 0 & 0 & 0 & 0 & 0 & 0 \\
0 & 0 & 0 & 0 & 0 & 0 & 0 & 0 & 0 & 0 & 0 & 0 \\
0 & 0 & -12\frac{EI_y}{L^3} & -6\frac{EI_y}{L^2} & 0 & 0 & 0 & 0 & 12\frac{EI_y}{L^3} & -6\frac{EI_y}{L^2} & 0 & 0 \\
0 & 0 & 6\frac{EI_y}{L^2} & 2\frac{EI_y}{L} & 0 & 0 & 0 & 0 & -6\frac{EI_y}{L^2} & 4\frac{EI_y}{L} & 0 & 0 \\
0 & 0 & 0 & 0 & -\frac{GI_t}{L} & 0 & 0 & 0 & 0 & 0 & \frac{GI_t}{L} & 0 \\
0 & 0 & 0 & 0 & 0 & 0 & 0 & 0 & 0 & 0 & 0 & 0
\end{bmatrix}
\begin{bmatrix}
u_{1X} \\ u_{1Y} \\ u_{1Z} \\ \varphi_{1X} \\ \varphi_{1Y} \\ \varphi_{1Z} \\ u_{2X} \\ u_{2Y} \\ u_{2Z} \\ \varphi_{2X} \\ \varphi_{2Y} \\ \varphi_{2Z}
\end{bmatrix} . \quad (7.54)
$$

For element II the local and global coordinates are identical. Figure 7.14 illustrates the state variables for element II in local coordinates. Therewith the single stiffness relation for element II results in:

$$
\begin{bmatrix}
0 & 0 & 0 & 0 & 0 & 0 & 0 & 0 & 0 & 0 & 0 & 0 \\
0 & 0 & 0 & 0 & 0 & 0 & 0 & 0 & 0 & 0 & 0 & 0 \\
0 & 0 & 12\frac{EI_y}{L^3} & 0 & -6\frac{EI_y}{L^2} & 0 & 0 & 0 & -12\frac{EI_y}{L^3} & 0 & -6\frac{EI_y}{L^2} & 0 \\
0 & 0 & 0 & 0 & 0 & 0 & 0 & 0 & 0 & 0 & 0 & 0 \\
0 & 0 & -6\frac{EI_y}{L^2} & 0 & 4\frac{EI_y}{L} & 0 & 0 & 0 & 6\frac{EI_y}{L^2} & 0 & 2\frac{EI_y}{L} & 0 \\
0 & 0 & 0 & 0 & 0 & 0 & 0 & 0 & 0 & 0 & 0 & 0 \\
0 & 0 & 0 & 0 & 0 & 0 & 0 & 0 & 0 & 0 & 0 & 0 \\
0 & 0 & 0 & 0 & 0 & 0 & 0 & 0 & 0 & 0 & 0 & 0 \\
0 & 0 & -12\frac{EI_y}{L^3} & 0 & 6\frac{EI_y}{L^2} & 0 & 0 & 0 & 12\frac{EI_y}{L^3} & 0 & 6\frac{EI_y}{L^2} & 0 \\
0 & 0 & 0 & 0 & 0 & 0 & 0 & 0 & 0 & 0 & 0 & 0 \\
0 & 0 & -6\frac{EI_y}{L^2} & 0 & 2\frac{EI_y}{L} & 0 & 0 & 0 & 6\frac{EI_y}{L^2} & 0 & 4\frac{EI_y}{L} & 0 \\
0 & 0 & 0 & 0 & 0 & 0 & 0 & 0 & 0 & 0 & 0 & 0
\end{bmatrix}
\begin{bmatrix}
u_{2X} \\ u_{2Y} \\ u_{2Z} \\ \varphi_{2X} \\ \varphi_{2Y} \\ \varphi_{2Z} \\ u_{3X} \\ u_{3Y} \\ u_{3Z} \\ \varphi_{3X} \\ \varphi_{3Y} \\ \varphi_{3Z}
\end{bmatrix}
=
\begin{bmatrix}
F_{2X} \\ F_{2Y} \\ F_{2Z} \\ M_{2X} \\ M_{2Y} \\ M_{2Z} \\ F_{3X} \\ F_{3Y} \\ F_{3Z} \\ M_{3X} \\ M_{3Y} \\ M_{3Z}
\end{bmatrix} .
$$

$$(7.55)$$

Fig. 7.15 Element III with
state variables in local
coordinates

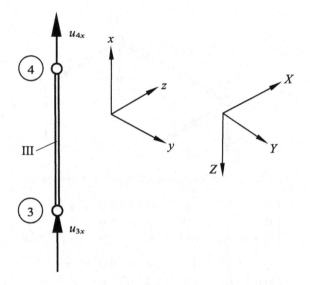

For element III the local coordinate system is rotated compared to the global one.
The state variables are illustrated in local coordinates in Fig. 7.15.

The single stiffness relations for element III in local coordinates:

$$
\begin{bmatrix}
\frac{EA}{L} & 0 & 0 & 0 & 0 & 0 & -\frac{EA}{L} & 0 & 0 & 0 & 0 & 0 \\
0 & 0 & 0 & 0 & 0 & 0 & 0 & 0 & 0 & 0 & 0 & 0 \\
0 & 0 & 0 & 0 & 0 & 0 & 0 & 0 & 0 & 0 & 0 & 0 \\
0 & 0 & 0 & 0 & 0 & 0 & 0 & 0 & 0 & 0 & 0 & 0 \\
0 & 0 & 0 & 0 & 0 & 0 & 0 & 0 & 0 & 0 & 0 & 0 \\
0 & 0 & 0 & 0 & 0 & 0 & 0 & 0 & 0 & 0 & 0 & 0 \\
-\frac{EA}{L} & 0 & 0 & 0 & 0 & 0 & \frac{EA}{L} & 0 & 0 & 0 & 0 & 0 \\
0 & 0 & 0 & 0 & 0 & 0 & 0 & 0 & 0 & 0 & 0 & 0 \\
0 & 0 & 0 & 0 & 0 & 0 & 0 & 0 & 0 & 0 & 0 & 0 \\
0 & 0 & 0 & 0 & 0 & 0 & 0 & 0 & 0 & 0 & 0 & 0 \\
0 & 0 & 0 & 0 & 0 & 0 & 0 & 0 & 0 & 0 & 0 & 0 \\
0 & 0 & 0 & 0 & 0 & 0 & 0 & 0 & 0 & 0 & 0 & 0
\end{bmatrix}
\begin{bmatrix}
u_{3x} \\ u_{3y} \\ u_{3z} \\ \varphi_{3x} \\ \varphi_{3y} \\ \varphi_{3z} \\ u_{4x} \\ u_{4y} \\ u_{4z} \\ \varphi_{4x} \\ \varphi_{4y} \\ \varphi_{4z}
\end{bmatrix}
=
\begin{bmatrix}
F_{3x} \\ F_{3y} \\ F_{3z} \\ M_{3x} \\ M_{3y} \\ M_{3z} \\ F_{4x} \\ F_{4y} \\ F_{4z} \\ M_{4x} \\ M_{4y} \\ M_{4z}
\end{bmatrix} .
\tag{7.56}
$$

The transformation rule of local to global coordinates for a vector regarding element
III is called:

$$
\begin{bmatrix}
\cos(-\frac{\pi}{2}) & \cos(\frac{\pi}{2}) & \cos(\pi) \\
\cos(\frac{\pi}{2}) & \cos(0) & \cos(\frac{\pi}{2}) \\
\cos(0) & \cos(-\frac{\pi}{2}) & \cos(-\frac{\pi}{2})
\end{bmatrix} .
\tag{7.57}
$$

The total transformation matrix T^{III} therefore results in:

$$
T^{\text{III}} = \begin{bmatrix}
0 & 0 & -1 & 0 & 0 & 0 & 0 & 0 & 0 & 0 & 0 & 0 \\
0 & 1 & 0 & 0 & 0 & 0 & 0 & 0 & 0 & 0 & 0 & 0 \\
1 & 0 & 0 & 0 & 0 & 0 & 0 & 0 & 0 & 0 & 0 & 0 \\
0 & 0 & 0 & 0 & 0 & -1 & 0 & 0 & 0 & 0 & 0 & 0 \\
0 & 0 & 0 & 0 & 1 & 0 & 0 & 0 & 0 & 0 & 0 & 0 \\
0 & 0 & 0 & 1 & 0 & 0 & 0 & 0 & 0 & 0 & 0 & 0 \\
0 & 0 & 0 & 0 & 0 & 0 & 0 & 0 & -1 & 0 & 0 & 0 \\
0 & 0 & 0 & 0 & 0 & 0 & 0 & 1 & 0 & 0 & 0 & 0 \\
0 & 0 & 0 & 0 & 0 & 0 & 1 & 0 & 0 & 0 & 0 & 0 \\
0 & 0 & 0 & 0 & 0 & 0 & 0 & 0 & 0 & 0 & 0 & -1 \\
0 & 0 & 0 & 0 & 0 & 0 & 0 & 0 & 0 & 0 & 1 & 0 \\
0 & 0 & 0 & 0 & 0 & 0 & 0 & 0 & 0 & 1 & 0 & 0
\end{bmatrix} . \tag{7.58}
$$

The single stiffness relation for element II in global coordinates reads:

$$
\begin{bmatrix}
0 & 0 & 0 & 0 & 0 & 0 & 0 & 0 & 0 & 0 & 0 & 0 \\
0 & 0 & 0 & 0 & 0 & 0 & 0 & 0 & 0 & 0 & 0 & 0 \\
0 & 0 & \frac{EA}{L} & 0 & 0 & 0 & 0 & 0 & -\frac{EA}{L} & 0 & 0 & 0 \\
0 & 0 & 0 & 0 & 0 & 0 & 0 & 0 & 0 & 0 & 0 & 0 \\
0 & 0 & 0 & 0 & 0 & 0 & 0 & 0 & 0 & 0 & 0 & 0 \\
0 & 0 & 0 & 0 & 0 & 0 & 0 & 0 & 0 & 0 & 0 & 0 \\
0 & 0 & 0 & 0 & 0 & 0 & 0 & 0 & 0 & 0 & 0 & 0 \\
0 & 0 & 0 & 0 & 0 & 0 & 0 & 0 & 0 & 0 & 0 & 0 \\
0 & 0 & -\frac{EA}{L} & 0 & 0 & 0 & 0 & 0 & \frac{EA}{L} & 0 & 0 & 0 \\
0 & 0 & 0 & 0 & 0 & 0 & 0 & 0 & 0 & 0 & 0 & 0 \\
0 & 0 & 0 & 0 & 0 & 0 & 0 & 0 & 0 & 0 & 0 & 0 \\
0 & 0 & 0 & 0 & 0 & 0 & 0 & 0 & 0 & 0 & 0 & 0
\end{bmatrix}
\begin{bmatrix}
u_{3X} \\ u_{3Y} \\ u_{3Z} \\ \varphi_{3X} \\ \varphi_{3Y} \\ \varphi_{3Z} \\ u_{4X} \\ u_{4Y} \\ u_{4Z} \\ \varphi_{4X} \\ \varphi_{4Y} \\ \varphi_{4Z}
\end{bmatrix}
=
\begin{bmatrix}
F_{3X} \\ F_{3Y} \\ F_{3Z} \\ M_{3X} \\ M_{3Y} \\ M_{3Z} \\ F_{4X} \\ F_{4Y} \\ F_{4Z} \\ M_{4X} \\ M_{4Y} \\ M_{4Z}
\end{bmatrix} . \tag{7.59}
$$

Now all single stiffness relations in global coordinates are known. The global stiffness relation can be established by arranging the single stiffness relations adequately. The dimensions of the global stiffness relation results in 24×24, respectively 6 parameters are considered on the 4 nodes.

Only the rows and columns with non-zero entries are considered in the illustration of the global stiffness relation. The column matrices of the state variables in global coordinates are called:

$$
[u_{1Z}, \varphi_{1X}, \varphi_{1Y}, u_{2Z}, \varphi_{2X}, \varphi_{2Y}, u_{3Z}, \varphi_{3Y}, u_{4Z}]^{\text{T}} \tag{7.60}
$$

and

$$
[F_{1Z}, M_{1X}, M_{1Y}, F_{2Z}, M_{2X}, M_{2Y}, F_{3Z}, M_{3Y}, F_{4Z}]^{\text{T}} \tag{7.61}
$$

Fig. 7.16 General structure
in the space with alternative
global coordinate system

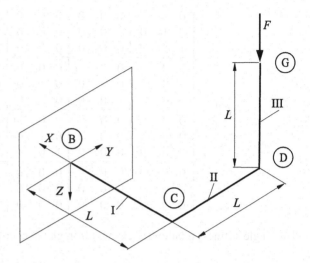

and the global stiffness matrix:

$$
\begin{bmatrix}
12\frac{EI_y}{L^3} & 6\frac{EI_y}{L^2} & 0 & -12\frac{EI_y}{L^3} & 6\frac{EI_y}{L^2} & 0 & 0 & 0 & 0 \\
6\frac{EI_y}{L^2} & 4\frac{EI_y}{L} & 0 & -6\frac{EI_y}{L^2} & 2\frac{EI_y}{L} & 0 & 0 & 0 & 0 \\
0 & 0 & k_t & 0 & 0 & -k_t & 0 & 0 & 0 \\
-12\frac{EI_y}{L^3} & -6\frac{EI_y}{L^2} & 0 & 24\frac{EI_y}{L^3} & -6\frac{EI_y}{L^2} & -6\frac{EI_y}{L^3} & -12\frac{EI_y}{L^3} & -6\frac{EI_y}{L^2} & 0 \\
6\frac{EI_y}{L^2} & 2\frac{EI_y}{L} & 0 & -6\frac{EI_y}{L^2} & 4\frac{EI_y}{L} & 0 & 0 & 0 & 0 \\
0 & 0 & -k_t & -6\frac{EI_y}{L^2} & 0 & k_t+4\frac{EI_y}{L} & 6\frac{EI_y}{L^2} & 2\frac{EI_y}{L} & 0 \\
0 & 0 & 0 & -12\frac{EI_y}{L^3} & 0 & 6\frac{EI_y}{L^2} & 12\frac{EI_y}{L^3}+k_z & 6\frac{EI_y}{L^2} & -k_z \\
0 & 0 & 0 & -6\frac{EI_y}{L^2} & 0 & 2\frac{EI_y}{L} & 6\frac{EI_y}{L^2} & 4\frac{EI_y}{L} & 0 \\
0 & 0 & 0 & 0 & 0 & 0 & -k_z & 0 & k_z
\end{bmatrix}
$$

$$(7.62)$$

with

$$
k_t = \frac{GI_t}{L} \quad , \quad k_z = \frac{EA}{L} .
$$

$$(7.63)$$

The boundary conditions are included in the global stiffness relation. The displacement u_{1Z} and the angles φ_{1X} and φ_{2Y} are zero at the fixed support. Thus, the corresponding rows and columns can be removed from the global stiffness relation. A reduced system remains.

$$
\begin{bmatrix}
24\dfrac{EI_y}{L^3} & -6\dfrac{EI_y}{L^2} & -6\dfrac{EI_y}{L^3} & -12\dfrac{EI_y}{L^3} & -6\dfrac{EI_y}{L^2} & 0 \\[2ex]
-6\dfrac{EI_y}{L^2} & 4\dfrac{EI_y}{L} & 0 & 0 & 0 & 0 \\[2ex]
-6\dfrac{EI_y}{L^2} & 0 & \dfrac{GI_t}{L}+4\dfrac{EI_y}{L} & 6\dfrac{EI_y}{L^2} & 2\dfrac{EI_y}{L} & 0 \\[2ex]
-12\dfrac{EI_y}{L^3} & 0 & 6\dfrac{EI_y}{L^2} & 12\dfrac{EI_y}{L^3}+k & 6\dfrac{EI_y}{L^2} & -k \\[2ex]
-6\dfrac{EI_y}{L^2} & 0 & 2\dfrac{EI_y}{L} & 6\dfrac{EI_y}{L^2} & 4\dfrac{EI_y}{L} & 0 \\[2ex]
0 & 0 & 0 & -k & 0 & k
\end{bmatrix}
\begin{bmatrix}
u_{2Z} \\ \varphi_{2X} \\ \varphi_{2Y} \\ u_{3Z} \\ \varphi_{3Y} \\ u_{4Z}
\end{bmatrix}
=
\begin{bmatrix}
0 \\ 0 \\ 0 \\ 0 \\ 0 \\ F
\end{bmatrix}
$$

$$(7.64)$$

with

$$
k = \frac{EA}{L}. \tag{7.65}
$$

Herefrom the unknown parameters can be determined:

$$
\begin{bmatrix}
u_{2Z} \\ \varphi_{2X} \\ \varphi_{2Y} \\ u_{3Z} \\ \varphi_{3Y} \\ u_{4Z}
\end{bmatrix}
=
\begin{bmatrix}
+\dfrac{F}{3\dfrac{EI_y}{L^3}} \\[3ex]
+\dfrac{F}{2\dfrac{EI_y}{L^2}} \\[3ex]
-\dfrac{F}{2\dfrac{GI_t}{L}} \\[3ex]
+\dfrac{2(GI_t+3EI_y)L^3F}{3EI_yGI_t} \\[2ex]
-\dfrac{L^2(GI_t+2EI_y)F}{2EI_yGI_t} \\[2ex]
+\dfrac{(3GI_tI_y+2GI_tAL^2)LF}{3EI_yAGI_t}
\end{bmatrix}
. \tag{7.66}
$$

By inserting the now known kinematic parameters into the global stiffness relation the clamping forces F_{1X} and fixed-end moments M_{1X} and M_{1Y} can be determined.

7.6 Supplementary Problems

7.1 Three-Dimensional Beam Structure

Determine the displacements and rotations for the above executed example for specific numerical values: $E = 210\,000\,\text{MPa}$, $G = 80\,707\,\text{MPa}$, $a = 20\,\text{mm}$, $I_t = 0.141\,a^4$, $F = 100\,\text{N}$.

7.2 Three-Dimensional Beam Structure, Alternative Coordinate System

Define a second global coordinate system for the above executed example. Figure 7.16 shows the definition of the coordinate axes.

The global Z-coordinate remained the same, the X- and Y-coordinate have changed positions.

Determine the kinematic parameters on the nodal points.

Reference

1. Stoer J (1989) Numerische Mathematik 1. Springer-Lehrbuch, Berlin

Chapter 8
Beam with Shear Contribution

Abstract This element describes the basic deformation mode of bending under the consideration of the shear influence. First, several basic assumptions for the modeling of the TIMOSHENKO beam will be introduced and the element used in this chapter will be differentiated from other formulations. The basic equations from the strength of materials, meaning kinematics, the equilibrium as well as the constitutive equation will be introduced and used for the derivation of a system of coupled differential equations. The section about the basics is ended with analytical solutions. Subsequently, the TIMOSHENKO bending element will be introduced, according to the common definitions for load and deformation parameters, which are used in the handling of the FE method. The derivation of the stiffness matrix is carried out through various methods and will be described in detail. Besides linear shape functions, a general concept for an arbitrary arrangement of the shape functions will be reviewed.

8.1 Introductory Remarks

The general differences regarding the deformation and stress distribution of a bending beam with and without shear influence have already been discussed in Chap. 5. In this chapter, the shear influence on the deformation is considered with the help of the TIMOSHENKO beam theory. Within the framework of the following introductive remarks, first the definition of the shear strain and the connection between shear force and shear stress is covered.

For the derivation of the equation for the shear strain in the x-y plane, the infinitesimal rectangular beam element $ABCD$, shown in Fig. 8.1, is considered, which deforms under exposure of shear stress. Here, a change of the angle of the original right angles as well as a change in the lengths of the edges occur.

The deformation of the point A can be described via the displacement fields $u_x(x, y)$ and $u_y(x, y)$. These two functions of *two* variables can be expanded in

© Springer International Publishing AG, part of Springer Nature 2018
A. Öchsner and M. Merkel, *One-Dimensional Finite Elements*,
https://doi.org/10.1007/978-3-319-75145-0_8

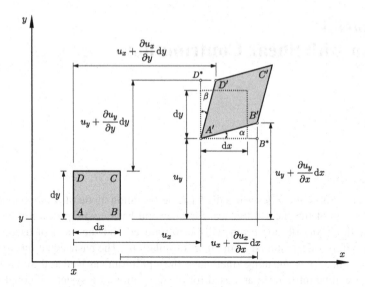

Fig. 8.1 For the definition of the shear strain γ_{xy} in the x-y plane at an infinitesimal beam element

a TAYLORs' series[1] of first order around point A to approximately calculate the deformations of the points B and D:

$$u_{x,B} = u_x(x + dx, y) \approx u_x(x, y) + \frac{\partial u_x}{\partial x} dx + \frac{\partial u_x}{\partial y} dy, \qquad (8.1)$$

$$u_{y,B} = u_y(x + dx, y) \approx u_y(x, y) + \frac{\partial u_y}{\partial x} dx + \frac{\partial u_y}{\partial y} dy, \qquad (8.2)$$

or alternatively

$$u_{x,D} = u_x(x, y + dy) \approx u_x(x, y) + \frac{\partial u_x}{\partial x} dx + \frac{\partial u_x}{\partial y} dy, \qquad (8.3)$$

$$u_{y,D} = u_y(x, y + dy) \approx u_y(x, y) + \frac{\partial u_y}{\partial x} dx + \frac{\partial u_y}{\partial y} dy. \qquad (8.4)$$

In Eqs. (8.1)–(8.4) $u_x(x, y)$ and $u_y(x, y)$ represent the so-called rigid-body displacements, which do not cause any deformation. If one considers that point B has the coordinates $(x + dx, y)$ and D the coordinates $(x, y + dy)$, the following results:

---

[1]For a function $f(x, y)$ of two variables usually a TAYLORs' series expansion of first order is assessed around the point (x_0, y_0) as follows: $f(x, y) = f(x_0 + dx, y_0 + dx) \approx f(x_0, y_0) + \left(\frac{\partial f}{\partial x}\right)_{x_0, y_0} \times (x - x_0) + \left(\frac{\partial f}{\partial y}\right)_{x_0, y_0} \times (y - y_0)$.

$$u_{x,B} = u_x(x, y) + \frac{\partial u_x}{\partial x} dx ,$$ (8.5)

$$u_{y,B} = u_y(x, y) + \frac{\partial u_y}{\partial x} dx ,$$ (8.6)

or alternatively

$$u_{x,D} = u_x(x, y) + \frac{\partial u_x}{\partial y} dy ,$$ (8.7)

$$u_{y,D} = u_y(x, y) + \frac{\partial u_y}{\partial y} dy .$$ (8.8)

The total shear strain γ_{xy} of the deformed beam element $A'B'C'D'$ results, according to Fig. 8.1, from the sum of the angles α and β, which can be identified at the rectangle, which is deformed as a rhombus. Under consideration of the two right-angled triangles $A'D^*D'$ and $A'B^*B'$ these two angles can be expressed via

$$\tan \alpha = \frac{\frac{\partial u_y}{\partial x} dx}{dx + \frac{\partial u_x}{\partial x} dx} \quad \text{and} \quad \tan \beta = \frac{\frac{\partial u_x}{\partial y} dy}{dy + \frac{\partial u_y}{\partial y} dy} .$$ (8.9)

It holds approximately for small deformations that $\tan \alpha \approx \alpha$ and $\tan \beta \approx \beta$ or alternatively $\frac{\partial u_x}{\partial x} \ll 1$ and $\frac{\partial u_y}{\partial y} \ll 1$, so that the following expression results for the shear strain:

$$\gamma_{xy} = \frac{\partial u_y}{\partial x} + \frac{\partial u_x}{\partial y} .$$ (8.10)

This total change of angle is also called the engineering definition. In contrast the expression $\varepsilon_{xy} = \frac{1}{2}\gamma_{xy} = \frac{1}{2}(\frac{\partial u_y}{\partial x} + \frac{\partial u_x}{\partial y})$ is known as the tensorial definition in the literature. Due to the general symmetry of the strain tensor, it applies that $\gamma_{ij} = \gamma_{ji}$.

The algebraic sign of the shear strain needs to be explained in the following with the help of Fig. 8.2 for the special case that only one shear force acts in parallel to the y-axis. If a shear force acts in direction of the positive y-axis at the right-hand face

Fig. 8.2 Definition of a **a** positive and **b** negative shear strain in the x-y plane

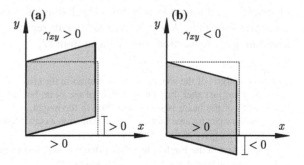

— hence a positive shear force distribution is assumed at this point —, according to Fig. 8.2a under consideration of Eq. (8.10) a positive shear strain results. Accordingly, a negative shear force distribution, according to Fig. 8.2b leads to a negative shear strain. It has already been mentioned in Chap. 5 that the shear stress distribution is alterable through the cross-section. As an example, the parabolic shear stress distribution for a rectangular cross-section was illustrated in Fig. 5.2. Via HOOKE's law for a one-dimensional shear stress state (for this see Sect. 4.1) it can be derived that the shear stress has to exhibit a corresponding parabolic course. From the shear stress distribution in the cross-sectional area at the location x of the beam[2] generally through integration, meaning

$$Q_y = \int_A \tau_{xy}(y, z) \mathrm{d}A \,, \tag{8.11}$$

the acting shear force results. However, the TIMOSHENKO beam theory assumes a simplification, i.e. an equivalent *constant* shear stress and strain:

$$\tau_{xy}(y, z) \rightarrow \tau_{xy} \,. \tag{8.12}$$

This constant shear stress results from the shear force, which acts in an equivalent cross-sectional area, the so-called shear area A_s:

$$\tau_{xy} = \frac{Q_y}{A_s}, \tag{8.13}$$

whereupon the relation between the shear area A_s and the actual cross-sectional area A is referred to as the shear correction factor k_s:

$$k_s = \frac{A_s}{A}. \tag{8.14}$$

Different assumptions can be made for the calculation of the shear correction factor [3]. As an example, it can be demanded [4] that the elastic strain energy of the equivalent shear stress has to be identical with the energy, which results from the acting shear stress distribution in the actual cross-sectional-area. Different characteristics of simple geometric cross-sections — including the shear correction factor[3] — are summarized in Table A.9 [5, 6]. Further details regarding the shear correction factor for arbitrary cross-sections can be taken from [7].

---

[2] A closer analysis of the shear stress distribution in the cross-sectional area shows that the shear stress does not just alter through the height of the beam but also through the width of the beam. If the width of the beam is small compared to the height, only a small change along the width occurs and one can assume in the first approximation a constant shear stress throughout the width: $\tau_{xy}(y, z) \rightarrow \tau_{xy}(y)$. See for example [1, 2].

[3] One notes that in the English literature often the so-called form factor for shear is stated. This results as the reciprocal of the shear correction factor.

Obviously, the equivalent constant shear stress can alter along the center line of the beam, in case that the shear force changes along the center line of the beam. The attribute 'constant' thus just refers to the cross-sectional area on the location x and the equivalent constant shear stress is therefore in general a function of the coordinate of length for the TIMOSHENKO beam:

$$\tau_{xy} = \tau_{xy}(x). \tag{8.15}$$

8.2 Basic Description of the Beam with Shear Contribution

The so-called TIMOSHENKO beam can be generated by superposing a shear deformation on a BERNOULLI beam according to Fig. 8.3.

One can see that the BERNOULLI hypothesis is partly no longer fulfilled for the TIMOSHENKO beam: plane cross-sections also remain plane after the deformation, however a cross-section, which stood at the right angle on the beam axis before the deformation is not at a right angle on the beam axis any longer after the deformation. If the demand for planeness of the cross-sections is also given up, one reaches theories of third-order [8–10], at which a parabolic course of the shear strain and stress in the displacement field are considered, see Fig. 8.4. Therefore, a shear correction factor is unnecessary for these theories of third-order.

8.2.1 Kinematics

According to the alternative derivation in Sect. 5.2.1, the kinematics relation can also be derived for the beam with shear action, by considering the angle ϕ_z instead of the angle φ_z, see Fig. 8.3c. Following an equivalent procedure as in Sect. 5.2.1, the following relationships are obtained:

$$\sin\phi_z = \frac{u_x}{-y} \approx \phi_z \text{ or } u_x = -y\phi_z, \tag{8.16}$$

wherefrom, via the general relation for the strain, meaning $\varepsilon_x = \mathrm{d}u_x/\mathrm{d}x$, the kinematics relation results through differentiation:

$$\varepsilon_x = -y\frac{\mathrm{d}\phi_z}{\mathrm{d}x}. \tag{8.17}$$

Note that $\phi_z \to \varphi_z = \frac{\mathrm{d}u_y}{\mathrm{d}x}$ results from neglecting of shear deformation and the relation according to Eq. (5.15) results as a special case. Furthermore, the following relation between the angles can be derived from Fig. 8.3c

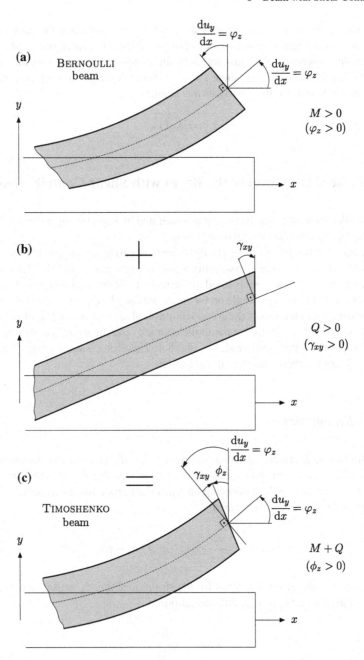

Fig. 8.3 Superposition of the BERNOULLI beam (**a**) and the shear deformation (**b**) to the TIM-OSHENKO beam (**c**) in the x-y plane. The marked orientations of the angles equal the positive definitions

Fig. 8.4 Deformation of originally plane cross-sections at the TIMOSHENKO beam (left) and at the theory of third-order (right) [11]

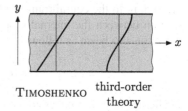

TIMOSHENKO   third-order theory

$$\phi_z = \frac{\mathrm{d}u_y}{\mathrm{d}x} - \gamma_{xy}, \tag{8.18}$$

which complements the set of the kinematics relations. It needs to be remarked that at this point that the so-called bending line was considered. Therefore, the displacement field u_y is just a function of *one* variable: $u_y = u_y(x)$.

8.2.2 Equilibrium

The derivation of the equilibrium condition for the TIMOSHENKO beam is identical with the derivation for the BERNOULLI beam according to Sect. 5.2.2:

$$\frac{\mathrm{d}Q_y(x)}{\mathrm{d}x} = -q_y(x), \tag{8.19}$$

$$\frac{\mathrm{d}M_z(x)}{\mathrm{d}x} = -Q_y(x). \tag{8.20}$$

8.2.3 Constitutive Equation

For the consideration of the constitutive relation, HOOKE's law for a one-dimensional normal stress state and for a shear stress state is used:

$$\sigma_x = E\varepsilon_x, \tag{8.21}$$

$$\tau_{xy} = G\gamma_{xy}, \tag{8.22}$$

whereupon the shear modulus G can be calculated through the elasticity modulus E and the POISSON's ratio ν via

$$G = \frac{E}{2(1+\nu)}. \tag{8.23}$$

According to the equilibrium condition shown in Fig. 5.8 and Eq. (5.29) the connection between the internal moment and the bending stress can be used for the TIMOSHENKO beam as follows:

$$\mathrm{d}M_z = (+y)(-\sigma_x)\mathrm{d}A , \tag{8.24}$$

or alternatively after integration under the use of the constitutive equation (8.21) and the kinematics relation (8.17):

$$M_z(x) = E I_z \frac{\mathrm{d}\phi_z(x)}{\mathrm{d}x} . \tag{8.25}$$

The connection between shear force and cross-section rotation results via the equilibrium relation (8.20) in:

$$Q_y(x) = -\frac{\mathrm{d}M_z(x)}{\mathrm{d}x} = -E I_z \frac{\mathrm{d}^2\phi_z(x)}{\mathrm{d}x^2} . \tag{8.26}$$

Before passing over to the differential equations of the bending line, the basic equations for the TIMOSHENKO beam are summarized in Table 8.1. Note that the normal stress and strain are functions of both coordinates of space x and y, however the shear stress and strain is only dependent on x, since an equivalent *constant* shear stress has been introduced via the cross-section as an approximation of the TIMOSHENKO beam approach.

8.2.4 Differential Equation of the Bending Line

Within the previous section the relation between the internal moment and the cross-section rotation was derived for the normal stress with the help of HOOKE's law. Differentiation of this relation according to Eq. (8.25) leads to the following connection

Table 8.1 Elementary basic equations for the TIMOSHENKO beam with shear contribution at the deformation in the x-y plane

Description	Equation
Kinematics	$\varepsilon_x(x, y) = -y\dfrac{\mathrm{d}\phi_z(x)}{\mathrm{d}x}$ and $\phi_z(x) = \dfrac{\mathrm{d}u_y(x)}{\mathrm{d}x} - \gamma_{xy}(x0)$
Equilibrium	$\dfrac{\mathrm{d}Q_y(x)}{\mathrm{d}x} = -q_y(x); \quad \dfrac{\mathrm{d}M_z(x)}{\mathrm{d}x} = -Q_y(x)$
Constitution	$\sigma_x(x, y) = E\varepsilon_x(x, y)$ and $\tau_{xy}(x) = G\gamma_{xy}(x)$

$$\frac{\mathrm{d}M_z}{\mathrm{d}x} = \frac{\mathrm{d}}{\mathrm{d}x}\left(EI_z\frac{\mathrm{d}\phi_z}{\mathrm{d}x}\right), \qquad (8.27)$$

which can be transformed with the help of the equilibrium relation (8.20) and the relation for the shear stress according to Eqs. (8.13) and (8.14) to

$$\frac{\mathrm{d}}{\mathrm{d}x}\left(EI_z\frac{\mathrm{d}\phi_z}{\mathrm{d}x}\right) = -k_s G A \gamma_{xy}. \qquad (8.28)$$

If the kinematics relation (8.18) is considered in the last equation, the so-called bending differential equation results in:

$$\frac{\mathrm{d}}{\mathrm{d}x}\left(EI_z\frac{\mathrm{d}\phi_z}{\mathrm{d}x}\right) + k_s G A \left(\frac{\mathrm{d}u_y}{\mathrm{d}x} - \phi_z\right) = 0. \qquad (8.29)$$

If now for the shear stress according to Eq. (8.22) the relation for the shear stress according to Eqs. (8.13) and (8.14) is considered in HOOKE's law, one obtains

$$Q_y = k_s A G \gamma_{xy}. \qquad (8.30)$$

Via the equilibrium relation (8.20) and the kinematics relation (8.18) the following results herefrom:

$$\frac{\mathrm{d}M_z}{\mathrm{d}x} = -k_s A G \left(\frac{\mathrm{d}u_y}{\mathrm{d}x} - \phi_z\right). \qquad (8.31)$$

After differentiation and the consideration of the equilibrium relation according to Eqs. (8.19) and (8.20) finally the so-called shear differential equation results in:

$$\frac{\mathrm{d}}{\mathrm{d}x}\left[k_s A G \left(\frac{\mathrm{d}u_y}{\mathrm{d}x} - \phi_z\right)\right] = -q_y(x). \qquad (8.32)$$

Therefore, the shear flexible TIMOSHENKO beam will be described through the following two coupled differential equations of second order:

$$\frac{\mathrm{d}}{\mathrm{d}x}\left(EI_z\frac{\mathrm{d}\phi_z}{\mathrm{d}x}\right) + k_s A G \left(\frac{\mathrm{d}u_y}{\mathrm{d}x} - \phi_z\right) = 0, \qquad (8.33)$$

$$\frac{\mathrm{d}}{\mathrm{d}x}\left[k_s A G \left(\frac{\mathrm{d}u_y}{\mathrm{d}x} - \phi_z\right)\right] = -q_y(x). \qquad (8.34)$$

Fig. 8.5 For the calculation
of the analytical solution of a
TIMOSHENKO beam under
distributed line load

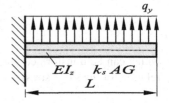

This system contains two independent unknown functions, namely the deflection
$u_y(x)$ and the cross-sectional rotation $\phi_z(x)$. Boundary conditions can be formulated
for both functions to be able to solve the system of differential equations.

8.2.5 Analytical Solutions

For the derivation of analytical solutions, the system of coupled differential equations
according to Eqs. (8.33) and (8.34) has to be solved. Through the use of a computer
algebra system (CAS) for the symbolic calculation of mathematical expressions,[4]
the general solution of the system results for constant EI_z, AG, and $q_y = q_0$ in:

$$u_y(x) = \frac{1}{EI_z}\left(\frac{q_0 x^4}{24} + c_1\frac{x^3}{6} + c_2\frac{x^2}{2} + c_3 x + c_4\right), \tag{8.35}$$

$$\phi_z(x) = \frac{1}{EI_z}\left(\frac{q_0 x^3}{6} + c_1\frac{x^2}{2} + c_2 x + c_3\right) + \frac{q_0 x}{k_s AG} + \frac{c_1}{k_s AG}. \tag{8.36}$$

The constants of integration $c_1, \ldots, c_4$ must be defined through appropriate boundary
conditions to calculate the special solution of a concrete problem, meaning under
consideration of the support and the load conditions of the beam.

As an example, the beam, which is illustrated in Fig. 8.5, needs to be considered
in the following. The loading occurs due to a constant distributed load q_y and the
boundary conditions are given as follows for this example:

$$u_y(x = 0) = 0 \ , \quad \phi_z(x = 0) = 0, \tag{8.37}$$

$$M_z(x = 0) = \frac{q_y L^2}{2} \ , \quad M_z(x = L) = 0. \tag{8.38}$$

The application of the boundary condition $(8.37)_1$ in the general analytical solution
for the deflection according to Eq. (8.35) immediately yields $c_4 = 0$. With the second
boundary condition in Eq. (8.37) the relation $c_3 = -c_1\frac{EI_z}{k_s AG}$ results with the general
analytical solution for the rotation according to Eq. (8.36). The further definition of

---

[4]Maple®, Mathematica® and Matlab® can be listed at this point as commercial examples.

the constants of integration demands that the bending moment is expressed with the help of the deformation. Via Eq. (8.25) the moment distribution results in

$$M_z(x) = EI_z \frac{d\phi_z}{dx} = \left(c_1 x + c_2 + \frac{3q_y x^2}{6} \right) + \frac{q_y E I_z}{k_s AG},$$ (8.39)

and the consideration of boundary conditions $(8.38)_1$ yields $c_2 = \frac{q_y L^2}{2} - \frac{q_y E I_z}{k_s AG}$. Accordingly, consideration of the second boundary condition in Eq. (8.38) yields the first constant of integration to $c_1 = -q_y L$ and finally $c_3 = \frac{q_y L E I_z}{k_s AG}$. Therefore, the deflection distribution results in

$$u_y(x) = \frac{1}{EI_z} \left(\frac{q_y x^4}{24} - q_y L \frac{x^3}{6} + \left[\frac{q_y L^2}{2} - \frac{q_y E I_z}{k_s AG} \right] \frac{x^2}{2} + \frac{q_y L E I_z}{k_s AG} x \right),$$ (8.40)

or alternatively the maximum deflection on the right-hand end of the beam, meaning for $x = L$, to:

$$u_y(x = L) = \frac{q_y L^4}{8 E I_z} + \frac{q_y L^2}{2 k_s AG}.$$ (8.41)

Further analytical solutions for the maximum deflection of a TIMOSHENKO beam are summarized in Table 8.2. Through comparison with the analytical solutions in Sect. 5.2.5 it becomes obvious that the analytical solutions for the maximum deflection compose additively from the classical solution for the BERNOULLI beam and an additional shear part.

Table 8.2 Maximum deflection of TIMOSHENKO beams at simple load cases for bending in the x-y plane

Load	Maximum deflection
F cantilever, $EI k_s AG$, L	$u_{y,\max} = u_y(L) = \dfrac{FL^3}{3EI_z} + \dfrac{FL}{k_s AG}$
q distributed load, $EI k_s AG$, L	$u_{y,\max} = u_y(L) = \dfrac{q_y L^4}{8EI_z} + \dfrac{q_y L^2}{2k_s AG}$
F simply supported, $EI k_s AG$, L	$u_{y,\max} = u_y\left(\frac{L}{2}\right) = \dfrac{FL^3}{48EI_z} + \dfrac{FL}{4k_s AG}$

Fig. 8.6 Comparison of the analytical solutions for the BERNOULLI and TIMOSHENKO beam for different boundary conditions: **a** one-sided fixed with end load, **b** one-sided fixed with distributed load, and **c** fixed with mid load

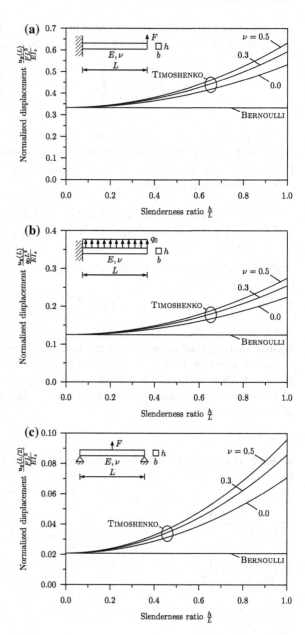

Table 8.3 Elementary basic equations for the TIMOSHENKO beam with shear contribution at deformations in the x-z plane

Notation	Equation
Kinematics	$\varepsilon_x(x, z) = -z \dfrac{d\phi_y(x)}{dx}$ and $\phi_y(x) = -\dfrac{du_z(x)}{dx} + \gamma_{xz}(x)$
Equilibrium	$\dfrac{dQ_z(x)}{dx} = -q_z(x); \quad \dfrac{dM_y(x)}{dx} = +Q_z(x)$
Constitution	$\sigma_x(x, z) = E\varepsilon_x(x, z)$ and $\tau_{xz}(x) = G\gamma_{xz}(x)$
Diff. Equations	$-\dfrac{d}{dx}\left(EI_y \dfrac{d\phi_y}{dx}\right) + k_s AG \left(\dfrac{du_z}{dx} + \phi_y\right) = 0$
	$\dfrac{d}{dx}\left[k_s AG \left(\dfrac{du_z}{dx} + \phi_y\right)\right] = -q_z(x)$

To highlight the influence of the shear part the maximum deflection needs to be presented in the following over the relation of beam height to beam length. As an example three different loading and support cases for a rectangular cross-section with the width b and the height h are presented in Fig. 8.6. It becomes obvious that the difference between the BERNOULLI and the TIMOSHENKO beam becomes smaller and smaller for a decreasing slenderness ratio, meaning for beams at which the length L is significantly larger compared to the height h.

The relative difference between the BERNOULLI and the TIMOSHENKO solution results, for example for a POISSON's ratio of 0.3 and a slenderness ratio of 0.1 – meaning for a beam, at which the length is ten times larger than the height – depending on the support and load in: 0.77% for the cantilever with point load, 1.03% for the cantilever with distributed load and 11.10% for the simply supported beam. Further analytical solutions for the TIMOSHENKO beam can be withdrawn, for example, from [12].

Conclusively it needs to be pointed out that for deformations in the x-z plane slightly modified equations occur compared to Table 8.1. The corresponding equations for bending in the x-z plane with shear part are summarized in Table 8.3.

8.3 The Finite Element of Plane Bending Beams with Shear Contribution

According to Sect. 5.3 the bending element is defined as a prismatic body with the x-axis along the center line and the y-axis orthogonally to the center line. Nodes are introduced at both ends of the bending element, at which displacements and rotations

Fig. 8.7 Definition of the positive direction for the bending element with shear contribution at deformation in the x-y plane: **a** deformation parameters; **b** load parameters

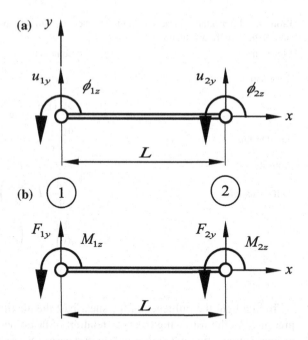

or alternatively forces and moments are defined, see Fig. 8.7. The deformation and load parameters are assumed to be positive in the drafted direction.

The two unknowns, meaning the deflection $u_y(x)$ and the herefrom independent cross-sectional rotation $\phi_z(x)$ will be approximated with the help of the following nodal approaches:

$$u_y(x) = N_{1u}(x)u_{1y} + N_{2u}(x)u_{2y}, \tag{8.42}$$

$$\phi_z(x) = N_{1\phi}(x)\phi_{1z} + N_{2\phi}(x)\phi_{2z}, \tag{8.43}$$

or alternatively in matrix notation as

$$u_y(x) = \begin{bmatrix} N_{1u}(x) & 0 & N_{2u}(x) & 0 \end{bmatrix} \begin{bmatrix} u_{1y} \\ \phi_{1z} \\ u_{2y} \\ \phi_{2z} \end{bmatrix} = \boldsymbol{N}_u \boldsymbol{u}_\mathrm{p}, \tag{8.44}$$

$$\phi_z(x) = \begin{bmatrix} 0 & N_{1\phi} & 0 & N_{2\phi}(x) \end{bmatrix} \begin{bmatrix} u_{1y} \\ \phi_{1z} \\ u_{2y} \\ \phi_{2z} \end{bmatrix} = \boldsymbol{N}_\phi \boldsymbol{u}_\mathrm{p}. \tag{8.45}$$

With these relations the derivative of the cross-sectional rotation in the coupled differential equations (8.33) and (8.34) results in

$$\frac{d\phi_z(x)}{dx} = \frac{dN_{1\phi}(x)}{dx}\phi_{1z} + \frac{dN_{2\phi}(x)}{dx}\phi_{2z} = \frac{dN_\phi}{dx}\boldsymbol{u}_p . \tag{8.46}$$

8.3.1 Derivation Through Potential

The elastic strain energy of a TIMOSHENKO beam with linear-elastic material behavior results in:

$$\Pi_{\text{int}} = \frac{1}{2}\int_\Omega \boldsymbol{\varepsilon}^{\mathrm{T}}\boldsymbol{\sigma}\,d\Omega = \frac{1}{2}\int_\Omega \begin{bmatrix} \varepsilon_x & \gamma_{xy} \end{bmatrix}\begin{bmatrix} \sigma_x \\ \tau_{xy} \end{bmatrix}d\Omega \tag{8.47}$$

$$= \frac{1}{2}\int_\Omega \varepsilon_x\sigma_x\,d\Omega + \frac{1}{2}\int_\Omega \gamma_{xy}\tau_{xy}\,d\Omega = \Pi_{\text{int,b}} + \Pi_{\text{int,s}} . \tag{8.48}$$

The bending and shear part of the elastic strain energy will be regarded separately in the following and subsequently be superposed.

The bending part of the elastic strain energy results through HOOKE's law according to Eq. (8.21) in:

$$\Pi_{\text{int,b}} = \frac{1}{2}\int_\Omega \varepsilon_x\sigma_x\,d\Omega = \frac{1}{2}\int_\Omega \varepsilon_x E\varepsilon_x\,d\Omega . \tag{8.49}$$

The kinematic relation (8.17) can be written as follows via Eq. (8.46)

$$\varepsilon_x = -y\frac{d\phi}{dx} = -y\left(\frac{dN_{1\phi}(x)}{dx}\phi_{1z} + \frac{dN_{2\phi}(x)}{dx}\phi_{2z}\right) \tag{8.50}$$

$$= -y\begin{bmatrix} 0 & \frac{dN_{1\phi}(x)}{dx} & 0 & \frac{dN_{2\phi}(x)}{dx} \end{bmatrix}\begin{bmatrix} u_{1y} \\ \phi_{1z} \\ u_{2y} \\ \phi_{2z} \end{bmatrix} = \boldsymbol{B}_b\boldsymbol{u}_p \tag{8.51}$$

whereupon a generalized $\boldsymbol{B}_b$ matrix as

$$\boldsymbol{B}_b = -y\frac{dN_\phi}{dx} \tag{8.52}$$

has been introduced. Therefore, the bending part of the elastic strain energy according to Eq. (8.49) results in:

$$\Pi_{\text{int,b}} = \frac{1}{2}\int_{\Omega} \left(B_b u_p\right)^T E \left(B_b u_p\right) d\Omega = \frac{1}{2}\int_{\Omega} u_p^T B_b^T E B_b u_p d\Omega$$

$$= \frac{1}{2} u_p^T \left[\int_{\Omega} B_b^T E B_b d\Omega\right] u_p = \frac{1}{2} u_p^T \left[\int_{\Omega} (-y)\frac{dN_\phi^T}{dx} E(-y)\frac{dN_\phi}{dx} d\Omega\right] u_p$$

$$= \frac{1}{2} u_p^T \left[\int_0^L \underbrace{\left(\int_A y^2 dA\right)}_{I_z} E \frac{dN_\phi^T}{dx}\frac{dN_\phi}{dx} dx\right] u_p .$$

$$= \frac{1}{2} u_p^T \underbrace{\left[\int_0^L E I_z \frac{dN_\phi^T}{dx}\frac{dN_\phi}{dx} dx\right]}_{k_b^e} u_p . \tag{8.53}$$

The element stiffness matrix was identified in the last equation with the help of the general formulation of the strain energy according to Eq. (5.91). Herefrom the element stiffness matrix for constant bending stiffness EI_z in components results in:

$$k_b^e = E I_z \int_0^L \begin{bmatrix} 0 & 0 & 0 & 0 \\ 0 & \frac{dN_{1\phi}}{dx}\frac{dN_{1\phi}}{dx} & 0 & \frac{dN_{1\phi}}{dx}\frac{dN_{2\phi}}{dx} \\ 0 & 0 & 0 & 0 \\ 0 & \frac{dN_{2\phi}}{dx}\frac{dN_{1\phi}}{dx} & 0 & \frac{dN_{2\phi}}{dx}\frac{dN_{2\phi}}{dx} \end{bmatrix} dx . \tag{8.54}$$

A further evaluation of Eq. (8.54) demands the introduction of the shape functions N_i. The shear part of the elastic strain energy results in the following with the help of Eqs. (8.11)–(8.14):

$$\Pi_{\text{int,s}} = \frac{1}{2}\int_{\Omega} \gamma_{xy}\tau_{xy} d\Omega = \frac{1}{2}\int_0^L \gamma_{xy}(x,y)\left(\int_A \tau_{xy}(x,y)dA\right) dx \tag{8.55}$$

$$= \frac{1}{2}\int_0^L \gamma_{xy} k_s G A \gamma_{xy} dx . \tag{8.56}$$

Via Eqs. (8.42) and (8.43) the kinematics relation (8.18) can be written as follows

$$\gamma_{xy} = \frac{du_y}{dx} - \phi = \frac{dN_{1u}}{dx}u_{1y} + \frac{dN_{2u}}{dx}u_{2y} - N_{1\phi}\phi_{1z} - N_{2\phi}\phi_{2z} \tag{8.57}$$

$$= \begin{bmatrix} \frac{\mathrm{d}N_{1u}}{\mathrm{d}x} & -N_{1\phi} & \frac{\mathrm{d}N_{2u}}{\mathrm{d}x} & -N_{2\phi} \end{bmatrix} \begin{bmatrix} u_{1y} \\ \phi_{1z} \\ u_{2y} \\ \phi_{2z} \end{bmatrix} = \boldsymbol{B}_{s} \boldsymbol{u}_{p}, \tag{8.58}$$

whereupon at this point a generalized $\boldsymbol{B}_s$-matrix for the shear part has been introduced. Therefore, the shear part of the elastic strain energy according to Eq. (8.55) results in:

$$\Pi_{\mathrm{int,s}} = \frac{1}{2} \int_{0}^{L} \left(\boldsymbol{B}_{s} \boldsymbol{u}_{p} \right)^{\mathrm{T}} k_{s} G A \left(\boldsymbol{B}_{s} \boldsymbol{u}_{p} \right) \mathrm{d}x \tag{8.59}$$

$$= \frac{1}{2} \boldsymbol{u}_{p}^{\mathrm{T}} \underbrace{\left[\int_{0}^{L} k_{s} G A \boldsymbol{B}_{s}^{\mathrm{T}} \boldsymbol{B}_{s} \mathrm{d}x \right]}_{k_{s}^{e}} \boldsymbol{u}_{p}. \tag{8.60}$$

The element stiffness matrix for constant shear stiffness GA results herefrom in components to:

$$\boldsymbol{k}_{s}^{e} = k_{s} G A \int_{0}^{L} \begin{bmatrix} \frac{\mathrm{d}N_{1u}}{\mathrm{d}x} \frac{\mathrm{d}N_{1u}}{\mathrm{d}x} & \frac{\mathrm{d}N_{1u}}{\mathrm{d}x}(-N_{1\phi}) & \frac{\mathrm{d}N_{1u}}{\mathrm{d}x} \frac{\mathrm{d}N_{2u}}{\mathrm{d}x} & \frac{\mathrm{d}N_{1u}}{\mathrm{d}x}(-N_{2\phi}) \\ (-N_{1\phi})\frac{\mathrm{d}N_{1u}}{\mathrm{d}x} & (-N_{1\phi})(-N_{1\phi}) & (-N_{1\phi})\frac{\mathrm{d}N_{2u}}{\mathrm{d}x} & (-N_{1\phi})(-N_{2\phi}) \\ \frac{\mathrm{d}N_{2u}}{\mathrm{d}x} \frac{\mathrm{d}N_{1u}}{\mathrm{d}x} & \frac{\mathrm{d}N_{2u}}{\mathrm{d}x}(-N_{1\phi}) & \frac{\mathrm{d}N_{2u}}{\mathrm{d}x} \frac{\mathrm{d}N_{2u}}{\mathrm{d}x} & \frac{\mathrm{d}N_{2u}}{\mathrm{d}x}(-N_{2\phi}) \\ (-N_{2\phi})\frac{\mathrm{d}N_{1u}}{\mathrm{d}x} & (-N_{2\phi})(-N_{1\phi}) & (-N_{2\phi})\frac{\mathrm{d}N_{2u}}{\mathrm{d}x} & (-N_{2\phi})(-N_{2\phi}) \end{bmatrix} \mathrm{d}x. \tag{8.61}$$

The two expressions for the bending and shear parts of the element stiffness matrix according to Eqs. (8.54) and (8.61) can be superposed for the principal finite element equation of the TIMOSHENKO beam on the element level

$$\boldsymbol{k}^{e} \boldsymbol{u}_{p} = \boldsymbol{F}^{e}, \tag{8.62}$$

whereupon the total stiffness matrix according to Eq. (8.63) is given.

$$
\underline{K}^e =
\begin{bmatrix}
k_s GA \int_0^L \dfrac{dN_{1u}}{dx}\dfrac{dN_{1u}}{dx}\,dx & k_s GA \int_0^L \dfrac{dN_{1u}}{dx}(-N_{1\phi})\,dx & k_s GA \int_0^L \dfrac{dN_{1u}}{dx}\dfrac{dN_{2u}}{dx}\,dx & k_s GA \int_0^L \dfrac{dN_{1u}}{dx}(-N_{2\phi})\,dx \\[2ex]
k_s GA \int_0^L (-N_{1\phi})\dfrac{dN_{1u}}{dx}\,dx & k_s GA \int_0^L (-N_{1\phi})(-N_{1\phi})\,dx + EI_z \int_0^L \dfrac{dN_{1\phi}}{dx}\dfrac{dN_{1\phi}}{dx}\,dx & k_s GA \int_0^L (-N_{1\phi})\dfrac{dN_{2u}}{dx}\,dx & k_s GA \int_0^L (-N_{1\phi})(-N_{2\phi})\,dx + EI_z \int_0^L \dfrac{dN_{2\phi}}{dx}\dfrac{dN_{1\phi}}{dx}\,dx \\[2ex]
k_s GA \int_0^L \dfrac{dN_{2u}}{dx}\dfrac{dN_{1u}}{dx}\,dx & k_s GA \int_0^L \dfrac{dN_{2u}}{dx}(-N_{1\phi})\,dx & k_s GA \int_0^L \dfrac{dN_{2u}}{dx}\dfrac{dN_{2u}}{dx}\,dx & k_s GA \int_0^L \dfrac{dN_{2u}}{dx}(-N_{2\phi})\,dx \\[2ex]
k_s GA \int_0^L (-N_{2\phi})\dfrac{dN_{1u}}{dx}\,dx & k_s GA \int_0^L (-N_{2\phi})(-N_{1\phi})\,dx + EI_z \int_0^L \dfrac{dN_{1\phi}}{dx}\dfrac{dN_{2\phi}}{dx}\,dx & k_s GA \int_0^L (-N_{2\phi})\dfrac{dN_{2u}}{dx}\,dx & k_s GA \int_0^L (-N_{2\phi})(-N_{2\phi})\,dx + EI_z \int_0^L \dfrac{dN_{2\phi}}{dx}\dfrac{dN_{2\phi}}{dx}\,dx
\end{bmatrix}
\tag{8.63}
$$

8.3.2 Derivation Through the Castigliano's Theorem

The elastic strain energy for a TIMOSHENKO beam according to Eq. (8.48) results via HOOKE's law (8.21) and the kinematics relation (8.17) or alternatively via the equation for the equivalent shear stress according to Eqs. (8.11)–(8.14) in:

$$
\begin{aligned}
\Pi_{\text{int}} &= \frac{1}{2} \int_{\Omega} \varepsilon_x \sigma_x d\Omega + \frac{1}{2} \int_{\Omega} \gamma_{xy}(x, y) \tau_{xy}(x, y) d\Omega \\
&= \frac{1}{2} \int_{\Omega} E \varepsilon_x^2 d\Omega + \frac{1}{2} \int_0^L \gamma_{xy}(x, y) \left(\int_A \tau_{xy}(x, y) dA \right) dx \\
&= \frac{1}{2} \int_{\Omega} E \left(\frac{d\phi}{dx} \right)^2 y^2 d\Omega + \frac{1}{2} \int_0^L \gamma_{xy} Q_y dx \\
&= \frac{1}{2} \int_0^L E \left(\frac{d\phi}{dx} \right)^2 \left(\int_A y^2 dA \right) dx + \frac{1}{2} \int_0^L \gamma_{xy} \tau_{xy} k_s A dx \\
&= \frac{1}{2} \int_0^L E I_z \left(\frac{d\phi}{dx} \right)^2 dx + \frac{1}{2} \int_0^L k_s G A \gamma_{xy}^2 dx \, .
\end{aligned}
\tag{8.64}
$$

Herefrom the elastic strain energy for a TIMOSHENKO beam with constant bending and shear stiffness results, via the approaches for the derivation of the cross-sectional rotation $\phi_z(x)$ according to Eq. (8.46) and the shear strain (8.57), in:

$$
\begin{aligned}
\Pi_{\text{int}} &= \frac{1}{2} E I_z \int_0^L \left(\frac{dN_{1\phi}(x)}{dx} \phi_{1z} + \frac{dN_{2\phi}(x)}{dx} \phi_{2z} \right)^2 dx \\
&+ \frac{1}{2} k_s G A \int_0^L \left(\frac{dN_{1u}}{dx} u_{1y} + \frac{dN_{2u}}{dx} u_{2y} - N_{1\phi}\phi_{1z} - N_{2\phi}\phi_{2z} \right)^2 dx \, .
\end{aligned}
\tag{8.65}
$$

Application of CASTIGLIANO's theorem on the strain energy in regards to the nodal displacement u_{1y} yields the external force F_{1y} on node 1:

$$
\begin{aligned}
\frac{d\Pi_{\text{int}}}{du_{1y}} &= F_{1y} \\
&= k_s G A \int_0^L \left(\frac{dN_{1u}}{dx} u_{1y} + \frac{dN_{2u}}{dx} u_{2y} - N_{1\phi}\phi_{1z} - N_{2\phi}\phi_{2z} \right) \frac{dN_{1u}}{dx} dx \, .
\end{aligned}
\tag{8.66}
$$

Accordingly the following results from the differentiation according to the other deformation parameters on the nodes:

$$
\frac{\mathrm{d}\Pi_{\mathrm{int}}}{\mathrm{d}\phi_{1z}} = M_{1z} = EI_z \int_0^L \left(\frac{\mathrm{d}N_{1\phi}(x)}{\mathrm{d}x}\phi_{1z} + \frac{\mathrm{d}N_{2\phi}(x)}{\mathrm{d}x}\phi_{2z} \right) \frac{\mathrm{d}N_{1\phi}(x)}{\mathrm{d}x}\mathrm{d}x +
$$

$$
+ k_s GA \int_0^L \left(\frac{\mathrm{d}N_{1u}}{\mathrm{d}x}u_{1y} + \frac{\mathrm{d}N_{2u}}{\mathrm{d}x}u_{2y} - N_{1\phi}\phi_{1z} - N_{2\phi}\phi_{2z} \right) (-N_{1\phi})\mathrm{d}x .
$$

$$(8.67)$$

$$
\frac{\mathrm{d}\Pi_{\mathrm{int}}}{\mathrm{d}u_{2y}} = F_{2y}
$$

$$
= k_s GA \int_0^L \left(\frac{\mathrm{d}N_{1u}}{\mathrm{d}x}u_{1y} + \frac{\mathrm{d}N_{2u}}{\mathrm{d}x}u_{2y} - N_{1\phi}\phi_{1z} - N_{2\phi}\phi_{2z} \right) \frac{\mathrm{d}N_{2u}}{\mathrm{d}x}\mathrm{d}x . \quad (8.68)
$$

$$
\frac{\mathrm{d}\Pi_{\mathrm{int}}}{\mathrm{d}\phi_{2z}} = M_{2z} = EI_z \int_0^L \left(\frac{\mathrm{d}N_{1\phi}(x)}{\mathrm{d}x}\phi_{1z} + \frac{\mathrm{d}N_{2\phi}(x)}{\mathrm{d}x}\phi_{2z} \right) \frac{\mathrm{d}N_{2\phi}(x)}{\mathrm{d}x}\mathrm{d}x +
$$

$$
+ k_s GA \int_0^L \left(\frac{\mathrm{d}N_{1u}}{\mathrm{d}x}u_{1y} + \frac{\mathrm{d}N_{2u}}{\mathrm{d}x}u_{2y} - N_{1\phi}\phi_{1z} - N_{2\phi}\phi_{2z} \right) (-N_{2\phi})\mathrm{d}x .
$$

$$(8.69)$$

The last four equations can be summarized as the principal finite element equation in matrix form, see Eqs. (8.62) and (8.63).

8.3.3   Derivation Through the Weighted Residual Method

According to the procedure in Sect. 5.3.2 one introduces approximate solutions into the differential equations (8.33) and (8.34) and demands that equations have to be fulfilled over a certain domain.

In the following first of all the shear differential equation (8.34) is considered, which is multiplied with a deflection weight function $W_u(x)$ to attain the following inner product:

$$
\int_0^L W_u^{\mathrm{T}}(x) \left\{ k_s AG \left(\frac{\mathrm{d}^2 u_y}{\mathrm{d}x^2} - \frac{\mathrm{d}\phi_z}{\mathrm{d}x} \right) + q_y(x) \right\} \mathrm{d}x \stackrel{!}{=} 0 . \quad (8.70)
$$

Partial integration of both expressions in the round brackets yields:

$$\int_0^L W_u^T k_s AG \frac{d^2 u_y}{dx^2} dx = \left[W_u^T k_s AG \frac{du_y}{dx} \right]_0^L - \int_0^L \frac{dW_u^T}{dx} k_s AG \frac{du_y}{dx} dx, \quad (8.71)$$

$$-\int_0^L W_u^T k_s AG \frac{d\phi_z}{dx} dx = -\left[W_u^T k_s AG \phi_z \right]_0^L + \int_0^L \frac{dW_u^T}{dx} k_s AG \phi_z \, dx. \quad (8.72)$$

Next, the bending differential equation (8.33) is multiplied with a rotation weight function $W_\phi(x)$ and is transformed in the inner product:

$$\int_0^L W_\phi^T(x) \left\{ \frac{d}{dx} \left(EI_z \frac{d\phi_z}{dx} \right) + k_s AG \left(\frac{du_y}{dx} - \phi_z \right) \right\} dx \overset{!}{=} 0 \quad (8.73)$$

Partial integration of the first expression yields

$$\int_0^L W_\phi^T EI_z \frac{d^2 \phi_z}{dx^2} dx = \left[W_\phi^T EI_z \frac{d\phi_z}{dx} \right]_0^L - \int_0^L \frac{dW_\phi^T}{dx} EI_z \frac{d\phi_z}{dx} dx \quad (8.74)$$

and the bending differential equations results in:

$$\left[W_\phi^T EI_z \frac{d\phi_z}{dx} \right]_0^L - \int_0^L \frac{dW_\phi^T}{dx} EI_z \frac{d\phi_z}{dx} dx + \int_0^L W_\phi^T(x) k_s AG \left(\frac{du_y}{dx} - \phi_z \right) dx = 0. \quad (8.75)$$

Summation of the two converted differential equations yields

$$\left[W_u^T k_s AG \frac{du_y}{dx} \right]_0^L - \int_0^L \frac{dW_u^T}{dx} k_s AG \frac{du_y}{dx} dx - \left[W_u^T k_s AG \phi_z \right]_0^L$$

$$+ \int_0^L \frac{dW_u^T}{dx} k_s AG \phi_z \, dx + \int_0^L W_u^T q_y dx + \int_0^L W_\phi^T(x) k_s AG \left(\frac{du_y}{dx} - \phi_z \right) dx$$

$$- \int_0^L \frac{dW_\phi^T}{dx} EI_z \frac{d\phi_z}{dx} dx + \left[W_\phi^T EI_z \frac{d\phi_z}{dx} \right]_0^L = 0, \quad (8.76)$$

or alternatively after a short conversion the weak form of the shear flexible bending beam:

$$
\int_0^L \frac{\mathrm{d}W_\phi^T}{\mathrm{d}x}EI_z\frac{\mathrm{d}\phi_z}{\mathrm{d}x}\mathrm{d}x + \int_0^L \underbrace{\left(\frac{\mathrm{d}W_u^T}{\mathrm{d}x} - W_\phi^T\right)}_{\delta\gamma_{xy}} k_s AG \underbrace{\left(\frac{\mathrm{d}u_y}{\mathrm{d}x} - \phi_z\right)}_{\gamma_{xy}}\mathrm{d}x
$$

$$
= \int_0^L W_u^T q_y \mathrm{d}x + \left[W_u^T k_s AG\left(\frac{\mathrm{d}u_y}{\mathrm{d}x} - \phi_z\right)\right]_0^L + \left[W_\phi^T EI_z\frac{\mathrm{d}\phi_z}{\mathrm{d}x}\right]_0^L. \qquad (8.77)
$$

One can see that the first part of the left-hand half represents the bending part and the second half the shear part. The right-hand side results from the external loads of the beam. In the following, first of all the left-hand half of the weak form will be considered to derive the stiffness matrix:

$$
\int_0^L \frac{\mathrm{d}W_\phi^T}{\mathrm{d}x}EI_z\frac{\mathrm{d}\phi_z}{\mathrm{d}x}\mathrm{d}x + \int_0^L \left(\frac{\mathrm{d}W_u^T}{\mathrm{d}x} - W_\phi^T\right)k_s AG\left(\frac{\mathrm{d}u_y}{\mathrm{d}x} - \phi_z\right)\mathrm{d}x. \qquad (8.78)
$$

In the next step the approaches for the deflection and rotation of the nodes or alternatively their derivatives according to Eqs. (8.44) and (8.45), meaning

$$
u_y(x) = N_u(x)u_p \ , \qquad \frac{\mathrm{d}u_y(x)}{\mathrm{d}x} = \frac{\mathrm{d}N_u(x)}{\mathrm{d}x}u_p , \qquad (8.79)
$$

$$
\phi_z(x) = N_\phi(x)u_p \ , \qquad \frac{\mathrm{d}\phi_z(x)}{\mathrm{d}x} = \frac{\mathrm{d}N_\phi(x)}{\mathrm{d}x}u_p , \qquad (8.80)
$$

have to be considered. The approaches for the weight functions are chosen analogous to the approaches for the unknowns:

$$
W_u(x) = N_u(x)\delta u_p , \qquad (8.81)
$$

$$
W_\phi(x) = N_\phi(x)\delta u_p , \qquad (8.82)
$$

or alternatively for the derivatives:

$$
\frac{W_u(x)}{\mathrm{d}x} = \frac{N_u(x)}{\mathrm{d}x}\delta u_p , \qquad (8.83)
$$

$$
\frac{W_\phi(x)}{\mathrm{d}x} = \frac{N_\phi(x)}{\mathrm{d}x}\delta u_p . \qquad (8.84)
$$

Therefore, the left-hand half of Eq. (8.78) – under consideration that the rotation or alternatively the virtual rotation can be considered as constant respective to the integration – results in:

$$\delta\boldsymbol{u}_{\text{p}}^{\text{T}} \int\limits_{0}^{L} EI_z \frac{\mathrm{d}\boldsymbol{N}_\phi^{\text{T}}}{\mathrm{d}x} \frac{\mathrm{d}\boldsymbol{N}_\phi}{\mathrm{d}x} \mathrm{d}x\, \boldsymbol{u}_{\text{p}} +$$

$$+ \delta\boldsymbol{u}_{\text{p}}^{\text{T}} \int\limits_{0}^{L} k_s AG \left(\frac{\mathrm{d}\boldsymbol{N}_u^{\text{T}}}{\mathrm{d}x} - \boldsymbol{N}_\phi^{\text{T}} \right) \left(\frac{\mathrm{d}\boldsymbol{N}_u}{\mathrm{d}x} - \boldsymbol{N}_\phi \right) \mathrm{d}x\, \boldsymbol{u}_{\text{p}} . \tag{8.85}$$

In the following, it remains to be seen that the virtual deformations $\delta\boldsymbol{u}^{\text{T}}$ can be eliminated with a corresponding expression on the right-hand side of Eq. (8.77). Therefore on the left-hand side there remains

$$\underbrace{\int\limits_{0}^{L} EI_z \frac{\mathrm{d}\boldsymbol{N}_\phi^{\text{T}}}{\mathrm{d}x} \frac{\mathrm{d}\boldsymbol{N}_\phi}{\mathrm{d}x} \mathrm{d}x\, \boldsymbol{u}_{\text{p}}}_{\boldsymbol{k}_{\text{b}}^{\text{e}}} + \underbrace{\int\limits_{0}^{L} k_s AG \left(\frac{\mathrm{d}\boldsymbol{N}_u^{\text{T}}}{\mathrm{d}x} - \boldsymbol{N}_\phi^{\text{T}} \right) \left(\frac{\mathrm{d}\boldsymbol{N}_u}{\mathrm{d}x} - \boldsymbol{N}_\phi \right) \mathrm{d}x\, \boldsymbol{u}_{\text{p}}}_{\boldsymbol{k}_{\text{s}}^{\text{e}}} \tag{8.86}$$

and the bending or alternatively the shear stiffness matrix can be identified, see Eqs. (8.54) and (8.61). Finally, the right-hand side of the weak form according to Eq. (8.77) is considered:

$$\int\limits_{0}^{L} \boldsymbol{W}_u^{\text{T}} q_y \mathrm{d}x + \left[\boldsymbol{W}_u^{\text{T}} k_s AG \left(\frac{\mathrm{d}u_y}{\mathrm{d}x} - \phi_z \right) \right]_0^L + \left[\boldsymbol{W}_\phi^{\text{T}} EI_z \frac{\mathrm{d}\phi_z}{\mathrm{d}x} \right]_0^L . \tag{8.87}$$

Consideration of the relations for the shear force and the internal moment according to Eqs. (8.30) and (8.25) in the right-hand side of the weak form yields

$$\int\limits_{0}^{L} \boldsymbol{W}_u^{\text{T}} q_y \mathrm{d}x + \left[\boldsymbol{W}_u^{\text{T}}(x) Q_y(x) \right]_0^L + \left[\boldsymbol{W}_\phi^{\text{T}}(x) M_z(x) \right]_0^L , \tag{8.88}$$

or alternatively after the introduction of the approaches for the deflection and rotation of the nodes according to Eqs. (8.44) and (8.45):

$$\delta\boldsymbol{u}_{\text{p}}^{\text{T}} \int\limits_{0}^{L} q_y \boldsymbol{N}_u^{\text{T}} \mathrm{d}x + \delta\boldsymbol{u}_{\text{p}}^{\text{T}} \left[Q_y(x) \boldsymbol{N}_u^{\text{T}}(x) \right]_0^L + \delta\boldsymbol{u}_{\text{p}}^{\text{T}} \left[M_z(x) \boldsymbol{N}_\phi^{\text{T}}(x) \right]_0^L . \tag{8.89}$$

$\delta\boldsymbol{u}_{\text{p}}^{\text{T}}$ can be eliminated with the corresponding expression in Eq. (8.85) and the following remains

$$\int_0^L q_y N_u^T dx + \left[Q_y(x) N_u^T(x)\right]_0^L + \left[M_z(x) N_\phi^T(x)\right]_0^L, \tag{8.90}$$

or alternatively in components:

$$\int_0^L q_y(x) \begin{bmatrix} N_{1u} \\ 0 \\ N_{2u} \\ 0 \end{bmatrix} dx + \begin{bmatrix} -Q_y(0) \\ 0 \\ +Q_y(L) \\ 0 \end{bmatrix} + \begin{bmatrix} 0 \\ -M_z(0) \\ 0 \\ +M_z(L) \end{bmatrix}. \tag{8.91}$$

One notes that the general characteristics of the shape function have been used during the evaluation of the boundary integrals:

$$\text{1st row:} \quad Q_y(L) \underbrace{N_{1u}(L)}_{0} - Q_y(0) \underbrace{N_{1u}(0)}_{1}, \tag{8.92}$$

$$\text{2nd row:} \quad M_z(L) \underbrace{N_{1\phi}(L)}_{0} - M_z(0) \underbrace{N_{1\phi}(0)}_{1}, \tag{8.93}$$

$$\text{3rd row:} \quad Q_y(L) \underbrace{N_{2u}(L)}_{1} - Q_y(0) \underbrace{N_{2u}(0)}_{0}, \tag{8.94}$$

$$\text{4th row:} \quad M_z(L) \underbrace{N_{2\phi}(L)}_{1} - M_z(0) \underbrace{N_{2\phi}(0)}_{0}. \tag{8.95}$$

8.3.4   Derivation Through Virtual Work

The derivation follows the course of reasoning as introduced for the bar element in Sect. 3.2.4 and the BERNOULLI beam element in Sect. 5.3.3. Let us consider a beam in bending as shown in Fig. 8.7 which is loaded by single forces and moments and which can freely deform. The equilibrium equations for the domain $\Omega =]0, L[$ are obtained according to Eqs. (5.22) and (5.26) as:

$$\frac{dQ_y(x)}{dx} + q_y(x) = 0, \tag{8.96}$$

$$\frac{dM_z(x)}{dx} + Q_y(x) = 0. \tag{8.97}$$

In a similar way, one can state the equilibrium conditions between the inner reactions and the external loads (F_{iy}, M_{iz}) at the boundaries, i.e. at $x = 0$ and $x = L$, as:

$$F_{1y} + Q_y(0) = 0, \qquad\qquad M_{1z} + M_z(0) = 0, \tag{8.98}$$

$$F_{2y} - Q_y(L) = 0, \qquad\qquad M_{2z} - M_z(L) = 0. \tag{8.99}$$

Following the procedure as outlined in Sect. 8.2.4, the equilibrium conditions according to Eqs. (8.96) and (8.97) give for constant bending and shear stiffnesses the following two coupled differential equations for the bending and shear behavior[5]:

$$EI_z\frac{d^2\phi_z}{dx^2} + k_sGA\left(\frac{du_y}{dx} - \phi_z\right) + m_z(x) = 0,\tag{8.100}$$

$$k_sAG\left(\frac{d^2u_y}{dx^2} - \frac{d\phi_z}{dx}\right) + q_y(x) = 0.\tag{8.101}$$

Under the influence of the virtual displacements δu_y and the virtual rotations $\delta\phi_z$, the beam virtually moves away from its state of equilibrium. We consider first the virtual work in the domain $\Omega =]0, L[$: The resulting forces and moments according to Eqs. (8.100)–(8.101) do some virtual work along the virtual displacements and rotations.

Multiplication of the bending differential equation (8.100) with the virtual rotation $\delta\phi_z$ gives

$$\int_0^L\left(EI_z\frac{d^2\phi_z}{dx^2} + k_sGA\left(\frac{du_y}{dx} - \phi_z\right) + m_z(x)\right)\delta\phi_z\,dx,\tag{8.102}$$

or after partial integration of the first expression under the integral:

$$\left[EI_z\frac{d\phi_z}{dx}\delta\phi_z\right]_0^L - \int_0^L EI_z\frac{d\phi_z}{dx}\frac{d\delta\phi_z}{dx}dx$$

$$+ \int_0^L k_sGA\left(\frac{du_y}{dx} - \phi_z\right)\delta\phi_z dx + \int_0^L m_z(x)\delta\phi_z dx.\tag{8.103}$$

In a similar war, the multiplication of the shear differential equation (8.101) with a virtual displacement δu_y gives

$$\int_0^L\left(k_sAG\left(\frac{d^2u_y}{dx^2} - \frac{d\phi_z}{dx}\right) + q_y(x)\right)\delta u_y\,dx,\tag{8.104}$$

or after integration by parts of the two expressions in the round brackets:

---

[5]Consider also the supplementary Problem 8.4.

$$\left[k_s A G \frac{\mathrm{d} u_y}{\mathrm{d} x} \delta u_y \right]_0^L - \int\limits_0^L k_s A G \frac{\mathrm{d} u_y}{\mathrm{d} x} \frac{\mathrm{d} \delta u_y}{\mathrm{d} x} \mathrm{d} x - \left[k_s A G \phi_z \delta u_y \right]_0^L$$

$$+ \int\limits_0^L k_s A G \phi_z \frac{\mathrm{d} \delta u_y}{\mathrm{d} x} \mathrm{d} x + \int\limits_0^L q_y(x) \delta u_y \mathrm{d} x . \qquad (8.105)$$

Application of the principle of virtual displacements by summation of Eqs. (8.103) and (8.104) and consideration of the contribution of the external loads (see Eq. (5.125)) at the boundaries gives after short calculation the following expression:

$$\int\limits_0^L E I_z \frac{\mathrm{d} \phi_z}{\mathrm{d} x} \frac{\mathrm{d} \delta \phi_z}{\mathrm{d} x} \mathrm{d} x + \int\limits_0^L k_s A G \left(\frac{\mathrm{d} u_y}{\mathrm{d} x} - \phi_z \right) \left(\frac{\mathrm{d} \delta u_y}{\mathrm{d} x} - \delta \phi_z \right) \mathrm{d} x =$$

$$\int\limits_0^L q_y(x) \delta u_y \mathrm{d} x + \int\limits_0^L m_z(x) \delta \phi_z \mathrm{d} x + \left[k_s A G \left(\frac{\mathrm{d} u_y}{\mathrm{d} x} - \phi_z \right) \delta u_y \right]_0^L$$

$$+ \left[E I_z \frac{\mathrm{d} \phi_z}{\mathrm{d} x} \delta \phi_z \right]_0^L + \left(F_{1y} + Q_y(0) \right)\big|_0 \delta u_y + \left(F_{2y} - \right.$$

$$\left. Q_y(L) \right)\big|_L \delta u_y + \left(M_{1z} + M_z(0) \right)\big|_0 \delta \phi_z + \left(M_{2z} - M_z(L) \right)\big|_L \delta \phi_z . \qquad (8.106)$$

Consideration of $Q_y(x) = k_s A G \left(\frac{\mathrm{d} u_y}{\mathrm{d} x} - \phi_z \right)$ and $M_z(x) = E I_z \frac{\mathrm{d} \phi_z}{\mathrm{d} x}$ in the boundary expressions in the last equation allows a further simplification.

Now we replace in the last equation the displacement and rotation distributions by the approximations according to Eqs. (8.42) and (8.43) and the virtual displacements and rotations according to the approach for the weight function in Eqs. (8.81) and (8.82). Elimination of $\delta \boldsymbol{u}_{\mathrm{p}}^{\mathrm{T}}$ and evaluation of the boundary expressions allows to express the weak form of the problem as:

$$\int\limits_0^L E I_z \frac{\mathrm{d} \boldsymbol{N}_\phi^{\mathrm{T}}}{\mathrm{d} x} \frac{\mathrm{d} \boldsymbol{N}_\phi}{\mathrm{d} x} \mathrm{d} x \boldsymbol{u}_{\mathrm{P}} + \int\limits_0^L k_s A G \left(\frac{\mathrm{d} \boldsymbol{N}_u^{\mathrm{T}}}{\mathrm{d} x} - \boldsymbol{N}_\phi^{\mathrm{T}} \right) \left(\frac{\mathrm{d} \boldsymbol{N}_u}{\mathrm{d} x} - \boldsymbol{N}_\phi \right) \mathrm{d} x \boldsymbol{u}_{\mathrm{p}}$$

$$= \int\limits_0^L q_y(x) \boldsymbol{N}_u^{\mathrm{T}} \mathrm{d} x + \int\limits_0^L m_z(x) \boldsymbol{N}_\phi^{\mathrm{T}} \mathrm{d} x + \begin{bmatrix} F_{1y} \\ M_{1z} \\ F_{2y} \\ M_{2z} \end{bmatrix} . \qquad (8.107)$$

8.3.5 Linear Shape Functions for the Deflection and Displacement Field

Only the first order derivatives of the shape functions appear in the element stiffness matrices k_b^e and k_s^e according to Eqs. (8.54) and (8.61). This demand on the differentiability of the shape functions leads to polynomials of minimum first order (linear functions) for the deflection and displacement field, so that in the approaches according to Eqs. (8.42) and (8.43) the following linear shape functions can be used:

$$N_{1u}(x) = N_{1\phi}(x) = 1 - \frac{x}{L},$$ (8.108)

$$N_{2u}(x) = N_{2\phi}(x) = \frac{x}{L}.$$ (8.109)

The necessary derivatives result in:

$$\frac{dN_{1u}}{dx} = \frac{dN_{1\phi}}{dx} = -\frac{1}{L},$$ (8.110)

$$\frac{dN_{2u}}{dx} = \frac{dN_{2\phi}}{dx} = \frac{1}{L}.$$ (8.111)

A graphical illustration of the shape function is given in Fig. 8.8. Additionally the shape functions in the natural coordinate $\xi \in [-1, 1]$ are given. This formulation is more beneficial for the numerical integration of the stiffness matrices.

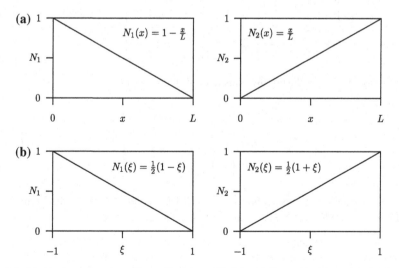

Fig. 8.8 Linear shape functions $N_1 = N_{1u}(x) = N_{1\phi}(x)$ and $N_2 = N_{2u}(x) = N_{2\phi}(x)$ for the TIMOSHENKO element in physical (x) and natural coordinates (ξ)

The integrals of the element stiffness matrices k_b^e and k_s^e according to Eqs. (8.54) and (8.61) are calculated analytically in the following. For the bending stiffness matrix the following results, using the linear approaches for the shape functions:

$$k_b^e = EI_z \int_0^L \begin{bmatrix} 0 & 0 & 0 & 0 \\ 0 & \frac{1}{L^2} & 0 & -\frac{1}{L^2} \\ 0 & 0 & 0 & 0 \\ 0 & -\frac{1}{L^2} & 0 & \frac{1}{L^2} \end{bmatrix} dx = EI_z \begin{bmatrix} 0 & 0 & 0 & 0 \\ 0 & \frac{x}{L^2} & 0 & -\frac{x}{L^2} \\ 0 & 0 & 0 & 0 \\ 0 & -\frac{x}{L^2} & 0 & \frac{x}{L^2} \end{bmatrix}_0^L , \qquad (8.112)$$

or alternatively under consideration of the integration boundaries:

$$k_b^e = EI_z \begin{bmatrix} 0 & 0 & 0 & 0 \\ 0 & \frac{1}{L} & 0 & -\frac{1}{L} \\ 0 & 0 & 0 & 0 \\ 0 & -\frac{1}{L} & 0 & \frac{1}{L} \end{bmatrix} . \qquad (8.113)$$

For the shear stiffness matrix the following results, using the linear approaches for the shape functions:

$$k_s^e = k_s AG \int_0^L \begin{bmatrix} \frac{1}{L^2} & \left(1-\frac{x}{L}\right)\frac{1}{L} & -\frac{1}{L^2} & \frac{x}{L^2} \\ \left(1-\frac{x}{L}\right)\frac{1}{L} & \left(1-\frac{x}{L}\right)^2 & -\left(1-\frac{x}{L}\right)\frac{1}{L} & \left(1-\frac{x}{L}\right)\frac{x}{L} \\ -\frac{1}{L^2} & -\left(1-\frac{x}{L}\right)\frac{1}{L} & \frac{1}{L^2} & -\frac{x}{L^2} \\ \frac{x}{L^2} & \left(1-\frac{x}{L}\right)\frac{x}{L} & -\frac{x}{L^2} & \frac{x^2}{L^2} \end{bmatrix} dx \qquad (8.114)$$

$$= k_s AG \begin{bmatrix} \frac{x}{L^2} & \frac{x(-2L+x)}{2L^2} & -\frac{x}{L^2} & \frac{x^2}{2L^2} \\ \frac{x(-2L+x)}{2L^2} & \frac{(-L+x)^3}{3L^2} & \frac{x(-2L+x)}{2L^2} & \frac{x^2(2x-3L)}{6L^2} \\ -\frac{x}{L^2} & \frac{x(-2L+x)}{2L^2} & \frac{x}{L^2} & -\frac{x^2}{2L^2} \\ \frac{x^2}{2L^2} & \frac{x^2(2x-3L)}{6L^2} & -\frac{x^2}{2L^2} & \frac{x^3}{3L^2} \end{bmatrix}_0^L \qquad (8.115)$$

and finally after considering the constants of integration:

$$k_s^e = k_s AG \begin{bmatrix} +\frac{1}{L} & +\frac{1}{2} & -\frac{1}{L} & +\frac{1}{2} \\ +\frac{1}{2} & +\frac{L}{3} & -\frac{1}{2} & +\frac{L}{6} \\ -\frac{1}{L} & -\frac{1}{2} & +\frac{1}{L} & -\frac{1}{2} \\ +\frac{1}{2} & +\frac{L}{6} & -\frac{1}{2} & +\frac{L}{3} \end{bmatrix} . \qquad (8.116)$$

The two stiffness matrices according to Eqs. (8.113) and (8.116) can be summarized additively to the total stiffness matrix of the TIMOSHENKO beam:

Fig. 8.9 Analysis of a TIMOSHENKO element under point load

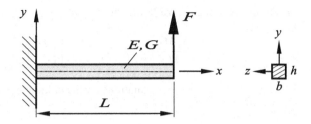

$$
k^{\mathrm{e}} =
\begin{bmatrix}
\frac{k_s AG}{L} & \frac{k_s AG}{2} & -\frac{k_s AG}{L} & \frac{k_s AG}{2} \\
\frac{k_s AG}{2} & \frac{k_s AGL}{3} + \frac{EI_z}{L} & -\frac{k_s AG}{2} & \frac{k_s AGL}{6} - \frac{EI_z}{L} \\
-\frac{k_s AG}{L} & -\frac{k_s AG}{2} & \frac{k_s AG}{L} & -\frac{k_s AG}{2} \\
\frac{k_s AG}{2} & \frac{k_s AGL}{6} - \frac{EI_z}{L} & -\frac{k_s AG}{2} & \frac{k_s AGL}{3} + \frac{EI_z}{L}
\end{bmatrix},
\tag{8.117}
$$

or alternatively via the abbreviation $\alpha = \frac{4EI_z}{k_s AG}$

$$
k^{\mathrm{e}} = \frac{k_s AG}{4L}
\begin{bmatrix}
4 & 2L & -4 & 2L \\
2L & \frac{4}{3}L^2 + \alpha & -2L & \frac{4}{6}L^2 - \alpha \\
-4 & -2L & 4 & -2L \\
2L & \frac{4}{6}L^2 - \alpha & -2L & \frac{4}{3}L^2 + \alpha
\end{bmatrix},
\tag{8.118}
$$

or alternatively via the abbreviation $\Lambda = \dfrac{EI_z}{k_s AGL^2}$

$$
k^{\mathrm{e}} = \frac{EI_z}{6\Lambda L^3}
\begin{bmatrix}
6 & 3L & -6 & 3L \\
3L & L^2(2 + 6\Lambda) & -3L & L^2(1 - 6\Lambda) \\
-6 & -3L & 6 & -3L \\
3L & L^2(1 - 6\Lambda) & -3L & L^2(2 + 6\Lambda)
\end{bmatrix}.
\tag{8.119}
$$

In the following, the deformation behavior of this analytically integrated[6] TIMO-SHENKO element is analyzed. For this, the configuration in Fig. 8.9 needs to be considered for which a beam has a fixed support on the left-hand side and a point load on the right-hand side. The displacement of the loading point has to be analyzed.

Through the stiffness matrix according to Eq. (8.118), the principal finite element equation for a single element results in

$$
\frac{k_s AG}{4L}
\begin{bmatrix}
4 & 2L & -4 & 2L \\
2L & \frac{4}{3}L^2 + \alpha & -2L & \frac{4}{6}L^2 - \alpha \\
-4 & -2L & 4 & -2L \\
2L & \frac{4}{6}L^2 - \alpha & -2L & \frac{4}{3}L^2 + \alpha
\end{bmatrix}
\begin{bmatrix}
u_{1y} \\
\phi_{1z} \\
u_{2y} \\
\phi_{2z}
\end{bmatrix}
=
\begin{bmatrix}
\cdots \\
\cdots \\
F \\
0
\end{bmatrix},
\tag{8.120}
$$

---

[6] A numerical GAUSS integration with two integration points yields the same results as the exact analytical integration.

or alternatively after considering the fixed support ($u_{1y} = 0$, $\phi_{1z} = 0$) of the left-hand side:

$$\frac{k_s A G}{4L} \begin{bmatrix} 4 & -2L \\ -2L & \frac{4}{3}L^2 + \alpha \end{bmatrix} \begin{bmatrix} u_{2y} \\ \phi_{2z} \end{bmatrix} = \begin{bmatrix} F \\ 0 \end{bmatrix}. \tag{8.121}$$

Solving this 2×2 system of equations for the unknown parameters on the right-hand end yields:

$$\begin{bmatrix} u_{2y} \\ \phi_{2z} \end{bmatrix} = \frac{4L}{k_s A G} \times \frac{1}{4(\frac{4}{3}L^2 + \alpha) - (-2L)(-2L)} \begin{bmatrix} \frac{4}{3}L^2 + \alpha & 2L \\ 2L & 4 \end{bmatrix} \begin{bmatrix} F \\ 0 \end{bmatrix}, \tag{8.122}$$

or alternatively solved for the unknown displacement on the right-hand end:

$$u_{2y}(L) = \frac{12 E I_z + 4 k_s A G L^2}{12 E I_z + k_s A G L^2} \times \left(\frac{FL}{k_s A G} \right). \tag{8.123}$$

Considering the rectangular cross-section, illustrated in Fig. 8.9, meaning $A = hb$ and $k_s = \frac{5}{6}$ and furthermore the relation for the shear modulus according to Eq. (8.23), after a short calculation the displacement on the right-hand end results:

$$u_{2y}(L) = \frac{12(1 + \nu)\left(\frac{h}{L}\right)^2 + 20}{60 + 25\left(\frac{L}{h}\right)^2 \frac{1}{1+\nu}} \times \left(\frac{FL^3}{E I_z} \right). \tag{8.124}$$

For very compact beams, meaning $h \gg L$, $\frac{L}{h} \to 0$ results and Eq. (8.124) converges against the analytical solution.[7] For very slender beams however, meaning $h \ll L$, a boundary value[8] of $\frac{4FL}{k_s A G}$ results from Eq. (8.123). This boundary value only contains the shear part without bending and runs against a wrong solution. This phenomenon is called *shear locking*. A graphical illustration of this behavior is given in Fig. 8.10 via the normalized deflection with the BERNOULLI solution. One can clearly see the different convergence behaviors for different domains of the slenderness ratio, meaning for slender and compact beams.

For the improvement of the convergence behavior, the literature suggests [13, 14] to conduct the integration via numerical GAUSS integration with only one integration point. Therefore, the arguments and the integration boundaries in the formulations of the element stiffness matrices for k_b^e and k_s^e according to Eqs. (8.54) and (8.61) have to be transformed into the natural coordinate $-1 \leq \xi \leq 1$. Furthermore, the shape functions need to be used according to Fig. 8.8. Via the transformation of the derivative to the new coordinate, meaning $\frac{dN}{dx} = \frac{dN}{d\xi} \frac{d\xi}{dx}$, and the transformation of the coordinate $\xi = -1 + 2\frac{x}{L}$ or alternatively $d\xi = \frac{2}{L}dx$, the bending stiffness matrix results in:

---

[7]For this see Fig. 8.6 and the supplementary Problem 8.6.

[8]One considers the definition of I_z and A in Eq. (8.123) and divides the fraction by h^3.

Fig. 8.10 Comparison of the analytical solution for a TIMOSHENKO beam and the corresponding discretization via one single finite element at analytical integration of the stiffness matrix

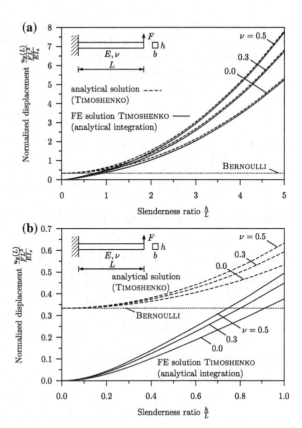

$$k_b^e = EI_z \int_0^L \frac{4}{L^2} \begin{bmatrix} 0 & 0 & 0 & 0 \\ 0 & \frac{dN_{1\phi}}{d\xi}\frac{dN_{1\phi}}{d\xi} & 0 & \frac{dN_{1\phi}}{d\xi}\frac{dN_{2\phi}}{d\xi} \\ 0 & 0 & 0 & 0 \\ 0 & \frac{dN_{2\phi}}{d\xi}\frac{dN_{1\phi}}{d\xi} & 0 & \frac{dN_{2\phi}}{d\xi}\frac{dN_{2\phi}}{d\xi} \end{bmatrix} \frac{L}{2} d\xi \,, \tag{8.125}$$

$$k_b^e = \frac{2EI_z}{L} \int_0^L \frac{4}{L^2} \begin{bmatrix} 0 & 0 & 0 & 0 \\ 0 & \frac{1}{4} & 0 & -\frac{1}{4} \\ 0 & 0 & 0 & 0 \\ 0 & -\frac{1}{4} & 0 & \frac{1}{4} \end{bmatrix} d\xi = \frac{EI_z}{2L} \sum_{i=1}^1 \begin{bmatrix} 0 & 0 & 0 & 0 \\ 0 & 1 & 0 & -1 \\ 0 & 0 & 0 & 0 \\ 0 & -1 & 0 & 1 \end{bmatrix} \times 2 \tag{8.126}$$

and after all in the final formulation in:

$$k_b^e = EI_z \begin{bmatrix} 0 & 0 & 0 & 0 \\ 0 & \frac{1}{L} & 0 & -\frac{1}{L} \\ 0 & 0 & 0 & 0 \\ 0 & -\frac{1}{L} & 0 & \frac{1}{L} \end{bmatrix} . \tag{8.127}$$

One can see that the same result for the bending stiffness matrix results as for the analytical integration. In the case of the bending stiffness matrix therefore the GAUSS integration with just one integration point is accurate.

The following expression results for the shear stiffness matrix under the use of the natural coordinate:

$$
\frac{2k_sGA}{L}\int_0^L
\begin{bmatrix}
\frac{dN_{1u}}{d\xi}\frac{dN_{1u}}{d\xi} & \frac{L}{2}\frac{dN_{1u}}{d\xi}(-N_{1\phi}) & \frac{dN_{1u}}{d\xi}\frac{dN_{2u}}{d\xi} & \frac{L}{2}\frac{dN_{1u}}{d\xi}(-N_{2\phi}) \\
\frac{L}{2}(-N_{1\phi})\frac{dN_{1u}}{d\xi} & \frac{L^2}{4}(N_{1\phi})(N_{1\phi}) & \frac{L}{2}(-N_{1\phi})\frac{dN_{2u}}{d\xi} & \frac{L^2}{4}(N_{1\phi})(N_{2\phi}) \\
\frac{dN_{2u}}{d\xi}\frac{dN_{1u}}{d\xi} & \frac{L}{2}\frac{dN_{2u}}{d\xi}(-N_{1\phi}) & \frac{dN_{2u}}{d\xi}\frac{dN_{2u}}{d\xi} & \frac{L}{2}\frac{dN_{2u}}{d\xi}(-N_{2\phi}) \\
\frac{L}{2}(-N_{2\phi})\frac{dN_{1u}}{d\xi} & \frac{L^2}{4}(N_{2\phi})(N_{1\phi}) & \frac{L}{2}(-N_{2\phi})\frac{dN_{2u}}{d\xi} & \frac{L^2}{4}(N_{2\phi})(N_{2\phi})
\end{bmatrix}
d\xi,
$$
$$(8.128)$$

or alternatively after the introduction of the shape functions

$$
\frac{2k_sGA}{L}\int_0^L
\begin{bmatrix}
\frac{1}{4} & \frac{L}{2}\left(\frac{1}{4}-\frac{x}{4}\right) & -\frac{1}{4} & \frac{L}{2}\left(\frac{1}{4}+\frac{x}{4}\right) \\
\frac{L}{2}\left(\frac{1}{4}-\frac{x}{4}\right) & \frac{L^2}{4}\left(\frac{(-1+x)^2}{4}\right) & \frac{L}{2}\left(-\frac{1}{4}+\frac{x}{4}\right) & \frac{L^2}{4}\left(\frac{1}{4}-\frac{x^2}{4}\right) \\
-\frac{1}{4} & \frac{L}{2}\left(-\frac{1}{4}+\frac{x}{4}\right) & \frac{1}{4} & \frac{L}{2}\left(-\frac{1}{4}-\frac{x}{4}\right) \\
\frac{L}{2}\left(\frac{1}{4}+\frac{x}{4}\right) & \frac{L^2}{4}\left(\frac{1}{4}-\frac{x^2}{4}\right) & \frac{L}{2}\left(-\frac{1}{4}-\frac{x}{4}\right) & \frac{L^2}{4}\left(\frac{(1+x)^2}{4}\right)
\end{bmatrix}
d\xi, \quad (8.129)
$$

or after the transition to the numerical integration

$$
\frac{2k_sGA}{L}
\begin{bmatrix}
\frac{1}{4} & \frac{L}{2}\frac{1}{4} & -\frac{1}{4} & \frac{L}{2}\frac{1}{4} \\
\frac{L}{2}\frac{1}{4} & \frac{L^2}{4}\frac{1}{4} & \frac{L}{2}\left(-\frac{1}{4}\right) & \frac{L^2}{4}\frac{1}{4} \\
-\frac{1}{4} & \frac{L}{2}\left(-\frac{1}{4}\right) & \frac{1}{4} & \frac{L}{2}\left(-\frac{1}{4}\right) \\
\frac{L}{2}\frac{1}{4} & \frac{L^2}{4}\frac{1}{4} & \frac{L}{2}\left(-\frac{1}{4}\right) & \frac{L^2}{4}\frac{1}{4}
\end{bmatrix}_{\xi_i=0}
\times 2 \qquad (8.130)
$$

and after all in the final formulation in:

$$
\boldsymbol{k}_s^e = k_sAG
\begin{bmatrix}
\frac{1}{L} & \frac{1}{2} & -\frac{1}{L} & \frac{1}{2} \\
\frac{1}{2} & \frac{L}{4} & -\frac{1}{2} & \frac{L}{4} \\
-\frac{1}{L} & -\frac{1}{2} & \frac{1}{L} & -\frac{1}{2} \\
\frac{1}{2} & \frac{L}{4} & -\frac{1}{2} & \frac{L}{4}
\end{bmatrix}.
\qquad (8.131)
$$

The two stiffness matrices according to Eqs. (8.127) and (8.131) can be summarized additively to the total stiffness matrix of the TIMOSHENKO beam and with the abbreviation $\alpha = \frac{4EI_z}{k_sAG}$ the following results:

$$
\boldsymbol{k}^e = \frac{k_sAG}{4L}
\begin{bmatrix}
4 & 2L & -4 & 2L \\
2L & L^2+\alpha & -2L & L^2-\alpha \\
-4 & -2L & 4 & -2L \\
2L & L^2-\alpha & -2L & L^2+\alpha
\end{bmatrix},
\qquad (8.132)
$$

or alternatively via the abbreviation $\Lambda = \frac{EI_z}{k_s AGL^2}$:

$$k^e = \frac{EI_z}{6\Lambda L^3} \begin{bmatrix} 6 & 3L & -6 & 3L \\ 3L & L^2(1,5+6\Lambda) & -3L & L^2(1,5-6\Lambda) \\ -6 & -3L & 6 & -3L \\ 3L & L^2(1,5-6\Lambda) & -3L & L^2(2+6\Lambda) \end{bmatrix}. \tag{8.133}$$

With the help of this formulation for the stiffness matrix the example according to Fig. 8.9 is analyzed once again in the following to investigate the differences to the analytical integration. Via the stiffness matrix according to Eq. (8.132) the principal finite element equation for a single element under consideration of the fixed support $(u_{1y} = 0, \ \phi_{1z} = 0)$ on the left-hand side results in:

$$\frac{k_s AG}{4L} \begin{bmatrix} 4 & -2L \\ -2L & L^2 + \alpha \end{bmatrix} \begin{bmatrix} u_{2y} \\ \phi_{2z} \end{bmatrix} = \begin{bmatrix} F \\ 0 \end{bmatrix}. \tag{8.134}$$

Solving this 2×2 system of equations for the unknown displacement on the right-hand side yields:

$$u_{2y}(L) = \left(1 + \frac{4EI_z}{k_s AGL^2}\right) \times \frac{FL^3}{4EI_z}. \tag{8.135}$$

Let us consider again at this point the rectangular cross-section as shown in Fig. 8.9. After a short calculation via $A = hb$, $k_s = \frac{5}{6}$ and the relation for the shear modulus according to Eq. (8.23), the displacement on the right-hand side results in:

$$u_{2y}(L) = \left(\frac{1}{4} + \frac{1}{5}(1+\nu)\left(\frac{h}{L}\right)^2\right) \times \left(\frac{FL^3}{EI_z}\right). \tag{8.136}$$

For very compact beams, meaning $h \gg L$, the solution converges against the analytical solution.[9] For very slender beams however, meaning $h \ll L$, a boundary value of $\frac{FL^3}{4EI_z}$ results from Eq. (8.136), whereupon the analytical solution yields a value of $\frac{FL^3}{3EI_z}$. However the phenomenon of shear locking does not occur and therefore, compared to the stiffness matrix based on the analytical integration, an improvement of the element formulation has been achieved.

A graphical illustration of this behavior via the normalized deflection is given in Fig. 8.11. One can clearly see the improved convergence behavior for small slenderness ratios. For large slenderness ratios the behavior remains according to the result of the analytical integration, since both approaches converge against the analytical solution.

When the differential equations according to Eqs. (8.33) and (8.34) are considered, it becomes obvious that the derivative $\frac{du_y}{dx}$ and the function ϕ_z itself are contained

---

[9]For this see Fig. 8.6 and the supplementary Problem 8.6.

Fig. 8.11 Comparison of the analytical solution for a TIMOSHENKO beam and the appropriate discretization via one single finite element at numerical integration of the corresponding matrix with the help of one integration point

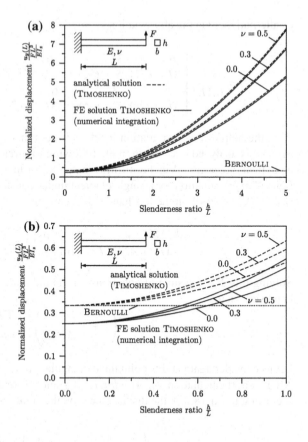

there. If linear shape functions are used for u_y and ϕ_z, the degree for polynomials for $\frac{du_y}{dx}$ and ϕ_z is different. In the limiting case of slender beams however the relation $\phi \approx \frac{du_y}{dx}$ has to be fulfilled and the consistency of the polynomials for $\frac{du_y}{dx}$ and ϕ_z is of importance. The linear approach for u_y yields for $\frac{du_y}{dx}$ a constant function and therefore also for ϕ_z a constant value would be desirable. However at this point it needs to be considered that the demand for the differentiability of ϕ_z at least results in a linear function. The one-point integration[10] in the case of the shear stiffness matrix with the expressions $N_{i\phi} N_{j\phi}$ causes however that the linear approach for ϕ_z is treated as a constant term, since two integration points would have to be used for an exact integration. A one-point integration can at most integrate a polynomial of first order exactly, meaning proportional to x^1, and therefore the following point of view results: $(N_{i\phi} N_{j\phi}) \sim x^1$. This however means that at most $N_{i\phi} \sim x^{0.5}$ or alternatively $N_{j\phi} \sim x^{0.5}$ holds. Since the polynomial approach solely allows integer values for the exponent of x, $N_{i\phi} \sim x^0$ or alternatively $N_{j\phi} \sim x^0$ results and the rotation needs

---

[10]The numerical integration according to the GAUSS–LEGENDRE method with n integration points integrates a polynomial, which degree is at most $2n - 1$, exactly.

to be seen as a constant term. This is consistent with the demand that the shear strain $\gamma_{xy} = \frac{du_y}{dx} - \phi_z$ has to be constant in an element for constant bending stiffness EI_z. Therefore in this case *shear locking* does not occur.

As another option for the improvement of the convergence behavior of linear TIMOSHENKO elements with numerical one-point integrations, references [13, 15] suggest to correct the shear stiffness $k_s AG$ according to the analytical correct solution.[11] To this the elastic strain energy is regarded, which results from Eqs. (8.49) and (8.56) for the energies and the kinematic relations (8.17) and (8.18) as follows:

$$
\Pi_{\text{int}} = \frac{1}{2} \int_0^L EI_z \left(\frac{d\phi_z(x)}{dx} \right)^2 dx + \frac{1}{2} \int_0^L k_s AG \left(\frac{du_y(x)}{dx} - \phi_z(x) \right)^2 dx . \quad (8.137)
$$

It is now demanded that the strain energy for the analytical solution and the finite element solution under the use of the corrected shear stiffness $(k_s AG)^*$ are identical. The analytical solution[12] for the problem in Fig. 8.9 results in

$$
u_y(x) = \frac{1}{EI_z} \left(-F\frac{x^3}{6} + FL\frac{x^2}{2} + \frac{EI_z F}{k_s AG} x \right) , \quad (8.138)
$$

$$
\phi_z(x) = \frac{1}{EI_z} \left(-F\frac{x^2}{2} + FLx \right) , \quad (8.139)
$$

and the elastic strain energy for the analytical solution therefore results in:

$$
\Pi_{\text{int}} = \frac{F^2}{2EI_z} \int_0^L (L - x)^2 dx + \frac{F^2 (EI_z)^2}{2k_s AG} \int_0^L dx = \frac{F^2 L^3}{6EI_z} + \frac{F^2 L}{2k_s AG} . \quad (8.140)
$$

Via Eq. (8.134) the finite element solution of the elastic strain energy results in:

$$
\Pi_{\text{int}} = \frac{EI_z}{2} \int_0^L \left(\frac{FL}{2EI_z} \right)^2 dx
$$

$$
+ \frac{(k_s AG)^*}{2} \int_0^L \left(\left(1 + \frac{4EI_z}{(k_s AG)^* L^2} \right) \frac{FL^2}{4EI_z} - \frac{FLx}{2EI_z} \right)^2 dx . \quad (8.141)
$$

This integral has to be evaluated numerically with a one-point integration rule and it is therefore necessary to introduce the natural coordinate via the transformation $x = \frac{L}{2}(\xi + 1)$:

---

[11] MACNEAL therefore uses the expression 'residual bending flexibility' [16, 17].
[12] For this see the supplementary Problem 8.5.

$$\Pi_{\text{int}} = \frac{F^2 L^3}{8EI_z}$$

$$+ \frac{(k_s AG)^*}{2} \int_0^L \left(\left(1 + \frac{4EI_z}{(k_s AG)^* L^2}\right) \frac{FL^2}{4EI_z} - \frac{FL}{2EI_z}(\xi + 1)\frac{L}{2}\right)^2 \frac{L}{2} d\xi$$

$$= \frac{F^2 L^3}{8EI_z} + \frac{(k_s AG)^*}{2} \left(\frac{4EI_z}{(k_s AG)^* L^2} \times \frac{FL^2}{4EI_z}\right)^2 \frac{L}{2} 2 \tag{8.142}$$

and finally

$$\Pi_{\text{int}} = \frac{F^2 L^3}{8EI_z} + \frac{F^2 L}{2(k_s AG)^*}. \tag{8.143}$$

Equalizing of the two energy expressions according to Eqs. (8.140) and (8.143) finally yields the corrected shear stiffness:

$$(k_s AG)^* = \left(\frac{L^2}{12EI_z} + \frac{1}{k_s AG}\right)^{-1}. \tag{8.144}$$

By inserting the with the 'residual bending flexibility' $\frac{L^2}{12EI_z}$ corrected shear stiffness into the finite element solution according to Eq. (8.135), the analytically exact solution results. The same result is derived in [15], starting from the general — meaning without considering a certain support of the beam — solution for the beam deflection, and in [13] the derivation for the equality of the deflection on the loading point according to the analytical and the corrected finite element solution takes place. It is to be considered that the derived corrected shear stiffness is not just valid for the cantilevered beam under point load, but yields the same value for arbitrary supports and loads on the ends of the beam. However, the derivation of the corrected shear stiffness for nonhomogeneous, anisotropic and non-linear materials remains problematic [13].

8.3.6 Higher-Order Shape Functions for the Beam with Shear Contribution

Within the framework of this subsection, first a general approach for a TIMOSHENKO element with an arbitrary amount of nodes will be reviewed [14]. Furthermore, the number of nodes, at which the deflection and the rotation are evaluated, can be different here. Therefore in the generalization of Eqs. (8.42) and (8.43) the following approach for the unknowns on the nodes results:

$$u_y(x) = \sum_{i=1}^{m} N_{iu}(x)u_{iy}, \tag{8.145}$$

$$\phi_z(x) = \sum_{i=1}^{n} N_{i\phi}(x)\phi_{iz}, \tag{8.146}$$

or alternatively in matrix notation as

$$u_y(x) = \begin{bmatrix} N_{1u} & \dots & N_{mu} & 0 & \dots & 0 \end{bmatrix} \begin{bmatrix} u_{1y} \\ \vdots \\ u_{my} \\ \phi_{1z} \\ \vdots \\ \phi_{nz} \end{bmatrix} = \boldsymbol{N}_u \boldsymbol{u}_p, \tag{8.147}$$

$$\phi_z(x) = \begin{bmatrix} 0 & \dots & 0 & N_{1\phi} & \dots & N_{n\phi} \end{bmatrix} \begin{bmatrix} u_{1y} \\ \vdots \\ u_{my} \\ \phi_{1z} \\ \vdots \\ \phi_{nz} \end{bmatrix} = \boldsymbol{N}_\phi \boldsymbol{u}_p. \tag{8.148}$$

With this generalized approach the deflection can be evaluated on m nodes and the rotation on n nodes. For the shape functions N_i usually LAGRANGE polynomials[13] are used, which in general are calculated as follows in the case of the deflection:

$$\begin{aligned} N_i &= \prod_{j=0 \wedge j \neq i}^{m} \frac{x_j - x}{x_j - x_i} \\ &= \frac{(x_1 - x)(x_2 - x) \cdots [x_i - x] \cdots (x_m - x)}{(x_1 - x_i)(x_2 - x_i) \cdots [x_i - x_i] \cdots (x_m - x_i)}, \end{aligned} \tag{8.149}$$

For the derivation of the general stiffness matrix we revert at this point to different methods. If, for example, the weighted residual method is considered, one can use the new approaches (8.147) and (8.148) in Eq. (8.86). Execution of the multiplication for the bending stiffness matrix yields

---

[13] At the so-called LAGRANGE interpolation, m points are approximated via the ordinate values with the help of a polynomial of the order $m - 1$. In the case of the HERMITE interpolation the slope of the regarded points is considered in addition to the ordinate value. For this see Chap. 6.

$$k_b^e = \int_0^L E I_z \begin{bmatrix} 0 & \cdots & 0 & 0 & \cdots & 0 \\ \vdots & (m \times m) & \vdots & \vdots & (m \times n) & \vdots \\ 0 & \cdots & 0 & 0 & \cdots & 0 \\ 0 & \cdots & 0 & \dfrac{dN_{1\phi}}{dx}\dfrac{dN_{1\phi}}{dx} & \cdots & \dfrac{dN_{1\phi}}{dx}\dfrac{dN_{n\phi}}{dx} \\ \vdots & (n \times m) & \vdots & \vdots & (n \times n) & \vdots \\ 0 & \cdots & 0 & \dfrac{dN_{n\phi}}{dx}\dfrac{dN_{1\phi}}{dx} & \cdots & \dfrac{dN_{n\phi}}{dx}\dfrac{dN_{n\phi}}{dx} \end{bmatrix} dx \quad (8.150)$$

and accordingly the execution of the multiplication for the shear stiffness matrix k_s^e yields

$$\int_0^L k_s AG \begin{bmatrix} \dfrac{dN_{1u}}{dx}\dfrac{dN_{1u}}{dx} & \cdots & \dfrac{dN_{1u}}{dx}\dfrac{dN_{mu}}{dx} & -N_{1\phi}\dfrac{dN_{1u}}{dx} & \cdots & -N_{n\phi}\dfrac{dN_{1u}}{dx} \\ \vdots & (m \times m) & \vdots & \vdots & (m \times n) & \vdots \\ \dfrac{dN_{mu}}{dx}\dfrac{dN_{1u}}{dx} & \cdots & \dfrac{dN_{mu}}{dx}\dfrac{dN_{mu}}{dx} & -N_{1\phi}\dfrac{dN_{mu}}{dx} & \cdots & -N_{m\phi}\dfrac{dN_{mu}}{dx} \\ -N_{1\phi}\dfrac{dN_{1u}}{dx} & \cdots & -N_{1\phi}\dfrac{dN_{mu}}{dx} & N_{1\phi}N_{1\phi} & \cdots & N_{1\phi}N_{n\phi} \\ \vdots & (n \times m) & \vdots & \vdots & (n \times n) & \vdots \\ -N_{n\phi}\dfrac{dN_{1u}}{dx} & \cdots & -N_{n\phi}\dfrac{dN_{mu}}{dx} & N_{n\phi}N_{1\phi} & \cdots & N_{n\phi}N_{n\phi} \end{bmatrix} dx.$$

$$(8.151)$$

These two stiffness matrices can be be superposed additively at this point and the following general structure for the total stiffness matrix yields:

$$k^e = \begin{bmatrix} k^{11} & k^{12} \\ k^{21} & k^{22} \end{bmatrix}, \quad (8.152)$$

with

$$k_{kl}^{11} = \int_0^L k_s AG \frac{dN_{ku}}{dx}\frac{dN_{lu}}{dx} dx , \quad (8.153)$$

$$k_{kl}^{12} = \int_0^L k_s AG \frac{dN_{ku}}{dx}(-N_{l\phi}) dx , \quad (8.154)$$

$$k_{kl}^{21} = k_{kl}^{12,\mathrm{T}} = \int\limits_0^L k_s A G (-N_{k\phi}) \frac{\mathrm{d}N_{lu}}{\mathrm{d}x} \, \mathrm{d}x \,, \tag{8.155}$$

$$k_{kl}^{22} = \int\limits_0^L \left(k_s A G N_{k\phi} N_{l\phi} + E I_z \frac{\mathrm{d}N_{k\phi}}{\mathrm{d}x} \frac{\mathrm{d}N_{l\phi}}{\mathrm{d}x} \right) \mathrm{d}x \,. \tag{8.156}$$

The derivation of the right-hand side can occur according to Eq. (8.91) and the following load vector results:

$$F^{\mathrm{e}} = \int\limits_0^L q_y(x) \begin{bmatrix} N_{1u} \\ \vdots \\ N_{mu} \\ 0 \\ \vdots \\ 0 \end{bmatrix} \mathrm{d}x + \begin{bmatrix} F_{1y} \\ \vdots \\ F_{my} \\ M_{1z} \\ \vdots \\ M_{nz} \end{bmatrix} \,. \tag{8.157}$$

In the following, a quadratic interpolation for $u_y(x)$ as well as a linear interpolation for $\phi_z(x)$ are chosen [14]. Therefore for $\frac{\mathrm{d}u_y(x)}{\mathrm{d}x}$ and $\phi_z(x)$ functions of the same order result and the phenomenon of *shear locking* can be avoided. Quadratic interpolation for the deflection means that the deflection will be evaluated on three nodes. The linear approach for the rotation means that the unknowns will be evaluated on only two nodes. Therefore, the illustrated configuration in Fig. 8.12 for this TIMOSHENKO element results.

Evaluation of the general LAGRANGE polynomial according to Eq. (8.149) for the deflection, meaning under consideration of three nodes, yields

$$N_{1u} = \frac{(x_2 - x)(x_3 - x)}{(x_2 - x_1)(x_3 - x_1)} = 1 - 3\frac{x}{L} + 2\left(\frac{x}{L}\right)^2, \tag{8.158}$$

$$N_{2u} = \frac{(x_1 - x)(x_3 - x)}{(x_1 - x_2)(x_3 - x_2)} = 4\frac{x}{L} - 4\left(\frac{x}{L}\right)^2, \tag{8.159}$$

$$N_{3u} = \frac{(x_1 - x)(x_2 - x)}{(x_1 - x_3)(x_2 - x_3)} = -\frac{x}{L} + 2\left(\frac{x}{L}\right)^2, \tag{8.160}$$

or alternatively for both nodes for the rotation:

$$N_{1\phi} = \frac{(x_2 - x)}{(x_2 - x_1)} = 1 - \frac{x}{L}, \tag{8.161}$$

$$N_{2\phi} = \frac{(x_1 - x)}{(x_1 - x_2)} = \frac{x}{L}. \tag{8.162}$$

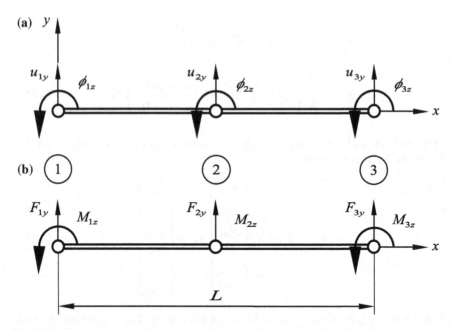

Fig. 8.12 TIMOSHENKO bending element with quadratic shape functions for the deflection and linear shape functions for the rotation: **a** deformation parameters; **b** load parameters

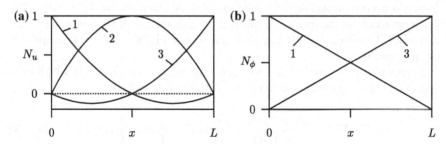

Fig. 8.13 Shape functions for a TIMOSHENKO element with **a** quadratic approach for the deflection and **b** linear approach for the rotation

A graphical illustration of the shape functions is given in Fig. 8.13. One can see that the typical characteristics for shape functions, meaning $N_i(x_i) = 1 \wedge N_i(x_j) = 0$ and $\sum_i N_i = 1$ are fulfilled.

With these shape functions the submatrices $\boldsymbol{k}^{11}, \ldots, \boldsymbol{k}^{22}$ in Eq. (8.152) result in the following via analytical integration:

$$
\boldsymbol{k}^{11} = \frac{k_s AG}{3L} \begin{bmatrix} 7 & -8 & 1 \\ -8 & 16 & -8 \\ 1 & -8 & 7 \end{bmatrix}, \tag{8.163}
$$

$$k^{12} = \frac{k_s AG}{6} \begin{bmatrix} 5 & 1 \\ -4 & 4 \\ -1 & -5 \end{bmatrix} = (k^{21})^{\mathrm{T}} , \tag{8.164}$$

$$k^{22} = \frac{k_s AGL}{6} \begin{bmatrix} 2 & 1 \\ 1 & 2 \end{bmatrix} + \frac{EI_z}{L} \begin{bmatrix} 1 & -1 \\ -1 & 1 \end{bmatrix} , \tag{8.165}$$

which can be put together for the principal finite element equation by making use of the abbreviation $\Lambda = \frac{EI_z}{k_s AGL^2}$:

$$\frac{k_s AG}{6L} \begin{bmatrix} 14 & -16 & 2 & 5L & 1L \\ -16 & 32 & -16 & -4L & 4L \\ 2 & -16 & 14 & -1L & -5L \\ 5L & -4L & -1L & 2L^2(1+3\Lambda) & L^2(1-6\Lambda) \\ 1L & 4L & -5L & L^2(1-6\Lambda) & 2L^2(1+3\Lambda) \end{bmatrix} \begin{bmatrix} u_{1y} \\ u_{2y} \\ u_{3y} \\ \phi_{1z} \\ \phi_{3z} \end{bmatrix} = \begin{bmatrix} F_{1y} \\ F_{2y} \\ F_{3y} \\ M_{1z} \\ M_{3z} \end{bmatrix} . \tag{8.166}$$

Since only one displacement is evaluated on the middle node, the number of unknowns is not the same on each node. This circumstance complicates the creation of the global system of equations for several of these elements. The degree of freedom u_{2y} however can be expressed via the remaining unknowns and therefore the possibility exists to eliminate this node from the system of equations. For this, the second row[14] of Eq. (8.166) has to be evaluated:

$$\frac{k_s AG}{6L} \left(-16u_{1y} + 32u_{2y} - 16u_{3y} - 4L\phi_{1z} + 4L\phi_{3z} \right) = F_{2y} , \tag{8.167}$$

$$u_{2y} = \frac{6L}{32k_s AG} F_{2y} + \frac{u_{1y} + u_{3y}}{2} + \frac{\phi_{1z} - \phi_{3z}}{8} L . \tag{8.168}$$

Furthermore, it can be demanded that no external force should have an effect on the middle node, so that the relation between the deflection on the middle node and the other unknowns yields as follows:

$$u_{2y} = \frac{u_{1y} + u_{3y}}{2} + \frac{\phi_{1z} - \phi_{3z}}{8} L . \tag{8.169}$$

This relation can be introduced into the system of equations (8.166) to eliminate the degree of freedom u_{2u}. Finally, after a new arrangement of the unknowns, the following principal finite element equation results, which is reduced by one column and one row:

---

[14]It needs to be remarked that the influence of distributed loads is disregarded in the derivation. If distributed loads occur, the equivalent nodal loads have to be distributed on the remaining nodes.

$$\frac{EI_z}{6\Lambda L^3}\begin{bmatrix} 6 & 3L & -6 & 3L \\ 3L & L^2(1,5+6\Lambda) & -3L & L^2(1,5-6\Lambda) \\ -6 & -3L & 6 & -3L \\ 3L & L^2(1,5-6\Lambda) & -3L & L^2(1,5+6\Lambda) \end{bmatrix}\begin{bmatrix} u_{1y} \\ \phi_{1z} \\ u_{3y} \\ \phi_{3z} \end{bmatrix}=\begin{bmatrix} F_{1y} \\ M_{1z} \\ F_{3y} \\ M_{3z} \end{bmatrix}.$$

(8.170)

This element formulation is identical with Eq. (8.133), which was derived with linear shape functions and numerical one-point integration. However, it is to be considered that the interpolation between the nodes during the use of Eq. (8.170) takes place with quadratic functions.

Further details and formulations regarding the TIMOSHENKO element can be found in the scientific literature [11, 18].

8.4 Sample Problems and Supplementary Problems

8.4.1 Sample Problems

8.1 Discretization of a Beam with 5 Linear Elements with Shear Contribution

The beam,[15] which is illustrated in Fig. 8.14 needs to be discretized equally with five linear TIMOSHENKO elements and the displacement of the loading point needs to be discussed as dependent on the slenderness ratio and the POISSON's ratio. One considers the case of the (a) analytical and (b) the numerical (one integration point) integrated stiffness matrix.

Solution (a) Stiffness matrix via analytical integration:

The element stiffness matrix according to Eq. (8.118) can be used for each of the five elements, whereupon it has to be considered that the single element length results in $\frac{L}{5}$. The resulting total stiffness matrix has the dimensions 12×12, which reduces to a 10×10 matrix due to the consideration of the fixed support on the left-hand boundary ($u_{1y} = 0$, $\phi_{1z} = 0$). Through inversion of the stiffness matrix, the reduced system of equations can be solved via $u = K^{-1}F$. The following extract shows the most important entries in this system of equations:

$$\begin{bmatrix} u_{2y} \\ \vdots \\ u_{6y} \\ \phi_{6z} \end{bmatrix}=\frac{\frac{4}{5}L}{k_s AG}\underbrace{\begin{bmatrix} \mathsf{x} & \cdots & & \mathsf{x} & \mathsf{x} \\ \vdots & & & \vdots & \vdots \\ \mathsf{x} & \cdots & \frac{125(3\alpha+4L^2)}{4(75\alpha+L^2)} & \mathsf{x} \\ \mathsf{x} & \cdots & & \mathsf{x} & \mathsf{x} \end{bmatrix}}_{10\times10 \text{ matrix}}\begin{bmatrix} 0 \\ \vdots \\ \vdots \\ F \\ 0 \end{bmatrix}.$$

(8.171)

---

[15] A similar example is presented in [19].

Fig. 8.14 Discretization of a beam structure with elements under consideration of the shear contribution

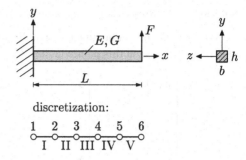

discretization:

Multiplication of the 9th row of the matrix with the load vector yields the displacement of the loading point as:

$$u_{6y} = \frac{25(3\alpha + 4L^2)}{75\alpha + L^2} \times \frac{FL}{k_s AG}, \tag{8.172}$$

or alternatively via $A = hb$, $k_s = \frac{5}{6}$ and the relation for the shear modulus according to Eq. (8.23) after a short calculation:

$$u_{6y} = \frac{12(1+\nu)\left(\frac{h}{L}\right)^2 + 20}{60 + \left(\frac{L}{h}\right)^2 \frac{1}{1+\nu}} \times \frac{FL^3}{EI_z}. \tag{8.173}$$

A graphical illustration of the displacement dependent on the slenderness ratio can be seen in Fig. 8.15. A comparison with Fig. 8.10 shows that the convergence behavior in the lower domain of the slenderness ratio for $0.2 < \frac{h}{L} < 1.0$ has significantly improved through the fine discretization, the phenomenon of the *shear lockings* for $\frac{h}{L} \to 0$ however still occurs.

Fig. 8.15 Discretization of a beam via 5 linear TIMOSHENKO elements at analytical integration of the stiffness matrix

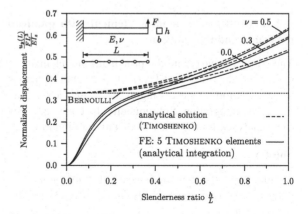

Solution (b) Stiffness matrix via numerical integration with one integration point:

According to the procedure in part (a) of this problem, the following 10×10 system of equations results at this point via the stiffness matrix according to Eq. (8.132)

$$
\begin{bmatrix} u_{2y} \\ \vdots \\ u_{6y} \\ \phi_{6z} \end{bmatrix} = \frac{\frac{4}{5}L}{k_s AG} \underbrace{\begin{bmatrix} \times & \cdots & \times & \times \\ \vdots & & \vdots & \vdots \\ \times & \cdots & \frac{25\alpha+33L^2}{20\alpha} & \times \\ \times & \cdots & \times & \times \end{bmatrix}}_{10 \times 10 \text{ matrix}} \begin{bmatrix} 0 \\ \vdots \\ F \\ 0 \end{bmatrix},
\tag{8.174}
$$

from which the displacement on the right-hand boundary can be defined as the following

$$
u_{6y} = \frac{4}{5}\left(\frac{5}{4} + \frac{33L^2}{20\alpha}\right) \times \frac{FL}{k_s AG}.
\tag{8.175}
$$

With the use of $A = hb$, $k_s = \frac{5}{6}$ and the relation for the shear modulus according to Eq. (8.23) results in the following after a short calculation:

$$
u_{6y} = \left(\frac{33}{100} + \frac{1}{5}(1+\nu)\left(\frac{h}{L}\right)^2\right) \times \frac{FL^3}{EI_z}.
\tag{8.176}
$$

The graphical illustration of the displacement in Fig. 8.16 shows that an excellent conformity with the analytical solution throughout the entire domain of the slenderness ratio results through the mesh refinement. Therefore, the accuracy at a TIMOSHENKO element with linear shape functions and reduced numerical integration can be increased considerably through mesh refinement.

8.2 TIMOSHENKO bending element with quadratic shape functions for the deflection and the rotation

The stiffness matrix and the principal finite element equation $k^e u_p = F^e$ are to be derived for the illustrated TIMOSHENKO bending element in Fig. 8.17 with quadratic shape functions. One distinguishes in the derivation between the analytical and numerical integration. Subsequently the convergence behavior of an element needs to be analyzed for the illustrated configuration in Fig. 8.9.

Solution

Evaluation of the general LAGRANGE polynomial according to Eq. (8.149) under consideration of 3 nodes yields the following shape functions for the deflection and the rotation:

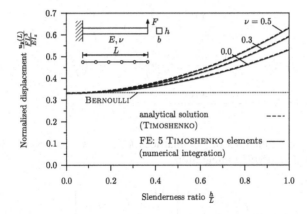

Fig. 8.16 Discretization of a beam via 5 linear TIMOSHENKO elements at numerical integration of the stiffness matrix with one integration point

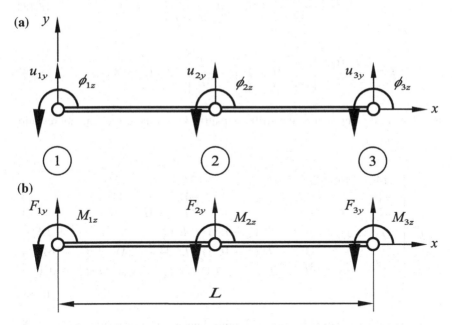

Fig. 8.17 TIMOSHENKO bending element with quadratic shape functions for the deflection and the rotation: **a** deformation parameters; **b** load parameters

$$N_{1u} = N_{1\phi} = \frac{(x_2 - x)(x_3 - x)}{(x_2 - x_1)(x_3 - x_1)} = 1 - 3\frac{x}{L} + 2\left(\frac{x}{L}\right)^2, \tag{8.177}$$

$$N_{2u} = N_{2\phi} = \frac{(x_1 - x)(x_3 - x)}{(x_1 - x_2)(x_3 - x_2)} = 4\frac{x}{L} - 4\left(\frac{x}{L}\right)^2, \tag{8.178}$$

$$N_{3u} = N_{3\phi} = \frac{(x_1 - x)(x_2 - x)}{(x_1 - x_3)(x_2 - x_3)} = -\frac{x}{L} + 2\left(\frac{x}{L}\right)^2. \tag{8.179}$$

With these shape functions the submatrices $k^{11}, \ldots, k^{22}$ in Eq. (8.152) result as follows through *analytical* integration:

$$k^{11} = \frac{k_s AG}{6L} \begin{bmatrix} 14 & -16 & 2 \\ -16 & 32 & -16 \\ 2 & -16 & 14 \end{bmatrix}, \tag{8.180}$$

$$k^{12} = \frac{k_s AG}{6L} \begin{bmatrix} 3L & 4L & -1L \\ -4L & 0 & 4L \\ 1L & -4L & -3L \end{bmatrix} = (k^{21})^{\mathrm{T}}, \tag{8.181}$$

$$k^{22} = \frac{k_s AGL}{30} \begin{bmatrix} 4 & 2 & -1 \\ 2 & 16 & 2 \\ -1 & 2 & 4 \end{bmatrix} + \frac{EI_z}{3L} \begin{bmatrix} 7 & -8 & 1 \\ -8 & 16 & -8 \\ 1 & -8 & 7 \end{bmatrix}, \tag{8.182}$$

which can be composed to the stiffness matrix k^e via the use of the abbreviation $\Lambda = \frac{EI_z}{k_s AGL^2}$:

$$\frac{k_s AG}{6L} \left[\begin{array}{ccc|ccc} 14 & -16 & 2 & 3L & 4L & -1L \\ -16 & 32 & -16 & -4L & 0 & 4L \\ 2 & -16 & 14 & 1L & -4L & -3L \\ 3L & -4L & 1L & L^2(\frac{4}{5}+14\Lambda) & L^2(\frac{2}{5}-16\Lambda) & L^2(-\frac{1}{5}+2\Lambda) \\ 4L & 0 & -4L & L^2(\frac{2}{5}-16\Lambda) & L^2(\frac{16}{5}+32\Lambda) & L^2(\frac{2}{5}-16\Lambda) \\ -1L & 4L & -3L & L^2(-\frac{1}{5}+2\Lambda) & L^2(\frac{2}{5}-16\Lambda) & L^2(\frac{4}{5}+14\Lambda) \end{array} \right] . \tag{8.183}$$

With this stiffness matrix the principal finite element equation results in $k^e u_p = F^e$, at which the deformation and load vector contains the following components:

$$u_p = \begin{bmatrix} u_{1y} & u_{2y} & u_{3y} & \phi_{1z} & \phi_{2z} & \phi_{3z} \end{bmatrix}^{\mathrm{T}}, \tag{8.184}$$

$$F^e = \begin{bmatrix} F_{1y} & F_{2y} & F_{3y} & M_{1z} & M_{2z} & M_{3z} \end{bmatrix}^{\mathrm{T}}. \tag{8.185}$$

For the analysis of the convergence behavior of an element for the illustrated beam in Fig. 8.9 with point load, the columns and rows for the entries u_{1y} and ϕ_{1z} in

Eq. (8.183) can be canceled due to the fixed support on this node. This reduced 4×4 stiffness matrix can be inverted and the following system of equations for the definition of the unknown degrees of freedom results:

$$
\begin{bmatrix} u_{2y} \\ u_{3y} \\ \vdots \\ \phi_{3z} \end{bmatrix} = \frac{6L}{k_s AG} \underbrace{\begin{bmatrix} \mathbf{x} & \cdots & & \cdots \mathbf{x} \\ \mathbf{x} & \frac{-3+340A+1200A^2}{8(-1-45A+900A^2)} & & \cdots \mathbf{x} \\ \vdots & & & \vdots \\ \mathbf{x} & & & \cdots \mathbf{x} \end{bmatrix}}_{4 \times 4 \text{ matrix}} \begin{bmatrix} 0 \\ F \\ \vdots \\ 0 \end{bmatrix} , \tag{8.186}
$$

from which, through evaluation of the second row, the displacement on the right-hand boundary can be defined as:

$$
u_{3y} = \underbrace{\frac{6L}{k_s AG}}_{\frac{6AL^3}{EI_z}} \times \frac{-3+340A+1200A^2}{8(-1-45A+900A^2)} \times F . \tag{8.187}
$$

For a rectangular cross-section $A = \frac{1}{5}(1+\nu)\left(\frac{h}{L}\right)^2$ results, and one can see that *shear locking* occurs also at this point for slender beams with $L \gg h$, since in the limit case $u_{3y} \to 0$ occurs.

In the following, the reduced numerical integration of the stiffness matrix is analyzed. For the definition of a reasonable amount of integration points one takes into account the following consideration:

If quadratic shape functions are used for u_y and ϕ_z, the degree of the polynomials for $\frac{du_y}{dx}$ and ϕ_z differs. The quadratic approach for u_y yields for $\frac{du_y}{dx}$ a linear function and thus a linear function would also be desirable for ϕ_z. The two-point integration however determines that the quadratic approach for ϕ_z is treated as a linear function. A two-point integration can exactly integrate a polynomial of third order, meaning proportional to x^3, at most and therefore the following view results: $(N_{i\phi}N_{j\phi}) \sim x^3$. This however means that $N_{i\phi} \sim x^{1.5}$ or alternatively $N_{j\phi} \sim x^{1.5}$ applies at most. Since the polynomial approach only allows integer values for the exponent, $N_{i\phi} \sim x^1$ or alternatively $N_{j\phi} \sim x^1$ results and the rotation needs to be regarded as a linear function.

The integration via numerical GAUSS integration with 2 integration points demands that the arguments and the integration boundaries in the formulations of the submatrices $k^{11}, \ldots, k^{22}$ in Eq. (8.152) have to be transformed to the natural coordinate $-1 \le \xi \le 1$. Via the transformation of the derivative onto the new coordinate, meaning $\frac{dN}{dx} = \frac{dN}{d\xi}\frac{d\xi}{dx}$ and the transformation of the coordinate $\xi = -1 + 2\frac{x}{L}$ or alternatively $d\xi = \frac{2}{L}dx$, the numerical approximation of the submatrices for two integration points $\xi_{1,2} = \pm\frac{1}{\sqrt{3}}$ results in:

$$k^{11} = \sum_{i=1}^{2} \frac{2k_s AG}{L} \begin{bmatrix} \frac{dN_{1u}}{d\xi}\frac{dN_{1u}}{d\xi} & \frac{dN_{1u}}{d\xi}\frac{dN_{2u}}{d\xi} & \frac{dN_{1u}}{d\xi}\frac{dN_{3u}}{d\xi} \\ \frac{dN_{2u}}{d\xi}\frac{dN_{1u}}{d\xi} & \frac{dN_{2u}}{d\xi}\frac{dN_{2u}}{d\xi} & \frac{dN_{2u}}{d\xi}\frac{dN_{3u}}{d\xi} \\ \frac{dN_{3u}}{d\xi}\frac{dN_{1u}}{d\xi} & \frac{dN_{3u}}{d\xi}\frac{dN_{2u}}{d\xi} & \frac{dN_{3u}}{d\xi}\frac{dN_{3u}}{d\xi} \end{bmatrix} \times 1 , \qquad (8.188)$$

$$k^{12} = \sum_{i=1}^{2} k_s AG \begin{bmatrix} \frac{dN_{1u}}{d\xi}(-N_{1\phi}) & \frac{dN_{1u}}{d\xi}(-N_{2\phi}) & \frac{dN_{1u}}{d\xi}(-N_{3\phi}) \\ \frac{dN_{2u}}{d\xi}(-N_{1\phi}) & \frac{dN_{2u}}{d\xi}(-N_{2\phi}) & \frac{dN_{2u}}{d\xi}(-N_{3\phi}) \\ \frac{dN_{3u}}{d\xi}(-N_{1\phi}) & \frac{dN_{3u}}{d\xi}(-N_{2\phi}) & \frac{dN_{3u}}{d\xi}(-N_{3\phi}) \end{bmatrix} \times 1 , \qquad (8.189)$$

$$k^{22} = \sum_{i=1}^{2} \frac{k_s AGL}{2} \begin{bmatrix} N_{1\phi}N_{1\phi} & N_{1\phi}N_{2\phi} & N_{1\phi}N_{3\phi} \\ N_{2\phi}N_{1\phi} & N_{2\phi}N_{2\phi} & N_{2\phi}N_{3\phi} \\ N_{3\phi}N_{1\phi} & N_{3\phi}N_{2\phi} & N_{3\phi}N_{3\phi} \end{bmatrix} \times 1 \qquad (8.190)$$

$$+ \sum_{i=1}^{2} \frac{2EI_z}{L} \begin{bmatrix} \frac{dN_{1\phi}}{d\xi}\frac{dN_{1\phi}}{d\xi} & \frac{dN_{1\phi}}{d\xi}\frac{dN_{2\phi}}{d\xi} & \frac{dN_{1\phi}}{d\xi}\frac{dN_{3\phi}}{d\xi} \\ \frac{dN_{2\phi}}{d\xi}\frac{dN_{1\phi}}{d\xi} & \frac{dN_{2\phi}}{d\xi}\frac{dN_{2\phi}}{d\xi} & \frac{dN_{2\phi}}{d\xi}\frac{dN_{3\phi}}{d\xi} \\ \frac{dN_{3\phi}}{d\xi}\frac{dN_{1\phi}}{d\xi} & \frac{dN_{3\phi}}{d\xi}\frac{dN_{2\phi}}{d\xi} & \frac{dN_{3\phi}}{d\xi}\frac{dN_{3\phi}}{d\xi} \end{bmatrix} \times 1 . \qquad (8.191)$$

The quadratic shape functions, which have already been introduced in Eqs. (8.158)–(8.160), still have to be transformed onto the new coordinates via the transformation $x = (\xi + 1)\frac{L}{2}$. Therefore for the shape functions or alternatively their derivatives the following results:

$$N_1(\xi) = -\frac{1}{2}(\xi - \xi^2) , \qquad \frac{dN_1}{d\xi} = -\frac{1}{2}(1 - 2\xi) , \qquad (8.192)$$

$$N_2(\xi) = 1 - \xi^2 , \qquad \frac{dN_2}{d\xi} = -2\xi , \qquad (8.193)$$

$$N_3(\xi) = \frac{1}{2}(\xi + \xi^2) , \qquad \frac{dN_3}{d\xi} = \frac{1}{2}(1 + 2\xi) . \qquad (8.194)$$

The use of these shape functions or alternatively their derivatives finally leads to the following submatrices

$$k^{11} = \frac{k_s AG}{6L} \begin{bmatrix} 14 & -16 & 2 \\ -16 & 32 & -16 \\ 2 & -16 & 14 \end{bmatrix} , \qquad (8.195)$$

$$k^{12} = \frac{k_s AG}{6L} \begin{bmatrix} 3L & 4L & -L \\ -4L & 0 & 4L \\ +L & -4L & -3L \end{bmatrix} , \qquad (8.196)$$

$$k^{22} = \frac{k_s AG}{6L} \begin{bmatrix} \frac{2}{3}L^2 & \frac{2}{3}L^2 & -\frac{1}{3}L^2 \\ \frac{2}{3}L^2 & \frac{8}{3}L^2 & \frac{2}{3}L^2 \\ -\frac{1}{3}L^2 & \frac{2}{3}L^2 & \frac{2}{3}L^2 \end{bmatrix} + \frac{EI_z}{L^3} \begin{bmatrix} \frac{7}{3}L^2 & -\frac{8}{3}L^2 & \frac{1}{3}L^2 \\ -\frac{8}{3}L^2 & \frac{16}{3}L^2 & -\frac{8}{3}L^2 \\ \frac{1}{3}L^2 & -\frac{8}{3}L^2 & \frac{7}{3}L^2 \end{bmatrix}, \quad (8.197)$$

which can be put together to the stiffness matrix k^e under the use of the abbreviation $\Lambda = \frac{EI_z}{k_s AGL^2}$:

$$\frac{k_s AG}{6L} \begin{bmatrix} 14 & -16 & 2 & & 3L & 4L & -1L \\ -16 & 32 & -16 & & -4L & 0 & 4L \\ 2 & -16 & 14 & & 1L & -4L & -3L \\ & & & & & & \\ 3L & -4L & 1L & L^2(\frac{2}{3}+14\Lambda) & L^2(\frac{2}{3}-16\Lambda) & L^2(-\frac{1}{3}+2\Lambda) \\ 4L & 0 & -4L & L^2(\frac{2}{3}-16\Lambda) & L^2(\frac{8}{3}+32\Lambda) & L^2(\frac{2}{3}-16\Lambda) \\ -1L & 4L & -3L & L^2(-\frac{1}{3}+2\Lambda) & L^2(\frac{2}{3}-16\Lambda) & L^2(\frac{2}{3}+14\Lambda) \end{bmatrix},$$
$$(8.198)$$

whereupon the deformation and load vectors contain the following components at this point:

$$u_p = \begin{bmatrix} u_{1y} & u_{2y} & u_{3y} & \phi_{1z} & \phi_{2z} & \phi_{3z} \end{bmatrix}^T, \quad (8.199)$$
$$F^e = \begin{bmatrix} F_{1y} & F_{2y} & F_{3y} & M_{1z} & M_{2z} & M_{3z} \end{bmatrix}^T. \quad (8.200)$$

For the analysis of the convergence behavior for the beam according to Fig. 8.9 the columns and rows for the entries u_{1y} and ϕ_{1z} in the present system of equations can be canceled. The inverted 4×4 stiffness matrix can be used for the definition of the unknown degrees of freedom:

$$\begin{bmatrix} u_{2y} \\ u_{3y} \\ \vdots \\ \phi_{3z} \end{bmatrix} = \frac{6L}{k_s AG} \underbrace{\begin{bmatrix} \mathbf{X} & \cdots & \cdots & \mathbf{X} \\ \mathbf{X} & \frac{1+3\Lambda}{18\Lambda} & \cdots & \mathbf{X} \\ \vdots & & & \vdots \\ \mathbf{X} & \cdots & \cdots & \mathbf{X} \end{bmatrix}}_{4\times 4 \text{ matrix}} \begin{bmatrix} 0 \\ F \\ \vdots \\ 0 \end{bmatrix}, \quad (8.201)$$

from which, through evaluation of the second row, the deformation on the right-hand boundary can be defined as:

$$u_{3y} = \underbrace{\frac{6L}{k_s AG}}_{\frac{6\Lambda L^3}{EI_z}} \times \frac{1+3\Lambda}{18\Lambda} \times F = \left(\frac{1}{3} + \Lambda \right) \frac{FL^3}{EI_z}. \quad (8.202)$$

For a rectangular cross-section $\Lambda = \frac{1}{5}(1+\nu)\left(\frac{h}{L}\right)^2$ results and one receives the exact solution[16] of the problem as:

---

[16] For this see the supplementary Problem 8.6.

$$u_{3y} = \left(\frac{1}{3} + \frac{1+\nu}{5} \left(\frac{h}{L} \right)^2 \right) \times \frac{FL^3}{EI_z} . \tag{8.203}$$

According to the procedure for the TIMOSHENKO element with quadratic-linear shape functions in Sect. 8.3.6, the middle node can be eliminated. Under the assumption that no forces or moments are acting on the middle node, the 2nd and 5th row of Eq. (8.198) yields the following relation for the unknowns on the middle node:

$$u_{2y} = \frac{1}{2}u_{1y} + \frac{1}{2}u_{3y} + \frac{1}{8}L\phi_{1z} - \frac{1}{8}L\phi_{3z} , \tag{8.204}$$

$$\phi_{2z} = \frac{-4u_{1y}}{L\left(\frac{8}{3}+32\Lambda\right)} + \frac{4u_{3y}}{L\left(\frac{8}{3}+32\Lambda\right)} - \frac{\left(\frac{2}{3}-16\lambda\right)\phi_{1z}}{\left(\frac{8}{3}+32\Lambda\right)} - \frac{\left(\frac{2}{3}-16\lambda\right)\phi_{3z}}{\left(\frac{8}{3}+32\Lambda\right)} . \tag{8.205}$$

These two relations can be considered in Eq. (8.198) so that the following principal finite element equation results after a short conversion:

$$\frac{2EI_z}{L^3(1+12\Lambda)}
\begin{bmatrix}
6 & 3L & -6 & 3L \\
3L & 2L^2(1+3\Lambda) & -3L & L^2(1-6\Lambda) \\
-6 & -3L & 6 & -3L \\
3L & L^2(1-6\Lambda) & -3L & 2L^2(1+3\Lambda)
\end{bmatrix}
\begin{bmatrix}
u_{1y} \\
\phi_{1z} \\
u_{3y} \\
\phi_{3z}
\end{bmatrix}
=
\begin{bmatrix}
F_{1y} \\
M_{1z} \\
F_{3y} \\
M_{3z}
\end{bmatrix} . \tag{8.206}$$

With this formulation the one-beam problem according to Fig. 8.9 can be solved a little bit faster since after the consideration of the boundary conditions only a 2×2 matrix needs to be inverted. In this case, for the definition of the unknown the following results:

$$\frac{L^3(1+12\Lambda)}{2EI_z}
\begin{bmatrix}
\frac{2(1+3\Lambda)}{3(1+12\Lambda)} & \frac{1}{L(1+12\Lambda)} \\
\frac{1}{L(1+12\Lambda)} & \frac{2}{L^2(1+12\Lambda)}
\end{bmatrix}
\begin{bmatrix}
F \\
0
\end{bmatrix}
=
\begin{bmatrix}
u_{3y} \\
\phi_{3z}
\end{bmatrix} , \tag{8.207}$$

which results from the exact solution for the deflection according to Eq. (8.203).

8.4.2 Supplementary Problems

8.3 Calculation of the Shear Correction Factor for Rectangular Cross-Section

For a rectangular cross-section with width b and height h, the shear stress distribution is given as follows [20]:

$$\tau_{xy}(y) = \frac{6Q_y}{bh^3}\left(\frac{h^2}{4} - y^2\right) \quad \text{with} \quad -\frac{h}{2} \le y \le \frac{h}{2}. \tag{8.208}$$

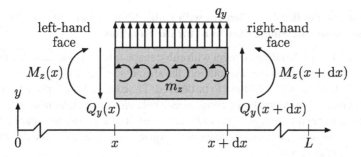

Fig. 8.18 Infinitesimal beam element with internal reactions and distributed loads

Compute the shear correction factor k_s under the assumption that the constant — in the surface A_s acting — equivalent shear stress $\tau_{xy} = Q_y/A_s$ yields the same shear strain energy as the actual shear stress distribution $\tau_{xy}(y)$, which acts in the actual cross-sectional area A of the beam.

8.4 Differential Equation under Consideration of Distributed Moment

For the derivation of the equilibrium condition, the infinitesimal beam element, illustrated in Fig. 8.18 needs to be considered, which is additionally loaded with a constant 'distributed moment' $m_z = \frac{\text{moment}}{\text{length}}$. Subsequently one needs to derive the differential equation for the TIMOSHENKO beam under consideration of a general moment distribution $m_z(x)$.

8.5 Analytical Calculation of the Distribution of the Deflection and Rotation for a Cantilever under Point Load

For the illustrated cantilever in Table 8.2, which is loaded with a point load F at the right-hand end, calculate the distribution of the deflection $u_y(x)$ and the rotation $\phi_z(x)$ under consideration of the shear influence. Subsequently the maximal deflection and the rotation at the loading point needs to be determined. Furthermore, the boundary value of the deflection at the loading point for slender ($h \ll L$) and compact ($h \gg L$) beams has to be determined.

8.6 Analytical Calculation of the Normalized Deflection for Beams with Shear Contribution

For the illustrated courses of the maximal normalized deflection $u_{y,\text{norm}}$ in Fig. 8.6 as a function of the slenderness ratio, the corresponding equations have to be derived.

8.7 Timoshenko Bending Element with Quadratic Shape Functions for the Deflection and Linear Shape Functions for the Rotation

For a TIMOSHENKO bending element with quadratic shape functions for the deflection and linear shape functions for the rotation, the stiffness matrix, after elimination of the middle node according to Eq. (8.170), is given. Derive the additional load vector on the right-hand side of the principal finite element equation which results from a distributed load $q_y(x)$. Subsequently the result for a constant load has to be simplified.

8.8 Timoshenko Bending Element with Cubic Shape Functions for the Deflection and Quadratic Shape Functions for the Rotation

For a TIMOSHENKO bending element with cubic shape functions for the deflection and quadratic shape functions for the rotation, the stiffness matrix and the principal finite element equation $k^e u_{\mathrm{p}} = F^e$ have to be derived. The exact solution has to be used for the integration. Subsequently the convergence behavior of an element configuration, which is illustrated in Fig. 8.9, has to be analyzed. The element deforms in the x-y plane. How does the principal finite element equation change, when the deformation in the x-z plane occurs?

References

1. Timoshenko SP, Goodier JN (1970) Theory of elasticity. McGraw-Hill, New York
2. Beer FP, Johnston ER Jr, DeWolf JT, Mazurek DF (2009) Mechanics of materials. McGraw-Hill, Singapore
3. Cowper GR (1966) The shear coefficient in Timoshenko's beam theory. J Appl Mech 33:335–340
4. Bathe K-J (2002) Finite-elemente-methoden. Springer, Berlin
5. Weaver W Jr, Gere JM (1980) Matrix analysis of framed structures. Van Nostrand Reinhold Company, New York
6. Gere JM, Timoshenko SP (1991) Mechanics of materials. PWS-KENT Publishing Company, Boston
7. Gruttmann F, Wagner W (2001) Shear correction factors in Timoshenko's beam theory for arbitrary shaped cross-sections. Comput Mech 27:199–207
8. Levinson M (1981) A new rectangular beam theory. J Sound Vib 74:81–87
9. Reddy JN (1984) A simple higher-order theory for laminated composite plate. J Appl Mech 51:745–752
10. Reddy JN (1997) Mechanics of laminated composite plates: theory and analysis. CRC Press, Boca Raton
11. Reddy JN (1997) On locking-free shear deformable beam finite elements. Comput Method Appl M 149:113–132
12. Wang CM (1995) Timoshenko beam-bending solutions in terms of Euler-Bernoulli solutions. J Eng Mech-ASCE 121:763–765
13. Cook RD, Malkus DS, Plesha ME, Witt RJ (2002) Concepts and applications of finite element analysis. Wiley, New York
14. Reddy JN (2006) An introduction to the finite element method. McGraw Hill, Singapore
15. MacNeal RH (1994) Finite elements: their design and performance. Marcel Dekker, New York
16. Russel WT, MacNeal RH (1953) An improved electrical analogy for the analysis of beams in bending. J Appl Mech 20:349
17. MacNeal RH (1978) A simple quadrilateral shell element. Comput Struct 8:175–183
18. Reddy JN (1999) On the dynamic behaviour of the Timoshenko beam finite elements. Sadhana-Acad P Eng S 24:175–198
19. Steinke P (2010) Finite-Elemente-Methode - Rechnergestützte Einführung. Springer, Berlin
20. Hibbeler RC (2008) Mechanics of materials. Prentice Hall, Singapore

Chapter 9
Beams of Composite Materials

Abstract The beam elements discussed so far consist of homogeneous, isotropic material. Within this chapter a finite element formulation for a special material type — composite materials — will be introduced. On the basis of plane layers the behavior for the one-dimensional situation on the beam will be developed. First, different description types for direction dependent material behavior will be introduced. Shortly a special type of composite material, the fiber reinforced materials, will be considered.

9.1 Composite Materials

In the previous chapters, homogeneous and isotropic material has been assumed. However, in practice, elements or components are made of different materials to fulfill the multiple operational demands through the combination of different materials with their specific characteristics. At this point, the treatment of these materials will be shown for bars and beams within the framework of the finite element formulation.

Figure 9.1a illustrates the assembly of a beam of composite material in the longitudinal cross-section. The single layers represent different materials with different material characteristics and can be variably thick. Figure 9.1b illustrates a quite simple composite beam. It consists of only two different materials. Figure 9.1c illustrates an often occurring special case. The assembly is symmetric. Figure 9.1d illustrates the assembly for a sandwich structure. The relatively thick core material and the relatively thin face sheets are typical.

The fiber reinforced materials stand for a composite material, at which the direction dependent behavior is predetermined via the structural assembly. Figure 9.2a illustrates a layer with fibers, which are embedded into a matrix.

In general, the fiber direction can be different for each layer (see Fig. 9.2b). In practice, one can often find a symmetric assembly.

© Springer International Publishing AG, part of Springer Nature 2018　　　　　205
A. Öchsner and M. Merkel, *One-Dimensional Finite Elements*,
https://doi.org/10.1007/978-3-319-75145-0_9

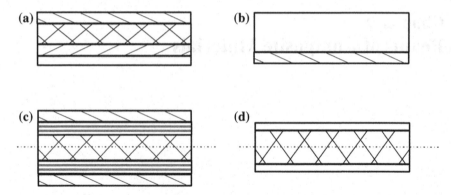

Fig. 9.1 Beams of composite materials: **a** general, **b** two materials, **c** symmetric assembly and
d sandwich with thick core material and thin face sheets

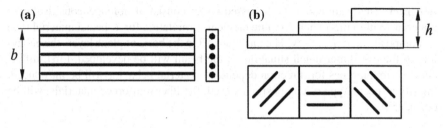

Fig. 9.2 a Composite layer with fibers and **b** composite with layers of different fiber directions

9.2 Anisotropic Material Behavior

Direction dependent behavior is a typical behavior of composite materials. As an
extension to an isotropic material, other description forms for the relation between
the strains and stress result. These will be introduced in the following. Regardless of
this, within this chapter a linear-elastic behavior is assumed for each material.

The general material description (constitutive description) for anisotropic bodies,
connects with

$$\sigma_{ij} = C_{ijpq}\,\varepsilon_{pq} \tag{9.1}$$

the strain tensor (2nd order) via a so-called elasticity tensor (4th order tensor) with
the stress tensor (2nd order tensor). Due to the symmetries of the stress and strain
tensors the first as well as the second index groups in the elasticity tensor

$$C_{jipq} = C_{ijpq} \quad ; \quad C_{ijqp} = C_{ijpq} \tag{9.2}$$

are invariant against permutation. Therewith only 36 from the originally 81 components of the elasticity tensor remain. Usually column matrices are introduced for the
symmetric stress tensor

$$
\begin{bmatrix} \sigma_{xx} & \sigma_{xy} & \sigma_{xz} \\ \sigma_{xy} & \sigma_{yy} & \sigma_{yz} \\ \sigma_{xz} & \sigma_{yz} & \sigma_{zz} \end{bmatrix} \Rightarrow \begin{bmatrix} \sigma_{xx} \\ \sigma_{yy} \\ \sigma_{zz} \\ \sigma_{yz} \\ \sigma_{zx} \\ \sigma_{xy} \end{bmatrix} \Rightarrow \begin{bmatrix} \sigma_1 \\ \sigma_2 \\ \sigma_3 \\ \sigma_4 \\ \sigma_5 \\ \sigma_6 \end{bmatrix}
\tag{9.3}
$$

and the symmetric strain tensor

$$
\begin{bmatrix} \varepsilon_{xx} & \varepsilon_{xy} & \varepsilon_{xz} \\ \varepsilon_{xy} & \varepsilon_{yy} & \varepsilon_{yz} \\ \varepsilon_{xz} & \varepsilon_{yz} & \varepsilon_{zz} \end{bmatrix} \Rightarrow \begin{bmatrix} \varepsilon_{xx} \\ \varepsilon_{yy} \\ \varepsilon_{zz} \\ \varepsilon_{yz} \\ \varepsilon_{zx} \\ \varepsilon_{xy} \end{bmatrix} \Rightarrow \begin{bmatrix} \varepsilon_1 \\ \varepsilon_2 \\ \varepsilon_3 \\ \varepsilon_4 \\ \varepsilon_5 \\ \varepsilon_6 \end{bmatrix} .
\tag{9.4}
$$

Therewith the stress strain relation (9.1) can be formulated in matrix notation as

$$
\begin{bmatrix} \sigma_1 \\ \sigma_2 \\ \sigma_3 \\ \sigma_4 \\ \sigma_5 \\ \sigma_6 \end{bmatrix} = \begin{bmatrix} C_{11} & C_{12} & C_{13} & C_{14} & C_{15} & C_{16} \\ C_{21} & C_{22} & C_{23} & C_{24} & C_{25} & C_{26} \\ C_{31} & C_{32} & C_{33} & C_{34} & C_{35} & C_{36} \\ C_{41} & C_{42} & C_{43} & C_{44} & C_{45} & C_{46} \\ C_{51} & C_{52} & C_{53} & C_{54} & C_{55} & C_{56} \\ C_{61} & C_{62} & C_{63} & C_{64} & C_{65} & C_{66} \end{bmatrix} \begin{bmatrix} \varepsilon_1 \\ \varepsilon_2 \\ \varepsilon_3 \\ \varepsilon_4 \\ \varepsilon_5 \\ \varepsilon_6 \end{bmatrix}
\tag{9.5}
$$

or in compact form as

$$
\sigma = C\,\varepsilon.
\tag{9.6}
$$

The specific elastic strain energy (related to the volume element) in matrix form

$$
\pi = \frac{1}{2}\varepsilon^{\mathrm{T}}\sigma ,
\tag{9.7}
$$

looks like this, which leads to the following, together with the constitutive equation according to Eq. (9.1)

$$
\pi = \frac{1}{2}\varepsilon^{\mathrm{T}}C\,\varepsilon.
\tag{9.8}
$$

Due to its energetic character this form has to be defined positively ($\pi \geq 0$). This however requires $C^{\mathrm{T}} = C$, thus the symmetry of the C-matrix. Because of this, only 21 components from the 36 components of the stiffness matrix are independent from each other ($C_{ij} = C_{ji}$). This material is also referred to as a *triclinic* material. In the strain-stress relation

$$
\varepsilon = S\sigma
\tag{9.9}
$$

the compliance matrix S connects the stress with the strains. The valid relation for the general three-dimensional case

$$\begin{bmatrix} \varepsilon_1 \\ \varepsilon_2 \\ \varepsilon_3 \\ \varepsilon_4 \\ \varepsilon_5 \\ \varepsilon_6 \end{bmatrix} = \begin{bmatrix} S_{11} & S_{12} & S_{13} & S_{14} & S_{15} & S_{16} \\ S_{21} & S_{22} & S_{23} & S_{24} & S_{25} & S_{26} \\ S_{31} & S_{32} & S_{33} & S_{34} & S_{35} & S_{36} \\ S_{41} & S_{42} & S_{43} & S_{44} & S_{45} & S_{46} \\ S_{51} & S_{52} & S_{53} & S_{54} & S_{55} & S_{56} \\ S_{61} & S_{62} & S_{63} & S_{64} & S_{65} & S_{66} \end{bmatrix} \begin{bmatrix} \sigma_1 \\ \sigma_2 \\ \sigma_3 \\ \sigma_4 \\ \sigma_5 \\ \sigma_6 \end{bmatrix} \tag{9.10}$$

can be simplified for various special cases. This will be introduced in the following section.

9.2.1 Special Symmetries

For further simplifications special symmetries will be considered. The following system can be regarded as an important selection. The stress-strain relation is represented in detail with the stiffness matrix C. The same derivation is valid for the strain-stress relation with the compliance matrix S.

Monoclinic Systems

Let us assume that the plane $z = 0$ is a symmetry plane. Then, all components of the C-matrix, which are related with the z-axis

$$\begin{bmatrix} \sigma_1 \\ \sigma_2 \\ \sigma_3 \\ \sigma_4 \\ \sigma_5 \\ \sigma_6 \end{bmatrix} = \begin{bmatrix} C_{11} & C_{12} & C_{13} & 0 & 0 & C_{16} \\ C_{12} & C_{22} & C_{23} & 0 & 0 & C_{26} \\ C_{13} & C_{23} & C_{33} & 0 & 0 & C_{36} \\ 0 & 0 & 0 & C_{44} & C_{45} & 0 \\ 0 & 0 & 0 & C_{45} & C_{55} & 0 \\ C_{16} & C_{26} & C_{36} & 0 & 0 & C_{66} \end{bmatrix} \begin{bmatrix} \varepsilon_1 \\ \varepsilon_2 \\ \varepsilon_3 \\ \varepsilon_4 \\ \varepsilon_5 \\ \varepsilon_6 \end{bmatrix} \tag{9.11}$$

are invariant against change of signs. Therewith 13 independent material constants remain.

Orthotropic Systems

Here, three mutually perpendicular planes of symmetry in the material exist. The corresponding invariancy against the change of signs yields for orthotropic systems in just 9 independent material constants:

$$\begin{bmatrix} \sigma_1 \\ \sigma_2 \\ \sigma_3 \\ \sigma_4 \\ \sigma_5 \\ \sigma_6 \end{bmatrix} = \begin{bmatrix} C_{11} & C_{12} & C_{13} & & & \\ C_{12} & C_{22} & C_{23} & & 0 & \\ C_{13} & C_{23} & C_{33} & & & \\ & & & C_{44} & & \\ & 0 & & & C_{55} & \\ & & & & & C_{66} \end{bmatrix} \begin{bmatrix} \varepsilon_1 \\ \varepsilon_2 \\ \varepsilon_3 \\ \varepsilon_4 \\ \varepsilon_5 \\ \varepsilon_6 \end{bmatrix} . \tag{9.12}$$

Transversely Isotropic Systems

This, for the fiber reinforced material important group, is characterized through an isotropic behavior in one plane (for example in the y-z plane). Therewith just 5 independent material constants are necessary for the description of the stress-strain relation for transversely isotropic systems:

$$
\begin{bmatrix} \sigma_1 \\ \sigma_2 \\ \sigma_3 \\ \sigma_4 \\ \sigma_5 \\ \sigma_6 \end{bmatrix} = \begin{bmatrix} C_{11} & C_{12} & C_{12} & & & \\ C_{12} & C_{22} & C_{23} & & 0 & \\ C_{12} & C_{23} & C_{22} & & & \\ & & & \frac{(C_{22}-C_{23})}{2} & & \\ & 0 & & & C_{66} & \\ & & & & & C_{66} \end{bmatrix} \begin{bmatrix} \varepsilon_1 \\ \varepsilon_2 \\ \varepsilon_3 \\ \varepsilon_4 \\ \varepsilon_5 \\ \varepsilon_6 \end{bmatrix} . \tag{9.13}
$$

The relation for the constant C_{44} comes from the equivalence of pure shear and a combined tension and compression load.

Isotropic Systems

If the material is isotropic, meaning invariant under all orthogonal transformations, just 2 independent material constants are needed for the stress-strain relation (HOOKE's material):

$$
\begin{bmatrix} \sigma_1 \\ \sigma_2 \\ \sigma_3 \\ \sigma_4 \\ \sigma_5 \\ \sigma_6 \end{bmatrix} = \begin{bmatrix} C_{11} & C_{12} & C_{12} & & & \\ C_{12} & C_{11} & C_{12} & & 0 & \\ C_{12} & C_{12} & C_{11} & & & \\ & & & \frac{(C_{11}-C_{12})}{2} & & \\ & 0 & & & \frac{(C_{11}-C_{12})}{2} & \\ & & & & & \frac{(C_{11}-C_{12})}{2} \end{bmatrix} \begin{bmatrix} \varepsilon_1 \\ \varepsilon_2 \\ \varepsilon_3 \\ \varepsilon_4 \\ \varepsilon_5 \\ \varepsilon_6 \end{bmatrix} . \tag{9.14}
$$

9.2.2 Engineering Constants

In the theory of isotropic continua usually the two material properties E (modulus of elasticity) and ν (lateral contraction number, POISSON's ratio) are used, which are easily defined experimentally. Likewise, based on the differential equation, the LAMÉ's coefficients λ and μ are used or the bulk modulus K, or the shear modulus G. The single parameters are dependent on each other and can be converted with

$$
\lambda = \frac{E\nu}{(1+\nu)(1-2\nu)} \quad , \quad \mu = \frac{E}{2(1+\nu)} = G \quad , \quad K = \frac{E}{2(1-2\nu)} \tag{9.15}
$$

$$
\nu = \frac{\lambda}{2(\lambda+\mu)} \quad , \quad E = \frac{\mu(3\lambda+2\mu)}{\lambda+\mu} \quad , \quad K = \lambda + \frac{2}{3}\mu .
$$

The meaning of these parameters can be read functionally from the compliance matrix.

Isotropic Systems

For isotropic materials the strain-stress relation looks as follows

$$
\begin{bmatrix}
\varepsilon_1 \\
\varepsilon_2 \\
\varepsilon_3 \\
\varepsilon_4 \\
\varepsilon_5 \\
\varepsilon_6
\end{bmatrix}
=
\begin{bmatrix}
\frac{1}{E} & \frac{-\nu}{E} & \frac{-\nu}{E} & & & \\
 & \frac{1}{E} & \frac{-\nu}{E} & & 0 & \\
 & & \frac{1}{E} & & & \\
 & & & \frac{1}{G} & & \\
 & \text{sym.} & & & \frac{1}{G} & \\
 & & & & & \frac{1}{G}
\end{bmatrix}
\begin{bmatrix}
\sigma_1 \\
\sigma_2 \\
\sigma_3 \\
\sigma_4 \\
\sigma_5 \\
\sigma_6
\end{bmatrix}
.
\tag{9.16}
$$

Through inversion of the so-called compliance matrix the stiffness matrix follows. The components of the stiffness matrix

$$
C_{11} = (1 - \nu)\overline{E} \tag{9.17}
$$
$$
C_{12} = \nu\overline{E}
$$
$$
\frac{(C_{11} - C_{12})}{2} = \frac{E}{2(1 + \nu)} = G
$$

with

$$
\overline{E} = E\frac{1}{(1 + \nu)(1 - 2\nu)} \tag{9.18}
$$

result from comparison with Eq. (9.14).

Transversely Isotropic Systems

As an example, the transversely isotropic behavior for the plane $x = 0$ is assumed. Of course the considerations can also be adopted to other directions in space. The compliance matrix for transversely isotropic system looks as follows

$$
\begin{bmatrix}
\varepsilon_1 \\
\varepsilon_2 \\
\varepsilon_3 \\
\varepsilon_4 \\
\varepsilon_5 \\
\varepsilon_6
\end{bmatrix}
=
\begin{bmatrix}
\frac{1}{E_1} & \frac{-\nu_{12}}{E_2} & \frac{-\nu_{12}}{E_2} & & & \\
 & \frac{1}{E_2} & \frac{-\nu_{23}}{E_2} & & 0 & \\
 & & \frac{1}{E_2} & & & \\
 & & & \frac{1}{2G_{23}} & & \\
 & \text{sym.} & & & \frac{1}{2G_{12}} & \\
 & & & & & \frac{1}{2G_{12}}
\end{bmatrix}
\begin{bmatrix}
\sigma_1 \\
\sigma_2 \\
\sigma_3 \\
\sigma_4 \\
\sigma_5 \\
\sigma_6
\end{bmatrix}
.
\tag{9.19}
$$

The shear modulus G_{23} can be calculated as with isotropic media from E_2 and ν_{23}.

The indexing of the lateral contraction ratio is carried out according to the following scheme:

- 1st index = direction of the contraction,
- 2nd index = direction of the load, which elicits this contraction.

Through inversion of the compliance matrix and comparison one obtains

$$C_{11} = (1 - \nu_{23}^2) \, \bar{\nu} \, E_1 \tag{9.20}$$

$$C_{22} = (1 - \nu_{12} \nu_{21}) \, \bar{\nu} \, E_2 \tag{9.21}$$

$$C_{12} = \nu_{12}(1 + \nu_{23}) \, \bar{\nu} \, E_1 = \nu_{21} \, (1 + \nu_1) \, \bar{\nu} \, E_2 \tag{9.22}$$

$$C_{23} = (\nu_{23} + \nu_{21} \nu_{12}) \, \bar{\nu} \, E_2 \tag{9.23}$$

$$C_{22} - C_{23} = (1 - \nu_{22} - 2\nu_{21} \nu_{12}) \, \bar{\nu} \, E_2 \tag{9.24}$$

$$C_{66} = G_{12} \tag{9.25}$$

with

$$\bar{\nu} = \frac{1}{(1 + \nu_{23}) \, (1 - \nu_{23} - 2\nu_{21} \nu_{12})} \tag{9.26}$$

the relation between the engineering constants and the components C_{ij} of the stiffness matrix. With the relation

$$\nu_{12} \, E_1 = \nu_{21} \, E_2 \tag{9.27}$$

the single material values are connected.

9.2.3 Transformation Behavior

In a composite material, materials for the single layers are used, which behave directionally dependent due to their structural configuration. This preferential direction — mostly it is one — is characteristic for one layer. The preferential directions can be defined variably in layers when consolidating in the network. For the macroscopic description of the composite material, transformation instructions are therefore necessary to consider the preferential direction of a single layer in the composite. An instruction has to be given on how the material equations transform in case of a change of the coordinate system. This instruction can be gained from the transformation behavior of tensors. Thereby however only against each other rotated Cartesian coordinate systems are necessary (so-called orthogonal transformations).

For an arbitrary tensor of 2nd order A_{ij} the following transformation is applicable in the case of a change from a Cartesian (ij) into another (kl') Cartesian system (herein are the c_{ij} the so-called direction cosines):

$$A'_{kl} = c_{ki} \, c_{lj} \, A_{ij} \tag{9.28}$$

and therefore especially for the strain and for the stress tensor

$$\varepsilon'_{kl} = c_{ki} \, c_{lj} \, \varepsilon_{ij} \,, \tag{9.29}$$

$$\sigma'_{kl} = c_{ki} \, c_{lj} \, \sigma_{ij} \,. \tag{9.30}$$

If the stresses and strains are each expressed as column matrices, the transformation can then be written as

$$\varepsilon' = T_\varepsilon \varepsilon \tag{9.31}$$
$$\sigma' = T_\sigma \sigma \tag{9.32}$$

or

$$\varepsilon = T_\varepsilon^{-1} \varepsilon' \tag{9.33}$$
$$\sigma = T_\sigma^{-1} \sigma'. \tag{9.34}$$

With the properties for these transformation matrices

$$T_\varepsilon^{-1} = T_\sigma^{\mathrm{T}} \ , \quad T_\sigma^{-1} = T_\varepsilon^{\mathrm{T}} \tag{9.35}$$

the relations for the transformations of the stiffness matrix follow based on

$$\sigma' = C' \varepsilon' \tag{9.36}$$

with the conversions

$$T_\sigma \sigma = C' T_\varepsilon \varepsilon \tag{9.37}$$
$$(T_\sigma)^{-1} T_\sigma \sigma = (T_\sigma)^{-1} C' T_\varepsilon \varepsilon$$
$$\sigma = (T_\sigma)^{\mathrm{T}} C' T_\varepsilon \varepsilon$$

or based on

$$\sigma = C \varepsilon \tag{9.38}$$

with the conversions

$$(T_\sigma)^{-1} \sigma' = C'(T_\sigma)^{-1} C' \tag{9.39}$$
$$(T_\sigma)(T_\sigma)^{-1} \sigma' = (T_\sigma) C(T_\sigma)^{-1} \varepsilon'$$
$$\sigma' = (T_\sigma) C(T_\varepsilon)^{\mathrm{T}} \varepsilon'$$

finally in

$$C' = T_\sigma C T_\sigma^{\mathrm{T}} \tag{9.40}$$
$$C = T_\varepsilon^{\mathrm{T}} C' T_\varepsilon. \tag{9.41}$$

Following the line of reasoning, one obtains the following for the transformation of the compliance matrix

$$S' = T_\varepsilon S (T_\varepsilon)^{\mathrm{T}} \tag{9.42}$$

$$S = T_\sigma^{\mathrm{T}} S' T_\sigma . \tag{9.43}$$

For the important group of transversal isotropic materials the following transformation matrices

$$T_\sigma = \begin{bmatrix} c^2 & s^2 & 0 & 0 & 0 & 2cs \\ s^2 & c^2 & 0 & 0 & 0 & -2cs \\ 0 & 0 & 1 & 0 & 0 & 0 \\ 0 & 0 & 0 & c & -s & 0 \\ 0 & 0 & 0 & s & c & 0 \\ -sc & -sc & 0 & 0 & 0 & c^2 - s^2 \end{bmatrix} \tag{9.44}$$

and

$$T_\varepsilon = \begin{bmatrix} c^2 & s^2 & 0 & 0 & 0 & cs \\ s^2 & c^2 & 0 & 0 & 0 & -ss \\ 0 & 0 & 1 & 0 & 0 & 0 \\ 0 & 0 & 0 & c & -s & 0 \\ 0 & 0 & 0 & s & c & 0 \\ -2cs & -2cs & 0 & 0 & 0 & c^2 - s^2 \end{bmatrix} \tag{9.45}$$

for a rotation around the z-axis with $s = \sin \alpha$ and $c = \cos \alpha$ result.

9.2.4 Plane Stress States

A crucial simplification of the stress-strain relation results from the reduction to two-dimensional, instead of spatial stress states [1]. A *thin* layer in the composite can be regarded under the assumption of the plane stress state. Stress components, which are not in the considered plane, are set equal to zero. It should be remarked that the plane stress state is only valid by approximation. Here the plane stress state for the x–y plane will be considered. For the stress states in the x–z or y–z plane similar formulations result.

The assumption of the plane stress states simplifies the stress-strain relation. Only three equations remain

$$\sigma_1 = C_{11}\varepsilon_1 + C_{12}\varepsilon_2 + C_{13}\varepsilon_3 + C_{16}\varepsilon_6 \tag{9.46}$$

$$\sigma_2 = C_{12}\varepsilon_1 + C_{22}\varepsilon_2 + C_{23}\varepsilon_3 + C_{26}\varepsilon_6 \tag{9.47}$$

$$\sigma_6 = C_{16}\varepsilon_1 + C_{26}\varepsilon_2 + C_{33}\varepsilon_3 + C_{66}\varepsilon_6 \tag{9.48}$$

for the three stress components σ_1, σ_2 and σ_6. The strain ε_3 at the right angle to the considered plane can be determined from the relation

$$\varepsilon_3 = -\frac{1}{C_{33}}(C_{13}\varepsilon_1 + C_{23}\varepsilon_2 + C_{36}\varepsilon_3). \tag{9.49}$$

If ε_3 is replaced in the above three equations, a modified form results

$$\sigma_i = \left(C_{ij} - \frac{C_{i3} \, C_{j3}}{C_{33}} \right) \varepsilon_j , \qquad i, j = 1, 2, 6 \tag{9.50}$$

which is usually described with

$$\sigma_i = Q_{ij} \, \varepsilon_j \tag{9.51}$$

or in matrix notation with

$$\boldsymbol{\sigma} = \boldsymbol{Q}\boldsymbol{\varepsilon}. \tag{9.52}$$

For the plane stress state the components of the compliance matrix S_{ij} in the strain-stress relation

$$\boldsymbol{\varepsilon} = \boldsymbol{S}\boldsymbol{\sigma} \tag{9.53}$$

remain the same. In further considerations the stress-strain relations and the strain-stress relations for the different lamina will be specified under a plane stress state.

In a practical application three different layers occur:

1. Layers, which are treated as quasi homogeneous and quasi isotropic. The elastic behavior does not show a preferential direction. A typical example are layers, whose matrix is supported with short fibers, whose direction is however arbitrary.
2. Layers, at which long fibers with preferential direction are embedded in a matrix, so-called *unidirectional lamina*. The load also occurs in this preferential direction. From a macroscopic point of view the material is regarded as quasi-homogeneous and orthotropic.
3. Layers as in (2). The load however can occur in any direction.

Isotropic Lamina

For isotropic lamina only two independent from each other material components are needed in the stress-strain relation

$$\begin{bmatrix} \sigma_1 \\ \sigma_2 \\ \sigma_6 \end{bmatrix} = \begin{bmatrix} Q_{11} & Q_{12} & 0 \\ Q_{12} & Q_{11} & 0 \\ 0 & 0 & Q_{66} \end{bmatrix} \begin{bmatrix} \varepsilon_1 \\ \varepsilon_2 \\ \varepsilon_6 \end{bmatrix} \tag{9.54}$$

with

$$Q_{11} = E/(1 - \nu^2) \tag{9.55}$$
$$Q_{12} = E\nu/(1 - \nu^2) \tag{9.56}$$
$$C_{66} = E/2(1 + \nu) \tag{9.57}$$

and in the strain-stress relation

$$\begin{bmatrix} \varepsilon_1 \\ \varepsilon_2 \\ \varepsilon_6 \end{bmatrix} = \begin{bmatrix} S_{11} & S_{12} & 0 \\ S_{12} & S_{11} & 0 \\ 0 & 0 & S_{66} \end{bmatrix} \begin{bmatrix} \sigma_1 \\ \sigma_2 \\ \sigma_6 \end{bmatrix} \tag{9.58}$$

with

$$S_{11} = 1/E \tag{9.59}$$
$$S_{12} = -\nu/E \tag{9.60}$$
$$S_{66} = 1/G = 2(1+\nu)/E. \tag{9.61}$$

The equations show that no coupling exists between the normal stress and the shear stress.

Unidirectional Lamina, Load in Direction of the Fiber

Usually a lamina related coordinate system (1', 2') is introduced for the description of the unidirectional lamina. The direction 1' equals the fiber direction (L), the direction 2' equals the direction perpendicular to the fiber direction (T). In the stress-strain relation

$$\begin{bmatrix} \sigma_1' \\ \sigma_2' \\ \sigma_6' \end{bmatrix} = \begin{bmatrix} Q_{11}' & Q_{12}' & 0 \\ Q_{12}' & Q_{22}' & 0 \\ 0 & 0 & Q_{66}' \end{bmatrix} \begin{bmatrix} \varepsilon_1' \\ \varepsilon_2' \\ \varepsilon_6' \end{bmatrix} \tag{9.62}$$

with

$$Q_{11}' = E_1'/(1 - \nu_{12}'\nu_{21}') \tag{9.63}$$
$$Q_{22}' = E_2'/(1 - \nu_{12}'\nu_{21}') \tag{9.64}$$
$$Q_{12}' = E_2'\nu_{12}'/(1 - \nu_{12}'\nu_{21}') \tag{9.65}$$
$$Q_{66}' = G_{12}' \tag{9.66}$$

and in the strain-stress relation

$$\begin{bmatrix} \varepsilon_1' \\ \varepsilon_2' \\ \varepsilon_6' \end{bmatrix} = \begin{bmatrix} S_{11}' & S_{12}' & 0 \\ S_{12}' & S_{22}' & 0 \\ 0 & 0 & S_{66}' \end{bmatrix} \begin{bmatrix} \sigma_1' \\ \sigma_2' \\ \sigma_6' \end{bmatrix} \tag{9.67}$$

with

$$S_{11}' = 1/E_1' \tag{9.68}$$
$$S_{22}' = 1/E_2' \tag{9.69}$$
$$S_{12}' = -\nu_{12}'/E_1' = -\nu_{21}'/E_2' \tag{9.70}$$
$$S_{66}' = G_{12}' \tag{9.71}$$

Fig. 9.3 Lamina with
angular misalignment
between preferential and
load direction

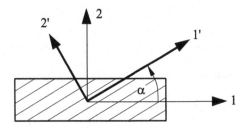

Four independent from each other material parameters are necessary. The valid parameters in the lamina own coordinate system can be formulated with the following

$$E_1' = E_L \quad , \quad E_2' = E_T \quad , \quad G_{12}' = G_{LT} \quad , \quad \nu_{12}' = \nu_{LT} \tag{9.72}$$

in the engineering constants.

Unidirectional Lamina, Arbitrary Load Direction in the Plane

In contrast to the above considerations the load cannot just occur in the preferential direction of the lamina, but in every direction of the plane (Fig. 9.3).

However, to be able to use the material values of the (1',2') coordinate system, a transformation of the stiffness and compliance matrix from the (1',2') system into the (1,2) system is necessary. The material equation for the plane stress state in the (1',2') system appears as follows:

$$
\begin{bmatrix} \varepsilon_1' \\ \varepsilon_2' \\ \varepsilon_6' \end{bmatrix}
=
\begin{bmatrix}
\frac{1}{E_1'} & \frac{-\nu_{12}}{E_1'} & 0 \\
\frac{-\nu_{12}}{E_1'} & \frac{1}{E_2'} & 0 \\
0 & 0 & \frac{1}{2G_{12}}
\end{bmatrix}
\begin{bmatrix} \sigma_1' \\ \sigma_2' \\ \sigma_6' \end{bmatrix}.
\tag{9.73}
$$

Through the application of the transformation relation of the transversal isotropic body with the 'plane' transformation matrix one obtains for the compliance matrix

$$
\begin{bmatrix} S_{11} \\ S_{12} \\ S_{16} \\ S_{22} \\ S_{26} \\ S_{66} \end{bmatrix}
=
\begin{bmatrix}
c^4 & 2c^2s^2 & s^4 & c^2s^2 \\
c^2s^2 & c^4+s^4 & c^2s^2 & -c^2s^2 \\
2c^3s & -2cs(c^2-s^2) & -2cs^3 & -cs(c^2-s^2) \\
s^4 & 2c^2s^2 & c^4 & c^2s^2 \\
2cs^3 & 2cs(c^2-s^2) & -2c^3s & cs(c^2-s^2) \\
4c^2s^2 & -8c^2s^2 & 4c^2s^2 & (c^2-s^2)^2
\end{bmatrix}
\begin{bmatrix} S_{11}' \\ S_{12}' \\ S_{22}' \\ S_{66}' \end{bmatrix}
\tag{9.74}
$$

and for the stiffness matrix

Table 9.1 Material models with the amount of none-zero entries and independent parameters

Material model	Three-dimensional		Two-dimensional	
	$\neq 0$	Independent parameter	$\neq 0$	Independent parameter
Isotropic	12	2	5	2
Transversely isotropic	12	5	–	–
Orthotropic	12	9	5	4
Monoclinic	20	13	–	–
Anisotropic	36	21	9	6

$$\begin{bmatrix} Q_{11} \\ Q_{12} \\ Q_{16} \\ Q_{22} \\ Q_{26} \\ Q_{66} \end{bmatrix} = \begin{bmatrix} c^4 & 2c^2s^2 & s^4 & 4c^2s^2 \\ c^2s^2 & c^4+s^4 & c^2s^2 & -4c^2s^2 \\ c^3s & -cs(c^2-s^2) & -cs^3 & -2cs(c^2-s^2) \\ s^4 & 2c^2s^2 & c^4 & 4c^2s^2 \\ cs^3 & cs(c^2-s^2) & -c^3s & 2cs(c^2-s^2) \\ c^2s^2 & -2c^2s^2 & c^2s^2 & (c^2-s^2)^2 \end{bmatrix} \begin{bmatrix} Q'_{11} \\ Q'_{12} \\ Q'_{22} \\ Q'_{66} \end{bmatrix} \tag{9.75}$$

with the abbreviations

$$s = \sin\alpha, \; c = \cos\alpha. \tag{9.76}$$

Therewith the stiffness and compliance matrix for transversely isotropic lamina in arbitrary (in the plane rotated) Cartesian coordinate systems can be illustrated (Assumption here: the rotation occurs in the plane around the z-axis, which is at right angles to the lamina plane).

In the following Table 9.1, the number of non-zero entries and the number of independent parameters are summarized for different material models. One distinguishes between the general three-dimensional and the plane stress state.

9.3 Introduction to the Micromechanics of the Fiber Composite Materials

Micromechanics serves to determine the properties of a composite from the properties of the single components. For the description of the fiber composite materials (FCM) so-called *unidirectinal lamina* are used as the model, which belongs to the group of transversely isotropic materials. The model is based on the following assumptions [2]:

1. the fibers are distributed equally in the matrix,
2. between the fibers and the matrix there are ideal contact conditions (consistency of the tangential component of the displacement)

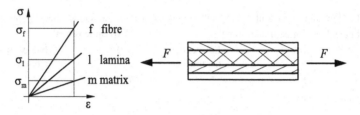

Fig. 9.4 Stress-strain relation at load in fiber direction

3. the matrix does not contain any cavities,
4. the external load acts in the direction of the fibers or perpendicular to that,
5. no eigenstress exists in the lamina,
6. the fiber as well as the matrix material are linear-elastic and
7. the fibers are infinitely long.

In regards to the loading of composite materials, one needs to distinguish between the load in and perpendicular to the fiber direction. Figure 9.4 illustrates the corresponding parameters for the load in the fiber direction.

Due to assumption (2) the following occurs

$$\varepsilon_f = \varepsilon_m = \varepsilon_1 \tag{9.77}$$

and according to assumption (6) the following is valid

$$\begin{aligned}
\sigma_f &= E_f\,\varepsilon_f = E_f\,\varepsilon_1 \\
\sigma_m &= E_m\,\varepsilon_m = E_m\,\varepsilon_1.
\end{aligned} \tag{9.78}$$

Since in general $E_f \geq E_m$ applies, $\sigma_f \geq \sigma_m$ follows.

From the equilibrium of forces

$$F = F_f + F_m \tag{9.79}$$

with A_f as the cross-section of the fibre and A_m as the cross-section of the matrix

$$\sigma_1 A_1 = \sigma_f A_f + \sigma_m A_m \tag{9.80}$$

the stress in the lamina results in

$$\sigma_1 = \sigma_f \frac{A_f}{A_1} + \sigma_m \frac{A_m}{A_1}. \tag{9.81}$$

Due to

$$A_1 = A_f + A_m \tag{9.82}$$

and the abbreviations

$$v_f = \frac{A_f}{A_1} \quad \text{and} \quad v_m = \frac{A_m}{A_1} \tag{9.83}$$

the following results

$$\frac{A_m}{A_1} = 1 - v_f = v_m \tag{9.84}$$

and therewith

$$\sigma_1 = \sigma_f v_f + \sigma_m (1 - v_f). \tag{9.85}$$

The division of this relation by ε_1 leads to the so-called *rule of mixtures*

$$E_1 = E_f v_f + E_m v_m = E_f v_f + E_m (1 - v_f). \tag{9.86}$$

With the relation

$$\frac{F_f}{F_1} = \frac{\sigma_f v_f}{\sigma_f v_f + \sigma_m (1 - v_f)} = \frac{E_f v_f}{E_f v_f + E_m (1 - v_f)} \tag{9.87}$$

the load fraction, which is transferred from the fibers from the total load is described.

9.4 Multilayer Composite

A composite material usually consists of various layers. These layers can differ in the geometric dimensions as well as in the material properties. In the following, a single layer will be analyzed at first, subsequently the entire composite. For the finite element formulation, the often occurring cases in practice will be introduced. The macromechanical behavior is described under the following assumptions:

- The single layers of the composite are perfectly connected with each other. There is *no intermediate layer*.
- Each layer can be regarded as quasi-homogeneous.
- The displacements and strains are continuously throughout the entire composite. Within one layer the displacements and strains can be described with a linear course.

9.4.1 One Layer in the Composite

For a single layer (lamina) it should be assumed that the thickness of the layer is much smaller than the length dimension. Therewith a plane stress state can be used for the description.

Furthermore two situations need to be distinguished. The stresses are

- constant or
- not constant

within a composite layer. In the first case a, from the stress resulting force vector for the kth layer in the composite results in

$$N^k = [N_1^k, N_2^k, N_6^k]^T , \tag{9.88}$$

which is defined through

$$N^k = \int_h \sigma \, dz . \tag{9.89}$$

N_i^k are forces, which are related to a unit width, the real normal forces one receives through multiplication with the width b^k of a composite layer. N_1^k, N_2^k are resulting normal forces, N_6^k stands for a shear force in the plane. For a constant stress over the cross-section the following results:

$$N^k = \sigma^k \, h^k . \tag{9.90}$$

In the reduced stiffness matrix Q the components are constant, too. With

$$\sigma^k = Q^k \, \varepsilon^0 \tag{9.91}$$

the following results

$$N^k = Q^k \, \varepsilon^0 \, h^k = A^k \varepsilon^0 . \tag{9.92}$$

For the case when the stresses are not constant over the layer thickness, a resulting momentum vector occurs

$$M^k = [M_1^k, M_2^k, M_6^k]^T , \tag{9.93}$$

which is defined via

$$M^k = \int_h \sigma^k(z) \, z \, dz . \tag{9.94}$$

M_i^k are moments, which are related to a unit width, whereupon M_1^k, M_2^k stand for bending moments and M_6^k for the torsional moment. According to the deformation model the strains run linearly over the cross-section and can be expressed via

$$\varepsilon^k(z) = z \, \kappa . \tag{9.95}$$

For the resulting momentum vector therefore the following results

$$M^k = \int_h Q^k \, \varepsilon^k \, z \, dz = Q^k \kappa \int_{-h/2}^{+h/2} z^2 dz = Q^k \kappa \frac{(h^k)^3}{12} = D^k \kappa. \tag{9.96}$$

If, for a layer, a constant as well as a linear fraction in the strains occur, the following is valid

$$\varepsilon(z) = \varepsilon^0 + z \, \kappa \tag{9.97}$$

and therefore

$$N^k = \int_h \sigma^k(z) \, dz = \int_h Q^k(\varepsilon^0 + z\kappa) dz \tag{9.98}$$

and

$$M^k = \int_h \sigma^k(z) \, z \, dz = \int_h Q^k(\varepsilon^0 + z\kappa) \, z \, dz. \tag{9.99}$$

Both a resulting force and momentum vector occur. Both derived formulations can be combined as

$$N^k = A^k \varepsilon_0 + B^k \kappa \tag{9.100}$$

$$M^k = B^k \varepsilon_0 + D^k \kappa \tag{9.101}$$

and compactly summarized as

$$\begin{bmatrix} N^k \\ M^k \end{bmatrix} = \begin{bmatrix} A^k & B^k \\ B^k & D^k \end{bmatrix} \begin{bmatrix} \varepsilon^0 \\ \kappa \end{bmatrix}. \tag{9.102}$$

For layers, which are symmetric to the center plane $z = 0$, the coupling matrix B^k disappears and it remains

$$\begin{bmatrix} N^k \\ M^k \end{bmatrix} = \begin{bmatrix} A^k & 0 \\ 0 & D^k \end{bmatrix} \begin{bmatrix} \varepsilon^0 \\ \kappa \end{bmatrix} \tag{9.103}$$

with

$$A^k = Q^k h^k \quad , \quad D^k = Q^k \frac{(h^k)^3}{12}. \tag{9.104}$$

A special case, for which solely one preferential direction is considered with

$$A_{11}^k = Q_{11}^k h^k \quad , \quad D_{11} = Q_{11}^k \frac{(h^k)^3}{12} \tag{9.105}$$

serves for the description of a composite layer in a beam.

9.4.2 The Multilayer Composite

The composite is built by various layers (laminate). In the determination of the resulting forces and moments one needs to integrate throughout the total heights. Since the stiffness matrices per lamina are independent of the z-coordinate, the integration can be substituted through a corresponding summation. Therewith the following results:

$$N = A\varepsilon^0 + Bk\,, \tag{9.106}$$

$$M = B\varepsilon^0 + Dk\,, \tag{9.107}$$

or summarized

$$\begin{bmatrix} N \\ M \end{bmatrix} = \begin{bmatrix} A & B \\ B & D \end{bmatrix} \begin{bmatrix} \varepsilon^0 \\ k \end{bmatrix}. \tag{9.108}$$

The matrices A, B and D represent abbreviations for:

$$A = \sum_{k=1}^{N} Q^k (z^k - z^{k-1})\,, \tag{9.109}$$

$$B = \frac{1}{2} \sum_{k=1}^{N} Q^k (z^k - z^{k-1})^2\,, \tag{9.110}$$

$$D = \frac{1}{3} \sum_{k=1}^{N} Q^k (z^k - z^{k-1})^3\,. \tag{9.111}$$

In the case of a layer construction of the composite, which is symmetric to the center plane ($z = 0$), the coupling matrix B disappears.
The general process to describe a composite is:

1. Calculation of the layer stiffness from the engineering constants for each layer in the corresponding layer coordinate system.
2. Potential transformation of each layer stiffness matrix into the layer coordinate system.
3. Calculation of the layer stiffnesses in the composite.
4. Through inversion of the stiffness matrix one can receive the compliance matrices of the composite from

$$\varepsilon^0 = \alpha N + \beta M\,, \tag{9.112}$$

$$k = \beta^{\mathrm{T}} N + \delta M\,, \tag{9.113}$$

whereupon the following matrices are introduced for the abbreviation

$$\alpha = A^{-1} + A^{-1} B \tilde{D}^{-1} B A^{-1}, \tag{9.114}$$

$$\beta = A^{-1} + B \tilde{D}^{-1}, \tag{9.115}$$

$$\beta^{\mathrm{T}} = \tilde{D}^{-1} B A^{-1}, \tag{9.116}$$

$$\delta = \tilde{D}^{-1}, \tag{9.117}$$

$$\tilde{D}^{-1} = D - B A^{-1} B. \tag{9.118}$$

From the above equations, layer deformations can be defined under given external loads. Through reverse transformation the layer deformations result in the corresponding lamina coordinate system and, through the stiffness matrices of the lamina, the *intralaminar* stress results.

9.5 A Finite Element Formulation

Within this chapter a finite element formulation for a *composite* element is derived [3]. The considerations about the general two-dimensional composite serve as the basis. Here the derivation concentrates on the one-dimensional situation, whereupon one needs to distinguish between the following two loading cases:

- The loading occurs in the direction of the center line of the beam. The beam can therefore be described as a bar. Tension and compression loads occur.
- The load occurs perpendicular to the center line of the beam. Bending and shear occur.

9.5.1 The Composite Bar

Figure 9.5 illustrates a composite bar under tensile loading. The general procedure for the determination of the stiffness matrix remains the same. The displacement distribution in the element is approximated via the nodal displacements and the shape functions. In the simplest case the approach is described with a linear approach. The stiffness matrix can be derived via various motivations, for example via the principle of virtual work or via the potential. For the tension bar with homogeneous, isotropic material, constant modulus of elasticity E and cross-sectional area A the stiffness matrix results in:

$$k^{\mathrm{e}} = \frac{EA}{L} \begin{bmatrix} 1 & -1 \\ -1 & 1 \end{bmatrix}. \tag{9.119}$$

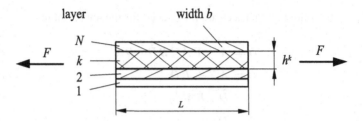

Fig. 9.5 Composite bar under tensile loading

If one also assumes for the derivation of the stiffness matrix for the composite bar that the material properties and the cross-sectional area are constant along the bar axis, a similar formulation results:

$$k^{\mathrm{e}} = \frac{(EA)^{\mathrm{V}}}{L} \begin{bmatrix} 1 & -1 \\ -1 & 1 \end{bmatrix}. \tag{9.120}$$

The expression $(EA)^{\mathrm{V}}$ stands for an axial stiffness, which refers to the unit length. From the comparison with the composite the following results

$$(EA)^{\mathrm{V}} = A_{11} b = b \sum_{k=1}^{N} Q_{11}^{k} h^{k}. \tag{9.121}$$

If the single composite layers each consist of a quasi-homogeneous, quasi-isotropic material, the relation simplifies

$$(EA)^{\mathrm{V}} = A_{11} b = b \sum_{k=1}^{N} E_{1}^{k} h^{k}. \tag{9.122}$$

The context can be simply but precisely interpreted as follows. The macroscopic axial stiffness which is represented of the composite material is composed of the weighted moduli of the elasticity of the single composite layers. At equal width the weights equate to the heights fractions.

Summary

For a composite bar with a symmetric composition throughout the thickness, the stiffness matrix can be derived similarly to the bar with homogeneous, isotropic material.

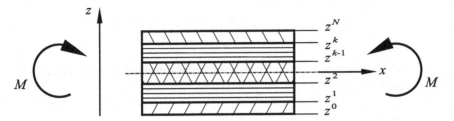

Fig. 9.6 Symmetric composite beam under bending conditions

9.5.2 The Composite Beam

Figure 9.6 illustrates a composite beam under bending condition.

First it needs to be assumed that only bending occurs as the loading condition. Therewith only the matrix D has to be provided in the relation, which reduces to D_{11} for the one-dimensional beam. The connection between bending moment and curvature results as

$$M_1 = D_{11}\,\kappa. \tag{9.123}$$

For a beam of homogeneous, isotropic material the relation between bending moment and curvature appears as follows:

$$M = EI\,\kappa. \tag{9.124}$$

From the comparison a similar formulation for the composite beam can be gained:

$$(EI)^{\mathrm{V}} = b\,D_{11}. \tag{9.125}$$

The expression $(EI)^{\mathrm{V}}$ represents macroscopically the bending stiffness of the composite beam. For a single composite layer the relation is

$$D_{11}^k = Q_{11}^k \frac{h^3}{12} = Q_{11}^k \frac{1}{12}(z^k - z^{k-1})^3, \tag{9.126}$$

whereupon as the absolute layer the $z = 0$ axis was assigned as the center plane. In the composite beam the cross-section dislocates from the 0-layer. Under consideration of the STEINER's fraction (parallel axis theorem)

$$\left(\frac{1}{2}(z^k + z^{k-1})\right)^2 b^k\,h^k = \frac{1}{4}(z^k + z^{k-1})^2\,b^k\,(z^k - z^{k-1}) \tag{9.127}$$

the following relation results:

$$(EI)^{\mathrm{V}} = D_{11}\, b = b\,\frac{1}{3}\sum_{k=1}^{N} Q_{11}^{k}\,((z^{k})^{3} - (z^{k-1})^{3}). \qquad (9.128)$$

If the single composite layers each consist of a homogeneous, isotropic material, the relation simplifies to

$$(EI)^{\mathrm{V}} = D_{11}\, b = b\,\frac{1}{3}\sum_{k=1}^{N} E_{1}^{k}\,((z^{k})^{3} - (z^{k-1})^{3}). \qquad (9.129)$$

The context can be simply but precisely interpreted as follows. The macroscopic bending stiffness which is representative of the composite material is composed of the weighted moduli of the elasticity of the single composite layers. At equal width the weights equate to the heights fractions under consideration of STEINER's fraction (parallel axis theorem) due to the eccentric position of a layer.

Summary

For a composite beam — designed symmetrically in respect to the thickness — the bending stiffness can be derived similarly to the homogeneous, isotropic beam.

9.6 Sample Problems and Supplementary Problems

9.1 Composite Bar with Three Layers

Given is a composite which is constructed symmetrically in height. The three layers are equally thick, which means the same heights h. Figure 9.7 illustrates the composite in the longitudinal section. Each layer consists of a homogeneous, isotropic material. For each layer the modulus of elasticity is given with $E^{(1)}$, $E^{(2)}$ and $E^{(3)} = E^{(1)}$. Furthermore $E^{(2)} = \frac{1}{10}\, E^{(1)}$ should be valid.

All layers in the composite have the same length L and the same width b.

Unknown are

1. the axial stiffness matrix for a load in the longitudinal direction of the composite and
2. the bending stiffness at a bending load. The bending moment is perpendicular to the x–z plane.

Fig. 9.7 Symmetric composite beam with three layers

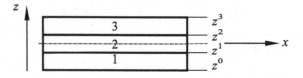

References

1. Altenbach H, Altenbach J, Naumenko K (1998) Ebene Flächentragwerke: Grundlagen der Modellierung und Berechnung von Scheiben und Platten. Springer, Berlin
2. Altenbach H, Altenbach J, Kissing W (2004) Mechanics of composite structural elements. Springer, Berlin
3. Kwon YW, Bang H (2000) The finite element method using MATLAB. CRC Press, Boca Raton

Chapter 10
Nonlinear Elasticity

Abstract The case of the nonlinear elasticity, meaning strain-dependent modulus of elasticity, will be considered within this chapter. The problem will be illustrated with the example of bar elements. First, the stiffness matrix or alternatively the principal finite element equation will be derived under consideration of the strain dependency. For the solving of the nonlinear system of equations three approaches will be derived, namely the direct iteration, the complete NEWTON–RAPHSON iteration and the modified NEWTON–RAPHSON iteration. These approaches will be demonstrated with the help of multiple examples. Within the framework of the complete NEWTON–RAPHSON iteration the derivation of the tangent stiffness matrix will be discussed in detail.

10.1 Introductory Remarks

In the context of the finite element method it is common to distinguish between the following kind of nonlinearities [1]:

- **Physical or material nonlinearities**: This relates to nonlinear material behavior, as, for example, in the elastic area (covered within this chapter) of rubber or elastoplastic behavior (covered in Chap. 11).
- **Nonlinear boundary conditions**: This is, for example, the case that in the course of the load application, a displacement boundary condition changes. Typical for this case are contact problems. This will not be covered within this book.
- **Geometric or kinematic nonlinearity**: This relates to large displacements and rotations at small strains. As examples structure elements such as wires and beams can be named. This will not be covered within this book.
- **Large deformations**: This relates to large displacements, rotations and large strains. This will not be covered within this book.
- **Stability problems**: Here, one has to distinguish between the geometric instability (as, for example, the buckling of bars and plates) and the material instability (as, for

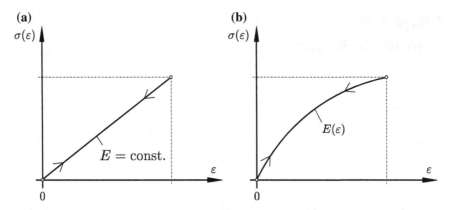

Fig. 10.1 Different behavior in the elastic range: **a** linear ; **b** nonlinear stress-strain diagram

example, the necking of tensile samples or the formation of shear bands). Within this book only the buckling of bars will be covered in Chap. 12.

The basic characteristic of elastic material behavior is that the strains go back to zero completely after unloading.[1] In the case of linear elasticity with a constant modulus of elasticity the loading and unloading takes places in the stress-strain diagram along a straight line, see Fig. 10.1a. The slope of this straight line equals exactly the constant modulus of elasticity E, according to HOOKE's law. In generalization of this linear-elastic behavior the loading and unloading can also take place along a nonlinear curve, and in this case one talks about nonlinear elasticity, see Fig. 10.1b. In this case HOOKE's law is only valid in an incremental or differential form:

$$\frac{\mathrm{d}\sigma(\varepsilon)}{\mathrm{d}\varepsilon} = E(\varepsilon). \tag{10.1}$$

One considers here that the denotation 'linear' or alternatively 'nonlinear' elasticity relates to the behavior of the stress-strain curve. Furthermore the modulus of elasticity can also be dependent on the coordinate. This is, for example, the case of functionally graded materials, the so-called gradient materials. Therefore the modulus of elasticity in general, under the consideration of the kinematic relation, can be indicated as

$$E = E(x, u). \tag{10.2}$$

However, a dependency from the x-coordinate can be treated as a variable cross-section[2] and demands no further analysis at this point. Therefore, in the following, the focus is on dependencies of the form $E = E(u)$ or alternatively $E = E(\frac{\mathrm{d}u}{\mathrm{d}x})$.

---

[1] At plastic material behavior remaining strains occur. This case will be covered in Chap. 11.
[2] For this, see the treatment of bar elements with variable cross-sectional areas $A = A(x)$ in Chap. 3.

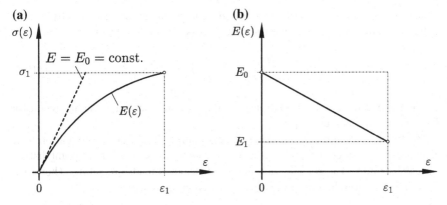

Fig. 10.2 a Nonlinear stress-strain diagram; **b** strain dependent modulus of elasticity

10.2 Element Stiffness Matrix for Strain Dependent Elasticity

The following derivations will be carried out for the example case that the modulus of elasticity is dependent linearly on the strain, see Fig. 10.2. Under this assumption, a nonlinear stress-strain diagram results, see Fig. 10.2a. The linear course of the modulus of elasticity can be defined in the following via the two sampling points $E(\varepsilon = 0) = E_0$ and $E(\varepsilon = \varepsilon_1) = E_1$.

Therefore, the following course of the function for the strain dependent modulus of elasticity results for the two sampling points:

$$E(\varepsilon) = E_0 - \frac{\varepsilon}{\varepsilon_1}(E_0 - E_1) = E_0\Big(1 - \varepsilon \times \underbrace{\frac{1 - E_1/E_0}{\varepsilon_1}}_{\alpha_{01}}\Big) = E_0(1 - \varepsilon\alpha_{01}). \qquad (10.3)$$

It needs to be remarked at this point that the principal route for the derivation does not change as long as the strain dependency of the modulus of elasticity can be described via a polynomial. This is often the case in practical applications, since experimental values are often approximated through a polynomial regression.

After the introduction of the kinematics relation for a bar, meaning $\varepsilon = \frac{du}{dx}$, herefrom the modulus of elasticity results in dependence of the displacement — or, to be precise, dependence of the derivative of the displacement — in:

$$E(u) = E_0\left(1 - \alpha_{01}\frac{du}{dx}\right). \qquad (10.4)$$

This strain dependent modulus of elasticity can be integrated analytically via the differential HOOKE's law and the following stress distribution results[3]:

$$\sigma(\varepsilon) = E_0\varepsilon - \frac{E_0 - E_1}{2E_0\varepsilon_1}\varepsilon^2 = E_0\varepsilon - \frac{1}{2}\alpha_{01}E_0\varepsilon^2 \,. \tag{10.5}$$

One notes that the classical relations for linear elastic material behavior result for $E_0 = E_1$ or alternatively $\alpha_{01} = 0$.

For the derivation of the element stiffness matrix, the differential equation for a bar has to be considered. For simplification reasons it is assumed at this point that the bar cross-section A is constant and that no distributed loads are acting. Therefore, the following formulation for the differential equation results:

$$A\frac{\mathrm{d}}{\mathrm{d}x}\left(E(u)\frac{\mathrm{d}u}{\mathrm{d}x}\right) = 0\,. \tag{10.6}$$

At first, the case is regarded that $E(u)$ is replaced by the expression according to Eq. (10.4):

$$A\frac{\mathrm{d}}{\mathrm{d}x}\left(E_0\left(1 - \alpha_{01}\frac{\mathrm{d}u}{\mathrm{d}x}\right)\frac{\mathrm{d}u}{\mathrm{d}x}\right) = AE_0\frac{\mathrm{d}}{\mathrm{d}x}\left(\frac{\mathrm{d}u}{\mathrm{d}x} - \alpha_{01}\left(\frac{\mathrm{d}u}{\mathrm{d}x}\right)^2\right) = 0\,. \tag{10.7}$$

After the completion of the differentiation the following expression results for the differential equation, which describes the problem:

$$AE_0\frac{\mathrm{d}^2u(x)}{\mathrm{d}x^2} - 2AE_0\alpha_{01}\frac{\mathrm{d}u(x)}{\mathrm{d}x}\frac{\mathrm{d}^2u(x)}{\mathrm{d}x^2} = 0\,. \tag{10.8}$$

Within the framework of the weighted residual method the inner product results herefrom through multiplication with the weight function $W(x)$ and subsequent integration via the bar length in:

$$\int_0^L W^{\mathrm{T}}(x)\left(AE_0\frac{\mathrm{d}^2u(x)}{\mathrm{d}x^2} - 2AE_0\alpha_{01}\frac{\mathrm{d}u(x)}{\mathrm{d}x}\frac{\mathrm{d}^2u(x)}{\mathrm{d}x^2}\right)\mathrm{d}x \overset{!}{=} 0\,. \tag{10.9}$$

Partial integration of the first expression in brackets yields:

---

[3] At this point it was assumed that for $\varepsilon = 0$ the stress turns 0. Therefore, for example no residual stress exists.

$$\int_0^L A E_0 \underbrace{W^{\mathrm{T}}}_{f} \underbrace{\frac{d^2 u}{dx^2}}_{g'} dx = A E_0 \Big[\underbrace{W^{\mathrm{T}}}_{f} \underbrace{\frac{du}{dx}}_{g} \Big]_0^L - \int_0^L A E_0 \underbrace{\frac{dW^{\mathrm{T}}}{dx}}_{f'} \underbrace{\frac{du}{dx}}_{g} dx . \qquad (10.10)$$

Accordingly, the second expression in brackets can be reformulated via partial integration:

$$\int_0^L 2 A E_0 \alpha_{01} \underbrace{\Big(W^{\mathrm{T}} \frac{du}{dx} \Big)}_{f} \underbrace{\frac{d^2 u}{dx^2}}_{g'} dx = 2 A E_0 \alpha_{01} \Big[\underbrace{W^{\mathrm{T}} \frac{du}{dx}}_{f} \underbrace{\frac{du}{dx}}_{g} \Big]_0^L$$

$$- \int_0^L 2 A E_0 \alpha_{01} \underbrace{\frac{d}{dx} \Big(W^{\mathrm{T}} \frac{du}{dx} \Big)}_{f'} \underbrace{\frac{du}{dx}}_{g} dx$$

$$= 2 A E_0 \alpha_{01} \Big[W^{\mathrm{T}} \Big(\frac{du}{dx} \Big)^2 \Big]_0^L - \int_0^L 2 A E_0 \alpha_{01} \Big(\frac{dW^{\mathrm{T}}}{dx} \frac{du}{dx} + W^{\mathrm{T}} \frac{d^2 u}{dx^2} \Big) \frac{du}{dx} dx$$

$$= 2 A E_0 \alpha_{01} \Big[W^{\mathrm{T}} \Big(\frac{du}{dx} \Big)^2 \Big]_0^L - \int_0^L 2 A E_0 \alpha_{01} \frac{dW^{\mathrm{T}}}{dx} \Big(\frac{du}{dx} \Big)^2 dx$$

$$- \int_0^L 2 A E_0 \alpha_{01} W^{\mathrm{T}} \frac{d^2 u}{dx^2} \frac{du}{dx} dx . \qquad (10.11)$$

Finally, the following results for the partial integration of the second expression:

$$\int_0^L 2 A E_0 \alpha_{01} W^{\mathrm{T}} \frac{du}{dx} \frac{d^2 u}{dx^2} dx$$

$$= A E_0 \alpha_{01} \Big[W^{\mathrm{T}} \Big(\frac{du}{dx} \Big)^2 \Big]_0^L - \int_0^L A E_0 \alpha_{01} \frac{dW^{\mathrm{T}}}{dx} \Big(\frac{du}{dx} \Big)^2 dx . \qquad (10.12)$$

The following expression results, when the expressions of the partial integrations according to Eqs. (10.10) and (10.12) are inserted into the inner product according to Eq. (10.9) and when the domain and boundary integrals are arranged:

$$\int\limits_0^L A E_0 \frac{\mathrm{d}W^{\mathrm T}}{\mathrm dx}\frac{\mathrm du}{\mathrm dx}\mathrm dx - \int\limits_0^L A E_0 \alpha_{01} \frac{\mathrm dW^{\mathrm T}}{\mathrm dx}\left(\frac{\mathrm du}{\mathrm dx}\right)^2 \mathrm dx$$

$$= A E_0 \left[W^{\mathrm T}\frac{\mathrm du}{\mathrm dx} - \alpha_{01}W^{\mathrm T}\left(\frac{\mathrm du}{\mathrm dx}\right)^2 \right]_0^L . \tag{10.13}$$

The introduction of the approaches for the displacement and the weight function, meaning $u(x) = N u_{\mathrm p}$ and $W(x) = N(x)\delta u_{\mathrm p}$, leads to the following expression, after elimination of the virtual displacement $\delta u_{\mathrm p}^{\mathrm T}$ and factoring out the displacement vector $u_{\mathrm p}$:

$$A E_0 \int\limits_0^L \left(\frac{\mathrm dN^{\mathrm T}(x)}{\mathrm dx}\frac{\mathrm dN(x)}{\mathrm dx} - \alpha_{01}\frac{\mathrm dN^{\mathrm T}(x)}{\mathrm dx}\left(\frac{\mathrm dN(x)}{\mathrm dx}u_{\mathrm p}\right)\frac{\mathrm dN(x)}{\mathrm dx}\right) \mathrm dx \times u_{\mathrm p}$$

$$= A E_0 \left[\frac{\mathrm dN^{\mathrm T}(x)}{\mathrm dx}\left(\frac{\mathrm du}{\mathrm dx} - \alpha_{01}\left(\frac{\mathrm du}{\mathrm dx}\right)^2\right)\right]_0^L . \tag{10.14}$$

Therefore, in dependence of the nodal displacement $u_{\mathrm p}$ the element stiffness matrix[4] results in:

$$k^{\mathrm e} = A E_0 \int\limits_0^L \left(\frac{\mathrm dN^{\mathrm T}(x)}{\mathrm dx}\frac{\mathrm dN(x)}{\mathrm dx} - \alpha_{01}\left(\frac{\mathrm dN^{\mathrm T}(x)}{\mathrm dx}\frac{\mathrm dN(x)}{\mathrm dx}\right)\left(u_{\mathrm p}\frac{\mathrm dN(x)}{\mathrm dx}\right)\right) \mathrm dx .$$

$$\tag{10.15}$$

If the shape functions are known, the stiffness matrix can be evaluated. The second expression in the outer brackets yields an additional symmetrical expression, which can be superposed to the classical stiffness matrix for linear elastic material behavior. For a constant modulus of elasticity $\alpha_{01} = 0$ results and one receives the classical solution. The following dimensions of the single matrix products results if the bar element has m nodes and therefore m shape functions:

$$\frac{\mathrm dN^{\mathrm T}(x)}{\mathrm dx}\frac{\mathrm dN(x)}{\mathrm dx} \rightarrow m \times m \text{ matrix}, \tag{10.16}$$

$$u_{\mathrm p}\frac{\mathrm dN(x)}{\mathrm dx} \rightarrow m \times m \text{ matrix}, \tag{10.17}$$

$$\left(\frac{\mathrm dN^{\mathrm T}(x)}{\mathrm dx}\frac{\mathrm dN(x)}{\mathrm dx}\right)\left(u_{\mathrm p}\frac{\mathrm dN(x)}{\mathrm dx}\right) \rightarrow m \times m \text{ matrix}. \tag{10.18}$$

---

[4]One considers that the associative law applies for matrix multiplications.

However, in the following an alternative strategy is illustrated, which leads slightly faster to the principal finite element equation. On the basis of the differential equation in the form (10.6), the inner product can be derived without replacing the expression for $E(u)$ a priori:

$$\int_0^L W^T(x) A \frac{d}{dx}\left(E(u(x))\frac{du(x)}{dx}\right) dx \stackrel{!}{=} 0. \tag{10.19}$$

Partial integration yields

$$\int_0^L \underbrace{W^T}_{f} \underbrace{A\frac{d}{dx}\left(E(u)\frac{du}{dx}\right)}_{g'} dx = \left[\underbrace{W^T}_{f} \underbrace{AE(u)\frac{du}{dx}}_{g}\right]_0^L - \int_0^L \underbrace{\frac{dW^T}{dx}}_{f'} \underbrace{AE(u)\frac{du}{dx}}_{g} dx = 0,$$

and the weak form of the problem appears as follows:

$$\int_0^L AE(u)\frac{dW^T}{dx}\frac{du}{dx} dx = \left[AE(u)W^T\frac{du}{dx}\right]_0^L. \tag{10.20}$$

Via the approaches for the displacement and the weight function, the following results herefrom:

$$A \underbrace{\int_0^L E(u)\frac{N^T N}{dx\ dx} dx}_{k^e} \times u_p = \left[AE(u)\frac{du}{dx}\frac{dN^T}{dx}\right]_0^L. \tag{10.21}$$

The right-hand side can be handled according to the procedure in Chap. 3 and yields the column matrix of the external loads. The left-hand side however requires that the modulus of elasticity $E(u)$ is considered appropriately. If the approach for the displacement, meaning $u(x) = N(x)u_p$, is considered in the formulation of the modulus of elasticity according to Eq. (10.4), the following results:

$$E(u_p) = E_0\left(1 - \alpha_{01}\frac{dN}{dx}u_p\right). \tag{10.22}$$

It can be considered at this point that the expression $\frac{dN}{dx} u_p$ yields a scalar parameter. Therefore, the stiffness matrix results in:

$$k^e = A E_0 \underbrace{\int_0^L \left(1 - \alpha_{01} \frac{dN}{dx} u_p \right)}_{\text{scalar}} \frac{dN^T}{dx} \frac{dN}{dx} dx \, . \tag{10.23}$$

This stiffness matrix is — as Eq. (10.15) — symmetric since the symmetric matrix $\frac{dN^T}{dx} \frac{dN}{dx}$ is multiplied by a scalar.

In the following, a bar element with two nodes, meaning linear shape functions, can be considered. Both shape functions and their derivatives in this case result in:

$$N_1(x) = 1 - \frac{x}{L} , \qquad\qquad \frac{dN_1(x)}{dx} = -\frac{1}{L} , \tag{10.24}$$

$$N_2(x) = \frac{x}{L} , \qquad\qquad \frac{dN_2(x)}{dx} = \frac{1}{L} . \tag{10.25}$$

Therefore, the stiffness matrix results in:

$$k^e = A E_0 \int_0^L \left(1 - \alpha_{01} \frac{dN_1}{dx} u_1 - \alpha_{01} \frac{dN_2}{dx} u_2 \right) \begin{bmatrix} \frac{dN_1}{dx} \frac{dN_1}{dx} & \frac{dN_1}{dx} \frac{dN_2}{dx} \\ \frac{dN_2}{dx} \frac{dN_1}{dx} & \frac{dN_2}{dx} \frac{dN_2}{dx} \end{bmatrix} dx , \tag{10.26}$$

or alternatively under consideration of the derivatives of the shape functions

$$k^e = \frac{A E_0}{L^2} \int_0^L \left(1 + \frac{\alpha_{01}}{L} u_1 - \frac{\alpha_{01}}{L} u_2 \right) \begin{bmatrix} 1 & -1 \\ -1 & 1 \end{bmatrix} dx . \tag{10.27}$$

After completion of the integration herefrom the element stiffness matrix results in

$$k^e = \frac{A E_0}{L^2} (L + \alpha_{01} u_1 - \alpha_{01} u_2) \begin{bmatrix} 1 & -1 \\ -1 & 1 \end{bmatrix} \tag{10.28}$$

or the principal finite element equation as:

$$\frac{A E_0}{L^2} (L + \alpha_{01} u_1 - \alpha_{01} u_2) \begin{bmatrix} 1 & -1 \\ -1 & 1 \end{bmatrix} \begin{bmatrix} u_1 \\ u_2 \end{bmatrix} = \begin{bmatrix} F_1 \\ F_2 \end{bmatrix} . \tag{10.29}$$

One considers that for the constant modulus of elasticity, meaning $\alpha_{01} = 0$, the classical solution from Chap. 3 results. For the variable modulus of elasticity the following system of equations results in matrix notation:

$$k^e(u_p) u_p = F^e , \tag{10.30}$$

or, alternatively, with various elements for the total system

$$K(u)u = F .\tag{10.31}$$

Since the stiffness matrix is dependent on the unknown nodal displacements u, a nonlinear system of equations results, which cannot be solved directly through inverting of the stiffness matrix.

10.3 Solving of the Nonlinear System of Equations

The solving of the nonlinear system of equations can be explained in the following for a bar, which is fixed on one side and loaded by a single force F on the other side, with the help of various methods, see Fig. 10.3. The modulus of elasticity is, according to Eq. (10.3), linearly dependent on the strain. First the discretization via one single element takes place, so that, under consideration of the fixed support, a system with one single degree of freedom results. The resulting equations are therefore solely dependent on one variable, the nodal displacement at the loading point. In the following step, one merges to the general case of a system with various degrees of freedom. The illustration takes place via a discretization of the problem according to Fig. 10.3a with two elements and therefore with two degrees of freedom.

For the example according to Fig. 10.3, the following values can be assumed: Geometry: $A = 100 \, \text{mm}^2$, $L = 400 \, \text{mm}$. Material characteristics: $E_0 = 70000 \, \text{MPa}$, $E_1 = 49000 \, \text{MPa}$, $\varepsilon_1 = 0.15$. Load: $F = 800 \, \text{kN}$.

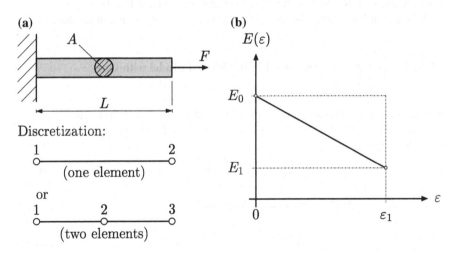

Fig. 10.3 Bar element under point load and strain dependent modulus of elasticity

Fig. 10.4 Schematic
illustration of the direct
iteration

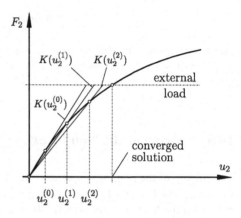

10.3.1 Direct Iteration

At the direct or PICARD's iteration [2, 3], the system of Eq. (10.31) is solved by evaluating the stiffness matrix in the previous and therefore known step. Through the selection of a reasonable initial value — for example from a linear elastic relation — the solution can be determined via the following formula through gradual inserting:

$$K(u^{(j)})u^{(j+1)} = F.$$ (10.32)

The schematic illustration of the direct iteration is shown in Fig. 10.4.

This method converges for modest nonlinearities with linear convergence rate.

10.3.1.1 Direct Iteration for a Finite Element Model with One Unknown

For the example corresponding to Fig. 10.3 and the principal finite element equation according to Eq. (10.29), under consideration of the fixed support, the iteration formula results in:

$$\frac{AE_0}{L^2}\left(L - \alpha_{01}u_2^{(j)}\right)u_2^{(j+1)} = F_2,$$ (10.33)

or alternatively solved for the new displacement:

$$u_2^{(j+1)} = \frac{F_2 L^2}{AE_0\left(L - \alpha_{01}u_2^{(j)}\right)}.$$ (10.34)

The evaluation of Eq. (10.34) for the example corresponding to Fig. 10.3 is summarized in Table 10.1 for an arbitrary initial value of $u_2^{(0)} = 20\,\text{mm}$. The normalized displacement difference was indicated as convergence criteria, whose fulfillment

Table 10.1 Numerical values for the direct iteration in the case of one element with an external load of $F_2 = 800\,\mathrm{kN}$ and an initial value of $u_2^{(0)} = 20\,\mathrm{mm}$. Geometry: $A = 100\,\mathrm{mm}^2$, $L = 400\,\mathrm{mm}$. Material characteristics: $E_0 = 70000\,\mathrm{MPa}$, $E_1 = 49000\,\mathrm{MPa}$, $\varepsilon_1 = 0.15$

Iteration j	$u_2^{(j)}$	$\varepsilon_2^{(j)}$	$\sqrt{\dfrac{\left(u_2^{(j)}-u_2^{(j-1)}\right)^2}{\left(u_2^{(j)}\right)^2}}$
	mm	–	–
0	20.000000	0.050000	–
1	50.793651	0.126984	0.606250
2	61.276596	0.153191	0.171076
3	65.907099	0.164768	0.070258
4	68.183007	0.170458	0.033379
5	69.360231	0.173401	0.016973
6	69.985252	0.174963	0.008931
7	70.321693	0.175804	0.004784
8	70.504137	0.176260	0.002588
9	70.603469	0.176509	0.001407
10	70.657668	0.176644	0.000767
11	70.687276	0.176718	0.000419
12	70.703461	0.176759	0.000229
13	70.712312	0.176781	0.000125
14	70.717152	0.176793	0.000068
15	70.719800	0.176800	0.000037
16	70.721248	0.176803	0.000020
17	70.722041	0.176805	0.000011
18	70.722474	0.176806	0.000006
19	70.722711	0.176807	0.000003
20	70.722841	0.176807	0.000002
21	70.722912	0.176807	0.000001
22	70.722951	0.176807	0.000001
23	70.722972	0.176807	0.000000
$\vdots$	$\vdots$	$\vdots$	$\vdots$
31	70.722998	0.176807	0.000000

requires 23 iterations for a value of 10^{-6}. Furthermore, one considers the absolute value of the displacement at the 31st increment, which is also consulted as a reference value in other methods.

10.3.1.2   Direct Iteration for a Finite Element Model with Various Unknowns

For the application of the direct iteration on a model with various unknowns, the bar, according to Fig. 10.3 can be considered in the following. The discretization should occur through two bar elements, which have the same length. Therefore, the following element stiffness matrix results for each of the two elements with length $\frac{L}{2}$:

$$\frac{4AE_0}{L^2}\left(\frac{L}{2}+\alpha_{01}u_1-\alpha_{01}u_2\right)\begin{bmatrix}1 & -1\\-1 & 1\end{bmatrix} \quad \text{(element I)}, \tag{10.35}$$

$$\frac{4AE_0}{L^2}\left(\frac{L}{2}+\alpha_{01}u_2-\alpha_{01}u_3\right)\begin{bmatrix}1 & -1\\-1 & 1\end{bmatrix} \quad \text{(element II)}. \tag{10.36}$$

The following reduced system of equations results, if the two matrices are summarized to the global principal finite element equation and if the boundary conditions are considered:

$$\frac{4AE_0}{L^2}\begin{bmatrix}(L-\alpha_{01}u_3) & -(\frac{L}{2}+\alpha_{01}u_2-\alpha_{01}u_3)\\-(\frac{L}{2}+\alpha_{01}u_2-\alpha_{01}u_3) & (\frac{L}{2}+\alpha_{01}u_2-\alpha_{01}u_3)\end{bmatrix}\begin{bmatrix}u_2\\u_3\end{bmatrix}=\begin{bmatrix}0\\F_3\end{bmatrix}. \tag{10.37}$$

Through inversion one obtains the following iteration formula of the direct iteration:

$$\begin{bmatrix}u_2\\u_3\end{bmatrix}_{(j+1)}=\frac{\frac{L^2}{4AE_0}}{DET^{(j)}}\begin{bmatrix}(\frac{L}{2}+\alpha_{01}u_2-\alpha_{01}u_3) & (\frac{L}{2}+\alpha_{01}u_2-\alpha_{01}u_3)\\(\frac{L}{2}+\alpha_{01}u_2-\alpha_{01}u_3) & (L-\alpha_{01}u_3)\end{bmatrix}_{(j)}\begin{bmatrix}0\\F_3\end{bmatrix}_{(j)}, \tag{10.38}$$

whereupon the determinant of the reduced stiffness matrix is given through the following equation:

$$DET=(L-\alpha_{01}u_3)\left(\frac{L}{2}+\alpha_{01}u_2-\alpha_{01}u_3\right)-\left(\frac{L}{2}+\alpha_{01}u_2-\alpha_{01}u_3\right)^2. \tag{10.39}$$

In general, the iteration instruction according to Eq. (10.38) can also be written as

$$u^{(j+1)}=\left(K\left(u^{(j)}\right)\right)^{-1}F. \tag{10.40}$$

The numerical results of the iteration for the example according to Fig. 10.3 with two elements are summarized in Table 10.2. A comparison with the direct iteration with one element, meaning Table 10.1, yields that the division in two elements has practically no influence on the convergence behavior. One considers that the displacements on node 2 and 3 are listed in Table 10.2 and that only in the converged situation the condition $u_2=\frac{1}{2}u_3$ results.

Table 10.2 Numerical values for the direct iteration in the case of two elements with an external load of $F_2 = 800\,\text{kN}$ and initial values of $u_2^{(0)} = 10\,\text{mm}$ and $u_3^{(0)} = 20\,\text{mm}$. Geometry: $A = 100\,\text{mm}^2$, $L_I = L_{II} = 200\,\text{mm}$. Material characteristics: $E_0 = 70000\,\text{MPa}$, $E_1 = 49000\,\text{MPa}$, $\varepsilon_1 = 0.15$

Iteration j	$u_2^{(j)}$	$u_3^{(j)}$	$\sqrt{\dfrac{\left(u_2^{(j)}-u_2^{(j-1)}\right)^2+\left(u_3^{(j)}-u_3^{(j-1)}\right)^2}{\left(u_2^{(j)}\right)^2+\left(u_3^{(j)}\right)^2}}$
	mm	mm	–
0	10.000000	20.000000	–
1	28.571429	49.350649	0.706244
2	32.000000	60.852459	0.174565
3	33.613445	65.739844	0.069707
4	34.430380	68.106422	0.032806
5	34.859349	69.3222247	0.016616
6	35.088908	69.9655414	0.008727
7	35.213000	70.3112206	0.004671
8	35.280446	70.4984992	0.002525
9	35.317213	70.6004116	0.001372
10	35.337288	70.6560035	0.000748
⋮	⋮	⋮	⋮
23	35.361489	70.7229715	0.000000
⋮	⋮	⋮	⋮
31	35.361499	70.7229976	0.000000

10.3.2 Complete Newton–Raphson Method

10.3.2.1 Newton's Method for a Function with One Variable

For the definition of the root of a function $f(x)$, meaning $f(x) = 0$, NEWTON's iteration is often used. For the derivation of the iteration method, one develops the function $f(x)$ around the point x_0 in a TAYLOR's series

$$f(x) = f(x_0) + \left(\frac{\mathrm{d}f}{\mathrm{d}x}\right)_{x_0} \cdot (x - x_0) + \frac{1}{2!}\left(\frac{\mathrm{d}^2 f}{\mathrm{d}x^2}\right)_{x_0} \cdot (x - x_0)^2 + \cdots + \frac{1}{k!}\left(\frac{\mathrm{d}^k f}{\mathrm{d}x^k}\right)_{x_0} \cdot (x - x_0)^k \,.$$

(10.41)

If the expressions of quadratic and higher order are disregarded, the following approximation results:

$$f(x) \approx f(x_0) + \left(\frac{\mathrm{d}f}{\mathrm{d}x}\right)_{x_0} \cdot (x - x_0) \,.$$

(10.42)

Fig. 10.5 Development of a function into a TAYLOR's series of first order

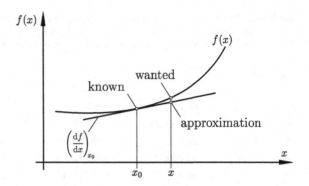

When considering that the derivative of a function equals the slope of the tangent line in the considered point and that the slope-intercept equation of a straight line is given via $f(x) - f(x_0) = m \times (x - x_0)$, one can see that the approximation via a TAYLOR's series of first order is given through the straight line through the point $(x_0, f(x_0))$ with slope $m = (df/dx)_{x_0}$, see Fig. 10.5.

For the derivation of the iteration formula for the definition of the roots, one sets Eq. (10.42) equal 0 and obtains the following calculation instruction via the substitutions $x_0 \to x^{(j)}$ and $x \to x^{(j+1)}$:

$$x^{(j+1)} = x^{(j)} - \frac{f(x^{(j)})}{\left(\frac{df}{dx}\right)_{x^{(j)}}}. \tag{10.43}$$

The principle course of action of a NEWTON's iteration is illustrated in Fig. 10.6. At the initial point of the iteration, the tangent is pictured on the graph of the function $f(x)$ and subsequently the root of this tangent will be defined. In the ordinate value of this root, the next tangent will be formed and the procedure will be continued according to the course of action in the initial point. If $f(x)$ is a continuous and monotonic function in the considered interval and if the initial point of the iteration lies 'close enough' to the unknown solution, the method converges quadratically against the root.

10.3.2.2 Newton–Raphson Method for a Finite Element Model with One Unknown

For the example according to Fig. 10.3, the problem reduces to locating the roots of the function, under consideration of the boundary conditions on the left-hand node

$$r(u_2) = \frac{AE_0}{L^2}(L - \alpha_{01}u_2)\,u_2 - F_2 = K(u_2)u_2 - F_2 = 0. \tag{10.44}$$

Fig. 10.6 Definition of the root of a function via NEWTON's iteration

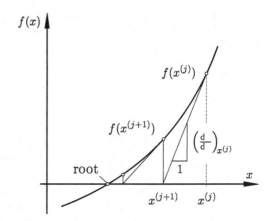

When applying the iteration instruction of the previous Sect. 10.3.2.1 on the residual function $r(u_2)$, the following NEWTON–RAPHSON iteration instruction[5] results in

$$u_2^{(j+1)} = u_2^{(j)} - \frac{r(u_2^{(j)})}{\frac{dr(u_2^{(j)})}{du_2}} = u_2^{(j)} - \left(K_T^{(j)}\right)^{-1} r(u_2^{(j)}), \qquad (10.45)$$

whereupon the parameter K_T is in general referred to as the tangent stiffness matrix.[6] In the example considered at this point, K_T however reduces to a scalar function. On the basis of Eq. (10.44) the tangent stiffness matrix for our example results in:

$$K_T(u_2) = \frac{dr(u_2)}{du_2} = K(u_2) + \frac{dK(u_2)}{du_2} u_2 \qquad (10.46)$$

$$= \frac{AE_0}{L^2}(L - \alpha_{01}u_2) - \frac{AE_0}{L^2}\alpha_{01}u_2$$

$$= \frac{AE_0}{L^2}(L - 2\alpha_{01}u_2) . \qquad (10.47)$$

When using the last result in the iteration instruction (10.45) and when considering the definition of the residual function according to (10.44), the iteration instruction for the regarded example finally results in:

$$u_2^{(j+1)} = u_2^{(j)} - \frac{\frac{AE_0}{L^2}\left(L - \alpha_{01}u_2^{(j)}\right)u_2^{(j)} - F_2^{(j)}}{\frac{AE_0}{L^2}\left(L - 2\alpha_{01}u_2^{(j)}\right)} . \qquad (10.48)$$

---

[5]In the context of the finite element method NEWTON's iteration is often referred to as the NEWTON–RAPHSON iteration [4].

[6]Alternative names in literature are HESSIAN, JACOBIAN or tangent matrix [1].

Table 10.3 Numerical values for the complete NEWTON–RAPHSON method at an external load of $F_2 = 800\,\text{kN}$. Geometry: $A = 100\,\text{mm}^2$, $L = 400\,\text{mm}$. Material behavior: $E_0 = 70000\,\text{MPa}$, $E_1 = 49000\,\text{MPa}$, $\varepsilon_1 = 0.15$

Iteration j	$u_2^{(j)}$	$\varepsilon_2^{(j)}$	$\sqrt{\dfrac{\left(u_2^{(j)}-u_2^{(j-1)}\right)^2}{\left(u_2^{(j)}\right)^2}}$
	mm	–	–
0	0	0	–
1	45.714286	0.114286	1
2	64.962406	0.162406	0.296296
3	70.249443	0.175624	0.075261
4	70.719229	0.176798	0.006643
5	70.722998	0.176807	0.000053
6	70.722998	0.176807	0.000000

The application of the iteration instruction according to Eq. (10.48) with $\alpha_{01} = 2$ leads to the summarized results in Table 10.3. One can see that only six iteration steps are necessary for the complete NEWTON–RAPHSON iteration, due to the quadratic convergence behavior, to achieve the convergence criteria $(<10^{-6})$ and the absolute value of $u_2 = 70.722998\,\text{mm}$. In the general case of the method however, the huge disadvantage arises that the tangent stiffness *matrix* has to be recalculated and inverted for each iteration step. For large systems of equations this leads to quite calculational intensive operations and can perhaps compensate the advantage of the quadratic convergence.

When increasing the external load F_2, a limit value results, however, from which no convergence can be achieved with the NEWTON–RAPHSON method any longer. A strain dependent modulus of elasticity according to Eq. (10.4) leads through integration to the illustrated parabolic stress distribution in Fig. 10.7. Based on this illustration the maximal stress to $\sigma_{\max} = \frac{E_0}{2\alpha_{01}}$ or alternatively the maximal force in a bar to $F_{\max} = \frac{E_0 A}{2\alpha_{01}}$ can be defined.

However, through gradual increasing of the external force F_2 in the regarded example, it results that the convergence limit is achieved clearly lower than the maximal force of $F_{\max} = 1750$ kN. Via a few iteration cycles it can be shown that starting with a value of about 900 kN, no convergence can be achieved any longer in the considered example. One also considers that a reasonable physical choice of the external force always has to meet the condition $F_2 \leq F_{\max}$.

To explain the loss of convergence, the residual function according to Eq. (10.44) has to be considered more closely, whereupon it has to be considered that the iteration method needs to define the roots of this function. The considered residual function is a quadratic function in u_2, which can be changed into the following equation of a parabola by completing the square:

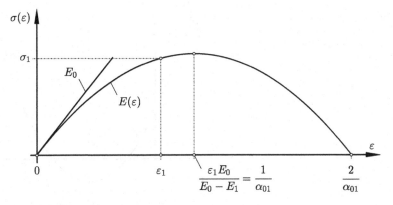

Fig. 10.7 Stress-strain course for a strain dependent modulus of elasticity according to Eq. (10.4)

$$\left(u_2 - \frac{L}{2\alpha_{01}}\right)^2 + \left(\frac{F_2}{E_0 A} - \frac{1}{4\alpha_{01}}\right)\frac{L^2}{\alpha_{01}} = 0. \tag{10.49}$$

Therefore, Eq. (10.44) represents an upward facing parabola with the vertex $\left(\frac{L}{2\alpha_{01}}, \left(\frac{F_2}{E_0 A} - \frac{1}{4\alpha_{01}}\right)\frac{L^2}{\alpha_{01}}\right)$. Dependent on the position of the vertex, a different number of roots results (see Fig. 10.8), so that the boundary value for the convergence of the iteration method is defined through the boundary point of the parabola with the u_2-axis:

$$\frac{F_2}{E_0 A} - \frac{1}{4\alpha_{01}} = 0. \tag{10.50}$$

Therefore, the NEWTON–RAPHSON iteration method for the considered case, that the modulus of elasticity according to Eq. (10.4) is dependent linearly on the strain, converges solely within the following boundaries:

$$F_2 \leq \frac{E_0 A}{4\alpha_{01}}, \quad \text{or alternatively} \quad \varepsilon \leq \frac{1}{2\alpha_{01}}. \tag{10.51}$$

The schematic process of the NEWTON–RAPHSON iteration is illustrated in Fig. 10.9. The tangent stiffness matrix $K_T^{(j)}$ is calculated in every single iteration point $u_2^{(j)}$, to conclude the follow-on value $u_2^{(j+1)}$ via a linearization. It is important at this point that the tangent stiffness matrix can be identified as the derivative in the force-displacement diagram, see Fig. 10.9a. To receive the illustration in a stress-strain diagram, one has to divide the residual Eq. (10.44) through the cross-sectional area and has to scale the displacement with the length, so that one obtains the following form:

Fig. 10.8 Illustration of the residual function according to Eq. (10.44) for different external loads F_2

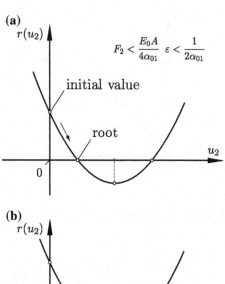

(a)

$$F_2 < \frac{E_0 A}{4\alpha_{01}} \quad \varepsilon < \frac{1}{2\alpha_{01}}$$

initial value

root

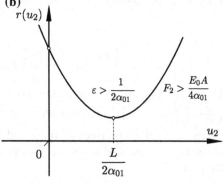

(b)

$$\varepsilon > \frac{1}{2\alpha_{01}} \quad F_2 > \frac{E_0 A}{4\alpha_{01}}$$

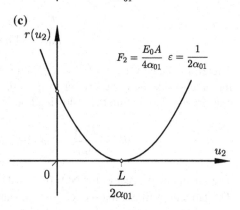

(c)

$$F_2 = \frac{E_0 A}{4\alpha_{01}} \quad \varepsilon = \frac{1}{2\alpha_{01}}$$

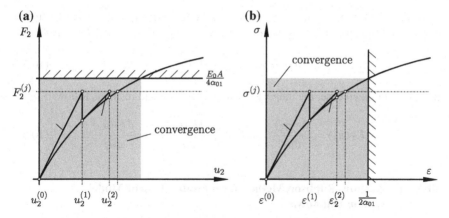

Fig. 10.9 Schematic illustration of the complete NEWTON–RAPHSON iteration

$$E_0\left(1 - \alpha_{01}\frac{u_2}{L}\right)\frac{u_2}{L} - \frac{F_2}{A} = 0 , \tag{10.52}$$

or alternatively in the variables stress and strain as

$$r(\varepsilon) = \underbrace{E_0\left(1 - \alpha_{01}\varepsilon\right)}_{E(\varepsilon)}\varepsilon - \sigma = 0 . \tag{10.53}$$

It is important at this point to note that the last equation is not confused with the stress-strain course according to Eq. (10.5), since the last equation deals with the outer and inner forces. Application of the iteration instruction according to Eq. (10.45) leads to the following formula at this juncture

$$\varepsilon^{(j+1)} = \varepsilon^{(j)} - \frac{r(\varepsilon^{(j)})}{\frac{\mathrm{d}r(\varepsilon^{(j)})}{\mathrm{d}\varepsilon}} , \tag{10.54}$$

whereupon

$$\frac{\mathrm{d}r(\varepsilon)}{\mathrm{d}\varepsilon} = E_{\mathrm{T}} = E(\varepsilon) + \frac{\mathrm{d}E}{\mathrm{d}\varepsilon}\varepsilon , \tag{10.55}$$

$$= E_0(1 - \alpha_{01}\varepsilon) - E_0\alpha_{01}\varepsilon , \tag{10.56}$$

$$= E_0(1 - 2\alpha_{01}\varepsilon) \tag{10.57}$$

is referred to as the consistent modulus E_{T} to the iteration formula. One considers the difference for the continuum mechanical modulus according to Eq. (10.3). Solely in the case of $\alpha_{01} = 0$, meaning for a constant modulus of elasticity, both moduli match.

At this point it needs to be remarked that the residual Eq. (10.44) can be further generalized by introducing a displacement dependent external load $F_2 = F_2(u_2)$:

$$r(u_2) = K(u_2)u_2 - F_2(u_2) = 0. \tag{10.58}$$

In this generalized case, the tangent stiffness matrix would result as follows:

$$K_T(u_2) = \frac{dr(u_2)}{du_2} = K(u_2) + \frac{dK(u_2)}{du_2}u_2 - \frac{dF_2(u_2)}{du_2}. \tag{10.59}$$

10.3.2.3 Newton–Raphson Method for a Finite Element Model with m Unknowns

The complete NEWTON–RAPHSON method [1, 5, 6] for a model with various unknowns is, in general, expressed through the following equation

$$\boldsymbol{u}^{(j+1)} = \boldsymbol{u}^{(j)} - \left(\boldsymbol{K}_T^{(j)}\right)^{-1} \boldsymbol{r}(\boldsymbol{u}^{(j)}), \tag{10.60}$$

whereupon the tangent stiffness matrix in general is defined as

$$\boldsymbol{K}_T = \frac{\partial \boldsymbol{r}(\boldsymbol{u})}{\partial \boldsymbol{u}}. \tag{10.61}$$

The vectorial function of the residuals is generally defined as

$$\boldsymbol{r}(\boldsymbol{u}) = \boldsymbol{K}\boldsymbol{u} - \boldsymbol{F} \tag{10.62}$$

and can be illustrated in components for a model with two linear bar elements as follows:

$$\begin{bmatrix} r_1(\boldsymbol{u}) \\ r_2(\boldsymbol{u}) \\ r_3(\boldsymbol{u}) \end{bmatrix} = \begin{bmatrix} K_{11} & K_{12} & K_{13} \\ K_{21} & K_{22} & K_{23} \\ K_{31} & K_{32} & K_{33} \end{bmatrix} \begin{bmatrix} u_1 \\ u_2 \\ u_3 \end{bmatrix} - \begin{bmatrix} F_1 \\ F_2 \\ F_3 \end{bmatrix}. \tag{10.63}$$

The JACOBIAN matrix $\frac{\partial \boldsymbol{r}}{\partial \boldsymbol{u}}$ of the residual function results in general from the partial derivatives r_i to:

$$\frac{\partial \boldsymbol{r}}{\partial \boldsymbol{u}}(\boldsymbol{u}) = \boldsymbol{K}_T(\boldsymbol{u}) = \begin{bmatrix} K_{T,11} & K_{T,12} & K_{T,13} \\ K_{T,21} & K_{T,22} & K_{T,23} \\ K_{T,31} & K_{T,32} & K_{T,33} \end{bmatrix} = \begin{bmatrix} \frac{\partial r_1}{\partial u_1} & \frac{\partial r_1}{\partial u_2} & \frac{\partial r_1}{\partial u_3} \\ \frac{\partial r_2}{\partial u_1} & \frac{\partial r_2}{\partial u_2} & \frac{\partial r_2}{\partial u_3} \\ \frac{\partial r_3}{\partial u_1} & \frac{\partial r_3}{\partial u_2} & \frac{\partial r_3}{\partial u_3} \end{bmatrix}. \tag{10.64}$$

The partial derivatives in Eq. (10.64) can be calculated the easiest, if the residual Eq. (10.63) are written in detail:

$$r_1(u_1, u_2, u_3) = K_{11}u_1 + K_{12}u_2 + K_{13}u_3 , \qquad (10.65)$$

$$r_2(u_1, u_2, u_3) = K_{21}u_1 + K_{22}u_2 + K_{23}u_3 , \qquad (10.66)$$

$$r_3(u_1, u_2, u_3) = K_{31}u_1 + K_{32}u_2 + K_{33}u_3 . \qquad (10.67)$$

As an example, two partial derivatives are given in the following:

$$\frac{\partial r_1}{\partial u_1} = \left(\frac{\partial K_{11}}{\partial u_1} u_1 + K_{11} \right) + \frac{\partial K_{12}}{\partial u_1} u_2 + \frac{\partial K_{13}}{\partial u_1} u_3 , \qquad (10.68)$$

$$\frac{\partial r_1}{\partial u_2} = \frac{\partial K_{11}}{\partial u_2} u_1 + \left(\frac{\partial K_{12}}{\partial u_2} u_2 + K_{12} \right) + \frac{\partial K_{13}}{\partial u_2} u_2 . \qquad (10.69)$$

Therefore, the tangent stiffness matrix results in the illustrated form in Eq. (10.75), which is composed from the stiffness matrix and a matrix with partial derivatives, which are multiplied with the nodal displacements. In general, the tangent stiffness matrix can therefore be formulated for a model with m degrees of freedom as

$$K_{T,ij} = K_{ij} + \sum_{k=1}^{m} \frac{\partial K_{ik}}{\partial u_j} u_k , \qquad (10.70)$$

or alternatively in matrix notation as

$$\boldsymbol{K}_T = \boldsymbol{K} + \frac{\partial \boldsymbol{K}}{\partial \boldsymbol{u}} \boldsymbol{u}. \qquad (10.71)$$

As a concluding remark, two important special cases need to be listed at this point:

• Scalar tangent stiffness matrix (see Sect. 10.3.2.2):

$$K_T(u) = K(u) + \frac{\mathrm{d}K}{\mathrm{d}u} u . \qquad (10.72)$$

• Two-dimensional tangent stiffness matrix (for example linear bar element without displacement boundary conditions):

$$\boldsymbol{K}_T(\boldsymbol{u}) = \begin{bmatrix} K_{11} & K_{12} \\ K_{21} & K_{22} \end{bmatrix} + \begin{bmatrix} \frac{\partial K_{11}}{\partial u_1} u_1 + \frac{\partial K_{12}}{\partial u_1} u_2 & \frac{\partial K_{11}}{\partial u_2} u_1 + \frac{\partial K_{12}}{\partial u_2} u_2 \\ \frac{\partial K_{21}}{\partial u_1} u_1 + \frac{\partial K_{22}}{\partial u_1} u_2 & \frac{\partial K_{21}}{\partial u_2} u_1 + \frac{\partial K_{22}}{\partial u_2} u_2 \end{bmatrix} . \qquad (10.73)$$

The general case with $\boldsymbol{u} = \begin{bmatrix} u_1, u_2, \ldots, u_m \end{bmatrix}^{\mathrm{T}}$ and $\dim(\boldsymbol{K}) = m \times m$ can easily be derived from the above considerations.

$$K_T = \begin{bmatrix} \dfrac{\partial K_{11}}{\partial u_1}u_1 + \dfrac{\partial K_{12}}{\partial u_1}u_2 + \dfrac{\partial K_{13}}{\partial u_1}u_3 + K_{11} & \dfrac{\partial K_{11}}{\partial u_2}u_1 + \dfrac{\partial K_{12}}{\partial u_2}u_2 + \dfrac{\partial K_{13}}{\partial u_2}u_3 + K_{12} & \dfrac{\partial K_{11}}{\partial u_3}u_1 + \dfrac{\partial K_{12}}{\partial u_3}u_2 + \dfrac{\partial K_{13}}{\partial u_3}u_3 + K_{13} \\[2ex] \dfrac{\partial K_{21}}{\partial u_1}u_1 + \dfrac{\partial K_{22}}{\partial u_1}u_2 + \dfrac{\partial K_{23}}{\partial u_1}u_3 + K_{21} & \dfrac{\partial K_{21}}{\partial u_2}u_1 + \dfrac{\partial K_{22}}{\partial u_2}u_2 + \dfrac{\partial K_{23}}{\partial u_2}u_3 + K_{22} & \dfrac{\partial K_{21}}{\partial u_3}u_1 + \dfrac{\partial K_{22}}{\partial u_3}u_2 + \dfrac{\partial K_{23}}{\partial u_3}u_3 + K_{23} \\[2ex] \dfrac{\partial K_{31}}{\partial u_1}u_1 + \dfrac{\partial K_{32}}{\partial u_1}u_2 + \dfrac{\partial K_{33}}{\partial u_1}u_3 + K_{31} & \dfrac{\partial K_{31}}{\partial u_2}u_1 + \dfrac{\partial K_{32}}{\partial u_2}u_2 + \dfrac{\partial K_{33}}{\partial u_2}u_3 + K_{32} & \dfrac{\partial K_{31}}{\partial u_3}u_1 + \dfrac{\partial K_{32}}{\partial u_3}u_2 + \dfrac{\partial K_{33}}{\partial u_3}u_3 + K_{33} \end{bmatrix}. \tag{10.74}$$

$$= \begin{bmatrix} K_{11} & K_{12} & K_{13} \\ K_{21} & K_{22} & K_{23} \\ K_{31} & K_{32} & K_{33} \end{bmatrix} + \begin{bmatrix} \dfrac{\partial K_{11}}{\partial u_1}u_1 + \dfrac{\partial K_{12}}{\partial u_1}u_2 + \dfrac{\partial K_{13}}{\partial u_1}u_3 & \dfrac{\partial K_{11}}{\partial u_2}u_1 + \dfrac{\partial K_{12}}{\partial u_2}u_2 + \dfrac{\partial K_{13}}{\partial u_2}u_3 & \dfrac{\partial K_{11}}{\partial u_3}u_1 + \dfrac{\partial K_{12}}{\partial u_3}u_2 + \dfrac{\partial K_{13}}{\partial u_3}u_3 \\[2ex] \dfrac{\partial K_{21}}{\partial u_1}u_1 + \dfrac{\partial K_{22}}{\partial u_1}u_2 + \dfrac{\partial K_{23}}{\partial u_1}u_3 & \dfrac{\partial K_{21}}{\partial u_2}u_1 + \dfrac{\partial K_{22}}{\partial u_2}u_2 + \dfrac{\partial K_{23}}{\partial u_2}u_3 & \dfrac{\partial K_{21}}{\partial u_3}u_1 + \dfrac{\partial K_{22}}{\partial u_3}u_2 + \dfrac{\partial K_{23}}{\partial u_3}u_3 \\[2ex] \dfrac{\partial K_{31}}{\partial u_1}u_1 + \dfrac{\partial K_{32}}{\partial u_1}u_2 + \dfrac{\partial K_{33}}{\partial u_1}u_3 & \dfrac{\partial K_{31}}{\partial u_2}u_1 + \dfrac{\partial K_{32}}{\partial u_2}u_2 + \dfrac{\partial K_{33}}{\partial u_2}u_3 & \dfrac{\partial K_{31}}{\partial u_3}u_1 + \dfrac{\partial K_{32}}{\partial u_3}u_2 + \dfrac{\partial K_{33}}{\partial u_3}u_3 \end{bmatrix} \tag{10.75}$$

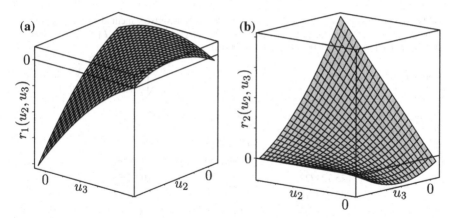

Fig. 10.10 Illustration of the residual functions according to Eq. (10.76)

In the following, the model with two bar elements according to Fig. 10.3 can be considered again. The discretization for two elements with the length $\frac{L}{2}$ leads the residual equation to:

$$
\begin{bmatrix} r_1 \\ r_2 \end{bmatrix} = \frac{4AE_0}{L^2} \begin{bmatrix} (L - \alpha_{01}u_3) & -(\frac{L}{2} + \alpha_{01}u_2 - \alpha_{01}u_3) \\ -(\frac{L}{2} + \alpha_{01}u_2 - \alpha_{01}u_3) & (\frac{L}{2} + \alpha_{01}u_2 - \alpha_{01}u_3) \end{bmatrix} \begin{bmatrix} u_2 \\ u_3 \end{bmatrix}
$$
$$
- \begin{bmatrix} 0 \\ F_3 \end{bmatrix} = 0 . \tag{10.76}
$$

A graphical illustration of the residual functions according to Eq. (10.76) is given in Fig. 10.10. Both functions are dependent on two variables, u_2 and u_3, in this case, and therefore at this point, surfaces in the space result, whose intersection curves have to be found via the u_2–u_3 planes. For this purpose, a tangent plane is built on the corresponding surface in every single point within the iteration scheme.

The application of the calculation instruction according to Eq. (10.73) leads to the tangent stiffness matrix as follows in this special case:

$$
K_\mathrm{T} = \frac{4AE_0}{L^2} \begin{bmatrix} (L - \alpha_{01}u_3) & -(\frac{L}{2} + \alpha_{01}u_2 - \alpha_{01}u_3) \\ -(\frac{L}{2} + \alpha_{01}u_2 - \alpha_{01}u_3) & (\frac{L}{2} + \alpha_{01}u_2 - \alpha_{01}u_3) \end{bmatrix}
$$
$$
+ \frac{4AE_0}{L^2} \begin{bmatrix} 0 - \alpha_{01}u_3 & -\alpha_{01}u_2 + \alpha_{01}u_3 \\ -\alpha_{01}u_2 + \alpha_{01}u_3 & \alpha_{01}u_2 - \alpha_{01}u_3 \end{bmatrix} . \tag{10.77}
$$

The two matrices in the last equation can still be summarized and one obtains the following illustration for the tangent stiffness matrix:

$$
K_\mathrm{T} = \frac{4AE_0}{L^2} \begin{bmatrix} L - 2\alpha_{01}u_3 & -\frac{L}{2} - 2\alpha_{01}u_2 + 2\alpha_{01}u_3 \\ -\frac{L}{2} - 2\alpha_{01}u_2 + 2\alpha_{01}u_3 & \frac{L}{2} + 2\alpha_{01}u_2 - 2\alpha_{01}u_3 \end{bmatrix} . \tag{10.78}
$$

Table 10.4 Numerical values for the complete NEWTON–RAPHSON method in the case of two elements with an external load of $F_2 = 800\,\text{kN}$. Geometry: $A = 100\,\text{mm}^2$, $L_\text{I} = L_\text{II} = 200\,\text{mm}$. Material behavior: $E_0 = 70000\,\text{MPa}$, $E_1 = 49000\,\text{MPa}$, $\varepsilon_1 = 0.15$

Iteration j	$u_2^{(j)}$	$u_3^{(j)}$	$\sqrt{\dfrac{\left(u_2^{(j)}-u_2^{(j-1)}\right)^2+\left(u_3^{(j)}-u_3^{(j-1)}\right)^2}{\left(u_2^{(j)}\right)^2+\left(u_3^{(j)}\right)^2}}$
	mm	mm	–
0	0	0	–
1	22.857143	45.714286	1
2	32.481203	64.962406	0.296296
3	35.124722	70.249443	0.075261
4	35.359614	70.719229	0.006643
5	35.361498	70.722998	0.000053
6	35.361499	70.722998	0.000000

The tangent stiffness matrix still has to be inverted[7] for the iteration scheme according to Eq. (10.60) and after a short calculation one obtains:

$$(\boldsymbol{K}_\text{T})^{-1} = \frac{L^2}{4AE_0\left(\frac{L}{2}-2\alpha_{01}u_2\right)}\begin{bmatrix} 1 & 1 \\ 1 & \frac{L-2\alpha_{01}u_3}{\frac{L}{2}+2\alpha_{01}u_2-2\alpha_{01}u_3} \end{bmatrix}. \tag{10.79}$$

Therefore, the iteration scheme $\boldsymbol{u}^{(j+1)} = \boldsymbol{u}^{(j)} - \left(\boldsymbol{K}_\text{T}^{(j)}\right)^{-1}\boldsymbol{r}(\boldsymbol{u}^{(j)})$ can be applied as follows for the example according to Fig. 10.3:

$$\begin{bmatrix} u_2 \\ u_3 \end{bmatrix}_{(j+1)} = \begin{bmatrix} u_2 \\ u_3 \end{bmatrix}_{(j)} - \frac{L^2(4AE_0)^{-1}}{\frac{L}{2}-2\alpha_{01}u_2^{(j)}}\begin{bmatrix} 1 & 1 \\ 1 & \frac{L-2\alpha_{01}u_3}{\frac{L}{2}+2\alpha_{01}u_2-2\alpha_{01}u_3} \end{bmatrix}_{(j)} \times$$

$$\left(\frac{4AE_0}{L^2}\begin{bmatrix} L-\alpha_{01}u_3 & -\frac{L}{2}-\alpha_{01}u_2+\alpha_{01}u_3 \\ -\frac{L}{2}-\alpha_{01}u_2+\alpha_{01}u_3 & \frac{L}{2}+\alpha_{01}u_2-\alpha_{01}u_3 \end{bmatrix}_{(j)}\begin{bmatrix} u_2 \\ u_3 \end{bmatrix}_{(j)} - \begin{bmatrix} 0 \\ F_3 \end{bmatrix}\right). \tag{10.80}$$

The numerical values of the iteration are summarized in Table 10.4. Due to a comparison with the values from Table 10.3 for a model with one single element, one can see that the convergence behavior is identical.

For practical applications however one would not calculate the tangent stiffness matrix of the global total system but the derivatives element by element. Subsequently the tangent stiffness matrices of the single elements — as in the case of the total stiffness matrix — can be assembled together for the tangent stiffness matrix of the global total system:

---

[7]One considers that the calculation of the inverse has to be carried out numerically in commercial programs.

$$K_{\mathrm{T}} = \sum K_{\mathrm{T}}^{\mathrm{e}}. \qquad (10.81)$$

For a linear element with a strain dependent modulus of elasticity according to Eq. (10.3) follows from the stiffness matrix according to Eq. (10.28), meaning

$$k^{\mathrm{e}} = \frac{A E_0}{L^2}(L + \alpha_{01}\, u_1 - \alpha_{01}\, u_2)\begin{bmatrix} 1 & -1 \\ -1 & 1 \end{bmatrix}, \qquad (10.82)$$

under application of the calculation instruction (10.73), the following tangent stiffness matrix for a single element with two nodes:

$$K_{\mathrm{T}}^{\mathrm{e}} = k^{\mathrm{e}} + \begin{bmatrix} \alpha_{01} u_1 - \alpha_{01} u_2 & -\alpha_{01} u_1 + \alpha_{01} u_2 \\ -\alpha_{01} u_1 + \alpha_{01} u_2 & \alpha_{01} u_1 - \alpha_{01} u_2 \end{bmatrix}$$

$$= \frac{A E_0}{L^2}(L + 2\alpha_{01}\, u_1 - 2\alpha_{01}\, u_2)\begin{bmatrix} 1 & -1 \\ -1 & 1 \end{bmatrix}. \qquad (10.83)$$

10.3.3 Modified Newton–Raphson Method

10.3.3.1 Modified Newton–Raphson Method for a Finite Element Model with One Unknown

The disadvantage of the complete NEWTON–RAPHSON method is that the tangent stiffness matrix has to be calculated and inverted subsequently at each iteration step. If the tangent stiffness matrix is only calculated once at the beginning, one attains the modified NEWTON–RAPHSON method [1, 5, 6]. From Eq. (10.45) the modified iteration scheme results in:

$$u_2^{(j+1)} = u_2^{(j)} - \frac{r(u_2^{(j)})}{\frac{\mathrm{d} r(u_2^{(0)})}{\mathrm{d} u_2}} = u_2^{(j)} - \left(K_{\mathrm{T}}^{(0)}\right)^{-1} r(u_2^{(j)}). \qquad (10.84)$$

A schematic illustration is given in Fig. 10.11. One can see that the same initial tangent is used in every iteration step, whereby in comparison with the complete method, more iteration steps result; the method does not converge quadratically anymore but only linearly. However the calculation intensive inversion of the tangent stiffness matrix in every step drops out and the calculation simplifies significantly.

If the iteration instruction of the modified method according to Eq. (10.84) is applied to the problem according to Fig. 10.3, the summarized results in Table 10.5 are obtained. 36 steps are necessary at this point for the fulfillment of the convergence criteria ($<10^{-6}$) and the reference value of $u_2 = 70.722998$ can only be achieved

Fig. 10.11 Schematic
illustration of the modified
NEWTON–RAPHSON iteration

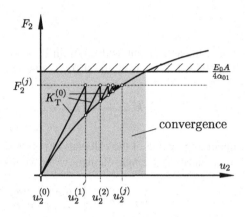

after 53 iteration steps. A comparison with the two other iteration schemes shows
that the modified NEWTON–RAPHSON method — with functions of one variable —
converges the slowest. However one considers that this conclusion does not have to
be valid for a system of equations anymore.

10.3.3.2   Modified Newton–Raphson Method for a Finite Element model with Various Unknowns

The modified NEWTON–RAPHSON method for a model with various unknowns is
generally given through the following equation

$$u^{(j+1)} = u^{(j)} - \left(K_{\mathrm{T}}^{(0)} \right)^{-1} r(u^{(j)}), \tag{10.85}$$

or alternatively for the example according to Fig. 10.3:

$$\begin{bmatrix} u_2 \\ u_3 \end{bmatrix}_{(j+1)} = \begin{bmatrix} u_2 \\ u_3 \end{bmatrix}_{(j)} - \frac{L^2 (4AE_0)^{-1}}{\frac{L}{2} - 2\alpha_{01} u_2^{(0)}} \begin{bmatrix} 1 & 1 \\ 1 & \frac{L - 2\alpha_{01} u_3}{\frac{L}{2} + 2\alpha_{01} u_2 - 2\alpha_{01} u_3} \end{bmatrix}_{(0)} \times$$

$$\left(\frac{4AE_0}{L^2} \begin{bmatrix} L - \alpha_{01} u_3 & -\frac{L}{2} - \alpha_{01} u_2 + \alpha_{01} u_3 \\ -\frac{L}{2} - \alpha_{01} u_2 + \alpha_{01} u_3 & \frac{L}{2} + \alpha_{01} u_2 - \alpha_{01} u_3 \end{bmatrix}_{(j)} \begin{bmatrix} u_2 \\ u_3 \end{bmatrix}_{(j)} - \begin{bmatrix} 0 \\ F_3 \end{bmatrix} \right).$$
$$\tag{10.86}$$

The numerical values of the iteration are summarized in Table 10.6. Due to a comparison with the values from Table 10.5 for the model with one single element, one
can see that the convergence behavior is identical.

Table 10.5 Numerical values for a modified NEWTON–RAPHSON method at an external load of $F_2 = 800$ kN. Geometry: $A = 100$ mm^2, $L = 400$ mm. Material behavior: $E_0 = 70000$ MPa, $E_1 = 49000$ MPa, $\varepsilon_1 = 0.15$

Iteration j	$u_2^{(j)}$	$\varepsilon_2^{(j)}$	$\sqrt{\dfrac{\left(u_2^{(j)} - u_2^{(j-1)}\right)^2}{\left(u_2^{(j)}\right)^2}}$
	mm	–	–
0	0	0	–
1	45.714286	0.114286	1
2	56.163265	0.140408	0.186047
3	61.485848	0.153715	0.086566
4	64.616833	0.161542	0.048455
5	66.590961	0.166477	0.029646
6	67.886066	0.169715	0.019078
7	68.756876	0.171892	0.012665
8	69.351825	0.173380	0.008579
9	69.762664	0.174407	0.005889
10	70.048432	0.175121	0.004080
11	70.248200	0.175621	0.002844
12	70.388334	0.175971	0.001991
13	70.486873	0.176217	0.001398
14	70.556282	0.176391	0.000984
15	70.605231	0.176513	0.000693
16	70.639779	0.176599	0.000489
17	70.664177	0.176660	0.000345
18	70.681416	0.176704	0.000244
19	70.693598	0.176734	0.000172
20	70.702210	0.176756	0.000122
⋮	⋮	⋮	⋮
35	70.722883	0.176807	0.000001
36	70.722916	0.176807	0.000000
⋮	⋮	⋮	⋮
53	70.722998	0.176807	0.000000

Table 10.6 Numerical values for a modified NEWTON–RAPHSON method in the case of two elements with an external load of $F_2 = 800\,\mathrm{kN}$. Geometry: $A = 100\,\mathrm{mm}^2$, $L = 400\,\mathrm{mm}$. Material behavior: $E_0 = 70000\,\mathrm{MPa}$, $E_1 = 49000\,\mathrm{MPa}$, $\varepsilon_1 = 0.15$

Iteration j	$u_2^{(j)}$	$u_3^{(j)}$	$\sqrt{\dfrac{\left(u_2^{(j)}-u_2^{(j-1)}\right)^2+\left(u_3^{(j)}-u_3^{(j-1)}\right)^2}{\left(u_2^{(j)}\right)^2+\left(u_3^{(j)}\right)^2}}$
	mm	mm	–
0	0	0	–
1	22.857143	45.714286	1
2	28.081633	56.163265	0.186046
3	30.742924	61.485848	0.086566
4	32.308416	64.616833	0.048455
5	33.295481	66.590961	0.029646
6	33.943033	67.886066	0.019078
7	34.378438	68.756876	0.012665
8	34.675913	69.351825	0.008579
9	34.881332	69.762664	0.005889
10	35.024216	70.048432	0.004080
⋮	⋮	⋮	⋮
36	35.361458	70.722916	0.000000
⋮	⋮	⋮	⋮
53	35.361499	70.722998	0.000000

10.3.4 Convergence Criteria

For the evaluation, if an iterative scheme converges, the following normalized displacement difference in the form

$$\sqrt{\frac{\left(u_2^{(j)} - u_2^{(j-1)}\right)^2 + \left(u_3^{(j)} - u_3^{(j-1)}\right)^2 + \cdots + \left(u_m^{(j)} - u_m^{(j)}\right)^2}{\left(u_2^{(j)}\right)^2 + \left(u_3^{(j)}\right)^2 + \cdots + \left(u_m^{(j)}\right)^2}} \tag{10.87}$$

was already used in previous chapters, whereupon m represents the number of unknown degrees of freedom. If this value is below a certain limit value, for example the computational accuracy in the program, the iteration can be regarded as converged.

Table 10.7 Calculation procedure in the linear and nonlinear elasticity (N–R = NEWTON–RAPHSON)

Procedure	Calculation instruction
Linear elasticity: $K\,u = F$	
• Inversion of the stiffness matrix	$u = (K)^{-1}\,F$
• ⋯	⋯
Nonlinear Elasticity: $K(u)u = F$	
• Direct iteration	$u^{(j+1)} = \left(K(u^{(j)})\right)^{-1} F$
• Complete N–R iteration	$u^{(j+1)} = u^{(j)} - \left(K_{\mathrm{T}}^{(j)}\right)^{-1} r(u^{(j)})$
• Modified N–R iteration	$u^{(j+1)} = u^{(j)} - \left(K_{\mathrm{T}}^{(0)}\right)^{-1} r(u^{(j)})$
• ⋯	⋯

Alternatively, the residual vector $r^{(j)} = K(u^{(j)})u^{(j)} - F^{(j)}$ can be regarded, whose norm can be indicated as follows

$$\sqrt{\sum_{i=1}^{m}\left(r_i^{(j)}\right)^2}. \tag{10.88}$$

If this norm is below a certain limit value, convergence is achieved.

At the end of this chapter, the discussed iteration instructions are summarized in Table 10.7, and those are opposed to the calculation procedure for linear elasticity.

It needs to be remarked at this point that the three listed procedures for linear elasticity simplify as the method of inversion of the stiffness matrix in the case of linear elasticity.

In the literature, a further series of methods are known, as for example the arc length method, with which the convergence range of the discussed methods here can be expanded significantly [7–9].

10.4 Sample Problems and Supplementary Problems

10.4.1 Sample Problems

10.1 Tension Bar with Quadratic Approach and Strain Dependent Modulus of Elasticity

One needs to derive the stiffness matrix for a bar element with quadratic shape functions for a strain dependent modulus of elasticity in the form

$$E(u) = E_0\left(1 - \alpha_{01}\frac{du}{dx}\right). \tag{10.89}$$

In this, the element has length L and the inner node is placed exactly in the middle of the element. Subsequently one needs to calculate the tangent stiffness matrix $\boldsymbol{K}_T$ based on the stiffness matrix.

Solution 10.1

Based on Eq. (10.23), meaning

$$\boldsymbol{k}^e = AE_0 \int_0^L \underbrace{\left(1 - \alpha_{01}\frac{d\boldsymbol{N}}{dx}\boldsymbol{u}_p\right)}_{\text{scalar}} \frac{d\boldsymbol{N}^T}{dx}\frac{d\boldsymbol{N}}{dx}\,dx\,, \tag{10.90}$$

and the shape functions for a quadratic bar element, or alternatively their derivatives

$$N_1(x) = 1 - 3\frac{x}{L} + 2\left(\frac{x}{L}\right)^2, \qquad \frac{dN_1(x)}{dx} = -\frac{3}{L} + 4\frac{x}{L^2}, \tag{10.91}$$

$$N_2(x) = 4\frac{x}{L} - 4\left(\frac{x}{L}\right)^2, \qquad \frac{dN_2(x)}{dx} = \frac{4}{L} - 8\frac{x}{L^2}, \tag{10.92}$$

$$N_3(x) = -\frac{x}{L} + 2\left(\frac{x}{L}\right)^2, \qquad \frac{dN_3(x)}{dx} = -\frac{1}{L} + 2\frac{x}{L^2}, \tag{10.93}$$

the stiffness matrix in general results in:

$$\boldsymbol{k}^e = AE_0 \int_0^L \left(1 - \alpha_{01}\frac{dN_1}{dx}u_1 - \alpha_{01}\frac{dN_2}{dx}u_2 - \alpha_{01}\frac{dN_3}{dx}u_3\right) \times$$

$$\begin{bmatrix} \frac{dN_1}{dx}\frac{dN_1}{dx} & \frac{dN_1}{dx}\frac{dN_2}{dx} & \frac{dN_1}{dx}\frac{dN_3}{dx} \\ \frac{dN_2}{dx}\frac{dN_1}{dx} & \frac{dN_2}{dx}\frac{dN_2}{dx} & \frac{dN_2}{dx}\frac{dN_3}{dx} \\ \frac{dN_3}{dx}\frac{dN_1}{dx} & \frac{dN_3}{dx}\frac{dN_2}{dx} & \frac{dN_3}{dx}\frac{dN_3}{dx} \end{bmatrix} dx\,. \tag{10.94}$$

After completion of the integration, the element stiffness matrix results herefrom to:

$$\boldsymbol{k}^e = \frac{AE_0}{3L}\begin{bmatrix} 7 & -8 & 1 \\ -8 & 16 & -8 \\ 1 & -8 & 7 \end{bmatrix}$$

$$+ \frac{AE_0\alpha_{01}}{3L^2}\begin{bmatrix} 15u_1 - 16u_2 + u_3 & -16u_1 + 16u_2 & u_1 - u_3 \\ -16u_1 + 16u_2 & 16u_1 - 16u_3 & -16u_2 + 16u_3 \\ ]u_1 - u_3 & -16u_2 + 16u_3 & -u_1 + 16u_2 - 15u_3 \end{bmatrix}. \tag{10.95}$$

Application of the calculation instruction for a (3×3) matrix according to Eq. (10.75) leads to the tangent stiffness matrix as:

$$\boldsymbol{K}_T = \boldsymbol{k}^e + \frac{AE_0\alpha_{01}}{3L^2}\begin{bmatrix} 15u_1 - 16u_2 + u_3 & -16u_1 + 16u_2 & u_1 - u_3 \\ -16u_1 + 16u_2 & 16u_1 - 16u_3 & -16u_2 + 16u_3 \\ u_1 - u_3 & -16u_2 + 16u_3 & -u_1 + 16u_2 - 15u_3 \end{bmatrix},$$

$$(10.96)$$

or, alternatively, after the summarization of the two matrices with the nodal displacements to:

$$\boldsymbol{K}_T^e = \frac{AE_0}{3L}\begin{bmatrix} 7 & -8 & 1 \\ -8 & 16 & -8 \\ 1 & -8 & 7 \end{bmatrix}$$

$$\frac{AE_0\alpha_{01}}{3L^2}\begin{bmatrix} 30u_1 - 32u_2 + 2u_3 & -32u_1 + 32u_2 & 2u_1 - 2u_3 \\ -32u_1 + 32u_2 & 32u_1 - 32u_3 & -32u_2 + 32u_3 \\ 2u_1 - 2u_3 & -32u_2 + 32u_3 & -2u_1 + 32u_2 - 30u_3 \end{bmatrix}.$$

$$(10.97)$$

10.2 One-sided Fixed Tension Bar with Quadratic Approach and Strain Dependent Modulus of Elasticity

With the derived bar element in example 10.1 with quadratic shape functions and strain dependent modulus of elasticity one can calculate a bar, which is fixed supported on the left-hand end and is loaded through a single force of 800 kN on the right-hand end. The material behavior is assumed as in example 10.1, whereupon the values $E_0 = 70000\,\text{MPa}$ and $\alpha_{01} = 2$ can be used. The length of the bar accounts $L = 400\,\text{mm}$ and the cross-sectional area is $A = 100\,\text{mm}^2$. For the solution one can make us of the complete NEWTON–RAPHSON method.

Solution 10.2

Under consideration of the boundary conditions, the principal finite element equation results as follows from Eq. (10.95)

$$\left(\frac{AE_0}{3L}\begin{bmatrix} 16 & -8 \\ -8 & 7 \end{bmatrix} + \frac{AE_0\alpha_{01}}{3L^2}\begin{bmatrix} -16u_3 & -16u_2 + 16u_3 \\ -16u_2 + 16u_3 & 16u_2 - 15u_3 \end{bmatrix} \right)\begin{bmatrix} u_2 \\ u_3 \end{bmatrix} = \begin{bmatrix} 0 \\ F_3 \end{bmatrix},$$

$$(10.98)$$

and from Eq. (10.97) the tangent stiffness matrix follows under consideration of the boundary conditions as

$$\boldsymbol{K}_T^e = \frac{AE_0}{3L}\begin{bmatrix} 16 & -8 \\ -8 & 7 \end{bmatrix} + \frac{AE_0\alpha_{01}}{3L^2}\begin{bmatrix} 32u_1 - 32u_3 & -32u_2 + 32u_3 \\ -32u_2 + 32u_3 & -2u_1 + 32u_2 - 30u_3 \end{bmatrix}.$$

$$(10.99)$$

The tangent stiffness matrix still has to be inverted for the iteration scheme according to Eq. (10.60) and one obtains the following representation after a short calculation:

$$(\boldsymbol{K_T})^{-1} = \frac{3L^2}{AE_0(3L^2 - 12\alpha_{01}u_3L + 64\alpha_{01}^2 u_2 u_3 - 4\alpha_{01}^2 u_3^2 - 64\alpha_{01}^2 u_2^2)}$$

$$\times \begin{bmatrix} \frac{7}{16}L + 2\alpha_{01}u_2 - \frac{15}{8}\alpha_{01}u_3 & \frac{1}{2}L + 2\alpha_{01}u_2 - 2\alpha_{01}u_3 \\ \frac{1}{2}L + 2\alpha_{01}u_2 - 2\alpha_{01}u_3 & L - 2\alpha_{01}u_3 \end{bmatrix}. \qquad (10.100)$$

The numerical results of the iteration are summarized in Table 10.8. A comparison with the results of the discretization with two linear elements in Table 10.4 shows that the results for the regarded case are identical.

10.3 Tension Bar with Three Different Elements for Strain Dependent Modulus of Elasticity and Force Boundary Condition

The illustrated finite element model in Fig. 10.12 of a one-sided fixed bar consists of three elements, which exhibit different characteristics. The bar is loaded with a point load F_0 on the right-hand end.

One considers the case that all three bars have a linear strain dependent modulus of elasticity according to Eq. (10.3) in the form

$$E^i(\varepsilon) = E_0^i(1 - \varepsilon\alpha_{01}), \quad i = \text{I, II, III}. \qquad (10.101)$$

Table 10.8 Numerical values for the complete NEWTON–RAPHSON method in the case of one element with quadratic shape function with one external load of $F_2 = 800\,\text{kN}$. Geometry: $A = 100\,\text{mm}^2$, $L = 400\,\text{mm}$. Material behavior: $E_0 = 70000\,\text{MPa}$, $E_1 = 49000\,\text{MPa}$, $\varepsilon_1 = 0.15$

Iteration j	$u_2^{(j)}$	$u_3^{(j)}$	$\sqrt{\dfrac{\left(u_2^{(j)}-u_2^{(j-1)}\right)^2+\left(u_3^{(j)}-u_3^{(j-1)}\right)^2}{\left(u_2^{(j)}\right)^2+\left(u_3^{(j)}\right)^2}}$
	mm	mm	–
0	0	0	–
1	22.857143	45.714286	1
2	32.481203	64.962406	0.296296
3	35.124722	70.249443	0.075261
4	35.359614	70.719229	0.006643
5	35.361498	70.722998	0.000053
6	35.361499	70.722998	0.000000

Fig. 10.12 Tension bar with three different elements for strain dependent modulus of elasticity and force boundary condition

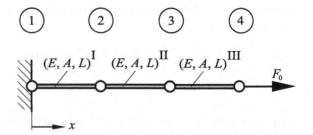

For the considered problem the following relations for the initial axial rigidity can be assumed:

$$(E_0 A)^I = 3 E_0 A, \tag{10.102}$$

$$(E_0 A)^{II} = 2 E_0 A, \tag{10.103}$$

$$(E_0 A)^{III} = 1 E_0 A. \tag{10.104}$$

As a numerical value one can use $F_0 = 800\,\text{kN}$, $A = 100\,\text{mm}^2$, $L^I = L^{II} = L^{III} = 400/3\,\text{mm}$, $E_0 = 70000\,\text{MPa}$, $E_1 = 49000\,\text{MPa}$, $\varepsilon_1 = 0.15$ and one can define the displacement of the nodes with the complete NEWTON–RAPHSON iteration procedure.

Solution 10.3

The element stiffness matrices according to Eq. (10.28) for the three elements result in

$$k^I = \frac{3 E_0 A}{L^2} (L + \alpha_{01} u_1 - \alpha_{01} u_2) \begin{bmatrix} 1 & -1 \\ -1 & 1 \end{bmatrix}, \tag{10.105}$$

$$k^{II} = \frac{2 E_0 A}{L^2} (L + \alpha_{01} u_2 - \alpha_{01} u_3) \begin{bmatrix} 1 & -1 \\ -1 & 1 \end{bmatrix}, \tag{10.106}$$

$$k^{III} = \frac{1 E_0 A}{L^2} (L + \alpha_{01} u_3 - \alpha_{01} u_3) \begin{bmatrix} 1 & -1 \\ -1 & 1 \end{bmatrix}, \tag{10.107}$$

which can be composed to the following reduced system of equations under consideration of the fixed support:

$$\frac{E_0 A}{L^2} \begin{bmatrix} 3L - 3\alpha_{01} u_2 + \\ 2L + 2\alpha_{01} u_2 & -2L - 2\alpha_{01} u_2 & \\ -2\alpha_{01} u_3 & +2\alpha_{01} u_3 & 0 \\ -2L - 2\alpha_{01} u_2 & 2L + 2\alpha_{01} u_2 - 2\alpha_{01} u_3 & -1L - 1\alpha_{01} u_3 \\ +2\alpha_{01} u_3 & +1L + 1\alpha_{01} u_3 - 1\alpha_{01} u_4 & +1\alpha_{01} u_4 \\ & -1L - 1\alpha_{01} u_3 & 1L + 1\alpha_{01} u_3 \\ 0 & +1\alpha_{01} u_4 & -1\alpha_{01} u_4 \end{bmatrix} \begin{bmatrix} u_2 \\ u_3 \\ u_4 \end{bmatrix} = \begin{bmatrix} 0 \\ 0 \\ F_0 \end{bmatrix}. \tag{10.108}$$

The tangent stiffness matrices for the three elements result in the following according to Eq. (10.83)

$$K_T^I = \frac{3 E_0 A}{L^2} (L + 2\alpha_{01} u_1 - 2\alpha_{01} u_2) \begin{bmatrix} 1 & -1 \\ -1 & 1 \end{bmatrix}, \tag{10.109}$$

$$K_T^{II} = \frac{2 E_0 A}{L^2} (L + 2\alpha_{01} u_2 - 2\alpha_{01} u_3) \begin{bmatrix} 1 & -1 \\ -1 & 1 \end{bmatrix}, \tag{10.110}$$

$$K_T^{III} = \frac{1 E_0 A}{L^2} (L + 2\alpha_{01} u_3 - 2\alpha_{01} u_3) \begin{bmatrix} 1 & -1 \\ -1 & 1 \end{bmatrix} \tag{10.111}$$

and can be combined to the following tangent stiffness matrix of the reduced system of equations under consideration of the fixed support:

$$K_T = \frac{E_0 A}{L^2} \begin{bmatrix} 3L - 6\alpha_{01} u_2 + \\ 2L + 4\alpha_{01} u_2 \\ -4\alpha_{01} u_3 & \begin{matrix} -2L - 4\alpha_{01} u_2 \\ +4\alpha_{01} u_3 \end{matrix} & 0 \\ \begin{matrix} -2L - 4\alpha_{01} u_2 \\ +4\alpha_{01} u_3 \end{matrix} & \begin{matrix} 2L + 4\alpha_{01} u_2 - 4\alpha_{01} u_3 \\ +1L + 2\alpha_{01} u_3 - 2\alpha_{01} u_4 \\ -1L - 2\alpha_{01} u_3 \\ +2\alpha_{01} u_4 \end{matrix} & \begin{matrix} -1L - 2\alpha_{01} u_3 \\ +2\alpha_{01} u_4 \end{matrix} \\ 0 & & \begin{matrix} 1L + 2\alpha_{01} u_3 \\ -2\alpha_{01} u_4 \end{matrix} \end{bmatrix} . \tag{10.112}$$

The iteration scheme $u^{(j+1)} = u^{(j)} - (K_T^{(j)})^{-1} r(u^{(j)})$ can be used via the reduced system of equations and the tangent stiffness matrix. The numerical results are summarized in Table 10.9.

10.4 Tension Bar with Three Different Elements for Strain Dependent Modulus of Elasticity and Displacement Boundary Condition

The finite element model of an one-sided fixed bar, which is illustrated in Fig. 10.13, consists of three elements, which exhibit different characteristics. A displacement u_0 is given on the right-hand end of the bar.

One can consider the case that all three bars exhibit a linear strain dependent modulus of elasticity according to Eq. (10.3) in the form

$$E^i(\varepsilon) = E_0^i (1 - \varepsilon \alpha_{01}), \quad i = I, II, III. \tag{10.113}$$

Table 10.9 Numerical values for the complete NEWTON–RAPHSON method in the case of three elements with an external load of $F_2 = 800\,\text{kN}$. Geometry: $A^i = 100\,\text{mm}^2$, $L^i = 400/3\,\text{mm}$. Material behavior: $E_0 = \beta^i \times 70000$ MPa, $E_1 = 49000\,\text{MPa}$, $\varepsilon_1 = 0.15$

Iteration j	$u_2^{(j)}$	$u_3^{(j)}$	$u_4^{(j)}$	$\sqrt{\dfrac{\sum_{i=1}^{3} \left(u_i^{(j)} - u_i^{(j-1)}\right)^2}{\sum_{i=1}^{3} \left(u_i^{(j)}\right)^2}}$
	mm	mm	mm	–
0	0	0	0	–
1	5.079365	12.698413	27.936508	1
2	5.535937	14.283733	35.937868	0.209121
3	5.539687	14.313393	37.729874	0.044001
4	5.539687	14.313407	37.886483	0.003831
5	5.539687	14.313407	37.887740	0.000030
6	5.539687	14.313407	37.887740	0.000000

Fig. 10.13 Tension bar with three different elements for strain dependent modulus of elasticity and displacement boundary condition

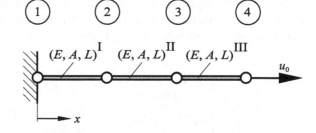

The following relations for the initial axial rigidity can be assumed for the considered problem:

$$(E_0 A)^{\mathrm{I}} = \beta^{\mathrm{I}} E_0 A \,, \tag{10.114}$$

$$(E_0 A)^{\mathrm{II}} = \beta^{\mathrm{II}} E_0 A \,, \tag{10.115}$$

$$(E_0 A)^{\mathrm{III}} = \beta^{\mathrm{III}} E_0 A \,, \tag{10.116}$$

whereupon two different cases need to be analyzed:

	β^{I}	β^{II}	β^{III}	u_0 in mm
Case a)	1	1	1	33
Case b)	3	2	1	37.887740

$$(10.117)$$

As further numerical values one can use $A = 100\,\mathrm{mm}^2$, $L^{\mathrm{I}} = L^{\mathrm{II}} = L^{\mathrm{III}} = 400/3\,\mathrm{mm}$, $E_0 = 70000\,\mathrm{MPa}$, $E_1 = 49000\,\mathrm{MPa}$, $\varepsilon_1 = 0.15$ and one can define the displacement of the nodes and the reaction force on the right-hand end via the complete NEWTON–RAPHSON iteration method.

Solution 10.4

According to the procedure in example 10.3, the total stiffness matrix results as follows, under consideration of the fixed support on the left-hand end:

$$\frac{E_0 A}{L^2}
\begin{bmatrix}
\begin{array}{l} \beta^{\mathrm{I}}L - \beta^{\mathrm{I}}\alpha_{01}u_2 + \\ \beta^{\mathrm{II}}L + \beta^{\mathrm{II}}\alpha_{01}u_2 \\ -\beta^{\mathrm{II}}\alpha_{01}u_3 \end{array} &
\begin{array}{l} -\beta^{\mathrm{II}}L - \beta^{\mathrm{II}}\alpha_{01}u_2 \\ +\beta^{\mathrm{II}}\alpha_{01}u_3 \end{array} & 0 \\[2em]
\begin{array}{l} -\beta^{\mathrm{II}}L - \beta^{\mathrm{II}}\alpha_{01}u_2 \\ +\beta^{\mathrm{II}}\alpha_{01}u_3 \end{array} &
\begin{array}{l} \beta^{\mathrm{II}}L + \beta^{\mathrm{II}}\alpha_{01}u_2 \\ -\beta^{\mathrm{II}}\alpha_{01}u_3 + \beta^{\mathrm{III}}L \\ +\beta^{\mathrm{III}}\alpha_{01}u_3 - \beta^{\mathrm{III}}\alpha_{01}u_4 \end{array} &
\begin{array}{l} -\beta^{\mathrm{III}}L - \beta^{\mathrm{III}}\alpha_{01}u_3 \\ +\beta^{\mathrm{III}}\alpha_{01}u_4 \end{array} \\[2em]
0 &
\begin{array}{l} -\beta^{\mathrm{III}}L - \beta^{\mathrm{III}}\alpha_{01}u_3 \\ +\beta^{\mathrm{III}}\alpha_{01}u_4 \end{array} &
\begin{array}{l} \beta^{\mathrm{III}}L + \beta^{\mathrm{III}}\alpha_{01}u_3 \\ -\beta^{\mathrm{III}}\alpha_{01}u_4 \end{array}
\end{bmatrix}.$$

$$(10.118)$$

If the known displacement is brought to the 'right-hand side' of the system of equations, the following reduced (2×2) system of equations results after canceling of the column and row, which belong to u_4:

$$\frac{E_0 A}{L^2}\begin{bmatrix} \beta^I L - \beta^I \alpha_{01} u_2 + & -\beta^{II} L - \beta^{II} \alpha_{01} u_2 \\ \beta^{II} L + \beta^{II} \alpha_{01} u_2 & +\beta^{II} \alpha_{01} u_3 \\ -\beta^{II} \alpha_{01} u_3 & \\ -\beta^{II} L - \beta^{II} \alpha_{01} u_2 & \beta^{II} L + \beta^{II} \alpha_{01} u_2 \\ +\beta^{II} \alpha_{01} u_3 & -\beta^{II} \alpha_{01} u_3 + \beta^{III} L \\ & +\beta^{III} \alpha_{01} u_3 - \beta^{III} \alpha_{01} u_4 \end{bmatrix}\begin{bmatrix} u_2 \\ u_3 \end{bmatrix}$$

$$= \frac{E_0 A}{L^2}\begin{bmatrix} 0 \\ -(-\beta^{III} L - \beta^{III} \alpha_{01} u_3 + \beta^{III} \alpha_{01} u_4)u_4 \end{bmatrix}. \tag{10.119}$$

According to the procedure in example 10.3 the tangent stiffness matrix results in (2×2) form in:

$$\boldsymbol{K}_\mathrm{T} = \frac{E_0 A}{L^2}\begin{bmatrix} \beta^I L - 2\beta^I \alpha_{01} u_2 + & -\beta^{II} L - 2\beta^{II} \alpha_{01} u_2 \\ \beta^{II} L + 2\beta^{II} \alpha_{01} u_2 & +2\beta^{II} \alpha_{01} u_3 \\ -2\beta^{II} \alpha_{01} u_3 & \\ -\beta^{II} L - 2\beta^{II} \alpha_{01} u_2 & \beta^{II} L + 2\beta^{II} \alpha_{01} u_2 - 2\beta^{II} \alpha_{01} u_3 \\ +2\beta^{II} \alpha_{01} u_3 & +\beta^I L + 2\beta^I \alpha_{01} u_3 - 2\beta^I \alpha_{01} u_4 \end{bmatrix}. \tag{10.120}$$

The iteration scheme $\boldsymbol{u}^{(j+1)} = \boldsymbol{u}^{(j)} - (\boldsymbol{K}_\mathrm{T}^{(j)})^{-1}\boldsymbol{r}(\boldsymbol{u}^{(j)})$ can be used due to the reduced system of equations and the tangent stiffness matrix. The reaction force F_{r4} on the right-hand end can be calculated after each iteration step by evaluating the 4th equation of the total system. The numerical results are summarized in Tables 10.10 and 10.11.

Case (a) with the results in Table 10.10 can be considered as a test case for the iteration scheme. Due to the displacement boundary condition on the right-hand end and the identical elements, the iteration needs to result in $u_2 = \frac{1}{3} u_0$ and $u_3 = \frac{2}{3} u_0$ at this point. As can be seen from Table 10.10, this is the case after five iterations for the chosen convergence criteria. The case (b) with the results in Table 10.11 represents the reversion of example 10.3. Since the result for the displacement from example 10.3 has been brought up as a boundary condition, the reaction force in the converged condition has to achieve a value of 800 kN. This is the case after four iteration steps.

Table 10.10 Numerical values for the complete NEWTON–RAPHSON method in the case of three elements with displacement boundary conditions of $u_0 = 33$ mm. Geometry: $A^i = 100$ mm^2, $L^i = 400/3$ mm. Material behavior: $E_0 = \beta^i \times 70000$ MPa, $E_1 = 49000$ MPa, $\varepsilon_1 = 0.15$, $\beta^I = \beta^{II} = \beta^{III} = 1$

Iteration j	$u_2^{(j)}$	$u_3^{(j)}$	$F_{r4}^{(j)}$	$\sqrt{\dfrac{\sum_{i=1}^{2}\left(u_i^{(j)}-u_i^{(j-1)}\right)^2}{\sum_{i=1}^{2}\left(u_i^{(j)}\right)^2}}$
	mm	mm	kN	–
0	0	0	0	–
1	16.338235	32.676471	16.902865	1
2	11.514910	23.029821	445.153386	0.418876
3	11.005802	22.011604	481.804221	0.046258
4	11.000001	22.000002	482.212447	0.000527
5	11.000000	22.000000	482.212500	0.000000

Table 10.11 Numerical values for the complete NEWTON–RAPHSON method in the case of three elements with displacement boundary conditions of $u_0 = 37.887740$ mm. Geometry: $A^i = 100$ mm^2, $L^i = 400/3$ mm. Material behavior: $E_0 = \beta^i \times 70000$ MPa, $E_1 = 49000$ MPa, $\varepsilon_1 = 0.15$, $\beta^I = 3$, $\beta^{II} = 2$, $\beta^{III} = 1$

Iteration j	$u_2^{(j)}$	$u_3^{(j)}$	$F_{r4}^{(j)}$	$\sqrt{\dfrac{\sum_{i=1}^{2}\left(u_i^{(j)}-u_i^{(j-1)}\right)^2}{\sum_{i=1}^{2}\left(u_i^{(j)}\right)^2}}$
	mm	mm	kN	–
0	0	0	0	–
1	6.152350	15.380875	782.695217	1
2	5.539025	14.319014	799.913803	0.079871
3	5.539687	14.313407	800.000003	0.000368
4	5.539687	14.313407	800.000000	0.000000

10.4.2 Supplementary Problems

10.5 Strain Dependent Modulus of Elasticity with Quadratic Course

The strain dependent modulus of elasticity, which is illustrated in Fig. 10.14 was defined by experiment. Approximate the course with a quadratic function of the form $E(\varepsilon) = a + b\varepsilon + c\varepsilon^2$ and define the constants $a, \ldots, c$. Subsequently, calculate the stress-strain course through integration and illustrate the course graphically.

In the next step, derive the element stiffness matrix for a linear bar element under consideration of the strain dependent modulus of elasticity. In the last step, define the tangent stiffness matrix.

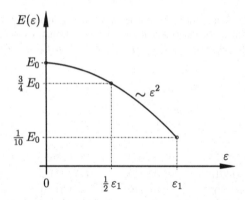

Fig. 10.14 Experimentally determined strain dependent modulus of elasticity

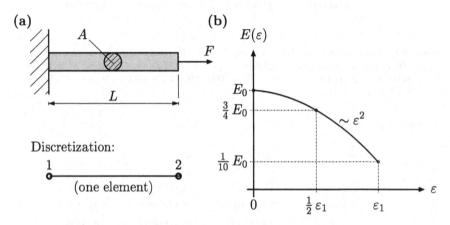

Fig. 10.15 Bar element under point load and quadratic strain dependency of the modulus of elasticity

10.6 Direct Iteration with Different Initial Values

Discretize the bar according to Fig. 10.3 with one single linear element and use the direct iteration for the solution at different initial values: $u_2^{(0)} = 0$ or 30 or 220 mm. Further data can be taken from Table 10.1.

10.7 Complete Newton–Raphson Scheme for a Linear Element with Quadratic Modulus of Elasticity

The beam illustrated in Fig. 10.15a can be discretized via one single linear element. The strain dependent modulus of elasticity exhibits a quadratic course according to Fig. 10.15b.

Based on the element stiffness matrix from Problem 10.5, solve the problem with the complete NEWTON–RAPHSON scheme for an external force of $F = 370$ kN. As convergence criteria use a relative displacement difference of $< 10^{-6}$. Subsequently,

Fig. 10.16 Experimentally determined strain dependent modulus of elasticity; general quadratic course

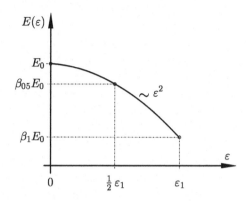

analyze the convergence range of the iteration scheme in general. For the geometry the concrete values $A = 100\,\text{mm}^2$ and $L = 400\,\text{mm}$ and for the material behavior the concrete values $E_0 = 70000\,\text{MPa}$ and $\varepsilon_1 = 0.15$ can be used.

10.8 Strain Dependent Modulus of Elasticity with General Quadratic Course

In extension of Problem 10.5 one can consider the illustrated course in Fig. 10.16 with the three sampling points $(0, E_0)$, $(\frac{1}{2}\varepsilon_1, \beta_{05}E_0)$ and $(\varepsilon_1, \beta_1 E_0)$. The form of the curve can be adapted more easily to the sampling points with the scale values β_{05} and β_1. The curve course can be approximated through a quadratic course in the form $E(\varepsilon) = a + b\varepsilon + c\varepsilon^2$. Define the constants $a, \ldots, c$ and derive the element stiffness matrix for a linear bar element under consideration of the strain dependent modulus of elasticity.

References

1. Wriggers P (2001) Nichtlineare finite-element-methoden. Springer, Berlin
2. Reddy JN (2004) An introduction to nonlinear finite element analysis. Oxford University Press, Oxford
3. Betten J (2004) Finite Elemente für Ingenieure 2: Variationsrechnung, Energiemethoden, Näherungsverfahren, Nichtlinearitäten. Numerische Integrationen, Springer, Berlin
4. Belytschko T, Liu WK, Moran B (2000) Nonlinear finite elements for continua and structures. Wiley, Chichester
5. Cook RD, Malkus DS, Plesha ME, Witt RJ (2002) Concepts and applications of finite element analysis. Wiley, New York
6. Bathe K-J (2002) Finite-elemente-methoden. Springer, Berlin
7. Riks E (1972) The application of newtons method to the problem of elastic stabilty. J Appl Mech 39:1060–1066
8. Crisfield MA (1981) A fast encremental/iterative solution procedure that handles snap through. Comput Struct 13:55–62
9. Schweizerhof K, Wriggers P (1986) Consitent linearization for path following methods in nonlinear fe-analysis. Comput Method Appl M 59:261–279

Chapter 11
Plasticity

Abstract The continuum mechanics basics for the one-dimensional bar will be compiled at the beginning of this chapter. The yield condition, the flow rule, the hardening law and the elasto-plastic modulus will be introduced for uniaxial, monotonic loading conditions. Within the scope of the hardening law the description is limited to isotropic hardening, which occurs for example for the uniaxial tensile test with monotonic loading. For the integration of the elasto-plastic constitutive equation, the incremental predictor-corrector method is generally introduced and derived for the fully implicit and semi-implicit backward-Euler algorithm. On crucial points the difference between one- and three-dimensional descriptions will be pointed out, to guarantee a simple transfer of the derived methods to general problems. Calculated examples and supplementary problems with short solutions serve as an introduction for the theoretical description.

11.1 Continuum Mechanics Basics

The characteristic feature of plastic material behavior is that a plastic strain ε^{pl} occurs after complete unloading, see Fig. 11.1a. Solely the elastic strains ε^{el} return to zero at complete unloading. An additive composition of the strains by their elastic and plastic parts

$$\varepsilon = \varepsilon^{\text{el}} + \varepsilon^{\text{pl}} \tag{11.1}$$

is permitted at restrictions to small strains. The elastic strains ε^{el} can hereby be determined via HOOKE's law, whereby the strain ε in Eq. (3.2) has to be substituted by the elastic strain ε^{el}.

Furthermore, no explicit correlation is given anymore for plastic material behavior in general between stress and strain, since the strain state is also dependent on the loading history. Due to this, rate equations are necessary and need to be integrated throughout the entire load history. Within the framework of the time-independent

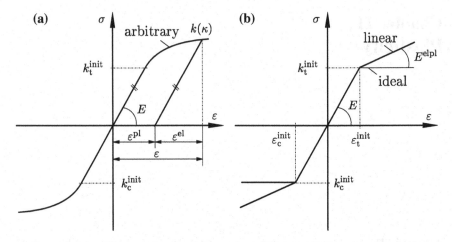

Fig. 11.1 Uniaxial stress-strain diagrams for different isotropic hardening approaches: **a** arbitrary hardening; **b** linear hardening and ideal plasticity

plasticity investigated here, the rate equations can be simplified to incremental relations. From Eq. (11.1) the additive composition of the strain increments results in:

$$\mathrm{d}\varepsilon = \mathrm{d}\varepsilon^{\mathrm{el}} + \mathrm{d}\varepsilon^{\mathrm{pl}} . \tag{11.2}$$

The constitutive description of plastic material behavior includes

- a yield condition,
- a flow rule and
- a hardening law.

In the following, solely the case of the monotonic loading[1] is considered, so that solely the isotropic hardening is considered in the case of the material hardening. This important case, for example, occurs in experimental mechanics at the uniaxial tensile test with monotonic loading. Furthermore, it is assumed that the yield stress is identical in the tensile and compressive regime: $k_{\mathrm{t}} = k_{\mathrm{c}} = k$.

11.1.1 Yield Condition

The yield condition enables one to determine whether the relevant material suffers only elastic or also plastic strains at a certain stress state at a point of the relevant body. In the uniaxial tensile test, plastic flow begins when reaching the initial yield

---

[1]The case of unloading or alternatively load reversal will not be regarded at this point due to simplification reasons.

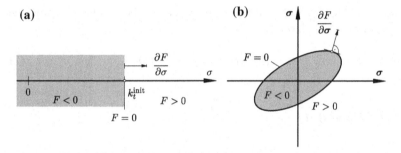

Fig. 11.2 Schematic illustration of the values of the yield condition and the direction of the stress gradient in the **a** one-dimensional and **b** multi-dimensional stress space. Here the σ-σ coordinate system represents a schematic illustration of the n-dimensional stress space

stress k^{init}, see Fig. 11.1. The yield condition in its general one-dimensional form can be set as follows ($\mathbb{R} \times \mathbb{R} \to \mathbb{R}$):

$$F = F(\sigma, \kappa), \tag{11.3}$$

whereupon κ represents the inner variable of the isotropic hardening. In the case of the ideal plasticity, see Fig. 11.1b, the following is valid: $F = F(\sigma)$. The values of F have the following mechanical meaning, see Fig. 11.2:

$$F(\sigma, \kappa) < 0 \quad \to \quad \text{elastic material behavior}, \tag{11.4}$$
$$F(\sigma, \kappa) = 0 \quad \to \quad \text{plastic material behavior}, \tag{11.5}$$
$$F(\sigma, \kappa) > 0 \quad \to \quad \text{invalid}. \tag{11.6}$$

A further simplification results under the assumption that the yield condition can be split into a pure stress fraction $f(\sigma)$, the so-called yield criterion,[2] and into an experimental material parameter $k(\kappa)$, the so-called flow stress:

$$F(\sigma, \kappa) = f(\sigma) - k(\kappa). \tag{11.7}$$

For a uniaxial tensile test (see Fig. 11.1) the yield condition can be noted in the following form:

$$F(\sigma, \kappa) = |\sigma| - k(\kappa) \leq 0. \tag{11.8}$$

If one considers the idealized case of the linear hardening (see Fig. 11.1b), Eq. (11.8) can be written as

$$F(\sigma, \kappa) = |\sigma| - (k^{\text{init}} + E^{\text{pl}}\kappa) \leq 0. \tag{11.9}$$

---

[2]If the unit of the yield criterion equals the stress, $f(\sigma)$ represents the equivalent stress or effective stress. In the general three-dimensional case the following is valid under consideration of the symmetry of the stress tensor $\sigma_{\text{eff}} : (\mathbb{R}^6 \to \mathbb{R}_+)$.

Parameter E^{pl} hereby is the plastic modulus (see Fig. 11.3), which becomes zero in the case of the ideal plasticity:

$$F(\sigma, \kappa) = |\sigma| - k^{\mathrm{init}} \leq 0. \tag{11.10}$$

11.1.2   Flow Rule

The flow rule serves as a mathematical description of the evolution of the infinitesimal increments of the plastic strain $\mathrm{d}\varepsilon^{\mathrm{pl}}$ in the course of the load history of the body. In its most general one-dimensional form, the flow rule can be set up as follows [1]:

$$\mathrm{d}\varepsilon^{\mathrm{pl}} = \mathrm{d}\lambda\, r(\sigma, \kappa), \tag{11.11}$$

whereupon the factor $\mathrm{d}\lambda$ is described as the consistency parameter ($\mathrm{d}\lambda \geq 0$) and $r : (\mathbb{R} \times \mathbb{R} \to \mathbb{R})$ as the function of the flow direction.[3] One considers that solely for $\mathrm{d}\varepsilon^{\mathrm{pl}} = 0$ $\mathrm{d}\lambda = 0$ results. Based on the stability postulate of DRUCKER [2] the following flow rule can be derived[4]:

$$\mathrm{d}\varepsilon^{\mathrm{pl}} = \mathrm{d}\lambda \frac{\partial F(\sigma, \kappa)}{\partial \sigma}. \tag{11.12}$$

Such a flow rule is referred to as the normal rule[5] (see Fig. 11.2a) or due to $r = \partial F(\sigma, \kappa)/\partial \sigma$ as the *associated* flow rule. Experimental results, among other things from the area of the granular materials [4] can however be approximated better if the stress gradient is substituted through a different function, the so-called plastic potential Q. The resulting flow rule is then referred to as the *non-associated* flow rule:

$$\mathrm{d}\varepsilon^{\mathrm{pl}} = \mathrm{d}\lambda \frac{\partial Q(\sigma, \kappa)}{\partial \sigma}. \tag{11.13}$$

In quite complicated yield conditions often the case occurs that a simpler yield condition is used for Q in the first approximation, for which the gradient can easily be determined.

---

[3]In the general three-dimensional case r hereby defines the direction of the vector $\mathrm{d}\varepsilon^{\mathrm{pl}}$, while the scalar factor $\mathrm{d}\lambda$ defines the absolute value.

[4]A formal alternative derivation of the associated flow rule can occur via the LAGRANGE multiplier method as extreme value with side-conditions from the principle of maximum plastic work [3].

[5]In the general three-dimensional case the image vector of the plastic strain increment has to be positioned upright and outside oriented to the yield surface, see Fig. 11.2b.

The application of the associated flow rule (11.12) to the yield conditions according to Eqs. (11.8)–(11.10) yields for all three types of yield conditions (meaning arbitrary hardening, linear hardening and ideal plasticity):

$$d\varepsilon^{\text{pl}} = d\lambda \, \text{sgn}(\sigma) \,, \tag{11.14}$$

whereupon $\text{sgn}(\sigma)$ represents the so-called sign function,[6] which can adopt the following values:

$$\text{sgn}(\sigma) = \begin{cases} -1 & \text{for } \sigma < 0 \\ 0 & \text{for } \sigma = 0 \\ +1 & \text{for } \sigma > 0 \end{cases} . \tag{11.15}$$

11.1.3 Hardening Law

The hardening law allows the consideration of the influence of material hardening on the yield condition and the flow rule. Within isotropic hardening the yield stress is described as dependent on an inner variable κ:

$$k = k(\kappa) \,. \tag{11.16}$$

If the equivalent plastic strain[7] is used for the hardening variable ($\kappa = |\varepsilon^{\text{pl}}|$), then one talks about strain hardening.

Another possibility is to describe the hardening in dependency of the specific[8] plastic work ($\kappa = w^{\text{pl}} = \int \sigma d\varepsilon^{\text{pl}}$). Then one talks about work hardening. If Eq. (11.16) is combined with the flow rule according to (11.14), the evolution equation for the isotropic hardening variable results in:

$$d\kappa = d|\varepsilon^{\text{pl}}| = d\lambda \,. \tag{11.17}$$

Figure 11.3 shows the flow curve, meaning the graphical illustration of the yield stress in dependence on the inner variable for different hardening approaches.

---

[6] Also signum function; from the Latin 'signum' for 'sign'.

[7] The effective plastic strain is in the general three-dimensional case the function $\varepsilon_{\text{eff}}^{\text{pl}} : (\mathbb{R}^6 \to \mathbb{R}_+)$. In the here regarded one-dimensional case the following is valid: $\varepsilon_{\text{eff}}^{\text{pl}} = \sqrt{\varepsilon^{\text{pl}}\varepsilon^{\text{pl}}} = |\varepsilon^{\text{pl}}|$. Attention: Finite element programs optionally use the more general definition for the illustration in the post processor, this means $\varepsilon_{\text{eff}}^{\text{pl}} = \sqrt{\frac{2}{3} \sum \Delta\varepsilon_{ij}^{\text{pl}} \sum \Delta\varepsilon_{ij}^{\text{pl}}}$, which considers the lateral contraction at uniaxial stress problems in the plastic area via the factor $\frac{2}{3}$. However in pure one-dimensional problems *without* lateral contraction, this formula leads to an illustration of the effective plastic strain, which is reduced by the factor $\sqrt{\frac{2}{3}} \approx 0.816$.

[8] This is the volume-specific definition, meaning $[w^{\text{pl}}] = \frac{\text{N}}{\text{m}^2}\frac{\text{m}}{\text{m}} = \frac{\text{kg}\,\text{m}}{\text{s}^2\text{m}^2}\frac{\text{m}}{\text{m}} = \frac{\text{kg}\,\text{m}^2}{\text{s}^2\text{m}^3} = \frac{\text{J}}{\text{m}^3}$.

Fig. 11.3 Flow curve for
different hardening
approaches

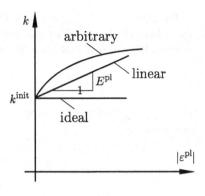

11.1.4 Elasto-Plastic Material Modulus

At plastic material behavior the stiffness of the material changes and the strain state is dependent on the loading history. Therefore, HOOKE's law for the linear-elastic material behavior according to Eq. (3.2) has to be replaced by the following infinitesimal incremental relation:

$$d\sigma = E^{\mathrm{elpl}}d\varepsilon . \tag{11.18}$$

In Eq. (11.18) E^{elpl} refers to the elasto-plastic material modulus (see Fig. 11.1b), which is derived in the following.[9] The total differential of the yield condition (11.8) gives:

$$dF = \left(\frac{\partial F}{\partial \sigma}\right)d\sigma + \left(\frac{\partial F}{\partial \kappa}\right)d\kappa = \mathrm{sgn}(\sigma)d\sigma + \left(\frac{\partial F}{\partial \kappa}\right)d\kappa = 0 . \tag{11.19}$$

If HOOKE's law (3.2) and the flow rule (11.14) are inserted in Eq. (11.2) for the additive composition of the elastic and plastic strain, one obtains:

$$d\varepsilon = \frac{1}{E}d\sigma + d\lambda\,\mathrm{sgn}(\sigma) . \tag{11.20}$$

Multiplication of Eq. (11.20) with $\mathrm{sgn}(\sigma)E$ and integration in Eq. (11.19) yields, under the use of the evolution equation of the hardening variables (11.17), the consistency parameter:

$$d\lambda = \frac{\mathrm{sgn}(\sigma)E}{E - \left(\frac{\partial F}{\partial \kappa}\right)}\,d\varepsilon . \tag{11.21}$$

Insertion of the consistency parameters in Eq. (11.20) and solving for $d\sigma$ finally yields the elasto-plastic material modulus:

---

[9]In the general three-dimensional case one talks about the elasto-plastic stiffness matrix $\boldsymbol{C}^{\mathrm{elpl}}$.

$$E^{\text{elpl}} = \frac{\mathrm{d}\sigma}{\mathrm{d}\varepsilon} = \frac{E \times \left(\frac{\partial F}{\partial \kappa}\right)}{\left(\frac{\partial F}{\partial \kappa}\right) - E}. \tag{11.22}$$

For the special case of linear hardening, meaning $\frac{\partial F}{\partial \kappa} = -E^{\text{pl}}$, Eq. (11.22) can be simplified as follows:

$$E^{\text{elpl}} = \frac{E \times E^{\text{pl}}}{E + E^{\text{pl}}}. \tag{11.23}$$

The different general definitions of the moduli are given comparatively in Table 11.1.

A comparison of the different equations and formulations of the one-dimensional plasticity with the general three-dimensional illustration (see for example [1, 5]) is given in Table 11.2.

Further details regarding plasticity can be taken from, for example, [6–10].

Table 11.1 Comparison of the different definitions of the stress-strain characteristics (moduli)

Range	Definition
Elastic	$E = \dfrac{\mathrm{d}\sigma}{\mathrm{d}\varepsilon^{\text{el}}}$
Plastic	$E^{\text{elpl}} = \dfrac{\mathrm{d}\sigma}{\mathrm{d}\varepsilon}$ for $\varepsilon > \varepsilon^{\text{init}}$
	$E^{\text{pl}} = \dfrac{\mathrm{d}\sigma}{\mathrm{d}|\varepsilon^{\text{pl}}|}$

Table 11.2 Comparison between general 3D plasticity and 1D plasticity with isotropic (arbitrary or ideal) strain hardening

General 3D plasticity	1D plasticity *arbitrary* hardening	1D plasticity *linear* hardening
Yield condition		
$F(\boldsymbol{\sigma}, \boldsymbol{q}) \leq 0$	$F = |\sigma| - k(\kappa) \leq 0$	$F = |\sigma| - (k^{\text{init}} + E^{\text{pl}}\kappa) \leq 0$
Flow rule		
$\boldsymbol{\varepsilon}^{\text{pl}} = \mathrm{d}\lambda \times \boldsymbol{r}(\boldsymbol{\sigma}, \boldsymbol{q})$	$\mathrm{d}\varepsilon^{\text{pl}} = \mathrm{d}\lambda \times \text{sgn}(\sigma)$	$\mathrm{d}\varepsilon^{\text{pl}} = \mathrm{d}\lambda \times \text{sgn}(\sigma)$
Hardening law		
$\boldsymbol{q} = [\kappa, \alpha]^{\text{T}}$	κ	κ
$\mathrm{d}\boldsymbol{q} = \mathrm{d}\lambda \times \boldsymbol{h}(\boldsymbol{\sigma}, \boldsymbol{q})$	$\mathrm{d}\kappa = \mathrm{d}\lambda$	$\mathrm{d}\kappa = \mathrm{d}\lambda$
Elasto-plastic material modulus		
$\boldsymbol{C}^{\text{elpl}} = \left(\boldsymbol{C} - \dfrac{(\boldsymbol{Cr}) \otimes \left(\boldsymbol{C}\frac{\partial F}{\partial \boldsymbol{\sigma}}\right)}{\left(\frac{\partial F}{\partial \boldsymbol{\sigma}}\right)^{\text{T}} \boldsymbol{Cr} - \left(\frac{\partial F}{\partial \boldsymbol{q}}\right)^{\text{T}} \boldsymbol{h}}\right)$	$E^{\text{elpl}} = \dfrac{E \times \left(\frac{\partial F}{\partial \kappa}\right)}{\left(\frac{\partial F}{\partial \kappa}\right) - E}$	$E^{\text{elpl}} = \dfrac{E \times E^{\text{pl}}}{E + E^{\text{pl}}}$

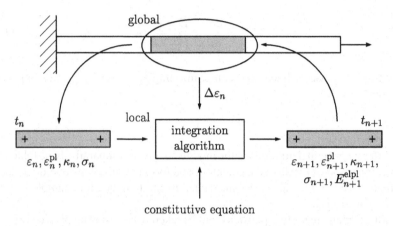

Fig. 11.4 Schematic illustration of the integration algorithm for plastic material behavior in the FEM; adopted from [11]. Integration points are marked schematically through the symbol '+'

11.2 Integration of the Material Equations

In comparison to a finite element calculation with pure linear-elastic material behavior, the calculation at a simulation of plastic material behavior cannot be conducted in one step any longer, since at this point in general no obvious connection between stress and strain exists.[10] The load is instead applied incrementally, whereupon in each increment a nonlinear system of equations has to be solved (for example with the NEWTON–RAPHSON algorithm). The principal finite element equation therefore has to be set in the following incremental form:

$$K \Delta u = \Delta F. \tag{11.24}$$

Additionally, the state variables — as for example the stress σ_{n+1} — have to be calculated for each increment $(n + 1)$ in each integration point (GAUSS point), based on the stress state at the end of the previous increment (n) and the given strain increment $(\Delta \varepsilon_n)$ (see Fig. 11.4).

To that, the explicit material law in infinitesimal form given has to be integrated numerically according to Eqs. (11.2) and (11.22). Explicit integration methods, as for example the EULER procedure[11] however are inaccurate and possibly unstable,

---

[10]In the general case with six stress and strain components (under consideration of the symmetry of the stress and strain tensor) an obvious relation only exists between effective stress and effective plastic strain. In the one-dimensional case however these parameters reduce to: $\sigma_{\text{eff}} = |\sigma|$ and $\varepsilon_{\text{eff}}^{\text{pl}} = |\varepsilon^{\text{pl}}|$.

[11]The explicit EULER procedure or polygon method (also EULER–CAUCHY method) is the most simple procedure for the numerical solution of an initial value problem. The new stress state results according to this procedure in $\sigma_{n+1} = \sigma_n + E_n^{\text{elpl}} \Delta \varepsilon$, whereupon the initial value problem can be named as $\frac{d\sigma}{d\varepsilon} = E^{\text{elpl}}(\sigma, \varepsilon)$ with $\sigma(\varepsilon_0) = \sigma_0$.

since a global error could accumulate [12]. Within the FEM one uses so-called predictor-corrector methods (see Fig. 11.5), in which, first a so-called predictor is explicitly determined and afterwards implicitly corrected instead of explicit integration procedures. In a first step, a test stress state (the so-called trial stress condition) is calculated under the assumption of pure linear-elastic material behavior via an elastic predictor[12]:

$$\sigma_{n+1}^{\text{trial}} = \sigma_n + \underbrace{E \Delta \varepsilon_n}_{\text{predictor } \Delta \sigma_n^{\text{el}}} \,. \tag{11.25}$$

The given hardening condition in this test stress state equals the condition at the end of the previous increment. Therefore, it is assumed that the load step occurs pure elastically, meaning without plastic deformation and therefore without hardening:

$$\kappa_{n+1}^{\text{trial}} = \kappa_n \,. \tag{11.26}$$

Based on the location of the test stress state in the stress space two elementary conditions can be distinguished with the help of the yield condition:

(a) The stress state is in the elastic area (see Fig. 11.5a) or on the yield surface boundary (valid stress state):

$$F(\sigma_{n+1}^{\text{trial}}, \kappa_{n+1}^{\text{trial}}) \leq 0 \,. \tag{11.27}$$

In this case, the test state can be taken on as the new stress/hardening state, since it equals the real state:

$$\sigma_{n+1} = \sigma_{n+1}^{\text{trial}} \,, \tag{11.28}$$

$$\kappa_{n+1} = \kappa_{n+1}^{\text{trial}} \,. \tag{11.29}$$

In conclusion, one proceeds to the next increment.

(b) The stress state is outside the yield surface boundary (invalid stress state), see Fig. 11.5b, c:

$$F(\sigma_{n+1}^{\text{trial}}, \kappa_{n+1}^{\text{trial}}) > 0 \,. \tag{11.30}$$

If this case occurs, a valid state ($F(\sigma_{n+1}, \kappa_{n+1}) = 0$) on the yield surface boundary is calculated in the second part of the procedure from the invalid test state. Therefore, the necessary stress difference

$$\Delta \sigma^{\text{pl}} = \sigma_{n+1}^{\text{trial}} - \sigma_{n+1} \tag{11.31}$$

is referred to as the plastic corrector.

---

[12]In the general three-dimensional case the relation is applied on the stress vector and the increment of the strain vector: $\sigma_{n+1}^{\text{trial}} = \sigma_n + C \Delta \varepsilon_n$.

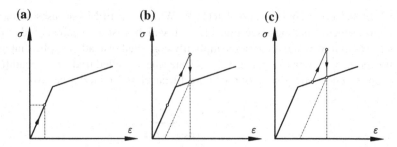

Fig. 11.5 Schematic illustration of the predictor-corrector method in the stress-strain diagram: **a** elastic predictor in the elastic area; **b** and **c** elastic predictor outside the yield surface boundary

For the calculation of the plastic corrector, the terms back projection, return mapping or catching up are used. Figure 11.6 compares the predictor-corrector method schematically in the one- and multi-dimensional stress space.

In the following the back projection is considered closely. Detailed descriptions can be found in [1, 5, 12–15].

The stress difference between initial and final state (stress increment)

$$\Delta\sigma_n = \sigma_{n+1} - \sigma_n \tag{11.32}$$

results, according to HOOKE's law from the elastic part of the strain increment, which results as the difference from the total strain increment and its plastic part:

$$\Delta\sigma_n = E\Delta\varepsilon_n^{el} = E(\Delta\varepsilon_n - \Delta\varepsilon_n^{pl}). \tag{11.33}$$

Figure 11.7 shows that the total strain increment in dependence on the trial stress state can be illustrated as follows:

$$\Delta\varepsilon_n = \varepsilon_{n+1} - \varepsilon_n = \frac{1}{E}(\sigma_{n+1}^{trial} - \sigma_n). \tag{11.34}$$

If the last equation as well as the flow rule[13] according to Eq. (11.14) are inserted in Eq. (11.33), the final stress state σ_{n+1} in dependence on the trial stress state σ_{n+1}^{trial} results in:

$$\sigma_{n+1} = \sigma_{n+1}^{trial} - \Delta\lambda_{n+1}E\,\mathrm{sgn}(\sigma). \tag{11.35}$$

Dependent on the location of the evaluation of the function $\mathrm{sgn}(\sigma)$ different methods result in the general case (see Table 11.3) to calculate the initial value for the plastic corrector or alternatively to define the final stress state iteratively. To obtain an initial value for the plastic corrector, $\mathrm{sgn}(\sigma)$ can either be evaluated in the trial stress state (backward-EULER) or on the yield surface (forward-EULER; at the transition from

---

[13]At this point within the notation it is formally switched from $d\lambda$ to $\Delta\lambda$. Therefore the transition from the differential to the incremental notation occurs.

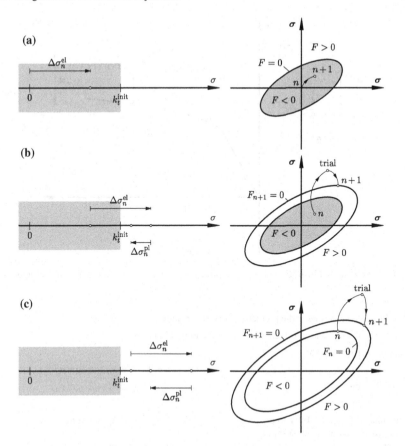

Fig. 11.6 Schematic illustration of the predictor-corrector method in the one-dimensional and multi-dimensional stress space. Here the σ-σ coordinate system represents a schematic illustration of the n-dimensional stress space. **a** elastic predictor in the elastic area; **b** and **c** elastic predictor outside of the yield surface

the elastic to the plastic area this is the initial yield surface, see Fig. 11.5b). If the function $\mathrm{sgn}(\sigma)$ is evaluated in the final state at the iterative calculation, the normal rule (see Sect. 11.1.2) is fulfilled in the final state. For this fully implicitbackward-EULER algorithm (also referred to as the closest point projection (CPP)) [1] however in the general three-dimensional case derivatives of higher order must be calculated. In the so-called cutting-plane algorithm [16] the function $\mathrm{sgn}(\sigma)$ is calculated in the stress state of the ith iteration step. The normal rule is not exactly fulfilled in the final state, however no calculations of derivatives of higher order are necessary. At the so-called mid-point rule [17] the function $\mathrm{sgn}(\sigma)$ is evaluated in the final state and on the yield surface as well as weighted in equal parts. If the function $\mathrm{sgn}(\sigma)$ is only evaluated on the yield surface, this leads to the semi-implicit backward-EULER algorithm [18], for which only derivatives of first order are necessary.

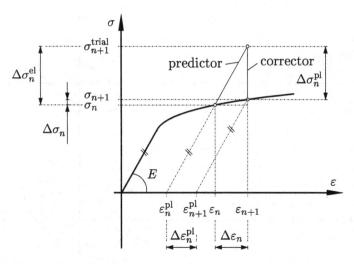

Fig. 11.7 Schematic illustration of the back projection in the stress-strain diagram. Adapted from [1]

Table 11.3 Overview over the predictor-corrector methods

Location of the evaluation of sgn(n)	Equation (11.35)
Initial value for corrector	
Trial condition	$\sigma_{n+1} = \sigma_{n+1}^{\text{trial}} - \Delta\lambda_{n+1} E \, \text{sgn}(\sigma_{n+1}^{\text{trial}})$
On the flow curve	$\sigma_{n+1} = \sigma_{n+1}^{\text{trial}} - \Delta\lambda_{n+1} E \, \text{sgn}(\sigma_n)$
During the iteration	
In the final stress state (fully impl. backward-EULER alg.) (closest point projection)	$\sigma_{n+1} = \sigma_{n+1}^{\text{trial}} - \Delta\lambda_{n+1} E \, \text{sgn}(\sigma_{n+1})$
On the flow curve (semi-impl. backward-EULER alg.)	$\sigma_{n+1} = \sigma_{n+1}^{\text{trial}} - \Delta\lambda_{n+1} E \, \text{sgn}(\sigma_n)$
In the final state and on the flow curve (mid-point rule)	$\sigma_{n+1} = \sigma_{n+1}^{\text{trial}} - \Delta\lambda_{n+1} E \frac{1}{2} \times$ $(\text{sgn}(\sigma_{n+1}) + \text{sgn}(\sigma_n))$
Stress state of ith iteration step (cutting-plane algorithm)	$\sigma_{n+1} = \sigma_{n+1}^{\text{trial}} - \Delta\lambda_{n+1} E \, \text{sgn}(\sigma^{(i)})$

If one considers the dependency of the yield condition from the hardening variable, one needs another equation, which describes the hardening. From the evolution equation of the hardening variables (11.17) the following incremental relation results

$$\kappa_{n+1} = \kappa_n + \Delta\lambda_{n+1} \qquad (11.36)$$

for the definition of the hardening variable.

Finally it can be remarked that three of the listed integration rules in Table 11.3 can be summarized via the following equation

$$\sigma_{n+1} = \sigma_{n+1}^{\text{trial}} - \Delta\lambda_{n+1} E \left([1 - \eta]\text{sgn}(\sigma_n) + \eta\,\text{sgn}(\sigma_{n+1}) \right). \tag{11.37}$$

The parameter η then becomes 1, 0 or $\frac{1}{2}$.

11.3 Derivation of the Fully Implicit Backward-Euler Algorithm

11.3.1 Mathematical Derivation

In this back projection method, the stress location on the yield surface, which is energetically closest to the test state (see Sect. 11.3.2) is calculated. Therefore this does not involve, as the name suggests, a calculation of the closest geometric point. The assumption that the plastic work takes on a maximum at a given strain serves as a basis for the method. Together with the elementary demand that the calculated stress state has to lie on the flow curve (yield surface), the CPP method can be interpreted in the mathematical sense as a solution of an extremal value problem (maximum of the plastic work) with side-conditions (the unknown stress condition has to be on the yield surface) [1]. The method is hereby implicit in the calculation of the function $\text{sgn}(\sigma)$ since the evaluation occurs in the final state $n + 1$. Because of this, the CPP algorithm is also referred to as the fully implicit backward-EULER algorithm. In the final state the following equations are therefore fulfilled:

$$\sigma_{n+1} = \sigma_{n+1}^{\text{trial}} - \Delta\lambda_{n+1} E\,\text{sgn}(\sigma_{n+1})\,, \tag{11.38}$$

$$\kappa_{n+1} = \kappa_n + \Delta\lambda_{n+1}\,, \tag{11.39}$$

$$F = F(\sigma_{n+1}, \kappa_{n+1}) = 0. \tag{11.40}$$

Outside of the final state, however, at each of these equations a residual[14] r remains:

$$
\begin{aligned}
r_\sigma(\sigma, \kappa, \Delta\lambda) &= \sigma - \sigma_{n+1}^{\text{trial}} + \Delta\lambda E\,\text{sgn}(\sigma) \neq 0 \quad \text{or} \\
&= E^{-1}\sigma - E^{-1}\sigma_{n+1}^{\text{trial}} + \Delta\lambda\,\text{sgn}(\sigma) \neq 0\,,
\end{aligned} \tag{11.41}
$$

$$
\begin{aligned}
r_\kappa(\kappa, \Delta\lambda) &= \kappa - \kappa_n - \Delta\lambda \neq 0 \quad \text{or} \\
&= -\kappa + \kappa_n + \Delta\lambda \neq 0\,,
\end{aligned} \tag{11.42}
$$

$$r_F(\sigma, \kappa) = F(\sigma, \kappa) = |\sigma| - k(\kappa) \neq 0\,. \tag{11.43}$$

---

[14]From the Latin 'residuus' for left or remaining.

The unknown stress/hardening state therefore represents the root of a vector function m, which consists of the single residual functions. Furthermore, it seems to make sense to also summarize the arguments for a single vector argument v:

$$m(v) \in (\mathbb{R}^3 \to \mathbb{R}^3) = \begin{bmatrix} r_\sigma(v) \\ r_\kappa(v) \\ r_F(v) \end{bmatrix} , \quad v = \begin{bmatrix} \sigma \\ \kappa \\ \Delta\lambda \end{bmatrix} . \tag{11.44}$$

The NEWTON method (iteration index: i) is used for the definition of the root[15]:

$$v^{(i+1)} = v^{(i)} - \left(\frac{\mathrm{d}m}{\mathrm{d}v}(v^{(i)}) \right)^{-1} m(v^{(i)}) , \tag{11.45}$$

whereupon the following

$$v^{(0)} = \begin{bmatrix} \sigma^{(0)} \\ \kappa^{(0)} \\ \Delta\lambda^{(0)} \end{bmatrix} = \begin{bmatrix} \sigma_{n+1}^{\mathrm{trial}} \\ \kappa_n \\ 0 \end{bmatrix} \tag{11.46}$$

has to be used as the initial value. The JACOBIan matrix $\frac{\partial m}{\partial v}$ of the residual functions results from the partial derivatives of the Eqs. (11.41)–(11.43) to:

$$\frac{\partial m}{\partial v}(\sigma, \kappa, \Delta\lambda) = \begin{bmatrix} \frac{\partial r_\sigma}{\partial\sigma} & \frac{\partial r_\sigma}{\partial\kappa} & \frac{\partial r_\sigma}{\partial\Delta\lambda} \\ \frac{\partial r_\kappa}{\partial\sigma} & \frac{\partial r_\kappa}{\partial\kappa} & \frac{\partial r_\kappa}{\partial\Delta\lambda} \\ \frac{\partial r_F}{\partial\sigma} & \frac{\partial r_F}{\partial\kappa} & \frac{\partial r_F}{\partial\Delta\lambda} \end{bmatrix} = \begin{bmatrix} E^{-1} & 0 & \mathrm{sgn}(\sigma) \\ 0 & -1 & 1 \\ \mathrm{sgn}(\sigma) & -\frac{\partial k(\kappa)}{\partial\kappa} & 0 \end{bmatrix} . \tag{11.47}$$

Next to the fulfillment of Eqs. (11.38)–(11.40), which are given due to plasticity, in each integration point also the global force equilibrium hast to be fulfilled. In order to make use of the NEWTON method, it is even necessary at small strains in the general three-dimensional case to define the elasto-plastic stiffness matrix,[16] which is consistent to the integration algorithm [11]. The consistent elasto-plastic modulus results from the following in the one-dimensional case:

$$E_{n+1}^{\mathrm{elpl}} = \frac{\partial\sigma_{n+1}}{\partial\varepsilon_{n+1}} = \frac{\partial\Delta\sigma_n}{\partial\varepsilon_{n+1}} . \tag{11.48}$$

With the inversion of the JACOBIan matrix $\frac{\partial m}{\partial v}$, which has to be evaluated in the converged condition of the above listed NEWTON iteration,

---

[15]The NEWTON method is usually used as follows for a one-dimensional function: $x^{(i+1)} = x^{(i)} - \left(\frac{\mathrm{d}f}{\mathrm{d}x}(x^{(i)}) \right)^{-1} \times f(x^{(i)})$.

[16]Also referred to as consistent elasto-plastic tangent modulus matrix, consistent tangent stiffness matrix or algorithmic stiffness matrix.

$$\left[\left(\frac{\partial \boldsymbol{m}}{\partial \boldsymbol{v}}\right)_{n+1}\right]^{-1} = \begin{bmatrix} \tilde{m}_{11} & \tilde{m}_{12} & \tilde{m}_{13} \\ \tilde{m}_{21} & \tilde{m}_{22} & \tilde{m}_{23} \\ \tilde{m}_{31} & \tilde{m}_{32} & \tilde{m}_{33} \end{bmatrix}_{n+1} \tag{11.49}$$

$$= \frac{E}{E + \frac{\partial k}{\partial \kappa}} \begin{bmatrix} \frac{\partial k}{\partial \kappa} & -\mathrm{sgn}(\sigma)\frac{\partial k}{\partial \kappa} & \mathrm{sgn}(\sigma) \\ \mathrm{sgn}(\sigma) & -1 & -E^{-1} \\ \mathrm{sgn}(\sigma) & E^{-1}\frac{\partial k}{\partial \kappa} & -E^{-1} \end{bmatrix}_{n+1} \tag{11.50}$$

the elasto-plastic modulus can be defined from

$$E_{n+1}^{\mathrm{elpl}} = \tilde{m}_{11}. \tag{11.51}$$

For this consider Eq. (11.22) and note that under the assumption of Eq. (11.7) the relation $\frac{\partial F}{\partial \kappa} = -\frac{\partial k}{\partial \kappa}$ results. As can be seen from Eq. (11.50), the consistent elasto-plastic modulus in the *one-dimensional* case does not depend on the chosen integration algorithm and equals the continuum form given in Eq. (11.22). However at this point it needs be considered that this identity does not have to exist at higher dimensions any longer.

For the special case of linear hardening, meaning $\frac{\partial k}{\partial \kappa} = E^{\mathrm{pl}} = \mathrm{const.}$, Eq. (11.45) is not to be solved iteratively and the unknown solution vector $\boldsymbol{v}_{n+1}$ results directly with the help of the initial value (11.46) in:

$$\boldsymbol{v}_{n+1} = \boldsymbol{v}^{(0)} - \left(\frac{\mathrm{d}\boldsymbol{m}}{\mathrm{d}\boldsymbol{v}}(\boldsymbol{v}^{(0)})\right)^{-1} \boldsymbol{m}(\boldsymbol{v}^{(0)}), \tag{11.52}$$

or in components as:

$$\begin{bmatrix} \sigma_{n+1} \\ \kappa_{n+1} \\ \Delta\lambda_{n+1} \end{bmatrix} = \begin{bmatrix} \sigma_{n+1}^{\mathrm{trial}} \\ \kappa_n \\ 0 \end{bmatrix} - \frac{E}{E + E^{\mathrm{pl}}} \times$$

$$\times \begin{bmatrix} E^{\mathrm{pl}} & -\mathrm{sgn}(\sigma_{n+1}^{\mathrm{trial}})E^{\mathrm{pl}} & \mathrm{sgn}(\sigma_{n+1}^{\mathrm{trial}}) \\ \mathrm{sgn}(\sigma_{n+1}^{\mathrm{trial}}) & -1 & -E^{-1} \\ \mathrm{sgn}(\sigma_{n+1}^{\mathrm{trial}}) & E^{-1}E^{\mathrm{pl}} & -E^{-1} \end{bmatrix} \begin{bmatrix} 0 \\ 0 \\ F_{n+1}^{\mathrm{trial}} \end{bmatrix}. \tag{11.53}$$

The third equation of (11.53) yields the consistency parameter in the case of the linear hardening to:

$$\Delta\lambda_{n+1} = \frac{F_{n+1}^{\mathrm{trial}}}{E + E^{\mathrm{pl}}}. \tag{11.54}$$

The insertion of the consistency parameter into the first equation of (11.53) yields the stress in the final stress state to:

$$\sigma_{n+1} = \left(1 - \frac{F_{n+1}^{\mathrm{trial}}}{E + E^{\mathrm{pl}}} \times \frac{E}{|\sigma_{n+1}^{\mathrm{trial}}|}\right)\sigma_{n+1}^{\mathrm{trial}}. \tag{11.55}$$

From the second equation of (11.53) the isotropic hardening variable in the final stress state is given in the following with the last two results:

$$\kappa_{n+1} = \kappa_n + \frac{F_{n+1}^{\text{trial}}}{E + E^{\text{pl}}} = \kappa_n + \Delta\lambda_{n+1} . \tag{11.56}$$

Finally, the plastic strain results into the following via the flow rule

$$\varepsilon_{n+1}^{\text{pl}} = \varepsilon_n^{\text{pl}} + \Delta\lambda_{n+1} \, \text{sgn}(\sigma_{n+1}) , \tag{11.57}$$

and the consistent elasto-plastic modulus can be defined according to Eq. (11.23).

To conclude, the calculation steps of the CPP algorithm are presented in compact form:

I. Calculation of the Test State

$$\sigma_{n+1}^{\text{trial}} = \sigma_n + E \Delta\varepsilon_n$$
$$\kappa_{n+1}^{\text{trial}} = \kappa_n$$

II. Test of Validity of the Stress State

$F(\sigma_{n+1}^{\text{trial}}, \kappa_{n+1}^{\text{trial}}) \leq 0$	
Y	N
$\sigma_{n+1} = \sigma_{n+1}^{\text{trial}}$ $\kappa_{n+1} = \kappa_{n+1}^{\text{trial}}$ End CPP	Back Projection necessary ($\Rightarrow$ Step III)

III. Back Projection

Initial Values:

$$\boldsymbol{v}^{(0)} = \begin{bmatrix} \sigma^{(0)} \\ \kappa^{(0)} \\ \Delta\lambda^{(0)} \end{bmatrix} = \begin{bmatrix} \sigma_{n+1}^{\text{trial}} \\ \kappa_n \\ 0 \end{bmatrix}$$

Root Finding with the NEWTON Method:

$$\boldsymbol{v}^{(i+1)} = \boldsymbol{v}^{(i)} - \left(\frac{d\boldsymbol{m}}{d\boldsymbol{v}}(\boldsymbol{v}^{(i)}) \right)^{-1} \boldsymbol{m}(\boldsymbol{v}^{(i)})$$

As long as $\| \boldsymbol{v}^{(i+1)} - \boldsymbol{v}^{(i)} \| < t_{\text{end}}^{v}$

In the termination criterion the vector norm (or length) of a vector has been used, which results in the following $\|x\| = \left(\sum_{i=1}^{n} x_i^2\right)^{0.5}$ for an arbitrary vector x.

IV. Actualization of the Parameters

$$\sigma_{n+1} = \sigma^{(i+1)}$$
$$\kappa_{n+1} = \kappa^{(n+1)}$$
$$E_{n+1}^{\text{elpl}} = \tilde{m}_{11}$$

The internally used calculation precision in the FE system appears perfect as termination precision t_{end}^v in the NEWTON method.

Figure 11.7 shows that the entire strain increment in dependence on the test stress state can be illustrated as

$$\Delta\varepsilon_n = \varepsilon_{n+1} - \varepsilon_n = E^{-1}(\sigma_{n+1}^{\text{trial}} - \sigma_n). \tag{11.58}$$

If one inserts the last equation and the flow rule[17] according to Eq. (11.14) in (11.33), the final stress state σ_{n+1} in dependence on the test stress state $\sigma_{n+1}^{\text{trial}}$ results in:

$$\sigma_{n+1} = \sigma_{n+1}^{\text{trial}} - \Delta\lambda_{n+1} E \, \text{sgn}(\sigma_{n+1}^{\text{trial}}). \tag{11.59}$$

The general procedure of an elasto-plastic finite element calculation is illustrated in Fig. 11.8.

One can see that the solution of an elasto-plastic problem occurs on two levels. On the global level, meaning for the global system of equations under consideration of the boundary conditions, the NEWTON–RAPHSON iteration scheme is made use of to define the incremental global displacement vector Δu_n. By summing the displacement increments, the total global displacement vector u_{n+1} of the unknown nodal displacement of a structure consisting of finite elements results. Via the strain-displacement relation the strain ε_{n+1} or, alternatively the strain increment $\Delta\varepsilon_n$ per element, can be determined from the nodal displacement.[18] The strain increment of an element is now used on the level of the integration points, to define the remaining state variables iteratively with the help of a predictor-corrector method.

---

[17] At this point it is switched from $d\lambda$ to $\Delta\lambda$.

[18] At this point for the considered linear bar elements a constant strain distribution per element results. In general, the strain results as a function of the element coordinates which is usually evaluated on the integration points. Therefore, one would in the general case normally define a strain *vector* ε per element, which combines the different strain values on the integration points. This however is unnecessary for a linear bar element. A scalar strain or alternative stress value is enough for the description.

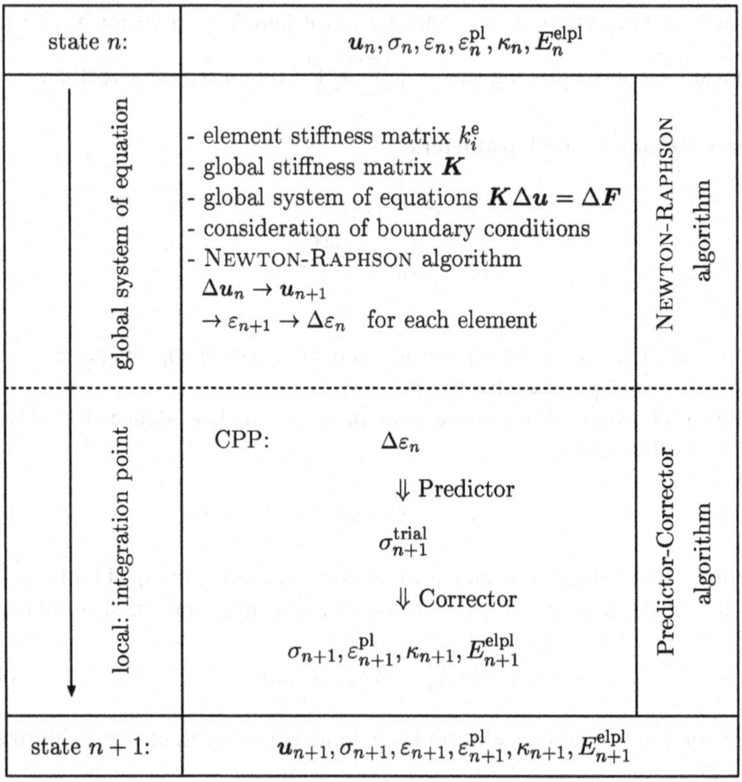

state n:	$\boldsymbol{u}_n, \sigma_n, \varepsilon_n, \varepsilon_n^{\mathrm{pl}}, \kappa_n, E_n^{\mathrm{elpl}}$	
global system of equation	- element stiffness matrix k_i^{e} - global stiffness matrix $\boldsymbol{K}$ - global system of equations $\boldsymbol{K}\Delta\boldsymbol{u} = \Delta\boldsymbol{F}$ - consideration of boundary conditions - NEWTON-RAPHSON algorithm $\Delta\boldsymbol{u}_n \to \boldsymbol{u}_{n+1}$ $\to \varepsilon_{n+1} \to \Delta\varepsilon_n$ for each element	NEWTON-RAPHSON algorithm
local: integration point	CPP: $\Delta\varepsilon_n$ $\Downarrow$ Predictor $\sigma_{n+1}^{\mathrm{trial}}$ $\Downarrow$ Corrector $\sigma_{n+1}, \varepsilon_{n+1}^{\mathrm{pl}}, \kappa_{n+1}, E_{n+1}^{\mathrm{elpl}}$	Predictor-Corrector algorithm
state $n+1$:	$\boldsymbol{u}_{n+1}, \sigma_{n+1}, \varepsilon_{n+1}, \varepsilon_{n+1}^{\mathrm{pl}}, \kappa_{n+1}, E_{n+1}^{\mathrm{elpl}}$	

Fig. 11.8 General procedure of an elasto-plastic finite element calculation

11.3.2 Interpretation as Convex Optimization Problem

The fully implicit backward-EULER algorithm can also be understood as a solution of a convex optimization problem. A general derivation of the following is given in [1]. As an objective function the complementary energy has to be minimized hereby under the side-condition that the yield condition is fulfilled.

In the following, the derivation of the optimization problem for the example of isotropic strain hardening is illustrated. The complementary energy[19] between the test stress state and an arbitrary state $(\sigma, |\varepsilon^{\mathrm{pl}}|)$ at an increment of the back projection can be split into its elastic and plastic parts according to

$$\bar{\pi}(\sigma, |\varepsilon^{\mathrm{pl}}|) = \bar{\pi}^{\mathrm{el}}(\sigma, |\varepsilon^{\mathrm{pl}}|) + \bar{\pi}^{\mathrm{pl}}(\sigma, |\varepsilon^{\mathrm{pl}}|). \tag{11.60}$$

At this point in the case of the considered *linear* elasticity, the complementary energy $\bar{\pi}^{\mathrm{el}}$ and the potential energy π^{el} are the same. Due to the assumption of the *linear*

---

[19]Hereby the energy per unit volume is considered.

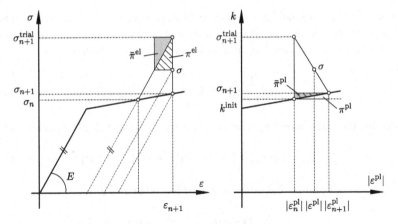

Fig. 11.9 Illustration of the elastic potential and plastic dissipative energy and the corresponding complementary energy between two states of the back projection in the case of the linear hardening

isotropic hardening accordingly it is valid that $\bar{\pi}^{\mathrm{pl}} = \pi^{\mathrm{pl}}$ (see Fig. 11.9). Therefore, the following occurs for the complementary energy

$$\bar{\pi} = \pi^{\mathrm{el}} + \bar{\pi}^{\mathrm{pl}}$$
$$= \int \left(\sigma_{n+1}^{\mathrm{trial}} - \sigma \right) \mathrm{d}\varepsilon^{\mathrm{el}} + \int \left(|\varepsilon^{\mathrm{pl}}| - |\varepsilon_n^{\mathrm{pl}}| \right) \mathrm{d}\sigma . \tag{11.61}$$

The assumption of linearity in the elastic and plastic area, meaning $\mathrm{d}\sigma = E\mathrm{d}\varepsilon^{\mathrm{el}}$ and $\mathrm{d}\sigma = E^{\mathrm{pl}}\mathrm{d}\varepsilon^{\mathrm{pl}}$, can be used in Eq. (11.61) so that the following finally results for the complementary energy

$$\bar{\pi}(\sigma, |\varepsilon^{\mathrm{pl}}|) = \frac{1}{2}(\sigma_{n+1}^{\mathrm{trial}} - \sigma)\frac{1}{E}(\sigma_{n+1}^{\mathrm{trial}} - \sigma) + \frac{1}{2}(|\varepsilon_n^{\mathrm{pl}}| - |\varepsilon^{\mathrm{pl}}|)E^{\mathrm{pl}}(|\varepsilon_n^{\mathrm{pl}}| - |\varepsilon^{\mathrm{pl}}|). \tag{11.62}$$

The fractions of $\bar{\pi}$ result as triangular areas in Fig. 11.9, which can also be used directly for the definition of the complementary energy. For the case that the flow curve k exhibits a certain course, the plastic energy parts for an arbitrary state $(\sigma, |\varepsilon^{\mathrm{pl}}|)$ can be calculated via

$$\pi^{\mathrm{pl}} = \int_{|\varepsilon_n^{\mathrm{pl}}|}^{|\varepsilon^{\mathrm{pl}}|} \left(k(|\varepsilon^{\mathrm{pl}}|) - \sigma_n \right) \mathrm{d}|\varepsilon^{\mathrm{pl}}| , \tag{11.63}$$

$$\bar{\pi}^{\mathrm{pl}} = \int_{|\varepsilon_n^{\mathrm{pl}}|}^{|\varepsilon^{\mathrm{pl}}|} \left(\sigma - k(|\varepsilon^{\mathrm{pl}}|) \right) \mathrm{d}|\varepsilon^{\mathrm{pl}}| . \tag{11.64}$$

The side-condition of the optimization problem is given through the yield condition and states that the final stress state has to be within or on the boundary of the elastic area. From Eq. (11.9) a limiting line results in a $\varepsilon^{\mathrm{pl}}$-$\sigma$ coordinate system:

$$\varepsilon^{\mathrm{pl}} \geq \frac{1}{E^{\mathrm{pl}}}\left(|\sigma| - k^{\mathrm{init}}\right) \quad \text{and} \quad \varepsilon^{\mathrm{pl}} \geq 0. \tag{11.65}$$

Generally, the side-condition, meaning the elastic area, can also be specified as

$$\mathbb{E}_\sigma := \left\{(\sigma, |\varepsilon^{\mathrm{pl}}|) \in \mathbb{R} \times \mathbb{R}_+ | F(\sigma, |\varepsilon^{\mathrm{pl}}|) \leq 0\right\} \tag{11.66}$$

and the convex optimization problem can be formulated as follows:

$$\text{Define}(\sigma_{n+1}, |\varepsilon^{\mathrm{pl}}_{n+1}|) \in \mathbb{E}_\sigma, \text{ so that}$$
$$\bar{\pi}(\sigma_{n+1}, |\varepsilon^{\mathrm{pl}}_{n+1}|) = \min\left\{\bar{\pi}(\sigma, |\varepsilon^{\mathrm{pl}}|)\right\}\Big|_{(\sigma, |\varepsilon^{\mathrm{pl}}|)\in\mathbb{E}_\sigma}.$$

Since $E > 0$ — and under the assumption $E^{\mathrm{pl}} > 0$ — it results that $\bar{\pi}$ is a convex function. The side-condition, meaning $F \leq 0$, also represents a convex function,[20] and the application of the LAGRANGE multiplier method leads to

$$\mathcal{L}(\sigma, |\varepsilon^{\mathrm{pl}}|, \mathrm{d}\lambda) := \bar{\pi}(\sigma, |\varepsilon^{\mathrm{pl}}|) + \mathrm{d}\lambda F(\sigma, |\varepsilon^{\mathrm{pl}}|). \tag{11.67}$$

The gradients of the LAGRANGE function $\mathcal{L}$ result in:

$$\frac{\partial}{\partial\sigma}\mathcal{L}\left(\sigma_{n+1}, |\varepsilon^{\mathrm{pl}}_{n+1}|, \mathrm{d}\lambda\right) = 0, \tag{11.68}$$

$$\frac{\partial}{\partial|\varepsilon^{\mathrm{pl}}|}\mathcal{L}\left(\sigma_{n+1}, |\varepsilon^{\mathrm{pl}}_{n+1}|, \mathrm{d}\lambda\right) = 0. \tag{11.69}$$

For Eqs. (11.68) and (11.69) the following results

$$\frac{\partial\mathcal{L}}{\partial\sigma} = (\sigma^{\mathrm{trial}}_{n+1} - \sigma)\left(-\frac{1}{E}\right) + \mathrm{d}\lambda\,\mathrm{sgn}(\sigma) = 0, \tag{11.70}$$

$$\frac{\partial\mathcal{L}}{\partial|\varepsilon^{\mathrm{pl}}|} = (|\varepsilon^{\mathrm{pl}}_n| - |\varepsilon^{\mathrm{pl}}|)(-E^{\mathrm{pl}}) + \mathrm{d}\lambda(-E^{\mathrm{pl}}) = 0. \tag{11.71}$$

The last two equations comply with the rules (11.56) and (11.35) of the previous section. A graphical interpretation of the implicit backward-EULER algorithm in the sense of a convex optimization problem is given in Fig. 11.10. If an invalid test state $(F > 0)$ results, the ellipsoid of the complementary energy lies outside of the valid

---

[20]The convexity of a yield condition can be derived from the DRUCKER's stability postulate [19, 20].

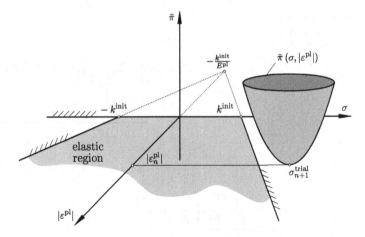

Fig. 11.10 Interpretation of the fully implicit backward-EULER algorithm as a convex optimization problem. Adapted from [1]

area of the elastic energy. It needs to be remarked at this point that the absolute minimum (without side-condition) of the complementary energy lies in the σ-$|\varepsilon^{\mathrm{pl}}|$ plane and therefore results in $\bar{\pi}(\sigma_{n+1}^{\mathrm{trial}}, |\varepsilon_n^{\mathrm{pl}}|) = 0$.

The minimum of the complementary energy under consideration of the side-condition, meaning $F \leq 0$, therefore has to be localized on the cutting curve between the ellipsoid of the complementary energy and the plane along the flow curve[21] (see Eq. (11.65)). If Eq. (11.65) in the complementary energy according to Eq. (11.62) is considered, the following results

$$\bar{\pi}(\sigma, |\varepsilon^{\mathrm{pl}}|) = \frac{1}{2E}(\sigma_{n+1}^{\mathrm{trial}} - \sigma)^2 + \frac{1}{2}E^{\mathrm{pl}}\left(|\varepsilon_n^{\mathrm{pl}}| - \frac{1}{E^{\mathrm{pl}}}(\sigma - k^{\mathrm{init}})\right)^2, \qquad (11.72)$$

this means a polynomial of 2nd order in the variable σ. The minimum of this function — and therefore the state $n + 1$ — results via partial derivative for $\sigma > 0$ (meaning for a tensile test) in:

$$\frac{\partial\bar{\pi}}{\partial\sigma} = -\frac{\sigma_{n+1}^{\mathrm{trial}} - \sigma_{n+1}}{E} - \left(|\varepsilon_n^{\mathrm{pl}}| - \frac{\sigma_{n+1} - k^{\mathrm{init}}}{E^{\mathrm{pl}}}\right) \qquad (11.73)$$

---

[21] This plane has to stand vertically on the σ-$|\varepsilon^{\mathrm{pl}}|$ plane. For a tensile test the plane has to go through the limit curve in the area $\sigma > 0$. For a compression test the according straight line from the area $\sigma < 0$ has to be chosen.

Fig. 11.11 Illustration of the complementary energy in a cutting plane along the flow curve

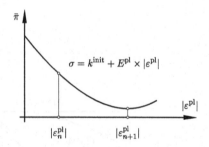

or

$$\sigma_{n+1} = \underbrace{\frac{E E^{\mathrm{pl}}}{E + E^{\mathrm{pl}}}}_{E^{\mathrm{elpl}}} \left(\frac{k^{\mathrm{init}}}{E^{\mathrm{pl}}} + \frac{\sigma_{n+1}^{\mathrm{trial}}}{E} + |\varepsilon_n^{\mathrm{pl}}| \right) . \qquad (11.74)$$

Via Eq. (11.65) the plastic strain in the final stress state results in:

$$\varepsilon_{n+1}^{\mathrm{pl}} = \underbrace{\frac{E E^{\mathrm{pl}}}{E + E^{\mathrm{pl}}}}_{E^{\mathrm{elpl}}} \left(\frac{k^{\mathrm{init}}}{E E^{\mathrm{pl}}} + \frac{\sigma_{n+1}^{\mathrm{trial}}}{E E^{\mathrm{pl}}} + \frac{|\varepsilon_n^{\mathrm{pl}}|}{E^{\mathrm{pl}}} \right) - \frac{k^{\mathrm{init}}}{E^{\mathrm{pl}}} . \qquad (11.75)$$

A graphical illustration of the cutting curve is given in Fig. 11.11. One can consider at this point that $|\varepsilon_n^{\mathrm{pl}}|$ equals the test stress state.

11.4 Derivation of the Semi-implicit Backward-Euler Algorithm

To avoid the higher derivatives in the JACOBIan matrix $\frac{\partial m}{\partial v}$ of the residual functions in the general three-dimensional case, the so-called semi-implicit backward-EULER algorithm can be made use of. This procedure is implicit in the consistency parameter (state $n + 1$), however explicit in the function $\mathrm{sgn}(\sigma)$ since the calculation occurs in the initial state n. Because of that, the normal rule in the final state $n + 1$ is not fulfilled exactly. To avoid a drift away from the flow curve, the yield condition in the final state $n + 1$ is fulfilled exactly. Therefore, the integration scheme results in:

$$\sigma_{n+1} = \sigma_{n+1}^{\mathrm{trial}} - \Delta\lambda_{n+1} E \, \mathrm{sgn}(\sigma_n) , \qquad (11.76)$$

$$\kappa_{n+1} = \kappa_n + \Delta\lambda_{n+1} , \qquad (11.77)$$

$$F = F(\sigma_{n+1}, \kappa_{n+1}) = 0 . \qquad (11.78)$$

Outside of the final state also at this point a residual r remains at each of these equations:

$$r_\sigma(\sigma, \kappa, \Delta\lambda) = E^{-1}\sigma - E^{-1}\sigma_{n+1}^{\text{trial}} + \Delta\lambda\,\text{sgn}(\sigma_n) \neq 0,$$
$$r_\kappa(\kappa, \Delta\lambda) = -\kappa + \kappa_n + \Delta\lambda \neq 0,$$
$$r_F(\sigma, \kappa) = F(\sigma, \kappa) = |\sigma| - k(\kappa) \neq 0. \tag{11.79}$$

The partial derivatives of the residual functions finally lead to the following JACOBIan matrix:

$$\frac{\partial \boldsymbol{m}}{\partial \boldsymbol{v}}(\sigma, \kappa, \Delta\lambda) = \begin{bmatrix} E^{-1} & 0 & \text{sgn}(\sigma_n) \\ 0 & -1 & 1 \\ \text{sgn}(\sigma) & -\frac{\partial k(\kappa)}{\partial \kappa} & 0 \end{bmatrix}. \tag{11.80}$$

If one compares the JACOBIan matrix according to Eqs. (11.80) and (11.47) one can see that the integration requirements for the fully implicit and semi-implicit algorithm are identical for the considered one-dimensional case at this point, as long as the stress state σ and σ_n lie in the same quadrant, meaning exhibiting the same algebraic sign. Similar conclusions can be drawn for the, in Table 11.3 summarized integration requirements.

To conclude, it can be remarked that the concept of the plastic material behavior, which was originally developed for the permanent deformation of metals, is also applied for other classes of material. Typically the macroscopic stress-strain diagram is regarded at this point, which exhibits a similar course as for classic metals. As an example, the following materials and disciplines can be listed:

- Plastics [21],
- Fibre-reinforced plastics [22],
- Soil mechanics [23, 24],
- Concrete [25].

11.5   Sample Problems and Supplementary Problems

11.5.1   Sample Problems

11.1   Back Projection at Linear Hardening – Continuum Bar

Figure 11.12a shows an idealized stress-strain diagram as it can, for example, be derived experimentally from an uniaxial tensile test. With the help of this material behavior the deformation of a tension bar (see Fig. 11.12b) can be simulated. Thereby the right-hand end is displaced by $u = 8 \times 10^{-3}$ m in total, whereupon the deformation is applied in 10 equal increments. The bar needs to be regarded as a continuum hereby and should not be discretized with finite elements.

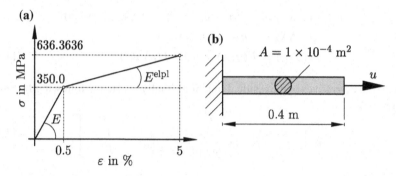

Fig. 11.12 Sample problem back projection at linear hardening: **a** stress-strain distribution; **b** geometry and boundary conditions

(a) Calculate the stress state with the help of the CPP algorithm in each increment and mark all values, which have to be updated.
(b) Graphically illustrate the stress distribution.

Solution 11.1

(a) For the back projection various material properties of the elastic and plastic regime are needed. The modulus of elasticity E results as a quotient from the stress and strain increment in the elastic regime in:

$$E = \frac{\Delta \sigma}{\Delta \varepsilon} = \frac{350 \text{ MPa}}{0.005} = 70000 \text{ MPa}. \tag{11.81}$$

The plastic modulus E^{pl} results as a quotient from the yield stress and the plastic strain increment in:

$$E^{\text{pl}} = \frac{\Delta k}{\Delta \varepsilon^{\text{pl}}} = \frac{636.3636 \text{ MPa} - 350 \text{ MPa}}{(0.05 - 636.3636 \text{ MPa}/E) - (0.005 - 350 \text{ MPa}/E)}$$
$$= 7000 \text{ MPa}. \tag{11.82}$$

Therefore, the elasto-plastic material modulus can be calculated via Eq. (11.23) in

$$E^{\text{elpl}} = \frac{E \times E^{\text{pl}}}{E + E^{\text{pl}}} = \frac{70000 \text{ MPa} \times 7000 \text{ MPa}}{70000 \text{ MPa} + 7000 \text{ MPa}} = 6363.636 \text{ MPa}. \tag{11.83}$$

Finally, the equation of the flow curve results in the following via the initial yield stress:

$$k(\kappa) = 350 \text{ MPa} + 7000 \text{ MPa} \times \kappa. \tag{11.84}$$

A graphical illustration of the flow curve is given in Fig. 11.13.

Fig. 11.13 Flow curve for
continuum bar

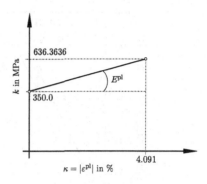

Table 11.4 Numerical values of the back projection for a continuum bar with linear hardening (10 increments, $\Delta\varepsilon = 0.002$)

Inc	ε	σ^{trial}	σ	κ	$d\lambda$	E^{elpl}
–	–	MPa	MPa	10^{-3}	10^{-3}	MPa
1	0.002	140.0	140.0	0.0	0.0	0.0
2	0.004	280.0	280.0	0.0	0.0	0.0
3	0.006	420.0	356.364	0.909091	0.909091	6363.636
4	0.008	496.364	369.091	2.727273	1.818182	6363.636
5	0.010	509.091	381.818	4.545455	1.818182	6363.636
6	0.012	521.818	394.545	6.363636	1.818182	6363.636
7	0.014	534.545	407.273	8.181818	1.818182	6363.636
8	0.016	547.273	420.000	10.000000	1.818182	6363.636
9	0.018	560.000	432.727	11.818182	1.818182	6363.636
10	0.020	572.727	445.455	13.636364	1.818182	6363.636

For the integration algorithm the strain increment is additionally necessary. At a total displacement of 8 mm and 10 equidistant steps, the strain increment can be determined via:

$$\Delta\varepsilon = \frac{1}{10} \times \frac{8\,\text{mm}}{400\,\text{mm}} = 0.002. \tag{11.85}$$

For the first two increments, trial stress states result in the elastic regime ($F < 0$), and the resulting stress can be calculated via HOOKE's law (11.25). From the third increment on an invalid test stress state ($F > 0$) results for the first time and the stress has to be calculated via Eq. (11.53), whereupon the constant matrix expression results in:

$$\begin{bmatrix} 7000 & -7000 & 1 \\ 1 & -1 & -(70000)^{-1} \\ 1 & 0,1 & -(70000)^{-1} \end{bmatrix}. \tag{11.86}$$

Table 11.4 summarizes the numerical results for the 10 increments.

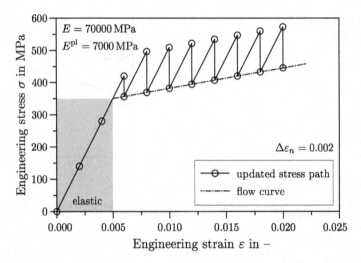

Fig. 11.14 Stress distribution in the case of the back projection for a continuum bar with linear hardening (10 increments, $\Delta\varepsilon = 0.002$)

(b) A graphical illustration of the stress distribution is given in Fig. 11.14. Due to the *linear* hardening behavior, the back projection for each increment occurs in one step.

Finally, it needs to be remarked at this point that for the special case of *linear* hardening at *uniaxial* stress states, the stress in the plastic area (inc ≥ 3) can directly be calculated via (see Fig. 11.1b)

$$
\begin{aligned}
\sigma(\varepsilon) &= k_t^{\text{init}} + E^{\text{elpl}} \times (\varepsilon - \varepsilon_t^{\text{init}}) \\
&= E\varepsilon_t^{\text{init}} + E^{\text{elpl}}\varepsilon - E^{\text{elpl}}\varepsilon_t^{\text{init}} \\
\sigma(\varepsilon) &= (E - E^{\text{elpl}}) \times \varepsilon_t^{\text{init}} + E^{\text{elpl}} \times \varepsilon.
\end{aligned}
\tag{11.87}
$$

The intention of this example however is to illustrate the concept of the back projection and not to define the stress according to the simplest method.

11.2 Back Projection at Linear Hardening – Discretization via one Finite Element, Displacement and Force Boundary Condition

The continuum bar from example 11.1 can be discretized within the frame of this example via one single finite element, see Fig. 11.15. The material behavior can be defined as shown in Fig. 11.12a. The load on the right-hand end of the bar can be applied in 10 equal increments, whereupon

(a) $u = 8\,\text{mm}$,
(b) $F = 100\,\text{kN}$

Fig. 11.15 Sample problem
back projection in the case of
linear hardening:
a displacement boundary
condition; **b** force boundary
condition

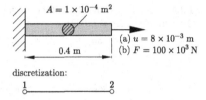

discretization:

can be applied. Calculate the stress state in each increment with the help of the CPP
algorithm and mark all values, which have to be updated. As convergence criteria an
absolute displacement difference of 1×10^{-5} mm can be assigned.

Solution 11.2

When using solely one element, the global system of equations results in the following
— without the consideration of the boundary conditions:

$$\frac{A\tilde{E}}{L} \begin{bmatrix} 1 & -1 \\ -1 & 1 \end{bmatrix} \begin{bmatrix} \Delta u_1 \\ \Delta u_2 \end{bmatrix} = \begin{bmatrix} \Delta F_1 \\ \Delta F_2 \end{bmatrix}. \tag{11.88}$$

Since in general this is a nonlinear system of equations, an incremental form has been
assigned. The modulus $\tilde{E}$ equals the elasticity modulus E in the elastic range and
the elasto-plastic material modulus E^{elpl} in the plastic range. Since a fixed support
is given on the left-hand node ($\Delta u_1 = 0$), Eq. (11.88) can be simplified to

$$\frac{A\tilde{E}}{L} \times \Delta u_2 = \Delta F_2. \tag{11.89}$$

Case (a) Displacement boundary condition $u = 8$ mm on the right-hand node:

In the case of the displacement boundary condition Eq. (11.89) must not be solved,
since for each increment $\Delta u_2 = 8\,\text{mm}/10 = 0.8$ mm is known. Via the equation for
the strain in the element, meaning $\varepsilon = \frac{1}{L}(u_2 - u_1)$, the strain increment results in
the following in the case of the fixed support on the left-hand node:

$$\Delta \varepsilon = \frac{1}{L} \times \Delta u_2 = \frac{0.8\,\text{mm}}{400\,\text{mm}} = 0.002. \tag{11.90}$$

The entire displacement or alternatively strain can be calculated through the summation of the incremental displacement or alternatively strain values, see Table 11.5. It
can be remarked at this point that for this case of displacement boundary condition
at one element, the calculation of the displacement or alternatively the strain for all
increments can be done without a stress calculation.

To calculate the stress and plastic strain in each increment, the calculation via the
CPP algorithm for each increment has to be conducted via the strain increment $\Delta \varepsilon$
from Table 11.5. This is exactly what is calculated in example 11.1 and the numerical
results can be taken from Table 11.4.

Table 11.5 Numerical values of the displacement and strain for displacement boundary condition (10 increments, $\Delta \varepsilon = 0.002$)

Inc	Δu_2	$\Delta \varepsilon$	u_2	ε
–	10^{-4} m	–	10^{-4} m	–
1	8.0	0.002	8.0	0.002
2	8.0	0.002	16.0	0.004
3	8.0	0.002	24.0	0.006
4	8.0	0.002	32.0	0.008
5	8.0	0.002	40.0	0.010
6	8.0	0.002	48.0	0.012
7	8.0	0.002	56.0	0.014
8	8.0	0.002	64.0	0.016
9	8.0	0.002	72.0	0.018
10	8.0	0.002	80.0	0.020

Case (b) Force boundary condition $F = 100$ kN on the right-hand node:

In the case of the force boundary condition, Eq. (11.89) can be solved via the NEWTON–RAPHSON method. To do so, this equation has to be written in the form of a residual r as

$$r = \frac{A\tilde{E}}{L} \times \Delta u_2 - \Delta F_2 = \tilde{E}(u_2) \times \frac{A}{L} \times \Delta u_2 - \Delta F_2 = 0. \tag{11.91}$$

If one develops the last equation into a TAYLOR's series and neglects the terms of higher order, the following form results

$$r(\Delta u_2^{(i+1)}) = r(\Delta u_2^{(i)}) + \left(\frac{\partial r}{\partial \Delta u_2}\right)^{(i)} \times \delta(\Delta u_2) + \cdots, \tag{11.92}$$

whereupon

$$\delta(\Delta u_2) = \Delta u_2^{(i+1)} - \Delta u_2^{(i)} \tag{11.93}$$

is valid and

$$\left(\frac{\partial r}{\partial \Delta u_2}\right)^{(i)} = K_{\mathrm{T}}^{(i)} \tag{11.94}$$

represents the tangent stiffness matrix[22] in the ith iteration step. Then Eq. (11.92) can also be written as

---

[22]In the considered example with linear hardening, $\tilde{E}$ is constant in the elastic range (increment 1 to 3) and in the plastic range (increment 4 to 10) and therefore not a function of u_2. In the general case however $\tilde{E}$ has to be differentiated as well.

$$\delta(\Delta u_2) K_{\mathrm{T}}^{(i)} = -r\left(\Delta u_2^{(i)}\right) = \Delta F^{(i)} - \frac{\tilde{E} A}{L} \Delta u_2^{(i)}. \tag{11.95}$$

Multiplication via $(K_{\mathrm{T}}^{(i)})^{-1}$ and the use of Eq. (11.93) finally leads to

$$\Delta u_2^{(i+1)} = \Delta F^{(i)} \times \frac{L}{\tilde{E} A}, \tag{11.96}$$

whereupon $\tilde{E} = E$ is valid in the elastic range (increment 1 to 3) and $\tilde{E} = E^{\mathrm{elpl}}$ in the plastic range (increment 4 to 10).

Application of Eq. (11.96) leads to a value of $\Delta u_2 = 0.571429$ mm in the elastic range (increment 1 to 3) and in the plastic range (increment 4 to 10) a displacement increment of $\Delta u_2 = 6.285715$ mm occurs. It can be remarked at this point that the calculation of the displacement increments (increments 1 to 3 and 4 to 10) does not need an iteration and the application of Eq. (11.96) directly yields the desired result. As soon as the displacement increments (Δu_2) are calculated, the entire displacement on node 2 results via summation of the incremental values. Subsequently the strain in the element can be calculated via the relation $\varepsilon = \frac{1}{L}(u_2 - u_1)$, and the strain increments result through subtraction of two consecutive strain values (see Table 11.7).

The calculation of the stress and the plastic strain now requires that the CPP algorithm is used in each increment, based on the strain increment $\Delta\varepsilon$. The graphical illustration of the back projection is given in Fig. 11.16. One can see clearly that the strain increments at a force boundary condition differ in the elastic and plastic range.

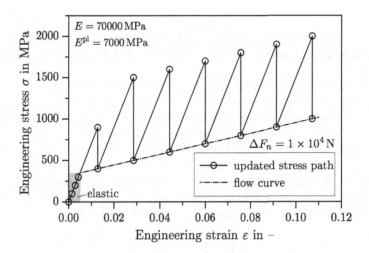

Fig. 11.16 Stress distribution in the case of the back projection for a continuum bar with linear hardening (10 increments, $\Delta F = 1 \times 10^4$)

Table 11.6 Numerical values for one element in the case of linear hardening (10 increments; $\Delta F = 1 \times 10^4$ N)

Inc	Ex. Force	Δu_2	u_2	ε	$\Delta\varepsilon$	σ	ε^{pl}
–	10^4 N	mm	mm	10^{-2}	10^{-2}	MPa	10^{-2}
1	1.0	0.5714	0.5714	0.1429	0.1429	1000	0.0
2	2.0	0.5714	1.1427	0.2857	0.1429	200.0	0.0
3	3.0	0.5714	1.7143	0.4286	0.1429	300.0	0.0
4	4.0	4.4286	5.1429	1.2857	0.8571	400.0	0.7143
5	5.0	6.2857	11.4286	2.8571	1.5714	500.0	2.1429
6	6.0	6.2857	17.7143	4.4286	1.5714	600.0	3.5714
7	7.0	6.2857	24.0000	6.0000	1.5714	700.0	5.0000
8	8.0	6.2857	30.2857	7.5714	1.5714	800.0	6.4286
9	9.0	6.2857	36.5714	9.1429	1.5714	900.0	7.8571
10	10.0	6.2857	42.8571	10.7143	1.5714	1000.0	9.2857

Fig. 11.17 Definition of the intermediate modulus $\tilde{E}$ in the case of the transition from the elastic to the plastic range

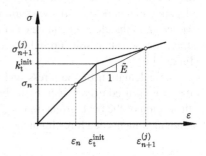

As a consequence, quite high values result for the test stress state in the plastic range. Thus for the increments 5 to 10 $\sigma_{n+1}^{\text{trial}} = \sigma_{n+1} + 1000$ MPa is valid (Fig. 11.16 and Table 11.6).

Particular attention is required at the transition from the elastic to the plastic regime, meaning from increment 3 to 4. Here the modulus $\tilde{E}$ is not defined clearly and Eq. (11.96) has to be solved iteratively. For the first cycle of the calculation (cycle $j = 0$) the arithmetic mean between the modulus of elasticity E and the elasto-plastic material modulus E^{elpl} can be applied. For the further cycles (cycle $j \geq 1$) $\tilde{E}$ can be approximated via an intermediate modulus (secant modulus, see Fig. 11.17). Therefore the following relation results as a calculation requirement for the intermediate modulus $\tilde{E}$ in the elasto-plastic transition regime (at this point in the considered example at the transition from increment 3 to 4):

$$\tilde{E} = \begin{cases} \frac{E + E^{\text{elpl}}}{2} & \text{for } j = 0 \\ \frac{\sigma_{n+1}^{(j)} - \sigma_n}{\varepsilon_{n+1}^{(j)} - \varepsilon_n} & \text{for } j > 0 \end{cases} . \tag{11.97}$$

Table 11.7 Numerical values for the transition from increment 3 to 4

Cycle	$\tilde{E}^{(i)}$	$u_2^{new} - u_2^{old}$
–	MPa	mm
0	38181.84	1.048×10^{-0}
1	23719.00	6.388×10^{-1}
2	17145.00	6.466×10^{-1}
3	14157.16	4.925×10^{-1}
4	12798.56	2.999×10^{-1}
5	12181.16	1.584×10^{-1}
6	11900.52	7.744×10^{-2}
7	11772.96	3.642×10^{-2}
8	11715.00	1.682×10^{-2}
9	11688.64	7.699×10^{-3}
10	11676.64	3.511×10^{-3}
11	11671.20	1.598×10^{-3}
12	11668.72	7.270×10^{-4}
13	11667.60	3.305×10^{-4}
14	11667.08	1.503×10^{-4}
15	11666.88	6.831×10^{-5}
16	11666.76	3.105×10^{-5}
17	11666.72	1.411×10^{-5}
18	11666.68	6.416×10^{-6}

The numerical values of the intermediate modulus $\tilde{E}$ at the transition from increment 3 to 4 and the differences of the displacement, which result herefrom on node 2 are summarized in Table 11.7. Since an absolute displacement difference of 1×10^{-5} mm was required as convergence criteria, 18 cycles are necessary to iterate the difference between the new and the old displacement on node 2 under these values. It can be considered at this point that the difference between the displacements on node 2 in the first cycle (cycle 0) result in $u_2^{new} - u_2^{old} = u_2^{(j=0)} - u_2|_{inc\ 3}$ and for the following cycles of the iteration via $u_2^{(j+1)} - u_2^{(j)}$ (Table 11.7).

The convergence behavior of the iteration rule is illustrated graphically in Fig. 11.18. An equidistant division was chosen in Fig. 11.18a and a logarithmic division (to be base 10) in Fig. 11.18b. One can see that quite a high convergence rate occurs at the beginning of the iteration, which flattens throughout the different cycles. At the chosen convergence criteria here of 10^{-5} the 18 iteration steps are therefore necessary, to finally reach the required absolute displacement difference. If one would require an absolute difference of 10^{-6} as convergence criteria, 21 iteration steps would be necessary.

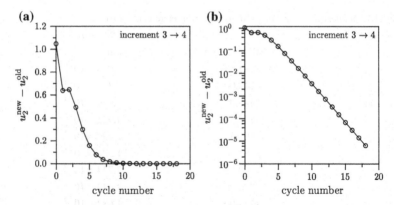

Fig. 11.18 Convergence behavior in the case of the transition from increment 3 to 4: **a** equidistant division; **b** logarithmic division of the absolute displacement difference

Fig. 11.19 Sample problem back projection in the case of a bar with different materials: **a** stress-strain distributions, **b** geometry and boundary conditions

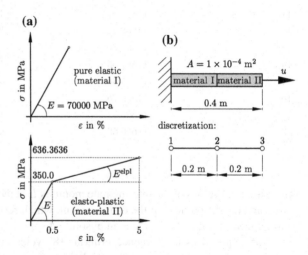

11.3 Back Projection for a Bimaterial Bar

Two different material behaviors (see Fig. 11.19a) can be considered in the following to model a bimaterial bar (see Fig. 11.19 b) via the FE method. The right-hand end is displaced by $u = 8$ mm hereby, whereupon the deformation is applied in 10 equal increments. The bar can be discretized with two finite elements hereby. Examine the following material combinations:

(a) Material I: pure elastic; Material II: pure elastic,
(b) Material I: elasto-plastic; Material II: elasto-plastic,
(c) Material I: pure elastic; Material II: elasto-plastic,

to calculate the displacement of the middle node. Furthermore, calculate the stress state in each element with the help of the CPP algorithm and mark all values, which have to be updated.

Solution 11.3

When using two finite elements, the global system of equations, without consideration of the boundary conditions, results in the following incremental form for this example:

$$\frac{A}{L} \times \begin{bmatrix} \tilde{E}^{\mathrm{I}} & -\tilde{E}^{\mathrm{I}} & 0 \\ -\tilde{E}^{\mathrm{I}} & \tilde{E}^{\mathrm{I}} + \tilde{E}^{\mathrm{II}} & -\tilde{E}^{\mathrm{II}} \\ 0 & -\tilde{E}^{\mathrm{II}} & \tilde{E}^{\mathrm{II}} \end{bmatrix} \begin{bmatrix} \Delta u_1 \\ \Delta u_2 \\ \Delta u_3 \end{bmatrix} = \begin{bmatrix} \Delta F_1 \\ \Delta F_2 \\ \Delta F_3 \end{bmatrix}. \tag{11.98}$$

The consideration of the boundary condition on the left-hand side, meaning $u_1 = 0$, yields the following reduced global system of equations:

$$\frac{A}{L} \times \begin{bmatrix} \tilde{E}^{\mathrm{I}} + \tilde{E}^{\mathrm{II}} & -\tilde{E}^{\mathrm{II}} \\ -\tilde{E}^{\mathrm{II}} & \tilde{E}^{\mathrm{II}} \end{bmatrix} \begin{bmatrix} \Delta u_2 \\ \Delta u_3 \end{bmatrix} = \begin{bmatrix} \Delta F_2 \\ \Delta F_3 \end{bmatrix}. \tag{11.99}$$

The consideration of the displacement boundary condition on the right-hand side, meaning $u_3 = u(t)$, and that $\Delta F_2 = \Delta F_3 = 0$ is valid, yields:

$$\frac{A}{L} \times (\tilde{E}^{\mathrm{I}} + \tilde{E}^{\mathrm{II}}) \Delta u_2 = \frac{\tilde{E}_2 A}{L} \Delta u_3, \tag{11.100}$$

or

$$\left(\frac{\tilde{E}^{\mathrm{I}}}{\tilde{E}^{\mathrm{II}}} + 1 \right) \Delta u_2 = \Delta u_3. \tag{11.101}$$

For the application of the NEWTON–RAPHSON method, Eq. (11.101) is written as a residual equation:

$$r = \left(\frac{\tilde{E}^{\mathrm{I}}}{\tilde{E}^{\mathrm{II}}} + 1 \right) \Delta u_2 - \Delta u_3 = 0. \tag{11.102}$$

If one develops the last equation into a TAYLOR's series and neglects the terms of higher order according to the procedure in example 11.2, finally the following iteration rule results for the definition of the displacement of the middle node:

$$\Delta u_2^{(i+1)} = \left(\frac{\tilde{E}^{\mathrm{I}}}{\tilde{E}^{\mathrm{II}}} + 1 \right)^{-1} \times \Delta u_3^{(i)}. \tag{11.103}$$

In Eq. (11.103) in the elastic regime the elasticity modulus E has to be used for $\tilde{E}$ and in the plastic regime the elasto-plastic material modulus E^{elpl}.

Case (a) Material I: pure elastic; Material II: pure elastic:

For the case that both sections exhibit pure elastic material behavior with $E^{\mathrm{I}} = E^{\mathrm{II}}$, Eq. (11.103) simplifies in:

Table 11.8 Numerical values for a bimaterial bar in the case of pure linear-elastic behavior (10 increments; $\Delta u_3 = 0.8\,\text{mm}$; $A = 100\,\text{mm}^2$)

Inc	Δu_2	u_2	$\Delta \varepsilon$	ε	σ	$F_{r,3}$
–	mm	mm	–	–	MPa	10^4 N
1	0.4	0.4	0.002	0.002	140.0	1.4
2	0.4	0.8	0.002	0.004	280.0	2.8
3	0.4	1.2	0.002	0.006	420.0	4.2
4	0.4	1.6	0.002	0.008	560.0	5.6
5	0.4	2.0	0.002	0.010	700.0	7.0
6	0.4	2.4	0.002	0.012	840.0	8.4
7	0.4	2.8	0.002	0.014	980.0	9.8
8	0.4	3.2	0.002	0.016	1120.0	11.2
9	0.4	3.6	0.002	0.018	1260.0	12.6
10	0.4	4.0	0.002	0.020	1400.0	14.0

$$\Delta u_2^{(i+1)} = (1+1)^{-1} \times \Delta u_3^{(i)} = \frac{1}{2} \times \Delta u_3^{(i)} = 4\,\text{mm}\,. \tag{11.104}$$

The strain in the left-hand element — which is identical with the strain in the right-hand element — can easily be defined via $\Delta \varepsilon = \frac{1}{200\,\text{mm}} \times \Delta u_2$, and the stress results from the strain through multiplication with the modulus of elasticity. The results of this pure elastic calculation are summarized in Table 11.8.

In addition to the displacement, strain and stress values[23] Table 11.8 also contains the reaction forces on node 3. These reaction forces result due to multiplication of the stiffness matrix with the resultant vector of the displacements and have to be in equilibrium with the resulting forces from the stress: $F_{r,3} = \sigma A$.

Case (b) Material I: elasto-plastic; Material II: elasto-plastic:

For the case that both sections exhibit the same elasto-plastic material behavior, $\tilde{E}^{\text{I}} = \tilde{E}^{\text{II}}$ is always valid, and Eq. (11.103) also at this point yields a displacement increment on the middle node from $\Delta u_2^{(i+1)} = \frac{1}{2} \times \Delta u_3^{(i)} = 4\,\text{mm}$ or alternatively a strain increment of $\Delta \varepsilon = 0.002$. For the calculation of the stress and the plastic strain for element II in the non-linear regime, the CPP algorithm has to be made use of, as in example 11.1. The corresponding values are summarized in Table 11.9.

Case (c) Material I: pure elastic; Material II: elasto-plastic:

At this point in the elastic regime of both elements also it occurs that the displacement at the middle node amounts to half of the displacement, which occurred on the right-hand node. As soon as the plastic material behavior occurs in the right-hand half of the bar, $\tilde{E}^{\text{I}} \neq \tilde{E}^{\text{II}}$ occurs and for the right-hand half of the bar, the elasto-plastic

---

[23] One considers that in both sections or alternatively elements, the stress and strain are identical.

Table 11.9 Numerical values for a bimaterial bar in the case of elasto-plastic behavior (10 Increments; $\Delta u_3 = 0.8$ mm; $A = 100$ mm^2)

Inc	Δu_2	u_2	$\Delta\varepsilon$	ε	σ	ε^{pl}	$F_{r,3}$
–	mm	mm	–	–	MPa	10^{-3}	10^4 N
1	0.4	0.4	0.002	0.002	140.0	0.0	1.4
2	0.4	0.8	0.002	0.004	280.0	0.0	2.8
3	0.4	1.2	0.002	0.006	356.364	0.909091	3.56364
4	0.4	1.6	0.002	0.008	369.091	2.727273	3.69091
5	0.4	2.0	0.002	0.010	381.818	4.545455	3.81818
6	0.4	2.4	0.002	0.012	394.545	6.363636	3.94545
7	0.4	2.8	0.002	0.014	407.273	8.181818	4.07273
8	0.4	3.2	0.002	0.016	420.000	10.000000	4.20000
9	0.4	3.6	0.002	0.018	432.727	11.818182	4.32727
10	0.4	4.0	0.002	0.020	445.455	13.636364	4.45455

material modulus has to be made use of. Therefore, the calculation requirement for the displacement increment can be summarized as follows:

$$\Delta u_2^{(i+1)} = \begin{cases} \frac{1}{2} \times \Delta u_3^{(i)} & \text{in the elastic region} \\ \left(\frac{E^{\mathrm{I}}}{E^{\mathrm{elpl,II}}} + 1\right)^{-1} \times \Delta u_3^{(i)} & \text{in the plastic region} \end{cases}. \tag{11.105}$$

The total displacement at the middle node results from the displacement increments through summation and the strain for each element can be defined via $\varepsilon = \frac{1}{L}(-u_1 + u_r)$ (Index 'l' for left-hand and index 'r' for right-hand node of the bar element). As soon as the right-hand part of the bar enters the plastic region, the predictor-corrector method has to be made use of to be able to calculate the state variables. The numerical values of the incremental solution method are summarized in Table 11.10. At this point it can be remarked that in the plastic region a similar relation as in the elastic region can be set to calculate the stress increment from the strain increment (see Eq. (11.106)). However thereby it has to be considered that the modulus of elasticity has to be substituted by the elasto-plastic material modulus:

$$\Delta\sigma = \begin{cases} \Delta\varepsilon \times E & \text{in the elastic region} \\ \Delta\varepsilon \times E^{\mathrm{elpl}} & \text{in the plastic region} \end{cases}. \tag{11.106}$$

The transition from the elastic to the plastic region, meaning from increment 2 to 3, demands a special consideration at this point. The intermediate modulus $\tilde{E}^{\mathrm{II}}$ hereby has to be calculated according to Eq. (11.97), to be able to define the displacement increment subsequently according to Eq. (11.105)$_2$. The absolute displacement on the middle node results from the summation, meaning $u_2^{(i)} = \Delta u_2^{(i)} + u_2|_{\mathrm{inc}\,2}$. The difference of the displacement on the middle node has to be determined in the first cycle (cycle 0) via $|u_2^{(i=0)} - u_2|_{\mathrm{inc}\,2}$ and for each further cycle (i) via $|u_2^{(i)} - u_2^{(i-1)}|$.

Table 11.10 Numerical values for a bimaterial bar in the case of different material behavior (10 Increments; $\Delta u_3 = 0.8$ mm; $A = 100$ mm^2)

Inc	Δu_2	u_2	ε^{I}	$\varepsilon^{\mathrm{II}}$	σ^{I}	σ^{II}	$\varepsilon^{\mathrm{pl,I}}$	$\varepsilon^{\mathrm{pl,II}}$	$F_{\mathrm{r},3}$
–	mm	mm	10^{-3}	10^{-3}	MPa	MPa	10^{-3}	10^{-3}	10^4 N
1	0.4	0.4	2.0	2.0	140.0	140.0	0.0	0.0	1.4
2	0.4	0.8	4.0	4.0	280.0	280.0	0.0	0.0	2.8
3	0.23333	1.03333	5.16667	6.83333	361.667	361.667	0.0	1.66667	3.61667
4	0.06667	1.10000	5.50000	10.50000	385.000	385.000	0.0	5.00000	3.85000
5	0.06667	1.16667	5.83333	14.16667	408.333	408.333	0.0	8.33333	4.08333
6	0.06667	1.23333	6.16667	17.83333	431.667	431.667	0.0	11.66667	4.31667
7	0.06667	1.30000	6.50000	21.50000	455.000	455.000	0.0	15.00000	4.55000
8	0.06667	1.36667	6.83333	25.16667	478.333	478.333	0.0	18.33333	4.78333
9	0.06667	1.43333	7.16667	28.83333	501.667	501.667	0.0	21.66667	5.01667
10	0.06667	1.50000	7.50000	32.50000	525.000	525.000	0.0	25.00000	5.25000

Table 11.11 Numerical values for transition from increment 2 to 3

Cycle	$\tilde{E}^{\mathrm{II},(i)}$	$\Delta u_2^{(i)}$	$u_2^{(i)}$	$\lvert u_2^{\mathrm{new}} - u_2^{\mathrm{old}} \rvert$	$\varepsilon^{\mathrm{II},(i)}$	$\Delta\varepsilon^{\mathrm{II},(i)}$	$\sigma^{\mathrm{II},(i)}$
–	MPa	mm	mm	10^{-2} mm	10^{-3}	10^{-3}	MPa
0	38181.82	0.282353	1.082353	28.235294	6.588235	2.588235	360.1070
1	30950.41	0.245272	1.045272	3.708073	6.773639	2.773639	361.2868
2	29306.91	0.236092	1.036092	0.918059	6.819542	2.819542	361.5789
3	28933.39	0.233962	1.033963	0.212904	6.830187	2.830187	361.6466
4	28848.50	0.233476	1.033476	0.0486115	6.832618	2.832618	61.6621
5	28829.20	0.233366	1.033366	0.0110598	6.833171	2.833171	361.6656
6	28824.82	0.233341	1.033341	0.0025142	6.833296	2.833296	361.6664
7	28823.82	0.233335	1.033335	0.0005714	6.833325	2.833325	361.6666

The calculation of the strain in the right-hand half of the bar can occur via the given boundary condition u_3 via $\varepsilon^{\mathrm{II},(i)} = \frac{1}{L}(-u_2^{(i)} + u_3)$. Finally, the stress results via the CPP algorithm, based on the strain increment $\Delta\varepsilon^{\mathrm{II},(i)} = \varepsilon^{\mathrm{II},(i)} - \varepsilon^{\mathrm{II}}|_{\mathrm{inc}\,2}$. If the stress and strains are known, the new intermediate modulus can be determined for the next cycle via Eq. (11.97). As convergence criteria for the absolute displacement difference a value of 10^{-5} was given within this example (Table 11.11).

11.5.2 Supplementary Problems

11.4 Plastic Modulus and Elasto-Plastic Material Modulus

Discuss the case (a) $E^{\mathrm{pl}} = E$ and (b) $E^{\mathrm{elpl}} = E$.

Fig. 11.20 Sample problem
back projection for a bar with
fixed supports at both ends

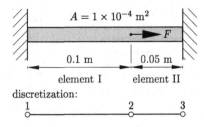

11.5 Back Projection at Linear Hardening

Calculate example 11.1 for the following linear flow curve of a steal:
$k(\kappa) = (690 + 21000\kappa)$ MPa. The modulus of elasticity amounts 210000 MPa. The
geometric dimensions of example 11.1 can be assumed.
(a) For 10 increments with $\Delta\varepsilon = 0.001$,
(b) For 20 increments with $\Delta\varepsilon = 0.0005$,
(c) For 20 increments with $\Delta\varepsilon = 0.001$.

Compare and interpret all the results.

11.6 Back Projection at Non-Linear Hardening

Calculate example 11.1 for the following non-linear flow curve: $k(\kappa) = (350 + 12900\kappa - 1.25 \times 10^5 \kappa^2)$ MPa. All other parameters can be taken as in example
11.1.

11.7 Back Projection for Bar at Fixed Support at Both Ends

Calculate for the illustrated bar in Fig. 11.20 with fixed support at both ends, the
displacement of the point of load application. The bar has an elasto-plastic material
behavior ($E = 1 \times 10^5$ MPa; $E^{\text{elpl}} = 1 \times 10^3$ MPa; $k_t^{\text{init}} = 200$ MPa) and a force
of $F = 6 \times 10^4$ N is applied in 3 increments equally. It can be assumed that the
material behavior is identical under tensile and compression loading. Calculate the
displacement of the point of load application and determine the stress and strain in
both elements. As convergence criteria an absolute displacement difference on the
point of load application of 10^{-5} mm can be given.

11.8 Back Projection for a Finite Element at Ideal-Plastic Material Behavior

Discuss the case of a single finite element with force boundary condition at ideal-
plastic material behavior. It can be assumed thereby that the applied force increases
linearly starting with zero. The problem and the material behavior are schematically illustrated in Fig. 11.21. Why is no convergence achieved in the plastic region
at the force boundary condition? What changes if the force boundary condition is
substituted by a displacement boundary condition?

Fig. 11.21 Sample problem back projection for a finite element in the case of ideal-plastic material behavior: **a** stress-strain diagram; **b** model representation

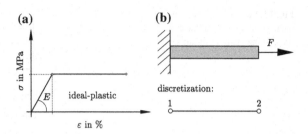

References

1. Simo JC, Hughes TJR (1998) Computational inelasticity. Springer, New York
2. Drucker DC (1952) A more fundamental approach to plastic stress-strain relations. In: Sternberg E et al (eds) (Hrsg) Proceedings of the 1st US National Congress of Applied Mechanics. Edward Brothers Inc, Michigan, pp 487–491
3. Betten J (2001) Kontinuumsmechanik. Springer, Berlin
4. de Borst R (1986) Non-linear analysis of frictional materials. Dissertation, Delft University of Technology
5. Belytschko T, Liu WK, Moran B (2000) Nonlinear finite elements for continua and structures. Wiley, Chichester
6. Mang H, Hofstetter G (2008) Festigkeitslehre. Springer, Wien
7. Altenbach H, Altenbach J, Zolochevsky A (1995) Erweiterte Deformationsmodelle und Versagenskriterien der Werkstoffmechanik. Deutscher Verlag für Grundstoffindustrie, Stuttgart
8. Jirásek M, Bazant ZP (2002) Inelastic analysis of structures. Wiley, Chichester
9. Chakrabarty J (2009) Applied plasticity. Springer, New York
10. Yu M-H, Zhang Y-Q, Qiang H-F, Ma G-W (2006) Generalized plasticity. Springer, Berlin
11. Wriggers P (2001) Nichtlineare finite-element-methoden. Springer, Berlin
12. Crisfield MA (2001) Non-linear finite element analysis of solids and structures. Bd. 1: Essentials. Wiley, Chichester
13. Crisfield MA (2000) Non-linear finite element analysis of solids and structures. Bd. 2: Advanced topics. Wiley, Chichester
14. de Souza Neto EA, Perić D, Owen DRJ (2008) Computational methods for plasticity: theory and applications. Wiley, Chichester
15. Dunne F, Petrinic N (2005) Introduction to computational plasticity. Oxford University Press, Oxford
16. Simo JC, Ortiz M (1985) A unified approach to finite deformation elastoplasticity based on the use of hyperelastic constitutive equations. Comput Method Appl M 49:221–245
17. Ortiz M, Popov EP (1985) Accuracy and stability of integration algorithms for elastoplastic constitutive equations. Int J Num Meth Eng 21:1561–1576
18. Moran B, Ortiz M, Shih CF (1990) Formulation of implicit finite element methods for multiplicative finite deformation plasticity. Int J Num Meth Eng 29:483–514
19. Betten J (1979) Über die Konvexität von Fließkörpern isotroper und anisotroper Stoffe. Acta Mech 32:233–247
20. Lubliner J (1990) Plasticity theory. Macmillan Publishing Company, New York
21. Balankin AS, Bugrimov AL (1992) A fractal theory of polymer plasticity. Polym Sci 34:246–248
22. Spencer AJM (1992) Plasticity theory for fibre-reinforced composites. J Eng Math 26:107–118
23. Chen WF, Baladi GY (1985) Soil plasticity. Elsevier, Amsterdam
24. Chen WF, Liu XL (1990) Limit analysis in soil mechanics. Elsevier, Amsterdam
25. Chen WF (1982) Plasticity in reinforced concrete. McGraw-Hill, New York

Chapter 12
Stability-Buckling

Abstract In common and technical parlance the term stability is used in many ways. Here it is restricted to the static stability of elastic structures. The derivations concentrate on elastic bars and beams. The initial situation is a loaded elastic structure. If the acting load remains under a critical value, the structure reacts 'simple' and one can describe the reaction with the models and equations of the preceding chapters. If the load reaches or exceeds the critical value, bars and beams begin to buckle. The situation becomes ambiguous, beyond the initial situation several equilibrium positions can exist. From the technical application the smallest load is critical for which buckling in either bars or beams appears.

12.1 Stability in Bar/Beam

The initial situation is a structure, which consists of bars and beams. Bars and beams are connected through nodes, through which forces and moments are introduced into each single element. As long as the loads on an element are lower than the critical limit, the element reacts linear elastic. If, however, a critical value is reached or exceeded, buckling occurs. Figure 12.1 illustrates various situations when buckling can occur.

For the analysis of the stability performance several description possibilities are available. In the following the energy-approach is used.

The total potential Π of a bar or beam can be generally described as

$$\Pi = \frac{1}{2} u^{\mathsf{T}} K u - u^{\mathsf{T}} F, \tag{12.1}$$

where u stands for the column matrix of deformation, K for the stiffness matrix and F for the column matrix of external loads. In an equilibrium state the entire potential energy Π of a system is stationary. For a stationary value of Π the first variation $\delta \Pi$ has to disappear:

$$\delta \Pi = \frac{\partial \Pi}{\partial u} \delta u \overset{!}{=} 0. \tag{12.2}$$

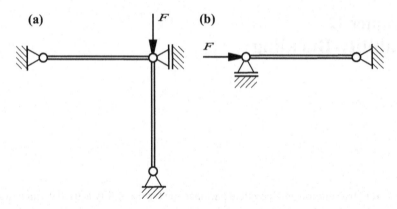

Fig. 12.1 Buckling of structures at **a** two elements **b** one element

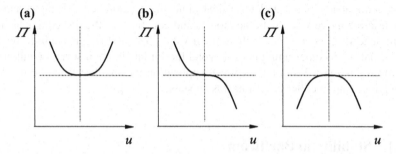

Fig. 12.2 States of equilibrium: **a** stable equilibrium, **b** indifferent or neutral equilibrium, **c** unstable equilibrium

For the clarification of the type of the stationary value the second variation of the potential has to be analyzed. Three states of equilibrium arise, see Fig. 12.2.

In the case of $\delta^2 \Pi > 0$ a stable equilibrium occurs. If the second variation disappears, one talks about an indifferent or neutral equilibrium. In the case of $\delta^2 \Pi < 0$ an unstable equilibrium exists. When buckling of bars and beams occurs, one assumes an indifferent balance. The second variation is called:

$$\delta^2 \Pi = \frac{\partial^2 \Pi}{\partial u^2} \delta^2 u = 0. \tag{12.3}$$

The demand for the second variation of Π to disappear can solely be fulfilled if the determinant of K becomes 0.

The stiffness matrix K for large deformations consists of an elastic and geometric fraction:

$$K = K^{\text{el}} + K^{\text{geo}}. \tag{12.4}$$

K^{el} stands for the stiffness matrix, which serves as a basis in the description of the linear elastic reaction. It is already known from the previous chapters. The assembly of K^{el} is independent of the axial load. In contrast, K^{geo} contains the axial load F as a prefactor. The detailed derivation of the geometric stiffness matrix K^{geo} follows later on.

If this force is scaled with a factor λ, one obtains:

$$K = K^{\text{el}} + \lambda \tilde{K}^{\text{geo}}. \tag{12.5}$$

The requirement that the determinant K has to disappear leads to:

$$\det(K) = \det\left[K^{\text{el}} + \lambda \tilde{K}^{\text{geo}} \right] \overset{!}{=} 0. \tag{12.6}$$

With this equation, an eigenvalue problem is formulated, where λ represents the unknown value. The formation of the determinant leads to a scalar function in λ, which is called the characteristic equation. It is obvious that this equation does not just possess a single eigenvalue. The roots of the characteristic equation correspond to the eigenvalues of the problem. The expression λF stands for the so-called buckling load. From a technical point of view, the smallest eigenvalue and therefore the smallest buckling load are interesting.

12.2 Large Deformations

Thus far, it was assumed that the occurring deformations are small. Equilibrium was established on the undeformed body. Within the discussion of nonlinear problems however, large deformations can occur. Those will now be described in more detail. The linear relation between the deformations and strains will be complemented by the nonlinear term in the strain-deformation relation

$$\varepsilon = \frac{1}{2}\left(\nabla u^{\text{T}} + u \nabla^{\text{T}} \right) + \frac{1}{2}\left(\nabla u^{\text{T}} \times u \nabla^{\text{T}} \right). \tag{12.7}$$

The second addend expresses the nonlinear term. During large deformations of the considered bars and beams a deformation in axial direction as well as another deformation occur. The complete strain matrix is called:

$$\varepsilon = \begin{pmatrix} \varepsilon_{xx} & \varepsilon_{xy} \\ \varepsilon_{yx} & \varepsilon_{yy} \end{pmatrix} \tag{12.8}$$

and results from:

$$\varepsilon = \frac{1}{2} \left[\begin{pmatrix} \dfrac{\partial}{\partial x} \\ \dfrac{\partial}{\partial y} \end{pmatrix} (u_x \, u_y) + \begin{pmatrix} u_x \\ u_y \end{pmatrix} \begin{pmatrix} \dfrac{\partial}{\partial x} & \dfrac{\partial}{\partial y} \end{pmatrix} \right] \qquad (12.9)$$

$$+ \frac{1}{2} \left[\begin{pmatrix} \dfrac{\partial}{\partial x} \\ \dfrac{\partial}{\partial y} \end{pmatrix} (u_x \, u_y) \times \begin{pmatrix} u_x \\ u_y \end{pmatrix} \begin{pmatrix} \dfrac{\partial}{\partial x} & \dfrac{\partial}{\partial y} \end{pmatrix} \right] . \qquad (12.10)$$

For further considerations only the elongation ε_{xx} towards the bar or beam axis is of relevance. This component can be extracted as

$$\varepsilon_{xx} = \frac{du_x}{dx} + \frac{1}{2} \left[\left(\frac{du_x}{dx} \right)^2 + \left(\frac{du_y}{dx} \right)^2 \right] \qquad (12.11)$$

from the complete strain matrix. Under the condition $du_x/dx \ll 1$ as well as $(du_x/dx)^2 \ll (du_y/dx)^2$ the entire term simplifies to

$$\varepsilon_{xx} = \frac{du_x}{dx} + \frac{1}{2} \left(\frac{du_y}{dx} \right)^2 . \qquad (12.12)$$

This relation for the strain can be used directly for bars. For beams, the complete deformation results from two parts

$$u_x = u_{xs} + u_{xb}. \qquad (12.13)$$

The first term represents the amount of deformation on the neutral axis of the beam. The second term represents the amount of pure bending and can be described as

$$u_{xb} = -y \frac{du_y}{dx} . \qquad (12.14)$$

With that said, the entire strain of the beam is represented in the following form:

$$\varepsilon_{xx} = \frac{du_{xs}}{dx} - y \frac{d^2 u_y}{dx^2} + \frac{1}{2} \left(\frac{du_y}{dx} \right)^2 . \qquad (12.15)$$

The elastic strain energy of the bar can be formulated as

$$\Pi_{\text{int}} = \frac{1}{2} \int_{\Omega} E \varepsilon_{xx}^2 \, d\Omega = \int_{\Omega} E \left[\frac{du_x}{dx} + \frac{1}{2} \left(\frac{du_y}{dx} \right)^2 \right]^2 d\Omega \qquad (12.16)$$

through the strains. After a few transformations the strain energy converts into the following:

$$\Pi_{\text{int}} = \frac{1}{2} AE \int\limits_L \left(\frac{du_x}{dx} \right)^2 dx + \frac{1}{2} F \int\limits_L \left(\frac{du_y}{dx} \right)^2 dx . \tag{12.17}$$

The elastic strain energy of the beam can be formulated as

$$\Pi_{\text{int}} = \frac{1}{2} \int\limits_\Omega E\varepsilon_{xx}^2 d\Omega = \frac{1}{2} \int\limits_\Omega E \left[\frac{du_{xs}}{dx} - y \frac{d^2 u_y}{dx^2} + \frac{1}{2} \left(\frac{du_y}{dx} \right)^2 \right]^2 d\Omega \tag{12.18}$$

through the strains. After a few transformations the strain energy converts into the following:

$$\Pi_{\text{int}} = \frac{1}{2} AE \int\limits_L \left(\frac{du_{xs}}{dx} \right)^2 dx + \frac{1}{2} EI \int\limits_L \left(\frac{d^2 u_y}{dx^2} \right)^2 dx + \frac{1}{2} F \int\limits_L \left(\frac{du_y}{dx} \right)^2 dx .$$

$$\tag{12.19}$$

The 1st and 3rd term is equivalent to the strain energy of the bar. The 2nd term is equivalent to the energy fraction of the bending.

12.3 Stiffness Matrices in Large Deformations

As for the small deformations, it should also be assumed that for large deformations the course through the nodal values and shape functions can be described. Figure 12.3 shows the kinematical quantities, which are relevant for buckling.

In general, various shape functions can be used for different directions of the displacement:

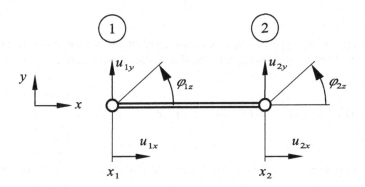

Fig. 12.3 State variables for buckling behavior

$$u_x(x) = N_x(x)\,u_{\mathrm{p}}\,,\tag{12.20}$$

$$u_y(x) = N_y(x)\,u_{\mathrm{p}}\,.\tag{12.21}$$

In the following, the stiffness matrices for large deformations for the bar and the beam will be derived.

12.3.1 Bar with Large Deformations

Within the strains according to Eq. (12.12) the first derivatives $\mathrm{d}u_x/\mathrm{d}x$ and $\mathrm{d}u_y/\mathrm{d}x$ occur. These lead to

$$\frac{\mathrm{d}u_x(x)}{\mathrm{d}x} = \frac{\mathrm{d}}{\mathrm{d}x}N_x(x)\,u_{\mathrm{p}} = N'_x(x)\,u_{\mathrm{p}}\,,\tag{12.22}$$

$$\frac{\mathrm{d}u_y(x)}{\mathrm{d}x} = \frac{\mathrm{d}}{\mathrm{d}x}N_y(x)\,u_{\mathrm{p}} = N'_y(x)\,u_{\mathrm{p}}\,.\tag{12.23}$$

In discretized form the entire potential can be formulated as

$$\Pi = \frac{1}{2}u_{\mathrm{p}}^{\mathrm{T}}\,AE\underbrace{\int_L N'^{\mathrm{T}}_x N'_x\,\mathrm{d}x}_{\boldsymbol{K}^{\mathrm{el,bar}}}\,u_{\mathrm{p}} + \frac{1}{2}u_{\mathrm{p}}^{\mathrm{T}}\,F\underbrace{\int_L N'^{\mathrm{T}}_y N'_y\,\mathrm{d}x}_{\boldsymbol{K}^{\mathrm{geo,bar}}}\,u_{\mathrm{p}} - u_{\mathrm{p}}^{\mathrm{T}}\boldsymbol{F}\,.\tag{12.24}$$

The stiffness matrices can be determined from Eq. (12.24). The submatrices result in:

$$\boldsymbol{K}^{\mathrm{el,\,bar}} = AE\int_L N'^{\mathrm{T}}_x N'_x\,\mathrm{d}x\,,\tag{12.25}$$

$$\boldsymbol{K}^{\mathrm{geo,\,bar}} = F\int_L N'^{\mathrm{T}}_y N'_y\,\mathrm{d}x\,.\tag{12.26}$$

Depending on the type of shape function, different stiffness matrices result. The shape functions for the displacement field $\boldsymbol{u}_x(x)$ are described in Chap. 3. With

$$\boldsymbol{u}_y(x) = N_1(x)\,u_{1y} + N_2(x)\,u_{2y} = \begin{bmatrix}0 & N_1(x) & 0 & N_2(x)\end{bmatrix}\begin{bmatrix}u_{1x}\\u_{1y}\\u_{2x}\\u_{2y}\end{bmatrix}\tag{12.27}$$

an appropriate approach for the displacement field $\boldsymbol{u}_y(x)$ is chosen. First of all, with

$$N_1(x) = \frac{1}{L}(x_2 - x) \quad \text{and} \quad N_2(x) = \frac{1}{L}(x - x_1) \tag{12.28}$$

a simple, linear approach for the deformation perpendicular to the principal bar axis is chosen. The derivatives of the shape functions N'_y are needed in Eq. (12.25). These result in:

$$\frac{\partial N_y(x)}{\partial x} = \begin{bmatrix} 0 & \dfrac{\partial N_1(x)}{\partial x} & 0 & \dfrac{\partial N_2(x)}{\partial x} \end{bmatrix} = \begin{bmatrix} 0 & -\dfrac{1}{L} & 0 & +\dfrac{1}{L} \end{bmatrix}. \tag{12.29}$$

With this, the integration $\int_L {N'_y}^T N'_y \, dx$ can be conducted. The geometric stiffness matrix, in dependence on the external load F results in:

$$K^{\text{geo, bar}} = F \int_L \frac{1}{L^2} \begin{bmatrix} 0 & 0 & 0 & 0 \\ 0 & 1 & 0 & -1 \\ 0 & 0 & 0 & 0 \\ 0 & -1 & 0 & -1 \end{bmatrix} dx = \frac{F}{L} \begin{bmatrix} 0 & 0 & 0 & 0 \\ 0 & 1 & 0 & -1 \\ 0 & 0 & 0 & 0 \\ 0 & -1 & 0 & -1 \end{bmatrix}. \tag{12.30}$$

With this, the overall stiffness matrix can be assembled through two submatrices

$$K^{\text{bar}} = \frac{EA}{L} \begin{bmatrix} 1 & 0 & -1 & 0 \\ 0 & 0 & 0 & 0 \\ -1 & 0 & 1 & 0 \\ 0 & 0 & 0 & 0 \end{bmatrix} + \frac{F}{L} \begin{bmatrix} 0 & 0 & 0 & 0 \\ 0 & 1 & 0 & -1 \\ 0 & 0 & 0 & 0 \\ 0 & -1 & 0 & 1 \end{bmatrix}. \tag{12.31}$$

12.3.2 Beams with Large Deformations

In the strains according to Eq. (12.15) the first derivatives du_{xs}/dx and du_y/dx as well as the second derivative $d^2 u_y/dx^2$ occur. These result in

$$\frac{du_{xs}(x)}{dx} = \frac{d}{dx} N_x(x) \, u_p = N'_x(x) \, u_p, \tag{12.32}$$

$$\frac{du_y(x)}{dx} = \frac{d}{dx} N_y(x) \, u_p = N'_y(x) \, u_p, \tag{12.33}$$

$$\frac{d^2 u_y(x)}{dx^2} = \frac{d^2}{dx^2} N_y(x) \, u_p = N''_y(x) \, u_p. \tag{12.34}$$

With this, the entire potential can be presented in discretized form as

$$\Pi = \frac{1}{2}\boldsymbol{u}_{\mathrm{p}}^{\mathrm{T}}\, AE \underbrace{\int_L \boldsymbol{N}_x'^{\mathrm{T}} \boldsymbol{N}_x'\mathrm{d}x}\, \boldsymbol{u}_{\mathrm{p}} + \frac{1}{2}\boldsymbol{u}_{\mathrm{p}}^{\mathrm{T}}\, EI \underbrace{\int_L \boldsymbol{N}_y''^{\mathrm{T}} \boldsymbol{N}_y''\mathrm{d}x}\, \boldsymbol{u}_{\mathrm{p}}$$

$$\underbrace{}_{\boldsymbol{K}^{\mathrm{el,\,bar}}} \qquad\qquad \underbrace{}_{\boldsymbol{K}^{\mathrm{el,\,bending}}}$$

$$+ \frac{1}{2}\boldsymbol{u}_{\mathrm{p}}^{\mathrm{T}}\, F \underbrace{\int_L \boldsymbol{N}_y'^{\mathrm{T}} \boldsymbol{N}_y'\mathrm{d}x}_{\boldsymbol{K}^{\mathrm{geo}}}\, \boldsymbol{u}_{\mathrm{p}} - \boldsymbol{u}_{\mathrm{p}}^{\mathrm{T}}\boldsymbol{F} \tag{12.35}$$

The elastic stiffness matrix $\boldsymbol{K}^{\mathrm{el}}$ consists of the fractions $\boldsymbol{K}^{\mathrm{el,\,bar}}$ and $\boldsymbol{K}^{\mathrm{el,\,bending}}$. The geometric stiffness matrix $\boldsymbol{K}^{\mathrm{geo}}$ is represented in the third term. The stiffness matrices can be calculated through Eq. (12.35). The submatrices result in:

$$\boldsymbol{K}^{\mathrm{el,\,bar}} = AE \int_L \boldsymbol{N}_x'^{\mathrm{T}} \boldsymbol{N}_x'\mathrm{d}x \,, \tag{12.36}$$

$$\boldsymbol{K}^{\mathrm{el,bending}} = EI \int_L \boldsymbol{N}_y''^{\mathrm{T}} \boldsymbol{N}_y''\mathrm{d}x \,, \tag{12.37}$$

$$\boldsymbol{K}^{\mathrm{geo}} = F \int_L \boldsymbol{N}_y'^{\mathrm{T}} \boldsymbol{N}_y'\mathrm{d}x \,. \tag{12.38}$$

According to the usual procedure the fraction from the elongation of the bar is approximately disregarded when describing the beam. For further considerations only the fraction from the bending is considered. Depending on the type of shape function different stiffness matrices occur. The general approach was already introduced in Chap. 5 and reads as follows:

$$\boldsymbol{u}_y(x) = N_1(x)\, u_{1y} + N_2(x)\, \varphi_1 + N_3(x)\, u_{2y} + N_4(x)\, \varphi_2 \,. \tag{12.39}$$

A cubic approach for the deformation perpendicular to the axial direction is chosen. The following shape functions are already known through Chap. 5

$$\begin{aligned}
N_1(x) &= 1 - 3\frac{x^2}{L^2} + 2\frac{x^3}{L^3}, \\
N_2(x) &= x - 2\frac{x^2}{L} + \frac{x^3}{L^2}, \\
N_3(x) &= 3\frac{x^2}{L^2} - 2\frac{x^3}{L^3}, \\
N_4(x) &= -\frac{x^2}{L} + \frac{x^3}{L^2}.
\end{aligned} \tag{12.40}$$

The derivatives of the shape functions N_y' are needed in Eq. (12.35). These result in:

$$\frac{\partial N_1(x)}{\partial x} = -\frac{6x}{L^2} + \frac{6x^2}{L^3},$$
$$\frac{\partial N_2(x)}{\partial x} = 1 - \frac{4x}{L} + \frac{3x^2}{L^2},$$
$$\frac{\partial N_3(x)}{\partial x} = \frac{6x}{L^2} - \frac{6x^2}{L^3},$$
$$\frac{\partial N_4(x)}{\partial x} = -\frac{2x}{L} + \frac{3x^2}{L^2}.$$

(12.41)

With this, the integration $\int_L N_y'^{\mathsf{T}} N_y' \, dx$ can be done. The integration is shown as an example using the matrix element $(1,1)$:

$$k_{11} = \int_0^L \left(-\frac{6x}{L^2} + \frac{6x^2}{L^3} \right)^2 dx = \frac{36}{L^4} \int_0^L \left(-x + \frac{x^2}{L} \right)^2 dx =$$
$$= \frac{36}{L^4} \left[\frac{1}{3}x^3 - \frac{1}{2L}x^4 + \frac{1}{5L^2}x^5 \right]_0^L = \frac{36}{L^4}\frac{L^3}{30} = \frac{36}{30L}.$$

(12.42)

The geometric stiffness matrix, depending on the external load F results in:

$$\boldsymbol{K}^{\text{geo}} = \frac{F}{30L} \begin{bmatrix} 36 & 3L & -36 & 3L \\ & 4L^2 & -3L & -L^2 \\ & & 36 & -3L \\ \text{sym.} & & & 4L^2 \end{bmatrix}.$$

(12.43)

The entire stiffness matrix therefore consists of the two submatrices

$$\boldsymbol{K} = \frac{EI}{L^3} \begin{bmatrix} 12 & 6L & -12 & 6L \\ 6L & 4L^2 & -6L & 2L^2 \\ -12 & -6L & 12 & -6L \\ 6L & 2L^2 & -6L & 4L^2 \end{bmatrix} + \frac{F}{30L} \begin{bmatrix} 36 & 3L & -36 & 3L \\ 3L & 4L^2 & -3L & -L^2 \\ -36 & -3L & 36 & -3L \\ 3L & -L^2 & -3L & 4L^2 \end{bmatrix}.$$

(12.44)

12.4 Examples of Buckling: The Four Euler's Buckling Cases

Given is a prismatic beam, which is loaded with a concentrated force F in axial direction on one end. The beam has a cross-sectional area A, the second moment of area I and the modulus of elasticity E. All factors are constant along the body axis. Required are the critical load F_{crit} and the buckling length L_{crit}, respectively (Fig. 12.4).

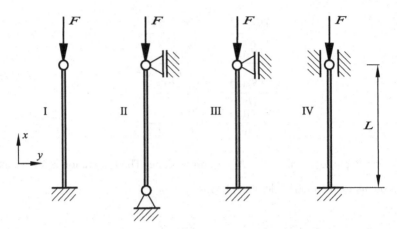

Fig. 12.4 The four EULER's buckling cases

The four EULER's buckling cases vary according to the boundary conditions on both ends.

12.4.1 Analytical Solutions for Euler's Buckling Loads

The differential equation of the buckling problem is [1]:

$$u_y'''' + \lambda^2 u_y'' = 0 \quad \text{with} \quad \lambda^2 = \frac{F}{EI}. \tag{12.45}$$

The general solution of this differential equation

$$u_y(x) = \bar{A}\,\cos(\lambda x) + \bar{B}\,\sin(\lambda x) + \bar{C}\,\lambda x + \bar{D} \tag{12.46}$$

consists of four constant terms.[1] The constant term $\bar{D}$ describes the translational rigid-body motion of the beam, the term $\bar{C}\,\lambda x$ describes the rigid-body rotation of the beam around the origin. The trigonometrical parts describe the deformation of the beam in the deformed position. The constant terms $\bar{A}, \bar{B}, \bar{C}$ and $\bar{D}$ can be determined from the boundary conditions. Required are the derivatives of the deformation from Eq. (12.46):

$$u_y'(x) = -\bar{A}\,\lambda\sin(\lambda x) + \bar{B}\,\lambda\cos(\lambda x) + \bar{C}\,\lambda, \tag{12.47}$$

$$u_y''(x) = -\bar{A}\,\lambda^2\cos(\lambda x) - \bar{B}\,\lambda^2\sin(\lambda x), \tag{12.48}$$

$$u_y'''(x) = +\bar{A}\,\lambda^3\sin(\lambda x) - \bar{B}\,\lambda^3\cos(\lambda x), \tag{12.49}$$

---

[1]In order to avoid conflicts with other variables the constants are headed by a bar.

Table 12.1 Buckling load and buckling length

EULER case	I	II	III	IV
Buckling load $F_{\text{crit}} = \pi^2 \dfrac{EI}{L^2} \times$	$\frac{1}{4}$	1	1.43^2	4
Buckling length $L_{\text{crit}} = L \times$	2	1	$\dfrac{1}{1.43}$	$\dfrac{1}{2}$

$$u_y^{IV}(x) = +\bar{A}\,\lambda^4 \cos(\lambda x) + \bar{B}\,\lambda^4 \sin(\lambda x)\,. \tag{12.50}$$

Table 12.1 shows the critical loads and buckling lengths for the EULER's buckling cases. Analogous to the critical load, the buckling lengths L_{crit} for EULER's buckling can be introduced. The critical load and the buckling length are, regardless of the boundary conditions, connected through

$$F_{\text{crit}} = \pi^2 \frac{EI}{L_{\text{crit}}^2}\,. \tag{12.51}$$

These values serve as a reference for the established solutions of the finite element method.

12.4.2 The Finite Element Method

The basis for the finite element analysis of the buckling behavior of beams is the stiffness matrix (12.44). With the abbreviations $e = \frac{EI}{L^3}$ and $f = \frac{F}{30L}$ the compact form of the entire stiffness matrix results in:

$$K = \begin{bmatrix} 12e - 36\lambda f & 6eL - 3\lambda f L & -12e + 36\lambda f & 6eL - 3\lambda f L \\ 6eL - 3\lambda f L & 4eL^2 - 4\lambda f L^2 & -6eL + 3\lambda f L & 2eL^2 + \lambda f L^2 \\ -12e + 36\lambda f & -6eL + 3\lambda f L & 12e - 36\lambda f & -6eL + 3\lambda f L \\ 6eL - 3\lambda f L & 2eL^2 + 3\lambda f L^2 & -6eL + 3\lambda f L & 4eL^2 - 4\lambda f L^2 \end{bmatrix}\,. \tag{12.52}$$

The four EULER's cases differ due to the boundary conditions. In the following the first EULER's buckling case will be described. Node 1 is clamped firmly. By this, the displacement u_{1x} and the rotation φ_{1z} vanish. The most simple finite element model consists of exactly one beam. The rows 1 and 2 as well as the columns 1 and 2 will be deleted in the system matrix. What remains is a reduced submatrix:

$$K^{\text{red}} = \begin{bmatrix} 12e - 36\lambda f L & -6eL + 3\lambda f \\ -6eL + 3\lambda f L & 4eL^2 - 4\lambda f L^2 \end{bmatrix}\,. \tag{12.53}$$

To define the eigenvalue λ_i, the determinant of the reduced system matrix has to be constituted. This leads to the characteristic equation. Two solutions result from the quadratic equation. For statements regarding the stability, the smallest eigenvalue is of relevance. With this, the following emerges for the buckling load:

$$F_{\text{crit}} = \lambda_{\min} F = \frac{4}{3}(13 - 2\sqrt{31})\frac{EI}{L^2} = 2.486\frac{EI}{L^2}. \qquad (12.54)$$

Compared to the analytical solution $F = \frac{\pi^2}{4}\frac{EI}{L^2}$ an error of 0.8% occurs.

12.5 Supplementary Problems

12.1 Entries of the geometric stiffness matrix

In the above descriptions, the integration $\int_L N_y'^{\mathrm{T}} N_y' \mathrm{d}x$ for an assembly of the geometric stiffness matrix of a bending beam has only been shown for the matrix element (1,1).

The geometric stiffness matrix can be defined for a cubic displacement approach perpendicular to the bending axis in all matrix elements.

12.2 Euler's buckling cases II, III and IV, one element

The above descriptions relate to EULER's buckling case I. Unknown are the finite element solution for the buckling load for EULER's buckling cases II, III and IV. Use *one* single beam element to discretize the buckling problem:

1. Set up of the system matrix consisting of elastic and geometric stiffness matrix.
2. Definition of the eigenvalues.

12.3 Euler's buckling cases, two elements

Unknown are finite element solutions for the critical buckling load of EULER's buckling cases I, II, III and IV. Use *two* beam elements to discretize the buckling beam.

12.4 Euler's buckling cases, error in regard to analytical solution

With the help of the finite element method the error from the determined solution of the critical buckling load in regard to the analytical solution can be discussed depending on the number of applied elements. Unknown are the finite element solution for critical buckling load of EULER's buckling cases I, II, III and IV. Thereby the bending beam can be discretized with various beam elements.

Reference

1. Gross D, Hauger W, Schröder J, Wall WA (2009) Technische Mechanik 2: Elastostatik. Springer, Berlin

Chapter 13
Dynamics

Abstract The transient behavior of the acting loads on the structure will be introduced additionally into the analysis within this chapter on dynamics. The procedure
for the analysis of dynamic problems depends essentially on the character of the
time course of the loads. At deterministic loads the column matrix of the external
loads is a given function of the time. The major amount of problems in engineering,
plant and vehicle construction can be analyzed under this assumption. In contrast to
that, randomness is relevant in the case of stochastic loads. Such cases will not be
regarded here. For deterministic loads a distinction is drawn between

- periodic and non-periodic,
- slow and fast changing load-time functions (relatively related to the dynamic eigen-
  behaviour of the structure).

In the following chapter linear dynamic processes will be considered, which can be
traced back to an external stimulation. The field of self-excited oscillation will not
be covered.

13.1 Principles of Linear Dynamics

The point of origin is an elastic continuum with mass which is, in contrast to previous
problems, stressed with time-dependent loads [1–3]. The mass with density ρ extends
over the volume Ω (see Fig. 13.1).

For dealing with dynamic problems in the context of the finite element method
various model assumptions can be discussed:

1. the distribution of the mass and
2. the treatment of the time dependency of all involved parameters.

Distribution of Mass

Within the framework of the FE method the continuum will be discretized. A first
model assumes that the distribution of the masses is not influenced by the discretization. The masses are also distributed continuously in the discretized condition. Figure 13.2a shows the continuously distributed mass for a bar. Another model

© Springer International Publishing AG, part of Springer Nature 2018   319
A. Öchsner and M. Merkel, *One-Dimensional Finite Elements*,
https://doi.org/10.1007/978-3-319-75145-0_13

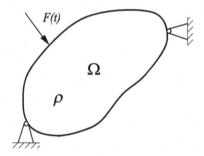

Fig. 13.1 Elastic continuum with mass under time-dependent load

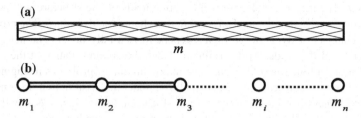

Fig. 13.2 Models of dynamic systems with **a** continuously and **b** discrete distributed masses

assumes that the originally continuously distributed mass can be concentrated on discrete points (see Fig. 13.2b).

The total mass

$$m = \sum_i^n m_i \qquad (13.1)$$

of the system remains. The connection between the points with mass will be applied with elements without mass, which may represent further physical properties, for example stiffness.

Time Dependency of Variables

Regarding the time dependency of the state variables, both the loads and the deformations as response of the system to the external loads are time changeable. Depending on the character of the external load, different problem areas are distinguished in dynamics (see Fig. 13.3) and pursue different strategies for the solution:

- Modal analysis
  Here the vibration behavior is considered without external loads. Eigenfrequency and eigenmodes are determined.
- Forced vibrations
  An external periodic force excites the component to resonate in the excitation frequency.

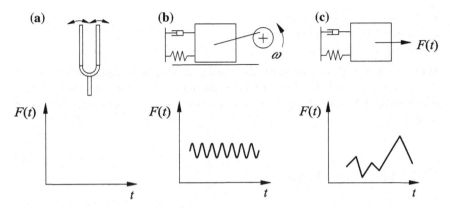

Fig. 13.3 Solution strategies structured according to the time course of the external loads: **a** impulse excitation, **b** sinusoidal excitation, **c** arbitrary excitation

- Transient analysis
  The external stimulating force $F(t)$ is an arbitrary non-periodic function of time.

Problem Definition

In addition to the elastic forces at pure static problems, inertia forces and frictional forces occur. According to the principle of d'ALEMBERT, these forces are at equilibrium with the external forces at all times:

$$F_m + F_c + F_k = F(t) . \tag{13.2}$$

In Eq. (13.2)

- F_m stands for the column matrix of the inertia forces,
- F_c stands for the column matrix of the damping forces, whereas in the following a velocity-related damping is assumed,
- F_k stands for the column matrix of the elastic reset forces and
- F stands for the column matrix of the external acting forces.

In static problems the deformation state on the inside of an element

$$u^e(x) = N(x)\, u_p \tag{13.3}$$

is expressed through shape functions and nodal displacements. This assumption also applies for dynamics. With $\ddot{u}$ as acceleration and second derivative of the displacement after the time and $\dot{u}$ as velocity and first derivative of the displacement after the time one obtains a differential equation

$$M\ddot{u} + C\dot{u} + Ku = F(t) \tag{13.4}$$

in the displacements u as basic equation of dynamics. Thereby

- M stands for the mass matrix,
- C stands for the damping matrix and
- K stands for the stiffness matrix, which is already known from statics.

In the continuum, this equation stands for a partial differential equation in space and time (wave equation). Resulting from the spatial discretization in the framework of the FE method, Eq. (13.4) only represents a system of common differential equations in time.

13.2   The Mass Matrices

The structure of the mass matrices is essentially defined through the assumption of the distribution of the masses. For an element, with continuously distributed mass, equivalent forces, which are acting in the nodal points can be determined via the principle of virtual work

$$\Delta u_p^T M u_p = \int_\Omega \rho (\Delta u)^T u \, d\Omega. \tag{13.5}$$

With the approaches for the displacements u and the accelerations $\ddot{u}$ one obtains the following as the mass matrix

$$M = \int_\Omega \rho N^T N d\Omega . \tag{13.6}$$

Consideration of the frictional forces F_c leads to the damping matrix

$$C = \int_\Omega N^T \mu N d\Omega . \tag{13.7}$$

For a distribution of the masses in discrete points, the mass matrix can be determined much easier. The proceeding will be shown in Sect. 13.6.2 for the example of the axial vibration of a bar.

13.3   Modal Analysis

An elastic, mass containing structure reacts to a time limited, external stimulation with a response in certain frequencies and modes of oscillation, whose entity is considered as an eigensystem of eigenfrequency and eigenmode. A basic assumption for the solution is that the changeable displacements in space and time are described in a separation approach

$$u(x, t) = \boldsymbol{\Phi}(x)\, q(t), \tag{13.8}$$

whereby with $\boldsymbol{\Phi}(x)$ the dependence of the displacement on space and with $q(t)$ the dependence of the displacement on the time are described.

Development of Eigenmodes and Eigenfrequencies

For low damped systems the eigenmodes can be established from the corresponding undamped system:

$$\boldsymbol{M\ddot{u}} + \boldsymbol{Ku} = \boldsymbol{0}. \tag{13.9}$$

With the approach for the displacement:

$$u(x, t) = \Phi(x)\, e^{i\omega t} \tag{13.10}$$

this leads to the generalized eigenvalue problem

$$(-\omega^2 \boldsymbol{M} + \boldsymbol{K})\, \boldsymbol{\Phi} = \boldsymbol{0}. \tag{13.11}$$

The nontrivial solutions (standing for the statics) one obtains from

$$\det(-\omega^2 \boldsymbol{M} + \boldsymbol{K}) = 0. \tag{13.12}$$

the ω_i, $i = 1, \ldots, n$, which fulfill the equation, are referred to as eigenfrequencies, and the corresponding $\boldsymbol{\Phi}_i$ are referred to as eigenmodes of the system with n degrees of freedom.[1] The single eigenmodes $\boldsymbol{\Phi}_i$ can be summarized in the modal matrix $\boldsymbol{\Phi}$

$$\boldsymbol{\Phi} = [\boldsymbol{\Phi}_1\ \boldsymbol{\Phi}_2\ \boldsymbol{\Phi}_3\ \ldots\ \boldsymbol{\Phi}_n]. \tag{13.13}$$

The eigenmodes possess essential characteristics:

1. The orthogonality of two eigenmodes:

$$\boldsymbol{\Phi}_i^{\mathrm{T}} \boldsymbol{\Phi}_j = 0\ ,\quad i \neq j. \tag{13.14}$$

2. The normalization regarding $\boldsymbol{M}$: The eigenmodes and therefore the eigenvectors can be $\boldsymbol{M}$-normalized. The eigenvectors can be stretched, so that an $\boldsymbol{M}$-orthonormality occurs. If the mass matrix $\boldsymbol{M}$ is multiplied with $\boldsymbol{\Phi}^{\mathrm{T}}$ from the left-hand and with $\boldsymbol{\Phi}$ from the right-hand, the modal mass matrix $\tilde{\boldsymbol{M}}$ results, which has entries exclusively on the main diagonal, to be precise a '1':

---

[1] With the eigenmodes the space dependent displacements are characterized. However, the absolute magnitude of any displacement cannot be determined. The reason is that the system (13.13) has always more unknowns than equations. For the illustration of eigenmodes one assigns a value for an arbitrary eigenmode and relates all other eigenmodes to that.

$$\boldsymbol{\Phi}^{\mathsf{T}}\boldsymbol{M}\boldsymbol{\Phi} = \tilde{\boldsymbol{M}} = \begin{bmatrix} 1 & 0 \\ 0 & 1 \end{bmatrix}. \tag{13.15}$$

If the stiffness matrix $\boldsymbol{K}$ is multiplied accordingly from the left-hand and the right-hand, the modal stiffness matrix $\tilde{\boldsymbol{K}}$ results, which has entries exclusively on the main diagonal, to be precise the squares of the eigenfrequencies ω_i:

$$\boldsymbol{\Phi}^{\mathsf{T}}\boldsymbol{K}\boldsymbol{\Phi} = \tilde{\boldsymbol{K}} = \begin{bmatrix} \omega_1^2 & 0 \\ 0 & \omega_2^2 \end{bmatrix}. \tag{13.16}$$

3. Modal damping: If the damping matrix $\boldsymbol{C}$ is multiplied accordingly from the left-hand and the right-hand, the modal damping matrix $\tilde{\boldsymbol{C}}$ results, which has entries exclusively on the main diagonal, to be precise the eigenfrequencies ω_i and the modal damping coefficients ζ_i:

$$\boldsymbol{\Phi}^{\mathsf{T}}\boldsymbol{C}\boldsymbol{\Phi} = \tilde{\boldsymbol{C}} = \begin{bmatrix} \omega_1\zeta_1 & 0 \\ 0 & \omega_2\zeta_2 \end{bmatrix}. \tag{13.17}$$

The damping approach is known under the name RAYLEIGH's damping and is possible, when the damping matrix is represented in the following form:

$$\boldsymbol{C} = \alpha\boldsymbol{M} + \beta\boldsymbol{K}. \tag{13.18}$$

4. Decoupling: In total, one receives an equivalent system of *decoupled* differential equations

$$\boldsymbol{\Phi}^{\mathsf{T}}\boldsymbol{M}\boldsymbol{\Phi} + \boldsymbol{\Phi}^{\mathsf{T}}\boldsymbol{C}\boldsymbol{\Phi} + \boldsymbol{\Phi}^{\mathsf{T}}\boldsymbol{K}\boldsymbol{\Phi} = 0, \tag{13.19}$$

which can be written in generalized displacements $\boldsymbol{q}$, also called modal coordinates, as

$$\ddot{q}_j + 2\omega_j\dot{q}_j + \omega_j^2 = \tilde{F}_j. \tag{13.20}$$

13.4 Forced Oscillation, Periodic Load

One talks about forced oscillation if a system suffers a periodic stimulation. Any eigenoscillations are decayed due to damping. Since every periodic stimulation can be analyzed via a FOURIER analysis, it is enough to assume single forces of the following kind

$$F(t) = F_0 e^{i\omega t}, \tag{13.21}$$

which take effect periodically with the frequency ω. In linear systems the total response comes from the superposition of the single responses. It can be assumed that in the equation of motion

$$M\ddot{u} + C\dot{u} + Ku = F_0 e^{i\omega t} \tag{13.22}$$

the deformation, the velocities and the accelerations can be illustrated as vectors of the following type

$$u(t) = u_0 e^{i(\omega t - \psi)}, \tag{13.23}$$

$$\dot{u}(t) = i\omega u_0 e^{i(\omega t - \psi)}, \tag{13.24}$$

$$\ddot{u}(t) = -\omega^2 u_0 e^{i(\omega t - \psi)}. \tag{13.25}$$

If the splitting of the complex displacement is inserted into the real and imaginary part

$$\mathbf{u}(t) = u_{\text{Re}} e^{i\omega t} + i u_{\text{Im}} e^{i\omega t} = \mathbf{u}_0 e^{i\omega t} \left(\cos \psi + i \sin \psi \right), \tag{13.26}$$

one obtains from

$$\left[-\omega^2 M \left(u_{\text{Re}} + i u_{\text{Im}} \right) + i\omega C \left(u_{\text{Re}} + i u_{\text{Im}} \right) + K \left(u_{\text{Re}} + i u_{\text{Im}} \right) \right] e^{i\omega t} = F_0 e^{i\omega t} \tag{13.27}$$

via

$$\left[\left(K - \omega^2 M \right) u_{\text{Re}} - \omega C u_{\text{Im}} \right] + i \left[\left(K - \omega^2 M \right) u_{\text{Im}} + \omega C u_{\text{Re}} \right] = F_0 \tag{13.28}$$

after the separation of the products of the real matrix with the complex vectors in the real and imaginary parts $2n$ equations of the following kind

$$\left(K - \omega^2 M \right) u_{\text{Re}} - \omega C u_{\text{Im}} = F_0, \tag{13.29}$$

$$\left(K - \omega^2 M \right) u_{\text{Im}} + \omega C u_{\text{Re}} = 0 \ . \tag{13.30}$$

With n degrees of freedom this is a solvable linear system of equations with the $2n$ unknowns of the respective n component of the real and imaginary part of the complex displacement $u = u_{\text{Re}} + i u_{\text{Im}}$. For each of the n degrees of freedom the amplitude is defined through

$$u_k = \sqrt{u_{k,\text{Re}}^2 + u_{k,\text{Im}}^2} \tag{13.31}$$

and the phase shift through

$$\psi_k = \arctan \left(\frac{u_{k,\mathrm{Im}}}{u_{k,\mathrm{Re}}} \right),\tag{13.32}$$

except the multiples of π.

13.5   Direct Methods of Integration, Transient Analysis

Transient dynamic analysis requires the integration of the equation of motion (13.4), which describes the correlation between acceleration, damping, deformation and the external force throughout the time interval of interest. Required are integration procedures, which identify the deformation in the regarded time interval from the equation of motion

$$\ddot{u}(t) = M^{-1}\left[F(t) - [C\dot{u}(t) + Ku(t)]\right].\tag{13.33}$$

An estimation of the displacement regarding time $t + \Delta t$ can be received through the expansion in series up to the 2nd element

$$u(t + \Delta t) \approx u(t) + \Delta t \dot{u}(t) + \frac{\Delta t^2}{2}\ddot{u}(t)\tag{13.34}$$

and an estimation of the velocity through an expansion of the series

$$\dot{u}(t + \Delta t) \approx \dot{u}(t) + \Delta t \ddot{u}(t),\tag{13.35}$$

which interrupts after the 1st element. To be able to make use of the integration instructions according to Eqs. (13.33) and (13.35), displacement and velocity from the initial moment t_0 have to be known. The necessity of two instructions follows from the fact that the equation of motion (13.4) represents a PDE of 2nd order in time (two time derivatives occur). With a sufficiently small Δt, the thus found displacement approximates the time course of the displacement $u(t)$ satisfactorily. The basic construction of the two mostly used integration procedures, which are ideally similar to the quadratic procedures (or procedures of 2nd order), will be described here.

13.5.1   Integration According to Newmark

In the time interval $[t, t + \Delta t]$ the constant averaged acceleration

$$\ddot{u}_m = \frac{1}{2}\left[\dot{u}(t) + \ddot{u}(t + \Delta t)\right]\tag{13.36}$$

is assumed. Therewith a quadratic course results for the displacement

$$u\,(t + \Delta t) = u\,(t) + \Delta t \dot{u}\,(t) + \frac{\Delta t^2}{4}\,[\ddot{u}\,(t) + \ddot{u}\,(t + \Delta t)] \tag{13.37}$$

and a linear course for the velocity $\dot{u}\,(t)$

$$\dot{u}\,(t + \Delta t) = \dot{u}\,(t) + \frac{\Delta t^2}{2}\,[\ddot{u}\,(t) + \ddot{u}\,(t + \Delta t)] \ . \tag{13.38}$$

Together with the equation of motion (13.4) at the point of time $t + \Delta t$

$$M\ddot{u}\,(t + \Delta t) + C\dot{u}\,(t + \Delta t) + Ku\,(t + \Delta t) = F\,(t + \Delta t) \tag{13.39}$$

three equations for three unknowns are available $u\,(t + \Delta t)$, $\dot{u}\,(t + \Delta t)$, $\ddot{u}\,(t + \Delta t)$. Setting $\Delta u = u\,(t + \Delta t) - u\,(t)$, for this increase of the displacement the following results

$$\Delta u = S^{-1}F\,(t + \Delta t) - Ku\,(t) + M\left[\ddot{u}\,(t) + \frac{4}{\Delta t}\dot{u}\,(t)\right] + C\dot{u}\,(t) \tag{13.40}$$

with

$$S = \frac{4}{\Delta t^2}M + \frac{2}{\Delta t}C + K. \tag{13.41}$$

The velocity $\dot{u}\,(t + \Delta t)$ and the acceleration $\ddot{u}\,(t + \Delta t)$ are calculated from Eqs. (13.35) and (13.33). The time integration according to NEWMARK in fact requires the often expensive calculation of these inverses, however allows relatively large time steps, so that this disadvantage is compensated for in many cases. In particular, for linear problems, in which the system matrices are not dependent on the actual displacements, this procedure can be used very effectively since the inverse S^{-1} only has to be calculated once.

13.5.2 Central Difference Method

The velocity $\dot{u}\,(t)$, as first derivative of the displacement according to time, can be approximated through the displacement to the times $t - \Delta t$ and $t + \Delta t$ at sufficiently small time step Δt through

$$\dot{u}\,(t) \approx \frac{u\,(t + \Delta t) - u\,(t - \Delta t)}{2\Delta t}\ . \tag{13.42}$$

The acceleration $\ddot{u}\,(t)$ as second derivative of the displacement according to time is approximated with

$$\ddot{u}\,(t) \approx \frac{u\,(t+\Delta t) - 2u\,(t) + u\,(t-\Delta t)}{\Delta t^2}. \tag{13.43}$$

If these relations are inserted into the equation of motion (13.4) at the point of time t, one obtains with the abbreviations $u_1 = u\,(t+\Delta t), u_0 = u\,(t)$ and $u_{-1} = u\,(t-\Delta t)$

$$M\frac{u_1 - 2u_0 + u_{-1}}{\Delta t^2} + C\frac{u_1 - u_{-1}}{2\Delta t} + Ku_0 = F\,(t) \tag{13.44}$$

a relation, from which the displacement $u_1 = u\,(t+\Delta t)$ can be calculated, if the displacement at the previous points of time t and $t - \Delta t$ are known:

$$u_1 = S^{-1}F\,(t) - \left(K - \frac{2M}{\Delta t^2}\right)u_0 - \left(\frac{M}{\Delta t^2} - \frac{C}{2\Delta t}\right)u_{-1} \tag{13.45}$$

with

$$S = \frac{1}{\Delta t^2}M + \frac{1}{2\Delta t}C. \tag{13.46}$$

To calculate the new displacement $u_1 = u(t+\Delta t)$, values of the displacement u at two points of time are necessary. Since for a transient problem, initial displacement and velocity and therefore according to Eq. (13.33) also the acceleration at the point of time $t = 0$ have to be known, a virtual displacement at the time $-\Delta t$ from the expansion of series is supplied

$$u_1 = u\,(t-\Delta t) \approx u\,(0) - \Delta t\dot{u}\,(0) + \frac{\Delta t^2}{2}\ddot{u}\,(0) \tag{13.47}$$

and, in the first time step, the displacement $u_1 = u\,(\Delta t)$ can be calculated.

The central difference method is *explicitly* named, since the displacement $u\,(t+\Delta t)$ is calculated from the conditions at the point of time t and not with an analysis of the equation of motion at the point of time $t + \Delta t$, while the *implicit* NEWMARK method considers the equilibrium of forces at the point of time $t + \Delta t$. This explicit method is of great importance for diagonal mass and damping matrices M and C, at which the inverses can easily be defined by

$$S = \begin{bmatrix} S_{1,1} & 0 & \cdots & 0 \\ 0 & S_{2,2} & \cdots & 0 \\ \cdots & \cdots & \cdots & \cdots \\ \cdots & \cdots & \cdots & \cdots \\ 0 & 0 & \cdots & S_{n,n} \end{bmatrix} \tag{13.48}$$

through

$$\mathbf{S}^{-1} = \begin{bmatrix} \frac{1}{S_{1,1}} & 0 & \cdots & 0 \\ 0 & \frac{1}{S_{2,2}} & \cdots & 0 \\ \cdots & \cdots & \cdots & \cdots \\ \cdots & \cdots & \cdots & \cdots \\ 0 & 0 & \cdots & \frac{1}{S_{n,n}} \end{bmatrix} \qquad (13.49)$$

with

$$S_{i,i} = \frac{M_i}{\Delta t^2} + \frac{C_i}{2\Delta t} \quad , \quad (i = 1 - n). \qquad (13.50)$$

The extremely fast, nonlinear crash programs, which conduct hundreds of thousands of integration steps during a calculation and in the process constantly calculate new matrices, make use of this or the herefrom derived methods. The time steps, with which the motion of a component can be calculated satisfactory, are clearly smaller than with the NEWMARK method, in return the calculations are done very easily and are outstandingly parallelizable, meaning very fast on computers with various or many processors. Furthermore only little storage space is needed, since the matrices in Eq. (13.50) never have to be derived completely.

13.6 Examples

So far, the introduced approaches can be discussed with the help of examples:

- Axial vibration of a bar and
- Bending vibration of a beam.

Essentially three models will be analysed:

1. The analytical solution, which results from the solution of the differential equation,
2. the solution according to the FE method, whereupon the masses are continuously distributed and
3. the solution according to the FE method, when the masses are concentrated on discrete points.

First of all the necessary mass and stiffness matrices will be given in general.

13.6.1 Definition of Mass and Stiffness Matrices

The general calculation instruction for the mass matrix

Fig. 13.4 Bar with degrees
of freedom for the dynamic
analysis

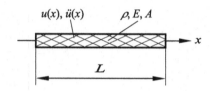

Fig. 13.5 Bar with linear
approach

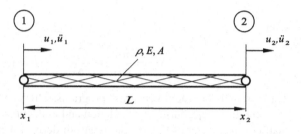

$$M^e = \int_\Omega \rho N^T N \, d\Omega \tag{13.51}$$

with continuously distributed mass and for the stiffness matrix

$$K^e = \int_\Omega B^T D B \, d\Omega \tag{13.52}$$

are known from previous chapters. In the following, the issue will be discussed with
the help of examples.

Axial Vibration of a Bar

In Fig. 13.4 the bar is drafted with degrees of freedom, which serve as a basis for
the analysis of the dynamic behavior. The names are closely connected with the
definition of the degrees of freedom in statics.

Besides the displacement $u(x)$ also the acceleration $\ddot{u}(x)$ is a state variable within
the considered system. Figure 13.5 illustrates the bar element with the degrees of
freedom for a linear approach. With linear shape functions the following mass matrix
results

$$M^e = \frac{\rho A L}{6} \begin{bmatrix} 2 & 1 \\ 1 & 2 \end{bmatrix} \tag{13.53}$$

and the following stiffness matrix results

$$K^e = \frac{EA}{L} \begin{bmatrix} 1 & -1 \\ -1 & 1 \end{bmatrix}. \tag{13.54}$$

The expression

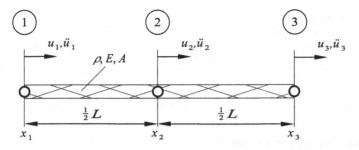

Fig. 13.6 Bar element with quadratic approach

$$M^e \ddot{u}^e + K^e u^e \tag{13.55}$$

can therefore be written as

$$\frac{\rho A L}{6} \begin{bmatrix} 2 & 1 \\ 1 & 2 \end{bmatrix} \begin{bmatrix} \ddot{u}_1 \\ \ddot{u}_2 \end{bmatrix} + \frac{EA}{L} \begin{bmatrix} 1 & -1 \\ -1 & 1 \end{bmatrix} \begin{bmatrix} u_1 \\ u_2 \end{bmatrix}. \tag{13.56}$$

Figure 13.6 shows the bar element with the degrees of freedom for a quadratic approach. With quadratic shape functions the mass matrix results in

$$M^e = \frac{\rho A L}{30} \begin{bmatrix} 4 & 2 & -1 \\ 2 & 16 & 2 \\ -1 & 2 & 4 \end{bmatrix} \tag{13.57}$$

and the stiffness matrix in:

$$K^e = \frac{EA}{3L} \begin{bmatrix} 7 & -8 & 1 \\ -8 & 16 & -8 \\ 1 & -8 & 7 \end{bmatrix}. \tag{13.58}$$

The expression

$$M^e \ddot{u}^e + K^e u^e \tag{13.59}$$

can therefore be written as

$$\frac{\rho A L}{30} \begin{bmatrix} 4 & 2 & -1 \\ 2 & 16 & 2 \\ -1 & 2 & 4 \end{bmatrix} \begin{bmatrix} \ddot{u}_1 \\ \ddot{u}_2 \\ \ddot{u}_3 \end{bmatrix} + \frac{EA}{3L} \begin{bmatrix} 7 & -8 & 1 \\ -8 & 16 & -8 \\ 1 & -8 & 7 \end{bmatrix} \begin{bmatrix} u_1 \\ u_2 \\ u_3 \end{bmatrix}. \tag{13.60}$$

Fig. 13.7 Bending beam
with degrees of freedom for
the dynamic analysis

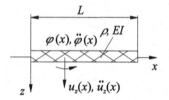

Bending Vibration of a Beam

In Fig. 13.7 the bending beam with the degrees of freedom is drafted, which serve as
a basis for the analysis of the dynamic behavior. The notation is closely connected
with the definition of the degrees of freedom in statics. First of all the influence
of the inertia in rotation can be disregarded. This issue will be introduced later on.
From static analysis the relation, how the deflection $u_z(x)$ at an arbitrary point x is
connected with the fixed nodal values u_{1z}, φ_{1y}, u_{2z} and φ_{2y} is already known. The
basis for this is an approach for the displacement in the form

$$u_z(x) = \sum_i^4 N_i(x)\, u_i \,. \tag{13.61}$$

With the four shape functions

$$
\begin{aligned}
N_1(x) &= 1 - \frac{3x^2}{L^2} + \frac{2x^3}{L^3}\,,\\
N_2(x) &= -x + \frac{2x^2}{L} - \frac{x^3}{L^2}\,,\\
N_3(x) &= \frac{3x^2}{L^2} - \frac{2x^3}{L^3}\,,\\
N_4(x) &= \frac{x^2}{L} - \frac{x^3}{L^2}
\end{aligned}
\tag{13.62}
$$

one obtains the following description for the deformation

$$
\begin{aligned}
u_z(x) &= \left(1 - \frac{3x^2}{L^2} + \frac{2x^3}{L^3}\right) u_{1z} + \left(-\frac{x}{L} + \frac{2x^2}{L^2} - \frac{x^3}{L^3}\right) L\,\varphi_{1y}\\
&\quad + \left(\frac{3x^2}{L^2} - \frac{2x^3}{L^3}\right) u_{2z} + \left(\frac{x^2}{L^2} - \frac{x^3}{L^3}\right) L\,\varphi_{2y}
\end{aligned}
\tag{13.63}
$$

regarding nodal values and shape functions.

Single entries can be determined with the shape functions from the calculation instruction for the mass matrix (13.51). For the bending beam 16 entries result for the mass matrix. The calculation will be illustrated as an example on the entries m_{11} and m_{12}. From the matrix element m_{11}

$$
\begin{aligned}
m_{11} &= \rho A \int_0^L \left(1 - \frac{3x^2}{L^2} + \frac{2x^3}{L^3}\right) dx \\
&= \rho A \int_0^L \left(1 - \frac{6x^2}{L^2} + \frac{4x^3}{L^3} + \frac{9x^4}{L^4} - \frac{12x^5}{L^5} + \frac{4x^6}{L^6}\right) dx \\
&= \rho A \left[x - \frac{2x^3}{L^2} + \frac{x^4}{L^3} + \frac{9x^5}{5L^4} - \frac{2x^6}{L^5} + \frac{4x^7}{7L^6}\right]\Big|_0^L \\
&= \frac{156}{420}\rho AL
\end{aligned}
$$
(13.64)

the matrix element m_{12}

$$
\begin{aligned}
m_{12} &= \rho A \int_0^L \left(1 - \frac{3x^2}{L^2} + \frac{2x^3}{L^3}\right) \times \left(-\frac{x}{L} + \frac{2x^2}{L^2} - \frac{x^3}{L^3}\right) L\, dx \\
&= \frac{22}{420}\rho AL^2,
\end{aligned}
$$
(13.65)

up to the matrix element m_{44}

$$
m_{44} = \rho A \int_0^L \left(\frac{x^2}{L^2} - \frac{x^3}{L^3}\right)^2 \times L^2 dx = \frac{4}{420}\rho AL^3
$$
(13.66)

the total mass matrix results

$$
M = \frac{\rho AL}{420}\begin{bmatrix} 156 & -22L & 54 & 13L \\ & 4L^2 & -13L & -3L \\ & & 156 & 22L \\ \text{sym.} & & & 4L^2 \end{bmatrix}
$$
(13.67)

for the description of the bending vibration in the bending beam.

So far, the influence of the cross-sectional rotation is disregarded. Besides the deflection u_z in z-direction the cross-section rotates around the y-axis. For the vibration behavior the inertia in rotation is additionally considered. The total mass matrix

$$M^e = \frac{\rho A L}{420} \begin{bmatrix} 156 & -22L & 54 & 13L \\ & 4L^2 & -13L & -3L^2 \\ & & 156 & 22L \\ \text{sym.} & & & 4L^2 \end{bmatrix}$$

$$+ \frac{\rho A L}{30} \left(\frac{I_y}{A \times L^2} \right) \begin{bmatrix} 36 & -3L & -36 & 3L \\ & 4L^2 & 3L & -L^2 \\ & & 36 & -3L \\ \text{sym.} & & & 4L^2 \end{bmatrix}$$

(13.68)

can be dispersed in a translational and a rotatory part. The expression I_y stands for the axial second moment of area of 2nd order around the y-axis. The first matrix corresponds with the already known matrix from consideration without the rotatory part.

13.6.2 Axial Vibration of a Bar

A prismatic tension bar serves as the point of origin, which has a continuously distributed mass (density ρ) and whose modulus of elasticity E and cross-sectional area A are constant. Unknown are the eigenfrequencies (Fig. 13.8).

From the differential equation for the axial vibration of a bar

$$\frac{\partial^2 u(x, t)}{\partial t^2} = \frac{E}{\rho} \frac{\partial^2 u(x, t)}{\partial x^2}$$

(13.69)

the eigenfrequencies result in:

$$\omega_n = \frac{2n - 1}{2} \pi \sqrt{\frac{E}{\rho L^2}} .$$

(13.70)

The first eigenfrequencies calculate for $n = 1, 2, 3, 4$ in the following:

$$\omega_1 = \frac{1}{2}\pi = 1.5708\sqrt{E/\rho L^2} ,$$

(13.71)

$$\omega_2 = \frac{3}{2}\pi = 4.7124\sqrt{E/\rho L^2} ,$$

(13.72)

$$\omega_3 = \frac{5}{2}\pi = 7.854\sqrt{E/\rho L^2} ,$$

(13.73)

$$\omega_4 = \frac{7}{2}\pi = 10.99\sqrt{E/\rho L^2} .$$

(13.74)

The following illustration shows a finite element discretization with continuously distributed mass. The total mass and stiffness matrix can be established by combining

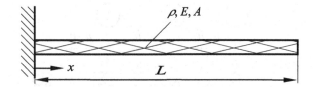

Fig. 13.8 Cantilever bar

the formulated matrices for a single element properly. Therefore one obtains the general equation of motion (Fig. 13.9):

$$
m
\begin{bmatrix}
2 & 1 & & & & \\
1 & 4 & 1 & & & \\
& 1 & 4 & 1 & & \\
& & \ddots & \ddots & \ddots & \\
& & & 1 & 4 & 1 \\
& & & & 1 & 2
\end{bmatrix}
\begin{bmatrix}
\ddot{u}_1 \\
\ddot{u}_2 \\
\ddot{u}_3 \\
\vdots \\
\ddot{u}_n
\end{bmatrix}
+ k
\begin{bmatrix}
1 & -1 & & & & \\
-1 & 2 & -1 & & & \\
& -1 & 2 & & & \\
& & \ddots & \ddots & \ddots & \\
& & & -1 & 2 & -1 \\
& & & & -1 & 2
\end{bmatrix}
\begin{bmatrix}
u_1 \\
u_2 \\
u_3 \\
\vdots \\
u_n
\end{bmatrix}
= 0
$$

(13.75)

with $m = \frac{\rho A \frac{1}{n} L}{6}$ and $k = \frac{EA}{\frac{1}{n}L}$. On the fixed support the boundary conditions $\ddot{u}_1 = 0$ (no acceleration) and $u_1 = 0$ (no displacement) apply. Hence the first row and the first column of a matrix in each case can be cancelled from the entire system of equations.

$$
\frac{\rho A \frac{1}{n} L}{6}
\begin{bmatrix}
4 & 1 & & & \\
1 & 4 & 1 & & \\
& \ddots & \ddots & \ddots & \\
& & 1 & 4 & 1 \\
& & & 1 & 2
\end{bmatrix}
\begin{bmatrix}
\ddot{u}_2 \\
\ddot{u}_3 \\
\vdots \\
\ddot{u}_N
\end{bmatrix}
+ \frac{EA}{\frac{1}{n}L}
\begin{bmatrix}
2 & -1 & & & \\
-1 & 2 & & & \\
& \ddots & \ddots & \ddots & \\
& & -1 & 2 & -1 \\
& & & -1 & 2
\end{bmatrix}
\begin{bmatrix}
u_2 \\
u_3 \\
\vdots \\
u_n
\end{bmatrix}
= 0.
$$

(13.76)

The matrices do not have a diagonal shape. Between two nodes (i) and $(i + 1)$ respectively there is a partial mass. This leads to secondary diagonals. In a nodal point (i) two *'finite masses'* bump against each other, in the endpoint only one. This becomes noticeable on the main diagonal in the last entry. Both matrices have a band structure with a band width of 3.

Lumped Mass Equivalent System (LMM)

For the lumped mass method (LMM), the continuously distributed mass will be concentrated on discrete points. When modeling the bar it needs to be considered that at the bar beginning and at the bar end only $m/2$ needs to be added respectively (see Fig. 13.10). One obtains the mass and stiffness matrix according to the above described procedure. The first row and the first column can be canceled in the equation of motion with $n + 1$ nodes since the following values for the displacement $u_0 = 0$ and the acceleration $\ddot{u}_0 = 0$ hold.

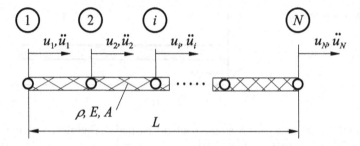

Fig. 13.9 FE discretization for axial vibration of a bar

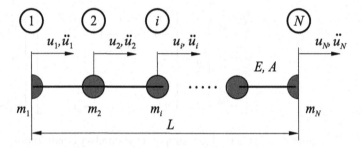

Fig. 13.10 FE discretization (concentrated mass) for tension vibration of a bar

$$
\rho A \frac{1}{n} L
\begin{bmatrix}
1 & & & & \\
& 1 & & 0 & \\
& & 1 & & \\
& & & \ddots & \\
& 0 & & 1 & \\
& & & & \frac{1}{2}
\end{bmatrix}
\begin{bmatrix}
\ddot{u}_2 \\
\ddot{u}_3 \\
\vdots \\
\ddot{u}_n
\end{bmatrix}
+ \frac{EA}{\frac{1}{n}L}
\begin{bmatrix}
2 & -1 & & & \\
-1 & 2 & -1 & & 0 \\
& -1 & 2 & -1 & \\
& & \ddots & \ddots & \ddots \\
0 & & -1 & 2 & -1 \\
& & & -1 & 1
\end{bmatrix}
\begin{bmatrix}
u_2 \\
u_3 \\
\vdots \\
u_n
\end{bmatrix}
= 0 .
$$

(13.77)

To be noted:

The mass matrix has a diagonal shape, the stiffness matrix has a band structure with a band width of 3.

13.6.2.1   Solutions with Linear Shape Functions

The number of finite elements has a decisive influence on the accuracy of the results. First of all the tension bar will be discretized with linear and later on with quadratic shape functions. The solution with continuously distributed masses as well as the solution with concentrated masses will be introduced.

Continuously Distributed Masses (CDM)

First of all the entire bar is regarded as a single element.

With the mass and stiffness matrix

$$M = \frac{\rho A L}{6} \begin{bmatrix} 2 & 1 \\ 1 & 2 \end{bmatrix} \quad \text{and} \quad K = \frac{EA}{L} \begin{bmatrix} 1 & -1 \\ -1 & 1 \end{bmatrix} \tag{13.78}$$

the equation of motion results as

$$\begin{bmatrix} 2 & 1 \\ 1 & 2 \end{bmatrix} \begin{bmatrix} \ddot{u}_1 \\ \ddot{u}_2 \end{bmatrix} + 6 \frac{E}{\rho L^2} \begin{bmatrix} 1 & -1 \\ -1 & 1 \end{bmatrix} \begin{bmatrix} u_1 \\ u_2 \end{bmatrix} = \begin{bmatrix} 0 \\ 0 \end{bmatrix}. \tag{13.79}$$

No acceleration ($\ddot{u} = 0$) and no displacement ($u = 0$) occur on the node 1 with the coordinate x_1. The first row of the system of equations can therefore be cancelled. Two equations result from the second line:

$$\ddot{u}_2 - 6 \frac{E}{\rho L^2} u_2 = 0 \quad , \quad 2\ddot{u}_2 + 6 \frac{E}{\rho L^2} u_2 = 0. \tag{13.80}$$

With the approach $u_2 = \hat{u}_2 e^{(i\omega t)}$ one obtains the following from the first equation:

$$\left(-\omega^2 - 6 \frac{E}{\rho L^2} \right) \hat{u}_2 = 0 \quad \Rightarrow \quad \omega = \sqrt{6 \frac{E}{\rho L^2}} i. \tag{13.81}$$

The second equation leads to:

$$\left(-2\omega^2 + 6 \frac{E}{\rho L^2} \right) \hat{u}_1 = 0 \quad \Rightarrow \quad \omega = \sqrt{3} \sqrt{\frac{E}{\rho L^2}} \tag{13.82}$$

or alternatively

$$\omega = 1.7321 \sqrt{\frac{E}{\rho L^2}}. \tag{13.83}$$

The result deviates significantly from that of the analytical solution ($+10.27\%$). An improvement can be achieved via a discretization with *two* finite elements (Fig. 13.11).

Two Elements

The system now consists of two finite elements with linear shape functions and three nodes 1, 2 and 3 on the coordinates x_1, x_2 and x_3 (Fig. 13.12). No accelerations ($\ddot{u} = 0$) and no displacements $u = 0$ occur on the fixing point. From the above considerations the reduced mass and stiffness matrix can be established

$$M^{\text{red}} = \frac{\rho A \frac{1}{2} L}{6} \begin{bmatrix} 4 & 1 \\ 1 & 2 \end{bmatrix} \quad \text{and} \quad K^{\text{red}} = \frac{EA}{\frac{1}{2} L} \begin{bmatrix} 2 & -1 \\ -1 & 1 \end{bmatrix} \tag{13.84}$$

so that the following characteristic equation

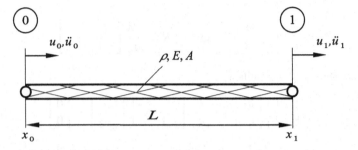

Fig. 13.11 An element with continuously distributed mass

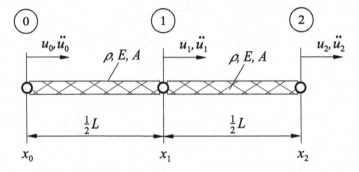

Fig. 13.12 Two elements with continuously distributed masses

$$\begin{vmatrix} -4\lambda^2 + 2 & -1\lambda^2 - 1 \\ -1\lambda^2 - 1 & -2\lambda^2 + 1 \end{vmatrix} \overset{!}{=} 0 \quad \Rightarrow \quad 2\left(1 - 2\lambda^2\right)^2 = \left(1 + \lambda^2\right)^2 \qquad (13.85)$$

with the abbreviations

$$\lambda^2 = \frac{1}{6}\frac{\rho\frac{1}{4}L^2}{E}\,\omega^2 \qquad (13.86)$$

results. Two solutions result

$$\lambda_1^2 = \frac{\sqrt{2} - 1}{1 + 2\sqrt{2}} \quad \text{and} \quad \lambda_2^2 = \frac{1 + \sqrt{2}}{2\sqrt{2} - 1}, \qquad (13.87)$$

which look as follows, written in detail:

$$\omega_1^2 = \frac{24\left(\sqrt{2} - 1\right)}{1 + 2\sqrt{2}}\frac{E}{\rho L^2} \quad \Rightarrow \quad \omega_1 = 1.61\sqrt{\frac{E}{\rho L^2}}, \qquad (13.88)$$

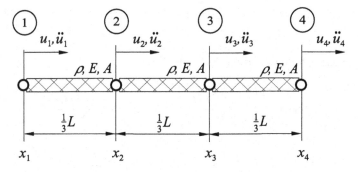

Fig. 13.13 Three elements with continuously distributed masses

$$\omega_2^2 = \frac{24\left(1+\sqrt{2}\right)}{2\sqrt{2}-1}\frac{E}{\rho L^2} \quad \Rightarrow \quad \omega_2 = 5.63\sqrt{\frac{E}{\rho L^2}}. \tag{13.89}$$

The values for the eigenfrequencies so far deviate significantly from the analytical solutions (+2.59%, +19.5%). The next simplification equals a division of the bar into *three* finite elements.

Three Elements

The system now consists of three finite elements with linear shape functions and four nodes 1, 2, 3 and 4 on the coordinates x_1, x_2, x_3 and x_4 (Fig. 13.13). No accelerations $(\ddot{u}_1 = 0)$ and no displacements $(u_1 = 0)$ occur at the fixing point. Therefore, the first row and the first column can be cancelled from the system of equations. The reduced mass and stiffness matrix remains

$$\boldsymbol{M}^{\mathrm{red}} = \frac{\rho A \frac{1}{3} L}{6}\begin{bmatrix} 4 & 1 & 0 \\ 1 & 4 & 1 \\ 0 & 1 & 2 \end{bmatrix} \quad \text{and} \quad \boldsymbol{K}^{\mathrm{red}} = \frac{EA}{\frac{1}{3}L}\begin{bmatrix} 2 & -1 & 0 \\ -1 & 2 & -1 \\ 0 & -1 & 1 \end{bmatrix}, \tag{13.90}$$

with which the following characteristic equation

$$\begin{vmatrix} -4\lambda^2 + 2 & -1\lambda^2 - 1 & 0 \\ -1\lambda^2 - 1 & -2\lambda^2 + 1 & -1\lambda - 1 \\ 0 & -1\lambda & 1 \end{vmatrix} \overset{!}{=} 0 \tag{13.91}$$

with the abbreviations

$$\lambda^2 = \frac{1}{6}\frac{\rho(\frac{1}{3}L)^2}{E}\omega^2 = \frac{1}{6}\frac{1}{9}\frac{\rho L^2}{E}\omega^2 \tag{13.92}$$

can be gained. Three solutions result

$$\lambda_1^2 = \frac{2 - \sqrt{3}}{4 + \sqrt{3}} \quad , \quad \lambda_2^2 = \frac{1}{2} \quad , \quad \lambda_3^2 = \frac{2 + \sqrt{3}}{4 - \sqrt{3}} \quad , \tag{13.93}$$

which appear as follows, written in detail

$$\omega_1 = \frac{54\left(2 - \sqrt{3}\right)}{4 + \sqrt{3}} \frac{E}{\rho L^2} \quad \Rightarrow \quad \omega_1 = 1.59 \sqrt{\frac{\rho L^2}{E}} \, , \tag{13.94}$$

$$\omega_2 = 27 \frac{E}{\rho L^2} \quad \Rightarrow \quad \omega_2 = 5.19 \sqrt{\frac{\rho L^2}{E}} \, , \tag{13.95}$$

$$\omega_3 = \frac{54\left(2 + \sqrt{3}\right)}{4 - \sqrt{3}} \frac{E}{\rho L^2} \quad \Rightarrow \quad \omega_3 = 9.43 \sqrt{\frac{\rho L^2}{E}} \, . \tag{13.96}$$

The deviations from the analytical solution are $+1.13$, $+8.23$ and $+20.1\%$.

Lumped Mass Method (LMM)

Within this method discretizations with one, two and three finite elements will be introduced. First of all, a discretization with just one element will be considered (Fig. 13.14). From Eq. (13.77) one obtains directly

$$-\omega^2 \rho A L \times \frac{1}{2} + \frac{EA}{L} = 0 \tag{13.97}$$

and herefrom the solution

$$\omega = \sqrt{2} \sqrt{\frac{E}{\rho L^2}} \tag{13.98}$$

for the eigenfrequency. This result deviates significantly from the analytical solution (-9.97%).

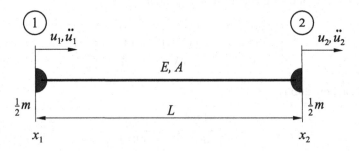

Fig. 13.14 One element with concentrated masses on the ends

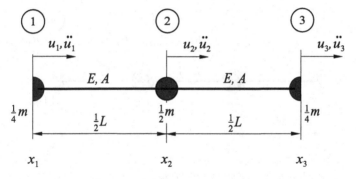

Fig. 13.15 Two elements with concentrated masses

Two Elements

Through a refining of the discretization with *two* elements a better solution can be achieved. The system consists of two finite elements with linear shape functions and three nodes 1, 2 and 3 on the coordinates x_1, x_2 and x_3 (Fig. 13.15). With the mass and stiffness matrix

$$\mathbf{M} = \rho A \frac{1}{2} L \begin{bmatrix} \frac{1}{2} & 0 & 0 \\ 0 & 1 & 0 \\ 0 & 0 & \frac{1}{2} \end{bmatrix} \text{ and } \mathbf{K} = \frac{EA}{\frac{1}{2}L} \begin{bmatrix} 1 & -1 & 0 \\ -1 & 2 & -1 \\ 0 & -1 & 1 \end{bmatrix} \qquad (13.99)$$

the equation of motion results in:

$$\begin{bmatrix} \frac{1}{2} & 0 & 0 \\ 0 & 1 & 0 \\ 0 & 0 & \frac{1}{2} \end{bmatrix} \begin{bmatrix} \ddot{u}_1 \\ \ddot{u}_2 \\ \ddot{u}_3 \end{bmatrix} + \frac{E}{\rho \frac{1}{4} L^2} \begin{bmatrix} 1 & -1 & 0 \\ -1 & 2 & -1 \\ 0 & -1 & 1 \end{bmatrix} \begin{bmatrix} u_1 \\ u_2 \\ u_3 \end{bmatrix} = \mathbf{0}. \qquad (13.100)$$

No acceleration ($\ddot{u}_1 = 0$) and no displacement ($u_1 = 0$) occur on the fixing point. The first row and the first column can be canceled from the system. From

$$\det\left(-\lambda^2 \begin{bmatrix} 1 & 0 \\ 0 & \frac{1}{2} \end{bmatrix} + \begin{bmatrix} 2 & -1 \\ -1 & 1 \end{bmatrix}\right) = 0 \qquad (13.101)$$

with

$$\lambda^2 = \frac{1}{4} \frac{\rho L^2}{E} \omega^2 \qquad (13.102)$$

one obtains via

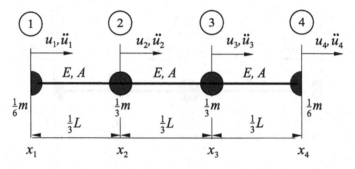

Fig. 13.16 Three elements with concentrated masses

$$\begin{vmatrix} -\lambda^2 + 2 & -1 \\ -1 & -\frac{1}{2}\lambda^2 + 1 \end{vmatrix} = 0 \quad \Rightarrow \quad (2 - \lambda^2)^2 = 2 \quad \Rightarrow \quad 2 - \lambda^2 = \pm\sqrt{2}$$

(13.103)

the solution $\lambda_1^2 = 2 - \sqrt{2}$ and $\lambda_2^2 = 2 + \sqrt{2}$. These can be written in detail as

$$\omega_1^2 = 4\left(2 - \sqrt{2}\right)\frac{E}{\rho L^2} \quad \Rightarrow \quad \omega_1 = 1.53\sqrt{\frac{E}{\rho L^2}}$$

(13.104)

and

$$\omega_2^2 = 4\left(2 + \sqrt{2}\right)\frac{E}{\rho L^2} \quad \Rightarrow \quad \omega_2 = 3.70\sqrt{\frac{E}{\rho L^2}}.$$

(13.105)

The solutions deviate significantly from the analytical solutions (-2.55 and -21.5%).

Three Elements

During the next refining the tension bar with three elements will be discretized. Therewith four nodes $(1, 2, 3, 4)$ are on the coordinates x_1, x_2, x_3 and x_4 in the system (Fig. 13.16). The tension bar is fixed on the node 1.

No acceleration ($\ddot{u}_1 = 0$) and no displacement ($u_1 = 0$) occur on the fixing point. Therefore the respective first rows and columns can be canceled from the mass and stiffness matrix. The reduced matrices remain

$$\mathbf{M}^{\text{red}} = \rho A \frac{1}{3} L \begin{bmatrix} 1 & 0 & 0 \\ 0 & 1 & 0 \\ 0 & 0 & \frac{1}{2} \end{bmatrix} \quad \text{and} \quad \mathbf{K}^{\text{red}} = \frac{EA}{\frac{1}{3}L} \begin{bmatrix} 2 & -1 & 0 \\ -1 & 2 & -1 \\ 0 & -1 & 1 \end{bmatrix}.$$

(13.106)

With the abbreviation

$$\lambda^2 = \frac{1}{9}\frac{\rho L^2}{E}\omega^2$$

(13.107)

Table 13.1 Relative errors in % respective to the analytically determined eigenfrequency on the basis of elements with linear shape functions

Number of elements	1	2		3		
Eigenfrequencies	1st	1st	2nd	1st	2nd	3rd
CDM	+10.27	+2.59	+19.5	+1.13	+8.23	+20.1
LMM	−9.97	−2.55	−21.5	−1.14	−9.96	−26.4

one obtains the determinant and the characteristic equation

$$\begin{vmatrix} -\lambda^2 + 2 & -1 & 0 \\ -1 & -\lambda^2 + 2 & -1 \\ 0 & -1 & -\frac{1}{2}\lambda^2 + 1 \end{vmatrix} \overset{!}{=} 0 \;\Rightarrow\; (2 - \lambda^2)\left[(2 - \lambda^2)^2 - 3\right] = 0$$

(13.108)

and herefrom the solutions

$$\lambda_1 = \sqrt{2 - \sqrt{3}} \;,\quad \lambda_2 = \sqrt{2} \;,\quad \lambda_3 = \sqrt{2 + \sqrt{3}} \;,$$ (13.109)

from which the eigenfrequency can be determined as follows

$$\omega_1 = 1.55\sqrt{\frac{\rho L^2}{E}} \;,\quad \omega_2 = 4.24\sqrt{\frac{\rho L^2}{E}} \;,\quad \omega_3 = 5.78\sqrt{\frac{\rho L^2}{E}}.$$ (13.110)

Table 13.1 summarizes all results. Therein are the relative errors in % for the FE solutions with continuously distributed and discretized masses (LMM). The errors relate to the analytical solution (Table 13.1).

Remarks: From the comparison in the above table one can see that the lumped mass method (LMM) delivers values which are *too low*, while one obtains values which are too high from the finite element method (CDM). Through the concentration of the continuously distributed mass model on the nodal points, the inertia effects are enlarged, whereby the eigenfrequencies become smaller. In contrast, the inertia effects are reduced when making use of a mass matrix M according to the CDM, which is based on a linear shape function matrix N. This leads to too high eigenfrequencies. Consequently a lower bound (LMM) and an upper bound (CDM) have been found for the limitation of the exact solution. If quadratic interpolation functions are used, the computing time of course increases. However a smaller number of elements is enough to achieve comparable results.

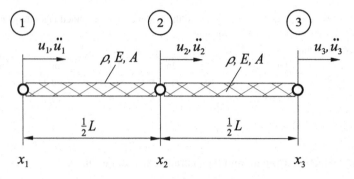

Fig. 13.17 One element with distributed mass and quadratic shape function

13.6.2.2 The Tension Bar with Quadratic Shape Functions

The problem will be described similarly to the previous section (see Fig. 13.11). In contrast to a linear approach, the element with three nodes is described in a quadratic approach. First of all, the entire bar with only one element is presented. The system consists of one finite element with quadratic shape function and three nodes 1, 2 and 3 on the coordinates x_1, x_2 and x_3 (Fig. 13.17).

With the mass and stiffness matrix for quadratic shape functions the following results for the equation of motion

$$
\begin{bmatrix} 4 & 2 & -1 \\ 2 & 16 & 2 \\ -1 & 2 & 4 \end{bmatrix} \begin{bmatrix} \ddot{u}_1 \\ \ddot{u}_2 \\ \ddot{u}_3 \end{bmatrix} + \frac{10E}{\rho L^2} \begin{bmatrix} 7 & -8 & 1 \\ -8 & 16 & -8 \\ 1 & -8 & 7 \end{bmatrix} \begin{bmatrix} u_1 \\ u_2 \\ u_3 \end{bmatrix} = \begin{bmatrix} 0 \\ 0 \\ 0 \end{bmatrix}, \quad (13.111)
$$

which can be simplified due to the boundary conditions on the fixing point $\ddot{u}_1 = 0$, $u_1 = 0$ to

$$
\begin{bmatrix} 16 & 2 \\ 2 & 4 \end{bmatrix} \begin{bmatrix} \ddot{u}_2 \\ \ddot{u}_3 \end{bmatrix} + \frac{10E}{\rho L^2} \begin{bmatrix} 16 & -8 \\ -8 & 7 \end{bmatrix} \begin{bmatrix} u_2 \\ u_3 \end{bmatrix} = \begin{bmatrix} 0 \\ 0 \end{bmatrix}. \quad (13.112)
$$

With the abbreviation

$$
\lambda^2 = \frac{\rho L^2}{10E} \omega^2 \quad (13.113)
$$

one obtains the characteristic equation

$$
\begin{vmatrix} -16\lambda^2 + 16 & -2\lambda^2 - 8 \\ -2\lambda^2 - 8 & -4\lambda^2 + 7 \end{vmatrix} \overset{!}{=} 0 \quad \Rightarrow \quad \lambda^4 - \frac{52}{15}\lambda^2 = -\frac{4}{5} \quad (13.114)
$$

with the solutions

$$\lambda^2 = \frac{26}{15} \pm \frac{1}{15}\sqrt{496} \ , \tag{13.115}$$

which can be written in detail as

$$\omega_1^2 = 2.486 \frac{E}{\rho L^2} \quad \Rightarrow \quad \omega_1 = 1.57 \sqrt{\frac{E}{\rho L^2}} \tag{13.116}$$

and

$$\omega_2^2 = 32.18 \frac{E}{\rho L^2} \quad \Rightarrow \quad \omega_2 = 5.67 \sqrt{\frac{E}{\rho L^2}}. \tag{13.117}$$

In contrast to the exact values ω_1 is affected by an error of $+0.38\%$ and ω_2 is affected by an error of $+20.4\%$. A slightly lower value for ω_1 with an error of $+1.13\%$ was not achieved until a division of the bar into *three* finite elements on the basis of a linear displacement approach took place. For a single element, a value of $\omega_1 = 1.7321\sqrt{E/\rho L^2}$ has been achieved, which is affected by an error of 10.27%. Therefore the value for ω_1 could be improved by $+9.89\%$ through a quadratic displacement approach. To receive a comparable value for ω_2, two finite elements are required.

Two Elements

With this modeling approach, the bar is divided into two elements with quadratic shape functions. The system consists of five nodes in total 1, 2, 3, 4 and 5 on the coordinates x_1, x_2, x_3, x_4 and x_5 (Fig. 13.18).

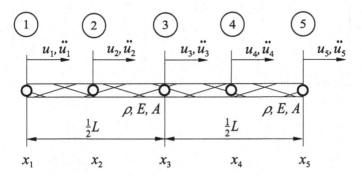

Fig. 13.18 Two elements with distributed masses and quadratic shape functions

The mass matrix

$$M = \frac{\rho A \frac{1}{2} L}{30} \begin{bmatrix} 4 & 2 & -1 & 0 & 0 \\ 2 & 16 & 2 & 0 & 0 \\ -1 & 2 & 8 & 2 & -1 \\ 0 & 0 & 2 & 16 & 2 \\ 0 & 0 & -1 & 2 & 4 \end{bmatrix} \tag{13.118}$$

and the stiffness matrix

$$K = \frac{EA}{3\frac{1}{2}L} \begin{bmatrix} 7 & -8 & 1 & 0 & 0 \\ -8 & 16 & -8 & 0 & 0 \\ 1 & -8 & 14 & -8 & 1 \\ 0 & 0 & -8 & 16 & -8 \\ 0 & 0 & 1 & -8 & 7 \end{bmatrix} \tag{13.119}$$

have the dimensions 5×5. No acceleration ($\ddot{u}_1 = 0$) and no displacement ($u_1 = 0$) occur on the fixing point. Therefore, the first row and the corresponding first column of the matrices can be canceled. From the equation of motion

$$\begin{bmatrix} 16 & 2 & 0 & 0 \\ 2 & 8 & 2 & -1 \\ 0 & 2 & 16 & 2 \\ 0 & -1 & 2 & 4 \end{bmatrix} \begin{bmatrix} \ddot{u}_2 \\ \ddot{u}_3 \\ \ddot{u}_4 \\ \ddot{u}_5 \end{bmatrix} + \frac{40E}{\rho L^2} \begin{bmatrix} 16 & -8 & 0 & 0 \\ -8 & 14 & -8 & 1 \\ 0 & -8 & 16 & -8 \\ 0 & 1 & -8 & 7 \end{bmatrix} \begin{bmatrix} u_2 \\ u_3 \\ u_4 \\ u_5 \end{bmatrix} = \begin{bmatrix} 0 \\ 0 \\ 0 \\ 0 \end{bmatrix} \tag{13.120}$$

the eigenvalue problem can be formulated

$$\det \left(-\omega^2 \begin{bmatrix} 16 & 2 & 0 & 0 \\ 2 & 8 & 2 & -1 \\ 0 & 2 & 16 & 2 \\ 0 & -1 & 2 & 4 \end{bmatrix} + \frac{40E}{\rho L^2} \begin{bmatrix} 16 & -8 & 0 & 0 \\ -8 & 14 & -8 & 1 \\ 0 & -8 & 16 & -8 \\ 0 & 1 & -8 & 7 \end{bmatrix} \right) = 0 \tag{13.121}$$

from which the solution for ω^2 or alternatively the eigenfrequency ω

$$\omega_1^2 = 2.468664757 \frac{E}{\rho L^2} \quad \Rightarrow \quad \omega_1 = 1.5712 \sqrt{\frac{E}{\rho L^2}}, \tag{13.122}$$

$$\omega_2^2 = 22.94616601 \frac{E}{\rho L^2} \quad \Rightarrow \quad \omega_2 = 4.7902 \sqrt{\frac{E}{\rho L^2}}, \tag{13.123}$$

$$\omega_3^2 = 77.06313717 \frac{E}{\rho L^2} \quad \Rightarrow \quad \omega_3 = 8.7786 \sqrt{\frac{E}{\rho L^2}}, \tag{13.124}$$

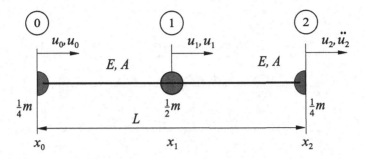

Fig. 13.19 One element with concentrated masses and quadratic shape function

$$\omega_4^2 = 198.6985027 \frac{E}{\rho L^2} \quad \Rightarrow \quad \omega_4 = 14.0961 \sqrt{\frac{E}{\rho L^2}} \tag{13.125}$$

can be determined.

The deviations from the analytical solutions are significantly lower in comparison with the approximations which were achieved with linear shape functions ($\omega_1 = 1.61\sqrt{\frac{E}{\rho L^2}}$, $w_2 = 5.63\sqrt{\frac{E}{\rho L^2}}$).

Method with Concentrated Masses (LMM)

In the first step of discretization the system consists of only one finite element with quadratic shape function and three nodes 1, 2 and 3 on the coordinates x_1, x_2 and x_3 (Fig. 13.19).

With the mass and stiffness matrix

$$M = \frac{\rho A L}{4} \begin{bmatrix} 1 & 0 & 0 \\ 0 & 2 & 0 \\ 0 & 0 & 1 \end{bmatrix} \quad \text{and} \quad K = \frac{EA}{3L} \begin{bmatrix} 7 & -8 & 1 \\ -8 & 16 & -8 \\ 1 & -8 & 7 \end{bmatrix} \tag{13.126}$$

the equation of motion appears as follows:

$$\begin{bmatrix} 1 & 0 & 0 \\ 0 & 2 & 0 \\ 0 & 0 & 1 \end{bmatrix} \begin{bmatrix} \ddot{u}_1 \\ \ddot{u}_2 \\ \ddot{u}_3 \end{bmatrix} + \frac{4}{3} \frac{E}{\rho L^2} \begin{bmatrix} 7 & -8 & 1 \\ -8 & 16 & -8 \\ 1 & -8 & 7 \end{bmatrix} \begin{bmatrix} u_1 \\ u_2 \\ u_3 \end{bmatrix} = \begin{bmatrix} 0 \\ 0 \\ 0 \end{bmatrix}. \tag{13.127}$$

No acceleration ($\ddot{u}_1 = 0$) and no displacement ($u_1 = 0$) occur on the fixing point. Therefore the first row and the respective first column of the matrices can be cancelled. With the abbreviation

$$\lambda^2 = \frac{3}{4} \frac{\rho L^2}{E} \omega^2 \tag{13.128}$$

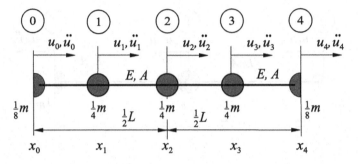

Fig. 13.20 Two elements with concentrated masses and quadratic shape functions

one obtains the characteristic equation

$$\begin{vmatrix} -2\lambda^2 + 16 & -8 \\ -8 & -\lambda^2 + 7 \end{vmatrix} \overset{!}{=} 0 \quad \Rightarrow \quad (\lambda^2 - 8)(\lambda^2 - 7) = 32 \qquad (13.129)$$

and therefore the solutions for λ_i

$$\lambda_{2,1}^2 = \frac{15}{2} \pm \frac{1}{2}\sqrt{129} \qquad (13.130)$$

and therefrom the eigenfrequencies

$$\omega_2 = 4.192\sqrt{\frac{E}{\rho L^2}} \quad \text{and} \quad \omega_1 = 1.558\sqrt{\frac{E}{\rho L^2}}. \qquad (13.131)$$

In comparison to the exact factor of 1.5708, the approximate value ω_1 is affected by an error of -8%, while ω_2 differs by -11.04%. One compares these results with the corresponding errors of $+0.38$ and $+20.4\%$ which result when using the equivalent mass matrix instead of the lumped-mass system.

Two Elements

In this step of discretization the system consists of two finite elements with quadratic shape functions and five nodes 1, 2, 3, 4 and 5 on the coordinates x_1, x_2, x_3, x_4 and x_5 (Fig. 13.20). With the mass and stiffness matrix

$$M = \frac{\rho A \frac{1}{2} L}{4} \begin{bmatrix} 1 & 0 & 0 & 0 & 0 \\ 0 & 2 & 0 & 0 & 0 \\ 0 & 0 & 2 & 0 & 0 \\ 0 & 0 & 0 & 2 & 0 \\ 0 & 0 & 0 & 0 & 1 \end{bmatrix} \quad \text{and} \quad K = \frac{EA}{3\frac{1}{2}L} \begin{bmatrix} 7 & -8 & 1 & 0 & 0 \\ -8 & 16 & -8 & 0 & 0 \\ 1 & -8 & 14 & -8 & 1 \\ 0 & 0 & -8 & 16 & -8 \\ 0 & 0 & 1 & -8 & 7 \end{bmatrix}$$

$$(13.132)$$

one obtains, under consideration of the boundary conditions ($\ddot{x}_1 = 0$, $x_1 = 0$) the equation of motion:

$$
\begin{bmatrix} 2 & 0 & 0 & 0 \\ 0 & 2 & 0 & 0 \\ 0 & 0 & 2 & 0 \\ 0 & 0 & 0 & 1 \end{bmatrix} \begin{bmatrix} \ddot{u}_2 \\ \ddot{u}_3 \\ \ddot{u}_4 \\ \ddot{u}_5 \end{bmatrix} + \frac{16}{3} \frac{E}{\rho L^2} \begin{bmatrix} 16 & -8 & 0 & 0 \\ -8 & 14 & -8 & 1 \\ 0 & -8 & 16 & -8 \\ 0 & 1 & -8 & 7 \end{bmatrix} \begin{bmatrix} u_2 \\ u_3 \\ u_4 \\ u_5 \end{bmatrix} = \begin{bmatrix} 0 \\ 0 \\ 0 \\ 0 \end{bmatrix}.
$$

$$(13.133)$$

From the solution of the eigenvalue problem

$$
\det\left(-\omega^2 \begin{bmatrix} 2 & 0 & 0 & 0 \\ 0 & 2 & 0 & 0 \\ 0 & 0 & 2 & 0 \\ 0 & 0 & 0 & 1 \end{bmatrix} + \frac{16}{3} \frac{E}{\rho L^2} \begin{bmatrix} 16 & -8 & 0 & 0 \\ -8 & 14 & -8 & 1 \\ 0 & -8 & 16 & -8 \\ 0 & 1 & -8 & 7 \end{bmatrix} \right) = 0
$$

$$(13.134)$$

one obtains the four real solutions

$$\omega_1^2 = 2.459021 \frac{E}{\rho L^2} \quad \Rightarrow \quad \omega_1 = 1.5681 \sqrt{\frac{E}{\rho L^2}}, \tag{13.135}$$

$$\omega_2^2 = 21.16383 \frac{E}{\rho L^2} \quad \Rightarrow \quad \omega_2 = 4.6004 \sqrt{\frac{E}{\rho L^2}}, \tag{13.136}$$

$$\omega_3^2 = 55.064934 \frac{E}{\rho L^2} \quad \Rightarrow \quad \omega_3 = 7.4206 \sqrt{\frac{E}{\rho L^2}}, \tag{13.137}$$

$$\omega_4^2 = 81.3122153 \frac{E}{\rho L^2} \quad \Rightarrow \quad \omega_4 = 9.0173 \sqrt{\frac{E}{\rho L^2}} \tag{13.138}$$

for ω_i^2 or alternatively the four eigenfrequencies ω_i.

Remark:

The eigenfrequencies approximated via the LMM are *smaller* as the analytically determined ones (*lower bound*). Table 13.2 summarizes all solutions. Given are the relative errors in % for the FE solutions with continuously distributed and discretized masses (LMM). The errors relate to the analytical solutions.

Table 13.2 Relative error in % respective to the analytically determined eigenfrequency on the basis of elements with quadratic shape functions

Number of elements	1		2			
Eigenfrequencies	1st	2nd	1st	2nd	3rd	4th
FEM	+0.38	+20.4	+0.03	+1.65	+11.77	+28.0
LMM	−0.8	−11.04	−0.17	−2.38	−5.52	−18.0

13.7 Supplementary Problems

13.1 Analytical Solutions for Bending Vibrations

Determine the first four eigenfrequencies for the one-sided fixed mass loaded beam with length L with constant bending stiffness EI.

Given: ρ, L, EI

13.2 FE Solution for Bending Vibrations

Determine the first four eigenfrequencies for the one-sided fixed mass loaded beam with length L with constant bending stiffness EI.

Given: ρ, L, EI

References

1. Betten J (2004) Finite Elemente für Ingenieure 1: Grundlagen, Matrixmethoden, Elastisches Kontinuum. Springer, Berlin
2. Gross D, Hauger W, Schröder J, Werner EA (2008) Hydromechanik, Elemente der Höheren Mechanik, Numerische Methoden. Springer, Berlin
3. Gross D, Hauger W, Schröder J, Wall WA (2009) Technische Mechanik 2: Elastostatik. Springer, Berlin

Chapter 14
Special Elements

Abstract Some elements for special applications are presented in the scope of this chapter. The first special element extends the classical Bernoulli element by consideration of an elastic foundation. In the scope of this element, the so-called Winkler foundation, which assumes that the distributed reaction forces of the foundation are proportional at every point to the deflection, is considered. The second special element considers the case of a stress singularity. A beam element with a particular mapping between the local and natural coordinates allows that the stress at one node converges to infinity. The third special element considers that the geometry can extend to infinity at a boundary. For the derivation of these elements, special shape functions for the interpolation from the local to the global positions are introduced.

14.1 Elastic Foundation

The case of elastic foundation [1–3] can be seen as an extension of the classical Bernoulli beam from Chap. 5. In particular, the beam is supported by an elastic foundation. This case is of interest in civil engineering when the loads on railway rails and the corresponding substructure are to be examined. This case is schematically shown in Fig. 14.1 where the foundation is represented by an arbitrarily dense number of springs. For the simplest modeling approach, the so-called Winkler foundation [4], it is assumed that the distributed reaction force q_k of the foundation is proportional to the local deflection:

$$q_k(x) = ku_y(x). \tag{14.1}$$

The parameter k in Eq. (14.1) is the foundation modulus and has in the case of a beam[1] the dimension force/length^2.

---

[1] In the general three-dimensional case, the dimension is force/length^3.

© Springer International Publishing AG, part of Springer Nature 2018
A. Öchsner and M. Merkel, *One-Dimensional Finite Elements*,
https://doi.org/10.1007/978-3-319-75145-0_14

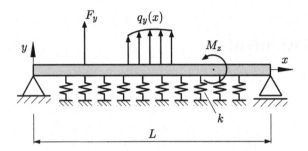

Fig. 14.1 Schematic representation of a beam with elastic foundation

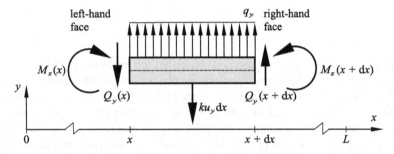

Fig. 14.2 Infinitesimal beam element of length $\mathrm{d}x$ with elastic foundation

The differential equation, which describes the problem from Fig. 14.1, will be derived in the following. To do this, consider the infinitesimal beam element shown in Fig 14.2 with internal reactions and the external load. The force equilibrium in the vertical direction gives:

$$-Q_y(x) + Q_y(x + \mathrm{d}x) + q_y\mathrm{d}x - ku_y(x + \tfrac{1}{2}\mathrm{d}x)\mathrm{d}x = 0. \qquad (14.2)$$

If the shear force on the right-hand face and the deflection in the middle are expanded in Taylor's series of first order, meaning

$$Q_y(x + \mathrm{d}x) \approx Q_y(x) + \frac{\mathrm{d}Q_y(x)}{\mathrm{d}x}\mathrm{d}x\,, \qquad (14.3)$$

$$u_y(x + \tfrac{1}{2}x) \approx u_y(x) + \frac{\mathrm{d}u_y(x)}{\mathrm{d}x}\tfrac{1}{2}\,\mathrm{d}x\,, \qquad (14.4)$$

Eq. (14.2) results in

$$-Q_y(x) + Q_y(x) + \frac{\mathrm{d}Q_y(x)}{\mathrm{d}x}\mathrm{d}x + q_y\mathrm{d}x - ku_y(x)\mathrm{d}x - k\frac{\mathrm{d}u_y(x)}{\mathrm{d}x}\tfrac{1}{2}\mathrm{d}x\,\mathrm{d}x = 0\,, \quad (14.5)$$

or alternatively after simplification and neglecting of the term with $(dx)^2$ finally to:

$$\frac{dQ_y(x)}{dx} = -q_y + ku_y(x).$$ (14.6)

The equilibrium of moments around the reference point at $x + dx$ gives:

$$M_z(x+dx) - M_z(x) - q_y dx(\tfrac{1}{2}x) + ku_y(x+\tfrac{1}{2}dx)dx(\tfrac{1}{2}dx) + Q_y(x)dx = 0.$$ (14.7)

If the bending moment on the right-hand face is expanded into a Taylor's series of first order similar to Eq. (14.3) and consideration that the terms $(dx)^2$ as infinitesimal small size of higher order can be disregarded, finally the following results:

$$\frac{dM_z(x)}{dx} = -Q_y(x).$$ (14.8)

The combination of Eqs. (14.6) and (14.8) leads to the relation between the bending moment, the distributed load and the spring force:

$$\frac{d^2 M_z(x)}{dx^2} = -\frac{dQ_y(x)}{dx} = q_y - ku_y(x).$$ (14.9)

The derivation of the differential equation, which describes the bending line, is carried out as in Sect. 5.2.4, i.e. two-time differentiation of Eq. (5.32) and consideration of Eq. (14.9):

$$\frac{d^2}{dx^2}\left(EI_z\frac{d^2 u_y(x)}{dx^2}\right) = q_y(x) - ku_y(x),$$ (14.10)

and accordingly for constant bending stiffness EI_z:

$$EI_z\frac{d^4 u_y(x)}{dx^4} = q_y(x) - ku_y(x).$$ (14.11)

Using the weighted residual method and the partial differential equation, the principal finite element equation can be derived according to Sect. 5.3.2. The inner product of a beam with elastic foundation — under the assumption that the foundation modulus is constant and that no distributed load is occurring — reads as follows:

$$\int_0^L W^T(x)\left(EI_z\frac{d^4 u_y(x)}{dx^4} + ku_y(x)\right)dx \overset{!}{=} 0.$$ (14.12)

The elemental stiffness matrix results, according to the procedure in Sect. 5.3.2, as:

$$k^{e} = \cdots + k \int_{0}^{L} N^{\mathrm{T}}(x)N(x)\mathrm{d}x .$$

(14.13)

Thus, an additive expression is added to the relation according to Eq. (5.111). After calculating the integral on the right-hand side of Eq. (14.13), the elemental stiffness matrix of a beam with constant elastic foundation finally becomes:

$$k^{e} = \frac{EI_{z}}{L^{3}} \begin{bmatrix} 12 & 6L & -12 & 6L \\ 6L & 4L^{2} & -6L & 2L^{2} \\ -12 & -6L & 12 & -6L \\ 6L & 2L^{2} & -6L & 4L^{2} \end{bmatrix} + \frac{kL}{420} \begin{bmatrix} 156 & 22L & 54 & -13L \\ 22L & 4L^{2} & 13L & -3L^{2} \\ 54 & 13L & 156 & -22L \\ -13L & -3L^{2} & -22L & 4L^{2} \end{bmatrix} .$$

(14.14)

It should be noted here that the classical stiffness matrix according to Eq. (5.82) is obtained for $k = 0$.

14.2 Stress Singularity

Within the framework for the modeling of mechanical members, stress singularities may occur in the case of sharp-edged geometries, such as, for example, in corners and edges, or in the case of concentrated point loads. Furthermore, singularities can be found in the modeling of cracks. Within the framework of fracture mechanics, a distinction is made between the basic types of crack separation (so-called modes), see Fig. 14.3:

- Mode I: Opening of the crack flanks due to normal loading to the crack front.
- Mode II: Opposite displacement of the crack flanks in the crack propagation direction.

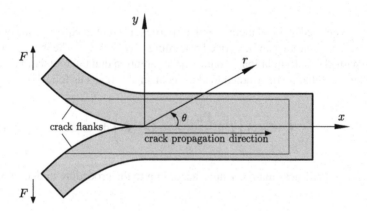

Fig. 14.3 Loading of a crack normal to the crack propagation direction (mode I)

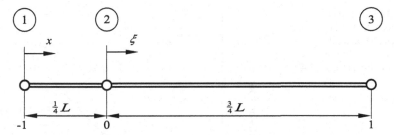

Fig. 14.4 Bar element with quadratic shape functions and excentric inner node

- Mode III: Displacement of the crack flanks across the crack propagation direction.

The stress distribution at the crack tip for mode I is given in polar coordinates r and θ as [5–7]:

$$
\begin{bmatrix} \sigma_x \\ \sigma_y \\ \tau_{xy} \end{bmatrix} = \frac{K_{\mathrm{I}}}{\sqrt{2\pi r}} \cos\left(\frac{\theta}{2}\right) \begin{bmatrix} 1 - \sin\left(\frac{\theta}{2}\right) \sin\left(\frac{3\theta}{2}\right) \\ 1 + \sin\left(\frac{\theta}{2}\right) \sin\left(\frac{3\theta}{2}\right) \\ \sin\left(\frac{\theta}{2}\right) \sin\left(\frac{3\theta}{2}\right) \end{bmatrix}, \tag{14.15}
$$

and accordingly along the x-axis ($\theta = 0, r \to x$) for each of the three stress components:

$$
\sigma_x = \sigma_y = \frac{K_{\mathrm{I}}}{\sqrt{2\pi x}}, \quad \tau_{xy} = 0. \tag{14.16}
$$

In the last two equations, K_{I} denotes the so-called stress intensity factor. This scalar quantity is dependent on the load, i.e. the crack opening mode, and the geometry of the crack [8].

Equation (14.16) indicates that both normal stress components reveal a stress singularity for $x \to 0$ — a so-called $1/\sqrt{r}$ singularity — which means that the stress tends to infinity.

To derive a special one-dimensional element[2], which approximates a $1/\sqrt{r}$ singularity, a bar element with quadratic shape functions is considered in the following, see Fig. 14.4. This bar element comprises three nodes where at the inner node is excentric located at $x_2 = \frac{L}{4}$. It is essential for the derivation that the origin of the natural coordinate ξ is located at the inner node. It should be noted here that the bar element with quadratic shape functions, as introduces in Sect. 6.4, used the standard approach whereat the origin of the natural coordinate ξ is located in the middle of the element.

The shape functions in the natural coordinate for such a quadratic element can be taken from Eq. (6.53) as follows:

---

[2]The following derivation is based on an idea, which is indicated in [9].

Fig. 14.5 Correlation
between the local coordinate
(x) and the natural
coordinate (ξ) for the
configuration in Fig. 14.4

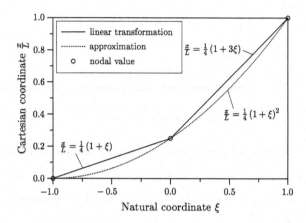

$$N_1(\xi) = \frac{\xi}{2}(\xi - 1) \; , \; N_2(\xi) = 1 - \xi^2 \, , \; N_3(\xi) = \frac{\xi}{2}(\xi + 1) \; . \qquad (14.17)$$

Thus, the transformation between the local and the natural coordinate can be stated for the left- and right-hand section of the interval as:

$$\begin{aligned} \frac{x}{L} &= \frac{1}{4}(1 + \xi) \;\; \text{für} \; -1 \le \xi \le 0 \\ \frac{x}{L} &= \frac{1}{4}(1 + 3\xi) \;\; \text{für} \;\; 0 \le \xi \le 1 \end{aligned} \qquad (14.18)$$

A graphical representation of this section-wise linear mapping between the local and natural coordinates is given in Fig. 14.5.

To avoid a distinction of cases, the section-wise linear correlation is approximated by a polynomial of order two, i.e., $\frac{x}{L} = a_0 + a_1\xi + a_2\xi^2$. Using the nodal values ($\frac{x}{L}(-1) = 0$, $\frac{x}{L}(0) = 0{,}25$, $\frac{x}{L}(1) = 1$) as sampling points, the following assignment rule is obtained:

$$\frac{x}{L} = \frac{1}{4}(\xi + 1)^2 \; , \qquad (14.19)$$

or rather solved for the natural coordinate:

$$\xi = 2\left(\frac{x}{L}\right)^{\frac{1}{2}} - 1 \; . \qquad (14.20)$$

The stress and strain distributions in a bar element can be generally expressed according to Eqs. (3.26) and (3.25) as $\sigma^{\mathrm{e}} = EBu_{\mathrm{p}}$ and $\varepsilon^{\mathrm{e}} = Bu_{\mathrm{p}}$. The B-matrix contains here the derivatives of the shape functions $N_i(x)$ with respect to the local coordinate x:

$$B = \frac{\mathrm{d}}{\mathrm{d}x} N(x) = \frac{\mathrm{d}\xi}{\mathrm{d}x}\frac{\mathrm{d}}{\mathrm{d}\xi} N(\xi) . \qquad (14.21)$$

Thus, the B-matrix is obtained in this case based on the shape functions (14.33) and the derivative $\frac{\mathrm{d}\xi}{\mathrm{d}x} = \frac{1}{\sqrt{Lx}}$ as:

$$B = \frac{1}{\sqrt{Lx}}\left[-\tfrac{1}{2}+\xi \quad -2\xi \quad \tfrac{1}{2}+\xi \right] \overset{(14.20)}{=}$$

$$\left[\tfrac{2}{L} - \tfrac{3}{2\sqrt{Lx}} \quad -\tfrac{4}{L} + \tfrac{2}{\sqrt{Lx}} \quad \tfrac{2}{L} - \tfrac{1}{2\sqrt{Lx}} \right] . \qquad (14.22)$$

The stress distribution in this special bar element can be expressed based on this result via the nodal displacements as:

$$\frac{\sigma_x^e}{E} = \left(\frac{2}{L} - \frac{3}{2\sqrt{Lx}} \right) u_1 + \left(-\frac{4}{L} + \frac{2}{\sqrt{Lx}} \right) u_2 + \left(\frac{2}{L} - \frac{1}{2\sqrt{Lx}} \right) u_3 . \qquad (14.23)$$

It results from the last equation that the stress tends to infinity for $x \to 0$.

Let us derive in the following the stiffness matrix for this special element. The general definition under consideration of the B-matrix results from Eq. (3.29) for constant material and geometric properties as:

$$k^e = \int_\Omega B^{\mathrm{T}} D B \, \mathrm{d}\Omega = EA \int_x B^{\mathrm{T}} B \, \mathrm{d}x = EA \int_0^L \frac{\mathrm{d}N^{\mathrm{T}}(x)}{\mathrm{d}x}\frac{\mathrm{d}N(x)}{\mathrm{d}x} \mathrm{d}x . \qquad (14.24)$$

It should be considered for the further derivation that the shape functions according to Eq. (14.33) are given in the natural coordinate ξ. Thus, a coordinate transformation $x \to \xi$ must be applied to Eq. (14.24) under utilization of $\frac{\mathrm{d}\xi}{\mathrm{d}x} = \frac{1}{\sqrt{Lx}}$ and $\mathrm{d}x = \sqrt{Lx}\,\mathrm{d}\xi$ respectively:

$$k^e = EA \int_{-1}^{1} \frac{\mathrm{d}\xi}{\mathrm{d}x}\frac{\mathrm{d}N^{\mathrm{T}}(\xi)}{\mathrm{d}\xi}\frac{\mathrm{d}\xi}{\mathrm{d}x}\frac{\mathrm{d}N(\xi)}{\mathrm{d}\xi} \sqrt{Lx}\,\mathrm{d}\xi . \qquad (14.25)$$

The last equation still requires that the local coordinate x is replaced by the natural coordinate ξ under consideration of Eq. (14.19). Thus, the following relationship is obtained:

$$k^e = \frac{EA}{L} \int_{-1}^{1} \frac{1}{\sqrt{\tfrac{1}{4}(\xi+1)^2}} \frac{\mathrm{d}N^{\mathrm{T}}(\xi)}{\mathrm{d}\xi}\frac{\mathrm{d}N(\xi)}{\mathrm{d}\xi} \mathrm{d}\xi , \qquad (14.26)$$

or with the derivatives of the single shape functions:

$$k^e = \frac{EA}{L} \int_{-1}^{1} \frac{1}{\sqrt{\frac{1}{4}(\xi+1)^2}} \begin{bmatrix} -\frac{1}{2}+\xi \\ -2\xi \\ \frac{1}{2}+\xi \end{bmatrix} \begin{bmatrix} -\frac{1}{2}+\xi & -2\xi & \frac{1}{2}+\xi \end{bmatrix} d\xi. \tag{14.27}$$

Thus, the following integration must be performed for the determination of the stiffness matrix:

$$k^e = \frac{EA}{L} \int_{-1}^{1} \begin{bmatrix} \frac{\text{sgn}(\xi+1)(-1+2\xi)^2}{2(\xi+1)} & -\frac{2\text{sgn}(\xi+1)\xi(-1+2\xi)}{\xi+1} & \frac{\text{sgn}(\xi+1)(-1+4\xi^2)}{2(\xi+1)} \\ -\frac{2\text{sgn}(\xi+1)\xi(-1+2\xi)}{\xi+1} & \frac{8\text{sgn}(\xi+1)\xi^2}{\xi+1} & -\frac{2\text{sgn}(\xi+1)\xi(1+2\xi)}{\xi+1} \\ \frac{\text{sgn}(\xi+1)(-1+4\xi^2)}{2(\xi+1)} & -\frac{2\text{sgn}(\xi+1)\xi(1+2\xi)}{\xi+1} & \frac{\text{sgn}(\xi+1)(1+2\xi)^2}{2(\xi+1)} \end{bmatrix} d\xi. \tag{14.28}$$

Analytical integration does not yield any results since a division by zero would occur. Numerical integration according to Gauss–Legendre (see Table 6.1) avoids this problem and the evaluation of two sampling points gives the following stiffness matrix:

$$k^e = \frac{EA}{L} \begin{bmatrix} 5.5 & -6 & 0.5 \\ -6 & 8 & -2 \\ 0.5 & -2 & 1.5 \end{bmatrix}. \tag{14.29}$$

Let us consider in the following a cantilevered bar (see Fig. 14.6), which has a stress singularity at the fixed support. Based on the previous derivations, the nodal displacements should be determined based on a single element.

The reduced system of equations is obtained under consideration of the boundary conditions as:

$$k^e = \frac{EA}{L} \begin{bmatrix} 8 & -2 \\ -2 & 1.5 \end{bmatrix} \begin{bmatrix} u_2 \\ u_3 \end{bmatrix} = \begin{bmatrix} 0 \\ F \end{bmatrix}, \tag{14.30}$$

Fig. 14.6 Bar with stress singularity at fixing point

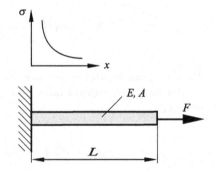

or rather solved for the unknown nodal displacements:

$$\begin{bmatrix} u_2 \\ u_3 \end{bmatrix} = \begin{bmatrix} 0.25 \\ 1 \end{bmatrix} \frac{FL}{EA}. \tag{14.31}$$

This result corresponds to the analytical solution. The stress distribution in the element can be calculated according to Eq. (14.23) based on the nodal displacements u_2 and u_3.

14.3 Infinite Extension

Despite the fact that no object can extend to infinity in the scope of classical mechanics, it may happen that the extension of a surrounding medium might be assumed infinite compared to the involved body. Classical examples for this are the foundation of a building, an airplane wing in the surrounding air or a boat in the surrounding sea. It might be meaningful in such cases to introduce elements with infinite extension to avoid that the surrounding medium must end at a certain distance to the considered body.

An introduction to the formulation of special elements with infinite extension can be found in [10–12].

Let us consider in the following a bar element with three nodes (see Fig. 14.7), whereat node 3 is located in infinity ($x_3 \to \infty$), see [12]. Furthermore, it is assumed that node 2 is located at a distance of a from the left-hand boundary of the element. It should be noted here that there is no geometry change on the level of the natural coordinate ξ, i.e. $-1 \leq \xi \leq +1$.

The interpolation of the filed variable $u^e(\xi)$ is based on the classical approach with interpolation functions and nodal values

$$u^e(\xi) = N_1(\xi)u_1 + N_2(\xi)u_2 + N_3(\xi)u_3 , \tag{14.32}$$

where the common interpolation functions N_i in natural coordinates according to Eq. (6.49) are used:

$$N_1(\xi) = \frac{\xi}{2}(\xi - 1) , \ N_2(\xi) = 1 - \xi^2 , \ N_3(\xi) = \frac{\xi}{2}(\xi + 1) . \tag{14.33}$$

For the interpolation of the local coordinate x^e, an approach based on the global coordinates of the nodes (x_i) and the shape functions M_i in natural coordinates is used

$$x^e(\xi) = M_1(\xi)x_1 + M_2(\xi)x_2 + M_3(\xi)x_3 , \tag{14.34}$$

where in this special case a different set of shape functions is introduced:

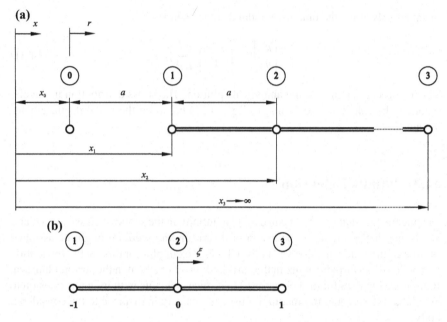

Fig. 14.7 One-dimensional infinite element in **a** local coordinate and **b** natural coordinate

$$M_1(\xi) = -\frac{2\xi}{1-\xi}, \; M_2(\xi) = \frac{1+\xi}{1-\xi}, \; M_3(\xi) = 0. \qquad (14.35)$$

The graphical representation in Fig. 14.8 indicates that the shape functions M_1 and M_2 take at their own node a value of 1 whereas the values tend to infinity for $\xi \rightarrow +1$. This combination of the shape functions causes that the local coordinate tends on the whole to infinity at its right-hand boundary:

$$\lim_{\xi \to 1} x^e(\xi) = \lim_{\xi \to 1} \frac{-2\xi x_1 + (1+\xi)x_2}{1-\xi} = \infty. \qquad (14.36)$$

In contrast to this, the field variable keeps a finite value at the right-hand boundary:

$$\lim_{\xi \to 1} u^e(\xi) = \lim_{\xi \to 1} \left(\frac{\xi}{2}(\xi - 1)u_1 + (1-\xi^2)u_2 + \frac{\xi}{2}(\xi + 1)u_3 \right) = u_3. \qquad (14.37)$$

Let us derive in the following the elemental stiffness matrix. The definition under consideration of the **B**-matrix results from Eq. (14.24) for constant material and geometric properties as:

Fig. 14.8 Graphical
representation of the shape
functions for the
interpolation of the local
coordinate

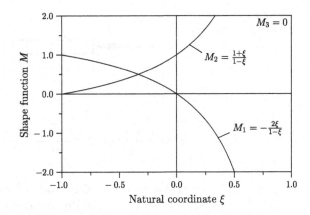

$$k^{\mathrm{e}} = E A \int\limits_{x} B^{\mathrm{T}} B \,\mathrm{d}x = E A \int\limits_{0}^{L} \frac{\mathrm{d}N^{\mathrm{T}}(x)}{\mathrm{d}x} \frac{\mathrm{d}N(x)}{\mathrm{d}x} \,\mathrm{d}x , \qquad (14.38)$$

and accordingly under the consideration that the shape functions are given in the
natural coordinate ξ:

$$k^{\mathrm{e}} = E A \int\limits_{-1}^{1} \frac{\mathrm{d}\xi}{\mathrm{d}x} \frac{\mathrm{d}N^{\mathrm{T}}(\xi)}{\mathrm{d}\xi} \frac{\mathrm{d}\xi}{\mathrm{d}x} \frac{\mathrm{d}N(\xi)}{\mathrm{d}\xi} f(\xi, x_i)\mathrm{d}\xi . \qquad (14.39)$$

The relation between the partial derivatives with respect to the local and natural
coordinate can be obtained via Eq. (14.49) by partial derivation with respect to the
natural coordinate:

$$\frac{\mathrm{d}x^{\mathrm{e}}(\xi)}{\mathrm{d}\xi} = \frac{\mathrm{d}M_1(\xi)}{\mathrm{d}\xi}x_1 + \frac{\mathrm{d}M_2(\xi)}{\mathrm{d}x}x_2 + \frac{\mathrm{d}M_3(\xi)}{\mathrm{d}x}x_3 ,$$

$$= \frac{-2}{(1-\xi)^2}x_1 + \frac{2}{(1-\xi)^2}x_2 . \qquad (14.40)$$

The global coordinates x_1 and x_2 can be alternatively expressed by the coordinate
x_0 and the characteristic length a via $x_1 = x_0 + a$ and $x_2 = x_0 + 2a$, see Fig. 14.7.
Thus, Eq. (14.40) can be expressed as:

$$\frac{\mathrm{d}x^{\mathrm{e}}(\xi)}{\mathrm{d}\xi} = \frac{2a}{(1-\xi)^2} .$$

Thus, the conditional equation for the elemental stiffness matrix results as:

Fig. 14.9 Infinitely extended bar with point load

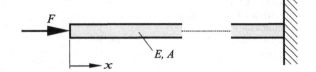

$$k^{\mathrm{e}} = \frac{EA}{2a} \int_{-1}^{1} (1 - \xi)^2 \frac{\mathrm{d} N^{\mathrm{T}}(\xi)}{\mathrm{d}\xi} \frac{\mathrm{d} N(\xi)}{\mathrm{d}\xi} \mathrm{d}\xi \,, \qquad (14.41)$$

or rather under consideration of the derivatives of the interpolation functions, i.e. $\frac{\mathrm{d} N(\xi)}{\mathrm{d}\xi} = \left[-\frac{1}{2} + \xi \quad -2\xi \quad \frac{1}{2} + \xi \right]$, finally via analytical or three-point Gauss–Legendre integration:

$$k^{\mathrm{e}} = \frac{EA}{a} \begin{bmatrix} \frac{23}{15} & -\frac{26}{15} & \frac{1}{5} \\ -\frac{26}{15} & \frac{32}{15} & -\frac{2}{5} \\ \frac{1}{5} & -\frac{2}{5} & \frac{1}{3} \end{bmatrix} . \qquad (14.42)$$

Let us consider in the following an infinitely extended bar, which is at its left-hand boundary loaded by a single force F and fixed at its right-hand end, see Fig. 14.9. Determine the displacement of the load application point based on a single element according to Eq. (14.42).

The reduced system of equations is obtained by canceling the third row and column in Eq. (14.42) as:

$$\frac{EA}{a} \begin{bmatrix} \frac{23}{15} & -\frac{26}{15} \\ -\frac{26}{15} & \frac{32}{15} \end{bmatrix} \begin{bmatrix} u_1 \\ u_2 \end{bmatrix} = \begin{bmatrix} F \\ 0 \end{bmatrix} . \qquad (14.43)$$

The solution of this system of equations results in the following column matrix of displacements:

$$\begin{bmatrix} u_1 \\ u_2 \end{bmatrix} = \frac{aF}{EA} \begin{bmatrix} 8 \\ 6.5 \end{bmatrix} . \qquad (14.44)$$

One can recognize that the solution of the problem is dependent on the characteristic length a and that an adequate choice is essential for the quality of the calculation. To enlighten a bit the meaning of the characteristic length a, consider Eq. (14.49) for the interpolation of the local coordinate and solve it for the natural coordinate:

$$\xi = \frac{x_2 - x}{-x + 2x_1 - x_2} . \qquad (14.45)$$

From this follows under consideration of the transformation (see Fig. 14.7) $x = x_0 + r$, $x_2 = x_0 + 2a$ and $x_1 = x_0 + a$ the following relation:

$$\xi = 1 - \frac{2a}{r}.$$ (14.46)

Insertion of the last relation in Eq. (14.32) for the interpolation of the displacement field gives:

$$u^{e}(r) = u_3 + (-u_1 + 4u_2 - 3u_3)\frac{a}{r} + (2u_1 - 4u_2 + 2u_3)\frac{a^2}{r^2}.$$ (14.47)

It can be seen from the last equation that for r towards infinity, the value u_3 is obtained. Furthermore, one can see that for r toward zero, the value of the field variable tends to infinity and thus, a pole is obtained at $x = x_0$, see Fig. 14.7. In the case that a problem reveals a singularity, one should place x_0 in the center of such a singularity [9].

14.4 Supplementary Problems

14.1 Beam with elastic foundation and point load

A beam has a fixed support at its left-hand boundary and is loaded at its right-hand boundary by a point load F, see Fig. 14.10. The elastic foundation is described by a constant foundation modulus k and it is furthermore assumed that the bending stiffness EI_z is constant. Calculate based on a single Bernoulli element the displacement and rotation of the load application point at $x = L$. Simplify afterwards the result for the case $k = 0$ and $EI_z = 0$ respectively.

14.2 Bar element with quadratic shape functions and excentric node: stress singularity

Derive for a bar element with quadratic shape functions and excentric node at $x = aL$ the B-matrix in such a way that at $x = 0$ a stress singularity is obtained. The correlation between the local and natural coordinate should be realized by a second-order polynomial (Fig. 14.11).

Fig. 14.10 Beam with elastic foundation and point load

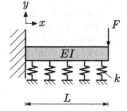

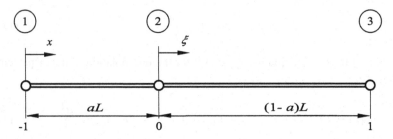

Fig. 14.11 Bar element with quadratic shape functions and excentric node (general case)

14.3 Infinite element with linear interpolation of the field variable

Derive according to the procedure in Sect. 14.3 an element formulation with linear interpolation of the displacement field, i.e.

$$u^e(\xi) = N_1(\xi)u_1 + N_2(\xi)u_2 = \frac{1-\xi}{2}u_1 + \frac{1+\xi}{2}u_2 , \qquad (14.48)$$

and quadratic interpolation of the local coordinate, i.e.

$$x^e(\xi) = M_1(\xi)x_1 + M_2(\xi)x_2 + M_3(\xi)x_3 . \qquad (14.49)$$

References

1. Czichos H, Hennecke M (2012) HÜTTE - Das Ingenieurwissen. Springer Vieweg, Berlin
2. Mourelatos ZP, Parsons MG (1987) A finite element analysis of beams on elastic foundation including shear and axial effects. Comput Struct 27:323–331
3. Wittenburg J (2001) Festigkeitslehre: Ein Lehr- und Arbeitsbuch. Springer, Berlin
4. Winkler E (1867) Die Lehre von der Elasticität und Festigkeit mit besonderer Rücksicht auf ihre Anwendung in der Technik. H. Dominicus, Prag
5. Goss D, Seelig T (2011) Bruchmechanik: Mit einer Einführung in die Mikromechanik. Springer, Berlin
6. Hertzberg RW (1996) Deformation and fracture mechanics of engineering materials. Wiley, Hoboken
7. Öchsner A (2016) Continuum damage and fracture mechanics. Springer, Singapore
8. Pilkey WD (2005) Formulas for stress, strain, and structural matrices. Wiley, Hoboken
9. Cook RD, Malkus DS, Plesha ME, Witt RJ (2002) Concepts and applications of finite element analysis. Wiley, New York
10. Bettes P (1977) Infinite elements. Int J Numer Methods Eng 11:53–64
11. Bettes P, Bettes JA (1984) Infinite elements for static problems. Eng Comput 1:4–16
12. Marques JMM, Owen DRJ (1984) Infinite elements in quasi-static materially nonlinear problems. Comput Struct 18:739–751

Appendix A

A.1 Mathematics

A.1.1 The Greek Alphabet (Table A.1)

A.1.2 Often Used Constants

$$\pi = 3.14159$$
$$e = 2.71828$$
$$\sqrt{2} = 1.41421$$
$$\sqrt{3} = 1.73205$$
$$\sqrt{5} = 2.23606$$
$$\sqrt{e} = 1.64872$$
$$\sqrt{\pi} = 1.77245$$

A.1.3 Special Products

$$(x + y)^2 = x^2 + 2xy + y^2, \tag{A.1}$$
$$(x - y)^2 = x^2 - 2xy + y^2, \tag{A.2}$$
$$(x + y)^3 = x^3 + 3x^2y + 3xy^2 + y^3, \tag{A.3}$$
$$(x - y)^3 = x^3 - 3x^2y + 3xy^2 - y^3, \tag{A.4}$$

© Springer International Publishing AG, part of Springer Nature 2018
A. Öchsner and M. Merkel, *One-Dimensional Finite Elements*,
https://doi.org/10.1007/978-3-319-75145-0

Table A.1 The Greek alphabet

Name	Lower case	Capital
Alpha	α	A
Beta	β	B
Gamma	γ	Γ
Delta	δ	Δ
Epsilon	ϵ	E
Zeta	ζ	Z
Eta	η	H
Theta	θ, ϑ	Θ
Iota	ι	I
Kappa	κ	K
Lambda	λ	Λ
My	μ	M
Ny	ν	N
Xi	ξ	Ξ
Omikron	o	O
Pi	π	Π
Rho	ρ, ϱ	P
Sigma	σ	Σ
Tau	τ	T
Ypsilon	υ	Υ
Phi	ϕ, φ	Φ
Chi	χ	X
Psi	ψ	Ψ
Omega	ω	Ω

$$(x + y)^4 = x^4 + 4x^3 y + 6x^2 y^2 + 4xy^3 + y^4, \tag{A.5}$$
$$(x - y)^4 = x^4 - 4x^3 y + 6x^2 y^2 - 4xy^3 + y^4. \tag{A.6}$$

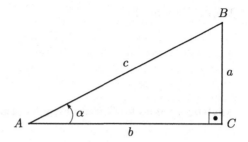

Fig. A.1 Triangle with a right angle at C

A.1.4 Trigonometric Functions

Definition of the Right-Angled Triangle

The triangle ABC has a right angle at C and the edge lengths a, b, c. The trigonometric functions of the angle α are defined in the following kind, see Fig. A.1:

$$\text{Sine of } \alpha = \sin\alpha = \frac{a}{c} = \frac{\text{Opposite}}{\text{Hypotenuse}}, \qquad (A.7)$$

$$\text{Cosine of } \alpha = \cos\alpha = \frac{b}{c} = \frac{\text{Adjacent}}{\text{Hypotenuse}}, \qquad (A.8)$$

$$\text{Tangent of } \alpha = \tan\alpha = \frac{a}{b} = \frac{\text{Opposite}}{\text{Adjacent}}, \qquad (A.9)$$

$$\text{Cotangent of } \alpha = \cot\alpha = \frac{b}{a} = \frac{\text{Adjacent}}{\text{Opposite}}, \qquad (A.10)$$

$$\text{Secant of } \alpha = \sec\alpha = \frac{c}{b} = \frac{\text{Hypotenuse}}{\text{Adjacent}}, \qquad (A.11)$$

$$\text{Cosecant of } \alpha = \csc\alpha = \frac{c}{a} = \frac{\text{Hypotenuse}}{\text{Opposite}}. \qquad (A.12)$$

Addition Theorem

$$\sin(\alpha \pm \beta) = \sin\alpha\cos\beta \pm \cos\alpha\sin\beta, \qquad (A.13)$$

$$\cos(\alpha \pm \beta) = \cos\alpha\cos\beta \mp \sin\alpha\sin\beta, \qquad (A.14)$$

$$\tan(\alpha \pm \beta) = \frac{\tan\alpha \pm \tan\beta}{1 \mp \tan\alpha\tan\beta}, \qquad (A.15)$$

$$\cot(\alpha \pm \beta) = \frac{\cot\alpha\cot\beta \mp 1}{\cot\beta \pm \cot\beta}. \qquad (A.16)$$

Mutual Presentation

$$\sin^2 \alpha + \cos^2 \alpha = 1, \tag{A.17}$$

$$\tan \alpha = \frac{\sin \alpha}{\cos \alpha}. \tag{A.18}$$

Analytical Values for Different Angles (Table A.2)

Table A.2 Analytical values of sine, cosine, tangent and cotangent for different angels

α in Degree	α in Radian	$\sin \alpha$	$\cos \alpha$	$\tan \alpha$	$\cot \alpha$
0°	0	0	1	0	$\pm\infty$
30°	$\frac{1}{6}\pi$	$\frac{1}{2}$	$\frac{\sqrt{3}}{2}$	$\frac{\sqrt{3}}{3}$	$\sqrt{3}$
45°	$\frac{1}{4}\pi$	$\frac{\sqrt{2}}{2}$	$\frac{\sqrt{2}}{2}$	1	1
60°	$\frac{1}{3}\pi$	$\frac{\sqrt{3}}{2}$	$\frac{1}{2}$	$\sqrt{3}$	$\frac{\sqrt{3}}{3}$
90°	$\frac{1}{2}\pi$	1	0	$\pm\infty$	0
120°	$\frac{2}{3}\pi$	$\frac{\sqrt{3}}{2}$	$-\frac{1}{2}$	$-\sqrt{3}$	$-\frac{\sqrt{3}}{3}$
135°	$\frac{3}{4}\pi$	$\frac{\sqrt{2}}{2}$	$-\frac{\sqrt{2}}{2}$	1	1
150°	$\frac{5}{6}\pi$	$\frac{1}{2}$	$-\frac{\sqrt{3}}{2}$	$-\frac{\sqrt{3}}{3}$	$-\sqrt{3}$
180°	π	0	-1	0	$\pm\infty$
210°	$\frac{7}{6}\pi$	$-\frac{1}{2}$	$-\frac{\sqrt{3}}{2}$	$\frac{\sqrt{3}}{3}$	$\sqrt{3}$
225°	$\frac{5}{4}\pi$	$-\frac{\sqrt{2}}{2}$	$-\frac{\sqrt{2}}{2}$	1	1
240°	$\frac{4}{3}\pi$	$-\frac{\sqrt{3}}{2}$	$-\frac{1}{2}$	$\sqrt{3}$	$\frac{\sqrt{3}}{3}$
270°	$\frac{3}{2}\pi$	-1	0	$\pm\infty$	0
300°	$\frac{5}{3}\pi$	$-\frac{\sqrt{3}}{2}$	$\frac{1}{2}$	$-\sqrt{3}$	$-\frac{\sqrt{3}}{3}$
315°	$\frac{7}{4}\pi$	$-\frac{\sqrt{2}}{2}$	$\frac{\sqrt{2}}{2}$	-1	-1
330°	$\frac{11}{6}\pi$	$-\frac{1}{2}$	$\frac{\sqrt{3}}{2}$	$-\frac{\sqrt{3}}{3}$	$-\sqrt{3}$
360°	2π	0	1	0	$\pm\infty$

Double Angle Functions

$$\sin(2\alpha) = 2\sin \alpha \cdot \cos \alpha, \tag{A.19}$$

$$\cos(2\alpha) = \cos^2 \alpha - \sin^2 \alpha$$

$$= 2\cos^2 \alpha - 1$$

$$= 1 - 2\sin^2 \alpha, \tag{A.20}$$

$$\tan(2\alpha) = \frac{2\tan\alpha}{1 - \tan^2\alpha}. \tag{A.21}$$

Reduction Formulae (Table A.3)

Table A.3 Reduction formulae for trigonometric functions

	$-\alpha$	$90° \pm \alpha$	$180° \pm \alpha$	$270° \pm \alpha$	$k(360°) \pm \alpha$
		$\frac{\pi}{2} \pm \alpha$	$\pi \pm \alpha$	$\frac{3\pi}{2} \pm \alpha$	$2k\pi \pm \alpha$
sin	$-\sin\alpha$	$\cos\alpha$	$\mp\sin\alpha$	$-\cos\alpha$	$\pm\sin\alpha$
cos	$\cos\alpha$	$\mp\sin\alpha$	$-\cos\alpha$	$\pm\sin\alpha$	$\cos\alpha$
tan	$-\tan\alpha$	$\mp\cot\alpha$	$\pm\tan\alpha$	$\mp\cot\alpha$	$\pm\tan\alpha$
csc	$-\csc\alpha$	$\sec\alpha$	$\mp\csc\alpha$	$-\sec\alpha$	$\pm\csc\alpha$
sec	$\sec\alpha$	$\mp\csc\alpha$	$-\sec\alpha$	$\pm\csc\alpha$	$\sec\alpha$
cot	$-\cot\alpha$	$\mp\tan\alpha$	$\pm\cot\alpha$	$\mp\tan\alpha$	$\pm\cot\alpha$

A.1.5 Basics for Linear Algebra

Vectors

With

$$\mathbf{a} = [a_1\, a_2\, a_i \ldots a_n] \tag{A.22}$$

a row vector and with

$$\mathbf{a} = \begin{bmatrix} a_1 \\ a_2 \\ a_i \\ \vdots \\ a_n \end{bmatrix} \tag{A.23}$$

a column vector[1] of dimension n is defined, whereupon the following is valid for all components: $a_i \in \mathbb{R} = 1, 2, \ldots, n$.

---

[1] The expression vector is used differently in mathematics and physics. In physics, a vector represents a physical dimension such as, for example, a force. A direction as well as an absolute value can be assigned to this vector. In mathematics, the expression vector is used for a positioning of components. Also here values can be defined, which however are without any physical meaning. Therefore at times vectors are also referred to as row or column matrices.

Matrices

The term matrix can be shown with the help of a simple example. The linear relation between a system of variables x_i and b_i

$$a_{11}x_1 + a_{12}x_2 + a_{13}x_3 + a_{14}x_4 = b_1 \tag{A.24}$$
$$a_{21}x_1 + a_{22}x_2 + a_{23}x_3 + a_{24}x_4 = b_2 \tag{A.25}$$
$$a_{31}x_1 + a_{32}x_2 + a_{33}x_3 + a_{34}x_4 = b_3 \tag{A.26}$$

can be summarized in a compact form as

$$\mathbf{Ax = b} \tag{A.27}$$

or

$$\begin{bmatrix} a_{11} & a_{12} & a_{13} & a_{14} \\ a_{21} & a_{22} & a_{23} & a_{24} \\ a_{31} & a_{32} & a_{33} & a_{34} \end{bmatrix} \begin{bmatrix} x_1 \\ x_2 \\ x_3 \\ x_4 \end{bmatrix} = \begin{bmatrix} b_1 \\ b_2 \\ b_3 \end{bmatrix}. \tag{A.28}$$

Thereby the following is valid for all coefficients a_{ij} and all components b_i and x_i: $a_{ij}, b_i, x_j \in \mathbb{R}, i = 1, 2, 3, j = 1, 2, 3, 4$.

Generally speaking a matrix $\mathbf{A}$ of the dimensions $m \times n$

$$\mathbf{A}^{m \times n} = \begin{bmatrix} a_{11} & a_{12} & \cdots & a_{1n} \\ a_{21} & a_{22} & \cdots & a_{2n} \\ \vdots & \vdots & \cdots & \vdots \\ a_{m1} & a_{m2} & \cdots & a_{mn} \end{bmatrix} \tag{A.29}$$

consists of m rows and n columns.

The **transpose** of a matrix results from interchanging of rows and columns:

$$\mathbf{A}^T = \begin{bmatrix} a_{11} & a_{21} & \cdots & a_{m1} \\ a_{12} & a_{22} & \cdots & a_{m2} \\ \vdots & \vdots & \vdots & \vdots \\ \vdots & \vdots & \vdots & \vdots \\ a_{1n} & a_{2n} & \cdots & a_{mn} \end{bmatrix}. \tag{A.30}$$

Quadratic matrices have equivalent rows and columns:

$$A^{n \times n} = \begin{bmatrix} a_{11} & a_{12} & \dots & a_{1n} \\ a_{21} & a_{22} & \dots & a_{2n} \\ \vdots & \vdots & \dots & \vdots \\ a_{n1} & a_{n2} & \dots & a_{nn} \end{bmatrix}. \tag{A.31}$$

If the following is valid additionally for a quadratic matrix

$$a_{ij} = a_{ji}, \tag{A.32}$$

a symmetric matrix results. As an example, a symmetric (3×3) matrix has the form

$$A^{3 \times 3} = \begin{bmatrix} a_{11} & a_{12} & a_{13} \\ a_{12} & a_{22} & a_{23} \\ a_{13} & a_{23} & a_{33} \end{bmatrix}. \tag{A.33}$$

Matrix Operations

The multiplication of two matrices looks as follows in index notation

$$c_{ij} = \sum_{k=1}^{m} a_{ik} b_{kj} \quad \begin{matrix} i = 1, 2, \dots, n \\ j = 1, 2, \dots, r \end{matrix} \tag{A.34}$$

or in matrix notation

$$C = AB. \tag{A.35}$$

Thereby the matrix $A^{n \times m}$ has n rows and m columns, matrix $B^{m \times r}$ m rows and r columns and the matrix product $C^{n \times r}$ n rows and r columns.

The multiplication of two matrices is not commutative, this means

$$AB \neq BA. \tag{A.36}$$

The product of two transposed matrices $A^{\mathrm{T}} B^{\mathrm{T}}$ results in

$$A^{\mathrm{T}} B^{\mathrm{T}} = (BA)^{\mathrm{T}}. \tag{A.37}$$

The transpose of a matrix product can be split with the help of

$$(AB)^{\mathrm{T}} = B^{\mathrm{T}} A^{\mathrm{T}} \tag{A.38}$$

into the product of the transposed matrices. When multiplying various matrices, the associative law

$$(AB)C = A(BC) = ABC \tag{A.39}$$

and the distributive law are valid

$$A\,(B + C) = A\,B + AC\,.\tag{A.40}$$

Determinant of a Matrix

The determinant of a quadratic matrix A of the dimension n can be determined recursively via

$$|A| = \sum_{i=1}^{n} (-1)^{i+1}\, a_{1i}\, |A_{1i}|\,.\tag{A.41}$$

The submatrix A_{1i} of dimensions $(n-1)(n-1)$ emerges due to canceling of the 1st row and the ith column of A.

Inverse of a Matrix

A is a quadratic matrix. The inverse A^{-1} is quadratic as well. The product of matrix and inverse matrix

$$A^{-1}A = I\tag{A.42}$$

yields the unit matrix. The inverse of a matrices product results as a product of the inverses of the matrices:

$$(AB)^{-1} = B^{-1}A^{-1}\,.\tag{A.43}$$

The inverse of the transposed matrix results as the transpose of the inverse matrix:

$$\left[A^{\mathrm{T}}\right]^{-1} = \left[A^{-1}\right]^{\mathrm{T}}\,.\tag{A.44}$$

Formally, with the inverse of a matrix, the system of equations

$$A\,x = b\tag{A.45}$$

can be solved. Thereby the quadratic matrix A and the vectors x and b have the same dimension. With the multiplication of the inverse from the left

$$A^{-1}Ax = A^{-1}b\tag{A.46}$$

one obtains the vector of the unknown to

$$x = A^{-1}b\,.\tag{A.47}$$

For a (2×2)- and a (3×3)-matrix the inverses are given explicitly. For a quadratic (2×2)-matrix

$$A = \begin{bmatrix} a_{11} & a_{12} \\ a_{21} & a_{22} \end{bmatrix} \tag{A.48}$$

the inverse results in

$$A^{-1} = \frac{1}{|A|} \begin{bmatrix} a_{11} & a_{12} \\ a_{21} & a_{22} \end{bmatrix} \tag{A.49}$$

with

$$|A| = a_{11}a_{22} - a_{12}a_{21} . \tag{A.50}$$

For the quadratic (3×3)-matrix

$$A = \begin{bmatrix} a_{11} & a_{12} & a_{13} \\ a_{21} & a_{22} & a_{23} \\ a_{31} & a_{32} & a_{33} \end{bmatrix} \tag{A.51}$$

the inverse results in

$$A^{-1} = \frac{1}{|A|} \begin{bmatrix} \tilde{a}_{11} & \tilde{a}_{12} & \tilde{a}_{13} \\ \tilde{a}_{21} & \tilde{a}_{22} & \tilde{a}_{23} \\ \tilde{a}_{31} & \tilde{a}_{32} & \tilde{a}_{33} \end{bmatrix} \tag{A.52}$$

with the coefficients of the inverses

$$\begin{aligned}
\tilde{a}_{11} &= +a_{22}a_{33} - a_{32}a_{23} \\
\tilde{a}_{12} &= -(a_{12}a_{33} - a_{13}a_{32}) \\
\tilde{a}_{13} &= +a_{12}a_{23} - a_{22}a_{13} \\
\tilde{a}_{21} &= -(a_{21}a_{33} - a_{31}a_{23}) \\
\tilde{a}_{22} &= +a_{11}a_{33} - a_{13}a_{31} \\
\tilde{a}_{23} &= -(a_{11}a_{23} - a_{21}a_{13}) \\
\tilde{a}_{31} &= +a_{21}a_{32} - a_{31}a_{22} \\
\tilde{a}_{32} &= -(a_{11}a_{32} - a_{31}a_{12}) \\
\tilde{a}_{33} &= +a_{11}a_{22} - a_{12}a_{21}
\end{aligned} \tag{A.53}$$

and with the determinant

$$\begin{aligned}
|A| = {} & a_{11}a_{22}a_{33} + a_{13}a_{21}a_{32} + a_{31}a_{12}a_{23} \\
& - a_{31}a_{22}a_{13} - a_{33}a_{12}a_{21} - a_{11}a_{23}a_{32} .
\end{aligned} \tag{A.54}$$

Equation Solving

The initial point is the system of equations

$$A x = b \tag{A.55}$$

with the quadratic matrix A and the vectors x and b, which both have the same dimension. The matrix A and the vectors b are both assigned with known values. It is the goal to determine the vector of the unknowns x.

The central operation at the direct equations solving is the partition of the system matrix

$$A = LU \tag{A.56}$$

into a lower and a upper triangular matrix. In detail this operation looks as follows:

$$LU = \begin{bmatrix} 1 & 0 & \ldots & 0 \\ L_{21} & 1 & \ldots & 0 \\ \vdots & & \ddots & \vdots \\ L_{n1} & L_{n2} & \ldots & 1 \end{bmatrix} \begin{bmatrix} U_{11} & U_{12} & \ldots & U_{1n} \\ 0 & U_{22} & \ldots & U_{2n} \\ \vdots & & \ddots & \vdots \\ 0 & 0 & \ldots & U_{nn} \end{bmatrix}. \tag{A.57}$$

The triangular decomposition is quite computationally intensive. Variants of the Gauss elimination are used as algorithms. Crucial is the structure of the system matrix A. If blocks with zero entries can be identified in advance in the system matrix the row and column operations can then be used for the blocks only with non-zero entries.

The equations solution is conducted with the paired solution of the two equations

$$Ly = b \tag{A.58}$$

and

$$Ux = y \tag{A.59}$$

whereupon y is solely an auxiliary vector. The single operations proceed as follows:

$$y_1 = b_1 \tag{A.60}$$

$$y_i = b_i - \sum_{j=1}^{i-1} L_{ij} y_j \quad i = 2, 3, \ldots, n \tag{A.61}$$

and

$$x_n = \frac{y_n}{U_{nn}} \tag{A.62}$$

$$x_i = \frac{1}{U_{ii}} \left(y_i - \sum_{j=i+1}^{n} U_{ij} x_j \right) \quad i = n - 1, n - 2, \ldots, 1. \tag{A.63}$$

The first two steps are referred to as forward partition and the last two steps as backward substitution.

The last equation is divided through the value of the diagonal of the upper triangular matrix. For very small and very large values this can lead to inaccuracies. An improvement can be achieved via a so-called pivoting, at which in the current row or column one has to search for the 'best' factor.

A.1.6 Derivatives

- $\dfrac{d}{dx}\left(\dfrac{1}{x}\right) = -\dfrac{1}{x^2}$

- $\dfrac{d}{dx} x^n = n \times x^{n-1}$

- $\dfrac{d}{dx}\sqrt[n]{x} = \dfrac{1}{n \times \sqrt[n]{x^{n-1}}}$

- $\dfrac{d}{dx}\sin(x) = \cos(x)$

- $\dfrac{d}{dx}\cos(x) = -\sin(x)$

- $\dfrac{d}{dx}\ln(x) = \dfrac{1}{x}$

- $\dfrac{d}{dx}|x| = \begin{cases} -1 \text{ if } x < 0 \\ 1 \text{ if } x > 0 \end{cases}$

A.1.7 Integration

A.1.7.1 Antiderivatives

- $\int e^x \, dx = e^x$

- $\int \sqrt{x} \, dx = \dfrac{2}{3} x^{\frac{3}{2}}$

- $\int \sin(x) dx = -\cos(x)$
- $\int \cos(x) dx = \sin(x)$

- $\int \sin(\alpha x) \cdot \cos(\alpha x) dx = \dfrac{1}{2\alpha} \sin^2(\alpha x)$

- $\int \sin^2(\alpha x) dx = \dfrac{1}{2}(x - \sin(\alpha x)\cos(\alpha x)) = \dfrac{1}{2}(x - \frac{1}{2\alpha}\sin(2\alpha x))$

- $\int \cos^2(\alpha x) dx = \dfrac{1}{2}(x + \sin(\alpha x)\cos(\alpha x)) = \dfrac{1}{2}(x + \frac{1}{2\alpha}\sin(2\alpha x))$

A.1.7.2   Partial Integration

One-Dimensional Case:

$$\int_a^b f(x)g'(x)dx = f(x)g(x)|_a^b - \int_a^b f'(x)g(x)dx \qquad (A.64)$$

$$= f(x)g(x)|_b - f(x)g(x)|_a - \int_a^b f'(x)g(x)dx . \qquad (A.65)$$

A.1.7.3   Integration and Coordinate Transformation

One-Dimensional Case:
$T : \mathbb{R} \to \mathbb{R}$ with $x = g(u)$ is a one-dimensional transformation of S to R. If g has a continuous partial derivative, so that the Jacobian matrix becomes nonzero, the following is valid

$$\int_R f(x)dx = \int_S f(g(u)) \left| \frac{dx}{du} \right| du , \qquad (A.66)$$

whereupon the Jacobian matrix in the one-dimensional case is given through $J = \left| \frac{dx}{du} \right| = x_u$.

A.1.7.4 One-Dimensional Integrals for the Calculation of the Stiffness Matrix

$$\int_{-1}^{1} (1-x)\mathrm{d}x = 2$$

$$\int_{-1}^{1} (1-x)^2(1+x)\mathrm{d}x = \frac{4}{3}$$

$$\int_{-1}^{1} (1+x)\mathrm{d}x = 2$$

$$\int_{-1}^{1} (1-x^2)\mathrm{d}x = \frac{4}{3}$$

$$\int_{-1}^{1} (1-x)(1+x)\mathrm{d}x = \frac{4}{3}$$

$$\int_{-1}^{1} (1+x^2)\mathrm{d}x = \frac{8}{3}$$

$$\int_{-1}^{1} (1-x)^2\mathrm{d}x = \frac{8}{3}$$

$$\int_{-1}^{1} (1-2x)x\mathrm{d}x = -\frac{4}{3}$$

$$\int_{-1}^{1} (1+x)^2\mathrm{d}x = \frac{8}{3}$$

$$\int_{-1}^{1} (1+2x)x\mathrm{d}x = \frac{4}{3}$$

$$\int_{-1}^{1} (1-x)^3\mathrm{d}x = 4$$

$$\int_{-1}^{1} (1-2x)^2\mathrm{d}x = \frac{14}{3}$$

$$\int_{-1}^{1} (1+x)^3\mathrm{d}x = 4$$

$$\int_{-1}^{1} (1+2x)^2\mathrm{d}x = \frac{14}{3}$$

$$\int_{-1}^{1} (1-x)(1+x)^2\mathrm{d}x = \frac{4}{3}$$

$$\int_{-1}^{1} (1-2x)(1+2x)\mathrm{d}x = -\frac{2}{3}$$

A.1.8   Expansion of a Function in a Taylor's Series

The expansion of a function $f(x)$ in a Taylor's series at the position x_0 yields:

$$f(x) = f(x_0) + \left(\frac{df}{dx}\right)_{x_0} \cdot (x - x_0) + \frac{1}{2!}\left(\frac{d^2 f}{dx^2}\right)_{x_0} \cdot (x - x_0)^2 + \cdots + \frac{1}{k!}\left(\frac{d^k f}{dx^k}\right)_{x_0} \cdot (x - x_0)^k .$$

(A.67)

An approximation of first order only takes into account the first derivative, and the approach for the function results in:

$$f(x) = f(x_0 + dx) \approx f(x_0) + \left(\frac{df}{dx}\right)_{x_0} \cdot (x - x_0) . \qquad (A.68)$$

If one considers from the analytical geometry that the first derivative of a function equals the slope of the tangent in the considered point and that the point-slope form of a straight line is given through $f(x) - f(x_0) = m \times (x - x_0)$, it results that the approximation of first order represents the equation of a straight line through the point $(x_0, f(x_0))$ with the slope $m = f'(x_0) = (df/dx)_{x_0}$, compare Fig. A.2.

Fig. A.2 Approximation of a function $f(x)$ via a Taylor's series of first order

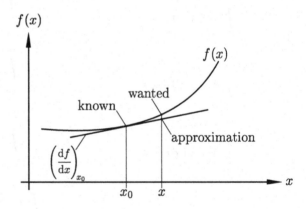

A.2 Units and Conversions

A.2.1 Consistent Units

In applying a finite element program usually there is no regulation on a specific physical mass or unit system. A finite element program retains *consistent* units throughout the analysis and requires the user to only enter absolute measure values without the information about a certain unit. Therewith the units, which are used by the user for the input, are also kept consistent in the output. The user therefore has to ensure for himself that his/her chosen units are consistent, meaning compatible with each other. The following Table A.4 shows an example of consistent units.

One can consider at this point the unit of the density. The following example shows the conversion of the density for steal:

$$\varrho_{St} = 7.8 \, \frac{kg}{dm^3} = 7.8 \times 10^3 \, \frac{kg}{m^3} = 7.8 \times 10^{-6} \, \frac{kg}{mm^3}. \tag{A.69}$$

Table A.4 Examples of consistent units

Size	Unit
Length	mm
Area	mm^2
Force	N
Stress	$MPa = \dfrac{N}{mm^2}$
Moment	Nmm
Moment of Inertia	mm^4
Elastic Modulus	$MPa = \dfrac{N}{mm^2}$
Density	$\dfrac{Ns^2}{mm^4}$
Time	s
Mass	$10^3 kg$

Table A.5 Example of consistent English units

Size	Unit
Length	in
Area	in^2
Force	lbf
Stress	$psi = \dfrac{lbf}{in^2}$
Moment	lbf in
Moment of Inertia	in^4
Elastic Modulus	$psi = \dfrac{lbf}{in^2}$
Density	$\dfrac{lbf\,sec^2}{in^4}$
Time	sec

With

$$1\,N = 1\,\frac{m\,kg}{s^2} = 1 \times 10^3\,\frac{mm\,kg}{s^2} \quad und \quad 1\,kg = 1 \times 10^{-3}\,\frac{Ns^2}{mm} \tag{A.70}$$

the consistent density results in:

$$\varrho_{St} = 7.8 \times 10^{-9}\,\frac{Ns^2}{mm^4}. \tag{A.71}$$

Since in the literature at one point or another also other units occur, the following Table A.6 shows an example of consistent English units (Table A.5):

One also considers at this point the conversion of the density:

$$\varrho_{St} = 0.282\,\frac{lb}{in^3} = 0.282\,\frac{1}{in^3} \times 0.00259\,\frac{lbf\,sec^2}{in} = 0.73038 \times 10^{-3}\,\frac{lbf\,sec^2}{in^4}. \tag{A.72}$$

A.2.2 Conversion of Important English Units (Table A.6)

Table A.6 Conversion of important English units

Type	English-speaking unit	Conversion
Length	Inch	$1 \text{ in} = 0.025400 \text{ m}$
	Foot	$1 \text{ ft} = 0.304800 \text{ m}$
	Yard	$1 \text{ yd} = 0.914400 \text{ m}$
	Mile (statute)	$1 \text{ mi} = 1609.344 \text{ m}$
	Mile (nautical)	$1 \text{ nm} = 1852.216 \text{ m}$
Area	Square inch	$1 \text{ sq in} = 1 \text{ in}^2 = 6.45160 \text{ cm}^2$
	Square foot	$1 \text{ sq ft} = 1 \text{ ft}^2 = 0.092903040 \text{ m}^2$
	Square yard	$1 \text{ sq yd} = 1 \text{ yd}^2 = 0.836127360 \text{ m}^2$
	Square mile	$1 \text{ sq mi} = 1 \text{ mi}^2 = 2589988.110336 \text{ m}^2$
	Acre	$1 \text{ ac} = 4046.856422400 \text{ m}^2$
Volume	cubic inch	$1 \text{ cu in} = 1 \text{ in}^3 = 0.000016387064 \text{ m}^3$
	Cubic foot	$1 \text{ cu ft} = 1 \text{ ft}^3 = 0.028316846592 \text{ m}^3$
	Cubic yard	$1 \text{ cu yd} = 1 \text{ yd}^3 = 0.764554857984 \text{ m}^3$
Mass	Ounce	$1 \text{ oz} = 28.349523125 \text{ g}$
	Pound (mass)	$1 \text{ lb}_m = 453.592370 \text{ g}$
	Short ton	$1 \text{ sh to} = 907184.74 \text{ g}$
	Long ton	$1 \text{ lg to} = 1016046.9088 \text{ g}$
Force	Pound-force	$1 \text{ lbf} = 1 \text{ lb}_F = 4.448221615260500 \text{ N}$
	Poundal	$1 \text{ pdl} = 0.138254954376 \text{ N}$
Stress	Pound-force per square inch	$1 \text{ psi} = 1 \frac{\text{lbf}}{\text{in}^2} = 6894.75729316837 \frac{\text{N}}{\text{m}^2}$
	Pound-force per square foot	$1 \frac{\text{lbf}}{\text{ft}^2} = 47.880258980336 \frac{\text{N}}{\text{m}^2}$
Energy	British thermal unit	$1 \text{ Btu} = 1055.056 \text{ J}$
	Calorie	$1 \text{ cal} = 4185.5 \text{ J}$
Power	Horsepower	$1 \text{ hp} = 745.699871582270 \text{ W}$

A.3   Mechanics

A.3.1   Second Moment of Area (Table A.7)

Table A.7 Axial second moment of area around the z- and y-axis

cross-section	I_z	I_y
	$\dfrac{\pi D^4}{64} = \dfrac{\pi R^4}{4}$	$\dfrac{\pi D^4}{64} = \dfrac{\pi R^4}{4}$
	$\dfrac{\pi b a^3}{4}$	$\dfrac{\pi a b^3}{4}$
	$\dfrac{b h^3}{12}$	$\dfrac{h b^3}{12}$
	$\dfrac{b h^3}{36}$	$\dfrac{h b^3}{36}$
	$\dfrac{b h^3}{36}$	$\dfrac{a h^3}{48}$

A.3.2 Equivalent Nodal Loads for Bending Elements (Table A.8)

Table A.8 Equivalent nodal loads for bending elements. Adapted from [1]

Load	Shear force	Bending moment
	$F_{1y} = -\dfrac{qL}{2}$ $F_{2y} = -\dfrac{qL}{2}$	$M_{1z} = -\dfrac{qL^2}{12}$ $M_{2z} = +\dfrac{qL^2}{12}$
	$F_{1y} = -\dfrac{qa}{2L^3}(a^3 - 2a^2L + 2L^3)$ $F_{2y} = -\dfrac{qa^3}{2L^3}(2L - a)$	$M_{1z} = -\dfrac{qa^2}{12L^2}(3a^2 - 8aL + 6L^2)$ $M_{2z} = +\dfrac{qa^3}{12L^2}(4L - 3a)$
	$F_{1y} = -\dfrac{3}{20}qL$ $F_{2y} = -\dfrac{7}{20}qL$	$M_{1z} = -\dfrac{qL^2}{30}$ $M_{2z} = +\dfrac{qL^2}{20}$
	$F_{1y} = -\dfrac{1}{4}qL$ $F_{2y} = -\dfrac{1}{4}qL$	$M_{1z} = -\dfrac{5qL^2}{96}$ $M_{2z} = +\dfrac{5qL^2}{96}$
	$F_{1y} = -\dfrac{Fb^2(3a + b)}{L^3}$ $F_{2y} = -\dfrac{Fa^2(a + 3b)}{L^3}$	$M_{1z} = -\dfrac{Fb^2a}{L^2}$ $M_{2z} = +\dfrac{Fa^2b}{L^2}$
	$F_{1y} = -6M\dfrac{ab}{L^3}$ $F_{2y} = +6M\dfrac{ab}{L^3}$	$M_{1z} = -M\dfrac{b(2a - b)}{L^2}$ $M_{2z} = -M\dfrac{a(2b - a)}{L^2}$

A.3.3 Characteristics of Different Cross-Sections in the y-x Plane (Table A.9)

Table A.9 Characteristics of different cross-sections in the y-z plane. I_z and I_y: axial second moment of area; A: cross-sectional area; k_s: shear correction factor. Adapted from [5]

cross-section	I_z	I_y	A	k_s
	$\dfrac{\pi R^4}{4}$	$\dfrac{\pi R^4}{4}$	πR^2	$\dfrac{9}{10}$
	$\pi R^3 t$	$\pi R^3 t$	$2\pi R t$	0.5
	$\dfrac{bh^3}{12}$	$\dfrac{hb^3}{12}$	hb	$\dfrac{5}{6}$
	$\dfrac{h^2}{6}(ht_w + 3bt_f)$	$\dfrac{b^2}{6}(bt_f + 3ht_w)$	$2(bt_f + ht_w)$	$\dfrac{2ht_w}{A}$
	$\dfrac{h^2}{12}(ht_w + 6bt_f)$	$\dfrac{b^3 t_f}{6}$	$ht_w + 2bt_f$	$\dfrac{ht_w}{A}$

A.3.4   Closed-Form Solutions of the Bending Line (Table A.10)

Table A.10   Closed-form solutions of the bending line at simple loading cases for statically determinate beams at bending in the x-y plane

Load	Bending line
	$u_y(x) = \dfrac{-F}{6EI_z} \times \left[3ax^2 - x^3 + \langle x - a \rangle^3 \right]$
	$u_y(x) = \dfrac{-M}{2EI_z} \times \left[-x^2 + \langle x - a \rangle^2 \right]$
	$u_y(x) = \dfrac{-q}{24EI_z} \times \left[6(a_2^2 - a_1^2)x^2 - 4(a_2 - a_1)x^3 + \right.$ $\left. + \langle x - a_1 \rangle^4 - \langle x - a_2 \rangle^4 \right]$
	$u_y(x) = \dfrac{-F}{6bEI_z} \times \left[(b - a)(b^2 x - x^3) - x \langle b - a \rangle^3 + \right.$ $\left. + b\langle x - a \rangle^3 - a \langle x - b \rangle^3 \right]$
	$u_y(x) = \dfrac{-M}{6bEI_z} \times \left[b^2 x - x^3 - 3x \langle b - a \rangle^2 + \right.$ $\left. + 3b \langle x - a \rangle^2 + \langle x - b \rangle^3 \right]$
	$u_y(x) = \dfrac{-q}{24bEI_z} \times \left[2\left(a_2^2 - a_1^2 - 2b(a_2 - a_1) \right) (x^3 - b^2 x) \right.$ $-x \langle b - a_1 \rangle^4 + x \langle b - a_2 \rangle^4 + b \langle x - a_1 \rangle^4 -$ $\left. - b \langle x - a_2 \rangle^4 - 2(a_2^2 - a_1^2) \langle x - b \rangle^3 \right]$

Reference

1. Buchanan GR (1995) Schaum's outline of theory and problems of finite element analysis. McGraw-Hill Book, New York

Short Solutions of the Exercises

Problems from Chap. 3

3.4: Tension bar with quadratic approximation

Three nodes are being introduced in a quadratic approach. The three shape functions are:

$$N_1(x) = 1 - 3\frac{x}{L} + 2\left(\frac{x}{L}\right)^2 = 1 + \frac{x}{L}\left(3 + 2\frac{x}{L}\right),$$

$$N_2(x) = 4\frac{x}{L} - 4\left(\frac{x}{L}\right)^2 = 4\frac{x}{L}\left(1 - \frac{x}{L}\right), \tag{A.73}$$

$$N_3(x) = -\frac{x}{L} + 2\left(\frac{x}{L}\right)^2 = \frac{x}{L}\left(-1 + 2\frac{x}{L}\right).$$

The derivatives of the three shape functions result in:

$$\frac{dN_1(x)}{dx} = N_1'(x) = -\frac{3}{L} + 4\frac{x}{L^2} = \frac{1}{L}\left(-3 + 4\frac{x}{L}\right),$$

$$\frac{dN_2(x)}{dx} = N_2'(x) = \frac{4}{L} - 8\frac{x}{L^2} = 4\frac{1}{L}\left(1 - 2\frac{x}{L}\right), \tag{A.74}$$

$$\frac{dN_3(x)}{dx} = N_3'(x) = -\frac{1}{L} + 4\frac{x}{L^2} = \frac{1}{L}\left(-1 + 4\frac{x}{L}\right).$$

With the general calculation rule for the stiffness matrix

$$k^e = \int_{\Omega} \boldsymbol{B}^{\mathrm{T}} \boldsymbol{D} \boldsymbol{B}\, d\Omega = EA \int_0^L \boldsymbol{B}^{\mathrm{T}} \boldsymbol{B}\, dx = EA \int_0^L \boldsymbol{N}'^{\mathrm{T}} \boldsymbol{N}'\, dx \tag{A.75}$$

the following results for a three-nodal element

© Springer International Publishing AG, part of Springer Nature 2018
A. Öchsner and M. Merkel, *One-Dimensional Finite Elements*,
https://doi.org/10.1007/978-3-319-75145-0

$$k^e = EA \int_0^L \begin{bmatrix} N_1'^2 & N_1'N_2' & N_1'N_3' \\ & N_2'^2 & N_2'N_3' \\ \text{sym.} & & N_3'^2 \end{bmatrix} dx \, . \tag{A.76}$$

After the integration

$$\int_0^L \begin{bmatrix} \left(-3+4\frac{x}{L}\right)^2 & \left(-3+4\frac{x}{L}\right)\left(4-8\frac{x}{L}\right) & \left(-3+4\frac{x}{L}\right)\left(-1+4\frac{x}{L}\right) \\ & \left(4-8\frac{x}{L}\right)^2 & \left(4-8\frac{x}{L}\right)\left(-1+4\frac{x}{L}\right) \\ \text{sym.} & & \left(-1+4\frac{x}{L}\right)^2 \end{bmatrix} dx \tag{A.77}$$

the stiffness matrix for a bar element with quadratic shape function results in:

$$k^e = \frac{EA}{3L} \begin{bmatrix} 7 & -8 & 1 \\ -8 & 16 & -8 \\ 1 & -8 & 7 \end{bmatrix} \, . \tag{A.78}$$

3.5: Tension Bar with variable Cross-Section and Quadratic Approximation

The three form function and the derivatives are the same as in the exercise before. The difference is the variable cross-section. This changes the calculation for the stiffness matrix to:

$$k^e = \int_\Omega B^T D B \, d\Omega = E \int_0^L B^T B A(x) dx = E \int_0^L N'^T N' A(x) dx \, . \tag{A.79}$$

The three nodal elements result in

$$k^e = E \int_0^L \begin{bmatrix} N_1'^2 & N_1'N_2' & N_1'N_3' \\ & N_2'^2 & N_2'N_3' \\ \text{sym.} & & N_3'^2 \end{bmatrix} A(x) dx \, . \tag{A.80}$$

For element k_{11} this means:

$$k_{11} = E\int_0^L (N_1'(x))^2 A(x)\,dx$$

$$= E\int_0^L \left(9\frac{1}{L^2} - 24x\frac{1}{L^3} + 16x^2\frac{1}{L^4}\right)\left(A_1 + A_2 x\frac{1}{L} - A_1 x\frac{1}{L}\right)dx$$

$$= E\frac{1}{L^2}\int_0^L 9A_1 + 9xA_2\frac{1}{L} - 9xA_1\frac{1}{L} - 24xA_1\frac{1}{L} - 24x^2 A_2\frac{1}{L^2}$$

$$+ 24xA_1\frac{1}{L^2} + 16x^2 A_1\frac{1}{L^2} + 16x^2 A_2\frac{1}{L^3} - 16x^2 A_1\frac{1}{L^3}dx$$

$$= E\frac{1}{L^2}\Big|\frac{1}{L^3}\Big(9xA_1 L^3 + 9x^2 A_2 L - 9x^2 A_1 L - 12x^2 A_1 L^2 - 8x^3 A_2 L$$

$$+ 8x^3 A_1 L + \frac{16}{3}x^3 A_1 L + 4x^4 A_2 - 4x^4 A_1\Big)\Big|_0^L$$

$$= E\frac{1}{L}\left[\left(9LA_1 + \frac{9}{2}LA_2 - \frac{9}{2}LA_1 - 12LA_1 - 8LA_2\right.\right.$$

$$+ \left.\left. 8LA_1 + \frac{16}{3}LA_1 + 4LA_2 - 4LA_1\right) - 0\right]$$

$$= E\frac{1}{6L}(11A_1 + 3A_2)$$

(A.81)

The other elements are calculated according to the same scheme. The complete stiffness matrix k^e results in:

$$k^e = \frac{E}{6L}\begin{bmatrix} 11A_1 + 3A_2 & -4(3A_1 + A_2) & A_1 + A_2 \\ -4(3A_1 + A_2) & 16(A_1 + A_2) & -4(3A_1 + A_2) \\ A_1 + A_2 & -4(3A_1 + A_2) & 3A_1 + 11A_2 \end{bmatrix}.$$

(A.82)

The stiffness matrix for a constant cross-section can be received for the case $A_1 = A_2 = A$ as follows:

$$k^e = \frac{E}{6L}\begin{bmatrix} 14A & -16A & 2A \\ -16A & 32A & -16A \\ 2A & -16A & 14A \end{bmatrix}$$

$$= \frac{EA}{3L}\begin{bmatrix} 7 & -8 & 1 \\ -8 & 16 & -8 \\ 1 & -8 & 7 \end{bmatrix}.$$

(A.83)

Problems from Chap. 4

4.2: Torsion Bar with linear Approximation

For the derivation of the linear shape functions one receives

$$B = \frac{1}{L}\begin{bmatrix} -1 & 1 \end{bmatrix} .$$
(A.84)

Since the bar has a rotational cross-section the torsional moment of inertia equals the polar moment of inertia and the torsional stiffness is defined through

$$D = GI_T = GI_p .$$
(A.85)

Now the stiffness matrix is calculated with:

$$
\begin{aligned}
k_t^e &= \int_L B^T D B \, dx = GI_p \int_L B^T B \, dx \\
&= GI_p \int_L \frac{1}{L}\begin{bmatrix} -1 \\ 1 \end{bmatrix} \frac{1}{L}\begin{bmatrix} -1 & 1 \end{bmatrix} \, dx = GI_p \left| \frac{1}{L^2} x \begin{bmatrix} 1 & -1 \\ -1 & 1 \end{bmatrix} \right|_0^L \\
&= \frac{GI_p}{L}\begin{bmatrix} 1 & -1 \\ -1 & 1 \end{bmatrix} .
\end{aligned}
$$
(A.86)

4.3: Heat conduction bar

The general equation for one-dimensional heat flux in the FE-Method is

$$\begin{bmatrix} \dot{Q}_1 \\ \dot{Q}_2 \end{bmatrix}^e = k_\Theta \begin{bmatrix} 1 & -1 \\ -1 & 1 \end{bmatrix}\begin{bmatrix} \Theta_1 \\ \Theta_2 \end{bmatrix} = -\frac{\lambda A}{L}\begin{bmatrix} 1 & -1 \\ -1 & 1 \end{bmatrix}\begin{bmatrix} \Theta_1 \\ \Theta_2 \end{bmatrix}$$
(A.87)

The heat fluxes at the ends can be calculated directly from the equation above:

$$\dot{Q}_1 = -\frac{\lambda A}{L}(\Theta_1 - \Theta_2) = -\frac{50\frac{W}{mK} \, 10^{-4}m^2}{0.5\,m} \times 100\,K = -1\,W$$
(A.88)

$$\dot{Q}_2 = -\frac{\lambda A}{L}(\Theta_2 - \Theta_1) = -\frac{50\frac{W}{mK} \, 10^{-4}m^2}{0.5\,m} \times -100\,K = 1\,W$$
(A.89)

Problems from Chap. 5

5.4: Equilibrium relation for infinitesimal beam element with variable load

For the setup of the equilibrium relation the changeable load is evaluated in the middle of the interval:

$$- Q_y(x) + Q_y(x + dx) + q_y\left(x + \frac{1}{2}dx\right)dx = 0. \qquad (A.90)$$

$$M_z(x + dx) - M_z(x) + Q_y(x)dx - \frac{1}{2}q_y\left(x + \frac{1}{2}dx\right)dx^2 = 0. \qquad (A.91)$$

5.5: Weighted residual method with variable distributed load

$$\int_0^L W^T(x)\left(EI_z\frac{d^4u_y(x)}{dx^4} - q_y(x)\right)dx = 0 \qquad (A.92)$$

$$\int_0^L EI_z\frac{d^2W^T}{dx^2}\frac{d^2u_y}{dx^2}dx = \int_0^L W^T q_y(x)dx + \left[-W^T\frac{d^3u_y}{dx^3} + \frac{dW^T}{dx}\frac{d^2u_y}{dx^2}\right]_0^L \qquad (A.93)$$

$$\cdots = \delta u_p^T \int_0^L N^T q_y(x)dx + \cdots \qquad (A.94)$$

$$\cdots = \int_0^L \begin{bmatrix} N_{1u} \\ N_{1\varphi} \\ N_{2u} \\ N_{2\varphi} \end{bmatrix} q_y(x)dx + \cdots \qquad (A.95)$$

The additional expression on the right-hand side yields the equivalent nodal loads for a load according to Eqs. (5.168) up to (5.171).

5.6: Stiffness matrix at bending in x-z plane

Bending in the x-z plane can be considered that the rotation is defined via $\varphi_y(x) = -\dfrac{\mathrm{d}u_z(x)}{\mathrm{d}x}$. Therefore, the following shape functions can be derived:

$$N_{1u}^{xz} = 1 - 3\left(\frac{x}{L}\right)^2 + 2\left(\frac{x}{L}\right)^3, \tag{A.96}$$

$$N_{1\varphi}^{xz} = -x + 2\frac{x^2}{L} - \frac{x^3}{L^2}, \tag{A.97}$$

$$N_{2u}^{xz} = 3\left(\frac{x}{L}\right)^2 - 2\left(\frac{x}{L}\right)^3, \tag{A.98}$$

$$N_{2\varphi}^{xz} = \frac{x^2}{L} - \frac{x^3}{L^2}. \tag{A.99}$$

A comparison with the shape functions in bending in the x-y plane according to Eqs. (5.60) up to (5.63) yields that the shape functions for the rotation have been multiplied with (-1).

5.7: Bending beam with changeable cross-section

The axial second moments of area result in:

$$I_z(x) = \frac{\pi}{64}\left(d_1 + (d_2 - d_1)\frac{x}{L}\right)^4 \quad \text{(circle)}, \tag{A.100}$$

$$I_z(x) = \frac{b}{12}\left(d_1 + (d_2 - d_1)\frac{x}{L}\right)^3 \quad \text{(rectangle)}. \tag{A.101}$$

The following results for the circular and rectangular cross-section:

$$
\kappa^e = \frac{\pi E}{64L^3}
\begin{bmatrix}
\frac{12(11d_2^4+11d_2^3d_1+5d_2^2d_1^2+3d_2d_1^3+5d_1^4)}{35} & \frac{2(19d_2^4+7d_2^4+8d_2^3d_1+9d_2^2d_1^2+22d_2d_1^3)L}{35} & \frac{12(11d_2^4+11d_2^4+5d_2^3d_1+3d_2^2d_1^2+5d_2d_1^3)}{35} & -\frac{2(47d_2^4+19d_2^4+22d_2^3d_1+9d_2^2d_1^2+8d_2d_1^3)L}{35} \\
\frac{2(19d_2^4+47d_2^4+8d_2^3d_1+9d_2^2d_1^2+22d_2d_1^3)L}{35} & \frac{4(3d_2^4+17d_2^4+2d_2^3d_1+9d_2^2d_1^2+9d_2d_1^3)L^2}{35} & \frac{2(19d_2^4+47d_2^4+8d_2^3d_1+9d_2^2d_1^2+22d_2d_1^3)L}{35} & -\frac{2(13d_2^4+13d_2^4+4d_2^3d_1+d_2^2d_1^2+4d_2d_1^3)L^2}{35} \\
\frac{12(11d_2^4+11d_2^4+5d_2^3d_1+3d_2^2d_1^2+5d_2d_1^3)}{35} & \frac{2(19d_2^4+47d_2^4+8d_2^3d_1+9d_2^2d_1^2+22d_2d_1^3)L}{35} & \frac{12(11d_2^4+11d_2^4+5d_2^3d_1+3d_2^2d_1^2+5d_2d_1^3)}{L} & \frac{2(47d_2^4+19d_2^4+22d_2^3d_1+9d_2^2d_1^2+8d_2d_1^3)L}{35} \\
\frac{2(47d_2^4+19d_2^4+22d_2^3d_1+9d_2^2d_1^2+8d_2d_1^3)L}{35} & \frac{2(13d_2^4+13d_2^4+4d_2^3d_1+d_2^2d_1^2+4d_2d_1^3)L^2}{35} & \frac{2(47d_2^4+19d_2^4+22d_2^3d_1+9d_2^2d_1^2+8d_2d_1^3)L}{35} & \frac{4(17d_2^4+3d_2^4+9d_2^3d_1+4d_2^2d_1^2+2d_2d_1^3)L^2}{35}
\end{bmatrix}
$$

$$(A.102)$$

$$
\kappa^e = \frac{bE}{12L^3}
\begin{bmatrix}
\frac{3(7d_2^3+3d_2^2d_1+3d_2d_1^2+7d_1^3)}{5} & \frac{3(2d_2^3+d_2^2d_1+2d_2d_1^2+5d_1^3)L}{5} & \frac{3(7d_2^3+3d_2^2d_1+3d_2d_1^2+7d_1^3)}{5} & -\frac{3(5d_2^3+2d_2^2d_1+d_2d_1^2+2d_1^3)L}{5} \\
\frac{3(2d_2^3+d_2^2d_1+2d_2d_1^2+5d_1^3)L}{5} & \frac{(2d_2^3+2d_2^2d_1+5d_2d_1^2+11d_1^3)L^2}{5} & \frac{3(2d_2^3+d_2^2d_1+2d_2d_1^2+5d_1^3)L}{5} & -\frac{(4d_2^3+d_2^2d_1+d_2d_1^2+4d_1^3)L^2}{5} \\
\frac{3(7d_2^3+3d_2^2d_1+3d_2d_1^2+7d_1^3)}{5} & \frac{3(2d_2^3+d_2^2d_1+2d_2d_1^2+5d_1^3)L}{5} & \frac{3(7d_2^3+3d_2^2d_1+3d_2d_1^2+7d_1^3)}{5} & \frac{3(5d_2^3+2d_2^2d_1+d_2d_1^2+2d_1^3)L}{5} \\
\frac{3(5d_2^3+2d_2^2d_1+d_2d_1^2+2d_1^3)L}{5} & \frac{(4d_2^3+d_2^2d_1+d_2d_1^2+4d_1^3)L^2}{5} & \frac{3(5d_2^3+2d_2^2d_1+d_2d_1^2+2d_1^3)L}{5} & \frac{(11d_2^3+5d_2^2d_1+2d_2d_1^2+2d_1^3)L^2}{5}
\end{bmatrix}
$$

$$(A.103)$$

5.8: Equivalent nodal load for quadratic load

$$q(x) = q_0 x^2 \qquad q(x) = q_0 \left(\frac{x}{L}\right)^2$$

$$F_{1y} = -\frac{q_0 L^3}{15} \qquad F_{1y} = -\frac{q_0 L}{15}$$

$$M_{1z} = -\frac{q_0 L^4}{60} \qquad M_{1z} = -\frac{q_0 L^2}{60}$$

$$F_{2y} = -\frac{4q_0 L^3}{15} \qquad F_{2y} = -\frac{4q_0 L}{15}$$

$$M_{1z} = \frac{q_0 L^4}{30} \qquad M_{1z} = \frac{q_0 L^2}{30}$$

5.9: Bending beam with changeable cross-section under point load

Analytical Solution:

$$EI_z(x)\frac{\mathrm{d}^2 u_y(x)}{\mathrm{d}x^2} = M_z(x),\tag{A.104}$$

$$\frac{E\pi h^4}{64}\left(2 - \frac{x}{L}\right)^4 \frac{\mathrm{d}^2 u_y(x)}{\mathrm{d}x^2} = -F(L-x).\tag{A.105}$$

$$u_y(x) = \frac{FL}{E\pi h^4}\left(\frac{64L^3}{2(-2L+x)} + \frac{64L^4}{6(-2L+x)}\right) + \frac{16L}{3}x + \frac{40L^2}{3}.\tag{A.106}$$

$$u_y(L) = -\frac{8}{3}\frac{FL^3}{E\pi h^4} \approx -2.666667\frac{FL^3}{E\pi h^4}.\tag{A.107}$$

Finite Element Solution:

$$u_y(L) = -\frac{7360}{2817}\frac{FL^3}{E\pi h^4} \approx -2.612709\frac{FL^3}{E\pi h^4}.\tag{A.108}$$

Problems from Chap. 6

6.1: Cubic displacement distribution in the tension bar

The natural coordinates of the four integration points are $\xi_1 = -1$, $\xi_2 = -1/3$, $\xi_3 = +1/3$ und $\xi_4 = +1$. The four shape functions

$$N_1 = \frac{(\xi - \xi_2)(\xi - \xi_3)(\xi - \xi_4)}{(\xi_1 - \xi_2)(\xi_1 - \xi_3)(\xi_1 - \xi_4)} = +\frac{9}{19}\left(\xi^2 - \frac{1}{9}\right)(\xi - 1),$$

$$N_2 = \frac{(\xi - \xi_1)(\xi - \xi_3)(\xi - \xi_4)}{(\xi_2 - \xi_1)(\xi_2 - \xi_3)(\xi_2 - \xi_4)} = -\frac{27}{16}\left(\xi - \frac{1}{3}\right)(\xi^2 - 1),$$

$$N_3 = \frac{(\xi - \xi_1)(\xi - \xi_2)(\xi - \xi_4)}{(\xi_3 - \xi_1)(\xi_3 - \xi_2)(\xi_3 - \xi_4)} = -\frac{27}{16}\left(\xi + \frac{1}{3}\right)(\xi^2 - 1),$$

$$N_4 = \frac{(\xi - \xi_1)(\xi - \xi_2)(\xi - \xi_3)}{(\xi_4 - \xi_1)(\xi_4 - \xi_2)(\xi_4 - \xi_3)} = +\frac{9}{19}\left(\xi^2 - \frac{1}{9}\right)(\xi + 1)$$

(A.109)

result via evaluation of Eq. (6.49) for $i = 1$ bis $i = n = 4$.

6.2: Coordinate transformation for tension bar in the plane

For the bar a normal force and a displacement in normal direction are defined on a node in local coordinates. In the plane the parameters each separate in the X- and Y-direction. Therefore, the transformation matrix has the dimensions 4×4. In the transformation matrix according to Eq. (6.16) it is the task to define the following expressions

$$\sin(30°) = \frac{1}{2} \quad \text{and} \quad \cos(30°) = \frac{1}{2}\sqrt{3}.$$

(A.110)

With this, the transformation matrix results in

$$\boldsymbol{T} = \begin{bmatrix} \frac{1}{2}\sqrt{3} & \frac{1}{2} \\ -\frac{1}{2} & \frac{1}{2}\sqrt{3} \end{bmatrix}.$$

(A.111)

The single stiffness relation in global coordinates

$$\begin{bmatrix} F_{1X} \\ F_{1Y} \\ F_{2X} \\ F_{2Y} \end{bmatrix} = \frac{1}{4}\frac{EA}{L} \begin{bmatrix} 3 & \sqrt{3} & -3 & -\sqrt{3} \\ \sqrt{3} & 1 & -\sqrt{3} & -1 \\ -3 & -\sqrt{3} & 3 & \sqrt{3} \\ -\sqrt{3} & -1 & \sqrt{3} & 1 \end{bmatrix} \begin{bmatrix} u_{1X} \\ u_{1Y} \\ u_{2X} \\ u_{2Y} \end{bmatrix}$$

(A.112)

results via the evaluation of Eq. (6.20).

Problems from Chap. 7

7.1: Short solution: Three-Dimensional Beam Structure

Via integration the following solution vector results:

$$
\begin{bmatrix} u_{2Z} \\ \varphi_{2X} \\ \varphi_{2Y} \\ u_{3Z} \\ \varphi_{3Y} \\ u_{4Z} \end{bmatrix} = \begin{bmatrix} +11.904 \\ +0.01785 \\ -0.05492 \\ +78.731 \\ -0.07277 \\ +78.732 \end{bmatrix} .
\tag{A.113}
$$

7.2: Three-Dimensional Beam Structure, Alternative Coordinate System

The column matrices of the state variables are the following in global coordinates:

$$
[u_{1Z}, \varphi_{1X}, \varphi_{1Y}, u_{2Z}, \varphi_{2X}, \varphi_{2Y}, u_{3Z}, \varphi_{3X}, u_{4Z}]^{\mathrm{T}}
\tag{A.114}
$$

and

$$
[F_{1Z}, M_{1X}, M_{1Y}, F_{2Z}, M_{2X}, M_{2Y}, F_{3Z}, M_{3X}, F_{4Z}]^{\mathrm{T}} .
\tag{A.115}
$$

The order of the entries on node 2 have changed in comparison to the original coordinate system. The angles for bending and torsion are exchanged:

$$
\begin{bmatrix} u_{2Z} \\ \varphi_{2X} \\ \varphi_{2Y} \\ u_{3Z} \\ \varphi_{3X} \\ u_{4Z} \end{bmatrix} = \begin{bmatrix} +\dfrac{F}{3\frac{EI_y}{L^3}} \\[2mm] +\dfrac{F}{2\frac{GI_t}{L}} \\[2mm] +\dfrac{F}{2\frac{EI_y}{L^2}} \\[2mm] +\dfrac{2(GI_t + 3EI_y)L^3 F}{3EI_y GI_t} \\[2mm] +\dfrac{L^2(GI_t + 2EI_y)F}{2EI_y GI_t} \\[2mm] +\dfrac{(3GI_t I_y + 2GI_t AL^2)LF}{3EI_y AGI_t} \end{bmatrix} .
\tag{A.116}
$$

The following solution vector results with the same numerical values as above

$$\begin{bmatrix} u_{2Z} \\ \varphi_{2X} \\ \varphi_{2Y} \\ u_{3Z} \\ \varphi_{3X} \\ u_{4Z} \end{bmatrix} = \begin{bmatrix} +11.904 \\ +0.05492 \\ +0.01785 \\ +78.731 \\ +0.07277 \\ +78.732 \end{bmatrix} \tag{A.117}$$

with the same values. The algebraic signs as well as the order of the entries on node 2 have changed. The angle for torsion and bending have changed positions.

Problems from Chap. 8

8.3: Calculation of the Shear Correction Factor for Rectangular Cross-Section

$$\int_{\Omega} \frac{1}{2G} \tau_{xy}^2 d\Omega \stackrel{!}{=} \int_{\Omega_s} \frac{1}{2G} \left(\frac{Q_y}{A_s}\right)^2 d\Omega_s, \tag{A.118}$$

$$k_s = \frac{Q_y}{A \int_A \tau_{xy}^2 dA} = \frac{5}{6}. \tag{A.119}$$

8.4: Differential Equation under Consideration of Distributed Moment

Shear force: no difference, meaning $\dfrac{dQ_y(x)}{dx} = -q_y(x)$.

Bending moment:

$$M_z(x + dx) - M_z(x) + Q_y(x)dx - \frac{1}{2}q_y dx^2 + m_z dx = 0. \tag{A.120}$$

$$\frac{dM_z(x)}{dx} = -Q_y(x) - m_z, \tag{A.121}$$

$$\frac{d^2 M_z(x)}{dx^2} + \frac{dm_z(x)}{dx} = q_y(x). \tag{A.122}$$

Differential Equations:

$$\frac{\mathrm{d}}{\mathrm{d}x}\left(EI_z\frac{\mathrm{d}\phi_z}{\mathrm{d}x}\right) + k_s A G\left(\frac{\mathrm{d}u_y}{\mathrm{d}x} - \phi_z\right) = -m_z(x)\,, \qquad \text{(A.123)}$$

$$\frac{\mathrm{d}}{\mathrm{d}x}\left[k_s A G\left(\frac{\mathrm{d}u_y}{\mathrm{d}x} - \phi_z\right)\right] = -q_y(x)\,. \qquad \text{(A.124)}$$

8.5: Analytical Calculation of the Distribution of the Deflection and Rotation for a Cantilever under Point Load

Boundary Conditions:

$$u_y(x = 0) = 0\,, \quad \phi_z(x = 0) = 0\,, \qquad \text{(A.125)}$$
$$M_z(x = 0) = FL\,, \quad Q_y(x = 0) = F\,. \qquad \text{(A.126)}$$

Integration Constants:

$$c_1 = -F\,;\ c_2 = FL\,;\ c_3 = \frac{EI_z}{k_s A G}F\,;\ c_4 = 0\,. \qquad \text{(A.127)}$$

Course of the Displacement:

$$u_y(x) = \frac{1}{EI_z}\left(-F\frac{x^3}{6} + FL\frac{x^2}{2} + \frac{EI_zF}{k_s A G}x\right)\,. \qquad \text{(A.128)}$$

Course of the Rotation:

$$\phi_z(x) = \frac{1}{EI_z}\left(-F\frac{x^2}{2} + FLx\right)\,. \qquad \text{(A.129)}$$

Maximum Bending:

$$u_y(x = L) = \frac{1}{EI_z}\left(\frac{FL^3}{3} + \frac{EI_zFL}{k_s A G}\right)\,. \qquad \text{(A.130)}$$

Bending at the Loading Point:

$$\phi_z(x = L) = \frac{FL^2}{2EI_z}\,. \qquad \text{(A.131)}$$

Limit Value:

$$u_y(x = L) = \frac{4F}{b}\left(\frac{L}{h}\right)^3 + \frac{F}{k_s b G}\left(\frac{L}{h}\right)\,. \qquad \text{(A.132)}$$

$$u_y(L)\big|_{h \ll L} \to \frac{4F}{b}\left(\frac{L}{h}\right)^3 = \frac{FL^3}{3EI_z}, \tag{A.133}$$

$$u_y(L)\big|_{h \gg L} \to \frac{F}{k_s bG}\left(\frac{L}{h}\right) = \frac{FL}{k_s AG}. \tag{A.134}$$

8.6: Analytical Calculation of the Normalized Deflection for Beams with Shear Contribution

$$I_z = \frac{bh^3}{12}, \ A = hb, \ k_s = \frac{5}{6}, G = \frac{E}{2(1+v)}. \tag{A.135}$$

$$u_{y,\,\text{norm}} = \frac{1}{3} + \frac{1+v}{5}\left(\frac{h}{L}\right)^2, \tag{A.136}$$

$$u_{y,\,\text{norm}} = \frac{1}{8} + \frac{1+v}{10}\left(\frac{h}{L}\right)^2, \tag{A.137}$$

$$u_{y,\,\text{norm}} = \frac{1}{48} + \frac{1+v}{5}\left(\frac{h}{L}\right)^2. \tag{A.138}$$

8.7: Timoshenko Bending Element with Quadratic Shape Functions for the Deflection and Linear Shape Functions for the Rotation

The nodal point displacement on the middle node as a function of the other unknown results in:

$$u_{2y} = \frac{u_{1y} + u_{3y}}{2} + \frac{\phi_{1z} - \phi_{3z}}{8}L + \frac{1}{32}\frac{6L}{k_s AG}\int_0^L q_y(x)N_{2u}(x)dx. \tag{A.139}$$

The additional load matrix on the right-hand side results in:

$$\cdots = \cdots + \begin{bmatrix} \int_0^L q_y(x)N_{1u}dx + \frac{1}{2}\int_0^L q_y(x)N_{2u}dx \\ +\frac{1}{8}L\int_0^L q_y(x)N_{2u}dx \\ \int_0^L q_y(x)N_{3u}dx + \frac{1}{2}\int_0^L q_y(x)N_{2u}dx \\ -\frac{1}{8}L\int_0^L q_y(x)N_{2u}dx \end{bmatrix}. \tag{A.140}$$

With $\int_0^L N_{1u} dx = \dfrac{L}{6}$, $\int_0^L N_{2u} dx = \dfrac{2L}{3}$ and $\int_0^L N_{3u} dx = \dfrac{L}{6}$ the following results for a constant line load q_y:

$$\cdots = \cdots + \begin{bmatrix} \dfrac{1}{2} q_y L \\ +\dfrac{1}{12} q_y L^2 \\ \dfrac{1}{2} q_y L \\ -\dfrac{1}{12} q_y L^2 \end{bmatrix}. \tag{A.141}$$

This result is identical with the equivalent line load for a Bernoulli beam. For this see Table A.8.

8.8: Timoshenko Bending Element with Cubic Shape Functions for the Deflection and Quadratic Shape Functions for the Rotation

The element is exact!

Deformation in the x-y plane:

$$\frac{2EI_z}{L^3(1+12\Lambda)} \begin{bmatrix} 6 & 3L & -6 & 3L \\ 3L & 2L^2(1+3\Lambda) & -3L & L^2(1-6\Lambda) \\ -6 & -3L & 6 & -3L \\ 3L & L^2(1-6\Lambda) & -3L & 2L^2(1+3\Lambda) \end{bmatrix} \begin{bmatrix} u_{1y} \\ \phi_{1z} \\ u_{2y} \\ \phi_{2z} \end{bmatrix} = \begin{bmatrix} F_{1y} \\ M_{1z} \\ F_{2y} \\ M_{2z} \end{bmatrix}. \tag{A.142}$$

Deformation in the x-z plane:

$$\frac{2EI_y}{L^3(1+12\Lambda)} \begin{bmatrix} 6 & -3L & -6 & -3L \\ -3L & 2L^2(1+3\Lambda) & 3L & L^2(1-6\Lambda) \\ -6 & 3L & 6 & 3L \\ -3L & L^2(1-6\Lambda) & 3L & 2L^2(1+3\Lambda) \end{bmatrix} \begin{bmatrix} u_{1z} \\ \phi_{1y} \\ u_{2z} \\ \phi_{2y} \end{bmatrix} = \begin{bmatrix} F_{1z} \\ M_{1y} \\ F_{2z} \\ M_{2y} \end{bmatrix}. \tag{A.143}$$

(See Tables A.9 and A.10)

Problems from Chap. 9

9.1: Solution for 1: Determination of the stiffness matrix

The stiffness matrix can directly be taken on from the above derivation:

$$k^e = \frac{(EA)^V}{L} \begin{bmatrix} 1 & -1 \\ -1 & 1 \end{bmatrix}. \tag{A.144}$$

The expression $(EA)^V$ has to be determined for the composite. Since each layer is homogeneous and isotropic and additionally all layers have the same thickness, the generally valid relation in Eq. (9.122) simplifies to

$$(EA)^V = A_{11} b = b \sum_{k=1}^{3} Q_{11}^k h^k = b \frac{1}{3} h \sum_{k=1}^{3} E^{(k)} \tag{A.145}$$

and furthermore to

$$(EA)^V = \frac{1}{3} b h (E^{(1)} + E^{(2)} + E^{(3)}) = \frac{1}{3} b h (2 E^{(1)} + E^{(2)}). \tag{A.146}$$

If one considers furthermore that $E^{(2)} = \frac{1}{10} E^{(1)}$ is valid, the relation simplifies to:

$$(EA)^V = \frac{1}{3} b h \, 2.1 \, E^{(1)} = 0.7 \, EA. \tag{A.147}$$

For the control of the result it can to be assumed that both moduli of elasticity are the same ($E^{(1)} = E^{(2)} = E^{(3)} = E$). Then the stiffness known for the homogeneous, isotropic tension bar with $(EA)^V = Ebh = EA$ results.

9.1: Solution for 2: Determination of the bending stiffness

The bending stiffness results according to Eq. (9.128) for three layers in the composite to

$$(EI)^V = b \frac{1}{3} \sum_{k=1}^{3} E^k \left((z^k)^3 - (z^{k-1})^3 \right). \tag{A.148}$$

The z-coordinates result in $z^{(0)} = -3/2h$, $z^{(1)} = -1/2h$, $z^{(2)} = +1/2h$ and $z^{(3)} = +3/2h$ if the layer thickness h is equal and if the construction is symmetric to the ($z = 0$)-axis. Via integration one obtains:

$$(EI)^V = b \frac{1}{3} h^3 \left[E^{(1)} \left(\left(-\frac{1}{2} \right)^3 - \left(-\frac{3}{2} \right)^3 \right) \right. \tag{A.149}$$

$$+ E^{(2)} \left(\left(+\frac{1}{2} \right)^3 - \left(-\frac{1}{2} \right)^3 \right) + E^{(1)} \left(\left(+\frac{3}{2} \right)^3 - \left(+\frac{3}{2} \right)^3 \right) \right] \quad \text{(A.150)}$$

$$= \frac{1}{3} b h^3 \left[E^{(1)} \left(-\frac{1}{8} + \frac{27}{8} + \frac{27}{8} - \frac{1}{8} \right) + E^{(2)} \left(+\frac{1}{8} + \frac{1}{8} \right) \right] \quad \text{(A.151)}$$

and finally

$$(EI)^V = \frac{1}{3} b h^3 \left[\frac{26}{4} E^{(1)} + \frac{1}{4} E^{(2)} \right] . \quad \text{(A.152)}$$

For the control of the result it can to be assumed that all moduli of elasticity are equal $(E^{(1)} = E^{(2)} = E^{(3)} = E)$. Then the bending stiffness EI for a homogeneous beam with the cross-section b and $3h$ results in $\frac{9}{4} E b h^3$.

Problems from Chap. 10

10.5: Strain Dependent Modulus of Elasticity with Quadratic Course (Fig. A.3) (Table A.11)

$$a = E_0 , b = -\frac{E_0}{10\varepsilon_1} , c = -\frac{4E_0}{5\varepsilon_1^2} , \quad \text{(A.153)}$$

Fig. A.3 Stress-strain diagram, based on quadratic modulus of elasticity

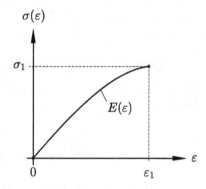

Table A.11 Numerical values for direct iteration at an external load of $F_2 = 800$ kN and different initial values. Geometry: $A = 100$ mm^2, $L = 400$ mm. Material properties: $E_0 = 70000$ MPa, $E_1 = 49000$ MPa, $\varepsilon_1 = 0.15$

Iteration j	$u_2^{(j)}$	$\varepsilon_2^{(j)}$	$\sqrt{\dfrac{\left(u_2^{(j)}-u_2^{(j-1)}\right)^2}{\left(u_2^{(j)}\right)^2}}$
	mm	–	–
Initial value: $u_2^{(0)} = 0$ mm			
0	0	0	–
1	45.714286	0.114286	1.000000
2	59.259259	0.148148	0.228571
⋮	⋮	⋮	⋮
23	70.722968	0.176807	0.000000
⋮	⋮	⋮	⋮
31	70.722998	0.176807	0.000000
Initial value: $u_2^{(0)} = 30$ mm			
0	30.000000	0.075000	–
1	53.781513	0.134454	0.442187
2	62.528736	0.156322	0.139891
⋮	⋮	⋮	⋮
22	70.722956	0.176807	0.000000
⋮	⋮	⋮	⋮
31	70.722998	0.176807	0.000000

(continued)

Table A.11 (continued)

Iteration j	$u_2^{(j)}$	$\varepsilon_2^{(j)}$	$\sqrt{\dfrac{\left(u_2^{(j)}-u_2^{(j-1)}\right)^2}{\left(u_2^{(j)}\right)^2}}$
	mm	–	–
Initial value: $u_2^{(0)} = 220$ mm			
0	220.000000	0.550000	–
1	−457.142857	−1.142857	1.481250
2	13.913043	0.034783	33.857143
$\vdots$	$\vdots$	$\vdots$	$\vdots$
25	70.722971	0.176807	0.000000
$\vdots$	$\vdots$	$\vdots$	$\vdots$
33	70.722998	0.176807	0.000000

$$E(\varepsilon) = E_0\left(1 - \underbrace{\frac{1}{10\varepsilon_1}}_{\alpha_1}\varepsilon - \underbrace{\frac{4}{5\varepsilon_1^2}}_{\alpha_2}\varepsilon^2\right), \tag{A.154}$$

$$\sigma(\varepsilon) = E_0\varepsilon\left(1 - \frac{\varepsilon}{20\varepsilon_1} - \frac{4\varepsilon^2}{15\varepsilon_1^2}\right), \tag{A.155}$$

$$k^e = \frac{AE_0}{L^2}\left(L + \alpha_1 u_1 - \alpha_1 u_2 - \frac{\alpha_2}{L}u_1^2 + \frac{2\alpha_2}{L}u_1 u_2 - \frac{\alpha_2}{L}u_2^2\right)\begin{bmatrix} 1 & -1 \\ -1 & 1 \end{bmatrix}, \tag{A.156}$$

$$K_T^e = \frac{AE_0}{L^2}\left(L + 2\alpha_1 u_1 - 2\alpha_1 u_2 - 3\frac{\alpha_2}{L}u_1^2 + 4\frac{\alpha_2}{L}u_1 u_2 - 3\frac{\alpha_2}{L}u_2^2\right)\begin{bmatrix} 1 & -1 \\ -1 & 1 \end{bmatrix}. \tag{A.157}$$

10.6: Direct Iteration with Different Initial Values (Table A.11)

10.7: Complete Newton–Raphson Scheme for a Linear Element with Quadratic Modulus of Elasticity

Table A.12 Numerical values for complete Newton–Raphson's method at an external load of $F_2 = 370\,\text{kN}$. Geometry: $A = 100\,\text{mm}^2$, $L = 400\,\text{mm}$. Material properties: quadratic course with $E_0 = 70000\,\text{MPa}$ and $\varepsilon_1 = 0.15$

Iteration j	$u_2^{(j)}$	$\varepsilon_2^{(j)}$	$\sqrt{\dfrac{\left(u_2^{(j)}-u_2^{(j-1)}\right)^2}{\left(u_2^{(j)}\right)^2}}$
	mm	–	–
0	0	0	–
1	21.142857	0.052857	1
2	25.648438	0.064121	0.175667
3	26.363431	0.065909	0.027121
4	26.384989	0.065962	0.000031
5	26.385009	0.065963	0.000001
6	26.385009	0.065963	0.000000

(See Table A.12)

Residual function:

$$r(u_2) = \frac{AE_0}{L^2}\left(L - \alpha_1 u_2 - \frac{\alpha_2}{L}u_2^2\right)u_2 - F_2 = K(u_2)u_2 - F_2 = 0. \quad (A.158)$$

Tangent stiffness:

$$K_{\text{T}}(u_2) = \frac{AE_0}{L^2}\left(L - 2\alpha_1 u_2 - \frac{3\alpha_2}{L}u_2^2\right). \quad (A.159)$$

Iteration scheme:

$$u_2^{(j+1)} = u_2^{(j)} - \frac{\dfrac{AE_0}{L^2}\left(L - \alpha_1 u_2^{(j)} - \dfrac{\alpha_2}{L}\left(u_2^{(j)}\right)^2\right)u_2^{(j)} - F_2^{(j)}}{\dfrac{AE_0}{L^2}\left(L - 2\alpha_1 u_2^{(j)} - \dfrac{3\alpha_2}{L}\left(u_2^{(j)}\right)^2\right)}. \quad (A.160)$$

Condition for convergence according to Fig. A.4:

$$r(u_{2,\text{max}}) \stackrel{!}{\geq} 0 \quad (A.161)$$

Fig. A.4 Illustration of the residual function according to Eq. (A.158)

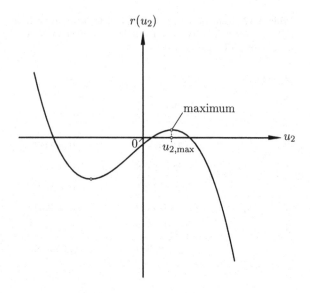

or

$$F \le \frac{AE_0}{27\alpha_2^2}\left(6\alpha_2 - \alpha_1\sqrt{\frac{\alpha_1^2 + 3\alpha_2}{\alpha_2^2}}\alpha_2 + \alpha_1^2\right)\left(\sqrt{\frac{\alpha_1^2 + 3\alpha_2}{\alpha_2^2}}\alpha_2 - \alpha_1\right). \quad \text{(A.162)}$$

The following results for the given numerical values, $F \le 410.803$ kN so that the iteration scheme converges.

10.8: Strain Dependent Modulus of Elasticity with General Quadratic Course

$$a = E_0 , b = \frac{E_0}{\varepsilon_1}(4\beta_{05} - \beta_1 - 3) , c = -\frac{4E_0}{\varepsilon_1^2}\left(\beta_{05} - \frac{1}{2}\beta_1 - \frac{1}{2}\right). \quad \text{(A.163)}$$

$$E(\varepsilon) = E_0\left(1 - \underbrace{\frac{(3 + \beta_1 - 4\beta_{05})}{\varepsilon_1}}_{\alpha_1}\cdot\varepsilon - \underbrace{\frac{4(-\frac{1}{2} - \frac{1}{2}\beta_1 + \beta_{05})}{\varepsilon_1^2}}_{\alpha_2}\cdot\varepsilon^2\right). \quad \text{(A.164)}$$

With the introduced definitions *here* of α_1 and α_2 a course according to Eq. (A.154) results. Thus under consideration of the here introduced definitions of the stiffness matrix according to Eq. (A.156) as well as the tangent stiffness matrix according to Eq. (A.157) can be used for α_1 and α_2.

Problems from Chap. 11

11.4: Plastic Modulus and Elasto-Plastic Material Modulus

(a) $E^{\text{elpl}} = \dfrac{E \times E^{\text{pl}}}{E + E^{\text{pl}}} = E \Rightarrow E = 0.$

Pure linear-plastic behavior without elastic part, meaning pure elastic behavior on macro-level.

(b) $E^{\text{pl}} = E \Rightarrow E^{\text{elpl}} = \dfrac{E \times E}{E + E} = \dfrac{1}{2} E.$

Linear hardening, whereupon the elasto-plastic modulus is half of the elastic modulus E.

11.5: Back Projection at Linear Hardening (Table A.13, A.14, A.15)

11.6: Back Projection at Non-Linear Hardening (Table A.16)

The flow curve results in:

$$k(\kappa) = 200 \text{ MPa} + 1010.\overline{10} \text{ MPa} \times \kappa. \tag{A.165}$$

The iteration scheme of the Newton–Raphson method can be used as follows (Table A.17):

$$\Delta u_2^{(i+1)} = \frac{\Delta F^{(i)}}{A\left(\frac{\tilde{E}^{\text{I}}}{L^{\text{I}}} + \frac{\tilde{E}^{\text{II}}}{L^{\text{II}}}\right)}. \tag{A.166}$$

Table A.13 Numerical values of the back projection for a continuum bar in the case of linear hardening (10 increments; $\Delta\varepsilon = 0.001$)

inc	ε	σ^{trial}	σ	κ	$\Delta\lambda$	E^{elpl}
–	–	MPa	MPa	10^{-3}		MPa
1	0.001	210.0	210.0	0.0	0.0	0.0
2	0.002	420.0	420.0	0.0	0.0	0.0
3	0.003	630.0	630.0	0.0	0.0	0.0
4	0.004	840.0	703.636	0.649	0.649	19090.909
5	0.005	913.636	722.727	1.558	0.909	19090.909
6	0.006	932.727	741.818	2.468	0.909	19090.909
7	0.007	951.818	760.909	3.377	0.909	19090.909
8	0.008	970.909	780.000	4.286	0.909	19090.909
9	0.009	990.000	799.091	5.195	0.909	19090.909
10	0.010	1009.091	818.182	6.104	0.909	19090.909

Table A.14 Numerical values of the back projection for a continuum bar in the case of linear hardening (20 increments; $\Delta\varepsilon = 0.0005$)

| inc | ε | σ^{trial} | σ | κ | $\Delta\lambda$ | E^{elpl} |
–	–	MPa	MPa	10^{-3}	10^{-3}	MPa
1	0.0005	105.0	105.0	0.0	0.0	0.0
2	0.0010	210.0	210.0	0.0	0.0	0.0
3	0.0015	315.0	315.0	0.0	0.0	0.0
4	0.0020	420.0	420.0	0.0	0.0	0.0
5	0.0025	525.0	525.0	0.0	0.0	0.0
6	0.0030	630.0	630.0	0.0	0.0	0.0
7	0.0035	735.0	694.091	0.195	0.195	19090.909
8	0.0040	799.091	703.636	0.649	0.455	19090.909
9	0.0045	808.636	713.182	1.104	0.455	19090.909
10	0.0050	818.182	722.727	1.558	0.455	19090.909
11	0.0055	827.727	732.273	2.013	0.455	19090.909
12	0.0060	837.273	741.818	2.468	0.455	19090.909
13	0.0065	846.818	751.364	2.922	0.455	19090.909
14	0.0070	856.364	760.909	3.377	0.455	19090.909
15	0.0075	865.909	770.455	3.831	0.455	19090.909
16	0.0080	875.455	780.000	4.286	0.455	19090.909
17	0.0085	885.000	789.545	4.740	0.455	19090.909
18	0.0090	894.545	799.091	5.195	0.455	19090.909
19	0.0095	904.091	808.636	5.649	0.455	19090.909
20	0.0100	913.636	818.182	6.104	0.455	19090.909

11.7: Back Projection for Bar at Fixed Support at Both Ends

For the fulfillment of the convergence criteria nine cycles are necessary for the second increment and four cycles for the third increment.

11.8: Back Projection for a Finite Element at Ideal-Plastic Material Behavior

Table A.15 Numerical values of the back projection for continuum bar at linear hardening (20 increments; $\Delta\varepsilon = 0.0005$)

inc −	ε −	σ^{trial} MPa	σ MPa	κ 10^{-3}	$\Delta\lambda$ 10^{-3}	E^{elpl} MPa
1	0.001	210.0	210.0	0.0	0.0	0.0
2	0.002	420.0	420.0	0.0	0.0	0.0
3	0.003	630.0	630.0	0.0	0.0	0.0
4	0.004	840.0	703.636	0.649	0.649	19090.909
5	0.005	913.636	722.727	1.558	0.909	19090.909
6	0.006	932.727	741.818	2.468	0.909	19090.909
7	0.007	951.818	760.909	3.377	0.909	19090.909
8	0.008	970.909	780.000	4.286	0.909	19090.909
9	0.009	990.000	799.091	5.195	0.909	19090.909
10	0.010	1009.091	818.182	6.104	0.909	19090.909
11	0.011	1028.182	837.273	7.013	0.909	19090.909
12	0.012	1047.273	856.364	7.922	0.909	19090.909
13	0.013	1066.364	875.455	8.831	0.909	19090.909
14	0.014	1085.455	894.545	9.740	0.909	19090.909
15	0.015	1104.545	913.636	10.649	0.909	19090.909
16	0.016	1123.636	932.727	11.558	0.909	19090.909
17	0.017	1142.727	951.818	12.468	0.909	19090.909
18	0.018	1161.818	970.909	13.377	0.909	19090.909
19	0.019	1180.909	990.000	14.286	0.909	19090.909
20	0.020	1200.000	1009.091	15.195	0.909	19090.909

In the elastic region, the calculation of the displacement on the node can take place. As soon as plastic material behavior occurs, no convergence can be achieved since no clear connection between load and strain exists. If the boundary force condition is being substituted by a boundary displacement condition, the stress in the bar is known and the stress can be calculated.

Table A.16 Numerical values of the back projection for a continuum bar in the case of nonlinear hardening hardening (10 increments; $\Delta\varepsilon = 0.002$)

inc	ε	σ^{trial}	σ	κ	$\Delta\lambda$	E^{elpl}
–	–	MPa	MPa	10^{-3}	10^{-3}	MPa
1	0.002	140.0	140.0	0.0	0.0	0.0
2	0.004	280.0	280.0	0.0	0.0	0.0
3	0.006	420.0	360.817	0.845469	0.845469	10741.553
4	0.008	500.817	381.995	2.542923	1.697454	10435.865
5	0.010	521.995	402.557	4.249179	1.706256	10125.398
6	0.012	542.557	422.494	5.964375	1.715196	9810.025
7	0.014	562.494	441.794	7.688654	1.724279	9489.616
8	0.016	581.794	460.449	9.422161	1.733507	9164.034
9	0.018	600.449	478.447	11.165046	1.742885	8833.140
10	0.020	618.447	495.778	12.917462	1.752416	8496.787

Table A.17 Numerical values for bar with fixed support on both sides (3 increments; $\Delta F_2 = 2 \times 10^4$ N)

inc	u_2	ε^{I}	ε^{II}	σ^{I}	σ^{II}	$\varepsilon^{\text{pl,I}}$	$\varepsilon^{\text{pl,II}}$
–	mm	10^{-3}	10^{-3}	MPa	MPa	10^{-3}	10^{-3}
1	0.0666667	0.666667	−1.33333	66.6667	−133.333	0.0	0.0
2	0.19806	1.9806	−3.9612	198.060	−201.938	0.0	−1.94182
3	6.88003	68.8003	−137.601	266.008	−333.992	66.1402	−134.261

Problems from Chap. 12

12.1: Entries of the geometric stiffness matrix

12.2: Short solution: Euler's buckling loads II, III and IV, one element

The elastic and geometric stiffness matrix is obtained as in the description to Euler's buckling load I. Due to the boundary conditions two eigenvalues result for case II and one eigenvalue for case III. Case IV cannot be modeled with just one element. The eigenvalues result for the Euler's buckling load II:

$$\lambda_{1/2} = (36 \pm 24)\frac{EI}{L^2} \tag{A.167}$$

and for the Euler's buckling load III:

$$\lambda = 30\frac{EI}{L^2}.$$ (A.168)

For the definition of the critical load respectively the smallest eigenvalues are of interest. The deviations from the analytical solutions are significant.

12.3: Short solution: Euler's buckling loads, two elements

The entire stiffness matrix consists of elastic and geometric stiffness matrix. Load IV can also be modeled with two elements. The eigenvalues can only be determined numerically, only for load I can an analytical solution with reasonable expense be named. The eigenvalues for Euler's buckling load I:

$$\frac{16}{17}\frac{EI}{L^2}\begin{bmatrix} 80 + 19\sqrt{2} + \sqrt{5847 + 3550\sqrt{2}} \\ 80 + 19\sqrt{2} - \sqrt{5847 + 3550\sqrt{2}} \\ 80 - 19\sqrt{2} + \sqrt{5847 + 3550\sqrt{2}} \\ 80 - 19\sqrt{2} - \sqrt{5847 + 3550\sqrt{2}} \end{bmatrix} = \frac{EI}{L^2}\begin{bmatrix} 198.69 \\ 2.4686 \\ 77.063 \\ 22.946 \end{bmatrix},$$ (A.169)

for Euler's buckling load II:

$$\frac{EI}{L^2}\begin{bmatrix} 48.0 \\ 128.72 \\ 9.9438 \\ 240.0 \end{bmatrix},$$ (A.170)

for Euler's buckling load III:

$$\frac{EI}{L^2}\begin{bmatrix} 197.52 \\ 20.708 \\ 75.101 \end{bmatrix},$$ (A.171)

Table A.18 Relative error in relation to analytical buckling load

Number of elements	Euler case			
	I	II	III	IV
1	7.52×10^{-3}	0.215854	0.485830	–
2	5.12×10^{-4}	7.52×10^{-3}	2.57×10^{-2}	1.32×10^{-2}
3	1.03×10^{-4}	1.58×10^{-3}	6.14×10^{-3}	2.19×10^{-2}
4	3.28×10^{-5}	5.12×10^{-4}	2.05×10^{-3}	7.52×10^{-3}
5	1.35×10^{-5}	2.12×10^{-4}	8.64×10^{-4}	3.21×10^{-3}
6	6.50×10^{-6}	1.03×10^{-4}	4.23×10^{-4}	1.58×10^{-3}
7	3.51×10^{-6}	5.58×10^{-5}	2.30×10^{-4}	8.66×10^{-4}
8	2.06×10^{-6}	3.28×10^{-5}	1.36×10^{-4}	5.12×10^{-4}
9	1.29×10^{-6}	2.05×10^{-5}	8.50×10^{-5}	3.22×10^{-4}
10	8.44×10^{-7}	1.35×10^{-5}	5.59×10^{-5}	2.12×10^{-4}

for Euler's buckling load IV:

$$\frac{EI}{L^2} \begin{bmatrix} 120.0 \\ 40.0 \end{bmatrix}. \tag{A.172}$$

For the determination of the critical load respectively the smallest eigenvalues are of interest. The deviations from the analytical solutions are significant.

12.4: Euler's buckling loads, errors in regard to analytical solution

The following Table A.18 shows the relative error of the solution of the critical buckling load, which was determined via the finite element method in regard to the analytical solution.

$$\text{error} = \frac{\text{FE solution - analytical solution}}{\text{analytical solution}} \tag{A.173}$$

The errors strongly differ for the different Euler's buckling loads. The error for the Euler's buckling load I is always the smallest, the one for the Euler's buckling load IV always the biggest. The difference in the single loads extends over two dimensions. At a cross-linking with four elements the error is already smaller 0.01 for all cases.

Problems from Chap. 13

13.1: Analytical Solutions for Bending Vibrations

With Eq. (13.70) one can determine the first four eigenfrequencies with $n = 1, \ldots, 4$. These result in

$$\omega_1 = 1.5708\sqrt{\frac{E}{\rho L^2}}$$

$$\omega_2 = 4.7124\sqrt{\frac{E}{\rho L^2}}$$

$$\omega_3 = 7.854\sqrt{\frac{E}{\rho L^2}}$$

$$\omega_4 = 10.99\sqrt{\frac{E}{\rho L^2}}\,.$$

13.2: FE Solution for Bending Vibration

One needs to solve Eq. (13.121). This leads to:

$$400\left(\frac{14745600E^4 - 6881280E^3L^2\rho w^2 + 380416E^2L^4\rho^2 w^4}{L^8 * \rho^4}\right.$$
$$\left. + \frac{-5120EL^6\rho^3 w^6 + 17 * L^8\rho^4 w^8}{L^8 * \rho^4}\right) = 0\,. \tag{A.174}$$

Solving this results in the eigenfrequencies

$$\omega_1 = \pm 1.5712\sqrt{\frac{E}{\rho L^2}}$$

$$\omega_2 = \pm 4.7902\sqrt{\frac{E}{\rho L^2}}$$

$$\omega_3 = \pm 8.7786\sqrt{\frac{E}{\rho L^2}}$$

$$\omega_4 = \pm 14.0961\sqrt{\frac{E}{\rho L^2}}\,,$$

where only the positive values are relevant, since there are no negative frequencies.

Problems from Chap. 14

14.1: Beam with elastic foundation and point load

From Eq. (14.14) the stiffness matrix k^e for a beam with elastic foundation is known. Inserting the known values k, EI_z, F leads to the following equation:

$$\left(\frac{EI_z}{L^3} \begin{bmatrix} 12 & 6L & -12 & 6L \\ 6L & 4L^2 & -6L & 2L^2 \\ -12 & -6L & 12 & -6L \\ 6L & 2L^2 & -6L & 4L^2 \end{bmatrix} + \frac{kL}{420} \begin{bmatrix} 156 & 22L & 54 & -13L \\ 22L & 4L^2 & 13L & -3L^2 \\ 54 & 13L & 156 & -22L \\ -13L & -3L^2 & -22L & 4L^2 \end{bmatrix} \right) \begin{bmatrix} u_{1y} \\ \phi_{1z} \\ u_{2y} \\ \phi_{2z} \end{bmatrix} = \begin{bmatrix} F \\ 0 \\ -F \\ 0 \end{bmatrix}.$$

$$\text{(A.175)}$$

Because of the fixed support on the left u_{1y} and ϕ_{1z} are zero and the system can be reduced to

$$\left(\frac{EI_z}{L^3} \begin{bmatrix} 12 & -6L \\ -6L & 4L^2 \end{bmatrix} + \frac{kL}{420} \begin{bmatrix} 156 & -22L \\ -22L & 4L^2 \end{bmatrix} \right) \begin{bmatrix} u_{2y} \\ \phi_{2z} \end{bmatrix} = \begin{bmatrix} -F \\ 0 \end{bmatrix} \qquad \text{(A.176)}$$

Solving this for the unknown deformations gives

$$\begin{bmatrix} u_{2y} \\ \phi_{2z} \end{bmatrix} = \left(\frac{L^3}{EI_z} \begin{bmatrix} 12 & -6L \\ -6L & 4L^2 \end{bmatrix}^{-1} + \frac{420}{kL} \begin{bmatrix} 156 & -22L \\ -22L & 4L^2 \end{bmatrix}^{-1} \right) \begin{bmatrix} -F \\ 0 \end{bmatrix}$$

$$= \left(\frac{L^3}{EI_z} \begin{bmatrix} \frac{1}{3} & \frac{1}{2L} \\ \frac{1}{2L} & \frac{1}{L^2} \end{bmatrix} + \frac{420}{kL} \begin{bmatrix} \frac{1}{35} & \frac{11}{70L} \\ \frac{11}{70L} & \frac{39}{35L^2} \end{bmatrix} \right) \begin{bmatrix} -F \\ 0 \end{bmatrix} \qquad \text{(A.177)}$$

Now the deformations can be calculated.

$$u_{2y} = \frac{-FL^3}{3EI_z} + \frac{-420F}{kL}$$

$$\phi_{2z} = \frac{-FL}{2EI_z} + 66 \frac{-F}{kL^2}$$

14.2: Bar element with quadratic shape functions and excentric node: stress singularity

14.3: Infinite element with linear interpolation of the field variable

Index

© Springer International Publishing AG, part of Springer Nature 2018
A. Öchsner and M. Merkel, *One-Dimensional Finite Elements*,
https://doi.org/10.1007/978-3-319-75145-0

Printed in the United States
By Bookmasters